Nanomaterials: Synthesis, Properties and Applications

Nanomaterials: Synthesis, Properties and Applications

Edited by
A S Edelstein

Naval Research Laboratory
Washington, DC

and

R C Cammarata

Department of Materials Science and Engineering
Johns Hopkins University
Baltimore, MD

Published in 1996 by
Taylor & Francis Group
270 Madison Avenue
New York, NY 10016

Published in Great Britain by
Taylor & Francis Group
2 Park Square
Milton Park, Abingdon
Oxon OX14 4RN

Printed in the United States of America on acid-free paper
10 9 8 7 6 5 4 3 2

International Standard Book Number-10: 0-7503-0578-9 (Softcover)
International Standard Book Number-13: 978-0-7503-0578-5 (Softcover)

Library of Congress Cataloging-in-Publication Data

Catalog record is available from the Library of Congress

Taylor & Francis Group
is the Academic Division of Informa plc.

**Visit the Taylor & Francis Web site at
http://www.taylorandfrancis.com**

Contents

Preface

Nanomaterials has recently become one of the most active research fields in the areas of solid state physics, chemistry, and engineering. Evidence of this interest is provided by the large number of recent conferences and research papers devoted to the subject. There are several reasons for this. One is the need to fabricate new materials on an ever finer scale to continue decreasing the cost and increasing the speed of information transmission and storage. Another is that nanomaterials display novel and often enhanced properties compared to traditional materials, which opens up possibilities for new technological applications.

This book grew out of the editors' realization, after attending several conferences on nanomaterials, of the breadth of this field and the lack of a text that covered its diverse subjects. Furthermore, workers in one area were largely unaware of work in other areas. This book is intended to satisfy this need for a broad coverage that will provide an introduction, background, and references to nearly all areas of nanomaterials research. It includes an extensive list of references at the end of nearly every chapter. As an illustration of the breadth of this subject, it is noted that even after the preparation of the book was at a fairly late stage, the editors found that important subjects had been overlooked and decided to add several new chapters.

The editors wish to thank the authors for their efforts in writing their chapters so that they would be accessible to a general audience, and Kathryn Cantley, Steve Clewer, Don Emerson and Pamela Whichard at Institute of Physics Publishing for their invaluable assistance.

The editors would like to acknowledge the important contributions made by many researchers in the field of nanomaterials whose work could not be cited because of lack of space.

A S Edelstein
R C Cammarata
February 1996

Authors' Addresses

R S Averback University of Illinois-Urbana, Department of Materials Science and Engineering, 104 West Green Street, Urbana, IL 61801, USA

R T Bate NanoFab Center, Texas Engineering Experiment Station, PO Box 830688, Mail Stop 32, University of Texas at Dallas, Richardson, TX 75083, USA

A Berkowitz Physics Department and Center for Magnetic Recording Research, University of California, San Diego, La Jolla, CA 92093, USA

F A Buot Electronics Science and Technology Division, Naval Research Laboratory, Washington, DC 20375, USA

R C Cammarata Department of Materials Science and Engineering, The Johns Hopkins University, Baltimore, MD 21218, USA

D-J Chen Department of Materials Science and Engineering, The Pennsylvania State University, University Park, PA 16802, USA

C L Chien Department of Physics and Astronomy, The Johns Hopkins University, Baltimore, MD 21218, USA

G M Chow Laboratory for Molecular Interfacial Interactions, Center for Biomolecular Science and Engineering, Naval Research Laboratory, Washington, DC 20375, USA

H G Craighead Applied and Engineering Physics, Cornell University, Ithaca, NY 14853, USA

D P E Dickson Department of Physics, University of Liverpool, Liverpool L69 3BX, UK

E A Dobisz Electronics Science and Technology Division, Naval Research Laboratory, Washington, DC 20375, USA

A S Edelstein Naval Research Laboratory, Washington, DC 20375, USA

M S El-Shall Department of Chemistry, Virginia Commonwealth University, VA 23284, USA

H J Fecht Technical University of Berlin, Institute of Metals Research, Hardenbergstrasse 36, PN 2-3, 10623 Berlin, Germany

K E Gonsalves Polymer Science Program, Institute of Materials Science and Department of Chemistry, University of Connecticut, Storrs, CT 06269, USA

G C Hadjipanayis Department of Physics and Astronomy, University of Delaware, Newark, DE 19716-2570, USA

D C Hague Department of Materials Science and Engineering, The Pennsylvania State University, University Park, PA 16802, USA

E Hanamura Department of Applied Physics, University of Tokyo, 7-3-1 Hongo, Bunkyou-ku, Tokyo

N Herron Central Research and Development, The Du Pont Company, PO Box 80356, Wilmington, DE 19880-0356, USA

D R Huffman Department of Physics, University of Arizona, Tucson, AZ 85721, USA

K J Klabunde Chemistry Department, Kansas State University, Manhattan, KS 66506, USA

L C Klein Rutgers—The State University of New Jersey, Ceramics Department, PO Box 909, Piscataway, NJ 08855-0909, USA

S W Koch Optical Sciences Center, University of Arizona, Tucson, AZ 85721, USA and Physics Department, University of Arizona, Tucson, AZ 85721, USA

C R K Marrian Electronics Science and Technology Division, Naval Research Laboratory, Washington, DC 20375, USA

Y Masumoto Institute of Physics, University of Tsukuba, Tsukuba, Ibaraki 305, Japan

M J Mayo Department of Materials Science and Engineering, The Pennsylvania State University, University Park, PA 16802, USA

J S Murday Code 6100, Naval Research Laboratory, Washington, DC 20375-5000, USA

J C Parker Nanophase Technologies Corporation, 453 Commerce Street, Burr Ridge, IL 60521, USA

N Peyghambarian Optical Sciences Center, University of Arizona, Tucson, AZ 85721, USA

S M Prokes Naval Research Laboratory, Washington, DC 20375, USA

B B Rath Associate Director of Research for Materials Science and Technology, Naval Research Laboratory, Washington DC 20375, USA

D E Rolison Surface Chemistry Branch, Code 6170, Naval Research Laboratory, Washington, DC 20375-5342, USA

H-E Schaefer Universität Stuttgart, Institut für Theoretische und Angewandte Physik, Pfaffenwaldring 57, 70550 Stuttgart, Germany

R W Siegel Materials Science Division, Argonne National Laboratory, Argonne, IL 60439, USA (Present address: Materials Science and Engineering Department, Rensselaer Polytechnic Institute, Troy, NY 12180-3590, USA)

C M Sorensen Department of Physics, Cardwell Hall, Kansas State University, Manhattan, KS 66506, USA

W M Tolles Executive Directorate, Code 1007, Naval Research Laboratory, Washington DC 20375, USA

K M Unruh Department of Physics and Astronomy, University of Delaware, Newark, DE 19716, USA

Y Wang Central Research and Development, The Du Pont Company, PO Box 80356, Wilmington, DE 19880-0356, USA

J R Weertman Materials Science and Engineering Department, Northwestern University, Evanston, IL 60208, USA

J Weissmüller National Institute of Standards and Technology, 223/B152, Gaithersburg, MD 20899, USA (Permanent address: Institut für Neue Materialen, Universität des Saarlandes im Stadtwald, D-66041 Saarbrücken, Germany)

E M Wright Optical Sciences Center, University of Arizona, Tucson, AZ 85721, USA and Physics Department, University of Arizona, Tucson, AZ 85721, USA

R Würschum Universität Stuttgart, Institut für Theoretische und Angewandte Physik, Pfaffenwaldring 57, 70550 Stuttgart, Germany

Acknowledgments

Chapter 3 GMC would like to thank the Office of Naval Research and the Naval Research Laboratory for supporting the work on nanostructured materials.
Chapter 4 The authors acknowledge their collaborators for contributions to this work during the past few years: A Suna, L T Cheng, W Mahler, J Calabrese, and W Farneth from Du Pont Co.; E Hilinski and his group at Florida State University; G D Stucky, H Eckert and their groups at UC Santa Barbara; as well as K Moller and T Bein at the University of New Mexico.
Chapter 5 The financial support by the Deutsche Forschungsgemeinschaft (Fe 313/1) and by the US Department of Energy (contract number DE-FG03-86ER45242) during the affiliation of the author with the California Institute of Technology, USA, where this work was initiated in collaboration with Professor W L Johnson, is gratefully acknowledged. The author would like to thank many colleagues, in particular at the California Institute of Technology and the Universität Augsburg, for collaboration and discussions over the past few years.
Chapter 6 The author thanks C L Chien, J Erlebacher, A L Greer, S M Prokes, K Sieradzki, F Spaepen, I K Schuller, and J W Wagner for stimulating discussions. Support during the preparation of this chapter by the National Science Foundation through grant number ECS-920222 and by the Office of Naval Research through grant number N00014-91-J-1169 is gratefully acknowledged.
Chapter 8 The authors gratefully acknowledge the support of the National Science Foundation through grant number DMR 9158098 in the preparation of this chapter.
Chapter 10 This chapter was prepared at the National Institute of Standards and Technology, Gaithersburg, MD, USA, while on leave from the Institut für Neue Materialen, Saarbrücken, Germany. The author would like to thank the Alexander von Humboldt Foundation for making the work at NIST possible through the award of a Feodor Lynen Fellowship. Helpful discussions with R Birringer, F Boscherini, J W Cahn, H Gleiter, C Krill, J Löffler, and R D Shull are gratefully acknowledged. The author thanks A J Allen and F W Gayle for critical reading of the manuscript and many helpful suggestions.

Chapter 11 The financial support of the Deutsche Forschungsgemeinschaft is appreciated.

Chapter 12 Stimulating arguments on the contents of this chapter with Professor George M Whitesides (Harvard University) are noted with pleasure. The support of the Office of Naval Research is gratefully acknowledged.

Chapter 14 The authors would like to acknowledge several of their students, including M Allitt, B M Patterson, and A Gavrin, who have contributed to many aspects of the work described in this chapter. In addition, the authors acknowledge a number of helpful conversations with S-T Chui, G C Hadjipanayis, P Sheng, A Tsoukatos (from whose thesis figures 14.1(*b*), 14.2(*b*), and 14.3 have been adapted), and L Withanawasam. Portions of this chapter have been supported by the Office of Naval Research under contract number N00014-91-J-1633 and the National Science Foundation under grant number DMR-9501195.

Chapter 15 This work was supported by NSF grants CHE 8706954 and CHE 9013930.

Chapter 16 The authors would like to thank the NSF, SDI/ONR/AFOSR, and NEDO for support of some of the work that is reported here. The work has been done in collaboration with B P McGinnis, K Kang, B Fleugel, A Kepner, Y Hu, V Esch, A Mysyrowicz, D Hulin, and S Gaponenko.

Chapter 18 The author is indebted to many co-workers, collaborators, and other scientists whose work in this field is discussed in this chapter. In particular the author would like to acknowledge the contributions of R B Frankel, S Mann, T G St Pierre, and J Webb.

Chapter 19 The author expresses appreciation to the Department of Energy for partial support of his work referenced in this paper through grant number DE-FG03-93ER12133. Special thanks to Mark Ross and Steven McElvany of the Naval Research Laboratory for their kind permission to use figure 19.2.

PART 1

INTRODUCTION

Chapter 1

Introduction

A S Edelstein and R C Cammarata

Early in the nineteenth century, scientific evidence was found that proved that matter is composed of discrete entities called atoms. It is likely that this discovery prompted a natural desire to be able to control the structure of matter atom by atom. Feynman [1] in his 1960 article *'There's plenty of room at the bottom'* discussed the advantages that could be provided by such control. For example, he pointed out that if a bit of information requires only 100 atoms, then all the books ever written could be stored in a cube with sides 0.02 in long. In recent years, researchers have been able to write bits of information in two dimensions using even fewer than 100 atoms by using a scanning tunneling microscope [2]. Economical fabrication of such structures remains a challenge [1–4]. Storage of information on an ever finer scale is just one aspect of the rapidly growing field of nanomaterials in which researchers are trying to control the fine-scale structure of materials.

In this book, we use the conventional definition of nanomaterials as materials having a characteristic length scale less than about a hundred nanometers. This length scale could be a particle diameter, grain size, layer thickness, or width of a conducting line on an electronic chip.

Several articles, books, and conference proceedings have covered one or more aspects of nanomaterials [5–17]. The subject of nanoelectronics has been discussed extensively in the proceedings of international symposia on the subject [6–8]. There were three NATO Advanced Study Institutes held between 1991 and 1993 devoted to different aspects of nanomaterials. The 1991 Institute had a broad scope with a particular emphasis on clusters [9]. The 1992 institute focused on the mechanical properties of materials with ultrafine microstructures [10]. The 1993 Institute returned to a broader scope but with very little coverage of clusters [11]. J W Gardner and H T Hingle [12] have considered the instrumentation and technology of controlling manufacturing on a nanoscale. Other selected aspects of nanotechnology were covered in the proceedings from the 1990 Warwick symposium [13] on this subject. The role of interphase boundaries in nanomaterials has been considered [14]. The synthesis and properties of metal cluster compounds was discussed in a book edited by de Jongh [15]. The synthesis and properties of nanoparticles were described in a book by Ichinose, Ozaki, and Kashu [16]. Some of the biological aspects of nanomaterials were discussed in the book by Mann, Webb, and Williams [17]. There is a journal devoted exclusively to nanostructured materials [18] and issues of other periodicals have from time to time focused on various areas within the field. One issue of *Science* [18] surveyed molecular self-assembly and nanochemistry, atomic and molecular manipulation with the scanning tunneling electron microscope, advances in quantum devices, and integrated sensors and microsystems. Other journals have also devoted special issues to areas such as nanotribology [19] and the optical behavior of nanostructures [20].

Despite these works, no previous book has offered the broad coverage necessary to provide a comprehensive introduction to nanomaterials research. Furthermore, because of the recent progress and interdisciplinary nature of the field, we felt there was a need for a book which provided some idea of the breadth and current status of this field. Our intent is that this book should provide access to most of the basic material on the synthesis, properties, and applications of nanomaterials. We believe it should be of particular use to graduate students and researchers in other fields seeking an extensive review of the subject, as well as researchers currently working in nanomaterials who want to learn about areas of the field beyond their current interests. Many references are provided from which the reader may obtain more detailed information. Although an attempt has been made to discuss most aspects of the broad subject of nanomaterials, certain fields are stressed more than others. Since thin films and bilayers are discussed at length elsewhere [21–23] they have not been included. Areas such as catalysis and nanoelectronics that have already been reviewed at some length in the recent past are covered succinctly, while other areas, such as isolated clusters, small particles, multilayers, fullerenes, biological nanomaterials, porous silicon, nanoelectronics, and assemblies of nanocrystals are discussed in greater detail.

It is appropriate to begin with a brief and selective history of the subject of nanomaterials. Nanomaterials are found in both biological systems and man-made structures. Nature has been using nanomaterials for millions of years. As Dickson has noted [24] 'Life itself could be regarded as a nanophase system.' Examples in which nanostructured elements play a vital role are magnetotactic bacteria [25], ferritin [26], and molluscan teeth [27]. Several species of aquatic bacteria use the Earth's magnetic field to orient themselves. They are able to do this because they contain chains of nanosized, single-domain magnetite (Fe_3O_4) particles. Because they have established their orientation, they are able to swim down to nutriments and away from what is lethal to them, oxygen. Another example of nanomaterials in nature is the storage of iron in a bioavailable form within the 8 nm protein cavity of ferritin [28]. Herbivorous mollusks use teeth attached to a tonguelike organ, the radula, to scrape their food. These teeth have a complex structure containing nanocrystalline needles of goethite [29]. We can utilize biological templates for making nanomaterials. Apoferritin has been used as a confined reaction environment for the synthesis of nanosized magnetite particles [30]. Some consider biological nanomaterials as model systems for developing technologically useful nanomaterials.

Scientific work on this subject can be traced back over 100 years. In 1861 the British chemist Thomas Graham coined the term colloid to describe a solution containing 1 to 100 nm diameter particles in suspension. Around the turn of the century, such famous scientists as Rayleigh, Maxwell, and Einstein studied colloids. In 1930 the Langmuir–Blodgett method for developing monolayer films was developed. By 1960 Uyeda had used electron microscopy and diffraction to study individual particles. At about the same time arc, plasma, and chemical flame furnaces were employed to produce submicron particles. Magnetic alloy particles for use in magnetic tapes were produced in 1970. By 1980, studies were made of clusters containing fewer than 100 atoms. In 1985, a team led by Smalley and Kroto found spectroscopic evidence that C_{60} clusters were unusually stable. In 1991, Iijima reported studies of graphitic carbon tube filaments.

Research on nanomaterials has been stimulated by their technological applications. The first technological uses of these materials were as catalysts [31, 32] and pigments [33]. The large surface area to volume ratio increases the chemical activity. Because of this increased activity, there are significant cost advantages in fabricating catalysts from nanomaterials. The properties of some single-phase materials can be improved by preparing them as nanostructures. For example, the sintering temperature can be decreased and the plasticity increased of single-phase, structural ceramics by reducing the grain size to several nanometers [34]. Multiphase nanostructured materials have displayed novel behavior resulting from the small size of the individual phases.

Technologically useful properties of nanomaterials are not limited to their structural, chemical, or mechanical behavior. Multilayers represent examples of materials in which one can modify or tune a property for a specific application by sensitively controlling the individual layer thickness. Examples of this microstructural control resulting in technologically useful materials include the development of multilayer x-ray [35] and neutron mirrors [36] as well as semiconductor superlattice devices [37]. It was discovered that the resistance of Fe–Cr multilayered thin films exhibited large changes in an applied magnetic field of several tens of kOe [38]. This effect was given the name giant magnetoresistance (GMR). More recently, suitably annealed magnetic multilayers have been developed that exhibit significant magnetoresistance effects even in fields as low as 5 to 10 Oe [39]. This effect may prove to be of great technological importance for use in magnetic recording read heads.

Confining electrons to small geometries gives rise to 'particle in a box' energy levels. This quantum confinement creates new energy states and results in the modification of the optoelectronic properties of semiconductors. In microelectronics, the need for faster switching times and ever larger integration has motivated considerable effort to reduce the size of electronic components. Increasing the component density increases the difficulty of satisfying cooling requirements and reduces the allowable amount of energy released on switching between states. It would be ideal if the switching occurred with the motion of a single electron. One kind of single-electron device is based on the change in the Coulombic energy when an electron is added or removed from a particle. For a nanoparticle this energy change can be large enough that adding a single electron effectively blocks the flow of other electrons. The use of Coulombic repulsion in this way is called Coulomb blockade.

In addition to technology, nanomaterials are also interesting systems for basic scientific investigations. For example, small particles display deviations from bulk solid behavior such as reductions in the melting temperature [40] and changes (usually reductions) in the lattice parameter [41]. The changes in the lattice parameter observed for metal and semiconductor particles result from the effect of the surface stress, while the reduction in melting temperature results from the effect of the surface free energy. Both the surface stress and surface free energy are caused by the reduced coordination of the surface atoms. By studying the size dependence of the properties of particles, it is possible to find the critical length scales at which particles behave essentially as bulk matter. Generally, the physical properties of a nanoparticle approach bulk values for particles containing more than a few hundred atoms.

Research on nanomaterials has led to the formation of new research fields, such as the one that developed from the discovery of the fullerene molecule C_{60} and of methods for producing fullerenes in large quantities. One highlight of this research was the discovery of superconductivity [42, 43] in K_xC_{60} and Rb_xC_{60}. Graphite carbon tubules have also been produced [44], and in later work, manganese-filled carbon tubules were grown [45]. In addition to the fullerenes, some metallo-carbohedrenes, M_8C_{12} (M = V, Zr, Hf, and Ti), called met-cars, have been discovered that also form stable clusters [46]. Studies have been made of clusters composed of fullerenes, i.e. clusters in which the individual building blocks are fullerenes. It was found [47] that clusters containing certain numbers of fullerenes (C_{60} or C_{70}) are more stable than others. This is similar to the case of atomic clusters where it was found [48] that clusters composed of certain numbers of atoms were more stable that others. The set of numbers for which clusters have greater stability is given the name 'magic numbers' . These magic numbers result from the increased stability of closed-shell electronic configurations for alkali metals or from geometric effects in rare gas clusters. Geometric effects are also responsible for the magic numbers in the case of fullerene clusters.

New techniques have been developed recently that have permitted researchers to produce larger quantities of other nanomaterials and to better characterize these materials. Each fabrication technique has its own set of advantages and disadvantages. Generally it is best to produce nanoparticles with a narrow size distribution. In this regard, free jet expansion techniques permit the study of very small clusters, all containing the same number of atoms. It has the disadvantage of only producing a limited quantity of material. Another approach involves the production of pellets of nanostructured materials by first nucleating and growing nanoparticles in a supersaturated vapor and then using a cold finger to collect the nanoparticles [49]. The nanoparticles are then consolidated under vacuum. Chemical techniques are very versatile in that they can be applied to nearly all materials (ceramics, semiconductors, and metals) and can usually produce a large amount of material. A difficulty with chemical processing is the need to find the proper chemical reactions and processing conditions for each material. Mechanical attrition, which can also produce a large amount of material, often makes less pure material. One problem common to all of these techniques is that nanoparticles often form micron-sized agglomerates. If this occurs, the properties of the material may be determined by the size of the agglomerate and not the size of the individual nanoparticles. For example, the size of the agglomerates may determine the void size in the consolidated nanostructured material.

The ability to characterize nanomaterials has been increased greatly by the invention of the scanning tunneling microscope (STM) and other proximal probes such as the atomic force microscope (AFM), the magnetic force microscope, and the optical near-field microscope. Near-field scanning optical microscopes overcome the limitation of conventional optics by making use of the evanescent field that exists within the first several nanometers of the surface of light pipes, and can have a resolution of 12 nm using 500 nm wavelength light [50]. STMs and AFMs [51] provide depth resolution and the AFM can be used as a very low-load mechanical properties microprobe. As mentioned earlier, the STM has been used to carefully place atoms on surfaces to write bits using a small number of atoms. It has also been employed to construct a circular arrangement of metal atoms on an insulating surface. Since electrons are confined to the circular path of metal atoms, it serves as a quantum 'corral' of atoms. This quantum corral was employed to measure the local electronic density of states of these circular metallic arrangements. By doing this, researchers were able to verify the quantum mechanical description of electrons confined in this way [52, 53].

Other new instruments and improvements of existing instruments are increasingly becoming important tools for characterizing surfaces of films, biological materials, and nanomaterials. The development of nanoindentors and the improved ability to interpret results from nanoindentation measurements have increased our ability to study the mechanical properties of nanostructured materials [54]. Improved high-resolution electron microscopes and modeling of the electron microscope images have improved our knowledge of the structure of the particles and the interphase region between particles in consolidated nanomaterials.

A variety of computational methods have been used to perform theoretical studies of the properties of nanomaterials. First-principles electronic structure calculations can be performed on clusters containing, at most, several hundred atoms. Treating larger clusters by first-principles calculations becomes very difficult because the time required generally increases as N^3 where N is the number of atoms in the cluster. For non-metals there are new, approximate methods for which the time required only increases as N or $N \log N$. Because the time required for first-principles calculations becomes prohibitively large, calculations on clusters containing more than several hundred atoms are usually performed using approximate methods, e.g. using embedded atom method potentials [55–58].

The reader should be aware that different points of view have been expressed on several matters, such as the atomic configuration in grain boundaries and the mechanism of photoluminescence in nanostructured silicon.

A brief outline of the subjects covered in the book is as follows. Part 2 begins with a review of classical nucleation theory, an understanding of which aids the subsequent discussion of the synthesis of nanomaterials. Several synthesis techniques for making nanomaterials are then described, including the production of nanomaterials from a supersaturated vapor, by chemical techniques, and by mechanical attrition. The fabrication and properties of multilayered materials are presented in part 3. Part 4 deals with the processing of nanostructured sol–gel materials and consolidation by compaction and sintering. Special attention is given to the problem of agglomeration. Techniques for microstructural and interfacial characterization are presented in part 5. The techniques discussed include x-ray diffraction, extended x-ray absorption fine structure, high-resolution electron microscopy, STM, AFM, and Raman spectroscopy, as well as density, porosity, diffusivity, and positron annihilation measurements. Part 6 discusses the chemical, electrical transport, optical, magnetic, and mechanical properties of nanomaterials. Some examples of special nanomaterials are described in part 7. The examples chosen are biological nanomaterials, fullerenes and carbon tubules, and porous silicon. Part 8 provides an introduction to nanoelectronics and nanofabrication of electronics. In part 9, several authors give their view on the development of different areas of nanotechnology and their assessment of the likely technological impact of these areas. The areas covered are electronic components, memories, special structural components, ceramics, adhesives, catalysts, magnetic recording, quantum dots, and sensors.

REFERENCES

[1] Feynman R P 1960 *Eng. Sci.* **23** 22; reprinted in 1992 *J. Micromech. Systems* **1** 60

[2] Eigler D M and Schweizer E K 1990 *Nature* **344** 524

[3] Drexler K E 1992 *Nanosystems: Molecular Machinery, Manufacturing, and Computation* (New York: Wiley)

[4] Krummenacker M and Lewis J (ed) 1995 Prospects in nanotechnology *Proc. 1st Gen. Conf. on Nanotechnology: Development, Applications, and Opportunities* (New York: Wiley)

[5] 1991 *Science* **254** November 29

[6] Drexler K E, Terson C and Pergamit G 1991 *Unbounding the Future: The Nanotechnology Revolution* (New York: Morrow)

[7] Reed M A and Kirk W P (ed) 1989 *Nanostructure Physics and Fabrication* (New York: Academic)

[8] Kirk W P and Reed M A (ed) 1992 *Nanostructures and Mesoscopic Systems* (Boston, MA: Academic)

[9] Jena P, Khanna S N and Rao B K (ed) 1992 *Physics and Chemistry of Finite Systems: From Clusters to Crystals (NATO ASI Series C 374)* (Dordrecht: Kluwer) (in two volumes)

[10] Natasi M, Parkin D M and Gleiter H (ed) 1993 *Mechanical Properties and Deformation Behavior of Materials Having Ulta-Fine Microstructures (NATO ASI Series E 233)* (Dordrecht: Kluwer)

[11] Hadjipanayis G C and Siegel R W (ed) 1994 *Nanostructured Materials: Synthesis, Properties, and Applications (NATO ASI Series E 260)* (Dordrecht: Kluwer)

[12] Gardner J W and Hingle H T (ed) 1991 *From Instrumentation to Nanotechnology* (Philadelphia, PA: Gordon and Breach)

[13] Whitehouse D J and Kawata K (ed) 1991 *Nanotechnology Proc. Joint Forum/ERATO Symp.* (Bristol: Hilger)

[14] Van Aken D C, Was G S and Ghosh A K (ed) 1991 *Microcomposites and Nanophase Materials* (Warrendale, PA: The Minerals, Metals and Materials Society)

[15] de Jongh L J (ed) 1994 *Physics and Chemistry of Metal Cluster Compounds* (Dordrecht: Kluwer)

[16] Ichinose N, Ozaki J and Kashu S 1992 *Superfine Particle Technology* (New York: Springer) Engl. Transl. by M James

[17] Mann S, Webb J and Williams R J P (ed) 1989 *Biomineralization* (New York: VCH)

[18] Kear B H, Siegel R W and Tsakalakos T (ed) *Journal of Nanostructured Materials* (New York: Pergamon)

[19] 1993 *Mater. Res. Soc. Bull.* **8** No 5

[20] 1993 *Phys. Today* **46** No 6

[21] Bunshah R F (ed) 1982 *Deposition Technologies for Films and Coatings* (Park Ridge, NJ: Noyes)

[22] Clemens B M and Johnson W L (ed) 1990 *Thin Film Structures and Phase Stability, Proc. Mater. Res. Soc. Symp.* vol 187 (Pittsburgh, PA: Materials Research Society)

[23] Thompson C V, Tsao J Y and Srolovitz D J (ed) 1991 *Evolution of Thin-Film and Surface Microstructure, Proc. Mater. Res. Soc. Symp.* vol 202 (Pittsburgh, PA: Materials Research Society)

[24] Dickson D P E page 459 of this volume.

[25] Blakemore R P, Frankel R B and Kalmijn A J 1980 *Nature* **286** 384

[26] Theil E C 1987 *Annu. Rev. Biochem.* **56** 289

[27] Lowenstam H A 1962 *Science* **137** 279

[28] Wade V J, Levi S, Arosio P, Treffry A, Harrison P M and Mann S 1991 *J. Mol. Biol.* **221** 1443

[29] Lawson D M, Treffry A, Artymiuk P J, Yewdall S J, Luzzago A, Cesareni G, Levi S and Arosio P 1989 *FEBS Lett.* **254** 207

[30] Meldrum F C, Heywood B R and Mann S 1992 *Science* **257** 522
[31] Ponec V 1975 *Catal. Rev. Sci. Eng.* **11** 41
[32] Sinfelt J H 1983 *Bimetallic Catalysts* (New York: Wiley)
[33] Martens C R (ed) 1968 *Technology of Paints, Varnishes and Lacquers* (New York: Reinhold) p 335
[34] Siegel R W, Ramasamy S, Hahn H, Gronsky R, Li Z Q and Lu T 1988 *J. Mater. Res.* **3** 1367
[35] Barbee T W Jr 1985 *Synthetic Modulated Structures* ed L L Chang and B C Giessen (Orlando, FL: Academic) p 313
[36] Mezei F 1988 *Physics, Fabrication, and Applications of Multilayered Structures (NATO ASI Series B)* vol 182, ed P Dhez and C Weisbuch (New York: Plenum) p 311
[37] Capasso F 1987 *Science* **235** 172
[38] Baibich M N, Broto J M, Fert A, Nguyen Van Dau F, Petroff F, Eitenne P, Creuzet B, Friederich A and Chazelas J 1988 *Phys. Rev. Lett.* **61** 2472
[39] Hylton T L, Coffey K R, Parker M A and Howard J K 1993 *Science* **261** 1021
[40] Borel J-P 1981 *Surf. Sci.* **106** 1
[41] Woltersdorf J, Nepijko A S and Pippel E 1981 *Surf. Sci.* **106** 64
[42] Hebard A F, Rosseinsky M J, Haddon R C, Murphy D W, Glarum S H, Palstra T T M, Ramirez A P and Kortan A R 1991 *Nature* **350** 600
[43] Rosseinsky M J, Ramirez A P, Glarum S H, Murphy D W, Haddon R C, Hebard A F, Palstra T T M, Kortan A R, Zahurak S M and Makhija A V 1991 *Phys. Rev. Lett.* **66** 2830
[44] Iijima S 1991 *Nature* **354** 56
[45] Ajayan P M, Colliex C, Lambert J M, Bernier P, Barbedette L, Tencé M and Stephan O 1994 *Phys. Rev. Lett.* **72** 1722
[46] Guo B C, Wei S, Purnell J, Buzza S and Castleman A W Jr 1992 *Science* **256** 515
[47] Martin T P, Näher U, Schaber H and Zimmermann U 1993 *Phys. Rev. Lett.* **70** 3079
[48] Knight W D, Clemenger K, de Heer W A, Saunders W A, Chou M Y and Cohen M L 1984 *Phys. Rev. Lett.* **52** 2141
[49] Birringer R and Gleiter H 1988 *Encylopedia of Materials Science and Engineering* suppl. vol 1, 1988 ed R W Cahn (Oxford: Pergamon) p 339
[50] Cassidy R 1993 *Res. Disc. Mag.* March 27
[51] Gratz A J, Manne S and Hansma P K 1991 *Science* **251** 1343
[52] Crommie M F, Lutz C P and Eigler D M 1993 *Science* **262** 218
[53] 1993 *Phys. Today* **46** No 11 17
[54] Wu T W 1991 *Mater. Res. Soc. Symp. Proc.* vol 226 (Pittsburgh, PA: Materials Research Society) p 165
[55] Daw M S, Foiles S M and Maskes M I 1993 *Mater. Sci. Rep.* **9** 251
[56] Gilmore C M 1989 *Phys. Rev.* B **40** 6402
[57] Sachdev A, Masel R I and Adams J B 1992 *J. Catal.* **136** 320
[58] Sachdev A and Masel R I 1993 *J. Mater. Res.* **8** 455

PART 2

SYNTHESIS

Chapter 2

Formation of clusters and nanoparticles from a supersaturated vapor and selected properties

M Samy El-Shall and A S Edelstein

2.1 INTRODUCTION

This chapter focuses on the physical methods commonly used to generate clusters and nanoparticles from a supersaturated vapor. The sizes of the nanoparticles made by these techniques cover the entire range from dimers to nanoparticles which are 100 nm in diameter. The term cluster, as used here, is mainly reserved for smaller nanoparticles containing fewer than 10^4 atoms or molecules.

The chapter consists of two major parts. The first is a discussion of the classical theory of nucleation and the methods for producing clusters by supersonic beam expansions, laser vaporization, and laser photolysis of organic compounds. The concepts of classical nucleation theory are discussed in some detail because they apply to many of the other synthesis techniques for producing nanomaterials. For example, classical nucleation theory can be applied to describe the synthesis of nanoparticles in liquids. Following this, some of the properties of clusters are reviewed. The second part, section 2.3, presents methods used for forming larger nanoparticles by sputtering, thermal evaporation, and laser methods. Section 2.3 also discusses particle coalescence and particle transport and collection. It concludes by describing the crystal structure, crystal habit, and size distribution of the nanoparticles.

2.2 CLUSTERS

2.2.1 Classical nucleation theory for cluster formation

In discussing clusters and nanoparticle formation, it is instructive to provide an abbreviated account of the nucleation process. Nucleation of new particles from a continuous phase can occur heterogeneously or homogeneously. Heterogeneous nucleation from a vapor phase can occur on foreign nuclei or dust particles, ions or surfaces. Homogeneous nucleation occurs in the absence of any foreign particles or ions when the vapor molecules condense to form embryonic droplets or nuclei. There are several reviews of the theory of nucleation from vapor [1–4], which was developed by Volmer, Becker and Döring and modified by Frenkel and Zeldovich [5, 6]. This theory is based on the assumption (known as the capillarity approximation) that embryonic clusters of the new phase can be described as spherical liquid drops with the bulk liquid density inside and the vapor density outside. The free energy of these clusters, relative to the vapor, is the sum of two terms: a positive contribution from the surface free energy and a negative contribution from the bulk free energy difference between the supersaturated vapor and the liquid. The surface free energy results from the reversible work in forming the interface between the liquid drop and the vapor. For a cluster containing n atoms or molecules, the interface energy is given by

$$\sigma A(n) = 4\pi\sigma(3v/4\pi)^{2/3}n^{2/3} \tag{2.1}$$

where σ is the interfacial tension or surface energy per unit area, $A(n)$ is the surface area of the cluster, and v is the volume per molecule in the bulk liquid. Since n molecules are transferred from the vapor to the cluster, the bulk contribution to the free energy of formation is $n(\mu_l - \mu_v)$ where μ_l and μ_v are the chemical potentials per molecule in the bulk liquid and vapor, respectively. Assuming an 'ideal' vapor it can be shown [7] that

$$(\mu_l - \mu_v)n = -nk_B T \ln S \tag{2.2}$$

where k_B is the Boltzmann constant, T is the temperature, and S, the supersaturation, is

$$S = P/P_e. \tag{2.3}$$

In equation (2.3), P is the vapor pressure and P_e is the equilibrium or 'saturation' vapor pressure at the temperature of the vapor. The sum of the contributions in equations (2.1) and (2.2) is the reversible work (free energy), $W(n)$, done in forming a cluster containing n atoms or molecules. This work is given by

$$W(n) = -nk_B T \ln S + 4\pi\sigma\left(\frac{3v}{4\pi}\right)^{2/3} n^{2/3}. \tag{2.4}$$

Equation (2.4) expresses the competition between 'bulk' and 'surface' free energy terms in determining the cluster stability and cluster concentration in the supersaturated vapor. Because of the positive interface contribution in equation (2.4), there is a free energy barrier which impedes nucleation. The smallest cluster of size n^* which can grow with a decrease in free energy is determined from the condition $\partial W/\partial n = 0$. It follows that

$$n^* = 32\pi\sigma^3 v^2/3(k_B T \ln S)^3 \tag{2.5}$$

$$r^* = 2\sigma v/(kT \ln S). \tag{2.6}$$

Substituting n^* into equation (2.4) yields the barrier height $W(n^*)$, given by

$$W(n^*) = 16\pi\sigma^3 v^2/3(k_B T \ln S)^2. \tag{2.7}$$

For $S > 1$, increasing S reduces the barrier height $W(n^*)$ and the critical size n^*, and increases the probability that fluctuations will allow some clusters to grow large enough to overcome the barrier and grow into stable droplets.

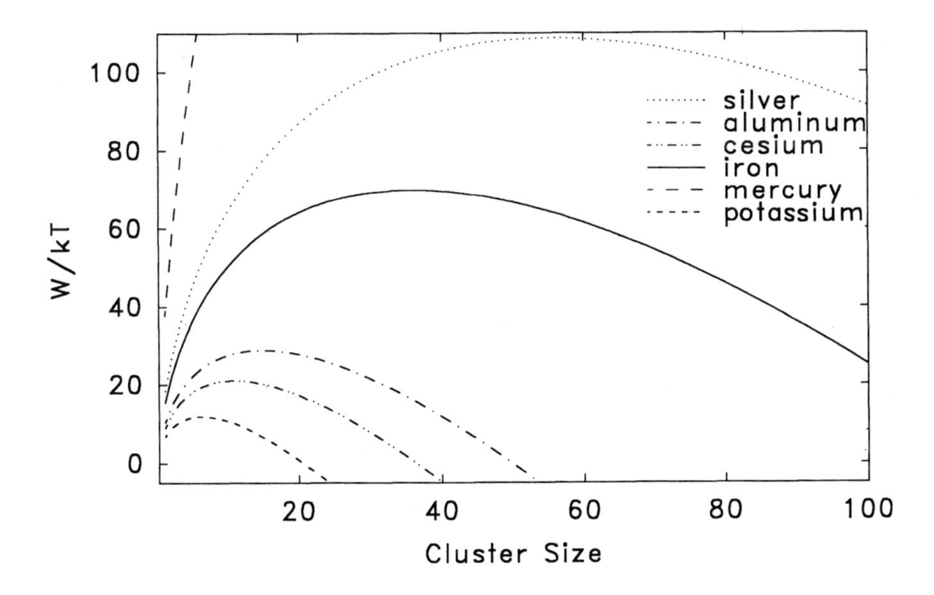

Figure 2.1. Free energy of formation as a function of size for several metals for a supersaturation ratio $S = 50$. Note there is an energy barrier that must be overcome in order to form a stable cluster.

Figure 2.1 shows plots of the dimensionless free energy of formation of different metal clusters as a function of size. The calculations were performed using equation (2.4) with $P_e = 0.01$ Torr and $P = 0.5$ Torr. For the metals Cs, K, Al, Ag, Fe, and Hg, the temperatures at which $P_e = 0.01$ Torr are 424, 464, 1472, 1262, 1678, and 328 K respectively. Literature values of the bulk surface tension and density were used in the calculations [8]. Note that S can be increased either by increasing P or decreasing P_e. The pressure P can be increased by increasing the rate at which atoms are placed in the vapor or decreasing the rate at which they leave the region where the particle nucleation and growth is occurring. The pressure P_e can be decreased by decreasing T since P_e is approximately given by

$$P_e = P_0 \, e^{L(0)/RT} \tag{2.8}$$

where the latent heat per mole has been approximated by its zero-temperature value $L(0)$, P_0 is a constant, and R is the gas constant.

The rate of homogeneous nucleation J, defined as the number of drops nucleated per cubic centimeter per second, is given by

$$J = K \exp[-W(n^*)/k_B T]. \tag{2.9}$$

The factor K incorporates both the effective collision rate of vapor molecules with a nucleus of size n^* and the departure of the cluster distribution from equilibrium [9, 10]. A critical supersaturation, S_c, can be defined as the supersaturation at which $J = 1\ \mathrm{cm}^{-3}\ \mathrm{s}^{-1}$. Setting $J = 1\ \mathrm{cm}^{-3}\ \mathrm{s}^{-1}$ in equation (2.9), S_c can be obtained using macroscopic values for the surface tension, liquid density, and P_e. Figure 2.2 plots S_c versus T for several metallic vapors. At lower temperatures, the critical supersaturations required for fixed nucleation rates are higher and the critical nuclei are smaller. However, one sees from equation (2.8) that higher values of S_c are obtained more easily at low temperatures.

Aside from the approximations made and the subtleties not considered in the classical theory [11–17] the theory is expected to fail in the case of high supersaturation. For high supersaturation, the change of state of the gas is faster than the time required to establish a local metastable equilibrium. Thus, the steady state nucleation theory is not applicable [18–21]. Further, since the nucleus could be less than ten atoms in the case of very high supersaturation, it is unreasonable to treat the nucleus as a macroscopic entity with macroscopic properties such as surface tension and density. Another problem in applying the nucleation theory arises when the clusters are crystalline. Although, in principle, the capillarity approximation can be used when the clusters are crystalline, the bulk properties, such as the surface tension at the relevant temperatures, are usually unavailable [22, 23].

2.2.2 Techniques for cluster formation

The generation of clusters or larger nanoparticles from the vapor phase requires achieving supersaturation by one of the following methods: (1) the physical cooling of a vapor by sonic or supersonic expansion techniques; (2) a gas phase chemical or photochemical reaction that can produce nonvolatile condensable

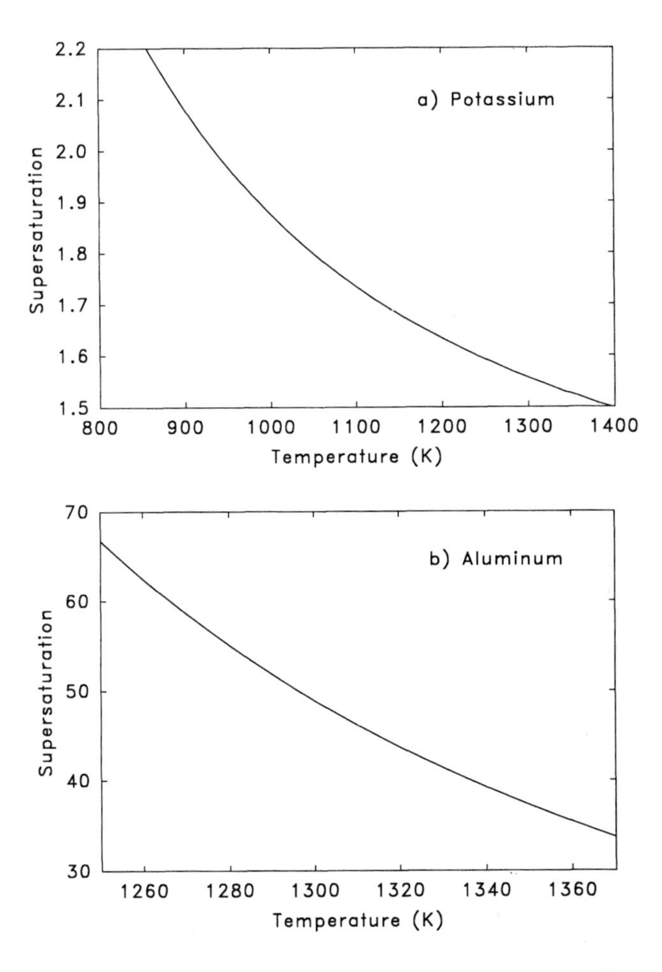

Figure 2.2. The critical supersaturation S_c versus T for (*a*) potassium and (*b*) aluminum metallic vapors.

products; (3) directly by thermal evaporation, sputtering, or laser ablation. Producing nanoparticles by thermal evaporation, sputtering, or laser methods is discussed in section 2.3.

In vapor expansion methods, the expansion takes place from a high-pressure gas source into a low-pressure ambient background. Our intent is not to present a review of these techniques which can be found in several sources [24–33] but to explain the mechanisms involved and the techniques utilized in producing clusters.

2.2.2.1 *Free jet expansion and scaling laws*

In free jet expansion, the vapor, mixed with an inert carrier gas (usually He or Ar), at a total pressure P_0 and a temperature T_0 is adiabatically expanded through a nozzle or orifice of diameter d into an ambient background pressure P_1. The gas starts from a negligibly small velocity defined by the stagnation state (P_0, T_0) and, due to the pressure difference $(P_0 - P_1)$, accelerates toward the source exit. The flow can achieve supersonic speeds. The vapor initially expands isentropically from the nozzle where it is continuum flow or collision dominated, to a region downstream where the flow becomes free molecular or collisionless and is no longer isentropic. The vapor cools during the expansion, crosses the gas/liquid coexistence line and becomes supersaturated. The density of clusters formed depends on the degree of supersaturation. Decreasing the nozzle diameter decreases the transit time through the free jet to the collisionless regime and produces smaller clusters. Typically decreasing d has a large effect on increasing the cooling rate. The residence time of the gaseous molecules in the collision zone is generally in the range of 0.1 to 100 μs. During this time, the temperature drops by about 10^2 K. Therefore, typical cooling rates in free jet expansions are 10^6–10^9 K s^{-1}. In general, the extent of clustering and the average cluster size increases with increasing the stagnation pressure and the aperture cross section, and with decreasing T_0.

Pulsed beam sources, reviewed by Gentry [34], can be used to produce a collimated beam of clusters. A conical aperture, called a skimmer, with a large diameter allows the center of an already formed jet to pass into a second vacuum chamber to form a practically collisionless flow field. By varying the nozzle to skimmer distance x, P_0, T_0, and d, it is possible to vary the cluster size from dimers and trimers to clusters containing 10^4 atoms or molecules per cluster. Typical experimental parameters are: P = 5–200 Torr diluted in an inert gas at a pressure for P_0 of 1–20 atm at a temperature T_0 of 80–1000 K. The background pressure is 10^{-3}–10^{-7} Torr. For substances that are liquids at ambient temperatures, the expansion can start with a saturated vapor by maintaining the stagnation mixture in contact with an ambient liquid.

Because of the absence of an exact theory for cluster nucleation and growth, scaling laws are useful for determining the onset conditions for cluster formation [35]. As indicated earlier, cluster formation and growth are favored by decreasing T_0 and increasing d. Hagena [36–38] introduced a reduced scaling parameter ζ^* defined as

$$\zeta^* = (r_{ch})^{3-q}(T_{ch})^{1.5-0.25q} \tag{2.10}$$

where q $(0.5 < q < 1)$ is a parameter determined experimentally from cluster beam measurements in which the nozzle diameter is varied at constant T_0, $r_{ch} = (m/\rho)^{1/3}$, $T_{ch} = \Delta H^0/k_B$, m is the atomic mass, ρ is the density of the solid, and ΔH_0^0 is the sublimation enthalpy per atom at 0 K. Empirically it has been found that for:

$\zeta^* < 200$	no clustering observed
$200 < \zeta^* < 1000$	clustering occurs
$1000 < \zeta^*$	massive condensation occurs with the cluster size > 100 atoms/cluster.

2.2.2.2 *Metal clusters from oven sources*

For low boiling point materials, an oven source can be used to produce a large vapor pressure. The sample is mixed with an inert carrier gas and then expanded into vacuum. The carrier gas serves as a heat sink for the seed and absorbs most of the heat of formation of the clusters. Because the heavier carrier gases are more efficient in removing the heat of condensation from the clusters, they are usually used to produce larger clusters [39]. Figure 2.3 illustrates the experimental apparatus for generating metal clusters from an oven source [40]. Figure 2.4 shows abundance spectra of potassium clusters produced in a supersonic jet beam from an oven source [41]. One sees that clusters containing eight, 20, 40, 58, and 92 potassium atoms are more prevalent. These values of n are called magic numbers and will be discussed later. Other examples of metal clusters produced from an oven source are reported in the literature [41, 42]. Figure 2.5 shows the antimony cluster distribution generated from an oven source ($T = 920$ K) and detected using 35 eV electron impact ionization [43]. It is interesting that for clusters with $N > 8$, only multiples of Sb_4 were observed suggesting that aggregation of Sb_4 was the main mechanism involved in the cluster growth at $T = 920$ K. Martin and coworkers used an inert gas condensation cell to generate large clusters containing up to 3000 Mg atoms [44, 45]. The time of flight (TOF) mass spectrum for Mg clusters is shown in figure 2.6. An oven source has also been used to generate mixed clusters such as PbS and PbSe [46], PS and AsS [47], PbAs [48], PbSb and BiSb [49], PbP, PbSr and PbIn [50], InP, GaAs and GaP [51, 52].

Although the free jet expansion using oven sources is a simple and reliable method of producing a wide range of cluster sizes, it is less useful for materials with high boiling points. For substances with high cohesive energies (e.g. refractory elements), very high temperatures are needed. Materials with cohesive energies of 0.5 to 1.5 eV per atom require about 100 Torr of atomic vapor to generate clusters in significant amounts in a continuous nonseeded expansion [53]. Also, the free-jet expansion technique has technical problems because of clogging, reactivity of the metal vapors with oven materials, and the need for large quantities of starting material.

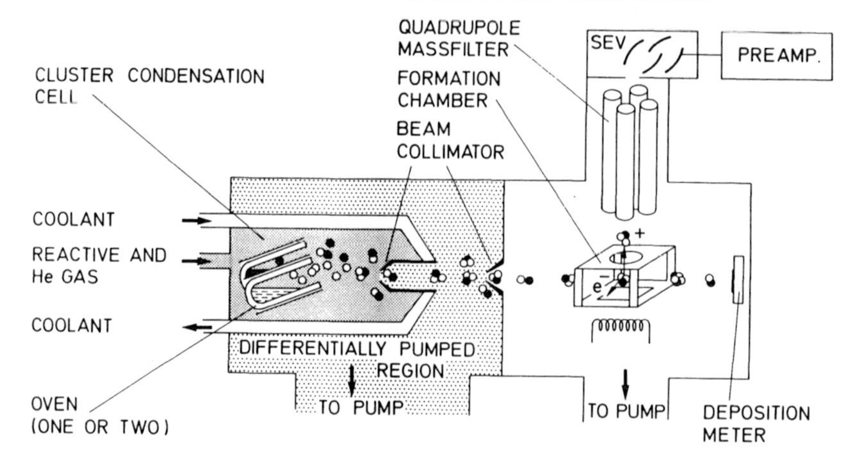

Figure 2.3. Experimental set-up for cluster generation from an oven source. (Reprinted from [40].)

2.2.2.3 Laser vaporization cluster beams

Laser vaporization cluster beams were introduced by Smalley and coworkers to overcome the limitations of oven sources [54–59]. In this method, a high-energy pulsed laser with an intensity flux exceeding 10^7 W cm^{-3} is focused on a target containing the material to be made into clusters. The resulting plasma causes highly efficient vaporization since with current, pulsed lasers one can easily generate temperatures at the target material greater than 10^4 K. This high temperature vaporizes all known substances so quickly that the rest of the source can operate at room temperature. Typical yields are 10^{14}–10^{15} atoms from a surface area of 0.01 cm^2 in a 10^{-8} s pulse. The local atomic vapor density can exceed 10^{18} atom cm^{-3} (equivalent to 100 Torr pressure) in the microseconds following the laser pulse. The hot metal vapor is entrained in a pulsed flow of a carrier gas (typically He) and expanded through a nozzle into a vacuum. The cool, high-density helium flowing over the target serves as a buffer gas in which clusters of the target material form, thermalize to near room temperature and then cool to a few K in the subsequent supersonic expansion. A typical laser vaporization source is shown in figure 2.7 and an example of TOF mass spectra measured for aluminum clusters [60] is shown in figure 2.8. Other studies of main group and transition metal clusters [60–69], mixed metal clusters and compound clusters [70, 71] have been performed using laser vaporization.

2.2.2.4 Laser photolysis of organometallic compounds

Photodissociation of organometallic precursor molecules can produce a complex pre-expansion mixture of metal atoms, ions, and unsaturated fragments [72]. During typical laser pulses of 10–20 ns of UV light, many organometallic

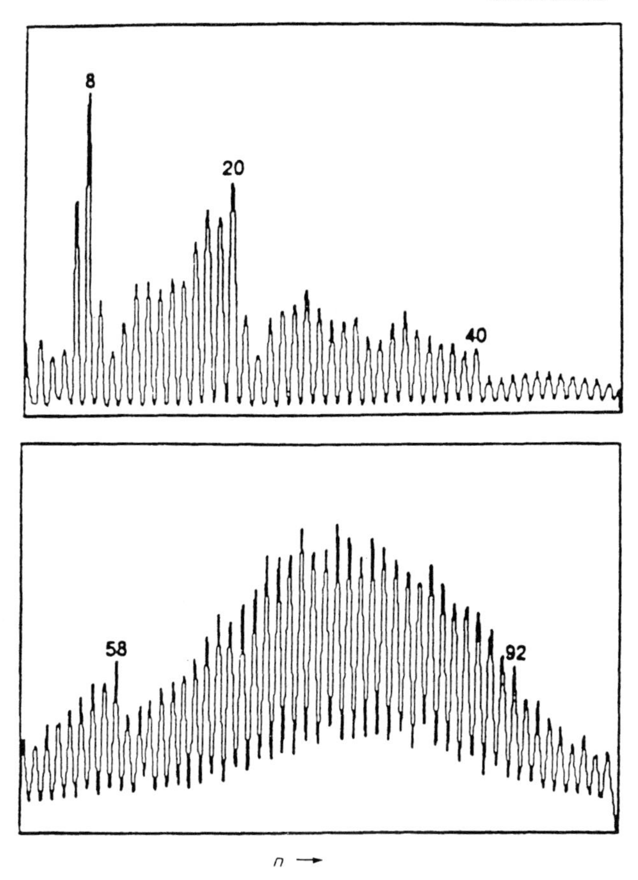

Figure 2.4. Mass spectra showing the abundance of potassium clusters from [41] produced in a supersonic beam with an oven source as a function of cluster size n.

compounds are reduced to their metal atoms. Immediately after the laser pulse, supersaturation ratios of refractory metal atoms of 10^6 or greater can exist. Laser photodissociation of organometallic precursor molecules was first observed by Tam and coworkers [73] in 1974 and was applied to generate metal clusters and ultrafine particles of ceramic and other materials [74–76]. Smalley and coworkers [77] showed that iron clusters, with $n = 1$–30, can be generated by laser photolysis of $Fe(CO)_5$ clusters using an excimer laser. The combination of laser vaporization and supersonic expansion of a metal carbonyl seeded carrier gas has been used to generate mixed metal clusters [78]. Also, the use of different hydrocarbons during the laser vaporization of metals results in the formation of metal–carbon clusters of different composition [79].

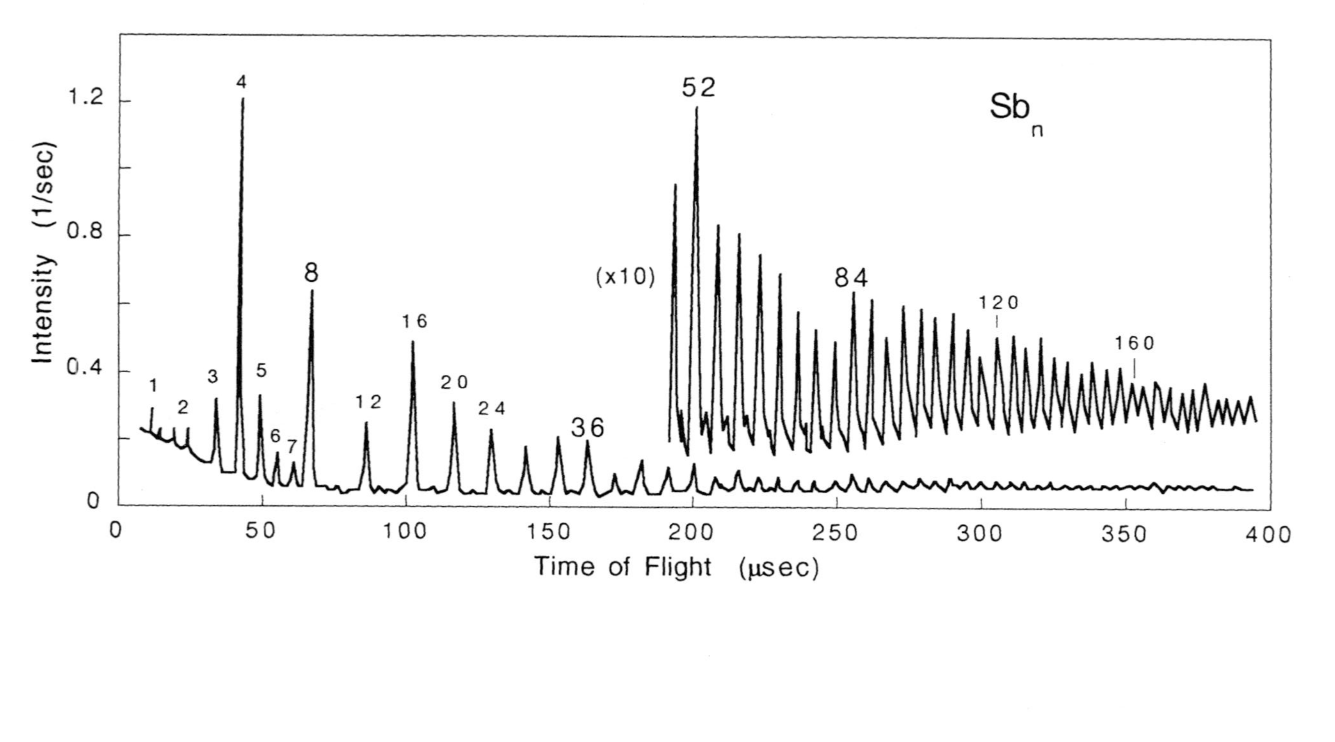

Figure 2.5. Mass spectra from [43] of antimony clusters produced in a supersonic beam with an oven source.

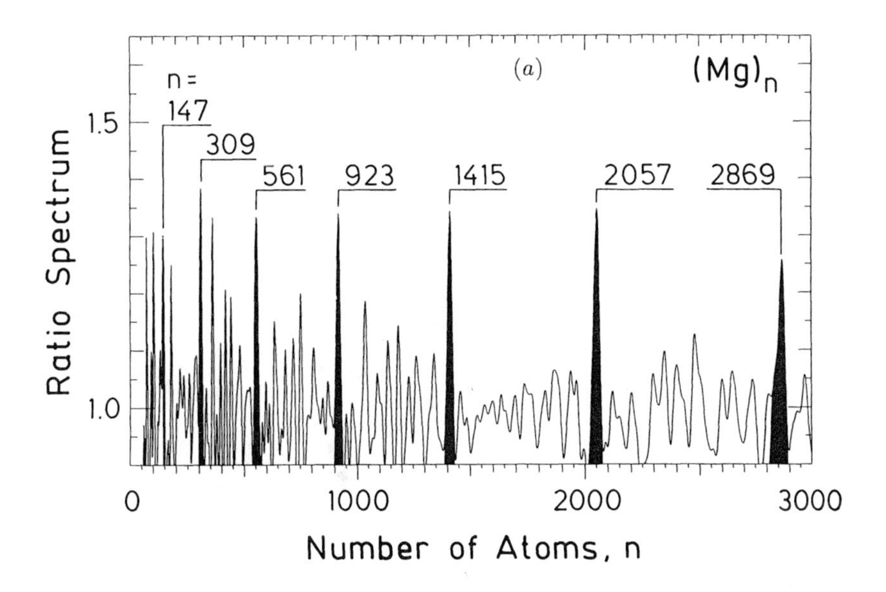

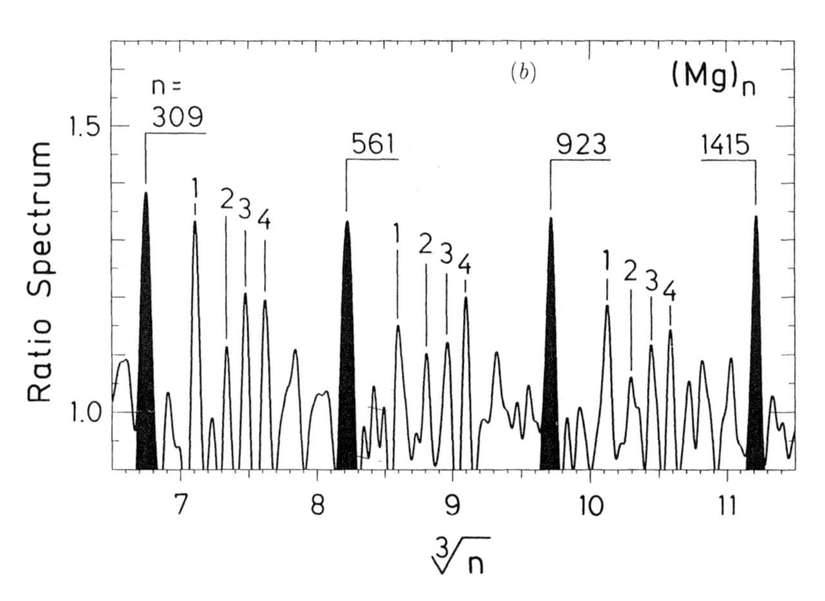

Figure 2.6. Mass spectra of large magnesium clusters from [45] produced in a supersonic beam with an oven source.

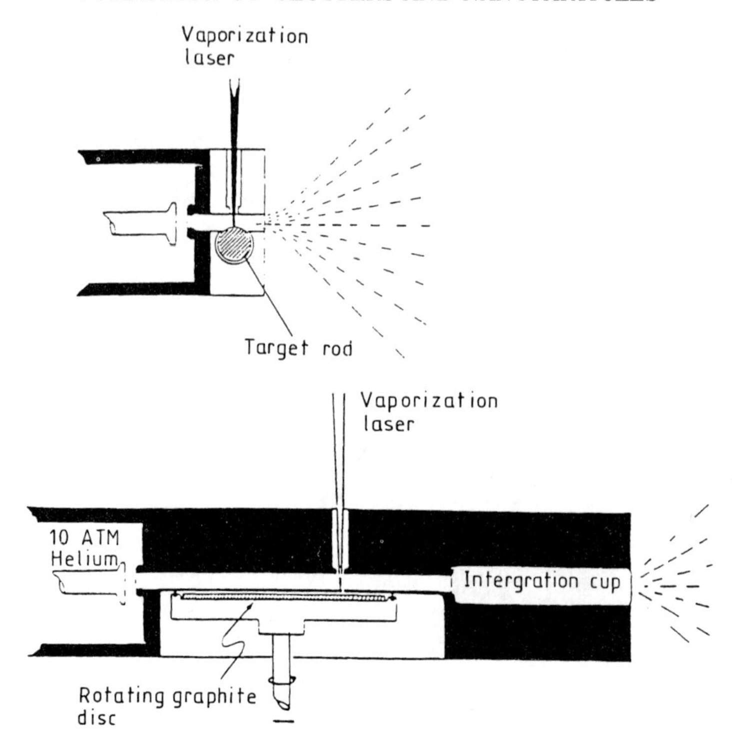

Figure 2.7. Typical laser vaporization source for the generation of metal clusters. (Reprinted from [216].)

2.2.3 Cluster assembled materials

Cluster beams can be used to deposit material on substrates. Cluster beam deposition falls into two classes: (1) depositions in which the clusters are ionized and accelerated so that they have keV of energy before reaching the substrate; (2) deposition of low-energy neutral cluster beams. If high-energy cluster beams are used, the clusters fragment on the substrate, thus creating additional nucleation sites and enhancing diffusion. Using a high-energy cluster beam, Yamada [80] was able to obtain an epitaxial coating of Al on Si (111) surfaces. If low-energy cluster beams are used, then the clusters do not fragment on striking the substrate and nanostructured thin films are produced.

Though clusters of low-energy beams do not fragment, they can diffuse on the surface. If two clusters meet by diffusion, they can 'touch' but still remain separate entities. The other possibility is that they fuse into a larger cluster. By fusing they decrease their surface free energy. There appears to be an energy barrier for coalescence which increases with cluster size. Thus, if both of the clusters are large, no coalescence will occur. As discussed below, metallic clusters containing approximately 1000 atoms may have reduced melting

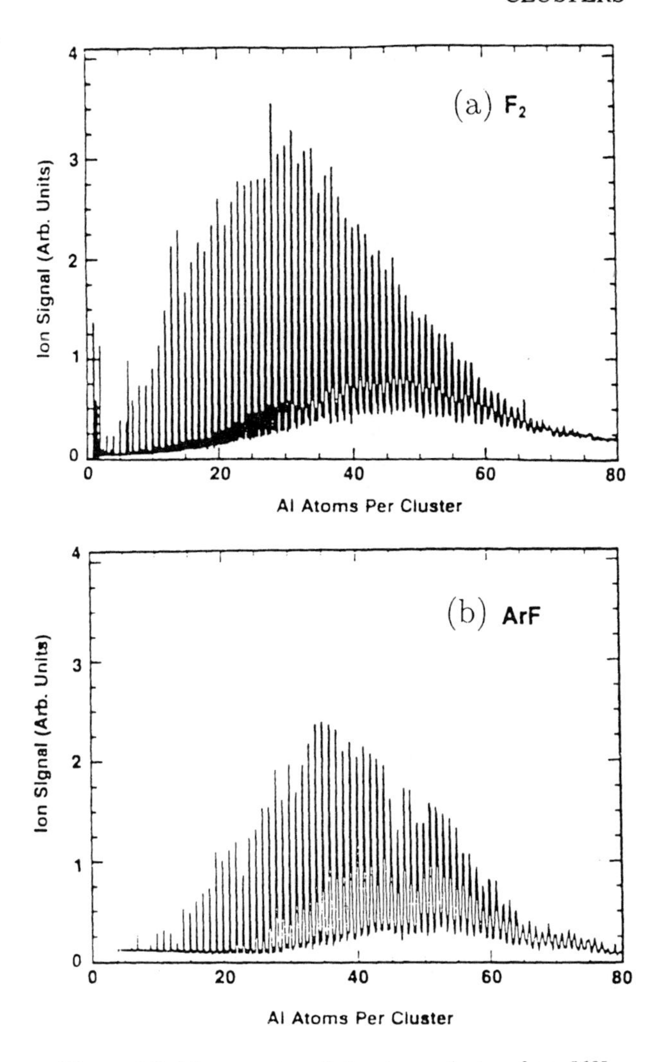

Figure 2.8. Mass spectra of aluminum clusters from [60].

temperatures. If the reduced melting temperature is comparable to the substrate temperature, then the clusters are 'liquid-like' and there is no barrier to fusion. In this case, coalescence will be unhindered until the melting temperature of the fused clusters is larger than the substrate temperature. Fusion of clusters of refractory metals and covalently bonded materials is sufficiently hindered that the average cluster size of refractory metals and covalently bonded materials on the substrate is likely to be similar to the average cluster size of the incident beam. The subject of cluster assembled materials has been reviewed by Melinon *et al* [81].

2.2.4 Physical and chemical properties of clusters

The study of the physical and chemical properties of clusters is an area of great current interest since it provides new ways to explore the gradual transition from atomic or molecular to condensed matter systems. Clusters provide unique systems for understanding certain complicated condensed phase phenomena such as nucleation, solvation, adsorption, and phase transitions. Studies of clusters are most easily performed *in situ* after they have been formed by a free jet expansion. In this case, they can be mass selected. It is not unusual for the properties of clusters to be different from those of the bulk material with the same composition. This is not too surprising in small clusters in which the atomic structure is different from that of the bulk material; however, even clusters large enough that their atomic structure is the same as the bulk structure can have modified properties. Several books as well as review articles have been devoted to detailed investigations of cluster properties [82–91]. In this section, we briefly describe some important properties of clusters with an emphasis on metal clusters.

2.2.4.1 *Surface to volume ratio of clusters*

Using a liquid drop model [1, 92], the volume V and surface area S of a cluster containing n atoms or molecules can be expressed as

$$V = \tfrac{4}{3}\pi R_0^3 n \qquad \text{and} \qquad S = 4\pi R_0^2 n^{2/3} \tag{2.11}$$

where R_0 is the radius of a single atom. The fraction of the atoms that are surface atoms $F = n_s/n$ is given by

$$F = 4/n^{1/3}. \tag{2.12}$$

In small clusters a large fraction of the atoms occupy surface states. For example, $F = 0.4$ for $n = 10^3$, $F = 0.2$ for $n = 10^4$ and $F = 0.04$ for $n = 10^6$. The large surface/volume ratio for clusters is responsible for a variety of electronic and vibrational surface excitations [92]. Jortner [92] has discussed the physical and chemical consequences of the large surface/volume ratio of clusters and has proposed cluster size equations (CSEs) which describe the gradual 'transition' from the large finite cluster to the infinite bulk system, with increasing cluster size. A hypothetical dependence of the value of a physical or chemical property $\chi(n)$ on n is shown in figure 2.9. The property $\chi(n)$ as a function of increasing n changes from $\chi(1)$, the atomic value, passes through a discrete region (clusters), and then approaches the thermodynamic limit $\chi(\infty)$. In the 'small'-cluster domain, an irregular size dependence of structural, energetic, electronic and electromagnetic properties is exhibited. In the 'large'-cluster domain, the property $\chi(n)$ smoothly approaches the bulk value $\chi(\infty)$. The CSEs proposed by Jortner have the form

$$\chi(n) = \chi(\infty) + An^{-\beta} \tag{2.13}$$

where A and β are constants and $0 \leq \beta \leq 1$. Equation (2.13) expresses the cluster property $\chi(n)$ in terms of the corresponding bulk value plus a correction term $An^{-\beta}$ [92].

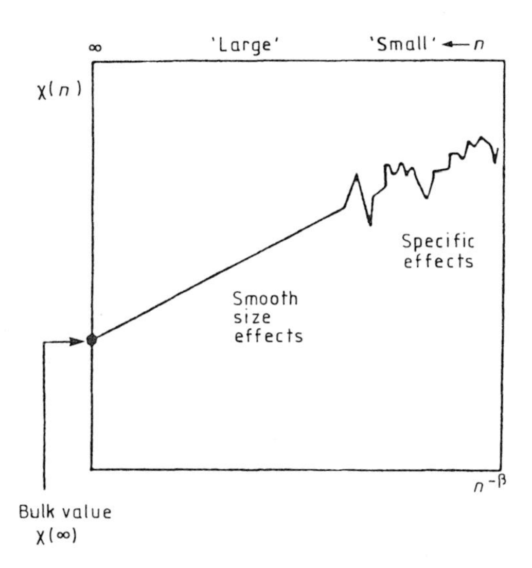

Figure 2.9. Schematic representation of the size dependence of the material's properties. (Reprinted from Jortner's article in [90] p 1.)

Similar to clusters, the properties of nanoparticles may also be size dependent. Because of the large surface to volume ratio of small clusters, the surface energy plays an important role in determining their properties and even their structure. For example, very small clusters of atoms containing fewer than approximately 20 atoms can have structures which are unique to these clusters. Somewhat larger particles have structures which are characteristic of the bulk but with reduced lattice parameters and reduced melting temperatures. The lattice constant is reduced as a result of changing the surface energy and elastic energy terms to minimize the free energy. This minimization gives rise to a Laplace pressure P given by

$$P = 2f/r \tag{2.14}$$

where f is the surface stress [93] and r is the radius. The lattice constant decrease for small nanoparticles [94] of Al is shown in figure 2.10. Once the cluster size exceeds approximately 1 nm, the particles usually have either the structure of the bulk materials or metastable structures with bulk values for the lattice constants and melting temperatures. In most cases, the particles produced directly from a supersaturated vapor are larger than 1 nm, and, consequently, most of their properties resemble those of the bulk material.

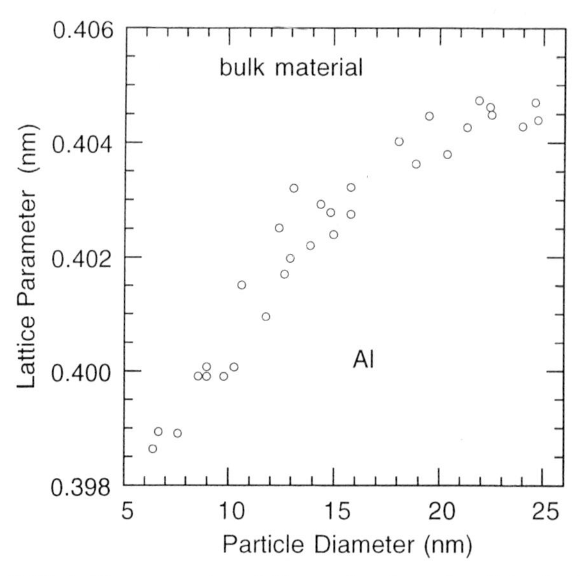

Figure 2.10. Lattice parameter of Al as a function of particle size from [94].

2.2.4.2 *Magic numbers*

Magic numbers refer to the unusually high mass spectral abundances that occur for certain cluster sizes which represent special electronic or geometric configurations. An example was mentioned earlier and shown in the case of potassium (see figure 2.4). For simple free-electron systems, such as alkali metal and the main group metal (I–IIIA) clusters, denoted by A_n, the magic numbers are often observed to follow the jellium-shell model [41, 95]. In this model, the valence electrons are highly delocalized and occupy spatially delocalized orbitals. There are nZ such electrons in an A_n cluster of valence Z ($Z = 1$ for Na, 2 for Mg, 3 for Al). The delocalized orbitals are classified according to their angular momentum as S, P, D, F, ... orbitals, just as in isolated atoms. Similar to the stability of atoms with filled atomic shells, clusters whose total number of valence electrons fill an electronic shell are especially stable. Therefore, the magic numbers observed as peaks in the abundance spectra (for example, for Na at $n = 8$, 20, 40, 58, 92) correspond to major electronic shell closings. The abundance spectra drop sharply above these magic numbers. These features are clearly evident in figure 2.4 which shows that abundances of K_n clusters obtained from free jet expansions.

For clusters of noble gas atoms and, generally, for large clusters ($n > 100$), the magic numbers are determined by geometric considerations and not by the filling of electronic shells. As suggested by Mackay [96], the magic numbers for inert gas clusters containing from 13 to 923 atoms arise from the closing of shells and sub-shells of icosahedral structures [97–102]. For example, Martin *et al* explained the magic numbers observed in the mass spectra of Na_n with $1500 < n < 22\,000$ by the completion of icosahedral or cuboctahedral shells of atoms [103]. Figure 2.11 shows the mass spectra of Na_n clusters obtained by photoionization with 415 and 423 nm light. The well defined minima are observed to occur at values of n corresponding to the total number of atoms in close-packed cuboctahedra and nearly close-packed icosahedra [44, 104]. Similar results have been obtained for Li_n clusters [105] and for Mg_n and Ca_n clusters [45]. In some other metal clusters, such as Pb_n, the stabilities are also dominated by geometric considerations [68, 106] and the strong magic numbers observed correlate with the icosahedral structures.

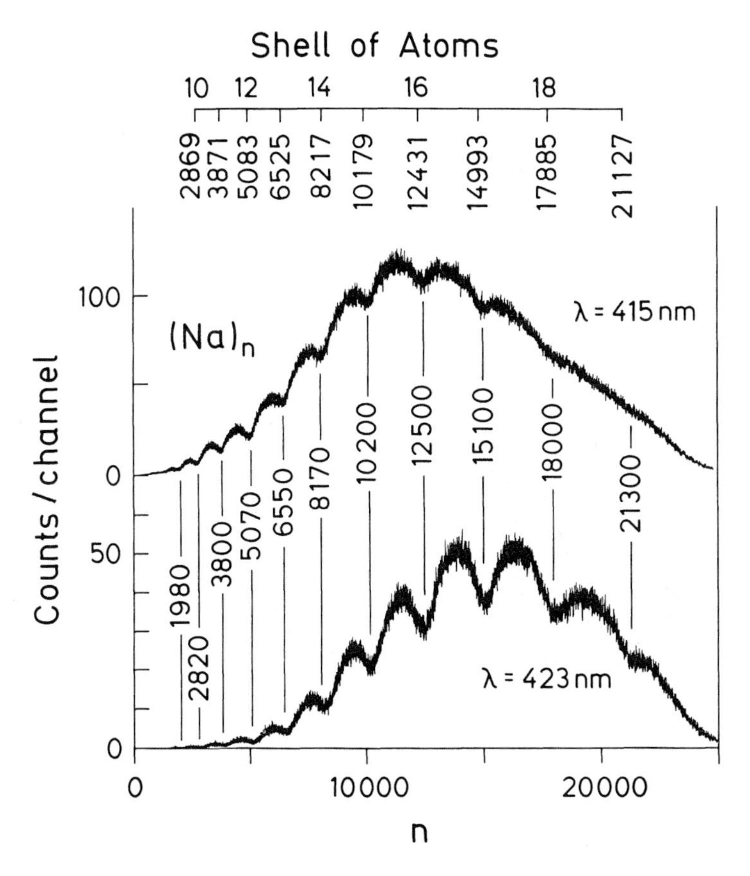

Figure 2.11. Mass spectrum of Na clusters condensed in Ar from [103].

The objects forming a cluster can be larger than individual atoms or molecules. For example, fullerenes can be the building blocks [107]. What is more interesting is that it was found that certain sizes of clusters ($n = 13$, 19, 23, 35, ...) of fullerenes were more prevalent. These magic numbers for fullerene clusters probably correspond to icosahedra similar to those found for Ar clusters [108]. Another example of larger building blocks is the self-assembly [109] of Mo cubes into larger cubes discussed below. Antimony clusters [43] exhibit a very distinctive distribution as shown in figure 2.5. For clusters containing more than eight atoms, only multiples of Sb_4 are seen. This suggests that the antimony tetramer is the building block for the clusters. Similarly, sulfur clusters produced in an oven gas aggregation source are characterized by a distribution containing maxima at multiples of eight [110]. This may be due to the condensation of clusters of S_8 which are known to be vaporized from orthorhombic sulfur.

2.2.4.3 *Ionization potentials and electron affinities*

The ionization potentials (IP) of metal clusters have been used to test the electronic shell model. Since the IP measures the orbital energy of the last electron, it should decrease abruptly at shell closings. Such decreases in IP have been observed for Na_n and K_n clusters [111–113]. However, as already mentioned, for Na_n clusters, the results are consistent with the atomic shell structures and the observation of magic numbers caused by an increase in the ionization energy each time a shell of atoms is completed. The results [114] are shown in figure 2.12. For Li_n clusters, the IP data support strong electronic shell effects at $n = 8$ and 20 as well as strong odd/even alternation as shown in figure 2.13 [115]. While the closed shell structures ($n = 8, 20$) can be explained in terms of the spherical shell model, this model cannot explain the odd/even alternation which is connected with ellipsoidal deformations [116]. Also, the IP of Al_n clusters are consistent with the electronic shell model [117]. However, the results for Fe_n are not in as good agreement with the electronic shell model because of the existence of many isomers with different IP values [118].

The IP data for Hg_n clusters have been interpreted in terms of a gradual transition from van der Waals to metallic bonding over the range $n = 20$–70, as evidenced by the drop in IP from high, atom-like values, to those near the predicted work function of a small metal droplet [119].

Ionization potentials were used to determine the energies of the ns and nd bands of Cu_n^- ($n = 1$–411) and Au_n^- ($n = 1$–233) clusters [120, 121]. The size dependence of these energies, shown in figure 2.14, resembles the behavior described by Jortner [92]. It is clear that the energies of the larger clusters converge to the corresponding bulk values but there are fluctuations for small values of n.

2.2.4.4 *Chemical reactivity*

The study of the chemical reactivity of metal clusters has attracted much attention in recent years because of its relevance to catalysis and materials processing [83, 122–124]. When the particles are suspended in a gas phase or in an inert matrix, they often show high reactivity compared to the conventional ligated

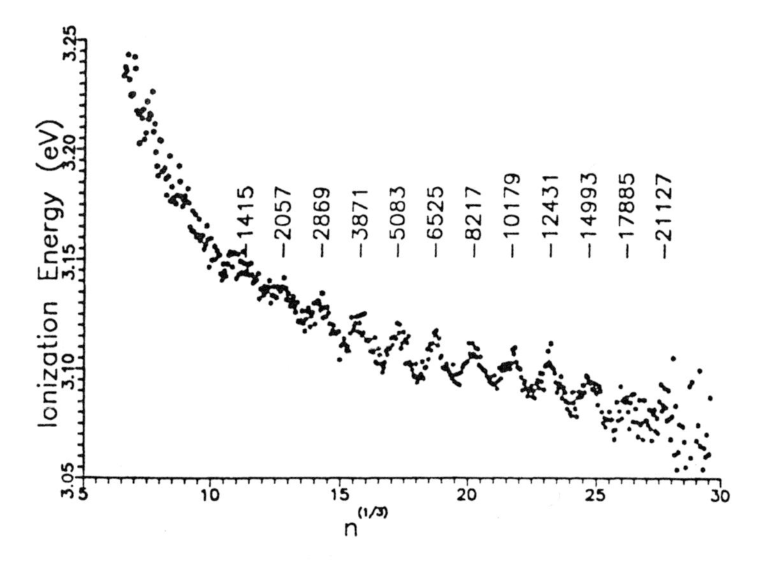

Figure 2.12. Ionization potential of large Na clusters versus $n^{1/3}$ from [114].

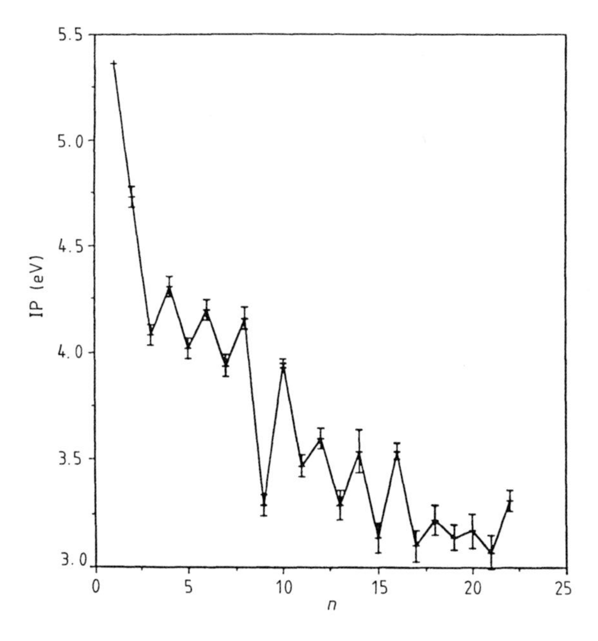

Figure 2.13. Ionization potential of Li clusters as a function of cluster size n from [115].

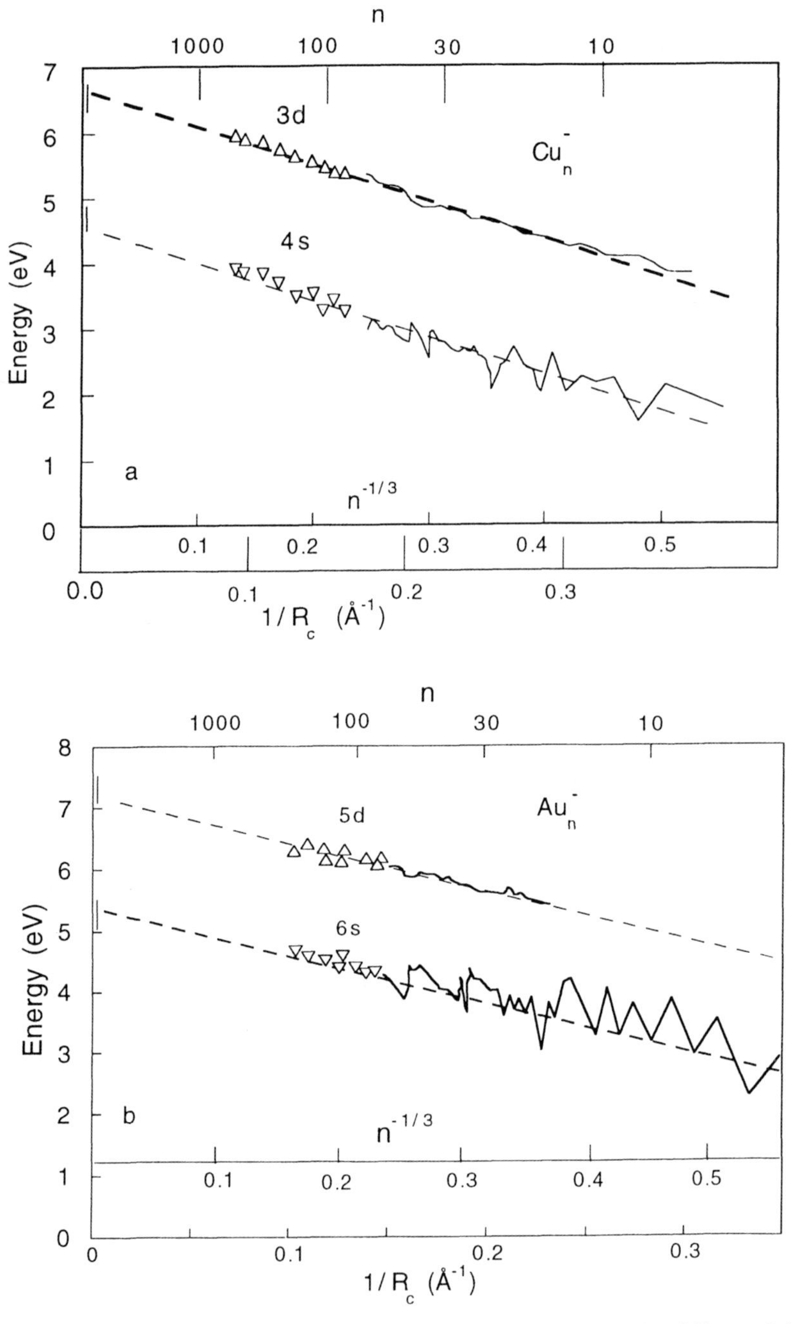

Figure 2.14. Size dependence of the energies of the *n*s and *n*d bands of Cu_n^- and Au_n^- clusters from [92]. Original data for Cu_n^- and Au_n^- clusters came from [120] and [121] respectively.

clusters where the metal centers are coordinated to different ligands. The reactivity of Al clusters toward H_2, D_2O, CO, CH_3OH, and O_2, as a function of the size of the Al cluster, varies strongly from one reactant to the next [60]. Overall the reactivity was roughly ordered as $O_2 > CH_3OH > CO > D_2O > D_2 > CH_4$ with CH_4 showing no reaction [60, 124].

Many reactions of transition metal clusters have been studied [62, 125–128]. For example, the reactions with hydrogen have been investigated with V [129], Fe [62, 63, 130], Co [130], Ni [130, 131], Cu [130], Nb [130, 132], and Pt [133] clusters. The reactions of transition metal clusters with a variety of other reagents have been studied. Examples include reactions of Pt and Nb clusters with hydrocarbons [127, 134–136], reactivity of Co, Fe, and Nb clusters with N_2 and CO [130, 137], and reactions of Fe clusters with O_2 [138]. The reader is referred to the literature [124, 139, 140] for details of the reactivity of metal clusters.

2.2.4.5 *Phase transitions*

Probably the most studied phase transition in nanomaterials is melting. Studies show that the melting temperature of nanomaterials decreases for clusters smaller than a few hundred ångströms [141, 142]. For example, the melting temperature of gold [143] decreases by approximately a factor of two when the cluster size is reduced from 10 nm to 2 nm. Figure 2.15 shows this effect [144]. The melting temperature of lead [145] decreases by approximately 7% as the particle size is reduced from 100 nm to 10 nm. Molecular dynamic simulations [146, 147] of melting of small rare gas atom clusters indicate that melting begins at the surface of the cluster. Couchman and Jesser [148] formulated a thermodynamic theory of the melting of small metal particles based on finding the criterion such that a liquid layer would spontaneously grow and consume the solid. Since melting originates on the surface, superheating of crystals without free surfaces may be possible. This probably explains why molecular dynamic simulations [149] show that clusters coated with higher-melting-point materials can be superheated above their thermodynamic melting point.

The phase and structure of small particles can fluctuate. For example, the coexistence of liquid-like and solid-like clusters of Ar_{13} has been calculated by microcanonical simulations [150]. High-resolution electron microscope studies found that small particles can fluctuate in their structure between different multiply twinned and single-crystal structures and that they remained in each state for about 1/30 s. Calculations indicate there is a quasi-molten phase below the melting curve as a function of particle size where the particles continuously fluctuate between different structures [151].

The nonmetal–metal transition in Hg_n clusters was discussed earlier in connection with the ionization potential of clusters. Tolbert and Alivisatos [152] found another example in which a phase transition depends on the sample size. They found that the solid–solid phase transition in CdSe nanocrystals from the wurtzite structure to the rock-salt structure required increasingly high pressures when the crystallite size decreased.

2.2.4.6 Magnetism

The magnetic properties of clusters can also differ from those of the corresponding bulk material. Table 2.1 lists some dramatic differences. For example, clusters of Rh [153, 154], Pd, Na, and K are ferromagnetic, whereas the bulk forms of these elements are paramagnetic. Other differences, such as the superparamagnetism of Fe, Co, and Ni clusters, are just a consequence of the small size of the clusters. Relatively few investigations have been performed on unsupported ferromagnetic clusters [155, 156]. The earliest results came from the Exxon group [155] who tried to use depletion in cluster beams to determine magnetic moments. The actual deflection profiles were, however, first observed by de Heer *et al* [156] on Fe_n clusters containing 56–256 atoms. Bucher *et al* [157] performed similar experiments on Co_n clusters. Further experiments on Fe, Co, and Ni clusters containing up to 700 atoms at temperatures between 80 and 1000 K have been reported [158, 159]. Ferromagnetism occurs even for the smallest sizes. For clusters with fewer than about 100–200 atoms the magnetic moments are enhanced and atom-like. As the size is increased up to 700 atoms, the magnetic moments decrease and approach the bulk limit, with oscillations possibly caused by surface induced spin-density waves or structural changes. The magnetism of nickel clusters converges to the bulk limit more rapidly than Co and Fe clusters. In particular, the moment of an Ni cluster with about three layers of atoms (Ni_{150}) is bulklike, whereas for Co and Fe

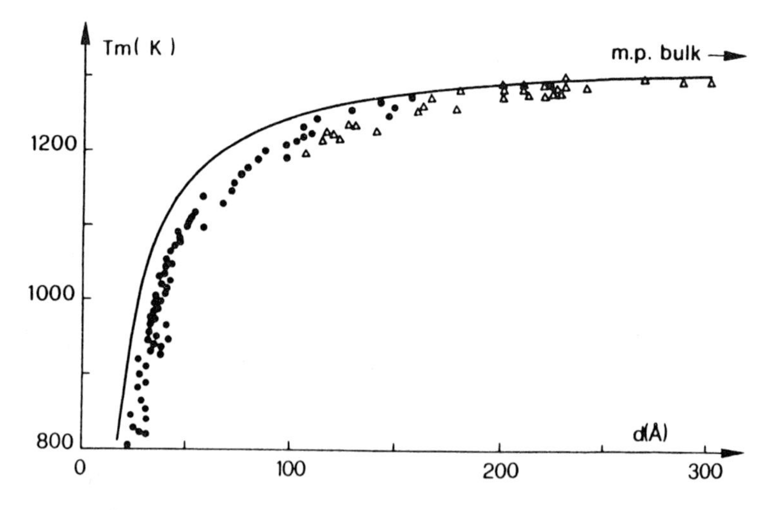

Figure 2.15. Melting temperature of gold particles versus size from [144]. △: Sambles experiment; ○: Buffat experiment; solid line: the Pawlow first-order theory.

about four to five layers are required (Co_{450}, Fe_{550}) before the same bulklike behaviors are observed. Compared with Ni and Co clusters, Fe clusters are anomalous. For $Fe_{120-140}$ near 600 K, a pronounced decrease in the average magnetic moment per atom in the cluster from the bulk value is observed. This effect has been interpreted [159] in terms of a crystallographic phase transition from the high-moment body-centered cubic phase to the low-moment face-centered cubic phase. This transition occurs in the bulk at $T = 1150$ K and it is suggested that in clusters the transition occurs at progressively lower temperatures as the cluster size is decreased.

Table 2.1. Comparison of the magnetic properties of clusters with those of bulk material.

System	Cluster[a]	Bulk
Na, K	Ferromagnetic	Paramagnetic
Fe, Co, Ni	Superparamagnetic	Ferromagnetic
Gd, Tb	Superparamagnetic	Ferromagnetic
Cr	Frustrated parmagnetic	Antiferromagnetic
Rh, Pd	Ferromagnetic	Paramagnetic

[a] Private communication from S N Khanna, cluster sizes 2 to 300 atoms.

Studies have also been performed on giant nickel carbonyl clusters [160] containing a core of $Ni_{38}Pt_6$ surrounded by a shell of CO ligands. The ligands quench the magnetism of the nickel atoms on the surface of the cluster but leave the magnetism of the inner-core atoms relatively unaffected.

2.3 NANOPARTICLES PRODUCED BY SPUTTERING AND THERMAL EVAPORATION AND LASER METHODS

2.3.1 Background

This section discusses the formation and collection of nanoparticles produced from a supersaturated vapor produced by thermal evaporation, sputtering, and laser methods. In the case of thermal evaporation, the technique was originated in 1930 by Pfund [161], by Burger and van Cittert [162], and by Harris *et al* [163]. The theory presented earlier on the nucleation and growth of clusters is applicable. However, there are two important differences between the nanoparticles and the clusters discussed earlier. First, nanoparticles produced directly from a supersaturated vapor are usually larger than the clusters discussed earlier. They range in size from 1 to 100 nm. Secondly, the nanoparticles are also usually produced in much larger quantities than the clusters produced in a free jet expansion. For example, it is not uncommon to produce gram quantities of material during a single run. These two properties are interrelated. The nanoparticles are this large because the particle density is so high that they collide and grow by coalescence. The nanoparticles produced by sputtering or thermal evaporation have been studied after being removed from the vacuum chamber. In contrast, the clusters produced in a free jet expansion are often studied by mass spectrometry and laser techniques while they are still in the vacuum chamber.

Most of the early works [161, 162] in preparing nanoparticles in the vapor were investigations of nanoparticles of elements. Oxides were also investigated at an early stage by the simple method of introducing some oxygen into the vapor [164–166]. More recent work has included investigations of alloys [167], and compounds [166, 168, 169], and mixtures of particles [170]. Probably the compound that has been studied the most is TiO_2 [169, 171]. In this case the oxide was formed by sputtering in a mixture of helium and oxygen. In many, if not most cases, the nanoparticles that are formed are single crystals. There have been several excellent reviews of this subject [172–174]. For example, Uyeda has reviewed the Japanese work [174]. In an appendix, Buckle *et al* [172] have tabulated the structure and morphology of elemental and alloy particles prepared by evaporation and condensation in an inert gas atmosphere.

Preparing nanoparticles directly from a supersaturated vapor was one of the earliest methods for producing nanoparticles. It has the advantages of being versatile, easy to perform and to analyze the particles, produces high-purity particles, and naturally produces films and coatings. It has the disadvantage that, despite preparing larger quantities of material than by a free jet expansion, the cost per gram of material is still very high. Further, it is difficult to produce as large a variety of materials and microstructures as one can produce by chemical means.

An early description of the process was given by Granqvist and Buhrman [175]. A more recent description was provided by Ichinose *et al* [176]. In its simplest form, the apparatus for producing nanoparticles from a supersaturated solution consists of a vapor source inside a vacuum chamber containing an inert gas, usually Ar or He. The vapor source can, for example, be an evaporation boat or a sputtering target. Supersaturation is achieved above the vapor source and nanoparticles are formed. Above the source is a collection surface which is often cooled to liquid nitrogen temperatures. A convective flow of the inert gas is set up between the warm region near the vapor source and the cold surface. The nanoparticles are carried by the convection flow to the cold surface where they are collected. Birringer, Gleiter and coworkers [173, 177] modified the method to make compacted pellets of the nanoparticles. Their apparatus is shown schematically in figure 2.16. The additions they made to the usual apparatus for producing nanoparticles were a scraper for removing the nanoparticles from the cold collection surface, a funnel for feeding the nanoparticles into a die, and a compaction unit for pressing the particles together to form a pellet. The consolidation of the particles and the properties of the consolidated particles are discussed in chapters 8–11 and 13. This section mainly focuses on how

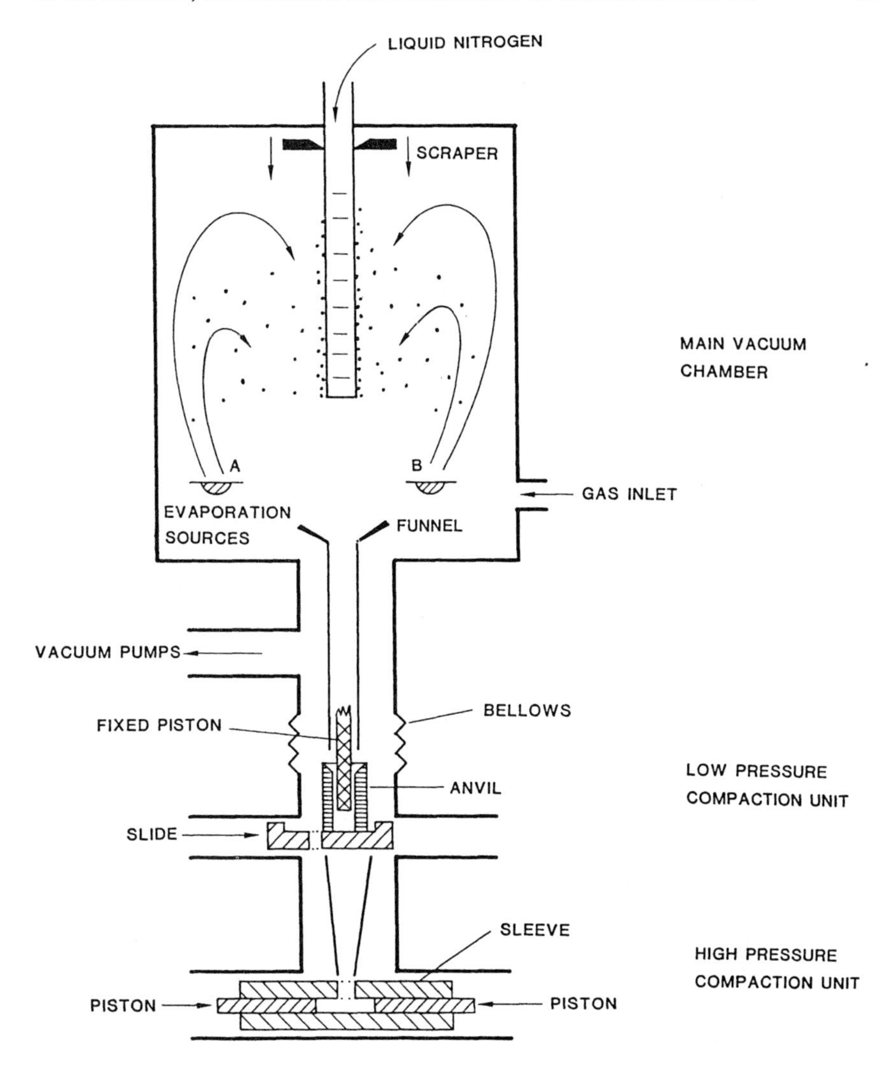

Figure 2.16. Typical apparatus for producing nanoparticles from a supersaturated vapor.

the nanoparticles are made and some of the properties of the unconsolidated particles.

The discussion of nanoparticles produced by sputtering and thermal evaporation is conveniently divided into discussing supersaturation, particle nucleation and growth, coalescence and coagulation, the transport of the particles, the collection of the particles, and their properties. The latter includes their structure, crystal habit, and size distribution.

2.3.2 Achieving supersaturation

Earlier we discussed jet expansion techniques for producing clusters. Other ways of achieving supersaturation are thermal evaporation [178], sputtering [179], electron beam evaporation, laser ablation [180], spark erosion [181], and flames [182].

Because spark erosion is discussed in connection with the synthesis of fullerenes in chapter 19 of this book, we will just briefly comment on this technique. Unlike the other methods, which employ an inert gas, small particles can be produced by spark erosion either in a dielectric liquid or an inert gas [183]. Spark erosion also differs from the other techniques in that particles can form from molten droplets that are produced by the spark. Interesting variants of the technique were used by Majetich *et al* [184] to produce carbon coated nanocrystallites and to separate the byproducts. In their separation technique they used a magnetic field gradient.

Thermal evaporation has the disadvantage that the operating temperature is limited by the choice of crucible material. Possible reactions with the crucible are also another concern. Sputtering has the advantage that it can be applied to almost any material, whereas thermal evaporation is largely limited to metals. To avoid charging effects for insulators, one must use radio frequency (RF) sputtering.

The reason for using a high pressure of an inert gas is that the frequent collisions with the gas atoms decrease the diffusion rate of atoms away from the source region. The collisions also cool the atoms. If the diffusion rate is not limited sufficiently, then supersaturation is not achieved and individual atoms or very small clusters of atoms are deposited on the collecting surface. Usually the gas that is used to limit the diffusion by shortening the mean free path is an inert gas, but mixtures of an inert gas and another gas that reacts with the sputtered atoms have also been used to produce molecular nanoparticles. For example, by adding oxygen to the inert gas Siegel and coworkers [169] have produced nanoparticles of TiO_2. The high pressure needed to limit the mean free path and thus confine the vapor to achieve supersaturation decreases the sputtering rate [185]. This occurs because the ions lose energy by collisions with the inert gas, and, thus, fewer of them have the necessary threshold energy to sputter atoms from the target. In most sputtering systems, there is a broad pressure range between the lowest pressure required to achieve supersaturation and the pressure at which the sputtering rate becomes unacceptably low.

The diffusion rate can be controlled by the choice of inert gas, the pressure, and the temperature difference between the source and a cold reservoir. This temperature difference drives the convection. We shall discuss how it is possible to have sufficiently regular flow so that supersaturation and particle nucleation and growth can occur at a cooler location several centimeters above the source. This is similar, but on a much slower time scale, to the formation of clusters in a free jet expansion. Heavier gas atoms are more effective at limiting the mean free path. Data by Granqvist and Buhrman [175] illustrating this effect are presented in figure 2.17. One sees that the particle sizes of Al and Cu increase when the mass of the inert gas increased by going from He to Ar and then to Xe. This occurs because the heavier gas atoms are more effective at confining the metal vapor.

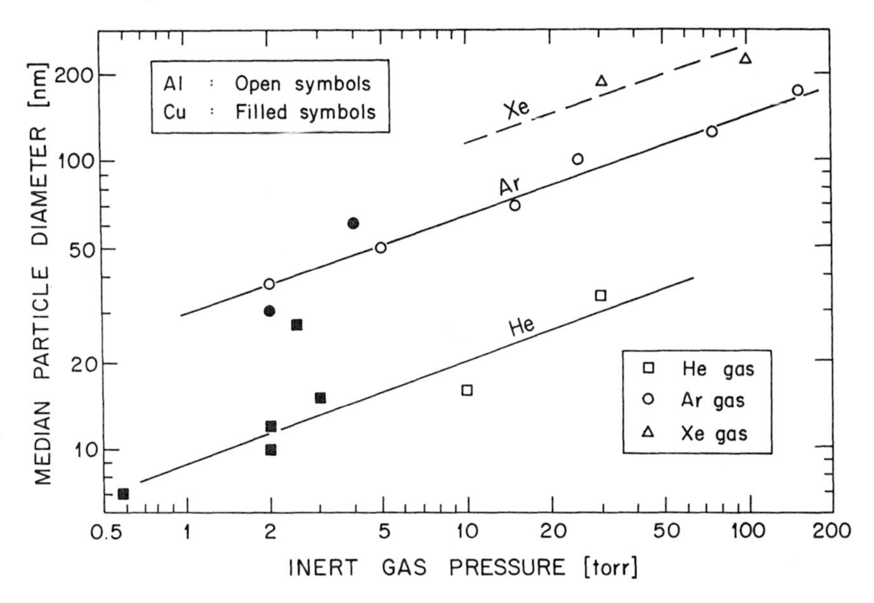

Figure 2.17. Effect of increasing the mass and pressure of the carrier gas on nanoparticle size. (Data from [175].)

Laser vaporization techniques provide several advantages as a source for producing nanoparticles. These advantages include the production of a high-density vapor of any metal within an extremely short time (10^{-8} s) and the generation of a directional, high-speed, metal vapor stream from the solid target [186]. These advantages have been demonstrated in the synthesis of ultrafine metal particles [187]. A variety of nanoscale metal oxides and nitrides have been synthesized using laser vaporization of metals followed by reactions between the metal vapor and oxygen or NH_3 to generate the oxide or nitride particles [188]. A novel version of this technique combines the laser vaporization process with controlled condensation in a diffusion cloud chamber in order to control the size distribution and the chemical compositions of the particles [189]. A detailed description of the chamber and its major components can be found in several references [10, 190]. The metal vapor is generated by pulsed laser vaporization using the second harmonic (532 nm) of a Nd-YAG laser (15–30 mJ/pulse). In the case of oxide syntheses, the hot metal atoms react with O_2 to form vapor phase metal oxide molecules and clusters. With the high total pressure of 800 Torr employed in these experiments, it is expected that the metal atoms and the oxide molecules approach the thermal energy of the ambient gas within several hundred microns from the vaporization target. The unreacted metal atoms and the less volatile metal oxide molecules and clusters are carried by convection to the nucleation zone near the top plate of the chamber. Figure 2.18 taken from the work of El-Shall *et al* [189] displays a scanning electron microscope (SEM) micrograph obtained for SiO_2 particles synthesized using 20% O_2 in He at a total pressure of 800 Torr and top and bottom plate temperatures of $-100\,^{\circ}$C and 20 °C, respectively. The particles exhibit an agglomerate pattern which appears as a weblike matrix. Based on the transmission electron microscope (TEM) analysis of the sample, the particle size is 10–20 nm. Decreasing the temperature gradient between the chamber plates results in larger particles. The agglomerate pattern is quite different from other nanoparticles synthesized using conventional oven or sputtering sources [175, 191, 192] and appears to be a result of the high cooling rate. This structure has potential for specific catalytic activity and for reinforcing agents for liquid polymers.

As discussed earlier, laser photolysis of volatile compounds can also be used to produce a supersaturated vapor. For example, the production of μm size crystalline particles of NaH and CsH from the photolysis of the metal vapor–H_2 mixture using a continuous wave (cw) Ar^+ laser has been reported [75]. Other metal particles and submicron-sized materials have been synthesized using similar laser induced clustering techniques [72, 193, 194].

In addition to laser photolysis, laser pyrolysis has been used for the production of nanoscale particles [195]. In this process, the heat generated through the absorption of CO_2 laser energy by one of the reactant species is used for the gas phase pyrolysis reaction of two or more molecular species. Nanoscale particles of Si, SiC, Si_3N_4 [195–198], ZrB_2 [199], TiO_2 [199, 200], and Fe_3C [201] have been produced using this technique.

2.3.3 Particle nucleation and growth

The classical theory of nucleation was presented in section 2.2. Here we only discuss those aspects of particle formation that are connected with particle formation directly from a vapor. For particles formed directly from the vapor, it is usually assumed that particle formation occurs via homogeneous nucleation. For large values of the scaled density ρ_s of atoms in the vapor, supersaturation occurs at all temperatures. For small values of the scaled density ρ_s supersaturation is only achieved at low temperatures. Using equation (2.8) from section 2.2, equation (2.6) for the critical radius r^* can be rewritten in scaled units as

$$\frac{L(0)k_b r^*}{2\sigma v R} = 1/[T_s \ln(\rho_s T_s) + 1] \tag{2.15}$$

where T_s is $RT/L(0)$ and ρ_s is $\rho L(0)/P_0$. The value of $L(0)k_b r^*/2\sigma v R$ from equation (2.15) is plotted in figure 2.19 as a function of T_s for various values of ρ_s. For low values of ρ_s, the values for r^* predicted by this equation increase

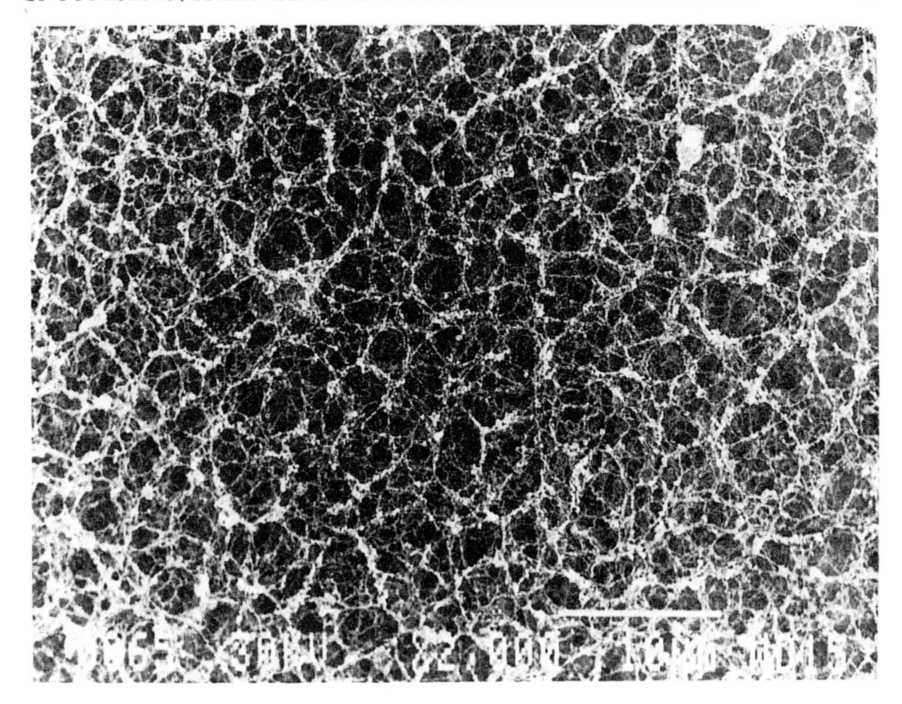

Figure 2.18. SEM micrograph of SiO_2 particles produced by laser ablation of Si in an oxygen/helium mixture. The primary spherical particles are aggregated into long chainlike strands. The strands are linked together to form a weblike network. See [217].

rapidly at high values of T_s. However, in the limit $T_s \to 0$, r^* is independent of ρ_s.

Typical values for the critical size predicted by this equation are of the order 1 nm or less. This is smaller than the usual sizes of nanoparticles produced in the vapor. Thus, the nanoparticles usually continue to grow after nucleation by acquiring more atoms from the vapor or by coalescence.

It is easy to obtain high evaporation rates in thermal evaporation. Because of these high evaporation rates, the large numbers of particles which are formed near the evaporation source are readily observable and have been given the name smoke. Usually the mean size of the particles increases as one increases the rate at which atoms are added to the vapor or if one increases the partial pressure of these atoms by increasing the inert gas pressure.

2.3.4 Coalescence, coagulation, and size distributions

The growth at short times has been modeled by Kim and Brock [202]. If the density of particles is relatively low and the time before the particles are collected is relatively short, then the particle agglomerates are small. At later

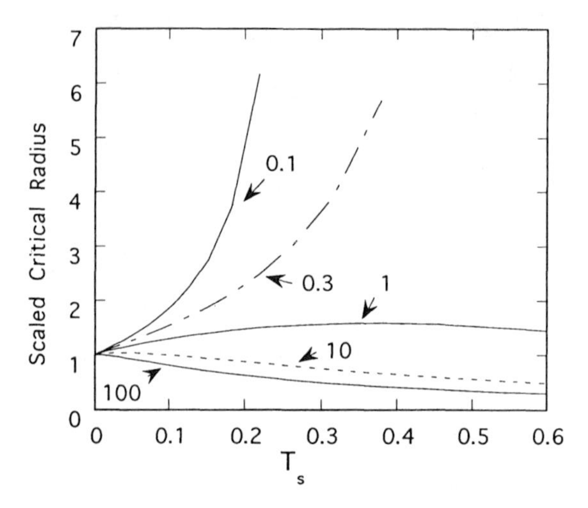

Figure 2.19. Scaled critical radius versus the scaled temperature $T_s = RT/L(0)$ for various scaled vapor densities. The curves are labeled according to the values of ρ_s. See the text for a more complete explanation.

times, after the particles are formed, they collide and either coalesce with one another to form a larger particle or coagulate. Which process will occur depends on the temperature and the available energy. Granqvist and Buhrman discuss how coalescence leads to a log-normal distribution for the particle radius r [175]. Granqvist and Buhrman state that coalescence is the dominant growth mechanism. The function describing this log-normal distribution [203] is

$$F(r) = \frac{C}{\sqrt{2\pi\sigma^2}}\frac{1}{r}\exp\left(\frac{-[\ln(r/r_M) - \sigma^2]^2}{2\sigma^2}\right). \qquad (2.16)$$

Note that this distribution has two independent variables, the mode or most probable radius r_m and the width σ. The mean $\bar{r}$ is related to the mode size by $\bar{r} = \exp(\frac{3}{2}\sigma^2)r_m$. It should be noted that equation (2.16) differs from the distribution function given by Granqvist and Buhrman by inclusion of the factor $1/r$. Both of these functions are discussed by Kurtz and Carpay [203]. Typical particle sizes are 2 to 30 nm and typical widths are 0.25 to 0.5 r_m. The main assumptions used in deriving the log-normal distribution are that particle growth occurs only by coalescence and that the change in volume that occurs at each coalescence event is a random fraction of the volume after coalescence. This distribution function is used below to fit the size distribution of Mo particles.

Kaito *et al* [165] present a simple model for the increase of the average radius, $\bar{r}$, of smoke particles due to coalescence. They describe the coalescence as taking place such that the particles unite with definite orientations so as to minimize their interface energy. In their model, the coalescence frequency, f, which is the number of coalescences per unit volume per unit time is given by

$$f = -\frac{\mathrm{d}N}{\mathrm{d}t} = kN^2\bar{r}^2 \tag{2.17}$$

where N is the number of particles per unit volume and k is a factor which includes the coalescence probability. Treating k as a constant, the result in the case of uniform convective flow with a constant cross section is

$$\bar{r} = \frac{Mk}{4\pi\rho\bar{v}}h + \overline{r_0} \tag{2.18}$$

where h is the height above the source, M the total mass of particles per unit volume, ρ is the density of particles, $\bar{v}$ is the mean vertical velocity of the particles, and $\overline{r_0}$ is the mean particle size at $h = 0$. Equation (2.18) predicts that the mean size should increase linearly in h; Kaito *et al*'s results [165] are in agreement with this prediction.

Magnetic particles can form intricate long strands when they agglomerate to minimize the magnetic energy. For nonmagnetic particles, the arrangement of clusters of particles is partially determined by surface energy considerations. This is the probable explanation for the aligning of the (100) faces of MgO crystals [164].

A very special case of collisions occurred when 5 nm cubes of Mo with a narrow size distribution collided with one another. In these experiments [109, 204] the pressure was adjusted to be near the threshold pressure for particle formation. At this pressure, the particles are formed in a cooler region several centimeters above the sputtering source. The narrow size distribution is apparently the result of collecting them from a small volume in the chamber soon after they have formed. In this case, when the 5 nm Mo cubes collided, they self-organized into larger cubes that were either $2 \times 2 \times 2$, $3 \times 3 \times 3$, or $4 \times 4 \times 4$ arrangements of the smaller 5 nm cubes. When the cubes are in contact with one another, the surface energy is lower. Thus, the self-organizing of the small cubes into the larger cubes probably occurs to decrease the surface energy of the system. To fit the experimental size distribution it was necessary to use a sum of four log-normal distributions, i.e. log-normal distribution for each of the four sizes of cubes. Some examples of TEM images of self-organized cubes, the individual log-normal distributions, and their sum are shown in figure 2.20.

2.3.5 Particle transport

The particles are transported to a surface where they are collected either by a convection current or by a combination of a forced gas flow [205] and a convection current. The convection current is set up by the temperature difference between the source and a cold surface as shown in figure 2.16. If there is a forced gas flow, the flow permits the particles to be collected at a considerable horizontal distance away from where they are formed. While

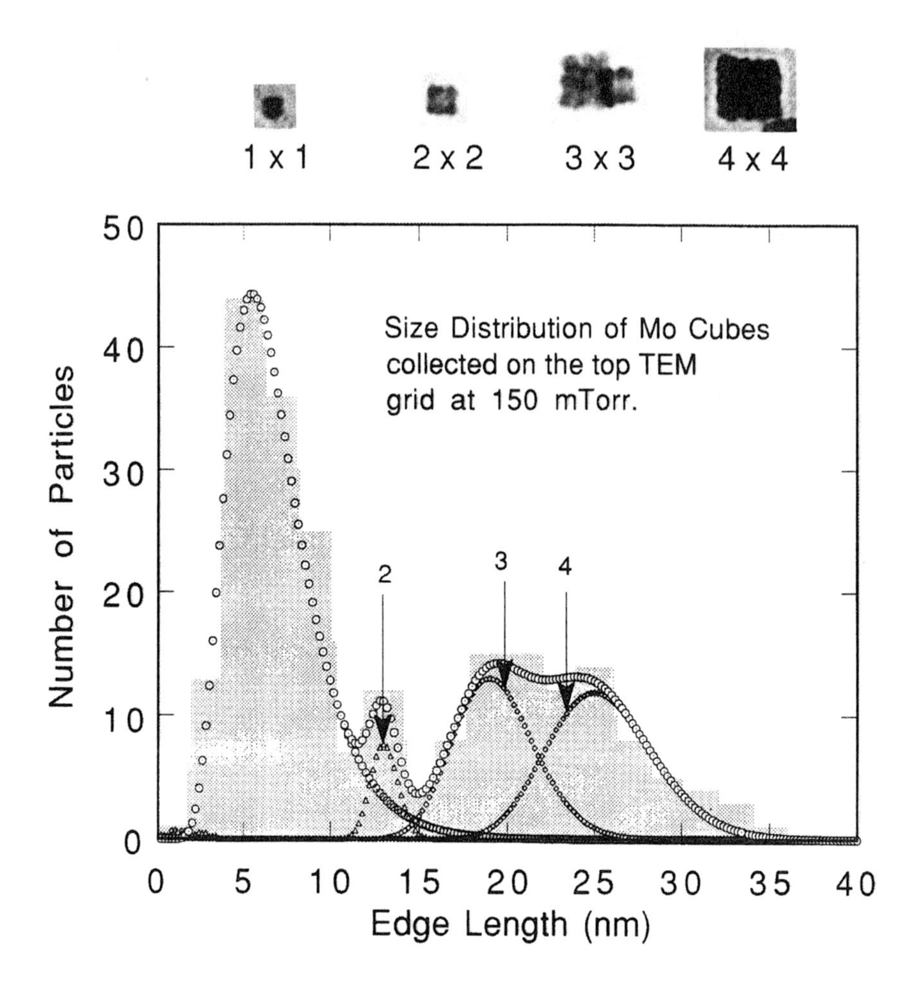

Figure 2.20. Shaded area is the histogram of edge lengths of Mo cubes produced by sputtering at high pressures. Also shown are: (1) individual log-normal distributions for $2 \times 2 \times 2$, $3 \times 3 \times 3$, and $4 \times 4 \times 4$ molybdenum cubes which are self-assembled from the smaller 5 nm molybdenum cubes; (2) the sum of these distributions denoted by (O); (3) insets showing TEM images taken with the same magnification of an individual cube and larger cubes which are $2 \times 2 \times 2$, $2 \times 2 \times 2$, and $3 \times 3 \times 3$, and a probable $4 \times 4 \times 4$ self-arrangements of small cubes like that shown in the first inset. (Reproduced from [204].)

the particles are moving in a forced gas flow, other processing steps can be performed on them. The equations governing convection are not simple [206, 207]. Solving these equations is difficult even for high-symmetry, idealized geometries [208] and would be especially difficult in the complicated geometries encountered in most deposition chambers. One way of getting an idea of the complexity of the convection currents is to look at the shape of the smoke. This complexity can be seen in the Mg smoke [209] shown in figure 2.21. The shape of the smoke depends on the convective currents and changes dramatically as a function of the evaporation source temperature and inert gas pressure.

2.3.6 Particle collection

The particles are usually collected on a cold surface located above the source. If one wants to examine the nanoparticles in a TEM, one can collect the particles on a thin substrate supported on a TEM grid located on or near this cold surface. If the number of particles deposited is small, one can examine the sample in a TEM without further processing. Thus, the processing is much simpler and does not require the thinning often required in preparing TEM samples. Thicker deposits can be investigated by small-angle x-ray scattering (SAXS). Thicker deposits of particles can also be deposited directly onto a cold cylindrical surface. After deposition they can be scraped off into a funnel, collected in a die, and finally compacted into a nanostructured solid. The apparatus for producing such compacts of nanosized particles is shown schematically in figure 2.16. The processing of these compacts is discussed in chapter 8 and their properties are discussed in chapters 9–11 and 13. If the particle agglomerates are large and have an irregular shape, it may be difficult to get rid of voids by compaction and to achieve a high density without substantial coarsening, i.e. increasing the grain size. The compaction of ceramic nanoparticles to high densities is especially difficult (see chapter 8).

2.3.7 Crystal structure and crystal habit

There have been many studies of the crystal structure, morphology, and habit of metal [178, 209–211] and oxide [165, 212] nanoparticles produced in vapor by evaporation. In these early studies, metals were evaporated and it was noted that metal smoke above the source consisted of three zones. The size of the particles and crystal habit of the particles varied depending upon the zone where the particles were collected. The differences are due to the fact that the three zones had different temperatures and metal vapor concentrations.

The large surface area of nanomaterials often makes them highly reactive. Thus, it is not surprising that metallic nanoparticles are sometimes coated with an oxide layer. It is not unusual for nanoparticles to oxidize completely after they are exposed to air. Eversole and Broida [213] found in making Zn particles that oxygen had the tendency of causing the particles to clump and form agglomerates

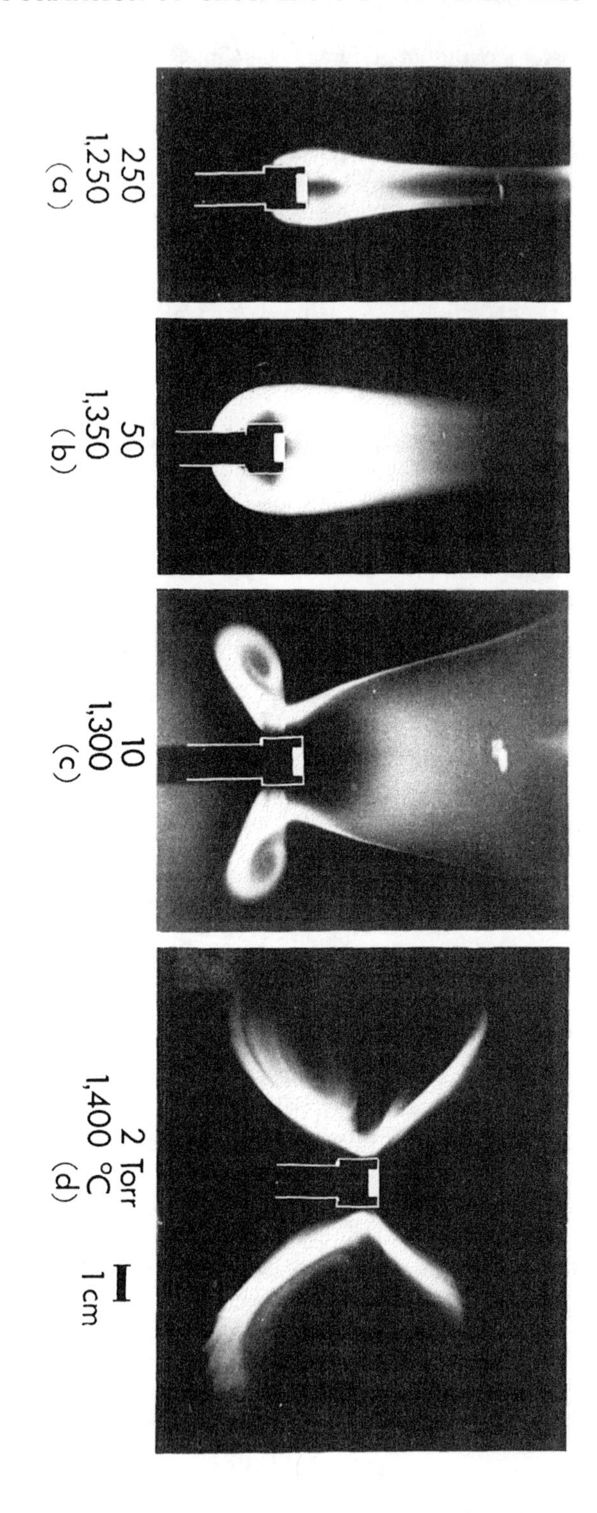

Figure 2.21. Examples from [208] of the convection current as seen by the shape of the smoke plume between the source and a cold surface.

approximately 0.5 μm in size.

In general, the crystal structure and crystal habit of the particles depends on the composition and temperature. Though the particles tend to be equiaxed in order to minimize their surface area, some of the MoO_3 particles of Kaito *et al* [165] were in the form of plates and needles. In many cases, the particles have the shape of polyhedra with varying degrees of truncation. These shapes occur either in order to minimize the surface energy of the particles or because of the kinetics of growth. If kinetics dominates, then the shape is determined by the rate at which the different crystal faces grow. If the particles are formed in thermal equilibrium, their shape or crystal habit results from minimizing the surface energy. In this case, the surface can be determined by performing a Wulff construction [22, 214]. It also should be noted that twinning may occur during coalescence [211].

Because of the combination of factors such as the temperature, kinetics, impurities, and surface energy effects, nanoparticles formed in the vapor can have unusual structures, shapes, and size distributions. For example, Saito *et al* [215] studied the formation of nanoparticles of some normally bcc elements produced by evaporation. They found that W nanoparticles had an A15 structure and that Cr and Mo particles could have either an A15 structure or a bcc structure.

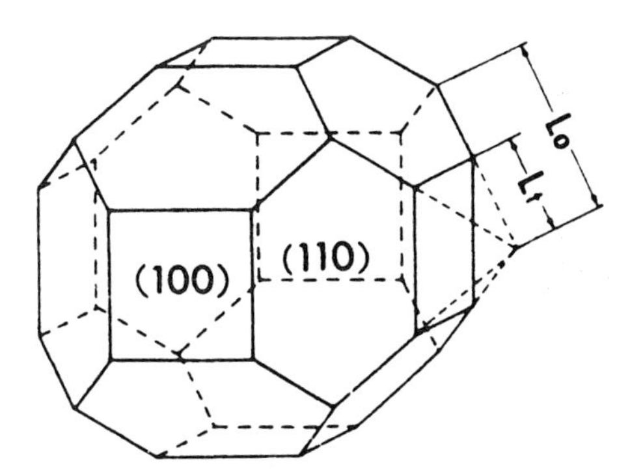

Figure 2.22. Truncated rhombic dodecahedron crystal habit of particles formed from bcc elements. The parameters L_t and L_0 which define the degree of truncation $R = L_t/L_0$ are shown in the figure.

The crystal habit of Fe, V, Nb, Cr, and Mo particles that have a bcc crystal structure is a truncated rhombic dodecahedron [215]. The geometry of a general truncated rhombic dodecahedron can be described by the lengths L_t and L_0 shown in figure 2.22. We define a parameter $R = L_t/L_0$. If $R = 1$, the dodecahedron is not truncated. If $R = 0$, the truncation is complete and the particle is a cube. For most crystals, if one uses a Wulff construction, the minimum radial distance to each crystal face is proportional to the surface energy of that face. Using this relationship one can show that

$$\begin{aligned} R &= 1 \text{ for } \gamma_{100}/\gamma_{110} \le 1/\sqrt{2} \rightarrow \text{cube} \\ R &= \sqrt{2}(\sqrt{2} - \gamma_{100}/\gamma_{110}) \text{ for } 1/\sqrt{2} < \gamma_{100}/\gamma_{110} < \sqrt{2} \\ &\qquad \rightarrow \text{truncated dodecahedron} \\ R &= 0 \text{ for } \gamma_{100}/\gamma_{110} \ge \sqrt{2} \rightarrow \text{dodecahedron} \end{aligned} \tag{2.19}$$

where γ_{100} and γ_{110} are the surface energies of the (100) and (110) surfaces.

Alternatively, as mentioned earlier, the shape may be determined by the growth kinetics. For example, if the growth of the (110) surface is much faster than that of the (100), then the resulting particle will be a cube with (100) crystal faces.

REFERENCES

[1] Abraham F F 1974 *Homogeneous Nucleation Theory* (New York: Academic)
[2] Zettlemoyer A C 1969 *Nucleation Phenomena* (New York: Dekker)
[3] LaMer V K and Dinegar R H 1950 *J. Am. Chem. Soc.* **72** 4847
[4] Fukuta N and Wagner P E 1992 *Nucleation and Atmospheric Aerosols* (Hampton, VA: Deepak)
[5] Frenkel J 1955 *Kinetic Theory of Liquids* (New York: Dover) ch 7
[6] Zeldovich Ja B 1943 *Acta Physicochim. (USSR)* **18** 1
[7] Reiss H 1985 *Advances in Chemical Reaction Dynamics (NATO ASI Series 184)* ed P M Rentzepis and C Capellos p 115
[8] Weast R C 1973 *Handbook of Chemistry and Physics* (Boca Raton, FL: Chemical Rubber Company)
[9] El-Shall M S 1989 *J. Chem. Phys.* **90** 6533
[10] Wright D, Caldwell R, Moxley C and El-Shall M S 1993 *J. Chem. Phys.* **98** 3356
[11] Ruth V, Hirth J P and Pound G M 1988 *J. Chem. Phys.* **88** 7079
[12] Pound G M 1972 *J. Phys. Chem. Ref. Data* **1** 119
[13] Langer J S and Turski L A 1973 *Phys. Rev.* A **8** 3230
[14] Reiss H, Katz J L and Cohen E R 1968 *J. Chem. Phys.* **48** 5553
[15] Reiss H and Katz J L 1967 *J. Chem. Phys.* **46** 2496
[16] Lothe J and Pound G M 1962 *J. Chem. Phys.* **36** 2080
[17] Feder J, Russell K C, Lothe J and Pound G M 1966 *Adv. Phys.* **15** 117
[18] Abraham F F 1969 *J. Chem. Phys.* **51** 1632
[19] Beylich A E and Robben F 1974 *Z. Angew. (Math.) Phys.* **25** 443
[20] Abraham O, Kim S S and Stein G D 1981 *J. Chem. Phys.* **75** 402
[21] Stein G D 1985 *Surf. Sci.* **156** 44
[22] Herring C 1953 *Structure and Properties of Solid Surfaces* (London: Cambridge University Press)
[23] Dunning W J 1969 *Nucleation* ed A C Zettlemoyer (New York: Dekker) p 2
[24] Anderson J B 1974 *Molecular Beams and Low Density Gas Dynamics* ed P P Wagner (New York: Dekker) p 1
[25] Hagena O 1974 *Molecular Beams and Low Density Gas Dynamics* ed P P Wagner (New York: Dekker) p 93

[26] Hagena O 1981 *Surf. Sci.* **106** 101
[27] Fenn J B 1982 *Appl. At. Collision Phys.* **5** 349
[28] Campargue R 1984 *J. Phys. Chem.* **88** 4466
[29] Ryali A and Fenn J 1984 *Ber. Bunsenges. Phys. Chem.* **88** 245
[30] Miller D R 1988 *Atomic and Molecular Beam Methods* ed G Scoles (New York: Oxford University Press) pp 14–53
[31] Zucrow M J and Hoffman J D 1976 *Gas Dynamics* (New York: Wiley)
[32] McCay T D and Price L L 1983 *Phys. Fluids* **26** 2115
[33] Ashkenas H and Sherman F S 1966 *Rarified Gas Dynamics* ed J H DeLeewuw (New York: Academic)
[34] Gentry W R 1988 *Atomic and Molecular Beam Methods* ed G Scoles (New York: Oxford University Press) p 55
[35] Milne T A, Vandegrift A E and Greene F T 1970 *J. Chem. Phys.* **52** 1552
[36] Hagena O 1963 *Z. Angew. Phys.* **16** 183
[37] Hagena O and Henkes W 1965 *Z. Naturf.* a **20** 1344
[38] Hagena O and Obert W 1972 *J. Chem. Phys.* **56** 1793
[39] Kappes M M, Kunz R W and Schumacher E 1982 *Chem. Phys. Lett.* **91** 413
[40] Martin T P 1986 *Angew. Chem. Int. Ed. Engl.* **25** 197
[41] Cohen M, Chou M Y, Knight W D and de Heer W A 1987 *J. Phys. Chem.* **91** 3141
[42] Phillips J C 1986 *Chem. Rev.* **86** 619
[43] Sattler K, Muhlbach J, Pfau P and Recknagel E 1982 *Phys. Lett.* **87A** 418
[44] Gohlich H, Lange T, Bergmann T and Martin T P 1990 *Phys. Rev. Lett.* **65** 748
[45] Martin T P, Bergmann T, Gohlich H and Lange T 1991 *Chem. Phys. Lett.* **176** 343
[46] Saito Y, Mihama K and Noda T 1983 *Japan. J. Appl. Phys.* **22** L179
[47] Martin T P 1984 *J. Chem. Phys.* **80** 170
[48] Sattler K 1985 *Surf. Sci.* **156** 292
[49] Sattler K 1986 *Z. Phys.* D **3** 223
[50] Martin T P 1985 *J. Chem. Phys.* **83** 78
[51] Foxon C T, Harvey J A and Joyce B A 1973 *J. Phys. Chem. Solids* **34** 1693
[52] Farrow R F C 1974 *J. Phys. D: Appl. Phys.* **7** 2436
[53] Kappes M M 1988 *Chem. Rev.* **88** 369
[54] Dietz T, Duncan M, Liverman M and Smalley R 1980 *J. Chem. Phys.* **73** 4816
[55] Dietz T, Duncan M, Powers D and Smalley R 1981 *J. Chem. Phys.* **74** 6511
[56] Smalley R E 1983 *Laser Chem.* **2** 167
[57] Powers D E, Hansen S G, Geusic M E, Puiu A C, Hopkins J B, Dietz T G, Duncan M A, Langridge-Smith P R R and Smalley R E 1982 *J. Phys. Chem.* **86** 2556
[58] Powers D E, Hansen S G, Geusic M E, Michalopoulos D L and Smalley R E 1983 *J. Chem. Phys.* **78** 2866
[59] Heath J R, Liu Y, O'Brien S C, Zhang Q-L, Curl R F, Tittel F K and Smalley R E 1985 *J. Chem. Phys.* **83** 5520
[60] Cox D M, Trevor D J, Whetten R L and Kaldor A 1988 *J. Phys. Chem.* **92** 421
[61] Mandich M L and Reents W D Jr 1990 *Atomic and Molecular Clusters* ed E R Bernstein (Amsterdam: Elsevier) p 69
[62] Riley S J, Parks E K, Pobo L G and Wexler S 1984 *Ber. Bunsenges. Phys. Chem.* **88** 287
[63] Whetten R L, Cox D M, Trevor D J and Kaldor A 1985 *Surf. Sci.* **156** 8
[64] Loh S K, Hales D A and Armentrout P B 1986 *Chem. Phys. Lett.* **129** 527

[65] Gole J L 1986 *Metal Clusters* ed M Moskovits (New York: Wiley)
[66] Rohlfing E A, Cox D M and Kaldor A 1984 *J. Chem. Phys.* **81** 3322
[67] Bloomfield L A, Freeman R R and Brown W L 1985 *Phys. Rev. Lett.* **54** 2246
[68] LaiHing K, Wheeler R G, Wilson W L and Duncan M A 1987 *J. Chem. Phys.* **87** 3401
[69] Wheeler R G, Lai-Hing K, Wilson W L and Duncan M A 1986 *Chem. Phys. Lett.* **131** 8
[70] O'Brien S C, Liu Y, Zhang Q, Heath J R, Tittel F K, Curl R F and Smalley R E 1986 *J. Chem. Phys.* **84** 4074
[71] Wheeler R G, Lai-Hing K, Wilson W L, Allen J D, King R B and Duncan M A 1986 *J. Am. Chem. Soc.* **108** 8101
[72] Chaiken J, Casey M J and Villarica M 1991 *J. Phys. Chem.* **96** 3183
[73] Tam A, Moe G G and Happer W 1975 *Phys. Rev. Lett.* **35** 1630
[74] Kizaki Y, Kandori T and Fujiyama Y 1985 *J. Appl. Phys.* **24** 800
[75] Yabuzaki T, Sato T and Ogawa T 1980 *J. Chem. Phys.* **73** 2780
[76] Rice G W 1986 *J. Am. Ceram. Soc.* **69** C-183
[77] Duncan M A, Dietz T G and Smalley R E 1981 *J. Am. Chem. Soc.* **103** 5245
[78] Beck S M 1989 *J. Chem. Phys.* **90** 6306
[79] Harano A, Kinoshita J and Koda S 1990 *Chem. Phys. Lett.* **172** 219
[80] Yamada I, Inokawa H and Takagi T 1984 *J. Appl. Phys.* **56** 2746
[81] Melinon P, Paillard V, Dupuis V, Perez A, Jensen P, Hoareau A, Perez J P, Tuaillon J, Broyer M, Vialle J L, Pellarin M, Baguenard B and Lerme J 1995 *Int. J. Mod. Phys.* B **9** 339
[82] Sugano S, Nishina Y and Ohnishi 1987 *Microclusters* (New York: Springer)
[83] Jena P, Rao B K and Khanna S N 1987 *Physics and Chemistry of Small Clusters (NATO-ASI Series 158)* (New York: Plenum)
[84] Moskovits M 1986 *Metal Clusters* (New York: Wiley)
[85] Johnson B F G 1980 *Transition Metal Clusters* (New York: Wiley)
[86] The entire issue of 1987 *J. Phys. Chem.* **91** (10) is devoted to clusters.
[87] The entire issue of 1986 *Chem. Rev.* **86** (3) is devoted to clusters.
[88] The entire issue of 1986 *Z. Phys.* D **3** (2/3) 101–327 is devoted to clusters.
[89] Russell D H 1989 *Gas Phase Inorganic Chemistry* (New York: Plenum)
[90] Jena P, Khanna S N and Rao B K 1992 *Physics and Chemistry of Finite Systems: From Clusters to Crystals (NATO-ASI Series)* (Deventer: Kluwer) vols I and II
[91] The entire issue of 1993 *Z. Phys.* D **26** (1–4) is devoted to clusters.
[92] Jortner J 1992 *Z. Phys.* D **24** 247
[93] Mays C W, Vermaak J S and Kuhlmann-Wilsdorf D 1968 *Surf. Sci.* **12** 134
[94] Woltersdorf J, Nepijko A S and Pippel E 1981 *Surf. Sci.* **106** 64
[95] Knight W D, Clemenger K, de Heer W A, Saunders W A, Chou M Y and Cohen M L 1984 *Phys. Rev. Lett.* **52** 2141
[96] Mackay A C 1962 *Acta Crystallogr.* **15** 916
[97] The subject of icosahedral shells of atoms in inert gas clusters has a long and interesting history. For a recent review see 1990 *Proc. Faraday Symp. on Large Gas Phase Clusters, J. Chem. Soc. Faraday Trans.* **86** 13
[98] Echt O, Sattler K and Recknagel E 1981 *Phys. Rev. Lett.* **47** 1121
[99] Farges J, De Feraudy M F, Raoult B and Torchet G 1986 *J. Chem. Phys.* **84** 3491
[100] Northby J A 1987 *J. Chem. Phys.* **86** 6166
[101] Lethbridge P G and Stace A J 1989 *J. Chem. Phys.* **91** 7685

[102] Schriver K E, Hahn M Y, Persson J L, LaVilla M E and Whetten R L 1989 *J. Phys. Chem.* **93** 2869

[103] Martin T P, Bergmann T, Gohlich H and Lange T 1990 *Chem. Phys. Lett.* **172** 209

[104] Bjornholm S, Borggreen J, Echt O, Hansen K, Pederson J and Rasmussen H D 1990 *Phys. Rev. Lett.* **65** 1627

[105] Brechignac C, Cahuzac Ph, de Frutos M, Roux J Ph and Bowen K 1992 *Physics and Chemistry of Finite Systems: From Clusters to Crystals* vol 1, ed P Jena, S N Khana and B K Rao (Deventer: Kluwer) p 369

[106] Farley R W, Ziemann P and Castleman A W Jr 1989 *Z. Phys.* **14** 353

[107] Martin T P, Näher U, Schaber H and Zimmermann U 1993 *Phys. Rev. Lett.* **70** 3079

[108] Harris I A, Kidwell R S and Northby J A 1984 *Phys. Rev. Lett.* **53** 2390

[109] Edelstein A S, Chow G M, Altman E I, Colton R J and Hwang D M 1991 *Science* **251** 1590

[110] Martin T P 1984 *J. Chem. Phys.* **81** 4426

[111] Kappes M M, Schär M, Radi P and Schumacher E 1986 *J. Chem. Phys.* **84** 1863

[112] Knight W D, de Heer W A and Saunders W A 1986 *Z. Phys.* D **3** 109

[113] Saunders W A, Clemenger K, de Heer W A and Knight W D 1985 *Phys. Rev.* B **32** 1366

[114] Gohlich H, Lange T, Bergmann T, Näher U and Martin T P 1992 *Physics and Chemistry of Finite Systems: From Clusters to Crystals* vol 1, ed P Jena, S N Khana and B K Rao (Deventer: Kluwer) p 581

[115] Dugourd Ph, Rayane D, Labastie P, Vezin B, Chevaleyer J and Broyer M 1992 *Physics and Chemistry of Finite Systems: From Clusters to Crystals* vol 1, ed P Jena, S N Khanna and B K Rao (Deventer: Kluwer) p 555

[116] Clemenger K 1985 *Phys. Rev.* B **32** 1359

[117] Pellarin M, Vialle J L, Lerme J, Valadier F, Baguenard B, Blanc J and Broyer M 1992 *Physics and Chemistry of Finite Systems: From Clusters to Crystals* vol 1, ed P Jena, S N Khanna and B K Rao (Deventer: Kluwer) p 633

[118] Yang S and Knickelbein M B 1990 *J. Chem. Phys.* **93** 1533

[119] Rademann K, Kaiser B, Even U and Hensel F 1987 *Phys. Rev. Lett.* **59** 2319

[120] Cheshnovsky O, Taylor K J, Conceicao J and Smalley R E 1990 *Phys. Rev. Lett.* **64** 1785

[121] Taylor K J, Pettietle-Hall C L, Cheshnovsky O and Smalley R E 1992 *J. Chem. Phys.* **96** 3319

[122] Gates B C, Guczi L and Knozinger H 1986 *Metal Clusters in Catalysis* (Amsterdam: Elsevier)

[123] Yamada I, Usui H and Takagi T 1986 *Metal Clusters* ed F Trager and G zu Putlitz (Berlin: Springer)

[124] Jarrold M F 1989 *Biomolecular Collisions* ed M N R Ashford and J E Baggott (London: Royal Society of Chemistry)

[125] Richtsmeier S C, Parks E K, Liu K, Pobo L G and Riley S J 1985 *J. Chem. Phys.* **82** 3659

[126] Geusic M E, Morse M D and Smalley R E 1985 *J. Chem. Phys.* **82** 590

[127] Trevor D J, Whetten R L, Cox D M and Kaldor A 1985 *J. Am. Chem. Soc.* **107** 518

[128] Geusic M E, Morse M D, O'Brian S C and Smalley R E 1985 *Rev. Sci. Instrum.* **56** 2124

[129] Cox D M, Zakin M R and Kaldor A 1987 *Physics and Chemistry of Small Clusters* ed P Jena, B K Rao and S N Khanna (New York: Plenum) p 741
[130] Morse M D, Geusic M E, Heath J R and Smalley R E 1985 *J. Chem. Phys.* **83** 2293
[131] Hoffman W F, Parks E K, Nieman G C, Pobo L G and Riley S J 1987 *Z. Phys.* D **7** 83
[132] Whetten R L, Zakin M R, Cox D M, Trevor D J and Kaldor A 1986 *J. Chem. Phys.* **85** 1697
[133] Kaldor A, Cox D M, Trevor D J and Zakin M R 1986 *Metal Clusters* ed F Trager and G zu Putlitz (Berlin: Springer) p 15
[134] St Pierre R J and El-Sayed M A 1987 *J. Phys. Chem.* **91** 763
[135] St Pierre R J, Chronister E L and El-Sayed M A 1987 *J. Phys. Chem.* **91** 5228
[136] Zakin M R, Cox D M and Kaldor A 1987 *J. Phys. Chem.* **91** 5224
[137] Cox D M, Reichmann K C, Trevo D J and Kaldor A 1988 *J. Chem. Phys.* **88** 111
[138] Riley S J, Parks E K, Nieman G C, Pobo L G and Wexler S 1984 *J. Chem. Phys.* **80** 1360
[139] Kaldor A, Cox D and Zakin M R 1987 *Springer Series Materials Science* vol 4 (Berlin: Springer) p 96
[140] Whetten R L and Schriver K E 1989 *Gas Phase Inorganic Chemistry* ed D H Russell (New York: Plenum) pp 193–226
[141] Goldstein A N, Echer C M and Alivisatos A P 1992 *Science* **256** 1425
[142] Coombes C J 1972 *J. Phys. F: Met. Phys.* **2** 441
[143] Buffat Ph and Borel J-P 1976 *Phys. Rev.* A **13** 2287
[144] Borel J-P 1981 *Surf. Sci.* **106** 1
[145] Peppiatt S J and Sambles J R 1975 *Proc. R. Soc.* A **345** 387
[146] Briant C L and Burton J J 1975 *J. Chem. Phys.* **63** 2045
[147] Cotterill R M J 1975 *Phil. Mag.* **32** 1283
[148] Couchman P R and Jesser W A 1977 *Nature* **269** 481
[149] Broughton J 1991 *Phys. Rev. Lett.* **67** 2990
[150] Davis H L, Jellinek J and Berry R S 1987 *J. Chem. Phys.* **86** 6456
[151] Ajayan P M and Marks L D 1988 *Phys. Rev. Lett.* **60** 585
[152] Tolbert S H and Alivisatos A P 1994 *Science* **265** 373
[153] Reddy B V, Khanna S N and Dunlap B I 1993 *Phys. Rev. Lett.* **70** 3323
[154] Cox A J, Louderback J G and Bloomfield L A 1993 *Phys. Rev. Lett.* **71** 923
[155] Cox D M, Trevor D J, Whetten R L, Rohlfing E A and Kaldor A 1985 *Phys. Rev.* B **32** 7290
[156] de Heer W A, Milani P and Châtelain A 1990 *Phys. Rev. Lett.* **65** 488
[157] Bucher J P, Douglass D C and Bloomfield L A 1991 *Phys. Rev. Lett.* **66** 3052
[158] Billas I M L, Becker J A, Châtelain A and de Heer W A 1993 *Phys. Rev. Lett.* **71** 4067
[159] Billas I M L, Châtelain A and de Heer W A 1994 *Science* **265** 1682
[160] van Leeuwen D A, van Ruitenbeek J M, de Jongh L J, Ceriotti A, Pacchioni G, Häberlen O D and Rösch N 1994 *Phys. Rev. Lett.* **73** 1432
[161] Pfund A H 1930 *Phys. Rev.* **35** 1434
[162] Burger H C and van Cittert P H 1930 *Z. Phys.* **66** 210
[163] Harris L, Jeffries D and Siegel B M 1948 *J. Appl. Phys.* **19** 791
[164] Chaudhari P and Matthews J W 1970 *Appl. Phys. Lett.* **17** 115
[165] Kaito C, Fujita K, Shibahara H and Shiojiri M 1977 *Japan. J. Appl. Phys.* **16** 697

[166] Kaito C 1978 *Japan. J. Appl. Phys.* **17** 601
[167] Yukawa N, Hida M, Imura T, Kawamura M and Mizuno Y 1972 *Metall. Trans.* **3** 887
[168] Hirayama T 1987 *J. Am. Ceram. Soc.* **70** C-122
[169] Siegel R W, Ramasamy S, Hahn H, Zongquan L, Ting L and Gronsky R 1988 *J. Mater. Res.* **3** 1367
[170] Kashu S, Fuchita E, Manabe T and Hayashi C 1984 *Japan. J. Appl. Phys.* **23** L910
[171] Mayo M J, Siegel R W, Narayanasamy A and Nix W D 1990 *J. Mater. Res.* **5** 1073
[172] Buckle E R, Tsakiropoulos P and Pointon K C 1986 *Int. Met. Rev.* **31** 258
[173] Gleiter H 1989 *Prog. Mater. Sci.* **33** 223
[174] Uyeda R 1991 *Prog. Mater. Sci.* **35** 1
[175] Granqvist C G and Buhrman R A 1976 *J. Appl. Phys.* **47** 2200
[176] Ichinose N, Ozaki Y and Kashu S 1992 *Superfine Particle Technology* (London: Springer) (Engl. Transl.)
[177] Birringer R, Herr U and Gleiter H 1986 *Trans. Japan. Inst. Met. Suppl.* **27** 43
[178] Uyeda R 1974 *J. Cryst. Growth* **24** 69
[179] Hahn H and Averback R S 1990 *J. Appl. Phys.* **67** 1113
[180] Mandich M L, Bondybey V E and Reents W D 1987 *J. Chem. Phys.* **86** 4245
[181] Berkowitz A E and Walter J L 1987 *J. Mater. Res.* **2** 277
[182] Hung C H and Katz J L 1992 *J. Mater. Res.* **7** 1861
[183] Berkowitz A E 1994 private communication
[184] Majetich S A, Artman J O, McHenry M E, Nuhfer N T and Staley S W 1993 *Phys. Rev.* B **48** 16845
[185] Edelstein A S, Kaatz F, Chow G M and Peritt J M 1992 *Studies of Magnetic Properties of Fine Particles & Their Relevance to Materials Science* ed J L Dormann and D Fiorani (Amsterdam: Elsevier) p 47
[186] Matsunawa A and Katayama S 1985 *Proc. ICALEO, Laser Welding, Machining and Materials Processing* ed C Abright (Berlin: IFS) pp 41, 205
[187] Matsunawa A and Katayama S 1985 *Trans. JWRI (Japan)* **14** 197
[188] Huibin X, Shusong T and Ngal L T 1992 *Trans. NF Soc.* **2** 58
[189] El-Shall M S, Slack W, Vann W, Kane D and Hanley D 1994 *J. Phys. Chem.* **98** 3067
[190] Katz J L 1970 *J. Chem. Phys.* **52** 4733
[191] Iwama S, Hayakawa K and Arizumi T 1982 *J. Cryst. Growth.* **56** 265
[192] Baba K, Shohata N and Yonezawa 1989 *Appl. Phys. Lett.* **54** 2309
[193] Zafiropulos V, Kollia Z, Fotakis C and Stockdale J A D 1993 *J. Chem. Phys.* **98** 5079
[194] Matsuzaki A, Horita H and Hamada Y 1992 *Chem. Phys. Lett.* **190** 331
[195] Haggerty J S 1981 *Laser-Induced Chemical Processes* ed J I Steinfeld (New York: Plenum)
[196] Fantoni R, Borsella E, Piccirillo S and Enzo S 1990 *Proc. SPIE* **1279** 77
[197] Buerki P R, Troxler T and Leutwyler S 1990 *High-Temp. Sci.* **27** 323
[198] Curcio F, Ghiglione G, Musci M and Nannetti C 1989 *Appl. Surf. Sci.* **36** 52
[199] Rice G W and Woodin R L 1988 *J. Am. Ceram. Soc.* **71** C181
[200] Curcio F, Musci M, Notaro N and De Michele G 1990 *Appl. Surf. Sci.* **46** 225
[201] Xiang Xin Bi, Ganguly B, Huffman G P, Huggins F E, Endo M and Eklund P C 1993 *J. Mater. Res.* **8** 1666

[202] Kim S G and Brock J R 1986 *J. Appl. Phys.* **60** 509
[203] Kurtz S K and Carpay F M A 1980 *J. Appl. Phys.* **51** 5725
[204] Kaatz F H, Chow G M and Edelstein A S 1993 *J. Mater. Res.* **8** 995
[205] Haas V, Gleiter H and Birringer R 1993 *Scr. Metall. Mater.* **28** 721
[206] Koschmieder E L 1993 *Bénard Cells and Taylor Vortices* (Cambridge: Cambridge University Press)
[207] Jaluria Y 1980 *Natural Convection: Heat and Mass Transfer* (Oxford: Pergamon)
[208] Kawamura K 1973 *Japan. J. Appl. Phys.* **12** 1685
[209] Kasukabe S, Yatsuya S and Uyeda R 1974 *Japan. J. Appl. Phys.* **13** 1714
[210] Yatsuya S, Kasukabe S and Uyeda R 1974 *J. Cryst. Growth* **24/25** 319
[211] Hayashi T, Ohno T, Yatsuya S and Uyeda R 1977 *Japan. J. Appl. Phys.* **16** 705
[212] Hayashi C 1987 *J. Vac. Sci. Technol.* A **5** 1375
[213] Eversole J D and Broida H P 1974 *J. Appl. Phys.* **45** 596
[214] Wulff G 1901 *Z. Kristallogr.* **34** 449
[215] Saito Y, Mihama K and Uyeda R 1980 *Japan. J. Appl. Phys.* **19** 1603
[216] Smalley R E 1990 *Atomic and Molecular Clusters* ed E R Bernstein (Amsterdam: Elsevier) p 1
[217] El-Shall M S, Li S, Turkki T, Graiver D, Pernisz U C and Baraton M I 1995 *J. Phys. Chem.* **99** 17 805

Chapter 3

Particle synthesis by chemical routes

G M Chow and K E Gonsalves

3.1 INTRODUCTION

In the chemical preparation of nanoscale particles with desired properties, the structural properties (crystalline or amorphous structure, size, shape, morphology), and chemical properties (composition of the bulk, interface, and surface) are important factors to be considered.

Because of its advantages, the role of chemistry in materials science has been rapidly growing [1]. The strength of chemistry in materials science is its versatility in designing and synthesizing new materials, which can be processed and fabricated into final components. Chemical synthesis permits the manipulation of matter at the molecular level. Because of mixing at the molecular level, good chemical homogeneity can be achieved. Also, by understanding the relationship between how matter is assembled on an atomic and molecular level and the material macroscopic properties, molecular synthetic chemistry can be tailor designed to prepare novel starting components. Better control of the particle size, shape, and size distribution can be achieved in particle synthesis. To benefit from the advantages of chemical processing, an understanding of the principles of crystal chemistry, thermodynamics, phase equilibrium, and reaction kinetics is required [2].

There are also potential difficulties in chemical processing. In some preparations, the chemistry is complex and hazardous. Entrapment of impurities in the final product needs to be avoided or minimized to obtain desired properties. Scaling up for the economical production of a large quantity of material may be relatively easy for some but not all systems. Another problem is that undesirable agglomeration at any stage of the synthesis process can change the properties.

Many liquid phase chemistry methods exist for synthesizing nanoscale or ultrafine particles. The scope of this chapter is limited to describing some selected examples of these methods for making nanoscale particles or powders, and is not intended to provide a comprehensive review. Chemical

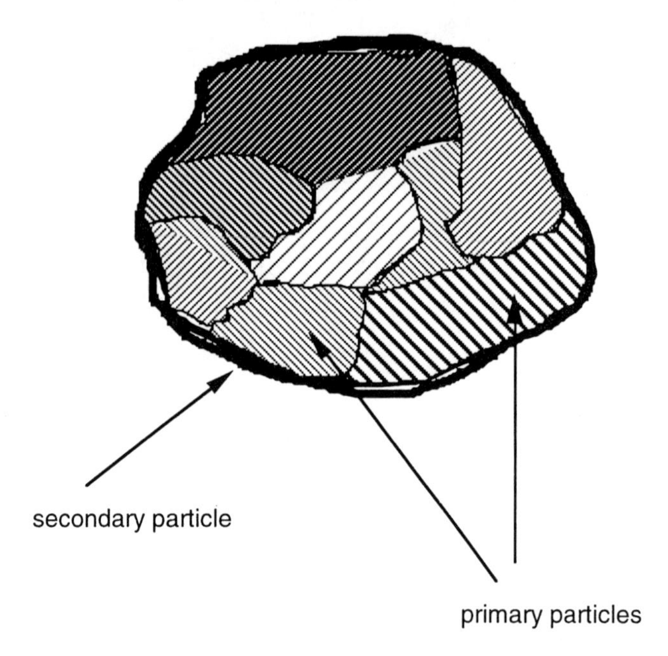

Figure 3.1. A schematic diagram showing the primary and secondary particles.

techniques for making films and modifying surfaces with nanoscale structures, such as electrochemical methods [3], are not discussed here. For further details concerning the procedures of the examples in this section and for descriptions of other related methods, the readers are encouraged to consult the references. The examples include both conventional and more recent methods. In some examples, liquid phase chemistry is used to prepare the precursor, which is subsequently converted to nanoscale particles by non-liquid phase chemical reactions. In the following, the size of a powder refers to the particle dimension as observed by imaging techniques such as scanning electron microscopy (SEM). The particle may be a single unit, e.g. a single crystal, or it may consist of subunits. The small subunits are defined as the primary particles and the agglomerates of these primary particles are called secondary particles (figure 3.1). The measurement of particle size by SEM often can only determine the size of the secondary particles. For crystalline materials, the size of primary nanoparticles can be estimated by the amount by which the x-ray line is broadened, or determined from dark-field imaging by transmission electron microscopy (TEM) or from lattice imaging by high-resolution transmission electron microscopy (HRTEM). Use of dark-field TEM and HRTEM for determining the primary particle size is preferred over x-ray line broadening. These techniques are more direct and less likely to be affected by experimental errors and/or other properties of the particles such as strain or a distribution in the size of the lattice parameter. For amorphous particles, the size of primary particles can also be estimated by bright-field imaging using TEM or HRTEM. We stress the importance of carefully defining the term 'particle size' to avoid confusion.

3.2 NUCLEATION AND GROWTH FROM SOLUTIONS

Precipitation of a solid from a solution is a common technique for the synthesis of fine particles. The general procedure involves reactions in aqueous or non-aqueous solutions containing the soluble or suspended salts. Once the solution becomes supersaturated with the product, a precipitate is formed by either homogeneous or heterogeneous nucleation. Homogeneous and heterogeneous nucleation refer to the formation of stable nuclei with or without foreign species respectively. After the nuclei are formed, their growth usually proceeds by diffusion. In diffusion controlled growth, concentration gradients and the temperature are important in determining the growth rate. To form monodispersed particles, i.e. unagglomerated particles with a very narrow size distribution, all the nuclei must form at nearly the same time, and subsequent growth must occur without further nucleation [4] or agglomeration of the particles.

In general, the particle size and particle-size distribution, the amount of crystallinity, the crystal structure, and the degree of dispersion can be affected by reaction kinetics. Factors influencing the rate of reactions include the concentration of reactants, the reaction temperature, the pH, and the order in which the reagents are added to the solution. A multi-element material is often made by coprecipitation of the batched ions. However, it is not always easy to simultaneously coprecipitate all the desired ions, since different species may only precipitate at different pH. Thus, special attention is required to control chemical homogeneity and stoichiometry. Phase separation may be avoided during liquid precipitation and the homogeneity at the molecular level improved by converting the precursor to powder form by using spray drying [5] and freeze drying [6]. Details of nucleation and growth of particles have been discussed in chapter 2.

3.3 STABILIZATION OF FINE PARTICLES AGAINST AGGLOMERATION

Fine particles, particularly nanoscale particles, since they have large surface areas, often agglomerate to form either lumps or secondary particles to minimize the total surface or interfacial energy of the system. When the particles are strongly stuck together, these hard agglomerates are called aggregates. Many materials containing fine particles, some examples including paints, pigments, electronic inks, and ferrofluids, are useful if the particles in the fluid suspension remain unagglomerated or dispersed. For instance, the desirable magnetic properties caused by single-magnetic-domain behavior cannot be realized if the ferromagnetic nanoscale particles are not isolated from each other. In the processing of ceramic materials, if the starting powders are adversely agglomerated with entrapped large pores, the green body formed by compacting the powders may have low density. The green body will fail to shrink and densify during sintering for pores above a certain size. Further details regarding the adverse effects of agglomeration of powders on consolidation may be found in chapter 8. Agglomeration of fine particles can occur at the synthesis stage, during drying and subsequent processing of the particles. Thus it is very important to stabilize the particles against adverse agglomeration at each step of particle production and powder processing. Surfactants are used to produce dispersed particles in the synthesis process or disperse as-synthesized agglomerated fine particles. The dispersion of fine particles in liquid media by surfactants has been studied intensively [7].

Many technologies use surfactants [8, 9]. A surfactant is a surface-active agent that has an amphipathic structure in that solvent, i.e. a lyophobic (solvent repulsive) and lyophilic group (solvent attractive). Depending on the charges at the surface-active portions, surfactants are classified as either anionic, cationic, zwitterionic (bearing both positive and negative charges), or non-ionic (no charges). At low concentrations, the surfactant molecules adsorb on the surfaces or interfaces in the system, and can significantly alter the interfacial energies.

Agglomeration of fine particles is caused by the attractive van der Waals force and/or the driving force that tends to minimize the total surface energy of the system. Repulsive interparticle forces are required to prevent the agglomeration of these particles. Two methods are commonly used. The first method provides the dispersion by electrostatic repulsion. This repulsion results from the interactions between the electric double layers surrounding the particles. An unequal charge distribution always exists between a particle surface and the solvent. Electrostatic stabilization of a dispersion occurs when the electrostatic repulsive force overcomes the attractive van der Waals forces between the particles. This stabilization method is generally effective in dilute systems of aqueous or polar organic media. This method is very sensitive to the electrolyte concentration since a change in the concentration may destroy the electric double layer, which will result in particle agglomeration.

The second method of stabilization involves the steric forces. Surfactant molecules can adsorb onto the surfaces of particles and their lyophilic chains will then extend into the solvent and interact with each other. The solvent–chain interaction, which is a mixing effect, increases the free energy of the system and produces an energy barrier to the closer approach of particles. When the particles come into closer contact with each other, the motion of the chains extending into the solvent become restricted and produce an entropic effect. Steric stabilization can occur in the absence of the electric barriers. Steric stabilization is effective in both aqueous and non-aqueous media, and is less sensitive to impurities or trace additives than electric stabilization. The steric stabilization method is particularly effective in dispersing high concentrations of particles.

Surfactants are evaluated by their efficiency and effectiveness [7]. Efficiency is a measure of the equilibrium concentration of a surfactant required to produce a given amount of effect on the interfacial process, and is related to the free energy change. An example of such an effect is the lack of agglomeration as determined by visual inspection. Effectiveness is the maximum effect that can be obtained when the surface is saturated with the surfactant, regardless of its concentration. A list of values for the effectiveness and efficiency of adsorption of surfactants at aqueous solution–air and aqueous solution–hydrocarbon interfaces is given in [7].

3.4 MATERIALS

3.4.1 Metals and intermetallics

3.4.1.1 Aqueous methods

Fine metal powders have applications as electronic and magnetic materials, explosives, catalysts, pharmaceuticals, and in powder metallurgy. Precious metal powders for electronic applications can be prepared by adding liquid reducing agents to the aqueous solutions of respective salts at adjusted pH [10]. The reducing agent need not be a liquid. Metal ions in aqueous solutions have been reduced by hydrogen to produce metal powders [11]. At sufficiently high concentrations of metal ions or high pH values, almost any metal can be reduced by hydrogen, provided that the formation of stable hydroxides can be avoided. Conventional aqueous precipitation provides powders with particle size in the micron range as seen by SEM. However, these micron-sized particles are sometimes secondary particles formed by the agglomeration of nanoscale primary particles.

Although amorphous particles are conventionally produced by non-chemical methods such as rapid quenching of a high-temperature melt, amorphous particles can also be made by solution chemistry, if the reaction temperature is less than the glass transition temperature. Nanoscale ferromagnetic particles can be obtained by the reduction of metal salts with aqueous sodium or potassium borohydride [12]. Using this technique, nanoscale particles of amorphous ferromagnetic alloys have been obtained at low reaction temperatures. In addition, the coprecipitation of boron in these alloys helped stabilize the metastable amorphous structure. This example shows that the incorporation of impurities can sometimes be beneficial. By reducing the boron content, crystalline metastable ferromagnetic alloy particles such as Fe–Cu [13] and metastable Co–Cu alloy particles [14] were made by the aqueous borohydride method. The x-ray diffraction (XRD) data of as-synthesized powders indicated that metastable alloys may have formed [13, 14]. Upon annealing these powders, phase separation occurred and composites with distinct phases were formed.

Figure 3.2 shows an HRTEM micrograph of the Fe–Cu particles. Generally, the precipitated fine powders are agglomerated and the secondary particles are typically microns in size. Stabilization of the powders against agglomeration can often be provided by adding surfactants [15] as discussed above.

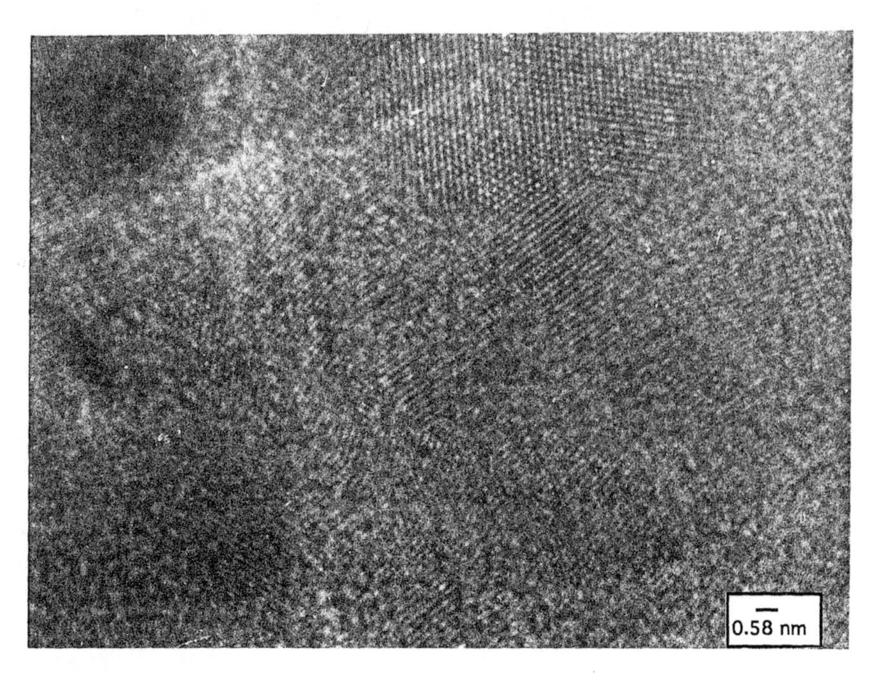

Figure 3.2. An HRTEM micrograph of Fe–Cu metastable alloy particles (after reference [13]).

3.4.1.2 Non-aqueous methods

Fine particles can also be synthesized using organic and organometallic reagents. Metal colloids, which can be used in applications requiring dispersed particles, for example, ferrofluids, have been made by the thermolysis of transition metal carbonyls in an inert atmosphere. Crystalline iron colloids with particle size between 5 and 20 nm have been produced by thermolysis of $Fe(CO)_5$ in polymer solutions [16]. However, the surface of these nanoscale particles oxidized when they were exposed to the room atmosphere. In general, the post-synthesis handling of nanoparticles requires special attention. Thermal decomposition of $Fe(CO)_5$ in an organic surfactant liquid also produced ferrofluids of amorphous iron particles with a 8.5 nm median diameter [17]. It was suggested that carbon atom impurities stabilized the amorphous structure of the particles. The colloidal stability of this ferrofluid was achieved by the long-chain surfactant molecules that prevented the close approach of particles because of entropic repulsion, thus preventing agglomeration. Monodispersed pure or coated iron submicron particles consisting of nanocrystallites were obtained by reducing the colloidal α-Fe_2O_3 (uncoated or coated with silica or cobalt oxide) in a hydrogen atmosphere [18]. The size of the iron crystallites increased with increasing reduction temperature and was approximately 80 nm at 450 °C.

Thermal reduction of inorganic metallic compounds in a polyol such as ethylene glycol or diethylene glycol was used to produce monodispersed particles of Co, Ni, Cu, and precious metals [19]. Reaction temperatures as high as the boiling point of the polyol can be used for metals that are not reduced easily. The polyol acts as a solvent for the starting metallic compounds and, subsequently, is also used to reduce them to metals. The reaction is a dissolution process and not a solid-state phase transformation. The number of nuclei formed and the rate of reduction increases with increasing temperature, resulting in a decrease in the particle size. Although micron-sized particles usually result from homogeneous nucleation in this process, nanoscale particles can be synthesized by the addition of impurities to promote heterogeneous nucleation. The particle size decreases as the concentration of foreign nuclei is increased.

Bimetallic PdCu colloids have been prepared by the thermal decomposition of copper acetate and palladium acetate in high-boiling-point organic solvents such as bromobenzene and xylenes [20]. An alternative method for producing PdCu colloids is by reducing the palladium acetate and copper acetate in a boiling alcohol such as 2-ethoxyethanol [21]. It was postulated [21] that the boiling alcohol rapidly reduced Pd(II) to Pd(0), which then reduced Cu(II) at the surface of the growing Pd particles. Poly(vinylpyrrolidone) was added to the solutions to stabilize the colloidal particles with sizes in the range of 3–5 nm.

The preparation of finely divided metal and alloy particles has been reported [22] via the reduction of metal salts of groups 6–12 and 14 in organic phases using hydroorganoborates of the general formal $M'H_\upsilon(BR_3)$ or $M'H_\upsilon[BR_n(OR')_{3-n}]_\upsilon$ (where M' = alkali or alkaline earth metal, $\upsilon = 1, 2$, R, R' = alkyl or aryl) at low temperatures (23–67 °C). The particle size, as determined by SEM, was between 10 and 100 nm, and the structure of the powders was found to be either microcrystalline or amorphous. The impurity content of boron was less than 1.5%. A simple co-reduction of different metal salts (metals of groups 6–12) can be used to prepare metallic alloys. For example, co-reduction of $FeCl_3$ and $CoCl_2$ using $LiBEt_3H$ in THF, where BEt_3 is triethyl boron and THF is tetrahydrofuran, yielded a boron free Fe/Co alloy powder with a particle size of approximately 100 nm [22]. Stable metal colloids of elements of groups 6–11 have also been synthesized in an organic phase by treating metal salts with tetraalkylammonium hydridoorganoborates [23]

$$MX_n + nNR_4BEt_3H \rightarrow M_{colloid} + nNR_4X + nBEt_3 + n/2H_2 \uparrow$$

M: metal of groups 6–11; X: Cl, Br; n: 2,3; R: alkyl C_4–C_{20}.

A TEM study of these particles indicated a particle size of about 6 nm. An analogous study was conducted [24] of the room-temperature reduction of group six metal chlorides $CrCl_3$, $MoCl_3$, $MoCl_4$, and WCl_4 in toluene with $NaBEt_3H$ to prepare agglomerates of 1–5 nm metallic crystallites.

Other approaches for the synthesis of highly reactive metal powders from the reduction of metal salts by reducing agents include the following [25]: (i) alkali metals in ethereal or hydrocarbon solvents; (ii) alkali metals in the presence of an electron carrier such as naphthalene; (iii) stoichiometric amounts of lithium naphthalide. An alternative approach is to rapidly reduce soluble compounds of transition metals and post-transition metals in dimethyl ether or tetrahydrofuran at temperatures of approximately $-50\,^{\circ}C$, by dissolved alkalides or electrides to produce metal particles with crystallite size from 3 to 15 nm [26]. This method can also be applied to the formation of finely dispersed metal particles on oxide supports.

The interest in intermetallic compounds arises from their attractive properties such as their high melting temperature, high-temperature strength, corrosion resistance, and low density. Powders of NiAl and Ni_3Al were prepared by heat treatment of the organometallic precursors synthesized by coprecipitation of constituent metallic salts in ammonium benzoate and hydradinium monochloride [27]. However, the particle size of these powders was quite large (1 to 3 microns). Nanoscale TiB_2 particles have been synthesized from the thermal conversion of the amorphous precursor $Ti(BH_4)_3(solvent)_n$ obtained by a wet chemical reaction of $TiCl_4$ with sodium borohydride [28]. The precipitated precursors were converted to a mixture of TiB_2 and $TiBO_3$ by heating them in a vacuum at $950\,^{\circ}C$. The TiB_2 phase existed both as rods with diameters ranging from 20 to 40 nm and with an aspect ratio of nine, and as equiaxed particles with diameters less than 300 nm.

3.4.2 Ceramics

Ceramic nanomaterials for functional or structural applications can be fabricated by ceramic powder processing. It has been pointed out [29] that it is very advantageous to produce ceramic submicron particles that are equiaxed, have a narrow size distribution, are dispersed or unagglomerated, and chemically homogeneous. For example, a green body consolidated from fine ceramic particles may be sintered at lower temperatures than that made of larger particles with the same packing density. Advances in the chemical synthesis of ceramics are discussed in several reviews [30]. Details of sol–gel processing [31], a method that can be used to make nanostructured powders, films, fibers, and monoliths, can be found in chapter 7 and are addressed here. A chemical method can be designed to improve the properties by controlling the fine-scale structure at an early stage of the fabrication [32].

Precipitation from solution is also an important conventional technique for the synthesis of ceramic particles such as oxides and hydroxides. The precipitated phase is often a precursor powder, and heating at a high temperature is necessary to obtain the final product. For example, precipitation reactions were used to obtain oxalates of lanthinum, barium, and copper from electrolyte solutions, which after sintering formed the high-T_c oxide superconductor $La_{1.85}Ba_{0.15}CuO_4$ [33]. Tetragonal ZrO_2 powders with particle size between 16 and 30 nm were obtained by hydrothermal treatment at 100 MPa of amorphous hydrous zirconia obtained by precipitation [34]. Nanoscale oxide ceramic powders can also be prepared by combustion synthesis. For example, a precursor was prepared by combining glycine with metal nitrates in aqueous solution [35]. The amino acid, glycine, had two functions. First, it complexed with metal ions, which increased their solubility and prevented phase separation during the removal of solvent; and second, it served as a fuel for the combustion synthesis. The combustion was self-sustained once the precursor solution was heated to autoignite. The combustion could produce either the final product or a precursor that required calcining to produce the final product.

Monodispersed particles were prepared by precipitation from homogeneous solutions using the following methods [36]: (i) deprotonation of hydrated cations; (ii) controlled release of precipitating anions; (iii) thermal decomposition of metal complexes such as organometallic compounds. It has been observed that submicron colloidal particles are formed by the aggregation of nanoscale spherical particles (figure 3.3) [37].

Low-temperature, non-aqueous phase reactions can be used to prepare non-oxide ceramics such as carbides, borides, and nitrides. For example, high-temperature materials such as molybdenum carbide and tungsten carbide particles have been synthesized by the room-temperature reduction of molybdenum and tungsten halides with $LiBEt_3H$ [38]. The micron-sized particles were agglomerates of 2 nm primary particles. Nanoscale silicon nitride powders were obtained by calcining the precursors derived from liquid phase reactions between silicon tetrachloride and ammonia [39]. Infrared transmitting ZnS powders consisting of primary particles 100 nm or smaller were obtained by passing H_2S (g) through a solution of $Zn(C_2H_5)_2$ and heptane [40]. Boride and carbide powders were synthesized from the reductive dechlorination of halide solutions [41]. The precursor derived from the reaction of NH_4Cl and $NaBH_4$ in benzene between 170 °C and 180 °C could be calcined in nitrogen at 1100 °C to yield aggregated crystalline BN powders consisting of 20–30 nm crystallites [42].

3.4.3 Composites

The traditional approach to making composite powders is to crush, grind, and blend the constituent powders. Great effort is required to produce submicron-size powder by this method. Chemical methods provide a more direct route for synthesizing composites. In chemical routes, chemical homogeneity is achieved at the molecular scale and the particle size can be reduced to a nanoscale regime. For example, ceramic/metal composite materials such as WC/Co have been synthesized by thermochemical conversion of precursor powders obtained by aqueous precipitation of tris(ethylenediamine)-cobalt (II) tungstate, tungstic

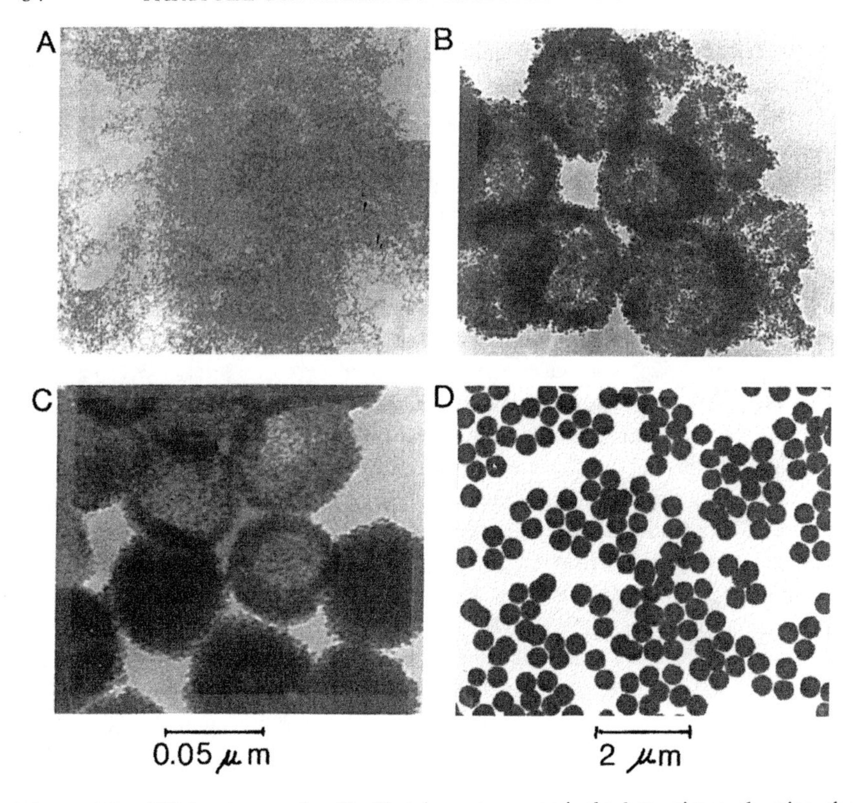

Figure 3.3. TEM micrographs (A–C) taken at successively later times showing the formation of colloidal CeO_2 particles of submicron size by aggregation of small spherical subunits. CeO_2 particles were formed by forced hydrolysis of an acidic (4.0×10^{-2} mol dm^{-3} H_2SO_4) solution of $Ce(SO_4)_2$ (1.0×10^{-3} mol dm^{-3}) heated at 90 °C during a 6 h period. (D) The dispersion at the completion of aging after 48 h. (After reference [37].) (Photo courtesy of E Matijevic.)

acids, and ammonium hydroxide. These materials have potential uses as fine drill bits, as cutting tools, and in high-wear applications. The precursor solution mixture was aerosolized and rapidly spray-dried to give extremely fine mixtures of tungsten and cobalt salts. This precursor powder was then reduced with hydrogen and reacted with carbon monoxide in a fluidized bed reactor to yield nanophase cobalt/tungsten carbide powder [43].

Preceramic polymers are inorganic or organometallic polymers that can be converted to ceramic materials by pyrolysis [44]. The following are two examples. Nanocomposite powders of Si_3N_4/SiC were obtained by thermochemical conversion of the preceramic powder by organometallic/polymer chemistry. A liquid organosilazane precursor $[CH_3SiHNH]_n$ ($n = 3, 4$) was synthesized by ammonolysis of methyl dichlorosilane [45]. Exposing aerosols of this organosilazane liquid precursor to a laser beam converted them into preceramic polymer particles [46]. Subsequent thermochemical treatment of the preceramic particles resulted in the formation of Si_3N_4/SiC nanoparticles. Nanocomposite powders of these materials have also been obtained by the conversion of the precursors in a hot-wall reactor [47]. Nanocomposite powders of AlN/BN were synthesized from the thermal conversion of a precursor gel derived from the reaction of tris-ethylaminoborane with diethyl aluminum amide [48]. Synthesis of this nanocomposite powder has also been recently achieved by a different approach [49] using boric acid, urea, and $AlCl_3 \cdot 6H_2O$ as the starting compounds. Ammonolysis of the aqueous solution of the above compound formed a precomposite gel. The latter was converted into the AlN/BN nanocomposite on further heat treatment. Thermal and compositional effects on the AlN/BN nanocomposite powders have also been recently investigated [50].

Magnetic nanocomposite powders of Fe_xN/BN ($x = 3, 4$) were prepared by the thermochemical conversion of a Fe–B–N containing precomposite synthesized through an aqueous chemical route [51]. The precomposite gel was converted by heating it at 400 °C in an NH_3 atmosphere to obtain the final composite material. Granular Co–Cu magnetic nanocomposite powders were fabricated by annealing the metastable Co–Cu alloys prepared by borohydride reduction [14], or by the polyol method [52].

3.4.4 Nanoparticles via organized membrane

As discussed above, the colloidal particles are often stabilized by the addition of surfactants or polymeric molecules to control the interparticle force. An alternative approach to synthesize stabilized, size controlled nanoscale particles is to utilize self-assembled membranes [53]. The self-organized biological and organic membrane assemblies used for this purpose include micelles, microemulsions, liposomes, and vesicles. The molecules of these assemblies have a polar headgroup with a nonpolar hydrocarbon tail, which self-assembles into membrane structures in an aqueous environment (figure 3.4). Aqueous and reverse micelles have diameters in the range of 3–6 nm. The diameters of the microemulsion are 5–100 nm. Liposomes and vesicles are closed bilayer aggregates formed from phospholipids and surfactants respectively. The multilamellar vesicles are 100–800 nm in diameter whereas the single-bilayer vesicles are 30–60 nm in diameter. Vesicles of uniform size can be prepared by sonication [54]. The membrane structures serve as reaction cages to provide a means to control supersaturation (thus nucleation and growth of the particle), particle size and distribution, and chemical homogeneity for composite material at the level of membrane dimension. They also act as agglomeration barriers.

Stabilized, colloidal nanoparticles such as semiconductor quantum crystallites of CdSe [55] have been made using reversed micelles. Phospholipid vesicles can also be used as reaction cages for intravesicular precipitation of nanoparticles. The general approach is to trap ions or molecules in

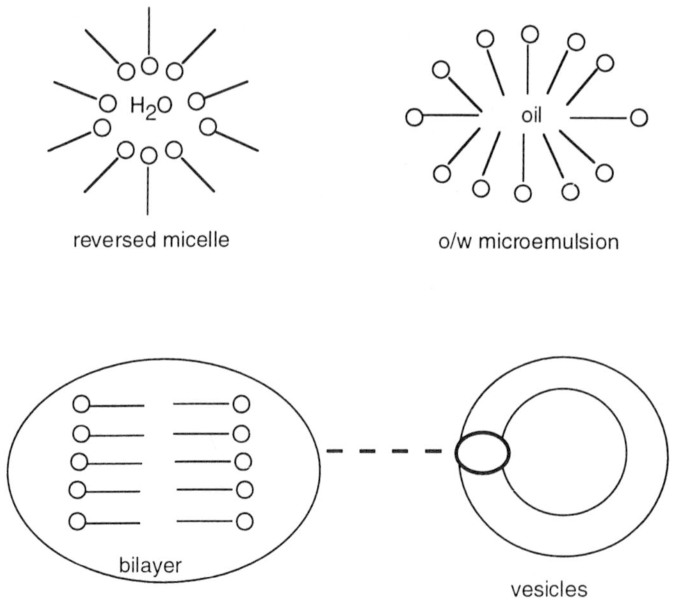

Figure 3.4. A schematic diagram showing different self-assembled membrane structures.

the aqueous compartment of lipid vesicles. This is done by forming the vesicles in the presence of these substances via sonication of the solution. The exogenous ions are removed from the solution by either gel filtration or dialysis. The permeability of anions across the membrane is several orders of magnitudes higher than that of cations. After the diffusion of appropriate anions, precipitation of inorganic particles occurs inside the vesicular compartment. For example, nanoparticles of silver oxide [56], iron oxides [57], aluminum oxide [58], and cobalt ferrite [59] have been synthesized by intravesicular precipitation. CdS and ZnS particles have also been deposited on extravesicular surfaces [60]. Particles are sterically stabilized when they are embedded within bilayer vesicles. However, particles deposited on the exterior surfaces of vesicles tend to be agglomerated. Metallic particles can also be made using the self-assembled vesicle membranes as the reaction cage. For example, nanoscale gold particles have been synthesized inside polymerized vesicles using electroless metallization [61]. Figure 3.5 shows a TEM micrograph of nanoscale gold particles synthesized using the polymerized vesicles. Nanoscale particles of Fe_3O_4 can also be synthesized in apoferritin which is used as a confined reaction environment [62] (see chapter 18 for further discussion). The method involves the removal of native ferrihydrite cores from horse spleens by dialysis, and then the reconstitution of apoferritin with Fe(II) solution under slow oxidative conditions. Transmission electron microscopy and electron diffraction confirmed the formation of 6 nm spherical single crystals of Fe_3O_4. The magnetic protein particles could have potential applications in biomedical imaging, cell labeling, and separation procedures.

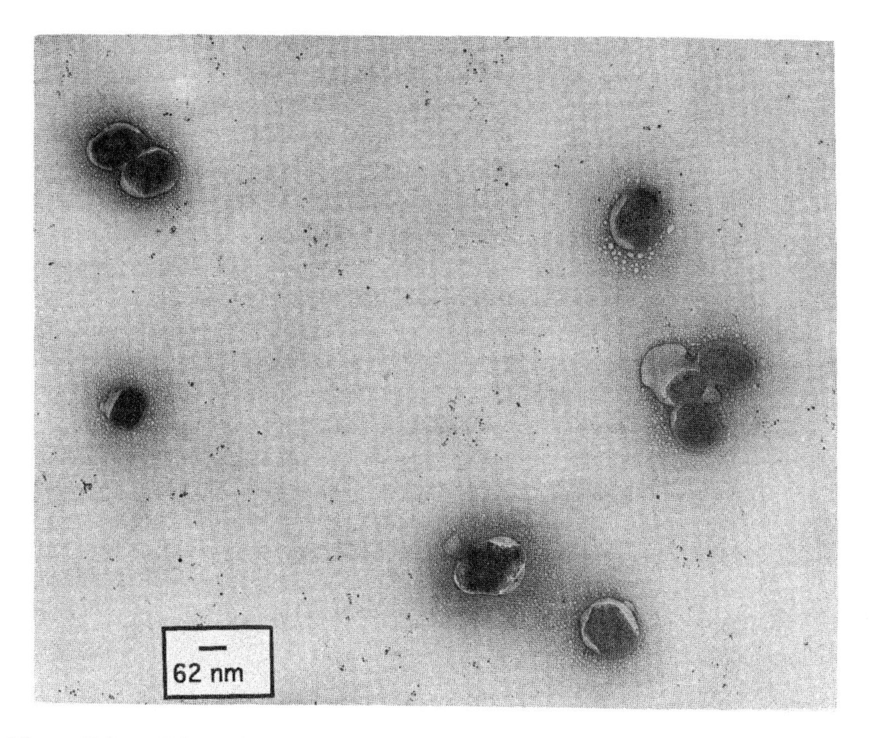

Figure 3.5. A TEM micrograph of polymerized vesicles in which gold particles have been deposited on the interior vesicle membrane surfaces. (After reference [61].)

3.4.5 Clusters

Clusters are finite aggregates containing 2–10^4 atoms, exhibiting unique physical and chemical properties. A detailed description of this field of research is beyond the scope of this chapter. Readers are referred to some examples discussed in chapters 2 and 4 and reviews in the literature [63–66]. Clusters are intermediates between molecular and condensed states of matters. The structural, energetic, electronic, electromagnetic, thermodynamic, and surface properties depend strongly on the size of the clusters [67]. Chemical methods can be used to synthesize ligand stabilized metallic clusters [68]. These clusters have a protecting ligand shell and can be handled like normal chemical compounds, in contrast to the unprotected, naked clusters produced by most other techniques. In addition, ligand stabilized clusters are also more uniform in their size distribution [69]. For example [69], chemically prepared Pd clusters were approximately 3.2–3.6 nm in diameter, with only about 10% of the particles outside this size range.

3.5 Conclusions

The use of chemistry for the synthesis of nanomaterials by a great diversity of routes has been discussed. Solution chemistry can either directly produce the desired particles or the precursors that are further treated by various reaction methods to obtain the final products. It should be pointed out that detailed work on characterization and property evaluation is required. Matters such as the yield and impurity content need to be critically addressed before a particular synthetic and processing method can be adopted for large-scale production. Though the problem of agglomeration of fine particles in solution can potentially be solved by controlling the interparticle forces, the subsequent processing must be designed carefully so that agglomeration can be minimized once the particles are removed from solution. The handling of fine particles after they are synthesized remains a serious problem. This problem is particularly acute for air-sensitive materials. The growing area of membrane mediated research, which addresses the stabilization and size control of nanoscale particles, is a potential practical approach to synthesize and process these materials. Chemistry synthesis of nanoparticles is a rapidly growing field with a great potential in making useful materials. The realization of this potential will require multidisciplinary interactions and collaborations between biologists, chemists, physicists, materials scientists, and processing engineers, to control the properties and to solve the problems described above,

REFERENCES

[1] Psaras P A and Langford H D (ed) 1987 *Advancing Materials Research, US National Academy of Engineering and National Academy of Sciences* (Washington, DC: National Academy Press) p 203

[2] Rao C N R 1993 *Mater. Sci. Eng.* B **18** 1

[3] Searson P C and Moffat T P 1994 *Crit. Rev. Surf. Chem.* **3** 171

[4] LaMer V K and Dinegar R H 1950 *J. Am. Chem. Soc.* **72** 4847
Overbeek J T G 1982 *Adv. Colloid Interface Sci.* **15** 251
Sugimoto T 1987 *Adv. Colloid Interface Sci.* **28** 65

[5] Nielsen F 1982 *Manufact. Chemist* **53** 38

[6] Real M W 1986 *Proc. Br. Ceram. Soc.* **38** 59

[7] Rosen M J 1989 *Surfactants and Interfacial Phenomena* 2nd edn (New York: Wiley)

[8] Rosen M J (ed) 1987 *Surfactants in Emerging Technologies* (New York: Dekker)

[9] McHale A E 1991 *Ceramics and Glasses, Engineered Materials Handbook* vol 4 (Metals Park, OH: ASM) p 115

[10] Yang K C and Rowan B D 1984 *Metals Handbook* 9th edn, vol 7 (Metals Park, OH: ASM) p 148

[11] Burkin A R and Richardson F D 1967 *Powder Metall.* **10** 32

[12] van Wonterghem J, Morup S, Koch C J W, Charles S W and Wells S 1986 *Nature* **322** 622
Inoue A, Saida J and Masumoto T 1988 *Metall. Trans.* A **19** 2315

Shen J, Hu Z, Hsia Y and Chen Y 1991 *Appl. Phys. Lett.* **59** 2510

Hadjipanayis G C, Gangopadhyay S, Yiping L, Sorensen C M and Klabunde K J 1991 *Science and Technology of Nanostructured Magnetic Materials* ed G C Hadjipanaysis and G A Prinz (New York: Plenum) p 497

[13] Chow G M, Ambrose T, Xiao J Q, Twigg M E, Baral S, Ervin A M, Qadri S B and Feng C R 1991 *Nanostruct. Mater.* **1** 361

[14] Chow G M, Ambrose T, Xiao J, Kaatz F and Ervin A 1993 *Nanostruct. Mater.* **2** 131

[15] Kunda W, Evans D and Mackiw V N 1966 *Modern Developments in Powder Metallurgy* vol 1 (New York: Plenum) p 15

[16] Griffiths C H, O'Horo M P and Smith T W 1979 *J. Appl. Phys.* **50** 7108

Smith T W and Wychick D 1980 *J. Phys. Chem.* **84** 1621

[17] van Wonterghem J, Morup S, Charles S, Wells S and Villadsen J 1985 *Phys. Rev. Lett.* **55** 410

[18] Ishikawa T and Matijevic E 1988 *Langmuir* **4** 26

[19] For a review, see Fievet F, Lapier J P and Figlarz M 1989 *MRS Bull.* **XIV** 29 (Pittsburgh, PA: Materials Research Society)

[20] Esumi K, Tano T, Torigoe K, Meguro K 1990 *Chem. Mater.* **2** 564

[21] Bradley J S, Hill E W, Klein C, Chaudret B and Dafeil A 1993 *Chem. Mater.* **5** 254

[22] Bönnemann H, Brijoux W and Joussen T 1990 *Angew. Chem. Int. Ed. Engl.* **29** 273

[23] Bönnemann H, Brijouk W, Brinkmann R, Dinjus E, Jouben T and Koraff B 1990 *Angew. Chem. Int. Ed. Engl.* **30** 1312

[24] Zeng D and Hampden-Smith M J 1993 *Chem. Mater.* **5** 681

[25] Riecke R D 1986 *Science* **246** 1260; 1991 *CRC Crit. Rev. Surf. Chem.* **1** 131

[26] Tsai K and Dye J L 1993 *Chem. Mater.* **5** 540

[27] Abe O and Tsuge A 1991 *J. Mater. Res.* **6** 928

[28] Axelbaum R L, Bates S E, Buhro W E, Frey C, Kelton K F, Lawton S A, Rosen L J and Sastry S M 1993 *Nanostruct. Mater.* **2** 139

[29] Bowen H K 1980 *Mater. Sci. Eng.* **44** 1

[30] For examples and reviews, see the series of *Better Ceramics through Chemistry*, the symposium proceedings series of the Materials Research Society from the early 1980s

Segal D 1989 *Chemical Synthesis of Advanced Ceramic Materials* (Cambridge: Cambridge University Press)

Gallagher P K 1991 *Ceramics and Glasses, Engineered Materials Handbook* vol 4 (Metals Park, OH: ASM) p 52

[31] Klein L C 1991 *Ceramics and Glasses, Engineered Materials Handbook* vol 4 (Metals Park, OH: ASM) p 209

[32] Robinson A L 1986 *Science* **233** 25

[33] Jorgensen J D, Schuttler H B, Hinks D G, Capone D W, Zhang K, Brodsky M B and Scalapino D J 1987 *Phys. Rev. Lett.* **58** 1024

[34] Tani E, Yoshimura M and Somiya S 1983 *J. Am. Ceram. Soc.* **66** 11

[35] Chick L A, Pederson L R, Maupin G D, Bates J L, Thomas L E and Exarhos G J 1990 *Mater. Lett.* **10** 6

Pederson L R, Maupin G D, Weber W J, McReday D J and Stephens R W 1991 *Mater. Lett.* **10** 437

[36] Matijevic E 1985 *Annu. Rev. Mater. Sci.* **15** 483

[37] Towe K M and Bradley W F 1967 *J. Colloid Interface Sci.* **24** 384
Hsu W P, Ronnquist L and Matijevic E 1988 *Langmuir* **4** 31
Matijevic E 1988 *Pure Appl. Chem.* **60** 1479
[38] Zeng D and Hampden-Smith M J 1992 *Chem. Mater.* **4** 968
[39] Mazdiyasni K S and Cooke C M 1973 *J. Am. Ceram. Soc.* **56** 628
[40] Johnson C E, Hickey D K and Harris D C 1986 *Mater. Res. Soc. Symp. Proc.* **73** 785
[41] Ritter J J 1986 *Mater. Res. Soc. Symp. Proc.* **73** 367
[42] Kalyoncu R S 1985 *Ceram. Eng. Sci. Proc.* **6** 1356
[43] McCandlish L E, Kear B H, Kim B K and Wu L W 1989 *Mater. Res. Soc. Symp. Proc.* **132** 67
[44] Wynne K J and Rice R W 1984 *Annu. Rev. Mater. Sci.* **14** 297
Seyferth D, Strohmann C, Tracy H J and Robison J L 1992 *Mater. Res. Soc. Symp. Proc.* **249** 3
Gonsalves K E and Xiao T D 1994 *Chemical Processing of Ceramics* ed B I Lee and E J A Pope (New York: Dekker) p 359
[45] Seyferth D and Wiseman G H 1984 *J. Am. Ceram. Soc.* **67** C-132
[46] Gonsalves K, Strutt P R and Xiao T D 1991 *Adv. Mater.* **3** 202
[47] Xiao T D, Gonsalves K, Strutt P R and Klemens P 1993 *J. Mater. Sci.* **28** 1334
[48] Kwon D, Schmidt W R, Interrante L V, Marchetti P and Maciel G 1991 *Inorganic and Organometallic Oligomers and Polymers* ed J F Harrod and R M Laine (Dordrecht: Kluwer) p 191
[49] Gonsalves K E, Xiao T D, Chow G M, Chen X and Strutt P R 1993 *Polymer Preprints, Am. Chem. Soc., Div. Polym. Chem.* **34** 362
Xiao T D, Gonsalves K E, Strutt P R, Chow G M and Chen X 1993 *Ceram. Eng. Sci. Proc.* **14** 1107
[50] Chow G M, Xiao T D, Chen X and Gonsalves K E 1994 *J. Mater Res.* **9** 168
[51] Xiao T D, Zhang Y D, Strutt P R, Budnick J I, Mohan K and Gonsalves K E 1993 *Nanostruct. Mater.* **2** 285
[52] Chow G M, Kurihara L K, Kemner K M, Schoen P E, Elam W T, Ervin A, Keller S, Zhang Y D, Budnick J and Ambrose T 1995 *J. Mater Res.* **10** 1546
[53] For a review see Fendler J H 1987 *Chem. Rev.* **87** 877
Heuer A H *et al* 1992 *Science* **255** 1098
[54] Huang C 1969 *Biochemistry* **8** 344
[55] Kortan A R, Hull R, Opila R L, Bawendi M G, Steigerwald M L, Carroll P J and Brus L E 1990 *J. Am. Chem. Soc.* **112** 1327
[56] Mann S and Williams R J P 1983 *J. Chem. Soc. Dalton Trans.* 311
[57] Mann S, Hannington J P and Williams R J P 1986 *Nature* **324** 565
[58] Bhandarkar S and Bose A 1990 *J. Colloid. Interface Sci.* **135** 531
[59] Bhandarkar S, Yaacob I and Bose A 1990 *Mater. Res. Soc. Symp. Proc.* **180** 637
[60] Heywood B R, Fendler J H and Mann S 1990 *J. Colloid. Interface Sci.* **138** 295
[61] Chow G M, Markowitz M A and Singh A 1993 *J. Miner. Met. Mater. Soc.* **45** 62
Markowitz M A, Chow G M and Singh A 1994 *Langmuir* **10** 4095
[62] Meldrum F C, Heywood B R and Mann S 1992 *Science* **257** 522
[63] 1977 *Proc. Int. Meeting on Small Particles and Inorganic Clusters, J. Physique Coll.* C2 199
[64] Jena P, Rao B K and Khanna N (ed) 1986 *The Physics and Chemistry of Small Clusters (NATO ASI Series)* (New York: Plenum)

[65] Echt O and Recknagel E (ed) 1991 *Proc. 5th Int. Symp. on Small Particles and Inorganic Clusters, Z. Phys.* D **19** 20

[66] Jena P, Khanna S N and Rao B K (ed) 1992 *Physics and Chemistry of Finite Systems: From Clusters to Crystals (NATO ASI Series)* (Dordecht: Kluwer)

[67] Jortner J 1992 *Physics and Chemistry of Finite Systems: From Clusters to Crystals (NATO ASI Series)* ed P Jena, S N Khanna and B K Rao (Dordecht: Kluwer) p 1

[68] Schmid G 1990 *Endeavour* **14** 172

[69] Schmid G, Harms M, Malm J-O, Bovin J-O, van Ruitenbeck J, Zandbergen H W and Fu W T 1993 *J. Am. Chem. Soc.* **115** 2046

Chapter 4

Synthesis of semiconductor nanoclusters

N Herron and Y Wang

The chemical synthesis of semiconductor nanoclusters from ~1 to ~20 nm in diameter (often called quantum dots, nanocrystals, or Q-particles) is a rapidly expanding area of research. In this size regime, the clusters possess short-range structures that are essentially the same as the bulk semiconductors, yet have optical and/or electronic properties which are dramatically different from the bulk [1–4]. They represent a relatively new class of materials and so have come under intensive investigation because of their quantum size effects [4], photocatalysis [3], nonlinear optical properties [5, 6], and more recently, photoconductivity [7].

Many methods have been developed for the synthesis of these kinds of material and it is our intention to attempt a broad overview of these methods in this section. Examples of such clusters, synthesized to date, include group II–VI, IV–VI, and III–V binary semiconductors [1–4, 8] as well as ternary diluted magnetic semiconductor clusters [9]. They can be prepared in the form of dispersed colloids or trapped and stabilized within micelles, polymers, zeolites, or glasses. In most cases, clusters prepared by these methods have poorly defined exterior surfaces and a relatively broad size distribution (~10–20%). The synthesis of single-sized (monodisperse) clusters with well defined surfaces remains a major goal in this field. We will, therefore, also review progress towards this goal, including our recent successful synthesis of a ~15 Å single-size tetrahedral CdS cluster [10].

While the bulk of this chapter is devoted to a selective review of the more common synthetic methods for nanocluster preparation, we begin with a brief discussion of some of the potential pitfalls which may be encountered in the synthesis and characterization of semiconductor clusters. In spite of the impressive progress made to date, problems still exist with the synthetic methods and these can easily lead to misinterpretation of the resultant characterizational data.

4.1 CHARACTERIZATION METHODS AND POTENTIAL PITFALLS IN THE SYNTHESIS

It was recognized very early [11] that many of the chemical synthetic routes, while designed to produce the desired semiconductor clusters, often form unexpected by-products. These by-products can interfere with the study of the semiconductor clusters and lead to misinterpretation of the data. For example, the absorption spectrum of the I_3^- ion was mistakenly assigned to a size-quantization affected spectrum of BiI_3 clusters [11–12]. More recently, the absorption spectrum of an unknown by-product was assigned to that of a GaAs cluster [13–14]. These problems are quite general and often stem from the ready reducibility or oxidizability of the components of the synthesis [11]. Although a wide variety of semiconductor clusters can now be synthesized, in many cases, there still exist unidentified by-products co-existing with the clusters in the sample. This is particularly true for III–V clusters. The technology for preparing II–VI semiconductor clusters such as CdS is more mature and relatively free of complications. In any event, it cannot be overemphasized that a thorough characterization of the sample is absolutely essential and one must always keep these complications in mind when interpreting the data.

One of the most useful characterizational tools is x-ray diffraction. In the ideal, but rare, case where single crystals of the cluster are available (an example is given in section 4.6), all structural information for the cluster can be obtained. More typically, the samples consist of crystalline semiconductor clusters randomly embedded in an amorphous matrix and, in such cases, powder x-ray diffraction data can be used to establish the identity, the phase, and the size of these clusters. Figure 4.1 shows an example of the x-ray diffraction pattern of PbS clusters in an ethylene–methacrylic acid co-polymer [15]. The phase (cubic) and the size ($\sim$40 Å) of the PbS clusters are established by a direct comparison with the computer simulated diffraction pattern (figure 4.1) [16, 17]. One may also calculate the average cluster size from the width of the diffraction peak using the Scherrer equation [18]. Based on direct comparisons with computer simulation [16] we have found that the Scherrer equation works quite well. However, several, more accurate, empirical formulae have also been derived based on simulation studies [4, 19]. We have found that the presence of size inhomogeneities and point defects does not significantly affect the accuracy of the size determination from the width of the x-ray diffraction peaks [19]. The presence of large-scale defects can be detected by the changes in the position of the diffraction peaks or, in some cases, the appearance of new diffraction peaks.

In most instances, the x-ray diffraction data can positively identify the presence of the desired semiconductor clusters in the sample. However, there is no guarantee that the measured absorption spectrum, which is of particular importance in the study of the quantum size effect and nonlinear optical properties, can be attributed to these semiconductor clusters. Other experiments are needed to establish this connection. One particularly valuable experiment is

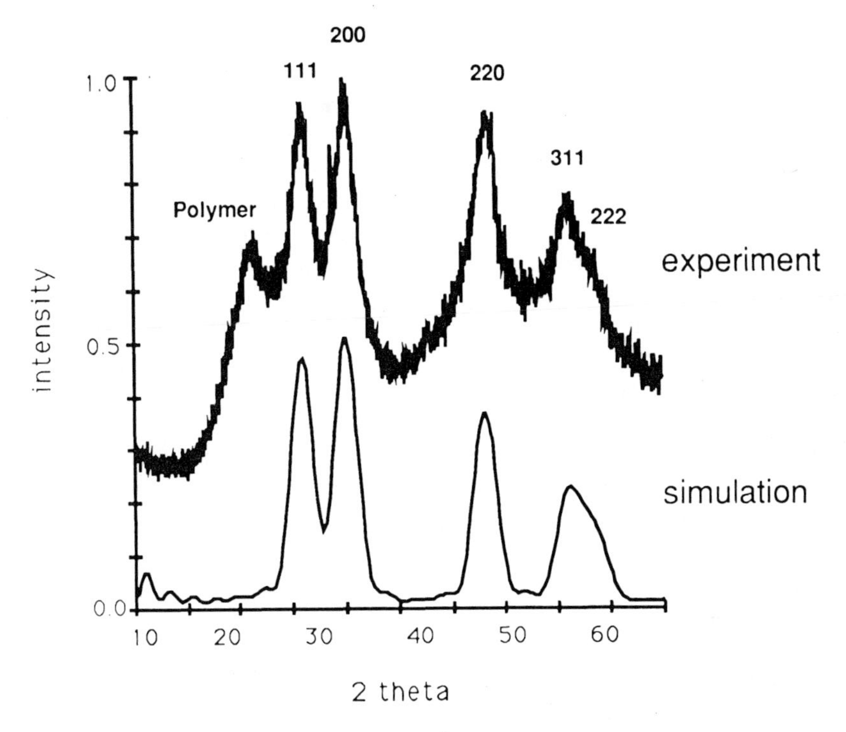

Figure 4.1. X-ray diffraction pattern of ~40 Å PbS clusters in ethylene–methacrylic acid co-polymer film compared to the simulated pattern of a 40 Å cubic PbS cluster.

the determination of the cluster absorption coefficients. These coefficients are obtained from the measured absorption spectrum and the volume concentration of the clusters in the sample as determined by chemical analysis. The magnitude of the absorption coefficient of the semiconductor cluster should be consistent with the known absorption coefficient of the bulk semiconductor after correction for the medium effect. To determine the band gap energy from the absorption spectrum, one must know the absorption coefficients. This is especially true if an exciton peak is absent in the spectrum. An absorption spectrum (optical density versus wavelength) with a featureless tail is useless for the determination of the band gap energy.

A useful technique for probing the local structure of the cluster is extended x-ray absorption fine structure (EXAFS) [9, 20, 21]. In certain ternary systems such as $Zn_xMn_{1-x}S$ [9] where the x-ray powder diffraction data cannot establish the location of Mn, this is a very powerful technique. Using EXAFS, the chemical identity of the nearest neighbors and the coordination number of the atom probed as well as its local bond lengths can all be determined. In principle, atomic resolution electron microscopy can also provide information about the local structure of the cluster. In reality, this has not yet been demonstrated. One major drawback of electron microscopy is that it measures only an extremely small portion of the sample. The observation of certain clusters or phases under the microscope does not mean that they are representative of the whole sample. When electron microscopy data are used alone, without confirmation by x-ray diffraction data, it cannot convincingly establish the identity of the sample and can lead to misinterpretation of the data [11–14].

Many other techniques such as nuclear magnetic resonance (NMR) [22, 23], Raman [24–26], hole-burning [27], and photoelectron spectroscopies [28] have been used for characterizing clusters. All can provide useful information when used in combination with other techniques.

In the following sections, we review various techniques developed for synthesizing nanometer-sized semiconductor clusters. We have divided these into categories according to the matrix in which the cluster is embedded or created. The final section is devoted to a discussion of means to prepare single-size clusters, the ultimate goal in this field. One subject we do not discuss is the aggregation of these nanoclusters. Such aggregation is absent in a zeolite matrix, less severe in polymers, and can be extensive in colloidal solutions. The extent of aggregation varies according to the nature and the concentration of the clusters as well as the formulation of the matrix. There is no systematic study on the subject but sometimes discussion on specific cases may be found in the references cited.

4.2 COLLOIDS/MICELLES/VESICLES

The preparation of colloidal semiconductor particles has a long history [29, 30]. Examples can be found in the literature from almost a hundred years ago [29]. To stabilize a colloid in the small-cluster-size regime, it is necessary to find an agent that can bind to the cluster surface and thereby prevent the uncontrolled growth into larger particles.

The simplest method for preparing such colloids involves the solvent itself acting as the stabilizer of the small clusters. This is illustrated by the generation of ZnO in alcohol solvents following base hydrolysis of a solution of a zinc salt [31]. This reaction produces a transparent colloid where the ZnO particle size increases slowly on standing. A variation on this approach [32] has led to extremely high concentrations and very stable solutions of highly luminescent ZnO particles in ethanol solution by a combined solvent/anion stabilization followed by hydrolysis using LiOH.

A more common approach to such colloids is the use of a polymeric surfactant/stabilizer which is added to a reaction designed to precipitate the bulk material. The polymer attaches to the surface of the growing clusters, usually electrostatically, and prevents their further growth. The most commonly used polymer is sodium polyphosphate (hexametaphosphate) and clusters of CdS [33], CdTe [34], and ZnTe [34] have been prepared with this surfactant.

A similar approach is the use of deliberately added capping agents to

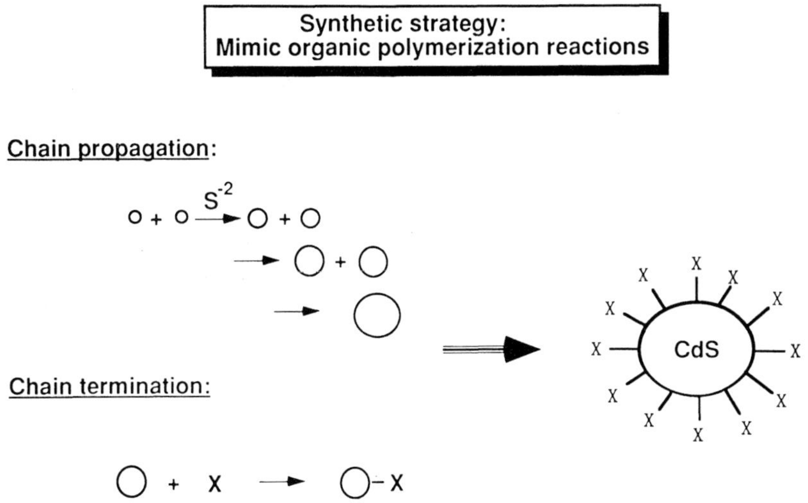

Synthetic strategy:
Mimic organic polymerization reactions

Chain propagation:

$\circ + \circ \xrightarrow{S^{-2}} \bigcirc + \bigcirc \rightarrow \bigcirc + \bigcirc \rightarrow \bigcirc$

Chain termination:

$\bigcirc + X \rightarrow \bigcirc\text{-}X$

CdS

Figure 4.2. Schematic diagram for the synthesis of thiophenolate capped CdS clusters (X represents thiophenolate ion) drawing analogy to organic polymerizations. In the chain propagation step, the size of the circle represents the size of the CdS clusters. The growth of the CdS clusters is analogous to the growth of a polymer chain as long as the surfaces are not covered by the terminating agent, X.

solutions of growing clusters. These agents, typically anionic, are added to a semiconductor precipitation reaction and intercept the growing clusters, preventing further growth by covalently binding to the cluster surface. Thiolates are the most commonly used capping agents [23, 35–38] and this method also forms the basis of the synthesis of monodispersed clusters (section 4.6). This approach can be thought of as mimicking an organic polymerization reaction (initiation, propagation, and termination phases) and is graphically depicted for CdS clusters in figure 4.2. In this analogy, mixing the cadmium and sulfide ions initiates the polymerization. The growth of the CdS clusters is viewed as a propagation step and is sustained by the presence of additional cadmium and sulfide ions. The growth of the clusters can be terminated by providing a capping agent such as thiophenolate ions which intercept the growing clusters by binding to the cluster surface [23]. The average cluster size can then be controlled by simply adjusting the sulfide to thiophenolate ratio in the solution, just as the average molecular weight of a polymer is controlled by adjusting the monomer to terminator (chain capper) ratio. We have also found that thiophenolate capped CdS clusters act somewhat like living polymers, i.e. they will continue to grow if fed more sulfide ions [23]. This living polymer property was used advantageously to produce monodisperse clusters which will be discussed in section 4.6.

Other capping agents have also been used. Glutathione peptides, produced by yeast, have been used to cap small CdS clusters [39]. The generation of GaAs in tetrahydrofuran (THF) solution [40] or glycol ether [41] may also be categorized as a reaction of this type. In the former case, pentamethylcyclopentadienyl ligands on the Ga ions and trimethylsilyl groups on the As ions regulate the cluster growth and maintain GaAs colloids in solution for extended periods. In the latter case, acetylacetonate ligands on the Ga ions and trimethylsilyl groups on the As ions control the cluster size.

The use of micelle forming reagents as a method of controlling cluster growth in semiconductor forming reactions is conceptually similar to the colloidal and capping approaches just described. In this case, however, a small region of physical space is defined by a micelle and the semiconductor is precipitated within this defined region. In contrast to the colloidal approach, the micellar reagent acts as a physical boundary rather than a surface capping agent. Normal vesicles using dihexadecyl phosphate [42, 43] or dioctadecyldimethylammonium chloride (DODAC) [44] can be generated in water so as to have diameters of the order of 150–300 nm. Dissolution of Cd or Zn ions in these vesicles followed by precipitation with H_2S leads to semiconductor clusters of up to 50 Å in diameter inside the micelles. Reversed micelles using bis(2-ethylhexyl)sulfosuccinate salts (AOT) allow formation of small water pools ($<$100 Å radius) in heptane solvent, and again incorporation of metal ions followed by chalcogenide treatment can precipitate semiconductor clusters within these pools [45–47]. A combination of the reversed micelle approach and surface capping of the resultant clusters has been developed to produce selenophenol capped CdSe clusters [35] or CdSe/ZnS [48].

A modification of the micelle technique has been reported where stabilized colloids of CdS have been entrained between layers of arachidic acid surfactant in a Langmuir–Blodgett film [49, 50].

4.3 POLYMERS

The colloidal preparation techniques discussed above have many inherent problems of irreproducibility in preparation, colloid instability, and the difficulty in establishing the definitive identity of the semiconductors by x-ray diffraction. For practical applications, it is also highly desirable to have a solid thin-film sample. By synthesizing small semiconductor clusters inside polymers, many of these problems have been solved. This was first achieved with CdS in Nafion® (perfluoroethylenesulfonic acid polymer) [51] and PbS in Surlyn® (ethylene–methacrylic acid co-polymer) [15, 51]. The definitive identification of semiconductor clusters by x-ray diffraction was first achieved with these stable solid samples.

Several approaches have been used to produce these composites. The first is an ion-exchange method. Polymers such as Nafion or ethylene–methacrylic acid co-polymer have cation exchange sites where Cd or Pb ions may be introduced into the polymer matrix. Treatment of such ion-exchanged films with chalcogenide sources results in *in situ* precipitation of the compound semiconductor within the hydrophilic regions of the polymer. In essence, these ionic polymers are the solid-state analogs of micelles. By appropriate control of the phase separated hydrophilic and hydrophobic regions of the polymer, the cluster size of the included material may be controlled [51, 52]. This method has been extended to synthesize layered semiconductor clusters such as PbI_2 [53] as well as magnetic particles such as Fe_2O_3 [54].

A somewhat different approach has been explored by Schrock *et al* [55] using direct incorporation of the metal ion Pb^{2+} into monomer units which eventually become part of a norbornene derived co-polymer following ring opening metathesis polymerization (ROMP). Again, the polymer is designed to have carefully phase separated and closely size controlled regions of high and low hydrophobicity. The semiconductor PbS is precipitated in the latter regions by treatment with H_2S from the gas phase.

The third approach is a hybrid of the polymer isolation and the surface capped cluster approaches. In this method, a well defined semiconductor cluster or colloid is prepared by the conventional capping techniques described above and dissolved in a solvent along with a soluble polymer. This mixed solution may then be spin-coated onto a substrate and dried to produce a polymer film doped with the semiconductor clusters. This simple approach has provided us with some new examples of interesting photoconductive composites by using a photoconductive polymer such as polyvinylcarbazole doped with semiconductor clusters such as CdS [7].

4.4 GLASSES

Small particles of CdS_xSe_{1-x} can be embedded in a borosilicate glass matrix and have long been used as color filters [56]. Such materials are prepared by traditional glass making technology where cadmium, sulfur, and selenium sources are added to a silicate [56, 57] or germanate [58] glass melt at elevated temperatures. After casting, the glass is struck (annealed at temperatures less than melting) and small particles of semiconductor form in the dense glass matrix. In this technique, the semiconductor must survive a very high-temperature oxidizing environment during the glass forming step. Thus, the technique is only applicable to a limited number of semiconductors. III–V semiconductor clusters such as GaAs cannot be prepared this way because of their thermal and oxidation sensitivity.

There exist alternate methods of incorporating oxidatively or thermally sensitive semiconductors into glass by avoiding the high-temperature processing steps. Examples of one such approach have utilized the internal void network of porous glass prepared by the selective acid leaching of borates from a borosilicate Vycor® glass. In these examples the semiconductor is essentially precipitated *in situ* within the pores of the glass at low temperatures and the physical restraint imposed by the pore size of the glass controls the size of the semiconductor clusters. Examples of II–VI [8, 59–61] and III–V [8, 62] semiconductors in such media have been reported. In related work we [8, 9, 63, 64] and others [65–68] have explored the use of sol–gel derived porous silica glass as a method for production of these materials. In this case we have found it desirable to assemble the semiconductor particles within the pore structure of the host glass matrix and then to back fill the residual porosity with a polymer such as PMMA. This reduces the light scattering which one frequently observes from the empty pore network and makes the composite more acceptable as a material for the construction of optical devices.

A novel technique of forming a glassy film of TiO_2 nanoparticles by electrostatic spraying of an alkoxide solution has been reported [69]. RF sputtering [70–74] and gas evaporation [75] techniques have also been used to produce small semiconductor clusters, which are sometimes doped in glasses. These techniques appear to be very useful for producing high-quality thin-film samples. Samples of Ge [70, 71], CdS [72], and CdTe [73] doped silica and CdS doped alumina [74] have all been prepared by RF sputtering. Gas evaporation has been used to produce small particles of Si, Ge, CdS, and ZnTe [75]. It remains to be seen whether these methods can be extended to include other semiconductors and whether they can be optimized to prepare monodisperse clusters.

4.5 CRYSTALLINE AND ZEOLITE HOSTS

The use of a crystalline host lattice for the isolation and stabilization of nanoclusters represents an attempt to impose crystalline order and physical size constraints upon the included semiconductor guests. CuCl crystallites of ~20–50 Å have been prepared within single crystals of NaCl or KCl by a melt procedure and were shown to exhibit exciton absorption features which were blue-shifted from the bulk material [76]. Another approach, one which we have explored, is the use of a porous crystalline host, zeolite, into which the semiconductor is either directly imbibed or else created *in situ*. Examples of the former technique are the sublimation of elemental selenium or tellurium into a variety of aluminosilicate zeolites [20, 77]. Characterization of the resultant selenium clusters and chains by EXAFS and optical spectroscopy have been reported by several groups [20, 77–79]. The second, 'ship-in-a-bottle' approach [80] has been used for the generation of CdS clusters and superclusters (cluster arrays) in zeolite Y [81, 82] which is discussed below.

Zeolite Y occurs naturally as the mineral faujasite and consists of a porous network of aluminate and silicate tetrahedra linked through bridging oxygen atoms. The structure consists of truncated octahedra called sodalite units arranged in a diamond net and linked through double six-rings [83]. This gives rise to two types of cavity within the structure—the sodalite cavity of ~5 Å diameter with access through ~2.5 Å windows and the supercage of ~13 Å diameter with access through ~7.5 Å windows (figure 4.3). These well defined and well ordered cavities provide an ideal enviroment for synthesizing single-size clusters and cluster arrays, while the windows provide access for transporting reagents to the cavities. Furthermore, each aluminum atom in the framework introduces one negative charge into the zeolite skeleton which is compensated for by loosely attached cations which, in turn, give rise to the well known ion-exchange properties of zeolites.

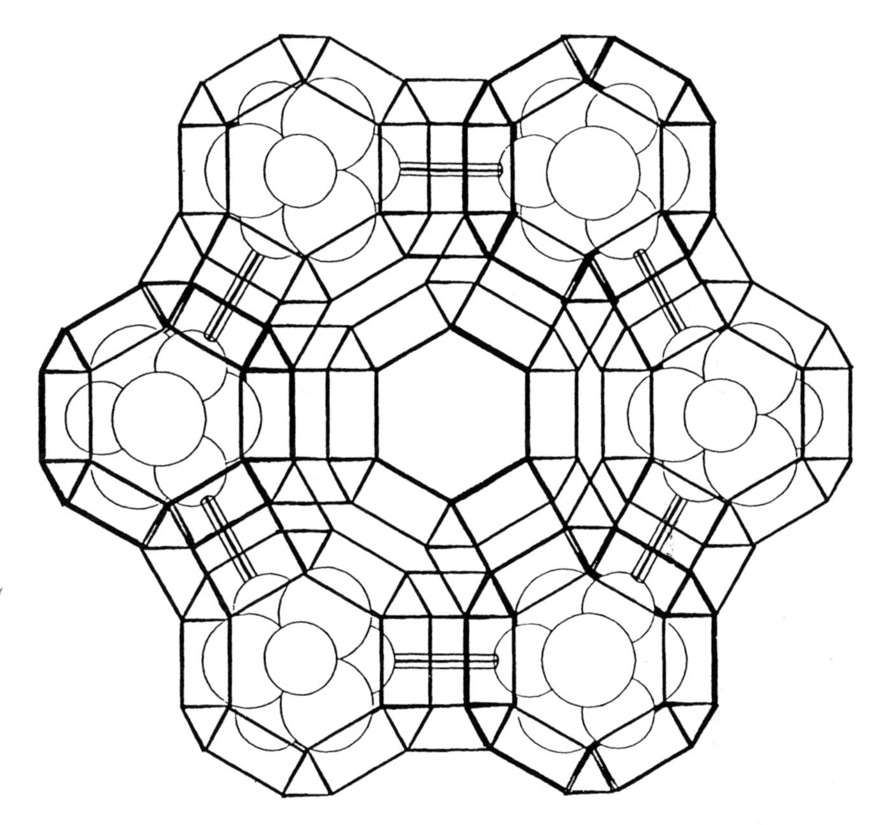

Figure 4.3. Cd_4S_4 cluster array in the sodalite cages of zeolite Y. Circles represent Cd_4S_4 clusters (smaller circles are Cd and larger circles are S). Six sodalite cages (occupied by Cd_4S_4) and one supercage (at the center, empty) are shown here. The sodalite cage has a diameter of ~5 Å and the supercage has a diameter of ~13 Å.

Based on the ion-exchange method, CdS clusters were successfully synthesized within zeolite Y [81, 82, 84, 85]. Detailed synchrotron x-ray, EXAFS, and optical absorption data reveal discrete $(CdS)_4$ cubes located within the small sodalite units of the structure (figure 4.3) [82]. At higher loadings, these cubes begin to occupy adjacent sodalite units where the Cd atoms point toward each other through the double six-rings linking the sodalite moeities with a Cd–Cd distance of ~6 Å (figure 4.3). These Cd_4S_4 cubes are not isolated. They interact with each other via through-bond interactions to form a supercluster (or cluster array) [82]. The absorption spectrum of the supercluster is shifted from ~290 nm for the isolated clusters to ~360 nm [81, 82]. There is a minimum number of clusters required in a supercluster to account for the change in the optical absorption spectrum, but this number is unknown at present.

The zeolite confinement approach has been extended to include many other semiconductor guests and sodalites as the host [86–89]. AgI in zeolite mordenite has been shown to have unusual optical behavior in terms of its photosensitivity [87] and, in a follow up on this work, Ozin *et al* [88] have generated silver halides in the sodalite zeolite family. PbI_2 in X, Y, A, and L type zeolites shows evidence of a strong size effect on the exciton absorption [89]. Cao *et al* [90] investigated a variant of this approach. They used layered crystalline hosts of the metal phosphonate series instead of zeolites.

In principle, zeolites offer a unique and exciting opportunity of preparing three-dimensional arrays of mutually interacting clusters with geometric structures imposed by the zeolite internal pore structure. The electronic properties of the cluster array are controlled by the different spatial arrangements of the clusters which in turn can be controlled by using different zeolites as the template. The work on CdS in zeolite Y represents the first step towards exploring this new class of materials. At present, imperfections do exist with these materials [91]. Alternative synthetic routes and better control in the fabrication process need to be explored in the future. The further success of this approach will depend on the availability of high-quality zeolite single crystals as well as zeolites with larger pore sizes.

4.6 TOWARDS SINGLE-SIZE CLUSTERS

Size inhomogeneity of as-synthesized clusters and the ambiguous nature of their surfaces are the problems that present the greatest barrier to future progress in this area. Quantitative understanding of the physics and chemistry of these clusters may best be achieved if single-size clusters can be prepared. The successful synthesis of C_{60} clusters provides the best illustration of this point [92]. In the following sections, we discuss several approaches developed in our laboratories to meet this challenge. One of the approaches has resulted in the first successful synthesis and structural determination of 15 Å single-size CdS clusters.

As mentioned in section 4.2, during the course of studying thiophenolate capped CdS clusters, we found that the clusters can be grown by simply adding extra sulfide to the cluster solution [36], an example of inorganic 'living polymerization'. Based on this observation, it seemed that, if one started with a well defined small molecular cluster, one might be able to grow larger single-size clusters by the careful addition of reagents selected to cement the smaller units together. The $(NMe_4)_4Cd_{10}S_4SPh_{16}$ cluster synthesized by Dance *et al* [93] provides an ideal candidate as a starting material. This compound belongs to a series of molecular clusters synthesized by the group of Dance *et al* [93, 94] who have produced the only instances of crystalline, well characterized semiconductor molecular clusters. These remarkable compounds, such as $(NMe_4)_4Cd_{10}S_4SPh_{16}$ [95] and $(NMe_4)_2Cd_{17}S_4SPh_{28}$ [96], contain a crystalline core having the same atomic arrangement as the bulk cubic phase of CdS and dimensions in the 7 to 9 Å range. These clusters remain very soluble in polar organic solvents and are true molecular species. With these well defined clusters as the starting materials, we have developed two approaches to the synthesis of single-size clusters.

4.6.1 Controlled cluster fusion in solution

The cluster synthesized by Dance *et al* [93] has the molecular formula $(Cd_{10}S_4(SPh)_{16})^{4-}$ (30 Cd and S atoms in the core), where Ph = C_6H_5, with a tetrahedral shape. The next largest cluster in the series has the formula $(Cd_{20}S_{13}(SPh)_{22})^{8-}$ (55 Cd and S atoms in the core), and is approximately 10 Å in size. To fuse two smaller clusters into a larger one requires the additon of five extra sulfide ions

$$2(Cd_{10}S_4(SPh)_{16})^{4-} + 5S^{2-} \longrightarrow (Cd_{20}S_{13}(SPh)_{22})^{8-} + 10Sph^{-}. \tag{4.1}$$

This cluster fusion idea has indeed been demonstrated [16]. Upon addition of sulfide ions to a 1.25×10^{-4} M dimethylformamide (DMF) solution of $(Cd_{10}S_4(SPh)_{16})^{4-}$, we observed a very sharp absorption feature at 351 nm with a shoulder at 330 nm at a precise ratio of 2.5 sulfide ions per original $(Cd_{10}S_4(SPh)_{16})^{4-}$ cluster. This absorption spectrum is assigned to that of $(Cd_{20}S_{13}(SPh)_{22})^{8-}$. Further addition of sulfide ions results in the disappearance of the 351 nm band and continuing growth of the clusters. The full width at half maximum of the absorption band is very small ($\leq$800 cm^{-1} at room temperature) indicating that the generated cluster is close to monodispersed. The clusters can be collected as stable solids and re-dissolved in solution. Based on the combined evidences of reaction stoichiometry, cluster chemical analysis data, x-ray powder diffractograms, and theoretical calculations of spectral properties, we have concluded that the resultant material has a discrete, tetrahedral $Cd_{20}S_{13}$ core surrounded by thiophenolate ion caps, consistent with the ideal structure of $(Cd_{20}S_{13}(SPh)_{22})^{8-}$ [16]. No single crystal is yet available for x-ray structural determination.

This approach has been extended to prepare even larger monodisperse CdS clusters. It was found that two 10 Å $(Cd_{20}S_{13}(SPh)_{22})^{8-}$ clusters may be fused together to form a 13 Å CdS cluster, which shows a sharp absorption peak at 370 nm and a luminescence peak at 390 nm [97]. This conclusion is supported by the concentration dependence of the absorption spectra, the luminescence–excitation spectra, the behavior in capillary zone electrophoresis, the electrochemical properties, and the powder x-ray diffraction data [97]. This cluster fusion process occurs naturally in solution at high concentrations of $(Cd_{20}S_{13}(SPh)_{22})^{8-}$. It can also be induced electrochemically by a two-electron oxidation process.

It has also been demonstrated that, even if the initially prepared clusters have a broad size distribution, it is possible to use techniques such as gel electrophoresis [98] or capillary zone electrophoresis [97] to separate them into monodispersed clusters.

4.6.2 Controlled thermolysis in the solid state

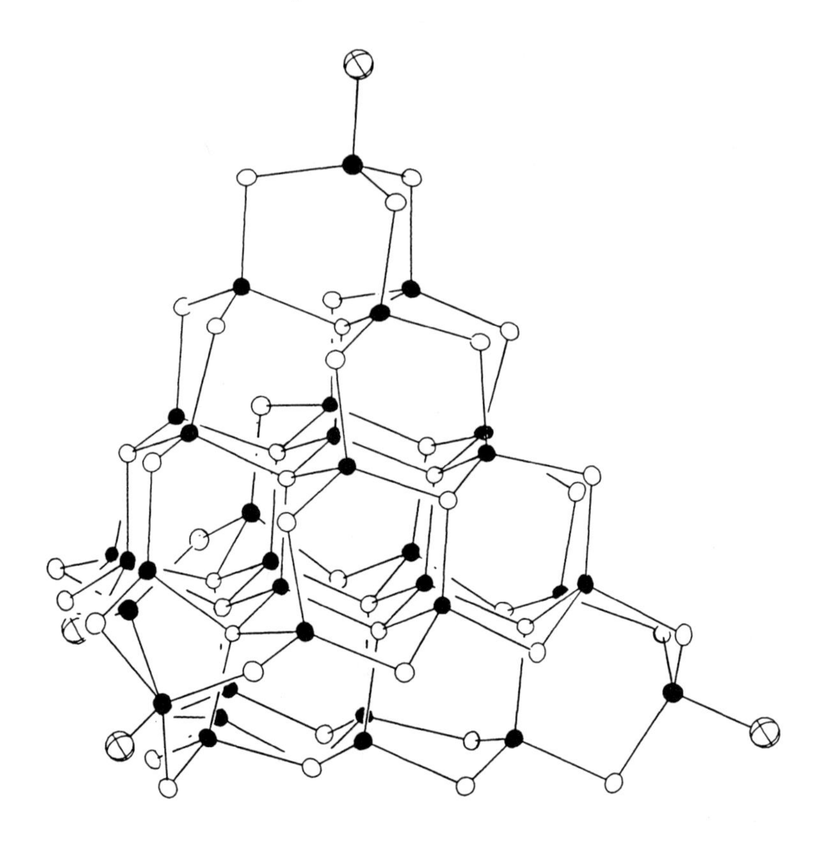

Figure 4.4. Single-crystal structure of the cluster $Cd_{32}S_{14}SPh_{36}DMF_4$. All phenyl groups have been omitted for clarity as have all but the nitrogen atoms of the terminally coordinated DMF molecules. All Cd atoms are marked as shaded ellipsoids, S atoms are open ellipsoids, and N atoms are large cross-hatched spheres. Selected bond lengths are: Cd–S^{2-} = 2.468(4) (triply bridging S in the center of cluster's tetrahedral face), 2.503(4) (central adamantyl core—see text), 2.538(8), 2.532(4), 2.537(5) Å (second shell out from central core); Cd–SPh = 2.495(5), 2.569(5), 2.546(5), 2.503(5), 2.554(5), 2.560(5) Å; Cd–N 2.33(4) Å.

Thermolysis of the compound $(NMe_4)_4Cd_{10}S_4SPh_{16}$ in inert atmospheres leads to the loss of thiophenolate capping groups and the eventual formation of bulk

CdS beginning at ~350 °C [99]. At an intermediate temperature of ~250 °C, where a mass loss corresponding to $4(NMe_4SPh)$ is observed, a material of stoichiometry $Cd_{10}S_4SPh_{12}$ can be isolated [99]. This material may be recrystallized into large cubic crystals from pyridine/dimethylformide (DMF) mixtures. Its structure and stoichiometry were established by x-ray diffraction measurements on a single crystal [10]. The material consists of a single-size cluster with the formula of $Cd_{32}S_{14}SPh_{36}DMF_4$. This cluster has an 82-atom tetrahedral core of cubic phase CdS with an overall tetrahedral shape (figure 4.4) [10]. This represents the largest crystallographically characterized semiconductor cluster ever and presents unique opportunities for the definitive assessment of optical and electronic properties for a monodispersed cluster size.

Both of these approaches, of taking well defined cluster species and expanding the cluster core by controlled synthesis, represent promising avenues for future research.

REFERENCES

[1] Efros Al L and Efros A L 1982 *Fiz. Tekh. Poluprovodn.* **16** 1209 (Engl. Transl. 1982 *Sov. Phys.–Semicond.* **16** 772)
Ekimov A I and Onushchenko A A 1984 *Zh. Eksp. Teor. Fiz.* **40** 337 (Engl. Transl. 1984 *JETP Lett.* **40** 1136)
[2] Steigerwald M L and Brus L E 1990 *Acc. Chem. Res.* **23** 183
[3] Henglein A 1988 *Topics Current Chem.* **143** 113
[4] Wang Y and Herron N 1991 *J. Phys. Chem.* **95** 525
[5] Wang Y, Herron N, Mahler W and Suna A 1989 *J. Opt. Soc. Am.* B **6** 808
[6] Wang Y 1991 *Acc. Chem. Res.* **24** 133
[7] Wang Y, Herron N, Harmer M and Suna A 1992 *Proc. MRS Spring Meetings* (Pittsburgh, PA: Materials Research Society)
[8] Wang Y and Herron N 1991 *Res. Chem. Intermed.* A **15** 17
[9] Wang Y, Herron N, Moller K and Bein T 1991 *Solid State Commun.* **77** 33
[10] Herron N, Calabrese J C, Farneth W E and Wang Y 1993 *Science* **259** 1426
[11] Wang Y and Herron N 1987 *J. Phys. Chem.* **91** 5005
[12] Sandroff C J, Kelty S P and Hwang D M 1986 *J. Chem. Phys.* **85** 5337
[13] Olshavsky M A, Goldstein A N, Alivasatos A P 1990 *J. Am. Chem. Soc.* **112** 9438
[14] Uchida H, Curtis C and Nozik A J 1991 *J. Phys. Chem.* **95** 5382
[15] Wang Y, Suna A, Mahler W and Kasowski R 1987 *J. Chem. Phys.* **87** 7315
[16] Herron N, Suna A and Wang Y 1992 *J. Chem. Soc. Dalton Trans.* 2329
[17] Bawendi M G, Kortan A R, Steigerwald M L and Brus L E 1989 *J. Chem. Phys.* **91** 7282
[18] Taylor A 1961 *X-ray Metallography* (New York: Wiley) p 674
[19] Wang Y and Sujna A 1995 unpublished results
[20] Parise J B, MacDougall J E, Herron N, Farlee R, Sleight A W, Wang Y, Bein T, Moller K and Moroney L M 1988 *Inorg. Chem.* **27** 210
[21] Marcus M A, Flood W, Steigerwald M, Brus L and Bawendi M 1991 *J. Phys. Chem.* **95** 1572

[22] Thayer A M, Steigerwald M L, Duncan T M and Douglass D C 1988 *Phys. Rev. Lett.* **60** 2673
[23] Herron N, Wang Y and Eckert H 1990 *J. Am. Chem. Soc.* **112** 1322
[24] Champagnon B, Andrianasolo B and Duval E 1991 *J. Chem. Phys.* **94** 5237
[25] Efros Al L, Ekimov A I, Kozlowski F, Petrova-Koch V, Schmidbaur H and Shumilov S 1991 *Solid State Commun.* **78** 853
[26] Alivisatos A P, Harris T D, Carrol P J, Steigerwald M L and Brus L E 1989 *J. Chem. Phys.* **90** 3463
[27] Alivisatos A P, Harris A L, Levinos N J, Steigerwald M L and Brus L E 1988 *J. Chem. Phys.* **89** 4001
[28] Colvin V L, Alivisatos A P and Tobin J G 1991 *Phys. Rev.* **66** 2786
[29] Lucas M 1896 *Bull. Soc. Chim.* **15** 40
[30] Berry C R 1967 *Phys. Rev.* **161** 848
[31] Bahnemann D W, Kormann C and Hoffmann M R 1987 *J. Phys. Chem.* **91** 989
[32] Spanhel L and Anderson M 1990 *J. Am. Chem. Soc.* **113** 2826
[33] Spanhel L, Haase M, Weller H and Henglein A 1987 *J. Am. Chem. Soc.* **109** 5649
Haase M, Weller H and Henglein A 1988 *J. Phys. Chem.* **92** 4706
Fischer Ch H, Weller H, Katsikas L and Henglein A 1989 *Langmuir* **5** 429
[34] Resch U, Weller H and Henglein A 1989 *Langmuir* **5** 1015
[35] Nosaka Y, Yamaguchi K, Miyama H and Hayashi H 1988 *Chem. Lett* 605
[36] Steigerwald M L, Alivisatos A P, Gibson J M, Harris T D, Kortan R, Muller A J, Thayer A M, Duncan T M, Douglass D C and Brus L E 1988 *J. Am. Chem. Soc.* **110** 3046
[37] Swayambunathan V, Hayes D, Schmidt K H, Liao Y X and Meisel D 1990 *J. Am. Chem. Soc.* **112** 3831
[38] Fischer C-H and Henglein A 1989 *J. Phys. Chem.* **93** 5578
[39] Dameron C T, Reese R N, Mehra R K, Kortan A R, Carroll P J, Steigerwald M L, Brus L E and Winge D R 1989 *Nature* **338** 596
[40] Byrne E K, Parkanyi L and Theopold K H 1988 *Science* **241** 332
[41] Uchida H, Curtis C, Kamat P V, Jones K M and Nozik A J 1992 *J. Phys. Chem.* **96** 1156
[42] Tricot Y M, Emeren Å and Fendler J H 1985 *J. Phys. Chem.* **89** 4721
[43] Youn H-C, Baral S and Fendler J H 1988 *J. Phys. Chem.* **92** 6320
[44] Watzke H J and Fendler J H 1987 *J. Phys. Chem.* **91** 854
[45] Meyer M, Wallberg C, Kurihara K and Fendler J H 1984 *J. Chem. Soc. Chem. Commun.* 90
[46] Lianos P and Thomas J K 1986 *Chem. Phys. Lett.* **125** 299
[47] Petit C, Lixon P and Pileni M P 1990 *J. Phys. Chem.* **94** 1598
[48] Kortan A R, Hull R, Opila R L, Bawendi M G, Steigerwald M L, Carroll P J and Brus L E 1990 *J. Am. Chem. Soc.* **112** 1327
[49] Xu S, Zhao X K and Fendler J H 1990 *Adv. Mater.* **2** 183
[50] Smotkin E S, Lee C, Bard A J, Campion A, Fox M A, Mallouk T E, Webber S E and White J M 1988 *Chem. Phys. Lett.* **152** 265
[51] Wang Y and Mahler W 1987 *Opt. Commun.* **61** 233
Hilinski E, Lucas P and Wang Y 1988 *J. Chem. Phys.* **89** 3435
Wang Y, Suna A, McHugh J, Hilinski E, Lucas P and Johnson R D 1990 *J. Chem. Phys.* **92** 6927
Mahler W 1988 *Inorg. Chem.* **27** 435

[52] Smotkin E S, Brown M R, Rabenberg L K, Salomon K, Bard A J, Campion A, Fox M A, Mallouk T E, Webber S E and White J M 1990 *J. Phys. Chem.* **94** 7543

[53] Goto T, Saito S and Tanaka M 1991 *Solid State Commun.* **80** 331

[54] Ziolo R F, Giannelis E P, Weinstein B A, O'Horo M P, Ganguly B N, Mehrotra V, Russell M W and Huffman D R 1992 *Science* **257** 219

[55] Sankaran V, Cummins C C, Schrock R R, Cohen R E and Silbey R J 1990 *J. Am. Chem. Soc.* **112** 6858

Cummins C C, Schrock R R and Cohen R E 1992 *Chem. Mater.* **4** 27

[56] Borelli N F, Hall D W, Holland H J and Smith D W 1987 *J. Appl. Phys.* **61** 5399

[57] Potter B G and Simmons J H 1988 *Phys. Rev.* **37** 10 838

[58] Arai T, Fujumura H, Umezu I, Ogawa T and Fujii A 1989 *Japan. J. Appl. Phys.* **28** 484

[59] Scoberg D J, Grieser F and Furlong D N 1991 *J. Chem. Soc. Chem. Commun.* 516

[60] Kimizuka N, Miyoshi T, Ichinose I and Kunitake T 1991 *Chem. Lett.* 2039

[61] Xu S, Zhao X K and Fendler J H 1990 *Adv. Mater.* **2** 183

[62] Luong J C and Borelli N F 1989 *Mater Res. Soc. Symp. Proc.* **144** 695

[63] Herron N and Wang Y *US Patent Application* 07/7630 filed 2/24/89

[64] Wang Y and Herron N 1992 *Int. J. Nonlinear Opt. Phys.* **1** 683

[65] Roy R, Komareni S and Roy D M 1984 *Mater Res. Soc. Symp. Proc.* **32** 347

[66] Rajh T, Vucemilovic M I, Dimitrijevic N M, Micic O I and Nozik A J 1988 *Chem. Phys. Lett.* **143** 305

[67] Minti H, Eytal M, Reisfeld R and Berkovic G 1991 *Chem. Phys. Lett.* **183** 277

[68] Kobayashi Y, Yamazaki S, Kurokawa Y, Miyakawa T and Kawaguchi H 1992 *J. Mater. Sci.: Mater. Electron.* **2** 20

[69] Park D G and Burlitch J M 1992 *Chem. Mater.* **4** 500

[70] Fujii M, Hayashi S and Yamamoto K 1990 *Appl. Phys. Lett.* **57** 2692

[71] Fujii M, Hayashi S and Yamamoto K 1991 *Japan. J. Appl. Phys.* **30** 687

[72] Tanahashi I, Tsujimura A, Mitsuyu T and Nishino A 1990 *Japan. J. Appl. Phys.* **29** 2111

[73] Potter B G and Simmons J H 1990 *J. Appl. Phys.* **68** 1218

[74] Tanahashi I, Yoshida M and Mitsuyu T 1991 *Bull. Chem. Soc. Japan.* **64** 2281

[75] Hayashi S, Tanimoto S, Fujii M and Yamamoto K 1990 *Superlatt. Microstruct.* **8** 13

Agata M, Kurase H, Hayashi S and Yamamoto K, 1990 *Solid State Commun.* **76** 1061

Hayashi S, Sanda H, Agata M and Yamamoto K 1989 *Phys. Rev.* **40** 5544

[76] Itoh T, Iwabuchi Y and Kataoka M 1988 *Phys. Status Solidi* b **145** 567

[77] Bogomolov V N, Lutsenko E L, Petranovskii V P and Kholodkevich S V 1976 *JETP Lett.* **23** 482

Bogomolov V N, Poborchii V V and Kholodkevich S V 1980 *JETP Lett.* **31** 434

Bogomolov V N, Poborchii V V, Kholodkevich S V and Shagin S I 1980 *JETP Lett.* **38** 533

Bogomolov V N, Poborchii V V and Kholodkevich S V 1985 *JETP Lett.* **42** 517

Bogomolov V N, Poborchii V V, Romanov S G, Shagin S I 1985 *J. Phys. C: Solid State Phys.* **18** L313

[78] Katayama Y, Yao M and Ajiro Y 1989 *J. Phys. Soc. Japan* **58** 1811

[79] Terasaki O, Yamazaki K, Thomas J M, Ohsuna T, Watanabe D, Sanders J V and Barry J C 1987 *Nature* **330** 58

[80] Herron N 1986 *Inorg. Chem.* **25** 4714
[81] Wang Y and Herron N 1987 *J. Phys. Chem.* **91** 257
[82] Herron N, Wang Y, Eddy M, Stucky G D, Cox D E, Bein T and Moller K 1989 *J. Am. Chem. Soc.* **111** 530
[83] Breck D W 1974 *Zeolite Molecular Sieves* (New York: Wiley)
[84] Stramel R D, Nakamura T and Thomas J K 1988 *J. Chem. Soc. Faraday Trans.* I **84** 1287
[85] Liu X and Thomas J K 1989 *Langmuir* **5** 58
[86] Stucky G D and MacDougall J E 1990 *Science* **247** 669
[87] Hirono T, Kawana A and Yamada T 1987 *J. Appl. Phys.* **62** 1984
[88] Ozin G A, Godber J P and Stein A 1988 *US Patent* Filed August 1988
Stein A 1988 *Thesis* University of Toronto
[89] Nozue Y, Tang Z K and Goto T 1990 *Solid State Commun.* **73** 531
[90] Cao G, Rabenberg L K, Nunn C M and Mallouk T E 1991 *Chem. Mater.* **3** 149
[91] Wang Y and Herron N 1988 *J. Phys. Chem.* **92** 4988
[92] Kratschmer W, Lamb L D, Fostiropoulos K and Huffman D R 1990 *Nature* **347** 354
[93] Dance I G, Choy A and Scudder M L 1984 *J. Am. Chem. Soc.* **106** 6285
[94] Dance I G 1986 *Polyhedron* **5** 1037
[95] Lee G S H, Craig D C, Ma I, Scudder M L, Bailey T D and Dance I G 1988 *J. Am. Chem. Soc.* **110** 4863
[96] Dance I G, Choy A and Scudder M L 1984 *J. Am. Chem. Soc.* **106** 6285
[97] Wang Y, Harmer M and Herron N 1993 *Isr. J. Chem.* **33** 31
[98] Eychmuller A, Katsikas L and Weller H 1990 *Langmuir* **6** 1605
[99] Farneth W E, Herron N and Wang Y 1992 *Chem. Mater.* **4** 917

Chapter 5

Formation of nanostructures by mechanical attrition

Hans J Fecht

5.1 INTRODUCTION AND BACKGROUND

Since the 1970s, mechanical attrition (MA) of powder particles as a method for materials synthesis has been developed as an industrial process to successfully produce new alloys and phase mixtures. For example, this powder metallurgical process allows the preparation of alloys and composites which can not be synthesized via conventional casting routes, e.g. uniform dispersions of ceramic particles in a metallic matrix and alloys of metals with quite different melting points with the goal of improved strength and corrosion resistance [1]. Furthermore, mechanical attrition has gained a lot of attention as a nonequilibrium process resulting in solid-state alloying beyond the equilibrium solubility limit and the formation of amorphous or nanostructured materials for a broad range of alloys, intermetallics, ceramics, and composites [2, 3]. In the case of mechanical attrition of a binary powder mixture, amorphous phase formation occurs by intermixing of the atomic species on an atomic scale, thus driving the crystalline solid solution outside of its stability range against 'melting' resulting in solid-state amorphization [4]. This process is considered to be a result of both mechanical alloying [5] and the incorporation of lattice defects into the crystal lattice [6].

However, in many cases it is experimentally difficult to clearly distinguish between a glassy structure (i.e. without translational symmetry, as in a liquid) or a nano- (or micro-) crystalline structure (i.e. an assembly of randomly oriented crystalline fragments of a bulk crystalline phase) (for a discussion see [7]). In order to avoid any confusion, the emphasis will be limited here to materials which are clearly nanocrystalline, i.e. characterized by the existence of grain boundaries between the nanometer-sized grains or interphase boundaries between the nanophase domains in multi-component systems.

As a consequence of the cold-working process occurring during mechanical attrition, a reduction of the average grain size is observed from typically 50 to 100 μm diameter (identical with the particle diameter) to sizes ranging from 2 to 20 nm. This internal refining process with a reduction of the average grain size by a factor 10^3–10^4 results from the creation and self-organization of large-angle grain boundaries within the powder particles during the milling process. Though the mechanism of microstructure formation is very different from other synthesis methods for nanostructured materials starting from clusters or ultrafine particles, the resulting microstructure is very similar. Furthermore, by this method the quantity limitations in preparing nanocrystalline materials can be overcome and the production of nanocrystalline powders can be scaled up to industrially relevant amounts. Although several problems still have to be solved in order to use these materials for technological applications, mechanical attrition offers interesting perspectives in preparing nanostructured powders with a number of different types of interface both in terms of structure (crystalline/crystalline, crystalline/amorphous) as well as atomic bonding (metal/metal, metal/semiconductor, metal/ceramic, etc). This opens exciting possibilities for the preparation of advanced materials with particular grain or interphase-boundary design.

5.2 HIGH-ENERGY BALL MILLING AND MECHANICAL ATTRITION

The milling of materials is of prime interest in the mineral, ceramic processing, and powder metallurgy industry [8]. Typical objectives of the milling process include particle size reduction (comminution), solid-state alloying, mixing or blending, and particle shape changes. These industrial processes are mostly restricted to relatively hard, brittle materials which fracture, deform, and cold weld during the milling operation. While oxide-dispersion strengthened superalloys have been the primary application of mechanical attrition, the technique has been extended to produce a variety of nonequilibrium structures including nanocrystalline [3], amorphous [2, 4, 9], and quasicrystalline [10] materials (for a review see [11]).

A variety of ball mills has been developed for different purposes including tumbler mills, attrition mills, shaker mills, vibratory mills, planetary mills, etc [12]. The basic process of mechanical attrition is illustrated in figure 5.1. Powders with typical particle diameters of about 50 μm are placed together with a number of hardened steel or tungsten carbide (WC) coated balls in a sealed container which is shaken or violently agitated. The most effective ratio for the ball to powder masses is five to 10.

High-energy milling forces can be obtained by using high frequencies and small amplitudes of vibration. Shaker mills (e.g. SPEX model 8000) which are preferable for small batches of powder (approximately 10 cm^3 is sufficient for research purposes) are highly energetic and reactions can take place one order of

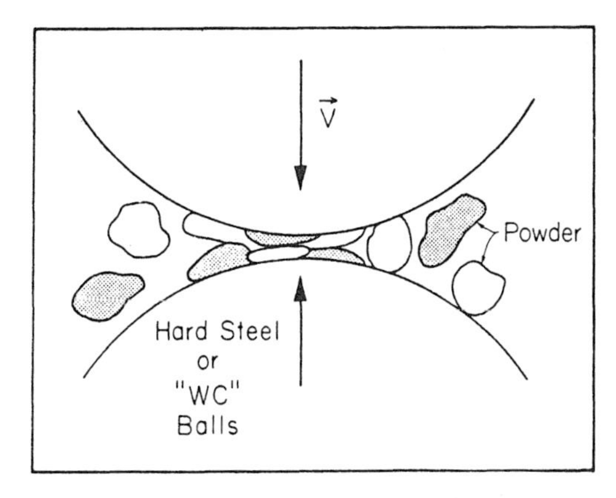

Figure 5.1. Schematic sketch of the process of mechanical attrition of metal powder.

magnitude faster than with other types of mill. Since the kinetic energy of the balls is a function of their mass and velocity, dense materials (steel or tungsten carbide) are preferable to ceramic balls. During the continuous severe plastic deformation associated with mechanical attrition, a continuous refinement of the internal structure of the powder particles to nanometer scales occurs during high-energy mechanical attrition. The temperature rise during this process is modest and is estimated to be $\leq$100 to 200 °C [4].

For all nanocrystalline materials prepared by a variety of different synthesis routes, surface and interface contamination is a major concern. In particular, during mechanical attrition contamination by the milling tools (Fe) and atmosphere (trace elements of O_2, N_2 in rare gases) can be a problem. By minimizing the milling time and using the purest, most ductile metal powders available, a thin coating of the milling tools by the respective powder material can be obtained which reduces Fe contamination tremendously. Atmospheric contamination can be minimized or eliminated by sealing the vial with a flexible 'O' ring after the powder has been loaded in an inert gas glove box. Small experimental ball mills can also be enclosed completely in an inert gas glove box. As a consequence, the contamination with Fe based wear debris can generally be reduced to less than 1–2 at.% and oxygen and nitrogen contamination to less than 300 ppm. However, milling of refractory metals in a shaker or planetary mill for extended periods of time (>30 h) can result in levels of Fe contamination of more than 10 at.% if high vibrational or rotational frequencies are employed. On the other hand, contamination through the milling atmosphere can have a positive impact on the milling conditions if one wants to prepare metal/ceramic nanocomposites with one of the metallic elements being chemically highly reactive with the gas (or fluid) environment.

5.3 PHENOMENOLOGY OF NANOSTRUCTURE FORMATION

5.3.1 Elements and intermetallics

During mechanical attrition the metal powder particles are subjected to severe mechanical deformation from collisions with the milling tools. Consequently, plastic deformation at high strain rates ($\sim 10^3$–10^4 s^{-1}) occurs within the particles and the average grain size can be reduced to a few nanometers after extended milling. This was first investigated in detail for a number of high-melting metals with bcc and hcp crystal structures [13, 14]. Metals with the fcc structure are inherently more ductile and often exhibit a stronger tendency to adhere to the container walls and to sinter to larger particles often several millimeters in diameter during the milling process. However, the successful preparation of nanocrystalline fcc metals has been described in a detailed study by Eckert *et al* [15].

By cold working, the metal is plastically deformed with most of the mechanical energy expended in the deformation process being converted into heat, but the remainder is stored in the metal, thereby raising its internal energy [16]. As model systems Ru, which is hcp, and the CsCl type AlRu intermetallic compound have been analyzed in detail with the relevant results summarized in the following. The microstructural changes as a result of mechanical attrition can be followed by x-ray diffraction methods, which average the structural information over the samples. The x-ray diffraction patterns exhibit an increasing broadening of the crystalline peaks as a function of milling time. As an example, figure 5.2 shows typical x-ray diffraction patterns (Cu Kα radiation) for (*a*) Ru and (*b*) AlRu powders before and after 32 h ball milling [17].

To separate the effects of grain size from internal strain the full widths at half maximum of the Bragg peaks ΔK as a function of their K value, the corresponding reciprocal space ($K = 2\sin\theta/\lambda$) variable, can be plotted [18]. A more sophisticated Warren–Averbach analysis basically gives the same results [19]. After corrections for Kα and instrumental broadening, the line broadening due to the small crystal size is constant in K space and is given by $\Delta K = 0.9\ (2\pi/d)$ where d is the average domain or grain diameter. The strain broadening corresponds to $\Delta K = A\langle e^2\rangle^{1/2}K$ with A being a constant depending on the strain distribution ($A \approx 1$ for a random distribution of dislocations [20]) and $\langle e^2\rangle^{1/2}$ the root mean square (RMS) strain. Here, additional defects which might contribute to the peak broadening, such as stacking faults, can be safely neglected in all cases discussed here. However, for some metals with very small stacking fault energies, e.g. Co, the contribution of stacking faults to the peak broadening is considerable [21]. See chapter 10 for a further discussion of these matters.

The change of the average coherently diffracting domain size (grain or crystal size) as a function of milling time is shown in figure 5.3. In the very beginning mechanical attrition leads to a fast decrease of the grain size to less

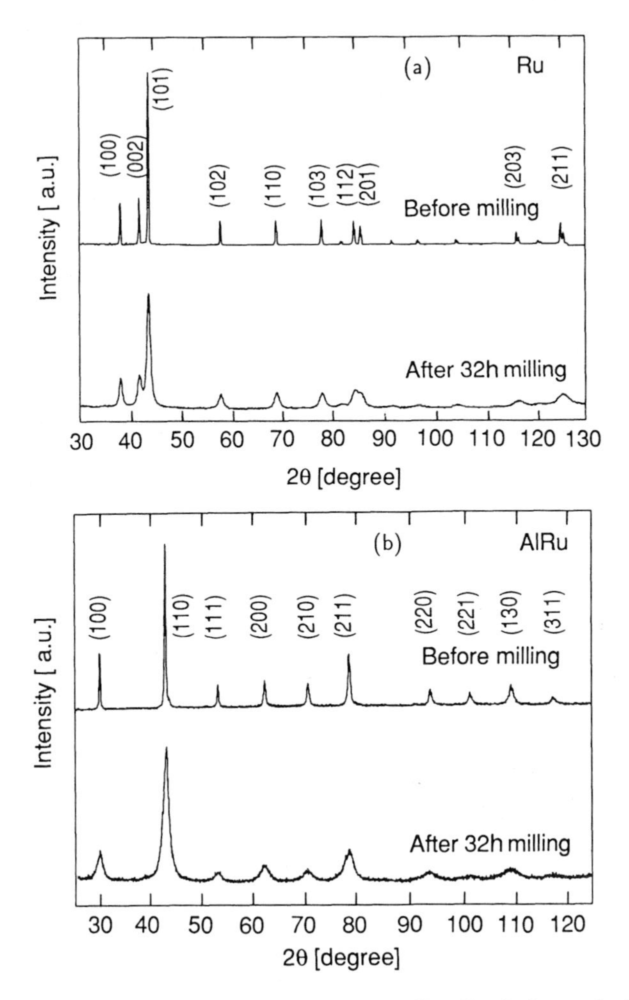

Figure 5.2. X-ray diffraction patterns of (*a*) Ru and (*b*) AlRu before and after MA for 32 h.

than 20 nm. Further refinement occurs slowly to about 10 nm after extended milling. In addition, the RMS strain as calculated from the x-ray broadening is shown in figure 5.4 as a function of reciprocal grain size. The deformation leads to an increase in atomic-level strains to about 1% (typical for metals) and 3% (typical for intermetallic compounds). The maximum value is reached at a grain size of about 12 nm. It is interesting to note that the strain decreases for grain sizes less than 12 nm.

Direct observation of the individual grains within the deformed powder particles by transmission electron microscopy (TEM) agrees well with the grain size determination by x-ray diffraction. Ultrathin sections (20–50 nm)

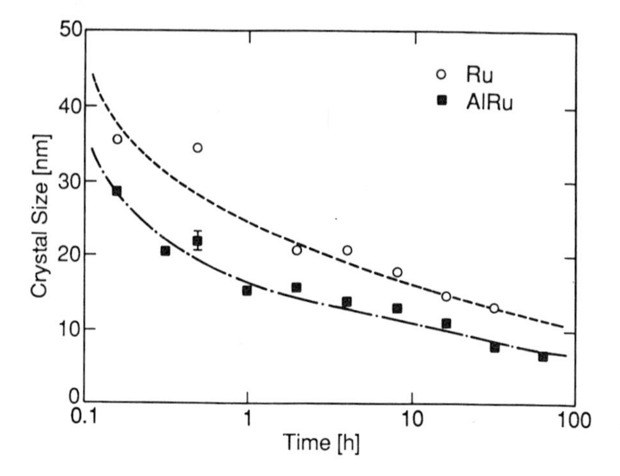

Figure 5.3. The average grain size as determined from x-ray line broadening as a function of milling time for Ru and AlRu.

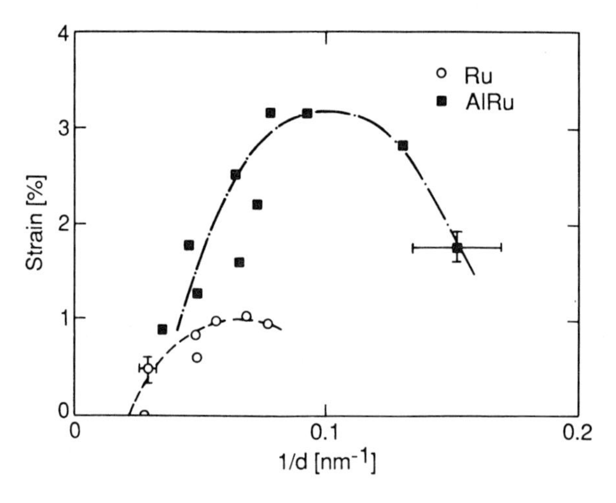

Figure 5.4. The measured RMS strain in Ru and AlRu as a function of reciprocal grain size.

of the powder particles are usually prepared with a diamond knife using an ultramicrotome. The sections are retrieved from water onto Cu grids coated with a holey carbon substrate. As an example, a TEM bright-field image of AlRu milled for 64 h together with the corresponding diffraction pattern is shown in figure 5.5. The AlRu has an estimated grain size of 5 to 7 nm. The distribution of grain sizes is relatively narrow, which indicates that extended ball milling produces a uniform comminution. The corresponding diffraction patterns and the lattice fringes of the crystals both show that the orientation of neighboring grains is completely random, i.e. the individual crystalline grains

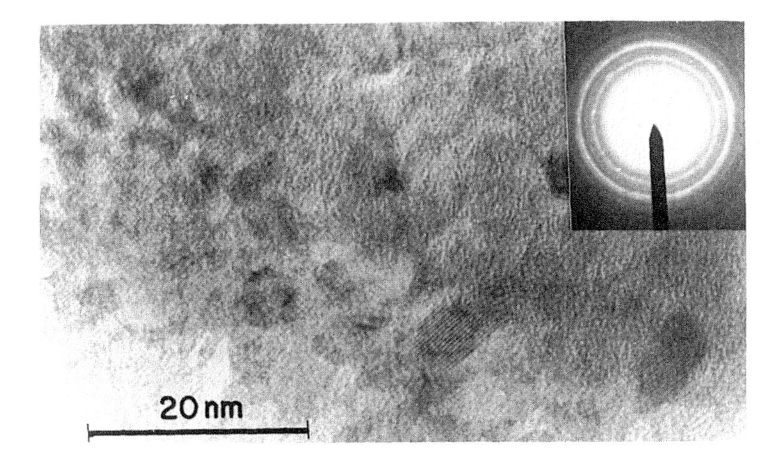

Figure 5.5. TEM high-resolution bright-field image of AlRu together with the corresponding diffraction pattern after MA for 64 h.

are separated by high-angle grain boundaries. It is interesting to note that by mechanical attrition plastic deformation can be introduced in nominally brittle materials such as the intermetallic compounds described here.

Basically all metals and compounds investigated so far exhibited similar behavior in terms of grain size reduction and an increase in atomic-level strains. Typical values for the average grain sizes of fcc metals vary between 22 nm for Al and 6 nm for Ir whereas a typical grain size of 8 nm is reached for most bcc metals and 13 nm for hexagonal metals [22].

The minimum grain or domain size for intermetallic compounds with CsCl structure has been found to vary between 12 nm for CuEr [23] and 2 nm (amorphous) for NiTi [24]. Furthermore, in the ordered intermetallic compounds anti-site disorder is introduced during mechanical attrition. Whereas for the CsCl compounds the reduction of the long-range chemical order parameter saturates at about 0.7, other intermetallic compounds exhibit complete disordering and the formation of a nanocrystalline solid solution. For example, the A-15 type compounds Nb_3Al [25], V_3Ga [25], and Nb_3Au [27] transform after extended milling to a bcc solid solution with nanometer-sized grains.

5.3.2 Nonequilibrium solid solutions

The mechanical attrition of multi-component powder mixtures results generally in the formation of solid solutions extended in composition far beyond their equilibrium solubility limit. The mechanical alloying process for binary alloys with a negative heat of mixing is a commonly observed process for mechanically driven systems and can reach extensions in compositions by a factor up to 10 beyond equilibrium solubility [28].

On the other hand, mechanical alloying is surprisingly achieved for powder mixtures having a positive enthalpy of mixing. Though in some cases, such as Ag–Fe, an intimate phase mixture of nanostructured Ag and Fe particles is produced [29], in other cases real miscibility on an atomic level can be obtained, e.g. for Cu–Fe [30], Cu–W [31], Cu–Ta [32], and Cu–V [33]. This apparent violation of the rules of equilibrium thermodynamics is a vital example of the potential of MA in synthesizing new materials under nonequilibrium conditions. For example, mechanical alloying can lead to the formation of nanocrystalline single-phase solid solutions of up to 60 at.% Fe in Cu and 20 at.% Cu in Fe [34, 35]. The steady-state grain sizes range from 20 nm for Cu to 8–10 nm for Fe-rich alloys as exhibited in figure 5.6.

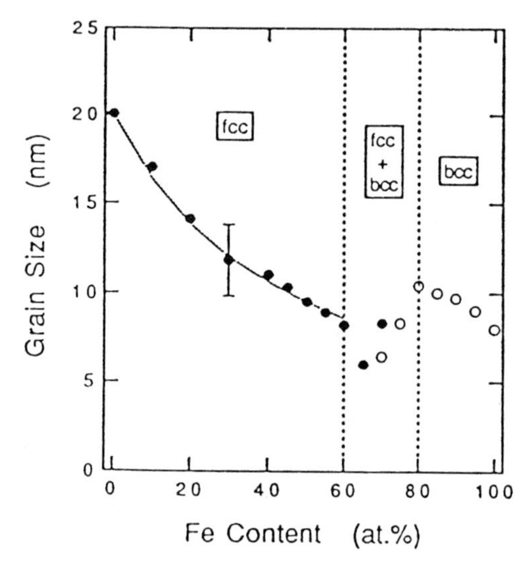

Figure 5.6. Average grain size for Fe_xCu_{100-x} powders after 24 h of milling versus Fe content (after [35]).

The enhanced solubility of alloys exhibiting spinodal behavior in coarse-grained systems has been attributed to the capillary pressure of the nanosized grains on the free energy due to their small radii of curvature. It has been found that during mechanical attrition of Fe–Cu powder mixtures agglomerates of multilayers are formed leading to microstructures very similar to those obtained by cold-rolling [36]. In this case, mechanical deformation proceeds by necking of one of the constituent phases with the surrounding matrix in the necking region undergoing heavy local deformation. This continuous process leads to the formation of small Fe particles embedded in the ductile Cu matrix. If the Gibbs free energy difference between the bcc and fcc structure is offset by the capillary pressure experienced by the Fe particle with radius r, i.e. $\Delta G_{Fe} = 2V_m\sigma_{FeCu}/r$, dissolution of Fe occurs. Here, V_m denotes the molar volume and σ_{FeCu} the Fe/Cu interfacial energy. σ_{FeCu} has been estimated to about 1.37 J m^{-2} [34].

Using established thermodynamic data, depending on composition the minimum diameter of Fe particles being dissolved was found to amount to 2–5 nm. Since the interfacial energy in nanostructured materials can be considerably larger in comparison with coarse-grained systems and the Gibbs free energy difference between bcc and fcc structures can be smaller by a factor of two at the prevalent conditions, the critical radius for the Fe particles is quite close to 10 nm, the typical grain diameter found during mechanical attrition of Fe–Cu mixtures. As such, for small particles of several nanometers in diameter, capillary pressures become important in the thermodynamic balance, thus changing solubility limits drastically as a function of grain size.

Furthermore, mechanical attrition can also produce ultra-fine-scaled phase mixtures if a brittle material is milled together with a more ductile material. For example, 10 nm Ge particles can be embedded in a ductile Sn or Pb matrix [37]. Similarly, very fine dispersions at the nanometer scale have been found in Fe–W, Cu–Ta, TiNi–C [38], and Ag–Fe [39]. In the case of Fe–Ag the grain size of the intermixed Ag and Fe domains is in the range of a few nanometers as shown in the high-resolution electron micrograph in figure 5.7 taken from [39]. By additional Mössbauer studies, the mutual solubility of Ag in Fe domains (and Fe in Ag domains) was shown. The resulting microstructure and chemical arrangement is very similar to nanocrystalline phase mixtures prepared by gas condensation methods [40].

5.3.3 Nanocomposites by mechano-chemistry

As mentioned earlier mechanical attrition is very sensitive to contamination resulting from the milling environment. As such, atmospheric control can be used to induce chemical reactions between the milled powders and their environment. By a proper choice of a reactive gas atmosphere (O_2, N_2, air, etc) or a milling fluid (organic fluid which also minimizes wear) the metal powder can be intentionally modified to a nanocrystalline metal–ceramic composite.

For example, ball milling of the intermetallic compound Fe_2Er (C15) in air resulted in the formation of a nanocrystalline Fe–ErN composite. By simply ball milling the powder in air in a container not hermetically sealed, a nanometer-scale phase separation into Fe-rich and Er-rich crystallites has been observed [41]. Chemical analysis of the samples revealed a large increase in nitrogen concentration of about 3 wt% whereas the oxygen concentration remained small (~0.05 wt%). X-ray diffraction results as shown in figure 5.8 indicated that after 10 h of milling the compound phase (figure 5.8(*a*)) vanished and was replaced by a mixture of α-Fe and fcc-ErN with an average grain size of 6 nm (figure 5.8(*b*)). With further milling the average grain size has been reduced further to about 2 nm (figure 5.8(*c*)). This result is consistent with TEM analysis revealing a grain size varying between 2 and 4 nm. This steady-state grain size distribution is considerably smaller than in the examples of MA mentioned above and is in the range of typical domain sizes of an 'amorphous' structure. However,

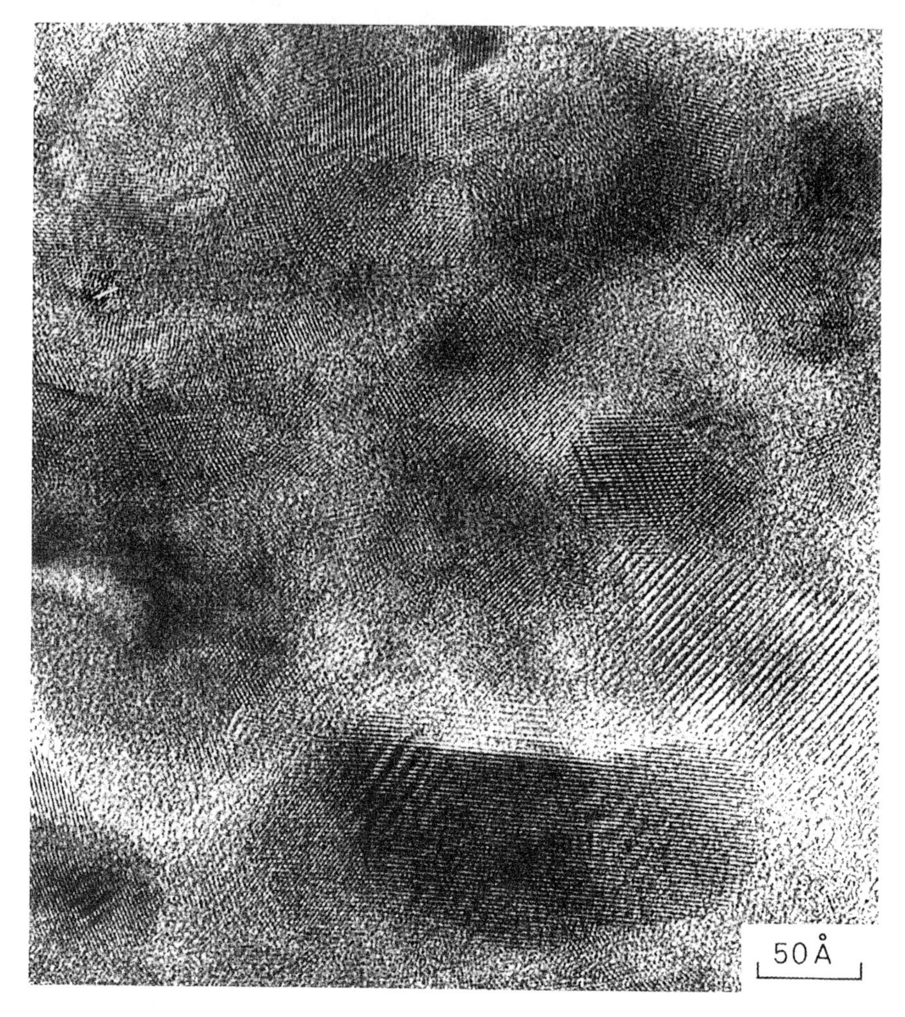

Figure 5.7. High-resolution electron microscopy image of a nanocrystalline two-phase composite of fcc Ag and bcc Fe with only small mutual solid solubility (after [39]).

the microstructures obtained from high-resolution TEM analysis clearly show crystalline regions remains within the nanocrystalline pattern. Similar results have been obtained by milling elemental Fe and Er powders at the same composition. However, milling of Er did not exhibit the formation of ErN; it turns out that Fe as a dissociation catalyst for the N_2 molecules is necessary to trigger the nitriding reaction.

A number of further examples demonstrate the potential of reactive milling in the preparation of metal nitrides and oxides. For example, the metal powders Ti, Fe, V, Zr, W, Hf, Ta, and Mo [42–44] transform to a nanocrystalline

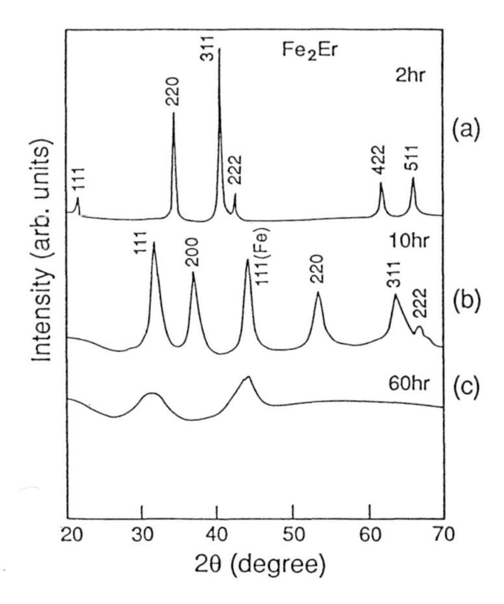

Figure 5.8. X-ray diffraction data (Cu Kα radiation) for the Laves phase Fe_2Er and the resulting nanostructured Fe–Er–N composites.

nitride by high-energy ball milling under nitrogen gas flow. This solid-state interdiffusion reaction during reactive ball milling is triggered by fragmentation of the starting powder thus creating new surfaces. These freshly created surfaces react with the flowing nitrogen gas to form a nitride surface layer over the unreacted core particle. With further milling this reaction continues and a homogeneous nitride phase is formed and the unreacted core of metal disappears resulting in a nanostructured (often metastable) metal nitride with a grain size of typically 5 nm.

By ball milling in organic fluids such as surfactants, which are sometimes used to prevent contamination by the milling tools, chemical reactions can be induced leading to the formation of fine carbides. For example, by milling AlTi, AlZr, and AlHf alloys in hexane an average grain size of 9 nm can be achieved with carbon being dissolved in the matrix [45]. During dynamic compaction at about 1300 K, grain growth occurs up to about 44 nm together with precipitation of ZrC particles, 7 nm in size. Such ultrafine-grained composites are expected to exhibit considerably improved ductility [46].

5.4 MECHANISM OF GRAIN-SIZE REDUCTION

For a better understanding of the refinement mechanism, the microstructure which develops in the powder particles has been investigated in more detail. Figure 5.9 shows a series of TEM micrographs of an AlRu particle after MA.

It can be seen that the crystal is heavily strained and the deformation occurs in a rather inhomogeneous way. The arrows in figure 5.9(*a*) indicate a highly deformed region of a width of about 1 μm which extends throughout the entire particle. These shear bands have been observed in rolled metals and are typical for deformation mechanisms that occur at high strain rates in contrast to slip and twinning mechanisms at low and moderate strain rates [47].

The observed shear bands are separated by areas of similar lateral dimensions in the micrometer range having low defect densities. High-resolution imaging of areas in the shear bands reveals a microstructure consisting of individual grains with a diameter of about 20 nm which are slightly rotated with respect to each other at a rotation angle of less than 20° as shown in figure 5.9(*b*). With longer durations of MA, the shear bands grow over larger areas and eventually (see figure 5.9(*c*)) the entire sample disintegrates into subgrains with a final grain size of 5–7 nm for AlRu after 64 h (see figure 5.5), thus ductilizing the originally brittle intermetallic compound.

The processes leading to the grain size refinement include three basic stages.

(i) Initially, the deformation is localized in shear bands consisting of an array of dislocations with high density.

(ii) At a certain strain level, these dislocations annihilate and recombine as small-angle grain boundaries separating the individual grains. The subgrains formed via this route are already in the nanometer range (~20–30 nm).

(iii) The orientations of the single-crystalline grains with respect to their neighboring grains become completely random. This can be understood in the following way. The yield stress σ required to deform a polycrystalline material by dislocation movement is related to the average grain size d by $\sigma = \sigma_0 + kd^{-1/2}$, where σ_0 and k are constants (Hall–Petch relationship). An extrapolation to nanocrystalline dimensions shows that very high stresses are required to maintain plastic deformation. Experimental values for k and σ_0 are typically $k = 0.5$ MN m$^{-3/2}$ and $\sigma_0 = 50$ MPa [48]. For a grain size of 10 nm the minimum yield stress is of the order of 5 GPa corresponding to 15% of the theoretical shear stress of a hexagonal metal, which sets a limit to the grain-size reduction achieved by plastic deformation during ball milling. Therefore, the reduction of grain size to a few nm is limited by the stresses applied during ball milling as long as no dramatic elastic softening of the crystal lattice occurs.

Further energy storage by mechanical deformation is only possible by an alternative mechanism. Grain boundary sliding has been observed in many cases at high temperatures leading to superplastic behavior. Alternatively, grain boundary sliding can also be achieved at very small grain size and low temperature by diffusional flow of atoms along the intercrystalline interfaces [49]. This provides a mechanism for the self-organization and rotation of the grains, thus increasing the energy of the grain boundaries proportional to their misorientation angle and excess volume.

(a)

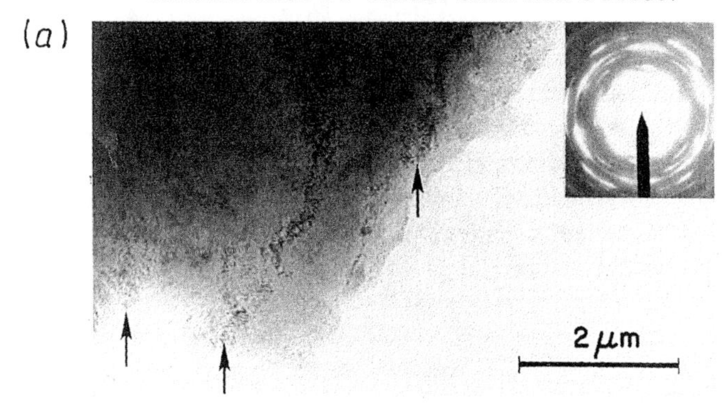

(b)

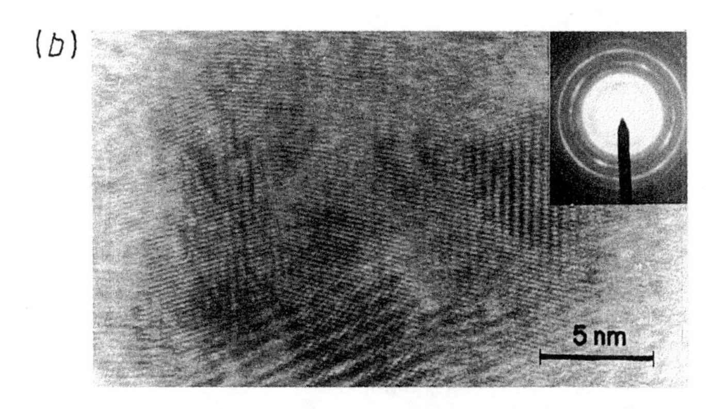

(c)

Figure 5.9. (*a*) TEM bright-field images at relatively low magnification of an AlRu powder particle after 10 min of MA. The arrows point to highly deformed regions (shear bands). The inset shows the corresponding diffraction pattern demonstrating the gradual smearing out of the initially sharp diffraction spots. (*b*) TEM high-resolution bright-field image of AlRu after 10 min of MA and its corresponding diffraction pattern. (*c*) TEM bright-field image and its corresponding diffraction pattern after 2 h of MA.

This behavior is typical for deformation processes of fcc and bcc metals at high strain rates. However, it is surprising that nominally brittle materials, such as intermetallics, develop considerable ductility under shear conditions as indicated by the formation of shear bands. Similar observations regarding the deformation mechanism have been reported in chips removed during machining [50] and in simple metal filings [51, 52]. In analogy to the mechanically attrited powder at the early stage, large inhomogeneities have been observed in the filings with the deformation process leading to the formation of small-angle grain boundaries. Here, the dislocation cell size dimensions are basically a function of the acting shear stress τ, resulting in an average cell size dimension L of $L = 10Gb/\tau$ with G being the shear modulus and b the Burgers vector [53].

More detailed studies are obtained from cold-rolling and torsion [54], wire drawing [55], and cyclic deformation [56] processes resulting in an asymptotic saturation of the flow stresses. This is considered to be a result of the simultaneous occurrence of dislocation multiplication and annihilation leading to a saturation of the dislocation density [57]. In particular, under cyclic deformation of, for example Cu, at strain amplitudes above $\gamma_{pl} \sim 10^{-4}$, slip becomes highly localized in so-called persistent slip bands (shear bands). These lie parallel to the primary glide plane and are separated by regions containing the original matrix structure [57]. These bands consist of dense walls of dislocations, largely screw dislocations having a density $\sim 10^{13}$ m^{-2}. The closest spacing between screw dislocations of opposite sign is $\sim$50 nm, the minimum distance before annihilation occurs. For edge dislocations, which are more relevant for the deformation of fcc crystals, this critical annihilation length is found to be 1.6 nm for Cu. It has been concluded that the annihilation of dislocations can set a natural limit to the dislocation densities which can be achieved by plastic deformation (typically less than 10^{13} m^{-2} for screw dislocations and 10^{16} m^{-2} for edge dislocations). Steady-state deformation is observed when the dislocation multiplication rate is balanced by the annihilation rate. This situation corresponds to the transition from stage (i) to stages (ii)/(iii) as described above. In this stage the role of dislocations becomes negligible and further deformation occurs via grain boundaries. It is expected that the shear modulus of the grain boundary regions is lowered by about 40% when the 'volume fraction' of the grain boundaries becomes comparable to that of the crystals [58, 59]. Localized deformation then proceeds by the dilatation of the grain boundary layers similar to superplastic behavior [60] with the undeformed crystallites moving in a 'sea' of dilated grain boundaries. Furthermore, the relative motion of the crystalline grains within the shear band leads to impingement which should give rise to large, locally inhomogeneous elastic stresses. As a consequence, in order to relax these strains, formation of nanovoids about 1 nm in diameter is expected to occur which inevitably leads to crack formation under tensile stress [61]. Such a deformation mode basically also provides a mechanism for the repeated fracturing and rewelding of the fresh surfaces during MA leading to a steady-state particle size.

5.5 PROPERTY–MICROSTRUCTURE RELATIONSHIPS

Decreasing the grain size of a material to the nanometer range leads to a drastic increase of the number of grain boundaries reaching typical densities of 10^{19} interfaces per cm^3. The large concentration of atoms located in the grain boundaries in comparison with the crystalline part scales roughly as the reciprocal grain-size $1/d$.

Consequently, due to their excess free volume the grain boundaries in the nanocrystallite cause large differences in the physical properties of nanocrystalline materials if compared with conventional polycrystals. In all cases discussed here the short-range order typical for an amorphous material is not observed as the characteristic structure of grain boundaries. As such, the grain boundary structure in these materials must be different from the structure of the single crystal as well as the amorphous structure of a glassy material. It turns out that the thermodynamic properties of nanostructured materials produced by MA can be realistically described on the basis of a free volume model for grain boundaries [62].

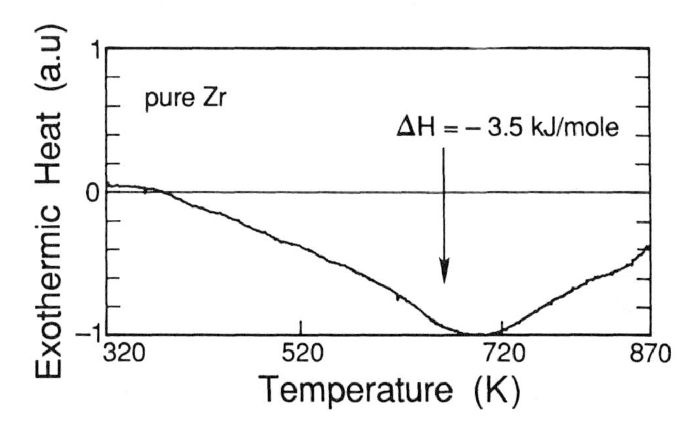

Figure 5.10. DSC heating scan at 20 K min^{-1} for Zr powder particles, ball milled for 24 h. Integration of the signal deviating from the baseline gives the stored enthalpy ΔH.

As a result of the cold work energy has been stored in the powder particles. This energy is released during heating to elevated temperatures due to recovery, relaxation processes within the boundaries, and recrystallization. For example, figure 5.10 exhibits a typical scan by a differential scanning calorimeter (DSC) for Zr after 24 h of MA. During heating in the DSC, a broad exothermic reaction is observed for all of the samples starting at about 370 K and being almost completed at 870 K. Integrating the exothermal signals gives the energy release ΔH during heating of the sample. ΔH values are listed in table 5.1 along with other characteristic values, such as the average grain size and excess specific heat after 24 h of MA for several metals and alloys.

Table 5.1. Structural and thermodynamic properties of metal and intermetallic powder particles after 24 h ball milling, including the melting temperature T_m, the average grain size d, the stored enthalpy or energy release ΔH as a percentage of the heat of fusion, and the excess heat capacity Δc_p.

Material	Structure	T_m (K)	d (nm)	ΔH (% of ΔH_f)	Δc_p (%)
Fe	bcc	1809	8	15	5
Cr	bcc	2148	9	25	10
Nb	bcc	2741	9	8	5
W	bcc	3683	9	13	6
Co	hcp	1768	(14)	6	3
Zr	hcp	2125	13	20	6
Hf	hcp	2495	13	9	3
Ru	hcp	2773	13	30	15
Al	fcc	933	22	43	—
Cu	fcc	1356	20	39	—
Ni	fcc	1726	12	25	—
Pd	fcc	1825	7	26	—
Rh	fcc	2239	7	18	—
Ir	fcc	2727	6	11	—
NiTi	CsCl	1583	5	25	2
CuEr	CsCl	1753	12	31	2
SiRu	CsCl	2073	7	39	10
AlRu	CsCl	2300	8	18	13

The stored enthalpy reaches values up to 7.4 kJ mol^{-1} (after 24 h) and 10 kJ mol^{-1} (after 32 h) for Ru, which corresponds up to 30–40% of the heat of fusion ΔH_f. These data are exhibited in figure 5.11 showing the energy release ΔH for different hcp and bcc metals with a grain size between 8 and 13 nm as a function of their respective melting temperatures. One would expect the recovery rates during the milling process to correlate with the melting point of the specific metal. With the exceptions of Co (due to a large number of stacking faults) and Hf, Nb, and W (possibly due to an increased level of Fe impurities from the milling tools stabilizing the nanostructure) such a relationship is indeed observed. Similar results have been obtained for metals with fcc structure as given in table 5.1 [15]. Consequently, most effective energy storage occurs for metals with melting points above 1500 K, resulting in average grain sizes between 6 (Ir) and 13 nm (Zr). For the compound phases similar high values for the stored energies are found ranging from 5 to 10 kJ mol^{-1} and corresponding

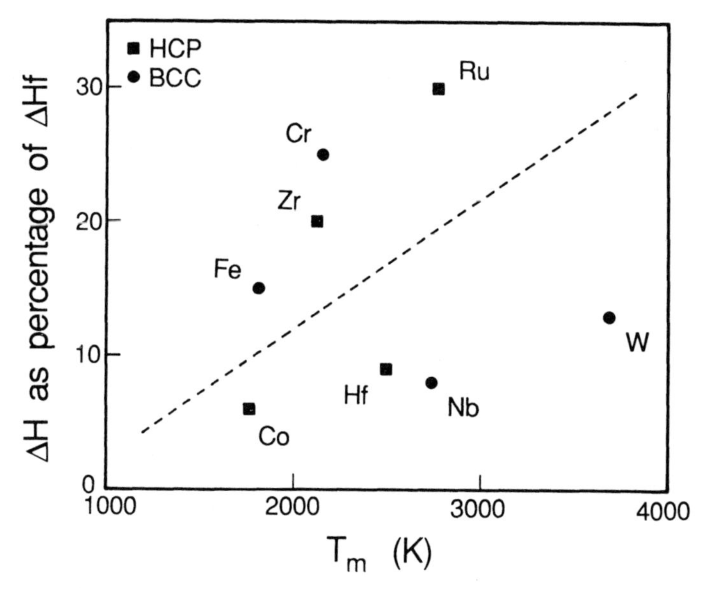

Figure 5.11. The released energy ΔH as a percentage of the heat of fusion ΔH_f measured by DSC up to 870 K. The data are shown as a function of the melting point T_m for several bcc and hcp metals.

to values between 18 and 39% of the heat of fusion for grain sizes between 5 and 12 nm.

Large differences generally also arise in the specific heat c_p at constant pressure. The specific heat of the heavily deformed powder particles was measured in the range from 130 K to 300 K, i.e. at low enough temperatures to prevent the recovery processes from taking place. For all samples, a considerable increase in c_p has been found experimentally after 24 h milling, reaching values up to 15% for Ru. These data are also included in table 5.1 given as a percentage of heat capacity increase in comparison to the unmilled state at 300 K. For pure metals, the heat capacity change Δc_p exhibits a linear dependence on the stored enthalpy ΔH given as a percentage of the heat of fusion $(\Delta H/\Delta H_f)$ after extended MA as presented in figure 5.12. Such a relationship is also predicted by the free volume model for grain boundaries.

The final energies stored during MA greatly exceed those resulting from conventional cold working of metals and alloys (cold-rolling, extrusion, etc). During conventional deformation, the excess energy is rarely found to exceed 1–2 kJ mol^{-1} and, therefore, is never more than a small fraction of the heat of fusion [4, 16]. In the case of MA, however, the energy can reach values typical for crystallization enthalpies of metallic glasses corresponding to about 40% ΔH_f. A simple estimate demonstrates that these energy levels cannot be achieved by the incorporation of defects which are found during conventional processing. In

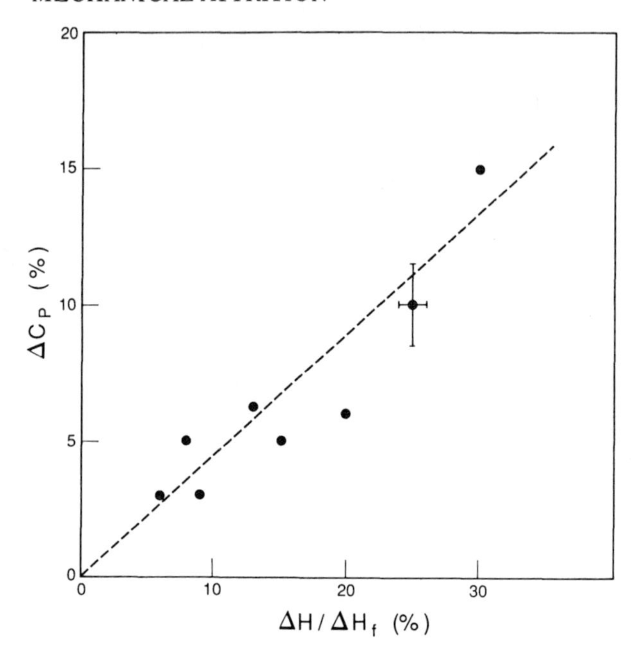

Figure 5.12. Specific heat increase Δc_p (%) in comparison to the unmilled state at room temperature as a function of the stored enthalpy ΔH (given as a percentage of ΔH_f) after 24 h ball milling of pure elemental powder samples.

the case of pure metals, the contribution of point defects (vacancies, interstitials) can be safely neglected because of the high recovery rate at the actual processing temperature [16]. Even taking nonequilibrium vacancies into account, which can form as a consequence of dislocation annihilation up to concentrations of 10^{-3} [57], such contributions are energetically negligible in comparison. On the other hand, for intermetallics point defects are relevant in describing the stability of the material [6].

The maximum dislocation densities that can be reached in heavily deformed metals are less than 10^{16} m^{-2} which would correspond to an energy of less than 1 kJ mol^{-1}. Therefore, it is assumed that the major energy contribution is stored in the form of grain boundaries and related strains within the nanocrystalline grains which are induced through grain boundary stresses as described above. Recent estimates suggest that the grain boundary energies in nanocrystalline metals are about twice as high as in high-energy grain boundaries in conventional polycrystals, close to about 1 J m^{-2} [15]. With the reduction of average grain size, an increase in hardness by a factor five as well as a decrease of Young's modulus by about 20% has been observed experimentally in nanocrystalline Nb with a grain size of about 12 nm [63].

The energetic microstructure–property relationship is further emphasized in figure 5.13. Here the stored enthalpy ΔH in Ru is shown as a function of average reciprocal grain size $1/d$. Since $1/d$ scales also with the volume density of grain boundaries in the nanocrystalline material ($\sim 3\delta/d$ with the thickness of the grain boundary about 1.2 nm [64]), this figure also represents the role of grain boundaries in energy storage.

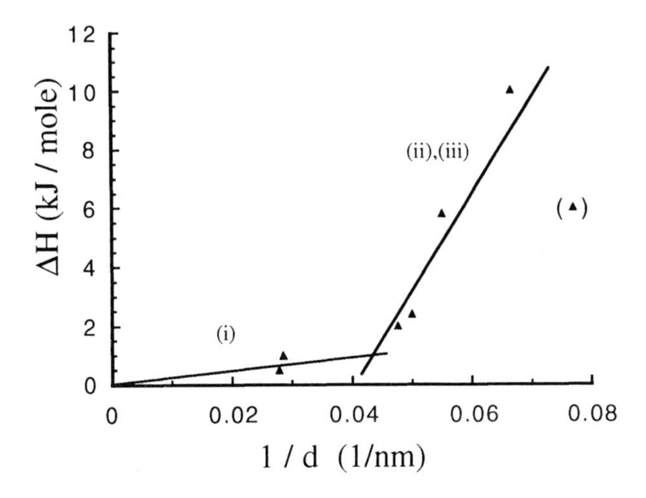

Figure 5.13. The stored enthalpy ΔH as a function of reciprocal grain size $1/d$ of Ru at different levels of MA. Two distinctively different stages can be observed: stage (i) which is dislocation controlled for $d > 25$ nm and stages (ii) and (iii) for $d < 25$ nm where deformation becomes controlled by grain boundaries.

Two different regimes can be clearly distinguished: for small grain size reductions at the early stages of MA, i.e. stage (i), the stored enthalpy shows only a weak grain size dependence typical for dislocation controlled deformation processes. After the average domain (grain) size is reduced below $d^* = 25$ nm, energy storage becomes more efficient. The critical grain size d^* as determined by figure 5.13 corresponds to the size of nanograins which are formed within the shear bands. Therefore, for $d < d^*$ a regime can be identified where deformation is controlled by the properties of the small-angle and later high-angle grain boundaries which are developing in stages (ii) and (iii). The slope of the corresponding strain versus $1/d$ relationship, which has been shown in figure 5.4, corresponds to the grain boundary regime of 0.1 nm, typical for atomic level strains [52].

It is important to note that the different deformation mechanisms, i.e. dislocation versus grain boundary mechanisms, result in a different microstructure which controls the physical properties of the material. In particular, in the nanocrystalline regime, i.e. for grain sizes less than $d^* \simeq 25$ nm, the thermodynamic and structural properties of the mechanically attrited material are controlled by the energy and structure of the grain boundaries developing. Such a transition from dislocation controlled properties to grain boundary controlled properties is expected for nanocrystalline materials synthesized by other methods as well [40, 65].

5.6 RELATED TOPICS

Many microscopic processes occurring during MA and mechanical alloying of powder particles exhibit common features with processes relevant in tribology and wear. For example, the effects of work hardening, material transfer, and erosion during wear situations result in similar microstructures of the wear surface as observed during MA [66]. In particular, during sliding wear, large plastic strains and strain gradients are created near the surface [67]. Similar to MA of powder particles, this is the consequence of the formation of dislocation cell networks, subgrains, and grain boundaries with the subgrains becoming smaller and sharper near the surface. Typical plastic shear strains occurring at the surface are estimated to be of order 10, corresponding to a strain rate of several 10^3 s^{-1}.

Cu nanocrystalline structures have been observed by high-resolution electron microscopy with an average grain size of 4–5 nm close to the surface of wear scars as well as in the wear debris [68]. Within the interiors of the grains no defects were observed suggesting that most of the defects are absorbed by the grain boundaries due to their proximity. However, this type of plastic deformation at high strain rates does not seem to be limited to metals and alloys [69], but has been observed in ceramics [70] and diamond [71] as well.

During sliding wear, transfer of material from one sliding component to another is found to occur in analogy to mechanical alloying. In such cases, a special tribo-layer develops on the surface of a sliding component being subjected to large plastic strains. This surface layer often is called the Beilby layer and for a long time was thought to be amorphous because its microstructure could not be resolved with the instruments commonly used [72]. There are indeed some systems in which truly amorphous layers are produced by sliding [73], but in most cases the sub-surface layer with a thickness of less than 1 μm is nanocrystalline. Except for limited microhardness data, very little is known about the mechanism of grain-size reduction and the related changes in the structural, mechanical, and thermodynamic properties of wear surfaces. As such, it is expected that the study of MA processes in the future will not only open new processing routes for a variety of advanced nanostructured materials but also improve the understanding of technologically relevant deformation processes on a nanostructure level.

REFERENCES

[1] Benjamin J S 1976 *Sci. Am.* **234** 40

[2] Schwarz R B and Johnson W L (ed) 1988 *Solid State Amorphization Transformation, J. Less-Common Met.* **140**

[3] Shingu P H (ed) 1992 *Mechanical Alloying, Mater. Sci. Forum* **88–90**
[4] Johnson W L 1986 *Prog. Mater. Sci.* **30** 81
[5] Fecht H J and Johnson W L 1988 *Nature* **334** 50
[6] Fecht H J 1992 *Nature* **356**
[7] Chen L C and Spaepen F 1991 *J. Mater. Sci. Eng.* A **133** 342
[8] Benjamin J S 1992 *Mater. Sci. Forum* **88** 1
[9] Koch C C, Cavin O B, McKamey C G and Scarbrough J O 1983 *Appl. Phys. Lett.* **43** 1017
[10] Eckert J, Schultz L and Urban K 1989 *Appl. Phys. Lett.* **55** 117
[11] Koch C C 1991 *Mechanical Milling and Alloying in Matrials Science and Technology (Processing of Metals and Alloys 15)* ed R W Cahn, P Haasen and E J Kramer (Weinheim: VCH) p 193
[12] Kuhn W E, Friedman I L, Summers W and Szegvari A 1985 *Powder Metallurgy (ASM Metals Handbook 7)* (Metals Park, OH: American Society of Metals) p 56
[13] Fecht H J, Hellstern E, Fu Z and Johnson W L 1990 *Metall. Trans.* A **21** 2333
[14] Fecht H J, Hellstern E, Fu Z and Johnson W L 1989 *Adv. Powder Metall.* **1** 111
[15] Eckert J, Holzer J C, Krill C E III and Johnson W L 1992 *J. Mater. Res.* **7** 1751
[16] Bever M B, Holt D L and Titchener A L 1973 *Prog. Mater. Sci.* **17** 5
[17] Hellstern E, Fecht H J, Fu Z and Johnson W L 1989 *J. Appl. Phys.* **65** 305
[18] Guinier A 1963 *X-ray Diffraction* (San Francisco: Freeman) p 124
[19] Wagner C N J and Boldrick M S 1991 *J. Mater. Sci. Eng.* A **133** 26
[20] Friedel J 1964 *Dislocations* (Oxford: Pergamon) p 418
[21] Moelle C, Schmauss M and Fecht H J 1992 unpublished
[22] Koch C C 1993 *Nanostruct. Mater.* **2** 109
[23] Hellstern E, Fecht H J, Fu Z and Johnson W L 1989 *J. Mater. Res.* **4** 1292
[24] Yamada K and Koch C C 1993 *J. Mater. Res.* **8** 1317
[25] Oehring M and Bormann R 1990 *J. Physique Coll.* **51** C4 169
[26] Bakker H, Di L M and Lo Cascio D M R 1992 *Solid State Phenom.* **23&24** 253
[27] Di L M and Bakker H 1992 *J. Appl. Phys.* **71** 5650
[28] Schwarz R B, Petrich R R and Saw C K 1985 *J. Non-Cryst. Solids* **76** 281
[29] Shingu P H, Huang B, Niskitani S R and Nasu S 1988 *Trans. JIM* **29** 3
[30] Yavari A R, Desré P J and Benameur T 1992 *Phys. Rev. Lett.* **68** 2235
[31] Gaffet E, Louison C, Harmelin M and Faudet F 1991 *J. Mater. Sci. Eng.* A **134** 1380
[32] Veltl G, Scholz B and Kunze H D 1991 *J. Mater. Sci. Eng.* A **134** 1410
[33] Fukunaga T, Mori M, Inou K and Mizutani U 1991 *J. Mater. Sci. Eng.* A **134** 863
[34] Yavari A R and Desré P J 1992 *Ordering and Disordering in Alloys* ed A R Yavari (Amsterdam: Elsevier) p 414
[35] Eckert J, Birringer R, Holzer J C, Krill C E III and Johnson W L 1993 *Preprint*
[36] Bordeaux F and Yavari A R 1990 *Z. Metall.* **81** 130
[37] Yang J S C and Koch C C 1990 *J. Mater. Res.* **5** 325
[38] Schlump W and Grewe H 1989 *New Materials by Mechanical Alloying Techniques* ed E Arzt and L Schultz (Oberursel: DGM) p 307
[39] Shingu P H, Ishihara K N and Kuyama J 1991 *Proc. 34th Japan Congress on Materials Research* (Kyoto: Society of Materials Science) p 19
[40] Gleiter H 1989 *Prog. Mater. Sci.* **33** 223
[41] Fu Z, Fecht H J and Johnson W L 1991 *Mater. Res. Soc. Symp. Proc.* **186** 169

[42] El-Eskandarany M S, Sumiyama K, Aoki K and Suzuki K 1992 *Mater. Sci. Forum* **88–90** 801
[43] Calka A 1991 *Appl. Phys. Lett.* **59** 1568
[44] Calka A and Williams J S 1992 *Mater. Sci. Forum* **88–90** 787
[45] Schwarz R B, Desch P B, Srinivasan S and Nash P 1992 *Nanostruct. Mater.* **1** 37
[46] Schulson E M and Barker D R 1983 *Scr. Metall.* **17** 519
[47] Cottrell A H 1972 *Dislocations and Plastic Flow in Crystals* (Oxford: Clarendon) p 162
[48] Meyers M A and Chawla K K 1984 *Mechanical Metallurgy* (Englewood Cliffs, NJ: Prentice-Hall) p 494
[49] Karch J, Birringer R and Gleiter H 1987 *Nature* **330** 556
[50] Turley D 1971 *J. Inst. Met.* **99** 271
[51] Wagner C N J 1957 *Acta Metall.* **5** 477
[52] Warren B E 1956 *Prog. Metals Phys.* **3** 147
[53] Kuhlmann-Wilsdorf D and Van der Merwe J H 1982 *J. Mater. Sci. Eng.* **55** 79
[54] Haessner F and Hemminger W 1978 *Z. Metall.* **69** 553
[55] Lloyd D J and Kenny D 1978 *Scr. Metall.* **12** 903
[56] Grosskreutz J C and Mughrabi H 1979 *Constitutive Equations in Plasticity* ed A S Argon (Cambridge, MA: MIT Press) p 251
[57] Essmann U and Mughrabi H 1979 *Phil. Mag.* A **40** 40
[58] Gilman J J 1975 *J. Appl. Phys.* **46** 1625
[59] Donovan P E and Stobbs W M 1983 *Acta Metall.* **31** 1
[60] Hatherly M and Malin A S 1984 *Scr. Metall.* **18** 449
[61] Goods S M and Brown L M 1979 *Acta Metall.* **27** 1
[62] Fecht H J 1990 *Phys. Rev. Lett.* **65** 610
[63] Kehrel A, Moelle C and Fecht H J 1994 *Proc. NATO ASI on Nanophase Materials* ed G C Hadjipanayis and R W Siegel (Dordrecht: Kluwer) p 125
[64] Kuwano H, Ouyang H and Fultz B 1992 *Nanostruct. Mater.* **1** 143
[65] Nieman G W, Weertman J R and Siegel R W 1991 *J. Mater. Res.* **6** 1012
[66] Rigney D A, Chen L H, Naylor M G S and Rosenfield A R 1984 *Wear* **100** 195
[67] Rigney D A 1988 *Annu. Rev. Mater. Sci.* **18** 141
[68] Ganapathi S K and Rigney D A 1990 *Scr. Metall.* **24** 1675
[69] Doyle F D and Aghan R L 1975 *Metall. Trans.* B **6** 143
[70] Mehrotra P K 1983 *Proc. Int. Conf. on Wear of Materials (Reston, VA, 1983)* (New York: ASME) p 194
[71] Humble P and Hannink R H J 1978 *Nature* **273** 37
[72] Beilby G 1921 *Aggregation and Flow of Solids* (London: Macmillan)
[73] Askenasy P 1992 *PhD Thesis* California Institute of Technology

PART 3

ARTIFICIALLY MULTILAYERED MATERIALS

Chapter 6

Artificially multilayered materials

Robert C Cammarata

6.1 INTRODUCTION

With the advent of advanced thin-film processing methods, it has become possible to produce artificially multilayered materials with precise control of the composition and the thickness of the layers. In contrast to conventional bulk laminate composites, the individual layer thickness in multilayered thin films can be reduced to atomic dimensions and therefore becomes the smallest and most important microstructural length scale. It is this length scale control that makes artificially multilayered thin films the epitome of microstructural engineering.

One of the earliest successful productions of an artificially multilayered thin film was reported by DuMond and Youtz in 1940 [1]. The film was prepared by alternate evaporation of copper and gold onto a glass substrate, and resulted in an x-ray mirror. However, it was unstable to room-temperature intermixing, probably a result of interdiffusion at moving grain boundaries of the small crystallites that made up the layers [2]. The first stable and well characterized metallic multilayered thin films were the result of the pioneering work of Hilliard and coworkers at Northwestern University in the late 1960s [3,4]. These materials, prepared by alternate shuttering of two evaporation sources, were composed of fcc metals such as Ag–Au and Cu–Pd. In the early 1970s, Esaki and coworkers at International Business Machines Corporation suggested that semiconductor multilayers with layer thicknesses smaller than the electron mean free path could be the basis of technologically useful devices [5,6]. Since that time there has been an exponential increase in the number of experimental and theoretical investigations concerning every aspect of multilayered materials.

Scientifically, artificially multilayered materials offer the possibility of investigating problems that previously had been mostly of academic interest. For example, the ability to produce layer thicknesses smaller than the mean free path for electrons allowed the one-dimensional particle-in-a-well problem from elementary quantum mechanics to be investigated experimentally. Artificially multilayered materials are also important from a technological point of view: as discussed throughout this chapter, these materials are involved in a variety of diverse engineering applications. Of course it is not possible to cover every aspect of the wide field of multilayered materials here. Reference [7] is an excellent source book, giving a thorough and detailed review of the subject up to the early 1980s. For advances since that time, the interested reader should consult the appropriate references cited at the end of the chapter.

6.2 STRUCTURE AND CHARACTERIZATION

6.2.1 Microstructure

Artificially structured materials composed of layers of different phases are known generically as *heterostructures*. Heterostructures can be composed of only a few layers or of many layers. Artificially multilayered materials are heterostructures composed of many alternating layers that are generally stacked in a periodic manner; such periodicity will henceforth be assumed unless otherwise noted. An artificially multilayered material is shown schematically in figure 6.1(a). The combined thickness of two adjacent layers is called the *bilayer repeat length* or *bilayer period*. The principal characteristic of multilayers is a *composition modulation*, that is, a periodic chemical variation. For this reason, multilayers are often referred to as *compositionally modulated materials* and the bilayer repeat length is often called the *composition modulation wavelength*. Many authors prefer to reserve the term 'compositionally modulated materials' for multilayers composed of mutually soluble layers separated by compositionally diffuse interfaces.

Multilayers composed of single-crystal layers that possess the same crystal structure and where the interfaces are in perfect atomic registry are called *superlattices*. An important example of this type of multilayer is the semiconductor superlattice thin film produced by molecular beam epitaxy (see section 6.3.1). In metallic superlattice thin films, the requirement that the layers are single crystals is often relaxed as long as they have an in-plane grain size many times larger than the layer thickness [2]. Also, single-crystal or large-grained metallic multilayers whose layers have different crystal structures but a well defined epitaxial relationship at the interfaces are often referred to as superlattices. It should be noted that some authors are very loose in their terminology and use the terms 'multilayered', 'compositionally modulated', and 'superlattice' interchangeably.

Since one of the chief characteristics of multilayered materials is the high density of interfaces, it is worth discussing the different types of interface structure. The most common types of interfaces found in solids are those involving chemical and/or structural changes. Almost all interfaces in multilayered thin films involve a chemical composition change. An 'ideal' multilayer would have sharp interfaces, and the composition profile would be

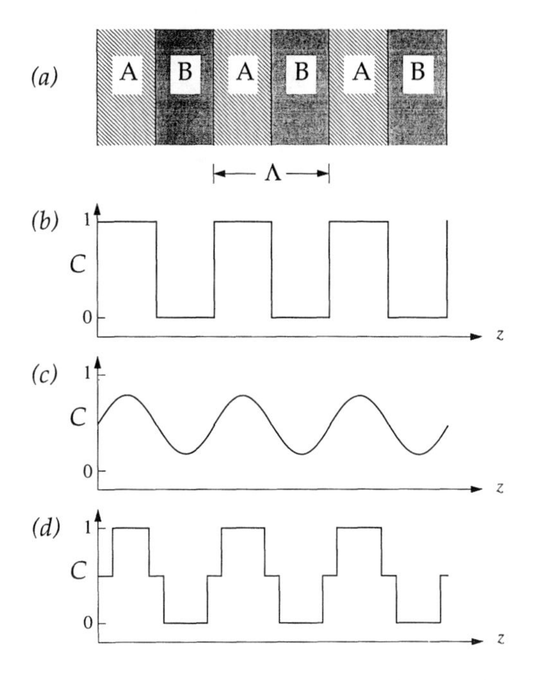

Figure 6.1. Schematic diagram and composition profiles for an artificially multilayered material of bilayer repeat length Λ. (*a*) Artificially layered materials composed of alternating layers of A and B; (*b*) ideal rectangular wave composition profile; (*c*) sinusoidal composition profile; (*d*) composition profile of a multilayer with a third interfacial phase.

a periodic rectangular wave (see figure 6.1(*b*)). If there is mutual solubility between the layer materials, then intermixing, for example, during deposition or during a post-deposition anneal, could lead to diffuse interfaces which in the extreme would result in a sinusoidal composition profile (see figure 6.1(*c*)). In this case, it is not possible to distinguish between an interface and the 'bulk' of a layer. If, instead of intermixing, a third interfacial phase such as an intermetallic is formed, then a composition profile as shown schematically in figure 6.1(*d*) could result.

In the case of an 'ideal' superlattice, there would be no structural variation at the interface. Interfaces in these materials are said to be *coherent*. Figure 6.2(*a*) schematically illustrates a coherent interface. If the layer materials have different equilibrium lattice parameters, there is said to be a *lattice parameter misfit*. In order to produce coherent interfaces in a multilayer where there is such a misfit, one or both of the layers must be elastically strained to produce atomic registry. A multilayer containing coherent interfaces between layers that have a lattice parameter misfit is called a *strained layer superlattice*.

Interfaces where some of the lattice parameter mismatch is accommodated

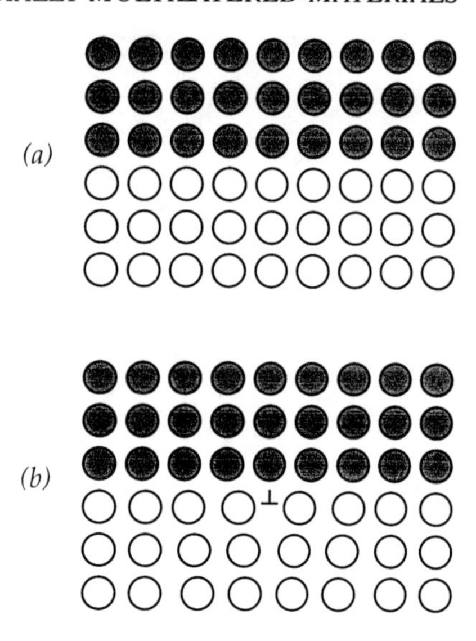

Figure 6.2. (*a*) Coherent interface having perfect lattice matching; (*b*) semicoherent interface where an edge dislocation (⊥) has accommodated some or all of the lattice misfit.

by an array of *misfit dislocations* but where large regions of lattice matching still remain are referred to as *semicoherent*. An example of a semicoherent interface is shown schematically in figure 6.2(*b*). Elastic strains in layers bounded by coherent or semicoherent interfaces are referred to as *coherency strains*. The dislocation spacing in a semicoherent interface is equal to the magnitude of the Burgers vector divided by the misfit strain being accommodated by the dislocations. Even if all of the lattice parameter misfit is accommodated by dislocations (i.e. there are no coherency strains), the interface can still be considered semicoherent as long as the misfit dislocations are not too closely spaced. However, if the dislocation spacing becomes very small, say less than about four or five Burgers vectors, or if there is little or no atomic registry at the interface, the interface is then said to be *incoherent*.

When a multilayer is composed of two materials with different crystal structures, a distinction between *ordered* and *disordered* interfaces can be made [2, 8]. An ordered interface is defined as one where there is strong epitaxial matching between the two layers. For all the interfaces in a multilayer to be ordered in this sense, any fluctuation in layer thickness must be a multiple of an interplanar crystal spacing. Such a multilayer can be considered to have long-range structural coherence. On the other hand, if there is no epitaxial matching, then the interfaces are said to be disordered. In this case, there can be a continuous distribution of layer thickness fluctuations, resulting in a lack of long-range structural coherence.

In some cases involving layer materials with two different bulk equilibrium crystal structures, it is possible to *pseudomorphically* grow one of the layers with the crystal structure of the other layer. Such a multilayer could be thermodynamically more favorable relative to one that has both layers retaining their bulk crystal structures if the reduction in interfacial free energy going from a semicoherent or incoherent interface to a coherent interface more than offsets the increase in volume free energy associated with the change in crystal structure. A striking example of pseudomorphism is bcc Ge formed in Mo–Ge multilayers [9].

In addition to crystalline materials, it is also possible to use amorphous solids as one or both of the layer materials. For example, crystalline Ag–amorphous $Fe_{0.7}B_{0.3}$ [10], amorphous $Fe_{0.8}B_{0.2}$–amorphous $Pd_{0.8}Si_{0.2}$ [11], and amorphous Si–amorphous Ge multilayered [12] films have been produced by sputtering (see section 6.3.1). Completely amorphous multilayered films are quite intriguing as they are amorphous solids that possess a well defined microstructural length scale, the bilayer period, that introduces translational symmetry.

6.2.2 Dislocation filters

An interesting application for artificial superlattices involves using them as 'dislocation filters' that improve the structural quality of epitaxial (i.e. single-crystal) films grown on top of them [13, 14]. In an epitaxial GaAs film on a Si substrate, the density of threading dislocations, which are grown-in dislocations that thread through a film from the film–substrate interface to the film surface, can be reduced by depositing a strained layer superlattice of InGaAs–GaAsP between the GaAs and Si. The threading dislocations are blocked and bent along the interfacial planes of the superlattice, resulting in a significant reduction in defect density in the GaAs film compared with one grown without the superlattice.

6.2.3 Characterization

The most common microstructural technique used to investigate the quality of layering in nanometer-scale multilayers is x-ray diffraction [15–17]. In superlattices with long-range structural coherence, satellite peaks will appear about the undiffracted beam (000) and the higher-angle Bragg peaks in the x-ray diffraction pattern. The reason for this can be illustrated in the following manner. Consider a single-crystal superlattice with completely coherent interfaces being investigated with a standard θ–2θ reflection x-ray diffractometer. For simplicity, assume each layer has the same number of atomic planes p and that the interplanar spacing d is constant along the direction perpendicular to the plane of the film (call this the z direction). The diffraction pattern can be taken as the square of the magnitude of the Fourier transform of the electron density along the z direction. The electron density variation along the z direction can be represented by a lattice function $L(z)$, which can be taken as an array of equally spaced delta functions, multiplied by a composition modulation function $C(z)$ of wavelength $\Lambda = 2pd$. The Fourier transform of the electron density will be proportional to the Fourier transform of the lattice $F[L(z)]$ convoluted with the Fourier transform of the composition modulation $F[C(z)]$. $F[L(z)]$ by itself would give the Bragg peaks of the x-ray diffraction pattern associated with a compositionally homogeneous single crystal. Owing to the convolution with $F[C(z)]$, around each Bragg peak and the undiffracted beam (000) there will be satellite peaks associated with the Fourier spectrum of the composition modulation $C(z)$. If $C(z)$ is a sine wave, as shown in figure 6.1(c), there will only be one satellite peak on either side of the main Bragg peaks; if $C(z)$ has sharper interfaces then there will be several orders of satellite peaks. Actual multilayers with long-range structural coherence may have a modulation in lattice parameter along the z direction, layers that are not of equal thickness, and/or interfaces that are not completely coherent. The analysis for these materials is more complicated but the general trend is still the same: the greater the number of peaks, the sharper the interface. Figure 6.3 [18] shows an example of an x-ray diffraction pattern obtained from a 12 nm Si–4 nm $Si_{0.7}Ge_{0.3}$ superlattice produced by molecular beam epitaxy (see section 6.3.1). The presence of several orders of satellite peaks indicates that the interfaces were relatively sharp.

Crystalline multilayered materials without long-range structural coherence

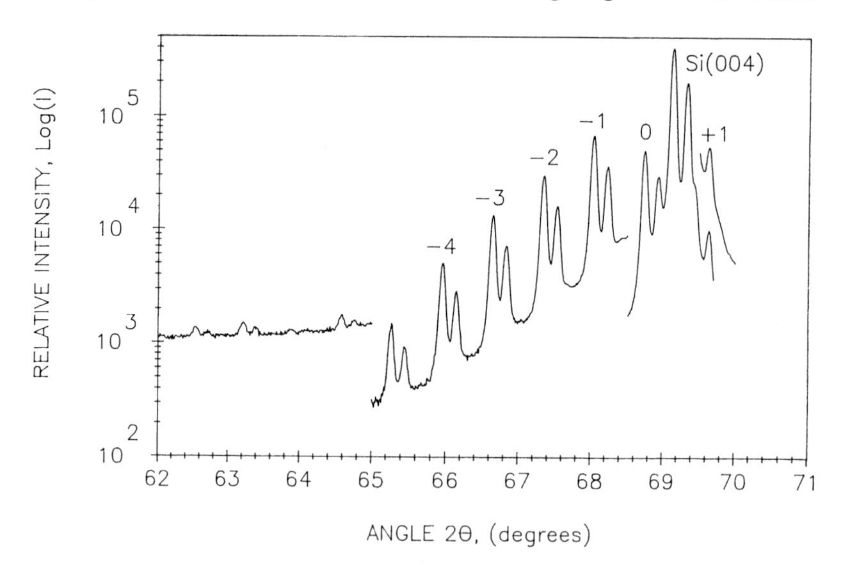

Figure 6.3. X-ray diffraction pattern (intensity versus angle 2θ) obtained from a Si–$Si_{0.7}Ge_{0.3}$ strained layer superlattice (taken from [18]). The splitting of the satellite peaks resulted from the diffraction of both K_{α_1} and K_{α_2} x-ray radiations.

will display weak or no satellite peaks around the high-angle Bragg peaks but will have satellites associated with the undiffracted beam (000). Also, there will only be satellite peaks from the undiffracted beam in amorphous multilayers since there are no high-angle Bragg peaks in the diffraction patterns of amorphous materials. An x-ray diffraction pattern obtained from an equal layer thickness amorphous Si–amorphous Ge multilayered thin film with a bilayer repeat length of 5.8 nm produced by sputtering (see section 6.3.1) is shown in figure 6.4 [12].

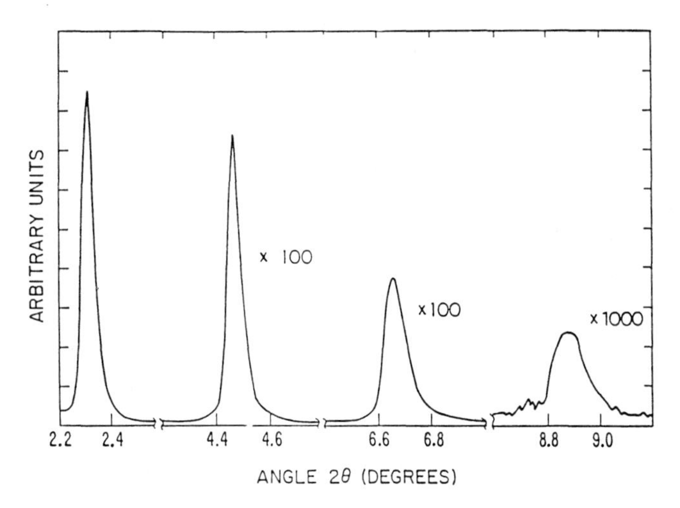

Figure 6.4. X-ray diffraction pattern (intensity versus angle 2θ) obtained from an amorphous Si–amorphous Ge multilayered film (taken from [12]).

X-ray diffraction patterns with satellite peaks can be used to extract quantitative structural information. The bilayer repeat length Λ of a multilayered film can be obtained from low-angle x-ray reflectivity measurements using Bragg's law for diffraction. However, it is necessary to take into account effects owing to refraction [19, 20]. The real part of the index of refraction n_r is generally slightly less than unity. Defining $\delta = 1 - n_r$, Bragg's law for a satellite peak associated with the undiffracted beam (000) can be expressed as

$$m\lambda = 2\Lambda \sin\theta_m [1 - (2\langle\delta\rangle / \sin^2\theta_m)]^{1/2} \tag{6.1}$$

where m is the order of the satellite, λ is the wavelength of the x-ray radiation, θ_m is the measured Bragg angle, and $\langle\delta\rangle$ represents a composite average value of δ appropriate for the multilayer. By measuring the Bragg angles for two or more satellites, both Λ and $\langle\delta\rangle$ can be obtained. It is possible to extract other structural features, such as roughness and diffuseness of the interfaces, from low- or high-angle diffraction patterns with the aid of x-ray scattering calculations [15–17, 19, 20].

Electron and neutron diffraction methods can also be used for microstructural characterization in a manner similar to that for x-ray diffraction.

A unique possibility offered by neutron diffraction is the ability to characterize multilayers composed of different isotopes of the same material [2, 21]. Such multilayers are useful in making highly sensitive self-diffusion coefficient measurements in a manner analogous to x-ray diffraction experiments performed to measure interdiffusivities (see section 6.4.2).

In addition to diffraction methods, measurements of certain physical properties that depend on the quality of the layering can be used to characterize the overall structure. For example, the width of optical absorption peaks obtained with quantum well semiconductor superlattices is often used as a measure of the quality of these materials (see section 6.5.1). The energies associated with optical absorption peaks depend on the thicknesses of the layers, and the sharper the absorption peaks, the higher the quality of the layering.

Diffraction studies and macroscopic physical property measurements can give information only about average layer quality. In order to investigate the structure on a more local scale, it is necessary to use techniques such as high-resolution transmission electron microscopy (HRTEM) and extended x-ray absorption fine structure (EXAFS). An example of a high-resolution cross-sectional transmission electron micrograph of an InAs–GaSb(100) superlattice grown by molecular beam epitaxy on a GaSb buffer layer (see section 6.3.1) is shown in figure 6.5.

Thin-film characterization methods such as Rutherford backscattering (RBS) and Auger electron spectroscopy (AES) can be used for chemical analysis of artificially multilayered thin films. AES depth profiling, where the chemical composition is continuously monitored during ion milling of the sample, can give a qualitative description of the composition modulation. Figure 6.6 shows an example of an AES profile for an amorphous metal $Fe_{0.8}B_{0.2}$–$Pd_{0.8}Si_{0.2}$ multilayered film with a composition modulation wavelength of 6.9 nm [11].

6.3 PROCESSING

Artificially multilayered materials are generally produced in thin-film form by standard thin-film processing methods modified to allow alternate deposition of two different materials. Examples of these processing techniques include Langmuir–Blodgett and related methods for producing organic films, electrodeposition, sputtering, evaporation (including molecular beam epitaxy), pulsed laser deposition, and chemical vapor deposition. Mechanical methods generally associated with the processing of bulk solids have also been developed that produce materials of lower quality in terms of the layering but in much larger quantities.

6.3.1 Thin-film deposition methods

For many years, Langmuir–Blodgett [22, 23] and self-assembly [24] methods of preparing organic multilayered films composed of monomolecular layers

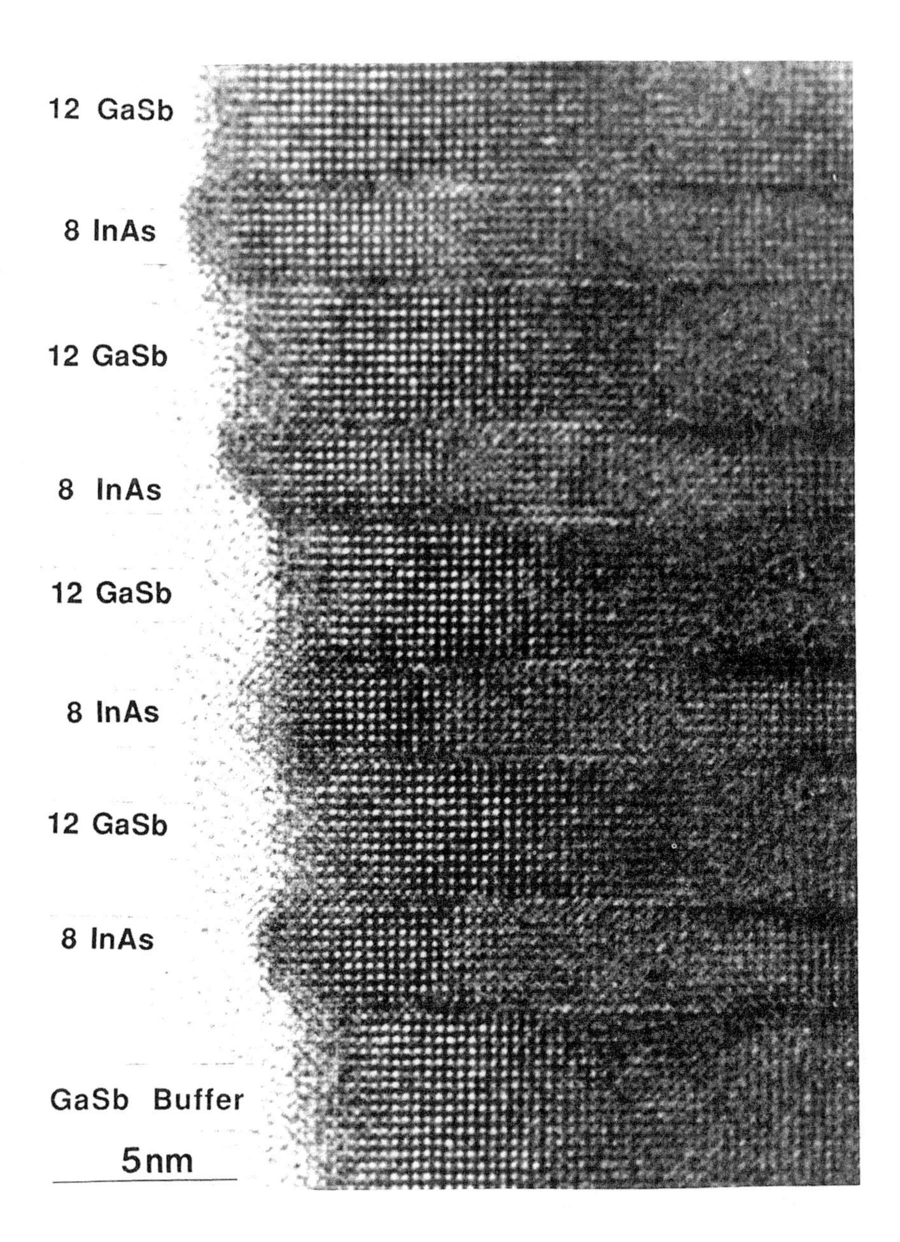

Figure 6.5. High-resolution cross-sectional transmission electron micrograph of an InAs–GaSb superlattice. Numbers indicate number of monolayers per layer. (Courtesy of M Twigg, L Ardis and B Bennett, Naval Research Laboratory, Washington DC, USA.)

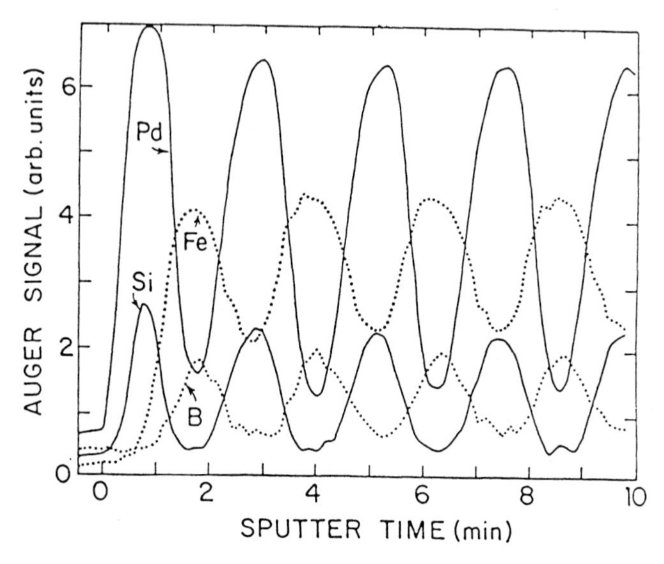

Figure 6.6. Auger electron spectroscopy depth profile of an amorphous metal $Fe_{0.8}B_{0.2}$–$Pd_{0.8}Si_{0.2}$ multilayered thin film with a composition modulation wavelength of 6.9 nm (taken from [11]).

have been investigated. Since Langmuir–Blodgett films are very fragile, their use in technological applications is somewhat limited. Mechanical stability is less of a problem in self-assembly systems, but it has sometimes proven difficult to build up multilayers with overall thicknesses of 100 nm or more that retain their long-range structural coherence. Very recently a related method of producing multilayered films composed of alternating layers of organic and inorganic macromolecules has been reported [25]. These materials were formed by sequential adsorption of polydiallyldimethylammonium chloride and exfoliated sheets of synthetic hectorite (a mica-type layered silicate) onto silicon substrates. X-ray diffraction characterization indicated that the multilayers remained structurally ordered even when the overall film thicknesses were relatively large (over 0.2 μm). It would appear that this approach can be extended to produce a wide variety of high-quality organic–inorganic layered systems.

Electrodeposition has proven to be a very successful and relatively inexpensive method of producing high-quality compositionally modulated materials, capable of depositing metallic, ceramic, semiconductor, and polymer multilayers [26, 27]. Generally speaking, electrodeposition methods for making multilayers can be divided into two types: alternately plating between two deposition potentials using a single electrolyte containing two different types of ion, or periodic transfer of the substrate from one electrolyte to another. In addition to low cost, an attractive feature of electrodeposition is the ability to deposit films over a large area and with a relatively large overall thicknesses (several tens of microns or more), dimensions much greater than those generally associated with vacuum deposition methods. Because of these features, electrodeposition may be the most promising approach with regard to scaling up thin-film processing methods to produce a multilayered structural material with nanoscale individual layer thicknesses but with a large overall thickness and in-plane area.

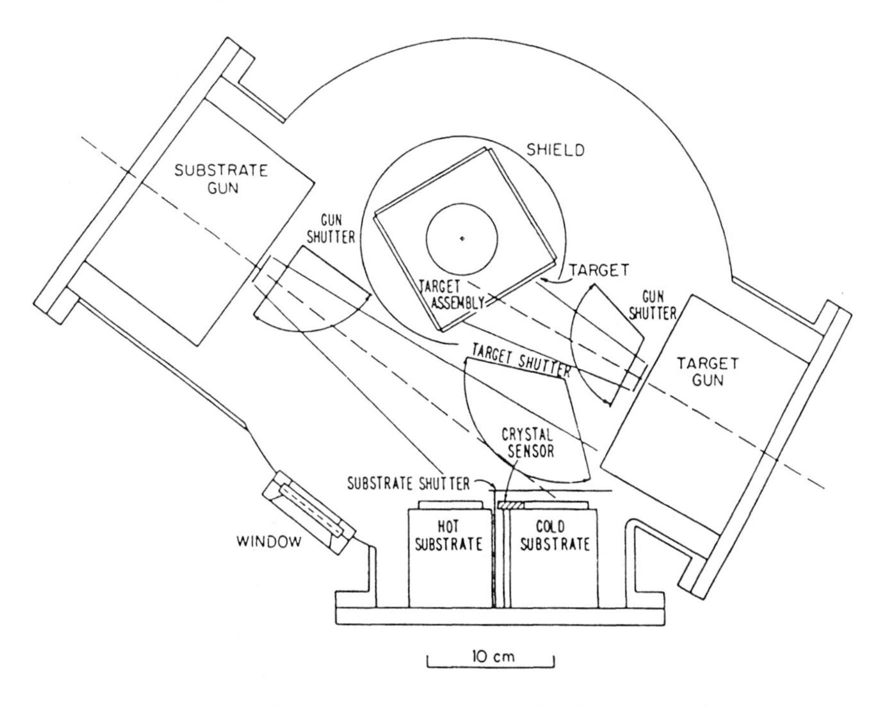

Figure 6.7. Schematic diagram of a multiple-target ion beam sputtering system used to produce artificially multilayered thin films (taken from [29]).

Sputtering has become a very popular preparation technique for the deposition of metallic and ceramic multilayers [28]. Sputtering involves the collisions of ions (usually of an inert gas such as Ar) with the surface of a target material, leading to the ejection of target atoms that are collected in thin-film form onto a substrate. A schematic diagram for an ion gun sputtering system capable of depositing multilayers is shown in figure 6.7 [29]. An Ar^+ ion beam from the target ion gun is used to sputter material from targets mounted on a rotating assembly. Two targets are alternately rotated in and out of the line of the target gun ion beam, resulting in the deposition of a multilayered film onto a stationary substrate. The substrate ion gun is used to sputter clean the substrate before deposition, and can also be used for ion beam assisted deposition processes [28]. Other sputtering systems employ schemes such as alternately exposing the substrate to deposition from two stationary targets, either by shuttering the targets or by rotating the substrate. In the preparation of ceramic multilayers with oxide or nitride layer materials, it is possible to sputter directly from ceramic targets using a neutralized ion beam or by radio frequency (RF) sputtering [28], or to bleed oxygen or nitrogen into the chamber and reactively sputter using metallic targets. Sputtering is often the best method of producing metal–ceramic multilayers (e.g. Al–Al_2O_3), either by alternately sputtering from metallic (Al) and ceramic (Al_2O_3) targets, or by alternately opening and closing a valve that bleeds in a gas (O_2) while continuously sputtering from one target (Al) [2]. Also, because of its very high effective quench rate, sputtering is often the best method of producing multilayers with amorphous metal or amorphous semiconductor layers.

Evaporation systems involving deposition from open sources in an evacuated chamber have been used extensively to produce multilayered films [28]. The deposited materials are evaporated or sublimated, either by an electron beam or in a resistively heated source. Molecular beam epitaxy (MBE) is often employed when it is desired to produce single-crystal superlattices [30]. An MBE system is essentially a fancy (and expensive) evaporator, where the deposition occurs under ultrahigh-vacuum (UHV) conditions, and where various characterization tools, such as reflection high-energy electron diffraction (RHEED) and AES analysis, are available for *in situ* characterization. Figure 6.8 is a schematic diagram of an MBE system that has been used to prepare $In_{1-x}Ga_xAs$–$GaSb_{1-y}As_y$ superlattices [31]. The various constituents are evaporated from Knudsen effusion cells. (Other MBE systems designed to deposit high-melting-temperature materials such as Si use an electron beam to evaporate the material.) A computer controlled shuttering system, using input from the mass spectrometer that monitors the flux rates, is used to create the layering. The ion sputtering gun is used for substrate cleaning.

Pulsed laser deposition (PLD), also referred to as laser ablation, involves the use of a short pulse (typically 20 to 30 nanoseconds) from the focused output of a laser (of order 0.1 to 1 mJ m^{-2}) to rapidly vaporize material from the surface of a bulk target that is then collected onto a substrate [32]. The deposition occurs in a vacuum chamber with a window that allows the laser pulse to enter the chamber and ablate (vaporize) the target surface. The thickness of material deposited per pulse is typically of order 0.1 nm, and the laser is pulsed at a rate of 1 to 100 Hz. The principal advantage of this process is that it allows deposition of a multicomponent film with the same composition as the target. This feature has made PLD a popular method for depositing stoichiometric compounds, especially high-temperature superconducting oxides. Multilayered materials can be produced by laser ablation from multiple targets.

Chemical vapor deposition (CVD) is a commonly used method for the preparation of semiconductor superlattices such as GaAs–$Al_xGa_{1-x}As$ [30, 33]. Also known as vapor phase epitaxy, CVD involves chemically reacting a volatile gaseous compound that has been adsorbed onto the substrate to produce

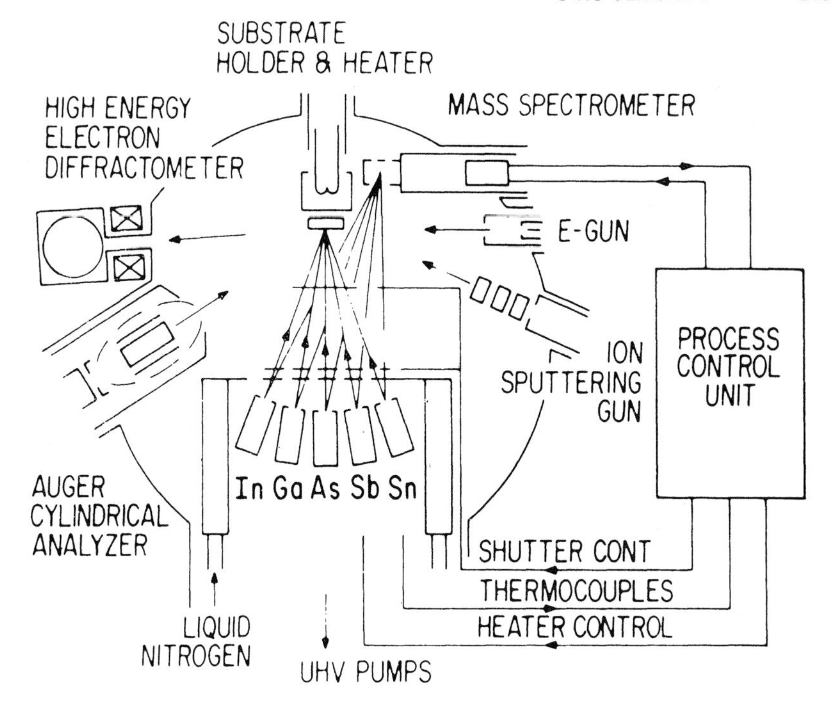

Figure 6.8. Schematic diagram of an MBE thin-film deposition system used to produce $In_{1-x}Ga_xAs$–$GaSb_{1-y}As_y$ superlattices (taken from [31]). Each Knudsen cell is labeled by its corresponding evaporant (In, Ga, As, Sb). The Sn cell is used for doping.

a nonvolatile thin solid film. A variety of reaction types (e.g. pyrolysis, reduction, oxidation) is used to produce semiconductors and metals, as well as compounds such as oxides and nitrides. Metallo-organic chemical vapor deposition (MOCVD) has become one of the principal processing techniques for depositing compound semiconductor films in general, and compound semiconductor superlattices in particular [30]. It involves the reaction of metal alkyls with a hydride of the nonmetal component. The following are examples of MOCVD reactions used to produce GaAs and $Al_xGa_{1-x}As$ layers in superlattices:

$$Ga(CH_3)_3(g) + AsH_3(g) \longrightarrow GaAs(s) + 3CH_4(g) \qquad (6.2a)$$

$$xAl(CH_3)_3(g) + (1-x)Ga(CH_3)_3(g) + AsH_3(g) \longrightarrow Al_xGa_{1-x}As(s) + 3CH_4(g). \qquad (6.2b)$$

By precisely controlling the temperature, pressure, and gas composition, high-quality superlattices can be produced.

The most sophisticated thin-film deposition systems are able to control deposition rates to better than 0.1% and can produce layers as small as one atom thick [2]. Thus, multilayered thin-film deposition represents the ultimate in microstructural control. In order to produce high-quality superlattice films, it is frequently necessary to first deposit a *buffer layer* on the substrate. This buffer layer, which is often (but not always) one of the layer materials that composes the multilayered film, is grown to a sufficient thickness that any islands formed during the initial stages of deposition have coalesced and transformed into a large-grained or single-crystal epitaxial film.

6.3.2 Mechanical processing methods

Although thin-film processing methods are by far the most popular choices for preparation of multilayered materials with nanoscale individual layer thicknesses, mechanical processing techniques have also been developed. In fact, mechanical methods of producing multilayered materials with very thin layers date back to medieval times to the forging of Damascus [34] and Japanese [35] swords composed of layers such as soft wrought iron and hardened steel that were repeatedly folded and rolled. Metallic multilayers of Ni–Er have been produced with individual layer thicknesses as small as 5 nm by repeated folding and rolling of initially thick (20 to 25 μm) foils of Ni and Er [36]. The quality of the layering is not as high as that obtained by thin-film deposition methods, but the total sample thickness can be quite large, up to 200 μm.

Multilayered polymer composites have been prepared by coextrusion of alternating layers of thermoplastic materials [37–40]. Examples of systems prepared in this manner include polystyrene–polypropylene, polystyrene–polyethylene, and poly(styrene-acrylonitrile)–polycarbonate. The multilayers so produced can be composed of thousands of layers, with the individual layer thicknesses as small as a few tens of nanometers [39].

6.4 THERMODYNAMICS AND KINETICS

6.4.1 Thermodynamics

Because of the high density of interfaces characteristic of multilayers (and of nanostructured materials in general), the thermodynamic nature of multilayered materials is intrinsically different from that of solids composed of bulk phases [2, 41]. For example, stresses in the layers owing to the presence of coherent interfaces can lead to violations of the Gibbs phase rule and the common tangent rule for determining multiphase equilibrium from free energy versus composition diagrams [42–44].

Associated with solid–solid interfaces are two types of thermodynamic quantity: interface free energy and interface stress. The interface free energy can be considered the reversible work per unit area needed to form new interface area by a process such as creep. Recently, zero-creep experiments using crystalline Ag–Ni(111) multilayered thin-film specimens have been performed to measure the Ag–Ni interface free energy [45, 46]. Many capillary effects resulting from the influence of the interface free energy (which can be positive or negative), such as variations in crystallization or melting temperatures, have been observed in multilayers. For example, melting point depressions of Pb layers a few nm thick in Pb–Ge multilayers have been attributed to the influence of the interface free energy [47].

Interface stress can be defined as the reversible work per unit area needed to elastically deform one or both of the layers on either side of the interface. There are two intrinsic interface stresses characteristic of a solid–solid interface: one can be related to elastic straining of one of the layers relative to the other layer, and the other is associated with uniform elastic straining of both layers [48, 49]. The strain associated with the former type of interface stress would lead to a change in the interfacial dislocation structure between two epitaxially matched crystalline phases. The latter type of interface stress could lead to uniform elastic strains in the plane of the interface as high as a few per cent relative to the equilibrium bulk state when the individual layer thickness is reduced to a value of order 1 nm [50–53]. A recent study of internal stresses in multilayered thin films has been performed to measure this type of interface stress in Ag–Ni(111) multilayered films [54].

Several novel metastable phases have been produced using multilayered thin films. One example has been cited previously: the formation of bcc Ge layers in Mo–Ge films (see section 6.2.1). Phase transformations can be induced in multilayered films by post-deposition anneals to produce other metastable phases. Perhaps the most remarkable example of this is the process known as *solid-state amorphization*, where polycrystalline metal layers in a multilayered thin film such as Au–La [55] and Ni–Zr [56] react during a post-deposition low-temperature anneal to form a single-phase amorphous alloy.

6.4.2 Kinetics

An often large thermodynamic driving force, coupled with relatively short diffusion distances, makes it possible to sensitively study the kinetics of atomic transport and phase transformations in multilayered thin films. In films where there is mutual solubility between the two types of layer material, intermixing can occur when the film is subjected to a relatively low-temperature anneal. It is possible to obtain the kinetics of the intermixing by monitoring the decay of x-ray satellite peaks that result from the layering to obtain an interdiffusion coefficient [11, 12, 17, 18]. This technique is the most sensitive method of determining interdiffusivities as it can measure diffusion coefficients as small as 10^{-27} m^2 s^{-1}. It has been of great use in the study of atomic transport in amorphous materials where it is necessary to use low-temperature anneals in order to avoid crystallization [17]. In addition, multilayered thin films mimic the microstructure of spinodally decomposed materials and early studies of atomic transport in multilayered films were performed in order to investigate the kinetics of this type of phase transformation [3, 4, 17].

The multilayer technique to measure interdiffusion is easily adapted to the study of the effects of pressure on diffusion by annealing films that are subjected to an external stress by, for example, stressing the film using a four-point bending configuration [57, 58]. It should noted that significant stresses can be generated in an initially stress-free multilayer if the components have different partial molar volumes and initially interdiffuse at different rates. If the stresses become large enough, they can inhibit chemical interdiffusion, and further intermixing would then be governed by the rate of stress relief [21, 59].

Neutron reflectivity measurements can be used to measure self-diffusivities in multilayers that are chemically homogeneous but where the alternating layers have different isotopes of the same element. The self-diffusion of nickel in ^{62}Ni–^{nat}Ni multilayers has been investigated in this manner [21]. For this type of study, it is desirable to choose an element such as nickel with readily available isotopes that have large differences in scattering length.

In multilayers where a third phase is formed by an interfacial reaction (see figure 6.1(*d*)), the kinetics of the interfacial reaction can be investigated by measuring the time it takes the product phase to grow and completely consume the adjacent reactant layers [21, 56, 60, 61]. This approach has been used in calorimetric studies to identify and quantitatively characterize both diffusion limited [60] and interface reaction limited kinetics [61] of silicide formation in transition metal–amorphous silicon multilayers. In certain multilayers, the interfacial reaction can be strongly exothermic, with the heat of reaction as high as tens of kJ mol^{-1} [2], which may be large enough to significantly raise the temperature of the film during the early stages of the reaction. This temperature increase may in turn significantly enhance the atomic transport of the layer materials, thereby accelerating the rate of intermetallic formation. For some systems, once the reaction is initiated (e.g. by a laser pulse) it is self-sustaining, and the multilayer completely transforms into the product phase. Self-sustaining reactions have been observed for several multilayered systems, the most common involving a transition metal (e.g. Ni [60]) and amorphous silicon.

6.5 ELECTRICAL AND OPTICAL PROPERTIES

6.5.1 Semiconductor superlattices

In the early 1970s, Esaki and coworkers [5, 6] proposed that in layered semiconductor heterostructures with layer thicknesses smaller than the electron mean free path of bulk three-dimensional semiconductors the motion of electrons and their interactions with photons would be modified. Such modifications would result in novel electronic behavior that could be exploited to produce new electronic and photonic devices. Since that time, semiconductor superlattices have become the basis of the most important technological applications of multilayers. Although there has been an active research effort concerning

$Si–Si_xGe_{1-x}$ superlattices [62], most of the interest has been focused on heterostructures based on compound semiconductor layer materials.

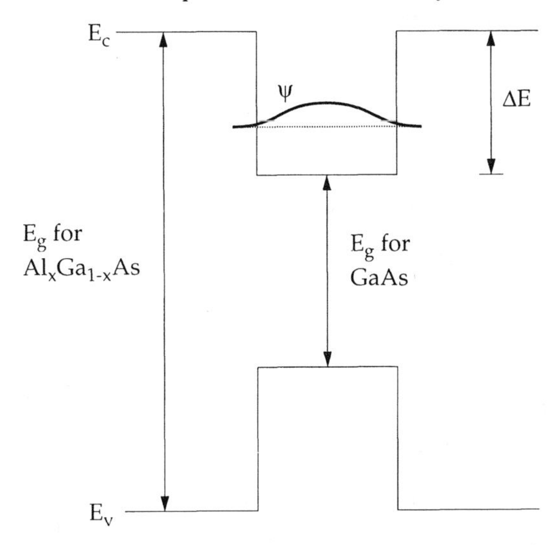

Figure 6.9. Quantum well composed of a GaAs layer between two $Al_xGa_{1-x}As$ layers. E_c is the energy of the bottom of the conduction band, E_v is the energy of the top of the valance band, and E_g is the band gap energy. The barrier height ΔE is the difference between the values of E_c for $Al_xGa_{1-x}As$ and GaAs.

Consider a superlattice composed of two semiconductor materials with a significant band gap difference, such as a $GaAs–Al_xGa_{1-x}As$ superlattice. Figure 6.9 schematically shows the energy band diagram for one of the GaAs layers between two of the $Al_xGa_{1-x}As$ layers. An electron in the GaAs layer, represented in the figure by its wavefunction ψ, can be considered to be partially confined in a *quantum well* of barrier height ΔE equal to the difference in the energies of the bottom of the conduction band E_c for the two layer materials. It is recalled from the one-dimensional particle-in-a-box problem of elementary quantum mechanics [63, 64] that in the limit as $\Delta E \rightarrow \infty$ all of the electron energy states are quantized, and can be expressed as

$$E_n = n^2h^2/8m^*L^2 \tag{6.3}$$

where h is Planck's constant, m^* is the effective mass of the electron, n is a quantum number $= 1, 2, 3, \ldots$, and L is the width of the well. It is noted that E_n is proportional to L^{-2}, and thus it is possible to tune the energy states by varying the width of the wells.

For the infinite-height well, an electron is completely confined for all values of n and there are an infinite set of discrete energy states. For a superlattice composed of a series of finite-height quantum wells, only a finite number of bound states exist, and the wells can interact [63]. Since the series of wells represents a periodic potential whose period is much larger than a crystal lattice spacing, folding of the Brillouin zones will occur, resulting in the creation of new energy bands that are often referred to as *subbands* or *minibands*. (The same kind of zone folding also occurs in phonon dispersion curves and is detectable by Raman spectroscopy.) As was discussed in section 6.2.2, the extra periodicity of a superlattice leads to satellite peaks in the x-ray diffraction pattern. These peaks can be considered as resulting from 'artificial' Bragg planes. It is recalled from elementary solid-state physics that associated with every Bragg plane is a Brillouin zone. Thus, the 'artificial' Bragg planes of a superlattice lead to 'artificial' Brillouin zones, which in turn result in the creation of new energy band gaps. The ability to produce superlattices of various layer thicknesses and composed of various layer materials offers the opportunity to manipulate the electronic structure of the film and perform what has been called 'band gap engineering'.

Another important way in which the electronic structure of a quantum well is different from a bulk three-dimensional solid is in the energy dependence of the electronic density of states [64]. For a three-dimensional free electron solid, the density of states is proportional to $E^{1/2}$, where E is the electron energy. A quantum well in a superlattice can be considered to be of finite width in one dimension and effectively of infinite width in the other two dimensions. In the limit of an infinite barrier height, the density of states N_{2D} is given by

$$N_{2D} = 4\pi m^*/h^2. \tag{6.4}$$

In contrast with the three-dimensional case, it is seen that the two-dimensional density of states is independent of the energy of the well within a given subband.

6.5.2 Microelectronics applications

Semiconductor superlattices have important technological applications in the area of high-speed microelectronics [63–68]. Consider a quantum well composed of a high-purity GaAs layer between two layers of the wide-band-gap semiconductor AlGaAs that is heavily doped with donors (see figure 6.9). The electrons of donor impurities in the AlGaAs layer can 'fall' into the smaller-band-gap GaAs well. Since the GaAs is high purity, there are no impurity scattering centers and the electrons have a mobility much higher than in doped GaAs. The same kind of effect can also be achieved using acceptor impurities where holes would be the dominant charge carriers. In fact, superlattices composed of a periodic sequence of n-doped and p-doped layers that have undoped (intrinsic) layers in between have been produced that are referred to as 'n–i–p–i' superlattices [66]. The introduction of a variation in dopant concentrations is called *modulation doping*. Modulation doping is used in field-effect transistors for high-frequency applications; such devices are commonly referred to by the acronym MODFET (modulation doped field-effect transistor).

At temperatures below about 4 K and in the presence of high magnetic fields above about 4 T applied perpendicular to the layers, high-mobility modulation doped GaAs-$Al_xGa_{1-x}As$ heterojunctions and multilayered structures display some remarkable electronic behavior [63, 67, 68]. The resistance measured parallel to the direction of a current passed in the plane of the layers exhibits broad minima as a function of the applied field. The apparent resistivity of the current carrying volume of some heterojunctions has been measured to be as low as 10^{-12} Ω cm, which is less than the resistivity of any known normal-state material. At the same fields where these parallel resistance minima occur, there are also wide flat plateaus in the Hall resistance, which is equal to the Hall voltage divided by the current along the layers. This latter phenomenon is known as the *quantized Hall effect*, and has been the subject of great scientific interest. The parallel resistance minima and Hall resistance plateaus can be understood, at least qualitatively, as occurring at fields where an integral number of Landau levels are filled [67].

6.5.3 Optoelectronics applications

Optical absorption measurements show that strong absorption occurs in bulk semiconductors such as GaAs at energies corresponding to the formation of a bound electron–hole pair called an exciton. If it is viewed as a hydrogen-like atom, an exciton has an effective 'Bohr radius' a_B in a bulk solid of order 5 to 10 nm. In a quantum well of width $L < a_B$, the exciton is confined to a region smaller than its equilibrium 'Bohr radius'. The smaller separation leads to an increase in the exciton 'ionization' energy compared to that for an exciton in a bulk semiconductor, which in turn leads to a much greater probability for exciton preservation and eventual radiative decay [64]. In superlattices containing direct-band-gap semiconductors such as GaAs, exciton absorption energies can be easily measured in a photoluminescence spectrum. These energies can be tuned by precise control of the width of the quantum wells (see equation (6.3)). The absorption spectrum can be used to characterize the quality of the layering since the more uniform the layering, the smaller the width of the emission peaks.

When the width of the layer separating semiconducting quantum wells is sufficiently thin (typically less than about 5 nm), electrons can tunnel between the wells. This ability of electrons to tunnel between quantum wells is the basis of many technologically important applications [63, 65]. For a double-well heterostructure, an injected carrier with an energy equal to one of the discrete levels (bound states) of the well can tunnel with unit transmittivity, an effect known as *resonant tunneling*. This phenomenon will lead to maxima in current–voltage curves, resulting in regimes of negative differential resistance that could be utilized for high-frequency device applications. Tunneling in multiple-quantum-well superlattices is also of practical interest. For example, the tunneling probability of a carrier is an exponentially decreasing function of the effective mass and therefore the superlattice can act as an effective mass filter. This effect can be used to produce a superlattice photoconductor [65]. In AlInAs–GaInAs superlattices, photogenerated heavy holes remain localized while the photoelectrons can be conducted through the superlattice until they recombine with the holes. Since the electron mobility depends exponentially on the barrier thickness, the photoconductor gain can be sensitively tuned by controlling this thickness. In MBE grown equal layer thickness AlInAs–GaInAs superlattices with a bilayer repeat length of 7 nm and 100 periods, current gains as high as 20 000 were obtained for wavelengths in the range 1.0 to 1.6 μm at low voltage (≤ 1 V).

The superlattice photoconductor is just one example of the many photonic device applications for semiconductor superlattices. Other examples include avalanche photodiodes, infrared lasers, and quantum confined Stark effect optical modulators; the interested reader should consult the reviews cited in the references [65, 69, 70].

6.5.4 Electronic transport in metallic multilayers

In contrast to the behavior of semiconductor superlattices, no evidence has been reported that has clearly demonstrated superlattice effects on the electronic transport properties of metallic superlattices. The principal experimental observation regarding normal-state transport in metallic superlattices is a contribution to the resistivity owing to electron scattering at the interfaces. This can be understood in terms of the classical Drude model for electrical conductivity [71], where the resistivity would be predicted to have a contribution from a term inversely proportional to the average layer thickness, the mean free path for the interface scattering mechanism. The resistivity as a function of the individual layer thickness in equal layer thickness Cu–Nb multilayers measured at a temperature of 20 K is shown in figure 6.10 [72, 73]. The measurements were made at this low temperature in order to greatly reduce the phonon scattering contribution to the resistivity. As seen in the figure, below a layer thickness of approximately 1 nm the resistivity saturated close to the value of 150 $\mu\Omega$ cm (the 'Ioffe–Regel limit' [63, 74]). Above this thickness it was found that the resistivity was inversely proportional to the layer thickness. It was also found that as the layer thickness was decreased, the temperature coefficient of resistivity became less positive and eventually zero at a layer thickness of about 1 nm. Multilayers with a vanishing temperature coefficient of resistivity could have applications as thin-film resistors that are stable to temperature variations.

A very few studies concerning the thermoelectric power in multilayered films have been conducted on Cu–Ni compositionally modulated thin films [73]. Although there are not enough data to interpret the behavior in detail, it appears that, for a given temperature, the thermopower depends on both the composition modulation amplitude and wavelength.

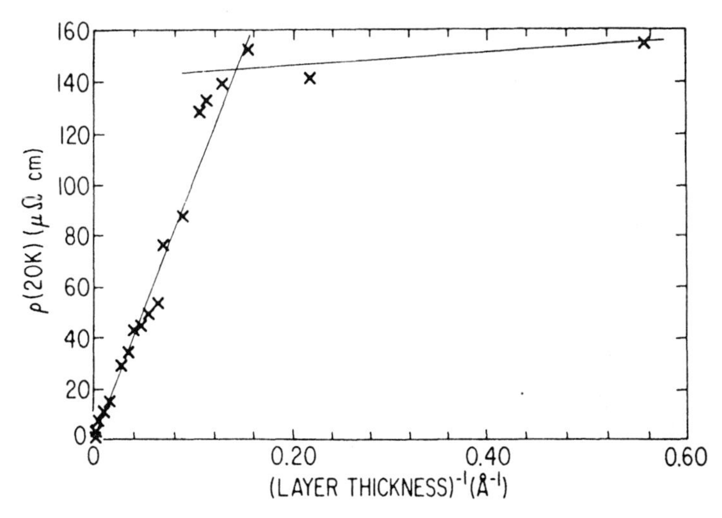

Figure 6.10. Electrical resistivity as a function of the individual layer thickness in equal layer thickness Cu–Nb multilayers measured at a temperature of 20 K (taken from [72]).

6.5.5 Bragg reflectors

As discussed in section 6.2.3, x-ray diffraction patterns of superlattices have satellite peaks about the (000) undiffracted beam and the main Bragg peaks (see figures 6.3 and 6.4). By controlling the bilayer repeat length, the wavelength of the x-rays needed to produce satellite peaks for a given incident angle can be tuned. Because of this ability to selectively diffract radiation of certain wavelengths, multilayered thin films have been used extensively as dispersion elements for x-rays, particularly soft x-rays [75–78]. When using multilayers as an x-ray mirror, optimal materials for the alternating layers should have a large difference in electron density, and common choices have been W–C and Mo–Si. Layer thicknesses can be chosen in order to maximize the reflectivity based on the optical properties of the constituent materials. It has been found that while it is generally best to maintain a constant bilayer repeat length, the ratio of individual layer thickness should be continuously varied in order to optimize the integrated reflectivity [75].

Multilayered x-ray Bragg reflectors have been used to form more complex structures. An x-ray Fabry–Perot etalon analogous to an optical Fabry-Perot etalon used for the visible spectral region has been developed using two identical multilayered Bragg reflectors separated by a low-absorption layer much thicker than one bilayer period of the multilayers [76]. An x-ray microscope has been constructed utilizing metallic multilayered x-ray mirrors [79].

Nonmetallic multilayers have also been used to produce Bragg reflectors. High-reflectivity GaAs–AlAs heterostructures can be used as mirrors in solid-state lasers [64], and multilayers composed of polymers with different refractive indices have been produced that reflect ultraviolet or near-infrared radiation as well as displaying iridescent color effects [38].

6.6 SUPERCONDUCTING PROPERTIES

6.6.1 Low-temperature superconductors

Several studies have been conducted on multilayered materials where one of the layer materials was a metallic (low-temperature) superconductor and where the individual layer thicknesses were less than the superconducting coherence length, typically 5 to 100 nm [80]. In systems involving one superconducting layer and one nonsuperconducting insulating layer, it is found that the critical temperature T_c in the superconducting layers generally decreases when the thickness of the nonsuperconducting layer is increased or the thickness of the superconducting layer is decreased. Also, there is a well defined transition from three-dimensional to two-dimensional behavior as the thickness of the nonsuperconducting layer is reduced below the coherence length. This transition has been observed in the temperature dependence of the upper critical field H_{c2} parallel to the layers near T_c in Nb–Ge multilayers, where the individual Nb layer thickness was greater than 3 nm. When the Nb layers were separated by a relatively large Ge thickness of 5 nm, H_{c2} varied as $(1-T/T_c)$, consistent with the two-dimensional behavior of the Nb predicted from the Ginzburg–Landau theory. When the Ge thickness was 0.7 nm, H_{c2} varied as $(1 - T/T_c)^{1/2}$, characteristic of three-dimensional behavior and indicating that there was strong coupling between the Nb layers. In addition to these effects which occur at temperatures below T_c, a change in dimensionality of the fluctuation superconductivity behavior above T_c was also observed.

6.6.2 High-temperature superconductors

Recently there has been a great deal of interest in producing multilayered thin films using high-temperature superconductors. Multilayered films with $YBa_2Cu_3O_7$ as one of the layers have been produced by sputtering [81] and pulsed laser deposition [82, 83]. A synthesis technique dubbed atomic layer-by-layer MBE has been developed to prepare superlattices with Bi–Sr–Ca–Cu–O type high-temperature superconducting layers [84]. This technique uses individual Knudsen cells for each of the cation components (Bi, Sr, Ca, Cu), whose deposition is controlled by a computer driven shuttering system. Ozone is introduced into the growth chamber to oxidize the films. It is believed that under proper operating conditions, only the top molecular layer is involved in the growth chemistry, so that precise atomic layer-by-layer growth can be achieved.

Superlattices composed of high-temperature superconducting layers have been used to investigate the dimensionality of the superconducting state. MBE prepared superlattices have been produced consisting of a monolayer of the high-temperature superconductor $Bi_2Sr_2CaCu_2O_8$ alternated with an integer number N of monolayers of the low-temperature superconductor $Bi_2Sr_2CuO_6$ acting as a spacer material [85]. Measurements of the critical temperature as a function of N were performed in order to determine whether any coupling between the high-temperature superconductor monolayers affected the critical temperature T_c. It was found that T_c was independent of N, strongly suggesting that the high-temperature superconductivity state was fundamentally two dimensional in nature.

6.7 MAGNETIC PROPERTIES

6.7.1 Magnetic superlattices

The magnetic behavior is the most heavily studied property of metallic superlattices, and has been covered in several excellent reviews [2, 73, 86–88]. Scientifically, magnetic superlattices offer the possibility of fundamental investigations of the physical origins of magnetism, and the potential for use in magnetic recording applications also makes them very promising materials technologically.

There was some early interest in the Cu–Ni system because of reports of an apparent enhanced magnetization per atom of Ni at temperatures below 180 K. However, more detailed studies have shown that the magnetization of the nickel is in fact reduced, and this behavior has been observed in other ferromagnetic–nonferromagnetic multilayered systems [73]. Generally, the magnetization decreases with decreasing thickness of the magnetic layer. One popular explanation for this behavior has been based on the effect of so-called *magnetically dead layers*, i.e. unmagnetized atomic planes at the interface. Similar dead layers at the free surfaces of ferromagnetic metals such as Fe and Co have also been proposed. However, recent experimental and theoretical results have called the dead-layer concept into question, at least in certain cases [89, 90]. It may be that intermixing or interfacial reactions are what lead to an overall reduced magnetization in multilayers.

Experimental studies of Fe–Cr, Co–Ru, and Co–Cr multilayers showed that the magnetic moments of the ferromagnetic layers could be spontaneously aligned either parallel or antiparallel depending on the thickness of the nonferromagnetic layer [88, 91]. These results indicated that the interaction between the ferromagnetic layers oscillated as a function of the nonferromagnetic layer thickness. This behavior may have resulted from Ruderman–Kittel–Kasuya–Yosida (RKKY) coupling mediated by spin polarization of the Cr or Ru layers [91].

In Ni–Mo multilayers, the dependence of the quantized spin-wave (magnon) frequency on the magnetic field could be altered by varying the bilayer repeat length [2]. Co–Pt and Co–Pd multilayers that display perpendicular anisotropy (that is, magnetization perpendicular to the plane of the film), high coercivity, and a square hysteresis loop have been studied with an eye toward magneto-optical recording applications [2]. Multilayers where one of the layer materials is a rare earth ferromagnet have also been widely investigated. In Gd–Y, variations in the Y layer thickness keeping the Gd layer thickness constant have resulted in parallel or antiparallel alignments [92], while in Dy–Y a coherent helical coupling between layers was observed [93].

The inherently complex nature of magnetic interactions betwen layers in superlattices may lead to dramatic changes in the properties. Bulk Cr is antiferromagnetic with a magnetic moment per atom of 0.59 μ_B (where μ_B is a Bohr magneton), but it has been predicted that in a coherent Cr–Au multilayer Cr can be made ferromagnetic with a moment per atom of 2.95 μ_B [94].

6.7.2 Giant magnetoresistance

One of the most interesting and potentially useful properties of metallic multilayers is the phenomenon known as *giant magnetoresistance* (GMR) which occurs in certain multilayers composed of alternating ferromagnetic and nonmagnetic layers such as Fe–Cr and Co–Cu [88, 95–98]. Discovered in 1988 [99], GMR refers to a significant change in the electrical resistance experienced by current flowing parallel to the layers when an external magnetic field H is applied. This effect occurs in a multilayer where the magnetic moments of the alternating ferromagnetic layers display an antiparallel alignment when $H = 0$. If a sufficiently strong magnetic field is applied, the magnetic moments of the ferromagnetic layers assume a parallel alignment. This change in orientation of the moments causes a change in resistance, the largest resistance occurring when the moments of the layers display an antiparallel alignment and the smallest resistance occurring when all the moments are parallel.

Plots of the resistivity change $\Delta\rho$ as a function of applied field measured at a temperature of 4 K of sputtered Fe–Cr(110) superlattices prepared using two different argon sputtering pressures are shown in figure 6.11 [100]. The difference in the maximum resistivity change for the two samples was attributed to differences in the degree of interfacial roughness, with the superlattice sputtered at an argon pressure of 12 mTorr displaying greater interfacial roughness and a greater value of $\Delta\rho$ at $H = 0$ compared to the superlattice sputtered at an argon pressure of 4 mTorr. The resistance change relative to the high-field value, $\Delta R/R$, was 14% and 6% for the 12 mTorr and 4 mTorr films respectively. In the original report of GMR [99], $\Delta R/R$ was measured to be 100% at room temperature, and effects as large as 220% at low temperatures have been observed [101].

Very recently GMR has been observed in Co–Cu [102–104] and (Fe, Ni)–Cu [103] multilayered nanowires. These multilayers were prepared by electrodeposition using nuclear track etched templates, and the resulting layered nanowires had diameters between 10 and 400 nm. In these materials, the resistance was measured for currents flowing perpendicular to the layers. Values

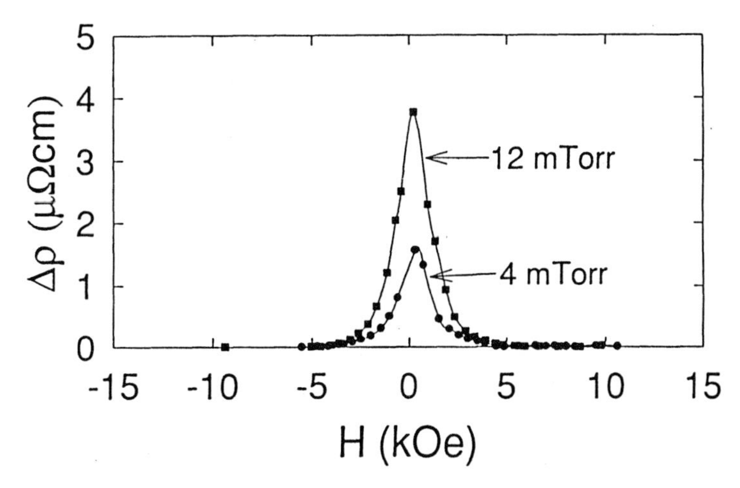

Figure 6.11. Resistivity change $\Delta\rho$ of sputtered Fe(3.0 nm)–Cr(1.8 nm) superlattices as a function of applied magnetic field H measured at a temperature of 4 K (taken from [100]). Circles: samples sputtered using 4 mTorr argon pressure; squares: samples sputtered using 12 mTorr argon pressure.

of $\Delta R/R$ of 10 to 14% measured at room temperature were observed.

If multilayers with room-temperature GMR can be prepared that have low magnetostriction and require low operating magnetic fields of about 10 gauss, then magnetoresistive random access memory could become an important memory technology for magnetic recording [95]. It should be mentioned that GMR has recently been observed in granular metal films [97] (see chapter 14).

6.8 MECHANICAL PROPERTIES

6.8.1 Elastic properties

In the earliest studies of the elastic properties of multilayered thin films, enhancements by factors of two or more were reported [105] in the elastic moduli of metallic multilayers such as Au–Ni and Cu–Pd when the bilayer repeat length was reduced to about 2 nm. Most of these early studies of this 'supermodulus effect' were performed using a bulge test apparatus, and subsequent investigations have suggested that the large apparent enhancements were in fact experimental artifacts [106]. Recent studies, using more precise dynamic techniques, have shown that certain metallic multilayers display modulus variations, both enhancements and reductions, of order 10 to 50% when the bilayer repeat length is reduced below about 4 nm. Generally associated with these modulus variations have been fairly large variations in the lattice spacings as measured by x-ray diffraction.

The most comprehensive study of the elastic behavior of multilayers was performed on Cu(111)–Nb(110) superlattices [107]. It was found that as the bilayer repeat length was decreased to 2.2 nm, the shear modulus C_{44} (referred to a coordinate system where the z axis is perpendicular to the layers) decreased by about 30%, the in-plane biaxial modulus increased by about 15%, and the flexural modulus (associated with a vibrating-reed type of deformation) remained approximately constant. X-ray diffraction results showed that concomitant with the elastic modulus variations was an increase in the average perpendicular interplanar spacing that reached a value of 2% when the modulation wavelength was reduced to 2.2 nm. The reasons for the modulus and lattice spacing variations are controversial. Perhaps the best model proposed so far is one based on higher-order elastic effects induced by interface stresses (see section 6.4.1) that can explain, at least in a semiquantitative manner, the observed modulus and lattice parameter variations [49, 50, 52, 53]. It appears that little if any dependence of the elastic behavior on the bilayer repeat length has been observed in the few studies conducted so far on ceramic or semiconductor superlattice films.

6.8.2 Damping capacity

Vibrating-reed measurements [108] on polycrystalline Cu–Ni multilayered films deposited on fused silica substrates revealed that above 300 °C these materials displayed an unusually large damping capacity (internal friction) [108]. Enhancements in the apparent damping capacity for the temperature range 50 °C to 200 °C were inferred from measurements on stressed polycrystalline Cu–Ni(100) vibrating membranes [109, 110]. (The stressed membranes also displayed a large apparent coefficient of thermal expansion that varied with the density of interfaces [111].) It has been proposed that the enhancements may be associated with grain-boundary sliding [108] or with energy dissipation mechanisms localized at the interfaces between the layers [109]. These effects need to be studied in more detail and in other multilayered systems before any definitive conclusions can be reached.

6.8.3 Plastic properties

Significant enhancements in the yield strength [112, 113], ultimate tensile strength [114], and hardness [105, 107, 115–117] of multilayered materials have been reported. Nanoindentation (low-load) hardness techniques [118, 119] have proven to be very useful when investigating the plastic properties of multilayered thin films. Figure 6.12 shows the relative nanoindentation hardness as a function of superlattice period (bilayer repeat length) for TiN–$V_xNb_{1-x}N$(100) strained layer ceramic superlattices produced by sputtering [115]. Similar hardness increases have also been reported for several metallic systems [105]. These hardness enhancements can be understood in a general way as resulting from interface strengthening effects owing to, for example, image forces associated with the interfaces that impede the motion and generation of dislocations [120,121], or from a Hall–Petch type effect as has been observed in fine-grained bulk alloys [122]. Low-load indentation studies of semiconductor superlattices such as Si–Si_xGe_{1-x} and GaAs–AlAs are just beginning to be conducted [123].

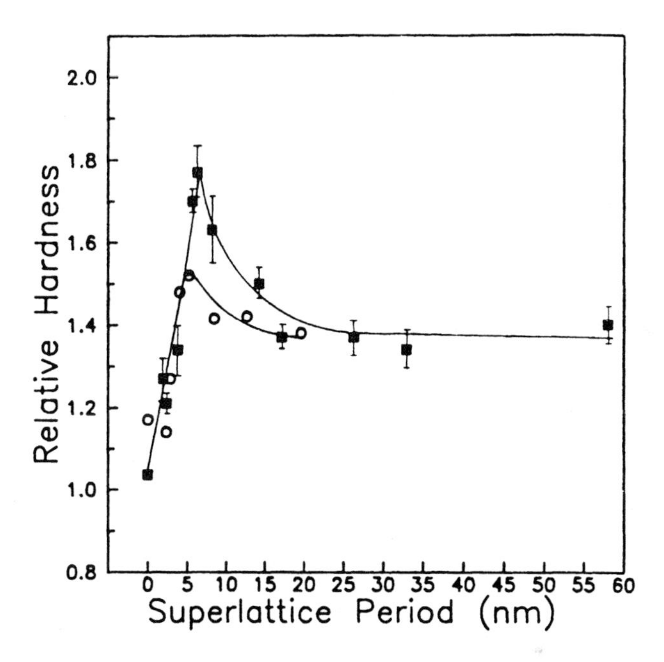

Figure 6.12. Nanoindentation hardness of TiN–VNbN superlattices as a function of superlattice period (bilayer repeat length) (taken from [115]). Squares: TiN–$V_{0.3}Nb_{0.7}N$ superlattices; circles: TiN–$V_{0.6}Nb_{0.4}N$ superlattices. A relative hardness value of one represents a rule of mixtures value.

Hardness experiments were recently performed on annealed electrodeposited Cu–Ni(100) compositionally modulated films with sine-wave composition profiles [124]. During the anneals the composition modulation wavelength Λ remained unchanged but the modulation amplitude A decreased due to interdiffusion. It was found that in films with different modulation wavelengths, the hardness enhancements owing to the composition modulation were linearly proportional to A. These results showed that it is necessary to take into account both of the important microstructural parameters, Λ and A, when interpreting the mechanical behavior of compositionally modulated thin films. Experimental [125] and theoretical [121, 122, 126, 127] studies of dislocation interactions with interfaces in multilayered thin films are being conducted in order to understand the microplasticity of these materials in greater detail.

6.8.4 Wear and friction

Tribology studies conducted on Cu–Ni multilayered coatings electrodeposited onto steel substrates indicated that these coatings displayed enhanced wear resistance, both to unlubricated sliding and to sliding using a paraffin oil lubricant, compared with the wear behavior of electrodeposited copper or nickel coatings [128,129]. The coefficients of unlubricated sliding friction of the multilayers were measured to be intermediate between those for copper and nickel. Further studies are needed to determine whether multilayered coatings are of practical use as tribological materials.

6.8.5 Fracture

Equal layer thickness Al–Cu multilayered films displayed significantly enhanced tensile fracture strength compared to the fracture strength of the constituent layer materials [112]. The tensile fracture stress increased as the individual layer thickness was reduced from 1000 to 70 nm, and then stayed approximately constant as the layer thickness was reduced from 70 to 20 nm. Qualitatively similar behavior was also observed for the yield stress dependence on layer thickness, suggesting that related mechanisms were responsible for both the enhanced plastic and fracture behavior of these materials. Theoretical models of fracture in multilayered films are currently being developed [130].

Coextruded polymer multilayered thin films composed of a high-modulus brittle layer and a low-modulus ductile layer have been prepared that have displayed synergistic improvements in the mechanical properties [38]. These brittle–ductile multilayers have a greater fracture toughness than the brittle material because crazes formed in a brittle layer will be blunted by the adjacent ductile layers. Such behavior has been observed in films composed of the brittle copolymer styrene–acrylonitrile (25% acrylonitrile by weight) and the ductile polymer polycarbonate [40]. In addition to displaying a greater fracture toughness than the brittle polymer, the multilayer will be stiffer than the lower-modulus ductile polymer. The net result is a film with a greater work-to-failure than either of the individual layer materials [38]. The improved mechanical behavior of brittle–ductile polymer multilayered films makes them attractive for use as heavy-duty wrapping and packaging materials.

REFERENCES

[1] DuMond J and Youtz J P 1940 *J. Appl. Phys.* **11** 357

[2] Greer A L and Somekh R E 1991 *Processing and Metals and Alloys (Materials Science and Technology 15)* ed R W Cahn (Weinheim: VCH) ch 15

[3] Cook H E and Hilliard J E 1969 *J. Appl. Phys.* **40** 2191

[4] Philofsky E M and Hilliard J E 1969 *J. Appl. Phys.* **40** 2198

[5] Esaki L and Tsu R 1970 *IBM J. Res. Dev.* **14** 61

[6] Chang L L, Esaki L, Howard W E and Ludeke R 1973 *J. Vac. Sci. Technol.* **10** 11

[7] Chang L L and Giessen B C (ed) 1985 *Synthetic Modulated Structures* (Orlando, FL: Academic)
[8] Clemens B M and Gay J G 1987 *Phys. Rev.* B **35** 9337
[9] Wilson L and Bienenstock A 1988 *Mater. Res. Soc. Symp. Proc.* **103** 69
[10] Chien C L, Liou S H and Xiao G 1988 *Metallic Multilayers and Epitaxy* ed M Hong, S Wolf and D C Gubser (Warrendale, PA: Metallurgical Society) p 245
[11] Cammarata R C and Greer A L 1984 *J. Non-Cryst. Solids* **61&62** 889
[12] Prokes S M and Spaepen F 1985 *Appl. Phys. Lett.* **47** 234
[13] El-Masry N A, Tarn J C L, Humphreys T P, Hamaguchi N, Karam N H and Bedair S M 1987 *Appl. Phys. Lett.* **51** 1608
[14] El-Masry N A and Karam N H 1988 *J. Appl. Phys.* **64** 3672
[15] McWhan D B 1985 *Synthetic Modulated Structures* ed L L Chang and B C Giessen (Orlando, FL: Academic) p 43
[16] Schuller I K 1980 *Phys. Rev. Lett.* **44** 1597
[17] Greer A L and Spaepen F 1985 *Synthetic Modulated Structures* ed L L Chang and B C Giessen (Orlando, FL: Academic) p 419
[18] Prokes S M and Wang K L 1990 *Appl. Phys. Lett.* **56** 2628
[19] Barbee T W Jr 1986 *Opt. Eng.* **25** 895
[20] Goldman L M, Atwater H A and Spaepen F 1990 *Mater. Res. Soc. Symp. Proc.* **160** 577
[21] Greer A L 1993 *J. Magn. Magn. Mater.* **126** 89
[22] Roberts G G 1985 *Adv. Phys.* **34** 475
[23] Swalen J D, Allara D L, Andrade J D, Chandross E A, Garoff S, Israelachvili J, McCarthy T J, Murray R, Pease R F, Rabolt J F, Wynne K J and Yu H 1987 *Langmuir* **3** 932
[24] Tillman N, Ulman A and Penner T L 1989 *Langmuir* **5** 101
[25] Kleinfeld E R and Ferguson G S 1994 *Science* **265** 370
[26] Searson P C and Moffat T F 1994 *Crit. Rev. Surf. Chem.* **3** 171
[27] Ross C 1994 *Annu. Rev. Mater. Sci.* **24** 159
[28] Ohring M 1991 *The Materials Science of Thin Films* (Boston, MA: Academic) ch 3
[29] Spaepen F, Greer A L, Kelton K F and Bell J L 1985 *Rev. Sci. Instrum.* **56** 1340
[30] Tu K N, Mayer J W and Feldman L C 1992 *Electronic Thin Film Science for Electrical Engineers and Materials Scientists* (New York: Macmillan) ch 6
[31] Chang L L and Esaki L 1980 *Surf. Sci.* **98** 70
[32] Hubler G K 1992 *MRS Bull.* **17** 26
[33] Ohring M 1991 *The Materials Science of Thin Films* (Boston: Academic) ch 3
[34] Slayter G 1962 *Sci. Am.* **206** 124
[35] Chikashige M 1936 *Alchemy and Other Chemical Achievements of the Ancient Orient* Engl. Transl. N Sasaki (Tokyo: Rokakuho Uchide) p 84
[36] Atzmon M, Unruh K M and Johnson W L 1985 *J. Appl. Phys.* **58** 3865
[37] Schrenk W J and Alfrey T Jr 1978 *Polymer Blends* vol 2, ed D R Paul and S Newman (New York: Academic) p 129
[38] Schrenk W J and Alfrey T Jr 1969 *Polym. Eng. Sci.* **9** 393 This and several other papers concerning the processing and properties of multilayered polymer films have been reprinted in Boyer R F and Mark H F (ed) 1986 *Selected Papers of Turner Alfrey* (New York: Dekker)

[39] Baer E, Hiltner A and Keith H D 1987 *Science* **235** 1015
[40] Gregory B L, Siegmann A, Im J, Hiltmer A and Baer E 1987 *J. Mater. Sci.* **22** 532
[41] Greer A L 1993 *Mechanical Properties and Deformation of Materials Having Ultrafine Microstructures* ed M Nastasi, D M Parkins and H Gleiter (Dordrecht: Kluwer) p 53
[42] Johnson W C 1987 *Metall. Trans.* A **18** 1093
[43] Johnson W C and Voorhees P W 1987 *Metall. Trans.* A **18** 1213
[44] Johnson W C 1988 *Mater. Res. Soc. Symp. Proc.* **103** 61
[45] Josell D and Spaepen F 1993 *Acta Metall. Mater.* **41** 3007
[46] Josell D and Spaepen F 1993 *Acta Metall. Mater.* **41** 3017
[47] Willens R H, Kornblit A, Testardi L R and Nakahara S 1982 *Phys. Rev.* B **25** 290
[48] Cahn J W and Larché F 1982 *Acta Metall.* **37** 51
[49] Cammarata R C 1994 *Prog. Surf. Sci.* **46** 1
[50] Cammarata R C and Sieradzki K 1989 *Phys. Rev. Lett.* **62** 2005
[51] Kosevich Yu A and Kosevich A M 1989 *Solid State Commun.* **70** 541
[52] Streitz F H, Cammarata R C and Sieradzki K 1994 *Phys. Rev.* B **49** 10 699
[53] Streitz F H, Cammarata R C and Sieradzki K 1994 *Phys. Rev.* B **49** 10 707
[54] Ruud J A, Witvrouw A and Spaepen F 1993 *J. Appl. Phys.* **74** 2517
[55] Schwarz R B and Johnson W L 1983 *Phys. Rev. Lett.* **51** 415
[56] Highmore R J, Evetts J E, Somekh R E and Greer A L 1987 *Appl. Phys. Lett.* **50** 566
[57] Prokes S M, Glembocki O J and Godby D J 1992 *Appl. Phys. Lett.* **60** 1087
[58] Prokes S M 1995 *Mater. Sci. Technol.* **11** 389
[59] Stephenson G B 1988 *Acta Metall.* **36** 2663
[60] Clevenger L A, Thompson C V, Cammarata R C and Tu K N 1988 *Appl. Phys. Lett.* **52** 795
[61] Schlesinger T E, Cammarata R C and Prokes S M 1991 *Appl. Phys. Lett.* **59** 449
[62] People R 1986 *IEEE J. Quantum Electron.* **QE-22** 1696
[63] Burns G 1985 *Solid State Physics* (San Diego: Academic) ch 18
[64] Tu K N, Mayer J W and Feldman L C 1992 *Electronic Thin Film Science for Electrical Engineers and Materials Scientists* (New York: Macmillan) ch 8
[65] Capasso F 1987 *Science* **235** 172
[66] Döhler G H and Ploog K 1985 *Synthetic Modulated Structures* ed L L Chang and B C Giessen (Orlando, FL: Academic) p 163
[67] Gossard A C and Pinczuk A 1985 *Synthetic Modulated Structures* ed L L Chang and B C Giessen (Orlando, FL: Academic) p 215
[68] Störmer H L, Haavasoja T, Narayanamurti V, Gossard A C and Wiegmann W 1983 *J. Vac. Sci. Technol.* B **1** 423
[69] Holonyak N Jr and Hess K 1985 *Synthetic Modulated Structures* ed L L Chang and B C Giessen (Orlando, FL: Academic) p 257
[70] Miller D A B 1990 *Opt. Photon. News* February p 257
[71] Burns G 1985 *Solid State Physics* (San Diego: Academic) ch 9
[72] Werner T R, Banerjee I, Yang Q S, Falco C M and Schuller I K 1982 *Phys. Rev.* B **26** 2224
[73] Falco C M and Schuller I K 1985 *Synthetic Modulated Structures* ed L L Chang and B C Giessen (Orlando, FL: Academic) p 339
[74] Ioffe A F and Regel A R 1960 *Prog. Semicond.* **4** 237

[75] Spiller E, Segmiller A and Haelbich R P 1980 *Ann. NY Acad. Sci.* **342** 188

[76] Barbee T W Jr 1985 *Synthetic Modulated Sructures* ed L L Chang and B C Giessen (Orlando, FL: Academic) p 313

[77] Barbee T W Jr 1986 *Opt. Eng.* **25** 954

[78] Houdy Ph and Boher P 1994 *J. Physique* III **4** 1589

[79] Underwood J H, Barbee T W Jr and Frieber C 1986 *Appl. Opt.* **25** 1730

[80] Ruggiero S T and Beasley M R 1985 *Synthetic Modulated Structures* ed L L Chang and B C Giessen (Orlando, FL: Academic) p 365

[81] Triscone J M, Fischer O, Bruner O, Antognaza L, Kent A D and Karkut M G 1990 *Phys. Rev. Lett.* **64** 804

[82] Li Q, Xi X X, Wu X D, Inam A, Vadlamannati S, McLean W L, Venkatesan T, Ramesh R, Hwang D M, Martinez J A and Nazar L 1990 *Phys. Rev. Lett.* **64** 3086

[83] Lowndes D H, Norton D P and Budai J D 1990 *Phys. Rev. Lett.* **65** 1160

[84] Schlom D G, Marshall A F, Harris J S Jr, Bozovic I and Eckstein J N 1991 *Advances in Superconductivity III* ed K Kajimura and H Hayakawa (Tokyo: Springer) p 1011

[85] Bozovic I, Eckstein J N, Klausmeier-Brown M E and Virshup G 1992 *J. Supercond.* **5** 19

[86] Schuller I K 1988 *Mater. Res. Soc. Symp. Proc.* **103** 335

[87] Jin B Y and Ketterson J B 1989 *Adv. Phys.* **38** 189

[88] Mathon J 1991 *Contemp. Phys.* **32** 143

[89] Freeman A J and Fu C L 1986 *Magnetic Properties of Low-Dimensional Systems* ed L M Falicov and J L Morán-López (Berlin: Springer) p 16

[90] Falicov L M, Victors R H and Tersoff J 1985 *The Structure of Surfaces* ed M A Van Hove and S Y Tong (Berlin: Springer) p 12

[91] Parkin S S P, More N and Roche K P 1990 *Phys. Rev. Lett.* **64** 2304

[92] Majkrzac C F, Cable J W, Hong J, McWhan D B, Yafet Y, Waszczak J V, Grimm H and Vettier C 1987 *J. Appl. Phys.* **61** 4055

[93] Hong M, Fleming R M, Kwo J, Schneemeyer L F, Waszczak J V, Mannaerts J P, Majkrzak C F, Gibbs D and Bohr J 1987 *J. Appl. Phys.* **61** 4052

[94] Fu C L, Freeman A J and Oguchi T 1985 *Phys. Rev. Lett.* **54** 2700

[95] Falicov L M 1992 *Phys. Today* October 46

[96] White R L 1992 *IEEE Trans. Magn.* **MAG-28** 2482

[97] Chien C L, Xiao J Q and Jiang J S 1993 *J. Appl. Phys.* **72** 5309

[98] Prinz G A 1995 *Phys. Today* April 58

[99] Baibich M N, Broto J M, Fert A, Nguyen Van Dau F, Petroff F, Etienne P, Greuzet G, Friederich A and Chazelas J 1988 *Phys. Rev. Lett.* **61** 2472

[100] Fullerton E E, Kelly D M, Guimpel J, Schuller I K and Bruynseràde Y 1992 *Phys. Rev. Lett.* **68** 859

[101] Schad R, Potter C D, Beliën P, Verbanck G, Moshchalkov V V and Bruynseraede Y 1994 *Appl. Phys. Lett.* **64** 3500

[102] Piraux L, George J M, Despres J F, Leroy C, Ferain E, Legras R, Ounadjela K and Fert A 1994 *Appl. Phys. Lett.* **65** 2484

[103] Blondel A, Meier J P, Doudin B and Ansermet J Ph 1994 *Appl. Phys. Lett.* **65** 3019

[104] Liu K, Nagodawithana K, Searson P C and Chien C L 1995 *Phys. Rev.* B **51** 7381

[105] Cammarata R C 1994 *Thin Solid Films* **248** 82

[106] Baker S P, Small M K, Vlassak J J, Daniels B J and Nix W D 1993 *Mechanical Properties and Deformation of Materials Having Ultrafine Microstructures* ed M Nastasi, D M Parkins and H Gleiter (Dordrecht: Kluwer) p 165

[107] Fartash A, Fullerton E E, Schuller I K, Bobbin S E, Wagner J W, Cammarata R C, Kumar S and Grimsditch M 1991 *Phys. Rev.* B **44** 13 760

[108] Berry B S and Pritchett W C 1976 *Thin Solid Films* **33** 19

[109] Su C M, Oberle R R, Wuttig M and Cammarata R C 1993 *Mater. Res. Soc. Symp. Proc.* **280** 527

[110] Oberle R R, Cammarata R C, Su C M and Wuttig M *J. Mater. Res.* submitted

[111] Su C M, Kim T, Oberle R R, Cammarata R C and Wuttig M 1994 *J. Appl. Phys.* **76** 4567

[112] Lehoczky S L 1978 *J. Appl. Phys.* **94** 5479

[113] Tench D and White J 1984 *Metall. Trans.* A **15** 2039

[114] Bunshah R F, Nimmaggada F, Doerr H J, Mochvan B A, Grechanuk N I and Dabizha E V 1980 *Thin Solid Films* **72** 261

[115] Mirkarimi P B, Barnett S A, Hubbard K M, Jervis T R and Hultman L 1994 *J. Mater. Res.* **9** 1456

[116] Barnett S M 1993 *Phys. Thin Films* **17** 1

[117] Barnett S M and Shinn M 1994 *Annu. Rev. Mater. Sci.* **24** 481

[118] Wu T W 1991 *Mater. Res. Soc. Symp. Proc.* **226** 165

[119] Pharr G M and Oliver W C 1992 *MRS Bull.* **17** 28

[120] Koehler J S 1970 *Phys. Rev.* B **2** 547

[121] Krzanowski J E 1991 *Scr. Metall. Mater.* **25** 1465

[122] Anderson P M and Li C 1995 *Nanostruct. Mater.* **5** 349

[123] Castell M R, Howie A, Petrovic D D, Whitehead A J, Ritchie D, Churchill A and Jones G A C 1993 *Mechanical Properties and Deformation of Materials Having Ultrafine Microstructures* ed M Nastasi, D M Parkins and H Gleiter (Dordrecht: Kluwer) p 489

[124] Oberle R R and Cammarata R C 1995 *Scr. Metall. Mater.* **32** 593

[125] Lashmore D S and Thomson R 1992 *J. Mater. Res.* **7** 2379

[126] Grilhé J 1993 *Mechanical Properties and Deformation of Materials Having Ultrafine Microstructures* ed M Nastasi, D M Parkins and H Gleiter (Dordrecht: Kluwer) p 255

[127] Embury J D and Hirth J P 1994 *Acta Metall. Mater.* **42** 2051

[128] Ruff A W and Myshkin N K 1989 *J. Tribol.* **111** 156

[129] Ruff A W and Wang Z X 1989 *Wear* **131** 259

[130] Anderson P M, Lin I-H and Thomson R 1992 *Scr. Metall. Mater.* **27** 687

PART 4

PROCESSING OF NANOMATERIALS

Chapter 7

Processing of nanostructured sol–gel materials

L C Klein

7.1 INTRODUCTION

The sol–gel process is a chameleon technology. Fifteen years ago, when the process enjoyed a resurgence in interest, the emphasis was on the duplication of ceramics and glasses which could be prepared by conventional means [1–3]. Scientific curiosity about the sol–gel process was based on the purity of the starting materials [4] and the generally lower temperatures for processing [5]. Along with other chemical methods for processing oxides, a large increase of conferences and reports occurred on this subject. The titles of these conferences tell the story of what was expected of sol–gel processing. On the one hand, sol–gel processing was a major part of the Materials Research Society Symposium series 'Better Ceramics through Chemistry', indicating no doubt that sol–gel processing leads to better ceramics [6–10]. On the other hand, sol–gel processing contributed to 'ultrastructure processing' which had the implication that sol–gel processing could be seen in the larger scheme of things where processing, properties, and performance are all linked [11–15]. There is mounting evidence that sol–gel processing makes better materials and that it will have a lasting role in advanced materials.

As a chameleon technology, sol–gel processing has taken on some other identities. In some respects, sol–gel processing can be viewed as a 'rapid solidification' because the resulting oxide is in a higher-energy, metastable state [16]. By other accounts, sol–gel processing is a 'near-net shape process' where replicating and miniaturizing of molds is possible [17]. One way to view sol–gel processing, which makes the distinction between precipitated powders and direct formation of porous preforms, is that it is a 'powder-free' process [18]. 'Powder-free' processing, which includes polymer pyrolysis and direct metal oxidation (DIMOX), is one of the more descriptive ways of classifying sol–gel processing, because it really highlights its possibilities.

Another rage of the last decade was fractals and fractal analysis, to which sol–gel processing conveniently lends itself. The scale of the sol–gel process and its amorphous nature require probing by scattering techniques. Neutron and x-ray studies reveal the fractal nature of sols, gels, and sol–gel transitions [19–22]. Interpretations of scattering data lead to surface area and density relationships which allow classification of gels and, at the same time, lead to predictions about porosity. Porosity in gels is the feature of interest both scientifically and technologically.

Finally, sol–gel processing is a form of nanostructure processing. Not only does the sol–gel process begin with a nanosized unit, it undergoes reactions on the nanometer scale resulting in a material with nanometer features [23]. When choosing a classification for processes according to scale, there is none more appropriate for sol–gel processing than the nanometer scale.

Now it should be clear why sol–gel processing is a chameleon technology. It has fitted in with each of the exciting new technologies which has surged in the last decade, and it is likely to adapt to new technologies ahead. Since the focus of this volume is nanostructured materials, those aspects of sol–gel processing having to do with synthesis, shaping, consolidating, and its use in composite fabrication will be described.

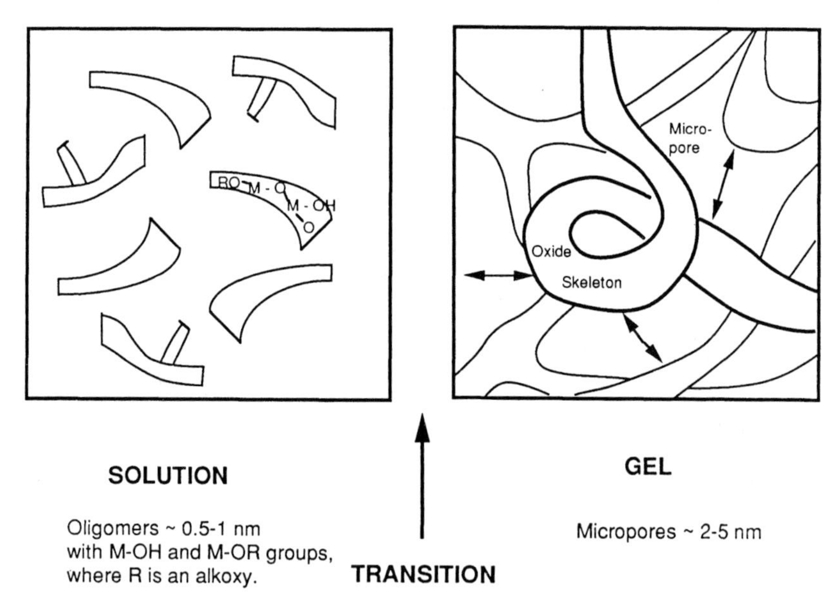

Figure 7.1. Schematic diagram of the oxide skeleton of a dried gel following the alkoxide solution route.

In particular, the sol–gel process is an appealing technology with respect to ceramic matrix composite fabrication where access to the nanostructure is the key. The concept is that the sol–gel process by a combination of chemical reactions turns a homogeneous solution of precursors and reactants into an infinite-molecular-weight oxide polymer. This polymer is a three-dimensional skeleton surrounding interconnected pores. A schematic of this skeleton is shown in figure 7.1 with the feature dimensions indicated. Ideally it is isotropic, homogeneous, uniform in microstructure, and it replicates its mold exactly and miniaturizes all features without distortion. Among the many names which have been associated with sol–gel processing, nanostructure processing is one which is likely to stick because of its precision in defining the scale of sol–gel processing.

7.2 SYNTHESIS OF OXIDES BY THE SOL–GEL PROCESS

The sol–gel process can be carried out either to form a powder or not to form a powder. The parallel powder and powder-free routes are shown in figure 7.2. The powder route follows the basic concepts which have been used for years in conventional ceramic processing. In conventional ceramics, powders are broken down in size and classified before combining in compacts. The alternative to breaking down the powders is building them from molecules. The innovation of the sol–gel process in forming powders is that the powder size, morphology, and surface chemistry are controlled simultaneously [24]. The sol–gel process is used to prepare monodisperse powders which are submicron in size. These monodisperse powders are said to avoid some of the flaws which occur in conventional powders. Nevertheless, they are still subject to ripening and coarsening, often placing them outside the realm of nanostructure processing.

The powder-free route in figure 7.2 is the heart of the concept of sol–gel processing as a nanostructure process. In the powder-free route, the nanometer-sized units which are the precursors link together in a continuous fashion, building ever larger structures without necessarily passing through a powder stage. The nanometer units generate the oxide skeleton. Because there are no discrete powder particles that link together to form necks, the problems of ripening and coarsening are avoided. In principle, the oxide skeleton is an infinite molecule taking on the dimensions of the mold in which it is reacting. The success of this continuous linking is then due in large part to the careful selection of precursors and synthetic methods. The precursors are alkoxides, soluble salts, and colloidal sols.

7.2.1 Alkoxide solution routes

The sol–gel process involves initially a homogeneous solution of one or more selected alkoxides [25]. In this case, the sol in the sol–gel process is shorthand for solution. Alkoxides are the organometallic precursors for silica, alumina, titania, zirconia, among others [26]. A catalyst is used to start reactions and control pH. The reactions are, first, hydrolysis to make the solution active (reaction I), followed by condensation polymerization (reaction II) along with

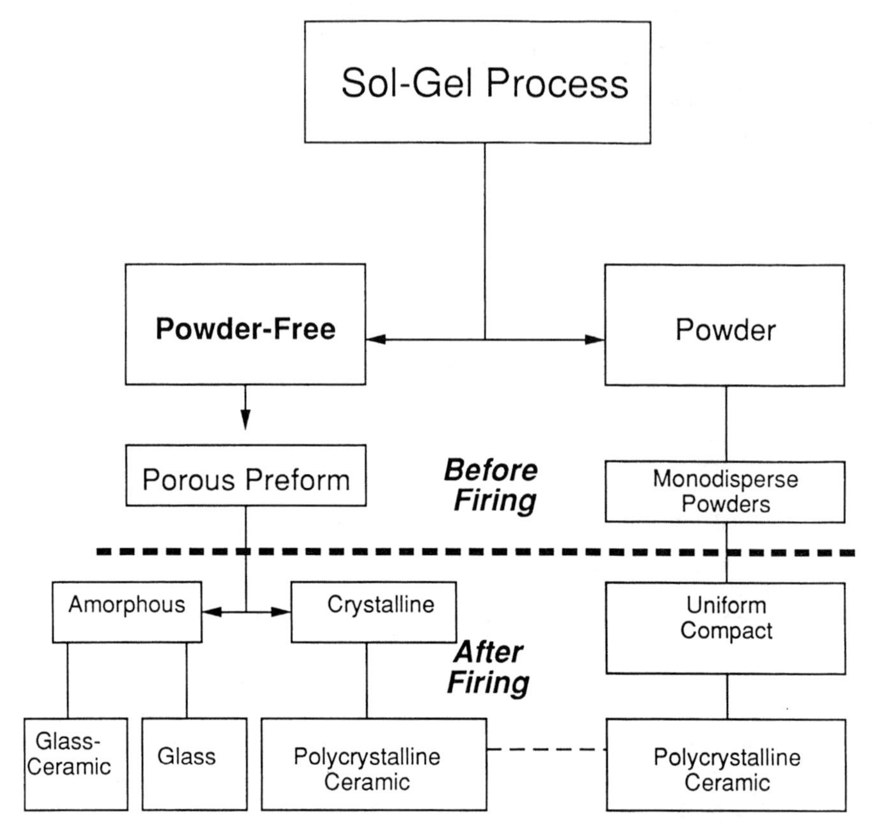

Figure 7.2. Flow chart of the sol–gel process following a powder route or a powder-free route.

further hydrolysis. These reactions increase the molecular weight of the oxide polymer (reactions III and IV). The pH of the water–alcohol mixture has an influence over the polymerization scheme such that acid catalyzed solutions typically remain transparent and base catalyzed solutions become opaque. Eventually, the solution reacts to a point where the molecular structure is no longer reversible. This point is known as the sol–gel transition. The gel is an elastic solid filling the same volume as the solution.

Over the years, the sol–gel process has been investigated widely for the synthesis of a variety of glasses and crystalline materials [27]. By far the most common system is one composed of tetraethyl orthosilicate (TEOS, $Si(OC_2H_5)_4$)–water–alcohol [28]. Various studies have elucidated the influence of the ratios of the starting species, the catalyst and other additives on the structure of the final gel or ceramic. Fortunately, all of these factors have been dealt with in a comprehensive volume [29].

7.2.2 Colloidal sols and suspensions

Colloidal sols are also referred to as precursors for sol–gel processing. In this case, the sol in sol–gel processing is not shorthand for solution, but, in fact, a colloidal solution. As long as the sol contains nanometer-sized particles, it is appropriate to include such precursors, recognizing that the mechanism for accomplishing the sol–gel transition is quite different. In sols such as Ludox™, the aggregation of sol particles is caused by changing the pH or changing the concentration [30]. When carried out in such a way that precipitates are generated, the product is no longer following a powder free route. Alternatively, the sols can be gelled in such a way that the process is powder free in the sense that the oxide skeleton is a continuous linking of sol particles, as shown schematically in figure 7.3. In comparison to figure 7.1, there are discrete features which make up the skeleton corresponding to the dimensions of the sol. The other features, pores within secondary particles and pores between secondary particles, are marked. The fundamental difference between alkoxide precursors and sol precursors is illustrated when comparing figures 7.1 and 7.3. The consequences of these differences may or may not be evidenced at later stages of the sol–gel process.

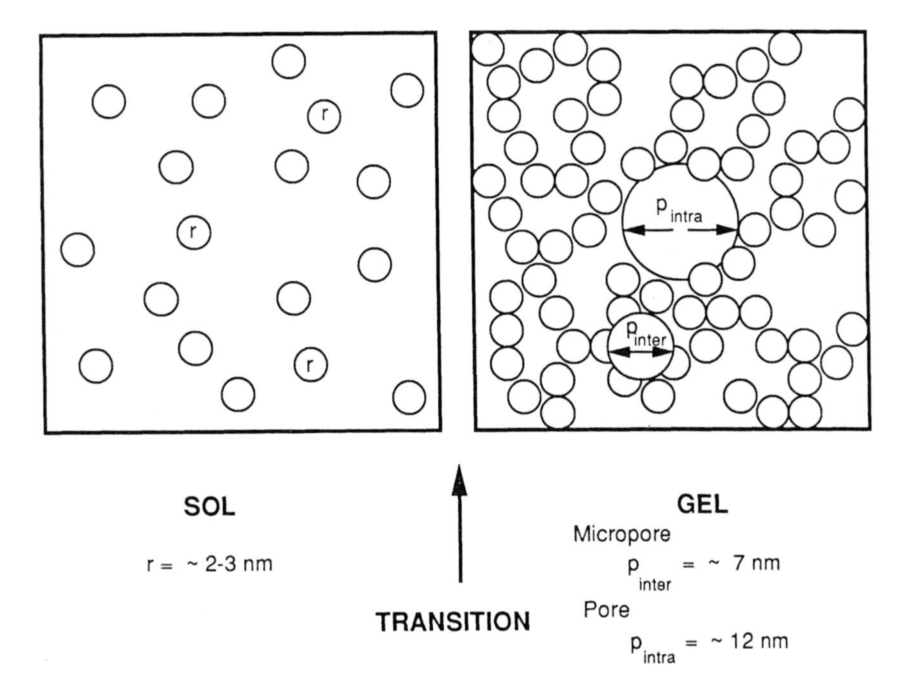

Figure 7.3. Schematic diagram of the linked particles of a dried gel following the colloidal route.

7.2.3 Single-component oxides

Most studies of the sol–gel process deal with a single alkoxide or sol. For anyone who has tried to study the reactions in a quantitative way, the reasons are obvious. Each precursor has its own reaction rates and complicated interdependences of pH, concentration, solvents, etc. Even in the relatively straightforward case of alumina, using aluminum-sec-butoxide (ASB), the expected reactions (I–IV) are

$$Al(OC_4H_9)_3 + H_2O = Al(OC_4H_9)_2(OH) + C_4H_9OH \quad (I)$$

$$2(Al(OC_4H_9)_2(OH) = 2(AlO(OH)) + yC_4H_9OH \quad (II)$$

$$2(Al(OC_4H_9)_2(OH)) + 2H_2O = 2Al(OH)_3 + 2C_4H_9OH \quad (III)$$

$$AlOOH \text{ or } Al(OH)_3 = Al_2O_3 + zH_2O. \quad (IV)$$

Either the monohydroxide AlOOH (boehmite) or the trihydroxide $Al(OH)_3$ (bayerite) can be transformed to gamma alumina [31]. In the production of transparent, activated alumina gels from ASB, one well studied composition for achieving these properties involves an excess of water in the initial solutions (100 moles water/ASB). Hydrolysis temperatures used were greater than 80 °C. This promoted formation of boehmite as opposed to bayerite.

The transition metal oxides have been formed using powder routes and powder-free routes. In systems with multiple valence, the intermediate species are oligomers which fall in the nanometer scale [32, 33].

7.2.4 Multi-component oxides

Despite the complexity of working with several alkoxides at once, each having its own reaction rates and complexing tendencies, there are many technologically important multi-component ceramics. The approaches to synthesizing these systems are largely empirical. While there is rarely a unique formulation for any given system, there is often a workable formulation which gives good results time after time. A recent example is yttrium aluminum garnet which was synthesized in garnet and monoclinic forms, and as the diphasic garnet with transition alumina [34].

The entire family of zeolites consists essentially of multi-component oxides from a sol–gel process. Although rarely grouped with sol–gel processing because of their single-crystal nature, zeolites and sol–gel processing share common precursors. Zeolites are certainly nanostructured materials. It is their nanoscaled microstructure that leads to their utility as catalysts. Their synthesis and characterization have been reported in detail [35].

7.3 POWDER-FREE PROCESSING OF GEL SHAPES

Figure 7.4 shows the options available for powder-free processing using the sol–gel process [27, 36, 37]. All of the options are more or less porous materials.

In all cases, it is important that the porosity remains interconnected, whether the form of the materials is essentially one dimensional such as a fiber, two dimensional such as a film, or three dimensional such as bulk monoliths. These applications can be divided into those which show isotropic shrinkage from the preform to the final form and those which show anisotropic shrinkage. Monoliths fall into the isotropic category. Monoliths formed from an alkoxide solution are achieved most often in dilute solutions. Thin films on substrates fall into the anisotropic category. Fibers also show anisotropic shrinkage, though it depends on whether or not the fiber is drawn from a preform many times its diameter or from a preform which is closer to its diameter. While monolithic pieces need a high water content, fibers and thin films can be drawn out of low-water-content solutions.

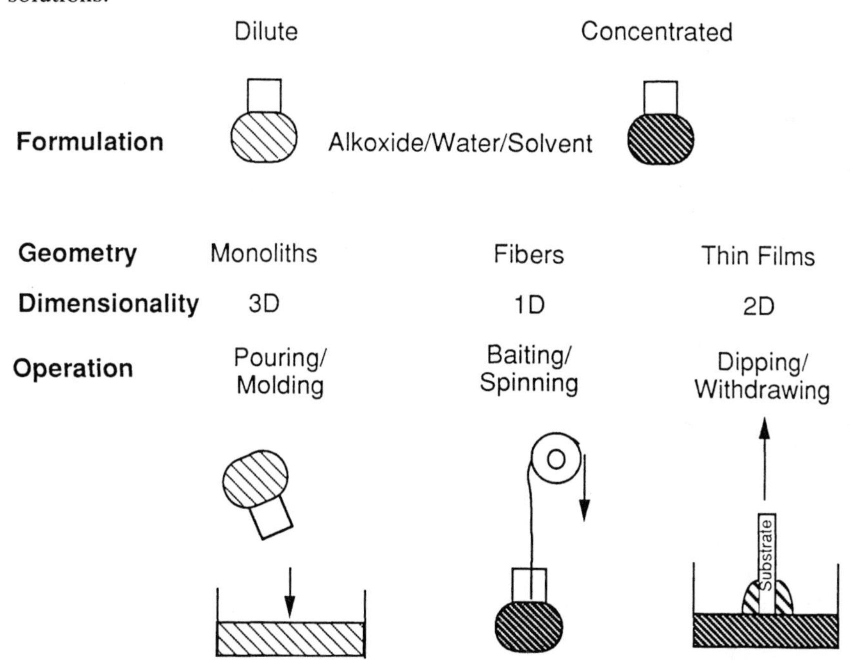

Figure 7.4. The options for gel geometry related to the concentration of the alkoxide solution: monoliths, fibers, thin films.

7.3.1 Thin films

Like sol–gel fibers, sol–gel thin films are prepared from low-water-content solutions. The technology of sol–gel thin films has been around for over 30 years [38]. The process is exceedingly simple. A solution containing the desired oxide precursors is applied to a substrate by spinning, dipping, or draining. The process is able to apply a coating to the inside and outside of complex shapes simultaneously. The equipment is inexpensive, especially in comparison to any deposition techniques that involve a vacuum. Typically, the coatings are applied at room temperature, though most need to be calcined and densified with heating. The basic steps for sol–gel thin-film depositions have been described elsewhere [39].

The physical properties of the coating solutions that are monitored are viscosity, surface tension, and time-to-gel. The time-to-gel is especially important when it comes to coatings because film formation, drying, and creation of pores must be rapid. Optimum film formulations correspond to those solutions that lose tackiness quickly. At present, the majority of sol–gel coatings are applied by dipping. This is a simple approach that allows the properties of the solution to control the deposition. A substrate is lowered into a vessel containing the solution. A meniscus develops at the contact of the liquid and the substrate. As the substrate is withdrawn, a continuous film is generated on the substrate [40]. Using appropriate assumptions, expressions relating film thickness to withdrawal rate and film thickness to oxide content can be derived. Two simple relationships are that the film thickness increases with increasing withdrawal rate and, for a given withdrawal speed, the film thickness increases with an increase in oxide content. When it comes to dip coatings, 50–500 nm coatings are easy to make but thicker coatings are more difficult. Repeated dipping builds up a thicker film, but drying the multiple coatings without peeling and cracking becomes a problem. Spin coating is a technique often used for planarizing films before photolithographic processing. In this case, the substrate, usually a silicon wafer, is placed on a spinner and rotated at perhaps 1000 RPM while the solution is dripped on the center of the substrate. As with dip coatings, a film between 50 and 500 nm results [41].

7.3.2 Fibers, sheets, and thick layers

As mentioned above, fibers can be drawn out of low-water-content solutions. The sol–gel process allows one to bait and draw a string of gel about the same diameter as the desired fiber directly from the solution [37]. The time and manner in which the system loses fluidity depend on composition and catalyst. Low-water solutions show a gradual increase in viscosity before losing fluidity, while high-water solutions lose fluidity more abruptly from a lower viscosity. This is visualized to mean that low-water solutions produce linear polymers and higher-water solutions produce crosslinked polymers or branched clusters. A small volume fraction of branched clusters will restrict flow when the same volume fraction of linear polymers will not. At a higher volume fraction the linear polymers will tangle and the low-water solution will gel. A problem with this method is that the shape of the fiber tends not to be cylindrical.

Another approach to making near-net-shaped fibers from a sol–gel process involves coating a shell or concentric shells of sol–gel glass on a filament of plastic [42]. This approach is referred to as the volatile host method, where a plastic filament, such as porous polypropylene, is used as a template. Due to the hydrophobic nature of the filament, it first has to be soaked in anhydrous alkoxides before subsequent coatings can be applied with partially hydrolyzed alkoxides. Different thicknesses can be obtained by multiple dipping or adjustment of the viscosity of the sol. Upon exposure to atmospheric moisture, the films gel. After gelling, a rigid, continuous shell runs the length of the filament. Once the shell has hardened, solvent and water can escape through interconnected pores. During heat treatment the interconnected pore structure remains open until the volatile host is eliminated and undesirable gases are removed.

The same concept of the volatile host can be used to make sheets of gel by coating gel onto sheets of plastic which can then be contoured. Other methods which have been tried include extruding gel through a slit [43] and floating gel on dense nonpolar organic liquids [44].

7.3.3 Microporous monoliths

Unlike the concentrated solutions used for fibers and thin films, dilute solutions are necessary for bulk objects based on casting and molding. The sol–gel process can be used to make a microporous preform that is nearly net shaped. This preform is called a monolith to refer to its continuity. Here monolith does not have the same sense as it does in monolithic ceramics where it is a term used to distinguish single-phase ceramics from composites. Monolithic gels can be formed from a colloidal sol or from an alkoxide solution. The main difference between colloidal gels and alkoxide gels is their pore structures. Alkoxide gels have small pores (<10 nm), while colloidal gels have bigger pores or voids between particles.

By either route, an almost necessary condition for sol–gel processing is the presence of water. With water come residual OH ions that have a deleterious effect primarily on optical properties and certainly on mechanical properties. That is why nonaqueous solvents, as well as fluorination or chlorination treatments, have been investigated to reduce the OH content [45]. Monolith fabrication is the most challenging demonstration of the sol–gel process yet. Despite obstacles, efforts persist in preparing larger and more complex sol–gel monoliths.

7.4 UNCONSOLIDATED GELS: AGING AND SYNERESIS

Having selected a geometry and designed the formulations accordingly, there are several further steps common to monoliths, films, and fibers. First of all, the gels must be dried. Drying of gels has been the subject of many studies [46–48], largely because it is viewed as a problem, even a barrier to the successful use of the sol–gel process. To illustrate why so much time and effort has been devoted to drying, consider the alkoxide route to silica. Starting with alkoxides, TEOS is less than 30% by weight silica and the hydrolyzed solution is even less, around 5 to 10%. From the time the solution is prepared to the time it gels, there is about 50% weight loss. Then, as the gel dries there is another reduction in weight by one half. To go along with the weight loss there is about a 70% volume reduction, most of which occurs by syneresis before any evaporation occurs. The weight loss and volume reduction are even more astounding in titania and alumina.

On the one hand, colloidal gels are said to be easily dried and less likely to crack, but their porosity may not be appropriate for nanocomposites. On the other hand, alkoxide gels are more difficult to dry than colloidal gels, largely because of the very same small pores that are ideal for nanocomposites. The goal is to have nanostructured materials and crack-free gels at the same time. Several drying treatments have been developed. One possibility is aerogels that are dried in an autoclave by hypercritical techniques [46]. That is the solvent is removed above its critical point. The resulting gel is about 10% dense and shows no shrinkage. The simpler case is xerogels that are dried by natural evaporation. Xerogels are 60% dense and have reductions 40 to 70% in volume [47]. Xerogels can be prepared with the addition of a drying control chemical additive (DCCA) to the solution (such as formamide or oxalic acid) [48]. In fact, drying crack-free monoliths is not an insurmountable problem. The efforts made to dry monoliths are rewarded when it comes to rigid hosts for nanocomposites (see section 7.5).

Unlike monoliths, films can be dried quickly in air because of the one thin dimension. It has been shown repeatedly that all shrinkage is taken up in the thin dimension and not in the plane of the substrate as long as the thickness is below 1 micron. Yet, the films remain adherent and continuous and maintain complete surface coverage.

Following drying and shrinkage comes the heat treatment. It would be inaccurate to say that ceramics have been formed at room temperature by merely reacting, gelling, and drying the solutions. In this state, the materials have many of the characteristics of the corresponding ceramic oxide, but they are more or less porous. Water and solvent escape through interconnected pores which remain open at the surface until the gels are fired to temperatures well above 600 °C. When needed to fulfill the requirements of a specific application, the hard gel is heated to various degrees of collapse. The microporosity in silica xerogels is not removed entirely until 1000 °C.

7.5 CONSOLIDATED GELS: SINTERING

When the goal is a pore-free dense oxide, the final stage of processing is sintering. Many studies of sintering and densification of dried gels have been carried out [28, 29]. Physical properties of gel derived ceramics have been compared to the properties of conventional polycrystalline ceramics. The properties in most cases are equivalent. In some rare cases where the properties are highly dependent on crystal size, such as the case for ferroelectricity, some reports are now probing the consequences of using nanostructured ceramics [49]. In cases where grain boundaries control the properties, there are important findings which are reported elsewhere in this book.

In the more general case, the high surface area of gels is looked at as a high driving force for sintering, so sintering is likely to occur at lower temperatures than in conventional powder compacts. Consider the two processing routes illustrated in figure 7.5 where the conventional powder is compacted and densified (left) and the powder-free preform is made pore-free (right). The question is whether or not the benefit of the powder-free route is evident in the end? In fact, the argument, at least for ceramics, for building from the molecule up as opposed to starting with comminuted powders is valid only when the molecule is so well designed that the product of building up the molecule cannot be achieved by conventional means. This argument takes on other components with many organic polymers being used for subsequent pyrolysis to ceramics [50].

Recapping the processing steps that are common to all geometries, they are: (I) dissolving and reacting the precursors; (II) forming and gelling the shapes; (III) drying; and (IV) firing to partial or full density. The dense gels should display all properties of the dense oxides—silica, titania alumina, or others—similar to conventional melted or sintered ceramics. The universal advantages of the sol–gel approach, according to this fundamental way of looking at the process, are simplicity and purity.

Considering that sintering is generally the last step in ceramic processing (excluding machining), it means that most gels are densified by sintering as well. This is the last step whether the compact was prepared from sol–gel powders or the compact was instead a continuous oxide skeleton resulting from a powder-free sol–gel process. This drive to remove porosity in the end from the dried gel preform produces a material similar to conventionally processed materials. The goals of high purity and uniform microstructure may be achieved in the sol–gel route, but the method with its many steps may seem somewhat indirect compared to well established conventional methods. The real challenge of the sol–gel process is to exploit the nanostructure aspects of the process to derive real benefits. For monolithic ceramics, the benefits may not be overwhelmingly evident at this time. For composites, there is a different story.

7.6 MATRICES FOR ACCESS TO THE NANOSTRUCTURE

In light of recent attention paid to nanostructure processing and nanostructured materials, a real opportunity exists for sol–gel processing in composite design and fabrication. The sol–gel process produces porous materials, with interconnected porosity, on a nanometer scale. The opportunity provided by sol–gel processing is that composites can be produced by filling the porosity using liquid infiltration, physical or chemical depositions, and chemical reactions

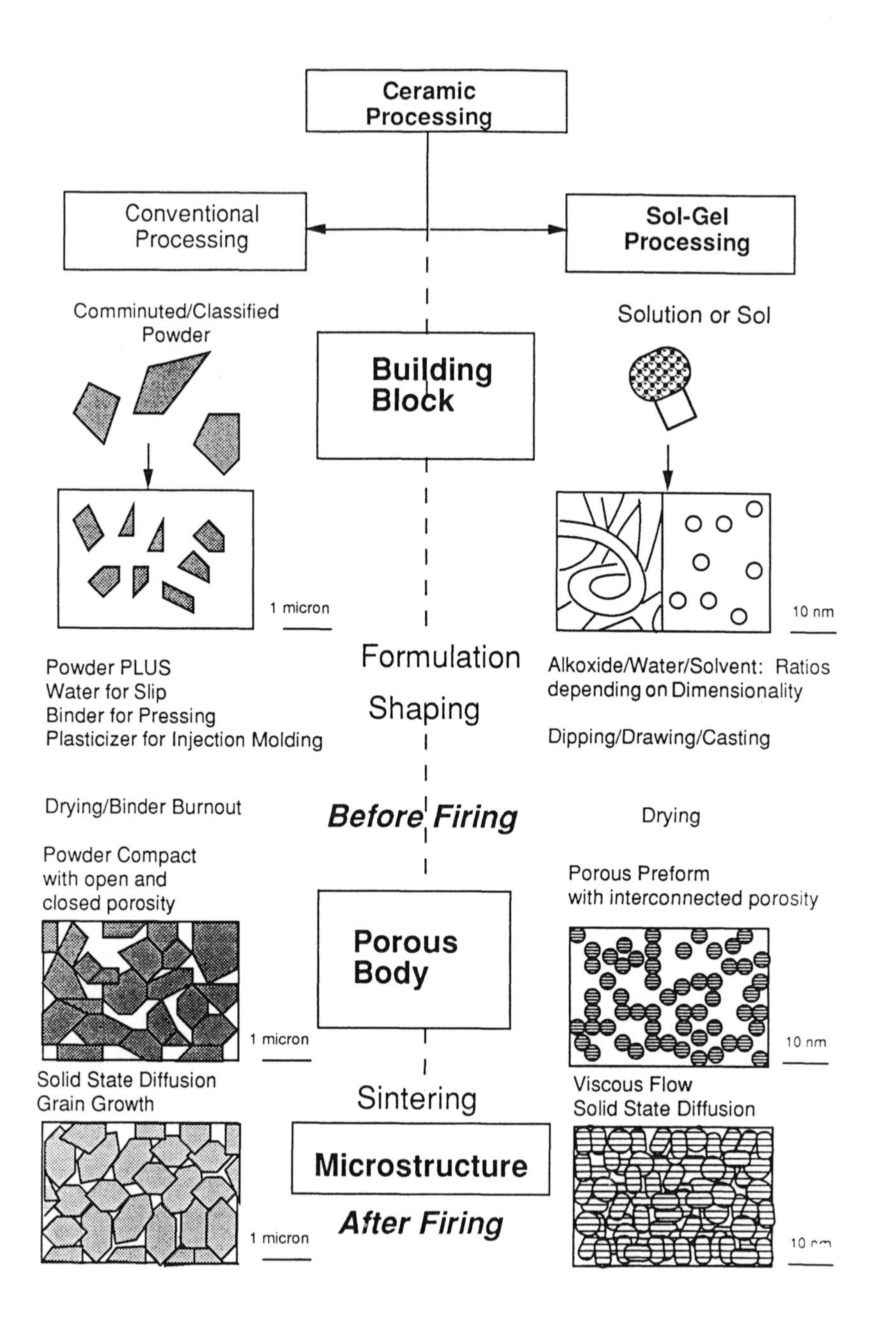

Figure 7.5. Comparison of steps in conventional processing of ceramics with those in sol–gel processing of ceramics.

such as pyrolysis or oxidation/reduction. Three possible approaches are shown schematically in figure 7.6 and described below.

7.6.1 *In situ* composites

The sol–gel process has been applied to *in situ* composites involving structural materials. In one case, short and long ceramic fibers were embedded in an alumina gel [51, 52]. In another case, a carbon containing composite was prepared *in situ* by pyrolyzing an organometallic gel precursor [53]. More recently, a class of materials called diphasic gels has been developed [54, 55]. These materials are a gel host for the precipitation of a second phase on an extremely fine scale.

The *in situ* composite may be formed in four steps. The first step of course is forming a homogeneous solution of the appropriate alkoxide precursors. The alkoxide solution is poured over the fibers and a thin film of liquid adheres to each fiber. Upon exposure to atmospheric moisture, the film gels. The solution is added repeatedly to build up a series of films. At this point, the gel is largely converted to oxide, and a rigid, continuous shell runs the length of the fiber. Enough solution then is added to fill in between the fibers.

The second step is drying of the matrix. Once the matrix has hardened, solvent and water must escape through interconnected pores. At the same time as the matrix is shrinking in all three dimensions, it must remain intact along the axis of the fiber. A third step is shrinkage of the matrix in the presence of reinforcements. Since densification by viscous sintering is only effective at the higher temperatures, it is conceivable that a stabilizing heat treatment can be designed which permits volume relaxation. After complete desiccation, little if any densification should occur at temperatures well below the glass transition and the interconnected pore structure remains open for toughness or thermal shock resistance.

The fourth step is complete densification. It is possible to densify by controlled heating, if necessary in vacuum. If the ultimate temperature is below the softening temperature for the matrix glass, the sintering can occur in essentially containerless conditions. That is the fiber composite does not need support during firing because of the high-viscosity matrix. This eliminates sources of contamination or surface abrasion. Hot pressing has also been used with remarkable success [56].

7.6.2 Volatile host method

Another approach for near-net-shaped processing requires an organic polymer as well as the inorganic polymer. The volatile host method mentioned previously for fibers involves a plastic filament dipped into an alkoxide solution such that a thin film of liquid adheres to the filament [42]. Upon exposure to atmospheric moisture, the film gels. The filament may be dipped repeatedly to build up a

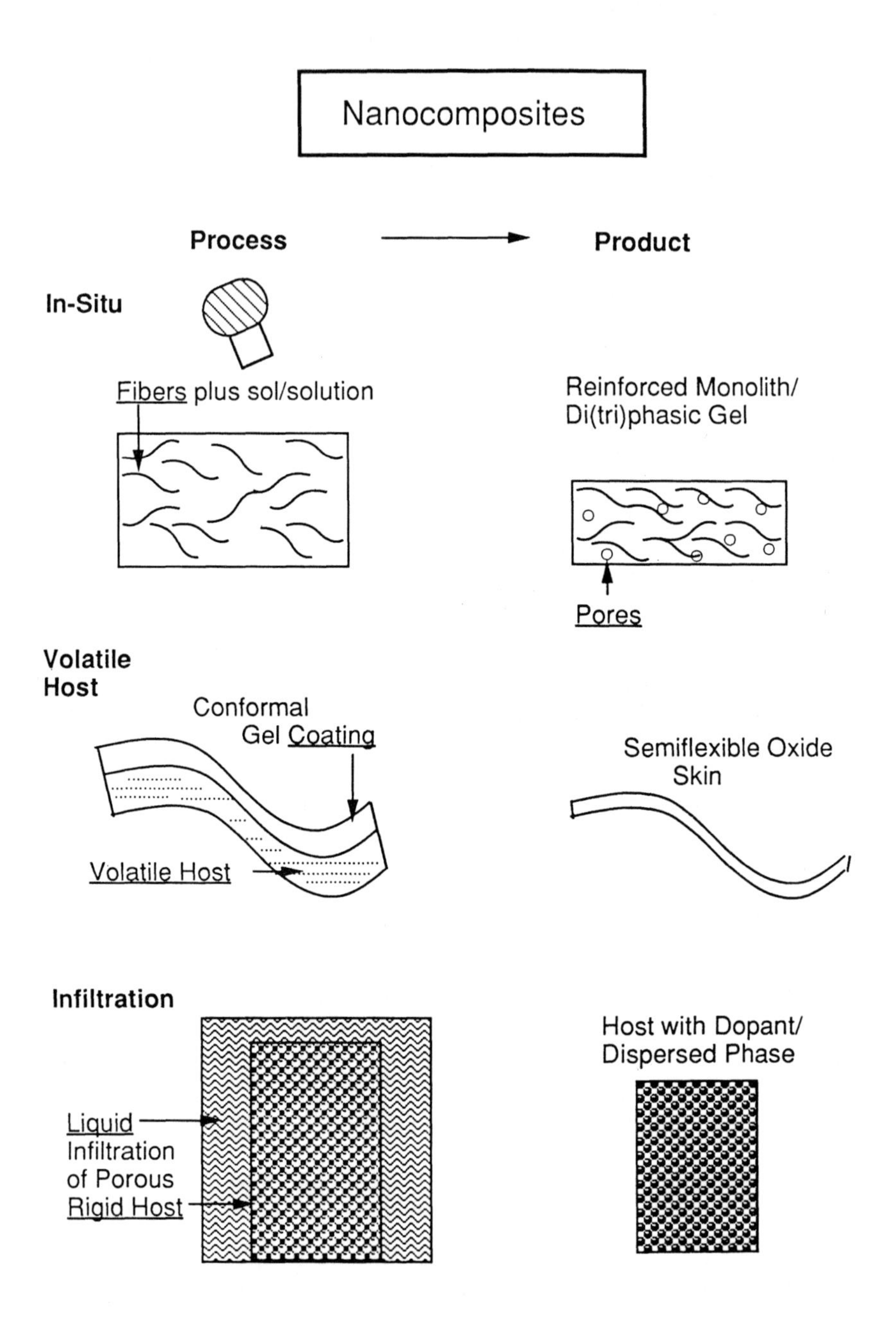

Figure 7.6. Schematic diagrams of nanocomposite fabrication schemes described in section 7.6.

series of films. Similar to the *in situ* composite fabrication, the second step is drying of the silica shell. The interconnected pore structure remains open until the volatile host is eliminated and the undesirable gases are removed. In the meantime, the open porosity can be used to facilitate exchange or infiltration treatments. The final step is the complete collapse of the shell at a temperature which is high enough to permit viscous sintering to full density. In addition, the volatile host, being flexible, can be used to facilitate conformal coatings on complex shapes.

In some cases, the organic polymer is left in the ceramic matrix when the composite is intended for use at low temperatures. Several name have been coined for such materials, including 'ceramers' [57] and 'ormosils' [58].

7.6.3 Infiltrated composites

The third approach shown in figure 7.6 involves infiltration [58]. In this case the objective is to use sol–gel processing to create a nanoporous matrix which can be infiltrated by a second phase. Building on the concept of the sol–gel process, the product of the sol–gel process is by its very nature two phase. At the time the oxide polymer condenses, the second phase is the rejected solvent. When the solvent is removed during drying, the second phase is the interconnected porosity. The channels where gas replaces liquid are typically 2 to 20 nm, uniform in cross section and narrowly distributed in size. In many dried sol–gel oxides, their porosity does not interfere with transmission of visible light despite measured surface areas in the range 200–500 $m^2 g^{-1}$. The porosity is on a scale which does not scatter visible light, allowing most of the light to be transmitted. This presents a new way of looking at sol–gel materials as a rigid transparent host [59]. The gel host may be a thin film or fiber, as well as a bulk shape. The appeal of the bulk shape is that it can be infiltrated with second phases, even metals [60, 61], molded to near-net shape, and can be cut and polished.

Another infiltration approach has been taken to prepare so-called hybrid gels incorporating organic and inorganic components. These methods which mix the components on the nanometer scale include the infiltration of previously formed oxide gels with monomers [62], reacting of alkoxide and organic monomers [63], or organic polymers in mutual solvents [64]. The products of these processes are truly nanocomposites, also referred to as interpenetrating networks (IPNs) [65]. Once again, sol–gel processing proves that it is a chameleon technology, able to play a role in the expanding field of IPNs, as well as in nanostructured materials.

7.7 PROSPECTS FOR THE SOL–GEL PROCESSING OF NANOSTRUCTURED MATERIALS

Among the various guises of sol–gel processing, the one that should stand out at this point is its use in nanocomposites. In the long run, the projected advantages of the sol–gel approach in nanocomposite fabrication are the simple processing steps, the flexibility of solution chemistry, the low-temperature treatments, and small investment in equipment. There is a need for innovative thinking to adapt sol–gel processing to composite fabrication. A few methods have been described here that may contribute to their eventual success.

REFERENCES

[1] Sakka S 1985 *Am. Ceram. Soc. Bull.* **64** 1463
[2] Johnson D W 1985 *Am. Ceram. Soc. Bull.* **64** 1597
[3] Colomban Ph 1989 *Ceram. Int.* **15** 23
[4] Dislich H 1986 *J. Non-Cryst. Solids* **80** 115
[5] Yoldas B E 1979 *J. Mater. Sci.* **14** 1843
[6] Brinker C J, Clark D E and Ulrich D R (ed) 1984 *Better Ceramics through Chemistry (Mat. Res. Soc. Symp. Proc. 32)* (New York: Elsevier)
[7] Brinker C J, Clark D E and Ulrich D R (ed) 1986 *Better Ceramics through Chemistry II (Mat. Res. Soc. Symp. Proc. 73)* (Pittsburgh, PA: Materials Research Society)
[8] Brinker C J, Clark D E and Ulrich D R (ed) 1988 *Better Ceramics through Chemistry III (Mat. Res. Soc. Symp. Proc. 121)* (Pittsburgh, PA: Materials Research Society)
[9] Zelinski B J J, Brinker C J, Clark D E, and Ulrich D R (ed) 1990 *Better Ceramics through Chemistry IV (Mat. Res. Soc. Symp. Proc. 180)* (Pittsburgh, PA: Materials Research Society)
[10] Hampden-Smith M J, Klemperer W G and Brinker C J (ed) 1992 *Better Ceramics through Chemistry V (Mat. Res. Soc. Symp. Proc. 271)* (Pittsburgh, PA: Materials Research Society)
[11] Hench L L and Ulrich D R (ed) 1984 *Ultrastructure Processing of Ceramics, Glasses and Composites* (New York: Wiley)
[12] Hench L L and Ulrich D R (ed) 1986 *Science of Ceramic Chemical Processing* (New York: Wiley)
[13] Mackenzie J D and Ulrich D R (ed) 1988 *Ultrastructure Processing of Advanced Ceramics* (New York: Wiley)
[14] Uhlmann D R and Ulrich D R (ed) 1992 *Ultrastructure Processing of Advanced Materials* (New York: Wiley)
[15] Hench L L and West J K (ed) 1992 *Chemical Processing of Advanced Materials* (New York: Wiley)
[16] Roy R 1987 *Science* **238** 1664
[17] Ohno M, Yamada T and Kurokawa T 1985 *J. Appl. Phys.* **57** 2951
[18] Klein L C 1990 *37th Sagamore Army Materials Research Conf. Proc.* ed D J Viechnicki (Watertown, MA: US Army) pp 142
[19] Winter R, Hua D-W, and Thiyagarajan P 1989 *J. Non-Cryst. Solids* **108** 137
[20] Long G G and Krueger S 1989 *J. Appl. Crystallogr.* **22** 539
[21] Brinker C J 1988 *J. Non-Cryst. Solids* **100** 31
[22] Lours T, Zarzycki J, Craievich A F and Aegerter M A 1990 *J. Non-Cryst. Solids* **121** 216
[23] Kumar K-N P, Keizer K, Burggraaf A J, Okubo T, Nagamoto H and Morooka S 1992 *Nature* **358** 48
[24] Barringer E A and Bowen H K 1982 *Commun. Am. Ceram. Soc.* **65** C199

[25] Mukherjee S P 1980 *J. Non-Cryst. Solids* **42** 477

[26] Bradley D C, Mehrotra R C and Gaur D P 1978 *Metal Alkoxides* (London: Academic)

[27] Klein L C (ed) *Sol–Gel Technology for Thin Films, Fibers, Preforms, Electronics and Speciality Shapes* (Park Ridge, NJ: Noyes)

[28] Klein L C 1985 *Annu. Rev. Mater. Sci.* **15** 227

[29] Brinker C J and Scherer G W 1990 *Sol–Gel Science* (Boston, MA: Academic)

[30] Shafer M W, Awschalom D D and Warnock J 1987 *J. Appl. Phys.* **61** 5438

[31] Yoldas B E 1975 *J. Mater. Sci.* **10** 1856

[32] Henry M, Jolivet J P and Livage J 1992 *Structure and Bonding 77* (Berlin: Springer) pp 153

[33] Livage J, Henry M and Sanchez C 1988 *Prog. Solid State Chem.* **18** 259

[34] Hay R S 1993 *J. Mater. Res.* **8** 578

[35] Klinowski J 1988 *Annu. Rev. Mater. Sci.* **18** 189

[36] Jones R W 1989 *Fundamental Principles of Sol–Gel Technology* (London: Institute of Metals)

[37] Kozuka H, Kuroki H and Sakka S 1988 *J. Non-Cryst. Solids* **100** 226

[38] Klein L C 1991 *Thin Film Processs II* ed J L Vossen and W Kern (New York: Academic) p 501

[39] Brinker C J, Frye G C, Hurd A J and Ashley C S 1991 *Thin Solid Films* **201** 97

[40] Brinker C J, Hurd A J, Schunk P R, Frye G C and Ashley C S 1992 *J. Non-Cryst. Solids* **147&148** 424

[41] Parill T M 1992 *J. Mater. Res.* **7** 2230

[42] de Lambilly H and Klein L C 1986 *SPIE* **683** 98

[43] Sakka S, Kamiya K, Makita K and Yamamoto Y 1988 *J. Non-Cryst. Solids* **63** 223

[44] Gallagher D and Klein L C 1986 *J. Colloid Interface Sci.* **109** 40

[45] Rabinovich E M and Wood D L 1986 *Better Ceramics through Chemistry II (Mat. Res. Soc. Symp. Proc. 73)* ed C J Brinker, D E Clark and D R Ulrich (Pittsburgh, PA: Materials Research Society) p 251

[46] Prassas M, Phalippou J and Zarzycki J 1984 *J. Mater. Sci.* **19** 1656

[47] Scherer G 1988 *J. Non-Cryst. Solids* **100** 77

[48] Wallace S and Hench L L 1984 *Better Ceramics through Chemistry (Mat. Res. Soc. Symp. Proc. 32)* ed C J Brinker, D E Clark and D R Ulrich (New York: Elsevier) p 47

[49] Dey S K and Zuleeg R 1990 *Ferroelectrics* **108** 1643

[50] Wynne K J and Rice R W 1984 *Annu. Rev. Mater. Sci.* **14** 297

[51] Lannutti J J and Clark D E 1984 *Better Ceramics through Chemistry (Mat. Res. Soc. Symp. Proc. 32)* ed C J Brinker, D E Clark and D R Ulrich (New York: Elsevier) p 369

[52] Lannutti J J and Clark D E 1984 *Better Ceramics through Chemistry (Mat. Res. Soc. Symp. Proc. 32)* ed C J Brinker, D E Clark and D R Ulrich (New York: Elsevier) p 375

[53] Chi F K 1983 *Ceram. Eng. Sci. Proc.* **4** 704

[54] Hoffman D, Roy R and Komarneni S 1984 *Mater. Lett.* **2** 245

[55] Hoffman D, Komarneni S and Roy S 1984 *J. Mater. Sci. Lett.* **3** 439

[56] Qi D and Pantano C G 1988 *Ultrastructure Processing of Advanced Ceramics* ed J D Mackenzie and D R Ulrich (New York: Wiley) p 635

[57] Huang H H, Orler B and Wilkes G L 1985 *Polym. Bull.* **14** 557

[58] Schmidt H and Wolter H 1990 *J. Non-Cryst. Solids* **121** 428
[59] Klein L C 1993 *Annu. Rev. Mater. Sci.* **23** 437
[60] Chatterjee A, Datta A, Giri A K, Das D and Chakravorty D 1992 *J. Appl. Phys.* **72** 3832
[61] Nogami M, Tohyama Y, Nagasaka K, Tokizaki T and Nakamura A 1991 *J. Am. Ceram. Soc.* **74** 238
[62] Abramoff B and Klein L C 1990 *SPIE* **1328** 241
[63] Giannelis E P 1992 *J. Miner. Met. Mater. Soc.* **44** 28
[64] Landry C J T, Coltrain B K and Brady B K 1992 *Polymer* **33** 1486
[65] Ellsworth M and Novak B M 1991 *J. Am. Chem. Soc.* **113** 2756

Chapter 8

Consolidation of nanocrystalline materials by compaction and sintering

M J Mayo, D-J Chen and D C Hague

8.1 INTRODUCTION

For many years nanocrystalline materials have been produced as curious by-products of certain manufacturing procedures. The fabrication of thin films by electron beam evaporation and other forms of physical vapor deposition [1–3], the melt spinning of ribbons [4], the electrodeposition of zinc coatings [5], the dehydration and firing of sols (liquid mixtures) to make ceramic fibers and bubbles [6], and the cold drawing and heat treatment of superconductor wires [7] have all yielded solids with sub-100 nm grain sizes. However, not until the recent surge of interest in the properties of such ultrafine-grained materials has a concerted effort been made to design manufacturing procedures specifically with the goal of producing nanocrystalline solids reliably in a variety of bulk sizes and geometries.

Several methods have emerged as promising candidates for the production of bulk, ultrafine-grained solids. One, which is applicable to metals and some intermetallics, is severe mechanical working. Numerous experiments which force metals to undergo equal channel angular pressing [8], high-energy ball milling [9, 10], or excessive friction and wear [11, 12] have been shown to yield nanocrystalline grain sizes in the heavily deformed workpiece. For a single phase material, the refinement of the grain size during severe working apparently occurs gradually, by the ordering and condensation of dislocation networks into ever smaller and tighter bands, or subgrain boundaries, that eventually become grain boundaries [9]. The use of mechanical working to produce nano-grained microstructures is described more fully in chapter 5.

Another commonly used technique for the production of ceramic as well as metallic solids is sintering. In this technique, a powder composed of individual metal or ceramic particles, each less than 50 nm in diameter, is first compacted into a raw shape (often called a green body, especially in ceramics terminology). This powder compact, or green body, is then exposed to elevated temperature in order for densification to occur by the diffusion of vacancies out of pores. As vacancies leave the pores and the pores shrink, the sample densifies. In concept, the pressureless sintering of fine particles is an easy method of preparing nanocrystalline metals and ceramics: one simply packs together an ultrafine powder and heats it. In practice, however, it is often very difficult to control grain growth at the same time as one is attempting to stimulate densification. Thus, the early attempts at producing nanocrystalline solids by the sintering of particles often resulted in materials that were fully dense, but with micron-sized grains, or, conversely, nano-grained but only partially dense [13]. This chapter attempts to identify some of the issues involved in the production of nanocrystalline solids by sintering, particularly with respect to the (sometimes incompatible) goals of minimizing grain growth while maximizing densification. As shall be seen, nanocrystalline ceramics are much more difficult to prepare by sintering than are nanocrystalline metals; thus, many of the intricacies of sintering protocol and theory are more relevant to the production of the former than the latter.

8.2 DRY COMPACTION OF NANOCRYSTALLINE PARTICLES

The first step in the sintering of nanocrystalline solids is the production of a suitable powder compact. For metals, which are capable of deforming plastically, the application of pressure during the powder compaction step will force metal particles to extrude into nearby pores and fill most of the available empty space [14]. Consequently, with high enough compaction pressures, it has been shown that one can produce 96–98% dense specimens directly from compacted nanocrystalline metal particles—that is, without subsequent high-temperature exposure [15–18]. Nevertheless, the mechanical strength of a three-dimensional structure of interlocking particles is inferior to the mechanical properties of a homogeneous solid with true grain boundaries, as several researchers have pointed out [15]. Thus, even though the density–grain size combination after compaction may appear ideal, some sintering is still necessary to achieve maximum mechanical strength.

Ceramic powder particles are not capable of plastically deforming during cold compaction. As a result, the compact densities that are observed in ceramics are far lower than in metals, and sintering is both required and well studied. An interesting observation is that the ultrafine size of nanocrystalline ceramic particles does not always reduce their ability to be compacted (see figure 8.1). It is generally accepted that as particles become progressively finer (and the total number of interparticle contacts inside the die becomes progressively larger), there arises a tremendous frictional resistance to the applied compaction force, generated by the need for many particles in the die to simultaneously slide past each other [19]. For some reason, this appears not always to be the case, and

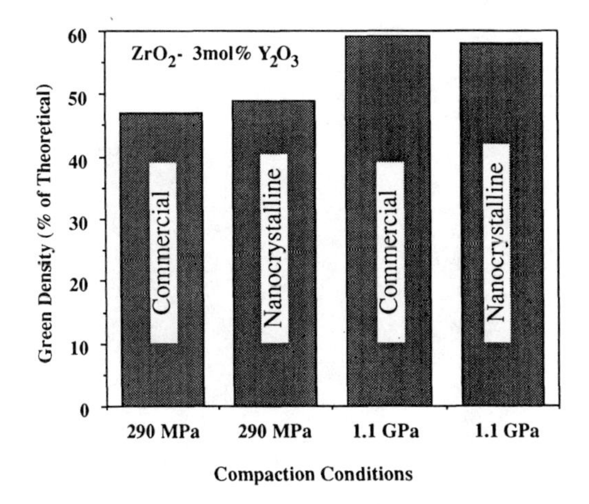

Figure 8.1. Post-compaction (green) densities of nanocrystalline (13 nm) and commercial submicron (Tosoh TZ-3Y brand, 0.17 μm) ZrO_2–3 mol% Y_2O_3 powders. Compaction pressures are indicated.

certain nanocrystalline ceramic powders appear to compact as easily as their submicron counterparts (figure 8.1).

It is interesting to note that an early concern of nanocrystalline researchers—surface contamination by atmospheric adsorbates—also appears not to hinder the compaction of the nanocrystalline powder. From geometry, a monolayer of 2 Å diameter surface adsorbates might be expected to contribute as much as an extra 24% to the effective volume of a 5 nm particle; this surface layer would then cause the total packing density to decrease noticeably (on the order of 20%) when comparing nanocrystalline (<100 nm) to submicron and micron-sized powders. As shown earlier (figure 8.1), this decrease is not observed. It is possible that adsorbates are locally displaced when adjoining particles make point-to-point contact, or even that adsorbates are not as ubiquitous as often assumed.

One parameter which dramatically affects packing density, and which cannot be overlooked in working with undeformable (e.g. ceramic) powders, is the arrangement of the particles within the starting powder. There is an enormous difference in the compaction (and sintering) behavior of powders in which each individual powder particle is free and independent of its neighbors, and those in which the primary particles, or crystallites, are chemically or physically bonded together to form larger units, termed agglomerates or aggregates. An example of an agglomerated powder is shown schematically in figure 8.2. Note that an agglomerated powder may not pack especially well, particularly if it contains loose arrangements of primary particles within the agglomerates as well as large interagglomerate pores. On the other hand, a powder that contains large, but quite dense agglomerates (such as those produced by the melting, welding, or coalescence of particles synthesized at high temperatures) can actually be

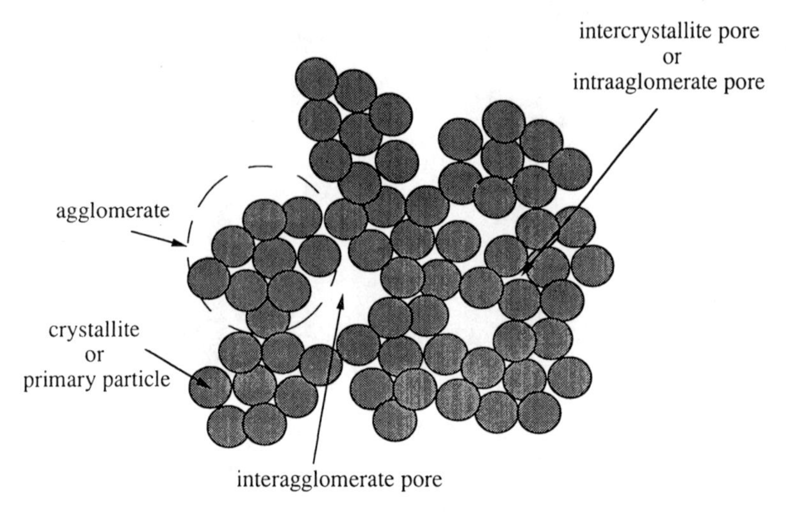

Figure 8.2. Schematic diagram of an agglomerated powder.

compacted to densities greater than one would predict based on the theoretical densest close packing of spheres (e.g. 74% for close packing [20] versus the observed value of 75% for strongly aggregated nanocrystalline TiO_2 [13]). Thus, the compaction behavior of a ceramic powder (for which the starting powder structure cannot be plastically deformed or easily rearranged) is heavily dictated by the degree and nature of agglomeration originally present in the powder. For the powders in figure 8.1, special effort was made to compare powders with morphologically similar particles. Future studies of the differences in compaction behavior between nanocrystalline and conventional powders will likewise require sets of powders which are morphologically similar and differ only in size.

A common method for detecting agglomeration is to measure the pore size distribution of the powder or its compact [21–24]. A non-agglomerated powder usually contains a single population of pores, all a certain fraction (typically $\frac{1}{2}$ to $\frac{1}{5}$) of the primary particle size. An agglomerated powder, on the other hand, typically displays a bimodal distribution of pores, since there are both small intraagglomerate pores and large interagglomerate pores contained within the powder [21–24]. By extension, a powder containing a wide range of agglomerate sizes, and thus a corresponding wide range of interagglomerate pore sizes, would exhibit a broad pore size distribution. A schematic diagram showing the pore size distributions of an agglomerated versus non-agglomerated powder is given in figure 8.3. Experimentally, pore size distributions can be obtained by either mercury intrusion or gas adsorption; the methodology and theory for each of these techniques is described in [25].

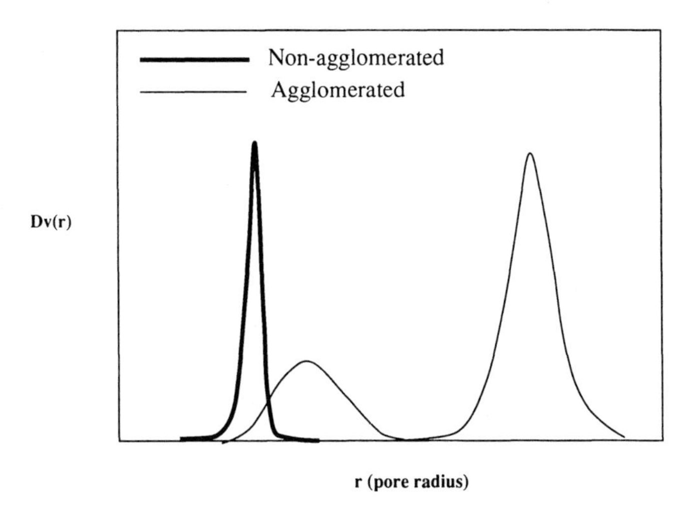

Figure 8.3. Schematic diagram of the pore size distribution for agglomerated (narrow line) versus non-agglomerated (thick line) powders. The agglomerated powder contains small intercrystallite pores as well as large interagglomerate pores and therefore has a bimodal pore size distribution. The non-agglomerated powder has a single population of pores, located at small pore radii, corresponding to the intercrystallite pores only. The ordinate axis, labelled as $Dv(r)$, represents the negative derivative of pore volume with respect to pore radius, calculated for a unit weight of powder. Thus, the area under the distribution represents the total volume of open pores measured in a sample of unit weight. Note that $Dv(r)$ scales with, but is not equal to, the number of pores present.

8.3 WET COMPACTION OF NANOCRYSTALLINE PARTICLES

Because particles (even non-agglomerated ones) can slide and rearrange much more easily in a liquid medium, it is often advantageous to fabricate powder compacts, particularly ceramic powder compacts (also known as green bodies) by wet-processing techniques. Extremely fine-structured, homogeneous green-body microstructures have been shown to result from pressure filtration, centrifugation, tape casting, and slip casting of submicron ceramic particles [26–29]. These procedures are beginning to be employed in the processing of nanocrystalline ceramics as well. (Note: metal particles are typically not wet processed due to their relatively good behavior in dry pressing and their tendency to form oxide layers during exposure to aqueous media.) As a dramatic illustration of the advantages offered by wet processing, a nanocrystalline zirconia powder suspended in a water–nitric acid solution of pH 2 and then centrifuged can achieve 100% density with an 80 nm grain diameter after sintering at 1100 °C [30]. Such a density–grain size combination can be achieved for equivalent dry-pressed zirconia samples only when they are compacted at 1.2 GPa prior to sintering [31–34].

Considering the magnitude of forces that would be required to achieve 1.2 GPa of pressure in any conventionally sized sample or component, one quickly sees the advantages of wet processing. Nevertheless, because dry pressing of nanocrystalline powders is the easier of the two procedures, it is the one for which we have the most information to date.

8.4 IDEAL DENSIFICATION DURING PRESSURELESS SINTERING

After compaction, a nanocrystalline powder can be densified at elevated temperatures without the application of an external pressure. This technique, known as pressureless sintering, has as its driving force the reduction of surface energy associated with the surfaces of the many internal pores. Thus densification can be thought of as the reduction of pore surface area (and, consequently, pore volume) in a powder compact. For metal powder compacts, which can be directly compacted to densities greater than 90% [15–18] (due to the ability of the metal particles to deform during compaction), the amount of porosity which has to be eliminated is very small. One simply has to sinter at the lowest possible temperature for a time sufficient to remove the residual porosity and establish coherent grain boundaries. The completion of the sintering process can be detected by monitoring the sample hardness. For metals with grain sizes $\geq$10 nm, the hardness should initially increase with sintering, as the microstructural characteristics of a true solid are established. If hardness begins to decrease with increased sintering, then only grain growth is occurring, and increased sintering will prove unproductive [15]. Unfortunately, more exact details of the sintering process have not yet been well studied for the case of nanocrystalline metals.

Because nanocrystalline ceramics compact to much lower densities than do nanocrystalline metals, substantial sintering is required to achieve reasonable densities in these materials. Consequently, the sintering process has been studied in more depth. To date, all evidence suggests that the densification of nanocrystalline ceramics during sintering follows rules previously established for the sintering of conventional ceramics. There are, for instance, three stages of densification [35, 36]: in the first stage, neck growth occurs at the contact points between adjacent particles. In the second stage, the ceramic has a sponge-like structure consisting of an extensive network of tubular pores open to the outside surface of the ceramic sample. Most of the actual densification takes place during this second stage, as the tubular pores shrink to smaller and smaller diameters. Once these pores are very narrow relative to their length, they become unstable and pinch off to form isolated, closed, spherical pores. The elimination of these closed pores then comprises the final stage of sintering. The three stages of sintering are shown schematically in figure 8.4.

Since most (70–80%) of the densification takes place during second stage sintering, by the shrinkage of open pores, it is instructive to examine this process in more detail. Coble [37] analyzed densification during second-stage sintering

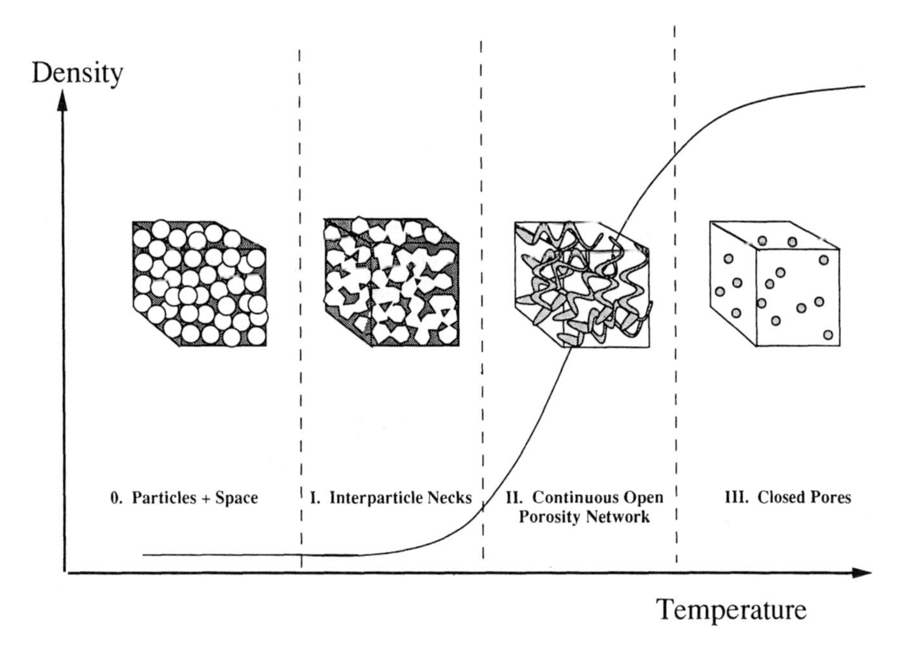

Figure 8.4. The three stages of sintering (I–III), starting from a powder compact (0).

in the 1960s based on the vacancy flux between a cylindrical pore (vacancy source) and the surrounding grain boundaries (vacancy sinks). As vacancies move from the pores to the grain boundaries (where they are either annihilated or transported to the specimen surface), the pores shrink in diameter and the sample densifies. For nanocrystalline ceramics, a modified version of Coble's original equation appears to hold [38]; it can be written as follows

$$\frac{1}{\rho(1-\rho)}\left(\frac{\partial\rho}{\partial t}\right)_T = A\frac{D_0\exp(-Q/RT)}{G^3}\frac{1}{r} \tag{8.1}$$

where ρ is the instantaneous (current) sample density, relative to the theoretical density (ρ is on a scale of 0 to 1), $(\partial\rho/\partial t)_T$ is the sample densification rate (slope of a density–time curve at constant temperature), A is a material constant, G is the instantaneous grain diameter, r is the instantaneous pore radius (it is assumed that all pores are of the same size in this simplified derivation), and, finally, $D_0\exp(-Q/RT)$ is the diffusivity, where Q is the activation energy for diffusion, R is the gas constant, and T is the absolute temperature.

As compared to Coble's expression for densification as it is usually written, equation (8.1) differs by the factor $1/\rho(1-\rho)$, which is introduced in order to convert the left-hand side from 'densification rate' to the somewhat more mathematically satisfying 'rate of elimination of pore volume per unit pore volume', since densification is, after all, the process of eliminating pores. (In other works, one often sees $(\partial\rho/\partial t)_T$ normalized by $1/\rho$, but this expression translates to 'rate of elimination of pore volume per unit volume of sample'.) Also, the $1/r$ term is also stated explicitly. Although the $1/r$ term does exist in Coble's original derivation [37], it is sometimes ignored and subsumed into the $1/G^3$ term when analyzing the densification of conventional ceramics. The rationale for incorporating $1/r$ into $1/G^3$ (and then compensating by raising the power of the latter term to $1/G^4$) has been that samples with smaller grains tend to have smaller pores anyway, so the two entities scale with each other and can be merged into a single term. This simplification may be true when comparing samples made of morphologically similar powders of dissimilar sizes. In that specific case, the packing arrangement of the two powders is identical, and the spaces between powder particles (which eventually become the pores) scale with the size of the particles themselves (which eventually become the grains). However, the assumption that r scales with G is not true, for instance, during the course of sintering a single powder. During sintering, the grains generally grow while the pores actually shrink. An inverse relationship between pore size and grain size therefore results in this case. The merging of the r and G terms also does a disservice to the densification analysis of a single powder consolidated by different methods (e.g. by dry pressing versus centrifugation) or even by the same method under somewhat different conditions (e.g. by dry pressing at two different pressures). In these cases, compacts containing particles or grains of the same size may well have pores of very different sizes; thus at the beginning of sintering, G may be the same in both cases, but r can be quite different. For all the above reasons, the modeling of densification in nanocrystalline ceramics appears to be more accurate when the $1/r$ term is explicitly included in equation (8.1).

Experimental evidence shows that equation (8.1) successfully describes the densification behavior of certain nanocrystalline ceramics during pressureless sintering. In particular, an analysis of a very weakly agglomerated ZrO_2–3 mol% Y_2O_3 powder with a starting particle size of $\approx$15 nm shows a densification rate that changes throughout sintering at 1050 °C in a manner consistent with the predictions of equation (8.1). For instance, figure 8.5 shows a reasonable correlation between $[1/\rho(1-\rho)](\partial\rho/\partial t)_T$ and $1/G^3$; however, only data from samples with similar pore sizes fall on the same line. A much stronger correlation can be obtained by considering the effect of pore size in addition to grain size and plotting $[1/\rho(1-\rho)](\partial\rho/\partial t)_T$ versus $(1/G^3)(1/r)$ (see figure 8.6). The latter relationship explains the differences in densification rates between submicron (commercial) and nanocrystalline ZrO_2–3 mol% Y_2O_3, as well as between nanocrystalline ZrO_2–3 mol% Y_2O_3 compacted to different green densities prior to sintering. For the commercial powder, its slow densification rate can be understood on the basis of its much larger grain and pore sizes. For the nanocrystalline powder, compaction to greater green densities causes no change in initial grain size but does reduce the starting pore radius. The smaller pores (e.g. 2.6 nm for a 58% dense green body versus 4 nm for a 50% dense green body) could be expected to accelerate the initial densification rates by about

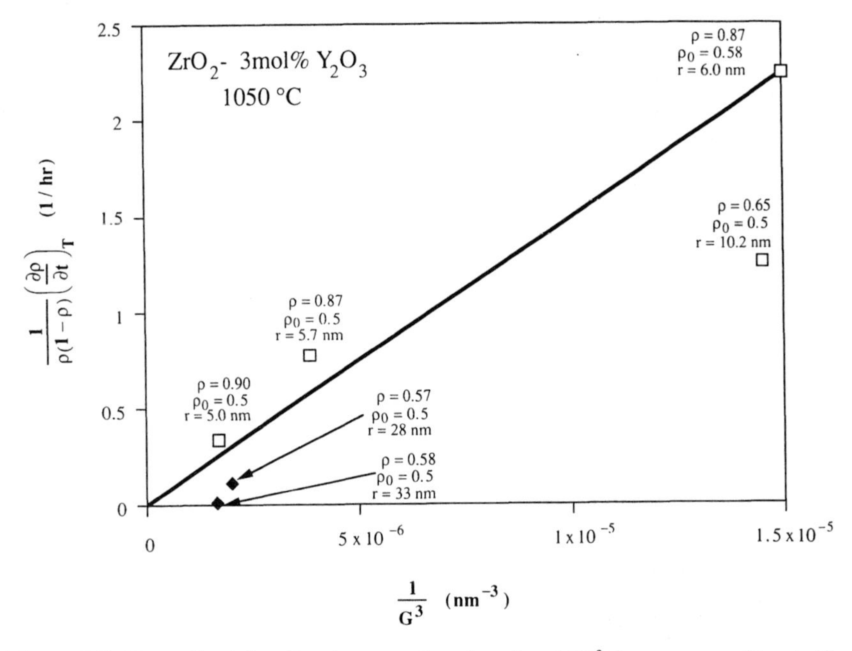

Figure 8.5. Normalized densification rate plotted against $1/G^3$ for nanocrystalline (white squares) and submicron (commercial Tosoh TZ-3Y brand, black diamonds) ZrO_2–3 mol% Y_2O_3 sintered to various densities at 1050 °C. Indicated are ρ, the density at which the data were taken, ρ_0, the initial (green) density of the sample, and r, the median pore radius of the sample when the data were taken. Here, G is the true grain diameter.

50% during initial sintering. However, the effect quickly becomes cumulative, as the faster densification rates allow one to achieve a given density at smaller grain sizes, and these smaller grain sizes in turn spur faster densification (at a given density). At later stages in the sintering process (compare the two data points for 87% dense samples in figure 8.6), it is primarily this difference in grain size (at a given density) that accounts for the difference in densification rates between samples which started at different green densities.

The impact of green density on densification rate can be seen more clearly in figure 8.7. Unfortunately, it is not yet possible to use equation (8.1) to predict densification curves like those in figure 8.7. Such an effort would require an integration of equation (8.1), which in turn would require knowing $G(t)$ and $r(t)$. The latter term is particularly difficult to assess, since the pore size (r) can coarsen, then shrink, during a typical sintering treatment. Given that pore size evolution is not yet well quantified, a thorough prediction of densification during sintering awaits further work.

Figure 8.7 exhibits another relevant feature of pressureless sintering, namely, that the densification rate decreases over time. This effect is largely

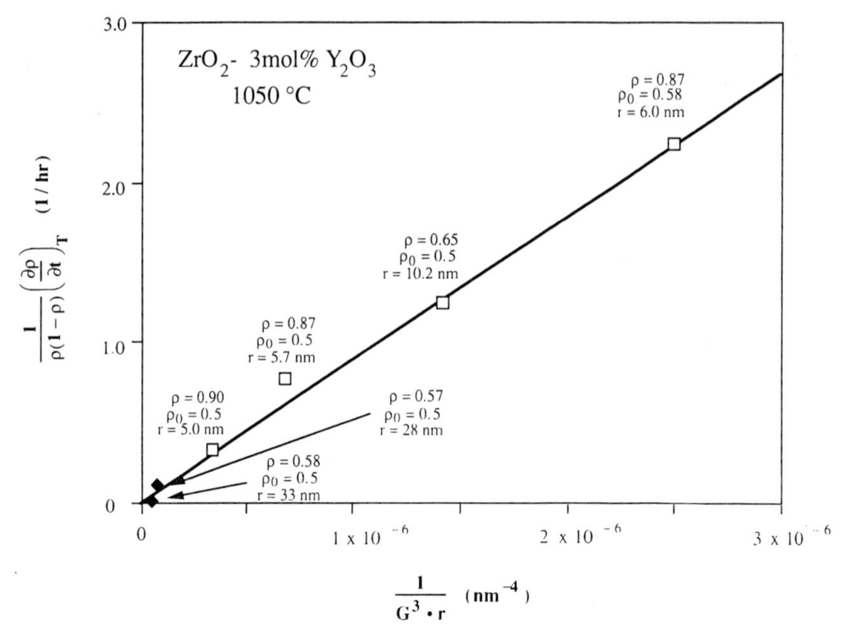

Figure 8.6. Normalized densification rate plotted against $(1/G^3)(1/r)$ for nanocrystalline and submicron (white squares) and submicron (commercial Tosoh TZ-3Y brand, black diamonds) ZrO_2–3 mol% Y_2O_3 sintered to various densities (ρ) at 1050 °C. Indicated are ρ, the density at which the data were taken, ρ_0, the initial (green) density of the sample, and r, the mean pore radius of the sample when the data were taken. Here, G is the true grain diameter.

due to grain growth: as the grains grow, the distance between vacancy sources (pores) and sinks (grain boundaries) increases, and it takes longer to move vacancies to their sinks and to achieve a given increment of densification. From equation (8.1) one also sees this fact mathematically: during sintering, the grains grow, and the $(1/G^3)$ term decreases quite rapidly. Even though the pores may shrink at the same time, leading to an increase in the $(1/r)$ term, this increase is never enough to offset the decrease in densification rate brought about by grain growth.

It should be noted that the densification rate can further decrease during sintering if there is a broad pore size distribution. While small pores are eliminated quickly (leading to an apparently fast densification rate), larger pores are eliminated more slowly (much slower densification rate), so the progressive elimination of smaller to larger pores will also manifest itself as a slowly decreasing densification rate.

From the above discussion and equation (8.1), it is clear that a small grain size, combined with a small pore size, is ideal for densifying at the fastest possible rates at a given temperature. As a result, nanocrystalline powders, with

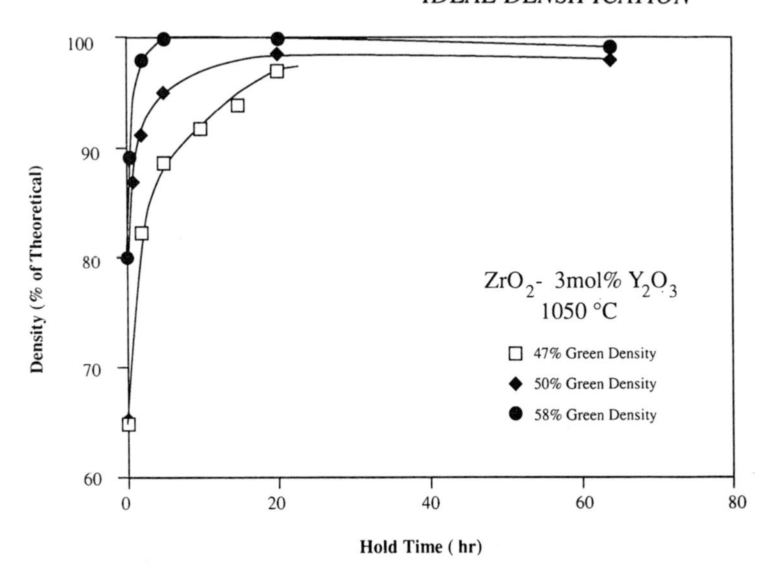

Figure 8.7. Density versus time for nanocrystalline ZrO_2–3 mol% Y_2O_3 during sintering at 1050 °C. A comparison between samples with differing green densities is shown (47% dense = compacted at 250 MPa; 50% dense = compacted at 300 MPa; 58% dense = compacted at 1.2 GPa). The slope of the above curves gives the densification rate at any point in time or density. Starting pore radii for the 50% dense and 58% dense samples are 4.0 and 2.8 nm, respectively. The reduction in initial pore size gives rise to an increase in the initial densification rate.

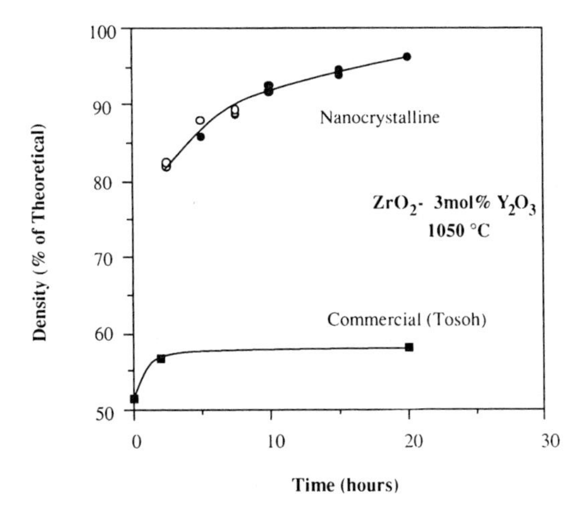

Figure 8.8. Comparison of the densification behavior of nanocrystalline (13 nm) ZrO_2–3 mol% Y_2O_3 and commercial (0.17 μm, Tosoh TZ-3Y brand) ZrO_2–3 mol% Y_2O_3 upon sintering at 1050 °C. Green densities of the powder compacts used for this comparison were 47% (nanocrystalline) and 50% (commercial) respectively.

their smaller grain and pore sizes, will sinter to much greater densities than coarser powders of the same composition, when subjected to the same sintering temperature (see figure 8.8). Conversely, the same density will generally be achieved by a nanocrystalline powder at much lower temperatures than its micron or submicron-sized cousins (see figures 8.9 and 8.10).

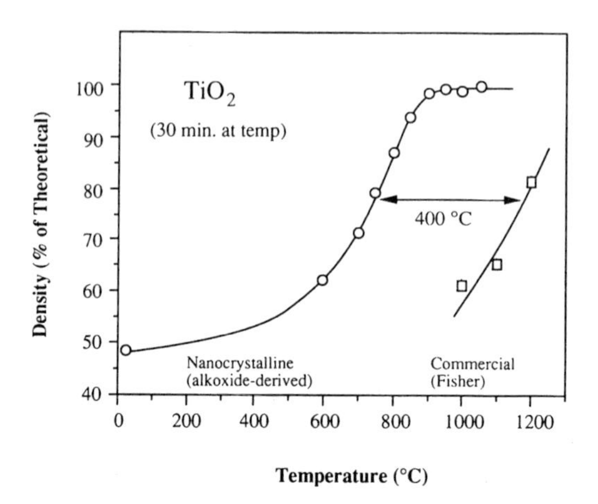

Figure 8.9. Comparison of the densification behavior of nanocrystalline (16 nm crystallites, 80 nm agglomerates) and commercial (Fisher Brand) TiO_2 as a function of temperature. Time spent at each temperature is 30 min.

It is obvious that the temperature has an enormous influence on densification rate, due to the exponential (Arrhenius) relationship between densification rate and temperature. The increase of density with temperature is often quite steep (see figures 8.9 and 8.10), so that a few degrees' miscalibration between furnace temperatures can yield quite different results from one study to the next. The activation energy for densification, obtained by plotting $\ln(\partial\rho/\partial t)_T$ versus $1/T$, yields some rather interesting results for the case of nanocrystalline TiO_2 [39]. In particular, there appear to be two stages of densification. At long times or low temperatures, the measured activation energy for densification (628 kJ mol^{-1}) is close to the known activation energy for grain boundary diffusion [39]. This result is expected, since the small grain size and many grain boundaries would tend to favor grain boundary diffusion over, say, lattice diffusion, as a way of moving vacancies from pores (or, equivalently, moving atoms *to* pores). However, in the initial stages of sintering—at low temperatures or short times—there appears to be another mechanism of densification, indicated by a different activation energy, $Q = 96$ kJ mol^{-1} [38]. Similarly low activation energies appear to control the initial stage densification of nanocrystalline ZrO_2–3 mol% Y_2O_3 [39]. To date, however, it is not clear what these activation energies represent, since the most likely explanation, surface diffusion, has not been reliably quantified for either the TiO_2 or ZrO_2–3 mol% Y_2O_3 system.

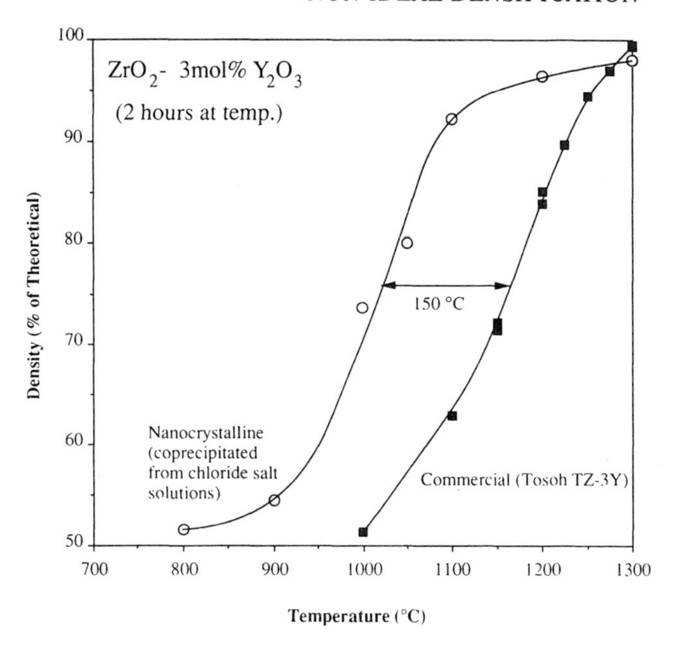

Figure 8.10. Comparison of the densification behavior of nanocrystalline (15 nm) commercial and (0.17 μm) ZrO_2–3 mol% Y_2O_3 as a function of temperature. Time spent at each temperature is 2 h.

8.5 NON-IDEAL DENSIFICATION DURING PRESSURELESS SINTERING

8.5.1 Agglomeration effects

In a non-agglomerated nanocrystalline powder, the pore sizes are quite small (typically $\frac{1}{2}$ to $\frac{1}{5}$ the crystallite size), so that densification proceeds quickly and at low temperatures. Sintering behavior can suffer dramatically if the nanocrystalline powder is agglomerated. Figure 8.11 shows the sintering behavior of three nanocrystalline TiO_2 powders, all of the anatase crystal structure, but with differing agglomerate (bold typeface) and crystallite (plain typeface) sizes. From figure 8.11 it is obvious that the agglomerate size, and not the crystallite size, has the dominant role in determining densification behavior. Recalling the earlier discussion of compaction behavior, this result is easily explained: the powders with larger mean agglomerate sizes contain larger interagglomerate pores at the beginning of sintering. These large pores have a small driving force for shrinkage (the large r yields a small $1/r$ term in equation (8.1). To compensate for the small driving force and to observe any reasonable rate of densification, the sintering temperature must be increased. The larger the agglomerates, the larger the interagglomerate pores, and the higher the sintering temperature required to reach a given density. The lowest sintering temperatures, as one would expect, are for powders which are nonagglomerated (NA in figure 8.11).

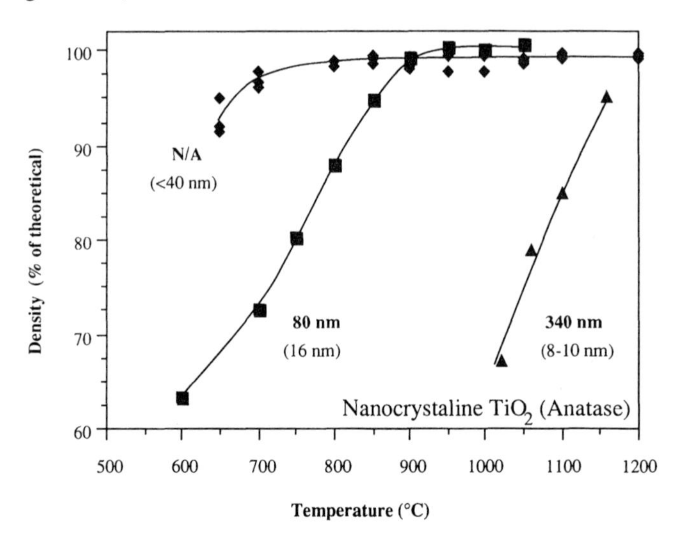

Figure 8.11. Densification behavior of nanocrystalline TiO_2 (anatase crystal structure) with three different agglomerate sizes. Note the larger the agglomerate size, the greater the sintering temperature. For the NA powder, the sintering time is 120 min (data from [66]); for the 80 nm agglomerate and 340 nm agglomerate powders, the sintering time is 30 min (data from [67] and [68] respectively). It should be noted that sintering time does not have a strong effect on the location of these curves [67].

Since grain growth is profoundly sensitive to temperature, the production of solids with <100 nm grains by pressureless sintering virtually requires NA starting powders and the low sintering temperatures they provide. This point cannot be overemphasized. For ceramic systems in which porosity controls grain growth, there is an additional reason why agglomerated powders undergo excessive grain growth during sintering. The small intercrystallite pores which pin the grain boundaries will disappear quickly, partway through the sintering process. When the constraint on grain growth has been relieved, the grains are free to grow until they encounter the remaining, more widely spaced large interagglomerate pores. In this manner, before sintering is completed, a grain size is achieved on the order of the starting agglomerate size. Ultimately, for reasons of both high sintering temperature and diminishment of pore number density during sintering, it is entirely possible for an agglomerated powder consisting of 10–20 nm crystallites to yield micron-sized grains [31, 40–42]. In previous sintering, experience has shown that it is extremely difficult to obtain a grain size at full density which is less than the starting agglomerate size.

It should be noted that one of the reasons why many ultrafine commercial ceramic powders (some of which have been available for many years) have not been successfully fabricated into solids with sub-100 nm grain sizes is this very problem of agglomeration. Agglomeration is not, however, as severe a problem for metal powders, since the large interagglomerate pores can be closed by plastic flow of the surrounding metal particles during compaction (i.e. prior to sintering). As shall be seen later, the use of plastic deformation *during* sintering can achieve the same result for nanocrystalline ceramic powders, and the temperature necessary to achieve complete densification can be dramatically reduced as a result.

8.5.2 Inhomogeneous sintering/differential densification

As has been shown earlier, the small grain sizes and pore sizes of nanocrystalline ceramics lead to fast densification rates, and these, in turn, lead to desirable, low sintering temperatures. However, the fast densification rates can themselves be a problem. Specifically, when a nanocrystalline ceramic sample is sintered in a furnace, it usually experiences a thermal gradient during furnace heat-up. If the outside of the sample is hotter than the inside (which is usually the case), it will densify at a faster rate and quickly turn into a hard, impervious shell. The hard shell constrains the inside of the sample from shrinking as it normally would, and large cracks or pores are formed as a result of the strain incompatibility. Figure 8.12 is a micrograph of such a sample. The thermal gradient/differential sintering phenomenon is not observed with ceramics of the same composition which have larger grain and pore sizes [43]. (Additionally, it should be mentioned that this problem of inhomogeneous sintering cannot be attributed to vapor transport or evolution of gases in the nanocrystalline materials, as explained in [43].) It is therefore the unfortunate characteristic of some nanocrystalline ceramics that their densification rates can be sufficiently fast, and their thermal conductivity sufficiently low, that the complete densification of the outside shell occurs much faster than the transport of heat to the inside of the ceramic body. (See [38] for an exact calculation of the magnitude of this effect for ZrO_2–3 mol% Y_2O_3 under different heating conditions.)

The problem of inhomogeneous densification intensifies as heating rates increase and sample dimensions become larger. Figure 8.13 shows the impact of increased heating rates on sintered density quite clearly; figure 8.14 shows that very small samples are not as prone to this problem as larger ones. The relatively minor impact of fast heating rates on small samples suggests one might be able to use fast heating rates for the densification of nanocrystalline thin films.

For moderate-to-large-sized samples, the problem of inhomogeneous densification can be dealt with in a number of ways. One can employ slow heating rates during sintering (2 °C min^{-1} for ZrO_2–3 mol% Y_2O_3, see figure 8.13) to prevent unequal heating of the sample. One can also reduce the sample's intrinsic densification rate by changing its starting microstructure. A 'precoarsening' treatment, in which the sample is heated at a low temperature

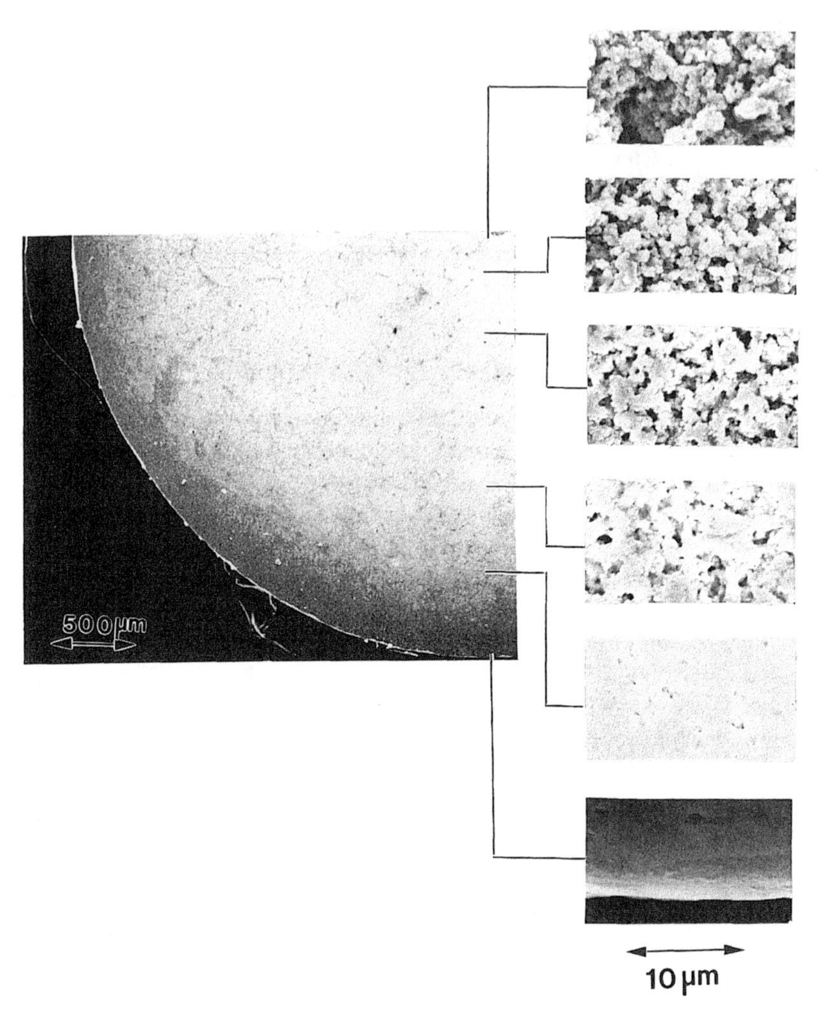

Figure 8.12. Scanning electron micrographs of a nanocrystalline ZrO_2–3 mol% Y_2O_3 sample which has experienced differential densification in response to thermal gradients. The sample was heated at 20 °C min^{-1} to 1300 °C and held for 2 h.

long enough for pores and grains to grow somewhat, should reduce the densification rate, according to equation (8.1). The slower densification rate would then reduce the amount of shrinkage taking place in the outer shell while heat is being transported to the inner regions of the sample. Smaller incompatibility strains would result. A precoarsening treatment should also lend greater mechanical strength to the compact, by allowing some densification and interparticle neck growth to take place before the sample experiences thermally induced stresses. As experimental verification of the precoarsening strategy,

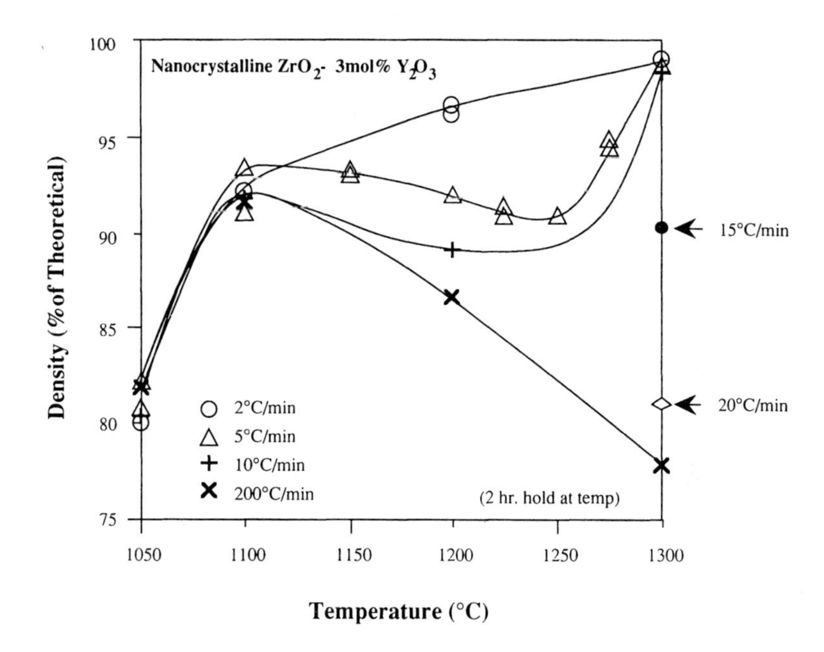

Figure 8.13. Effect of heating rate on the densification of nanocrystalline ZrO_2–3 mol% Y_2O_3. Fast heating rates lead to inhomogeneous densification, crack formation, and, ultimately, low sintered densities.

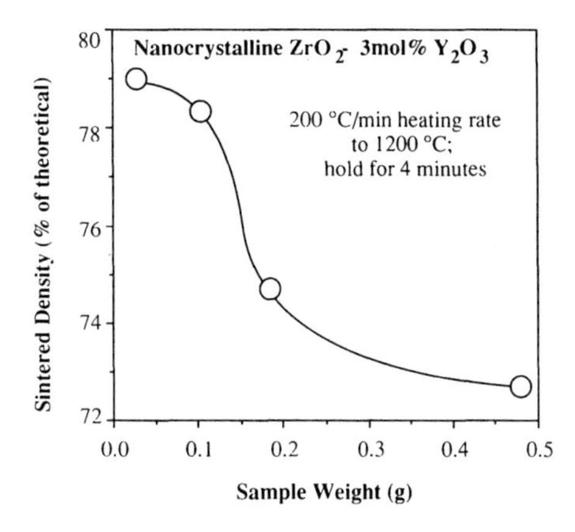

Figure 8.14. Influence of fast heating rates on the densification of various-sized samples. Very small samples are not as adversely affected as larger samples.

figure 8.15 shows that an improvement in sintered density does occur when a precoarsening treatment is used prior to rapid sintering of nanocrystalline ZrO_2–3 mol% Y_2O_3 at 1300 °C. Note that a precoarsening treatment which does not actually involve coarsening (e.g. presintering the same sample at 1000 °C, where little microstructural change occurs), has no beneficial effect upon subsequent sintering [39].

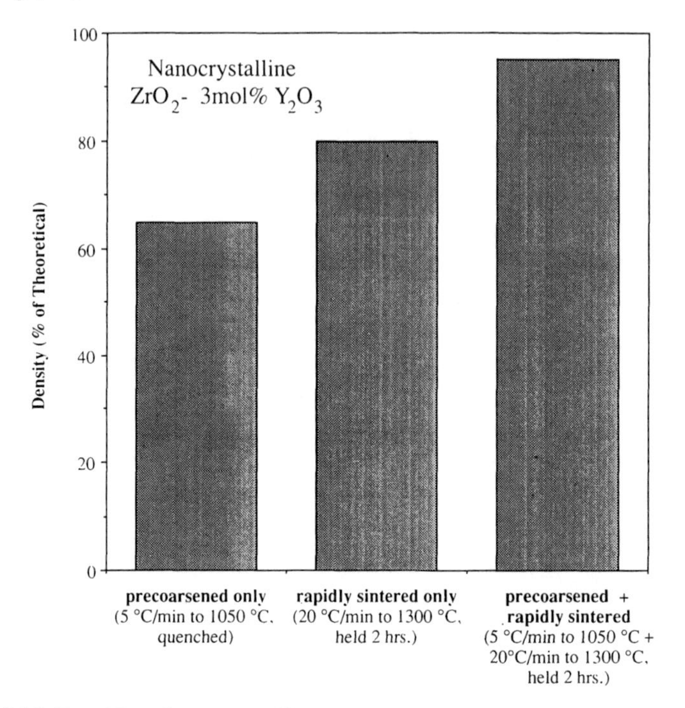

Figure 8.15. Densities of nanocrystalline ZrO_2–3 mol% Y_2O_3 samples showing the effect of precoarsening on the ultimate sintered density at fast heating rates.

Finally, it should be mentioned that the goal of decreasing the densification rate to avoid inhomogeneous sintering can also be met by choosing a sintering protocol of low temperatures in conjunction with long times. Thus, prolonged sintering of nanocrystalline ZrO_2–3 mol% Y_2O_3 at 1050 °C also results in crack-free, dense specimens [43].

8.6 GRAIN GROWTH DURING PRESSURELESS SINTERING

As was the case for densification, grain growth in nanocrystalline materials appears to follow the traditional phenomenological laws. A number of investigators have confirmed that the general relationship

$$G^N - G_0^N = kt \tag{8.2}$$

holds for both nanocrystalline metals [16] and ceramics (see [38] and [44] and figure 8.16). In equation (8.2), G is the instantaneous grain size, G_0 is the initial grain size, N is the grain size exponent (ranging from two to four), k is a rate constant, and t is time. The rate constant, k, is generally proportional to the diffusivity, $D_0 \exp(-Q/RT)$, and this gives the temperature dependence of grain growth. The exact value of k, however, is sensitive to the drag effects of solutes, pores, or second phases localized at the grain boundaries; k can therefore be substantially less than one would predict simply based on the diffusivity of the dominant chemical species. An interesting observation is that two different temperature regimes of grain growth have been observed for ZrO_2–3 mol% Y_2O_3, similar to the two densification regimes alluded to earlier. At intermediate and higher temperatures, the activation energy for grain growth is in the 300–400 kJ mol^{-1} range [38, 44], consistent with grain boundary diffusion [45], while at lower temperatures the activation energy is on the order of 100 kJ mol^{-1} [38, 44]. The different regimes of grain growth can be observed in figure 8.17.

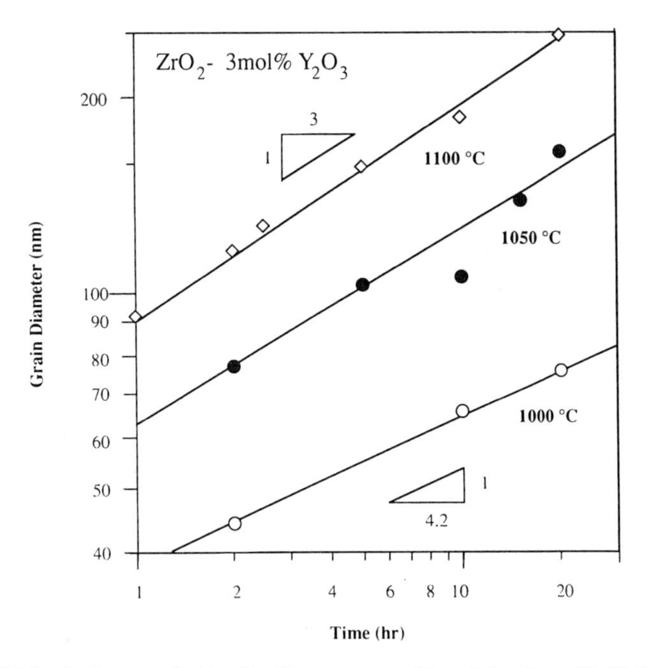

Figure 8.16. Grain growth kinetics for nanocrystalline ZrO_2–3 mol% Y_2O_3. The data appear to obey the law $G^N - G_0^N = kt$, where N is between 3 (for 1100 °C) and 4.2 (for 1050 °C). Note that the use of log–log plots to determine N is limited to cases where G_0 is small compared to G. Here, the values used for G are true grain diameters.

Equation (8.2) shows that, in the ideal case, grain growth depends on temperature and time, but, more importantly, it reveals what grain growth does *not* depend on (at least, not directly): microstructural parameters such

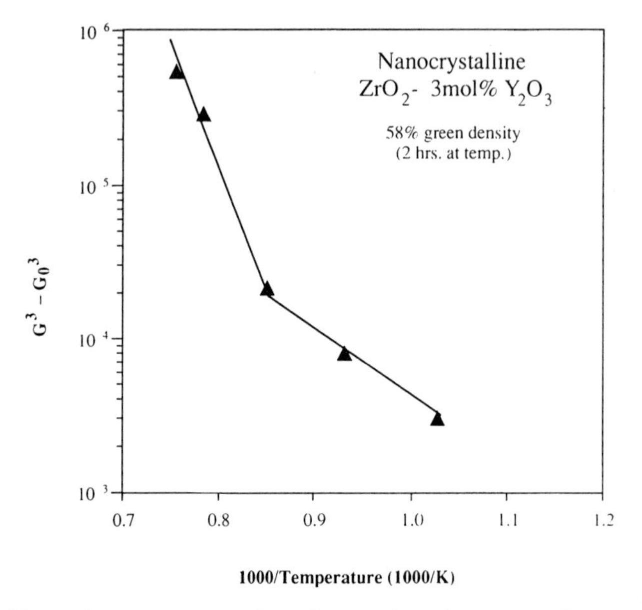

Figure 8.17. The temperature dependence of grain growth for nanocrystalline ZrO_2–3 mol% Y_2O_3.

as pore size, green density, or agglomerate size. The latter parameters can affect grain growth by altering the value of N in equation (8.2) [38], but the effect is a weak one. Figure 8.18 shows the effect of green density on grain growth in nanocrystalline ZrO_2–3 mol% Y_2O_3 at 1050 °C; note there is very little correlation between green density and grain growth. A more extensive survey of grain growth over a large range of temperatures [38] shows that higher green densities can occasionally lead to small increases in grain size after sintering at a given temperature, but again, the effect is very small.

Because the major impact of higher green density is to reduce the pore size, one can extrapolate that pore size also has little bearing on grain growth. The most striking example of how little grain growth depends on microstructural features can be taken from wet-processing experiments, in which the same ZrO_2–3 mol% Y_2O_3 powder is consolidated from aqueous suspensions formulated at pH values between 2 and 9 [30]. The effect of the pH is to change the structure of the consolidated green body from a well packed, ordered configuration of particles separated by small pores (pH 2; see figure 8.19(*a*)) to a loose structure of disorderly packed particles separated by larger pores (pH 9; see figure 8.19(*b*)). Despite these very different microstructures, the grain growth of these two samples is nearly identical: after 1 h at 1100 °C, both have a grain size between 70 and 80 nm, as do all the samples prepared from suspensions of intermediate pH values and morphologies [30]. Densities, on the other hand, differ tremendously after sintering at 1100 °C for 1 h: from 100% dense for the

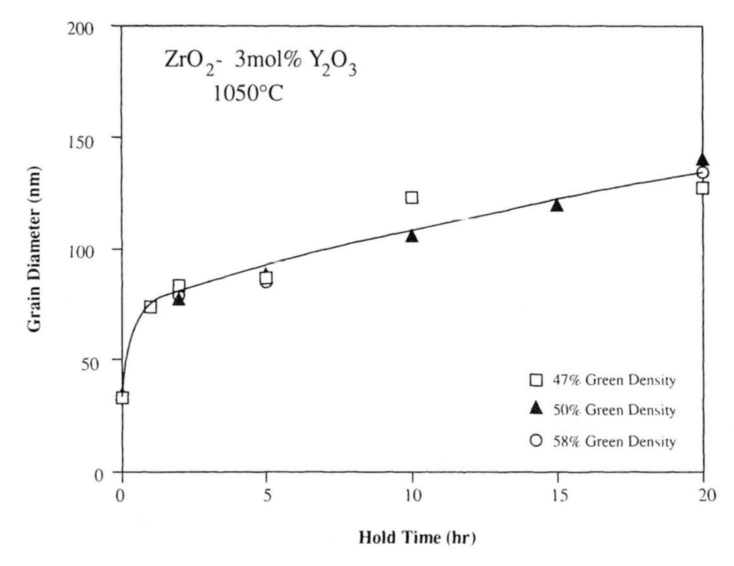

Figure 8.18. Effect of green density (or lack thereof) on grain growth at 1050 °C in ZrO_2–3 mol% Y_2O_3.

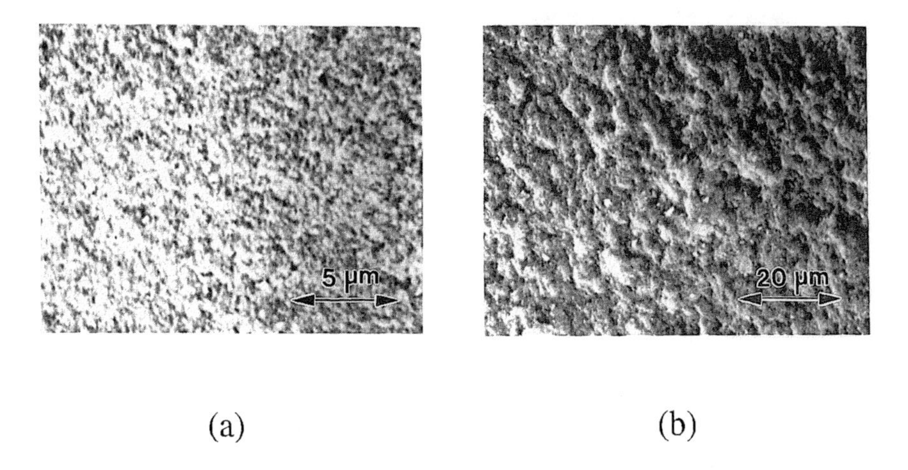

Figure 8.19. Particle packing and green body morphology in nanocrystalline ZrO_2–3 mol% Y_2O_3 centrifugally consolidated from aqueous suspensions adjusted to (*a*) pH 2, pore radius = 4.8 nm and (*b*) pH 9, pore radius = 9.0 nm.

well ordered pH 2 sample (initial pore size = 4.8 nm) to 67% dense for the loosely packed pH 9 sample (pore size = 9.0 nm) [29].

It might also be mentioned that other factors which can adversely affect densification, such as heating rate (which leads to differential sintering and hence incomplete densification) also have virtually no effect on grain growth;

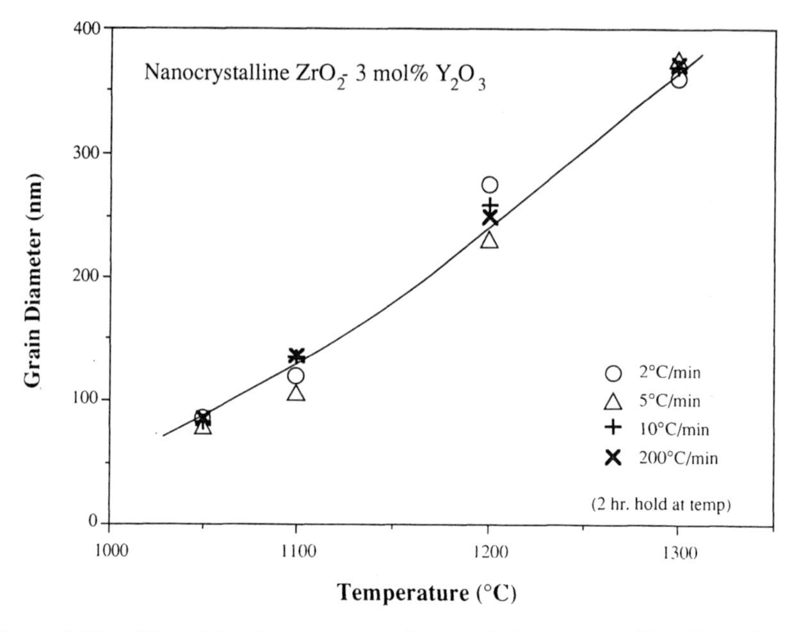

Figure 8.20. Effect of heating rate on grain growth in nanocrystalline ZrO_2–3 mol% Y_2O_3.

see figure 8.20. Thus grain size, in the ideal case, is dictated almost exclusively by time and temperature. Of these two parameters, time and temperature, temperature has by far the greater influence, since grain growth proceeds pseudo-exponentially in temperature and less than linearly in time.

8.7 GRAIN BOUNDARY PINNING BY PORES DURING PRESSURELESS SINTERING

In the ideal case, one would expect grain growth to go hand-in-hand with densification, since both are governed primarily by diffusion. Surprisingly, there are several examples of nanocrystalline ceramics where grain growth kinetics and densification kinetics do not appear to correlate. One example, shown in figure 8.21, is TiO_2, for which densification begins to accelerate long before grain growth. How the ceramic can be diffusionally active (as is necessary for densification) and yet avoid grain growth appears to be something of a mystery.

One hypothesis which may explain the observed lag between densification and grain growth is the influence of porosity on grain-boundary mobility. Closed pores can serve to pin the grain boundaries if they reside on the grain boundaries and their mobility (usually governed by surface diffusion) is slow relative to the mobility of the grain boundary (usually governed by grain-boundary diffusion) [46–48]. Work with submicron and larger-grained ceramics has previously

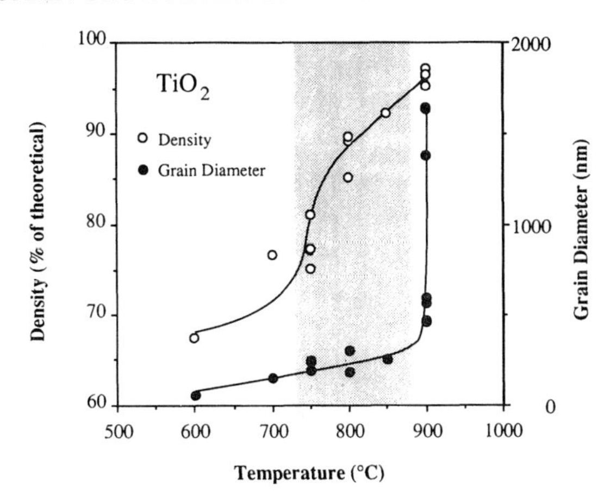

Figure 8.21. Comparison of the temperature dependence of grain growth and the temperature dependence of densification for nanocrystalline TiO_2. The shaded region indicates a regime in which significant densification occurs but grain growth is sluggish.

shown, both experimentally [49] and theoretically [50] that open pores are even more effective at pinning grain boundaries, primarily for geometrical reasons. The strong pinning ability of open pores then appears to explain the lack of grain growth during intermediate-stage sintering of many nanocrystalline ceramics, even though densification is active. As evidence, the point at which grain growth eventually accelerates in nanocrystalline TiO_2 corresponds almost exactly to the point at which the open porosity becomes closed (see figure 8.22). Furthermore, the grain size during intermediate-stage sintering appears to approximately equal the spacing between open pores, as shown in figure 8.23. Pore size may not have a strong influence on grain growth, but in select systems pore spacing can.

8.8 MINIMIZING GRAIN GROWTH AND MAXIMIZING DENSIFICATION DURING PRESSURELESS SINTERING

With an appreciation of the kinetics of both grain growth and densification, it becomes possible to postulate some simple principles for the production of fully dense materials with sub-100 nm grain size from ultrafine starting powders. The considerations for metals and ceramics are somewhat different and are enumerated separately below.

Metals

(i) Compact to the highest possible density prior to sintering.

(ii) Sinter at a low temperature (to avoid grain growth) for as long as necessary to remove residual porosity and establish equilibrium grain boundaries between neighboring particles. As porosity is removed and grain boundaries are established, hardness should increase. If hardness begins to decrease with

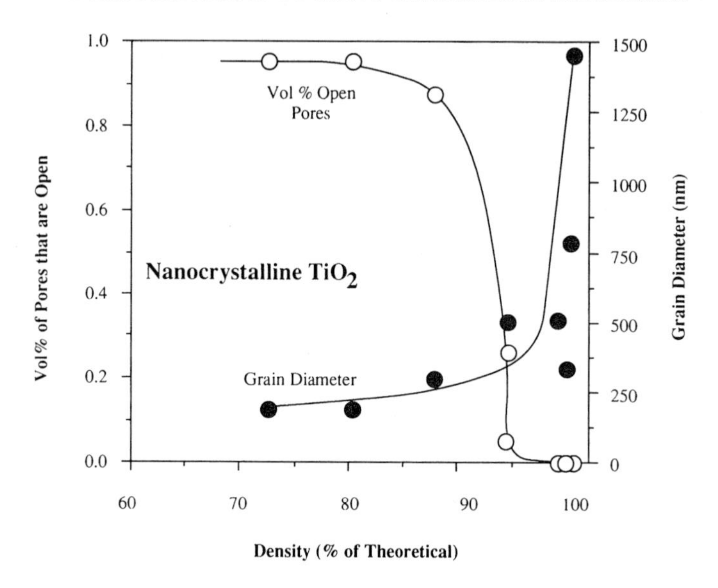

Figure 8.22. Correlation between the onset of accelerated grain growth and the closure of open pores, for nanocrystalline TiO_2.

increased sintering, then only grain growth is occurring, and increased sintering will prove unproductive. (An exception to this rule may occur at grain sizes <10 nm where the conventional relationship between hardness and grain size has not yet been proven.)

Ceramics

(i) Choose a non-agglomerated starting powder. For pressureless sintering, this is perhaps *the* most important step in synthesizing a sub-100 nm ceramic from ultrafine powders.

(ii) Maximize densification rates by minimizing the pore size within the starting compact. For small samples, narrow pores can be achieved by compacting at large pressures; for larger samples, fabricating the powder compact by wet-processing techniques is an option.

(iii) Minimize grain growth kinetics by choosing a low-sintering-temperature/long-sintering-time combination. Generally speaking, grain growth increases much more rapidly with temperature than with time, so a long time at a low sintering temperature will prevent 'overshooting'.

(iv) Avoid fast heating rates, especially in ceramics which are not thermally conductive. Rapid heating may be useful for small samples, such as thin films, but rapid heating can pose problems for larger samples, due to differential densification.

(v) For select ceramics in which grain growth is limited by the presence of pores, grain growth will be severely curtailed up to $\approx$90% density (the point at

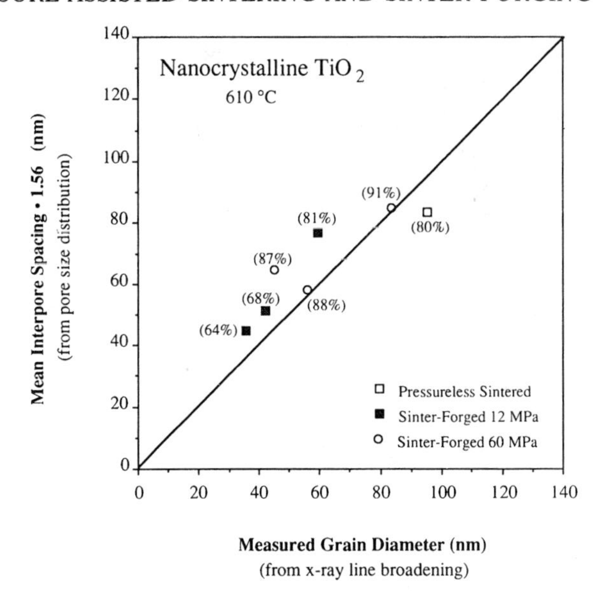

Figure 8.23. Correspondence between the mean spacing between open pores in a nanocrystalline titania (obtained by deconvoluting the ceramic's pore size distribution) and the grain size (measured by x-ray line broadening and cross-calibrated by transmission electron microscopy). The factor of 1.56 is necessary to convert a 2D mean linear intercept dimension to a 3D spatial dimension. Densities for each TiO_2 sample are shown in parentheses. Data represent sinter-forging times of 30 min, 60 min, and 6 h (for 12 MPa and 60 MPa) and pressureless sintering for a time of 6 h.

which open porosity usually becomes closed), even without special precautions. However, limiting grain growth beyond this point will require adherence to the above suggestions.

Using the above principles, it has been shown by several authors [21, 43, 44] that ZrO_2–3 mol% Y_2O_3 can be successfully sintered to densities between 97 and 100% with a grain size (true grain diameter) between 58 and 87 nm. In these cases, the powder was dry pressed to a high starting density, then sintered at 1050 °C, in air, for 5–6 h. A micrograph of such a sample is shown in figure 8.24.

8.9 PRESSURE ASSISTED SINTERING AND SINTER-FORGING

Because of the difficulty in finding non-agglomerated starting powders and, furthermore, the extensive experimentation involved in optimizing densification and grain growth kinetics, many experimentalists have turned to pressure assisted sintering, or sinter-forging, as an easy method of producing nanocrystalline ceramics. Pressure assisted sintering refers to any sintering operation in

Figure 8.24. Field emission scanning electron micrograph of a ZrO_2–3 mol% Y_2O_3 with a 85 nm diameter grain size, fabricated by compacting 13 nm diameter particles to 58% green density and pressureless sintering in air for 5 h at 1050 °C. The grains are not actually rounded, as indicated by the micrograph; rather, the rounded nature of the grains is due in part to their being blanketed by a deposited gold coating and in part to the resolution limitations of the machine used. (Micrograph courtesy of A H Carim, the Pennsylvania State University.)

which a pressure beyond atmospheric is externally applied to the specimen during sintering. Pressure assisted sintering thus includes such operations as hot pressing, hot isostatic pressing, hot extrusion, and sinter-forging. Sinter-forging itself is a more precise term and indicates an operation in which a uniaxial compressive stress is applied to the sample during sintering. The sample is therefore allowed to deform and densify simultaneously. Usually sinter-forging is performed without a die (in contrast to hot pressing). A number of experiments have shown that ultrafine-grained ceramics which are incapable of retaining a nanocrystalline grain size during pressureless sintering are successfully produced in nanocrystalline or near-nanocrystalline form during a pressure assisted sintering operation [51–54]. The reasons why pressure assisted sintering works for nanocrystalline ceramics are rather interesting.

Generally speaking, the application of pressure to accelerate densification is a well known tactic in the field of sintering. The applied pressure lowers the vacancy concentration at the grain boundaries (vacancy sinks) relative to the pores (vacancy sources) and, by thus increasing the concentration gradient of vacancies, serves to increase the vacancy flux. With a greater number of vacancies moving out of the pores per unit time, the pore shrinkage rate, and hence the densification rate, increases. This rationale works quite well for ceramics with conventional pore and grain sizes, but it does not explain the success of pressure assisted sintering for the case of nanocrystalline ceramics.

For nanocrystalline ceramics, there is already a tremendous driving force for densification by vacancy motion due to the pore curvature. When this driving force is expressed in terms of an effective stress (the so-called sintering stress), one sees that it is on the order of GPa

$$\sigma = \frac{-2\gamma}{r}. \tag{8.3}$$

Here, σ is the sintering stress, γ is the surface tension of the pore, and r is the pore radius. Assuming a 5 nm pore and a surface tension of 1 J m^{-2}, σ is 400 MPa. In contrast, a micron-grained ceramic with a pore size of 1 μm has a sintering stress of only 2 MPa. Thus, for a micron-grained ceramic, an externally applied pressure of even 10 MPa might be able to significantly assist the intrinsic sintering stress in driving diffusion; however, for a nanocrystalline ceramic, the applied stress would have to be greater than 0.5 GPa to have a noticeable effect. Interestingly, the applied stresses typically used for sinter-forging nanocrystalline ceramics are small by comparison—in the 50–100 MPa range [53–55]. The argument that these stresses accelerate densification rates by helping to drive diffusion therefore cannot hold. The diffusion rates for small pores are already rapid. Indeed, the experiments on nanocrystalline titania in [55] showed that all the small intercrystallite pores actually disappeared during furnace heat-up, well before the application of an applied stress. Clearly, another explanation for the impact of pressure on densification rates is required.

One explanation is that all the pores in nanocrystalline ceramics are not necessarily nanometer sized. This fact is certainly true of agglomerated powders, which may have submicron or micron-sized interagglomerate pores in addition to the nanometer-sized intercrystallite pores. With a lower intrinsic sintering stress, only a moderate applied stress would be necessary to achieve a visible increase in the shrinkage rates of these pores. It is worth noting that agglomerated powders were used in the sinter-forging studies of [53–55]. Furthermore, in [55], it was found that the disappearance rate of moderate-sized interagglomerate pores was indeed consistent with the kinetics of stress assisted diffusion.

There is evidence, however, that the primary advantage of sinter-forging lies largely in a unique pore shrinkage mechanism which is nondiffusional in nature and which is specific to fine-grained ceramics. While micron-grained ceramics are typically brittle, fine-grained ceramics have a singular ability to deform plastically at moderate temperatures (i.e. at temperatures around half the ceramic's melting point). This phenomenon has been observed in both nanocrystalline [56–58] and sub-micron-grained ceramics [59–62]. The mechanism by which plastic flow occurs in these fine-grained ceramics is thought to be superplastic deformation; the process of superplastic deformation, in turn, is envisaged as grains or groups of grains sliding past each other to take up new positions in the deforming solid. Although the shape of individual grains may not change, their relative positions do, in order to accommodate the imposed strains. An ability to deform plastically would allow pores to be squeezed shut

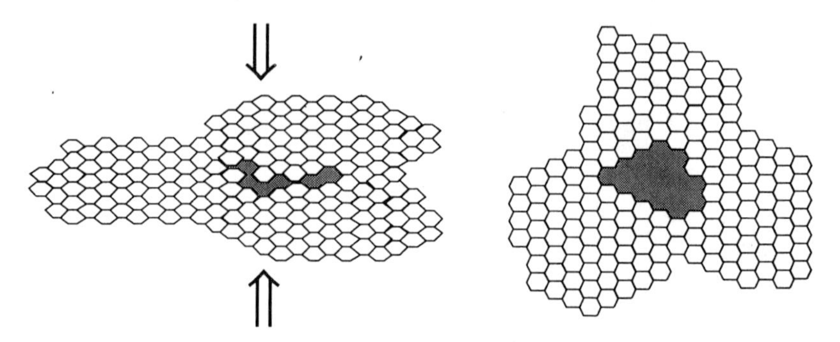

Figure 8.25. Schematic diagram of pore elimination by plastic flow/grain boundary sliding.

directly by plastic flow of the surrounding matrix (see figure 8.25), rather than isotropically shrinking by diffusion. Plastic strain induced pore closure is well documented and modeled in metals [63–65] but is not usually considered in the case of ceramics. By virtue of their excellent ability to deform, however, nanocrystalline ceramics can undergo this process [55]. A major consequence is that very large pores, which cannot possibly be closed by diffusion in any reasonable length of time, are able to be quickly eliminated. An example is shown in figure 8.26(*a*), in which a titania sample sinter-forged to a true strain of 0.22 shows no large pores upon examination by scanning electron microscopy. However, a similar sample sinter-forged to a much smaller true strain of 0.08 is riddled with such pores (figure 8.26(*b*)) and, in fact, looks very similar to a sample which has undergone only pressureless sintering (figure 8.26(*c*)). Note that small (sub-100 nm) pores are not visible at the magnifications of these micrographs.

The ability to close large pores with ease is a distinct advantage. With the elimination of the deleterious large interagglomerate pores, the processing of agglomerated powders into components with sub-100 nm grain sizes suddenly becomes possible. For the agglomerated TiO_2 powder used in figures 8.26(*a*)–(*c*), sinter-forging for 6 h at 60 MPa and 610 °C achieves a true strain of 0.27 and yields a 91% dense ceramic with an 87 nm grain diameter. By contrast, pressureless sintering requires temperatures of 800 °C to achieve the same density—resulting in a grain size of about 380 nm. Other researchers who have studied the sinter-forging of nanocrystalline TiO_2 (using a different starting powder) have been able to achieve 97% dense ceramics with a 60 nm grain size [53].

Despite such successes, it is worth noting that superplastic deformation, which provides the mechanism for strain induced pore closure, occurs in only a limited range of temperatures and stresses. Therefore one must be careful to select a stress and temperature combination for sinter-forging (or pressure assisted sintering) which is within the superplastic regime. If the stresses are too

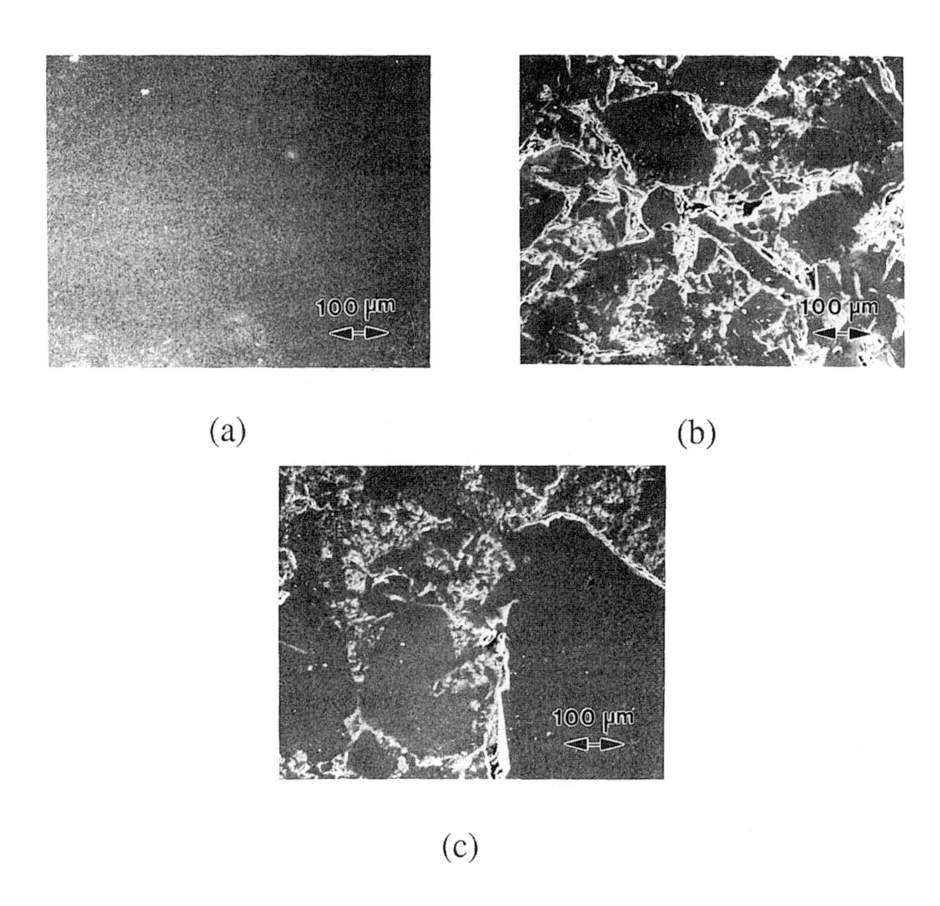

Figure 8.26. (*a*) Nanocrystalline TiO_2 sample sinter-forged at 610 °C to high strains (plastic strain = 0.22, initial stress = 60 MPa, and time under stress = 45 min). Total time at temperature is 1 h and 15 min. The polished surface of the sample is featureless under a scanning electron microscope, indicating a lack of large pores visible at this magnification. (*b*) Nanocrystalline TiO_2 sample sinter-forged at 610 °C to low strains (plastic strain = 0.08, initial stress = 12 MPa, and time under stress = 1 h). Total time at temperature is 1 h and 30 min. On the polished surface, large pores are seen as white areas; the white color actually outlines each pore surface and is due to electron charging of the pore surfaces while under observation in a scanning electron microscope. The large, dark featureless regions are densely sintered areas which arose from prior weak agglomerates (granules) in the starting powder. (*c*) Nanocrystalline TiO_2 sample pressureless sintered at 610 °C (plastic strain = 0, stress = 0 MPa, and time at temperature = 6.5 h). Microstructure is similar to (*b*) above, with many large pores (white areas) visible between the densely sintered agglomerates.

low, or the temperatures too high, one is likely to encounter a deformation regime in which diffusion—not grain boundary sliding—is the primary mechanism for the transport of matter. Examples of such regimes would include Coble creep or Nabarro–Herring creep. Theoretically, the strain required to close pores solely by plastic flow is large—about 0.62 [65]—and such large plastic strains are not easily achieved in diffusional creep, or elsewhere outside of the superplastic regime.

As the sinter-forging technique is expanded to different tool geometries and forming methods, it is worthwhile remembering that plastic deformation, which is the key to the success of this technique, occurs in response to shear stresses but not hydrostatic stresses. Thus, it is likely that the densification rates observed in different sinter-forging operations will correlate with the relative proportion of shear to hydrostatic stresses in the forming operation used. For example, one might expect (though it is not yet proven) that unconstrained, dieless sinter-forging will prove to be a superior densification method to hot pressing (in which shear deformation is constrained by the die), which in turn will be superior to hot isostatic pressing (in which no shear stresses exist, except in the immediate vicinity of the pores). Further work is required in this area.

8.10 SUMMARY

The production of nanocrystalline metals and ceramics by the compaction and sintering of ultrafine powders has been successfully achieved by several investigators. One of the greatest challenges, particularly for ceramics, is to simultaneously densify a nanocrystalline powder compact while preventing grain growth. For pressureless sintering, the fact that densification is highly sensitive to microstructural features (such as pore size and particle packing arrangements), while grain growth is much less so, allows one to choose microstructural configurations which favor densification while employing sintering conditions which discourage grain growth. For some powders, particularly agglomerated powders, pressureless sintering is completely unsuccessful in producing solids with sub-100 nm grain sizes. In these cases, pressure assisted sintering—and especially sinter-forging—have proven quite effective in eliminating large, interagglomerate pores and accelerating densification relative to grain growth.

REFERENCES

[1] Ortiz C, Rubin K A and Ajuria S 1988 *J. Mater. Res.* **3** 1196

[2] Sethurman A R, DeAngelis R J and Reucraft P J 1991 *J. Mater. Res.* **6** 749

[3] Lappalainen R and Raj R 1991 *Acta Metall. Mater.* **39** 3125

[4] Zaluska A, Xu Yan, Altounian Z, Ström-Olsen J O, Allem R and L'Espérance G 1991 *J. Mater. Res.* **6** 724

[5] Hansen P L and Jessen C Q 1989 *Scr. Metall.* **23** 1387

[6] Lange R W and Sowman H G 1979 Shaped and fired articles of TiO_2 *US Patent* 4 166 147
[7] De Moranville K, Yin D and Wong J 1993 *IEEE Trans. Appl. Supercond.* **AS-3** pt 3 982
[8] Valiev R Z, Krasilnikov N A and Tsenev N K 1991 *Mater. Sci. Eng.* A **137** 35
[9] Froes F H, Suryanarayana C, Chen Guo-Hao, Frefer A and Hyde G R 1992 *JOM* **44** 26
[10] Morris D G and Morris M A 1991 *Acta Metall. Mater.* **39** 1763
[11] Ganapathi S K and Rigney D A 1990 *Scr. Metall. Mater.* **24** 1675
[12] Ganapathi S K, Aindow M, Fraser H L and Rigney D A 1991 *Clusters and Cluster-Assembled Materials (Mater. Res. Soc. Symp. Proc. 206)* ed R S Averback, J Bernholc and D L Nelson (Pittsburgh, PA: Materials Research Society) p 593
[13] Hahn H, Logas J and Averback R S 1990 *J. Mater. Res.* **5** 609
[14] Hirschorn S 1969 *Introduction to Powder Metallurgy* (New York: American Powder Metallurgy Institute) p 121
[15] Fougere G E, Weertman J R, Siegel R W and Kim S 1992 *Scr. Metall. Mater.* **26** 1879
[16] Ganapathi S K, Owen D M and Chokshi A H 1991 *Scr. Metall. Mater.* **25** 2699
[17] Nieman G W, Weertman J R and Siegel R W 1991 *J. Mater. Res.* **6** 1012
[18] Nieman G W, Weertman J R and Siegel R W 1989 *Scr. Metall. Mater.* **23** 2013
[19] Boschi A O and Gilbart E 1990 *Advanced Ceramic Processing and Technology* vol 1, ed J G P Binner (Park Ridge, NJ: Noyes) p 73
[20] Evans J W and DeJonghe L C 1991 *The Production of Inorganic Materials* (New York: Macmillan) p 391
[21] Winnubst A J A, Thennissen G S A M and Burggraaf A J 1989 *Euro Ceramics (Proc. 1st Eur. Ceram. Soc. Conf.)* vol 1, ed G de With, R A Terpstra and R Metselaar (Amsterdam: Elsevier) p 1.391
[22] Whittemore O J 1981 *Powder Technol.* **29** 167
[23] Roosen A and Hausner H 1985 *Science and Technology of Zirconia (Advances in Ceramics 12)* vol 2, ed A H Heuer (Columbus, OH: American Ceramic Society) p 714
[24] Pampuch R and Haberko K 1983 *Ceramic Powders* ed P Vincenzini (Amsterdam: Elsevier) p 623
[25] Lowell S and Shields Joan E 1979 *Powder Surface Area and Porosity* (New York: Chapman and Hall) pp 55, 87, 139
[26] Hurst J B and Cutta S 1987 *J. Am. Ceram. Soc.* **70** C-303
[27] Evans J W and DeJonghe L C 1991 *The Production of Inorganic Materials* (New York: Macmillan) pp 135, 402
[28] Aksay I A and Schilling C H 1984 *Forming of Ceramics (Advances in Ceramics 9)* ed J A Mangels and G L Messing (Columbus, OH: American Ceramic Society) p 85
[29] Cowan R E 1976 *Ceramic Fabrication Processes (Treatise on Materials Science and Technology 9)* ed F F Y Wang (New York: Academic) p 153
[30] Mayo M J, Chen D-J, Cottom B A and Sharma M 1995 *J. Am. Ceram. Soc.* submitted
[31] Andrievski R A 1994 *J. Mater. Sci.* **29** 614
[32] Theunissen G S A M, Winnubst A J A and Burggraaf A J 1993 *J. Eur. Ceram. Soc.* **11** 315

[33] Winnubst A J A, Theunissen C S A M, Boutz M M R and Burggraaf A J 1990 *Structural Ceramics: Processing, Microstructure and Properties* ed J J Bentzen *et al* (Roskilde: Risø National Laboratory) p 523
[34] Mayo M J 1993 *Mater. Design* **14** 323
[35] Beere W 1975 *Acta. Metall.* **23** 139
[36] Kuhn W E 1963 *Ultrafine Particles* ed W E Kuhn (New York: Wiley) p 41
[37] Coble R L 1961 *J. Appl. Phys.* **32** 787
[38] Vergnon P, Astier M and Teichner S J 1974 *Fine Particles (Proc. Second Int. Conf. Fine Particles)* ed W E Kuhn (Princeton, NJ: Electrochemical Society)
[39] Chen D-J 1994 *MS Thesis* Pennsylvania State University
[40] Höfler H J and Averback R S 1990 *Scr. Metall. Mater.* **24** 2401
[41] Hahn H, Logas J and Averback R S 1990 *J. Mater. Res.* **5** 609
[42] Siegel R W, Ramasamy S, Hahn H, Zonghuan Z and Ting L 1988 *J. Mater. Res.* **3** 1367
[43] Chen D-J and Mayo M J 1993 *J. Nanostruct. Mater.* **2** 469
[44] Boutz M R, Theunissen G S A M, Winnubst A J A and Burgraaf A J 1990 *Superplasticity of Metals, Ceramics and Intermetallics (Mater. Res. Soc. Symp. Proc. 196)* ed M J Mayo, M Kobayashi and J Wadsworth (Pittsburgh, PA: Materials Research Society) p 87
[45] Helle A S 1986 *Lincentiate Thesis* Luleå, Sweden. As referenced in Bourell D L, Parimal and Kaysser W 1993 *J. Am. Ceram. Soc.* **76** 705
[46] Liu Y and Patterson B R 1993 *Acta Metall. Mater.* **41** 2651
[47] Nichols F A 1968 *J. Am. Ceram. Soc.* **51** 468
[48] Brook R J 1969 *J. Am. Ceram. Soc.* **52** 56
[49] Cameron C P and Raj R 1988 *J. Am. Ceram. Soc.* **71** 103
[50] Svoboda J and Riedel H 1992 *Acta Metall. Mater.* **40** 2829
[51] Owen D M and Chokshi A H 1993 *Nanostruct. Mater.* **2** 181
[52] Panda P C, Wang J and Raj R 1988 *J. Am. Ceram. Soc.* **71** C507
[53] Uchic M, Höfler H J, Flick W J, Tao R, Kurath P and Averback R S 1992 *Scr. Metall. Mater.* **26** 791
[54] Hague D C and Mayo M J 1993 *Mechanical Properties and Deformation Behavior of Materials Having Ultrafine Microstructures* ed M Nastasi, D Parkin and H Gleiter (Dordrecht: Kluwer) p 539
[55] Hague D C and Mayo M J *J. Mater. Res.* submitted
[56] Hahn H and Averback R S 1991 *J. Am. Ceram. Soc.* **74** 2918
[57] Mayo M J 1991 *Superplasticity in Advanced Materials* ed S Hori, M Tokizane and N Furoshiro (Osaka: Japan Society for Research on Superplasticity) p 541
[58] Mayo M J 1993 *Mechanical Properties and Deformation Behavior of Materials Having Ultrafine Microstructures* ed M Nastasi, D Parkin and H Gleiter (Dordrecht: Kluwer) p 361
[59] Wakai F, Sakaguchi S and Matsuno Y 1986 *Adv. Ceram. Mater.* **1** 259
[60] Chen I-Wei and Xue L A 1990 *J. Am. Ceram. Soc.* **73** 2585
[61] Maehara Y and Langdon T G 1990 *J. Mater. Sci.* **25** 2275
[62] Nieh T G, Wadsworth J and Wakai F 1991 *Int. Mater. Rev.* **36** 146
[63] Lionel F V and Ansell G S 1967 *Sintering and Related Phenomena* ed G C Kuczynski, N A Horton and C F Gibbon (New York: Gordon and Breach) p 351
[64] Duva J M and Crow P D 1992 *Acta Metall. Mater.* **40** 31

[65] Budiansky B, Hutchinson J W and Slutsky S 1982 *Mechanics of Solids, The Rodney Hill 60th Anniversary Volume* ed H G Hopkins and M J Sewell (Oxford: Pergamon) p 13

[66] Yan M F and Rhodes W W 1983 *Mater. Sci. Eng.* **61** 59

[67] Hague D C 1992 *MS Thesis* Pennsylvania State University

[68] Barringer E A, Brook R and Bowen H K 1984 *Sintering and Heterogeneous Catalysis* ed G C Kuczynski, A E Miller and G A Sargent (New York: Plenum) p 1

PART 5

CHARACTERIZATION OF NANOSTRUCTURED MATERIALS

Chapter 9

Nanostructures of metals and ceramics†

Richard W Siegel

9.1 INTRODUCTION

The ability to create nanophase materials, as discussed in earlier chapters, has developed rapidly over the past decade. This development has resulted in a new class of materials that, in contrast to conventional solids, have an appreciable fraction of their atoms residing in defect environments. For example, a nanophase material with a readily achievable average grain size of 5 nm has about 50% of its atoms within the first two nearest-neighbor planes of a grain boundary, in which significant atomic displacements from their normal lattice positions are exhibited. Since the properties of nanophase materials are so strongly related to their unique structures, this chapter will attempt to review what we know about the atomic scale structures of nanophase materials after almost a decade of research.

The structures of nanophase materials on a variety of length scales have an important bearing on their special chemical and physical properties [1–4]. They are dominated by their ultrafine grain sizes and by the large number of interfaces associated with their small grains, as indicated in figure 9.1. However, other structural features such as pores (and larger flaws), grain boundary junctions, and other crystal lattice defects that can depend upon the manner in which these materials are synthesized and processed also play a significant role. It has become increasingly clear during the past several years that all of these structural aspects must be carefully considered in trying to fully understand the properties of nanophase materials.

The atom clusters that make up the grains of cluster consolidated nanophase materials typically have rather narrow size distributions, whether they are created by the gas condensation method or by a variety of other physical or chemical methods discussed earlier. Most frequently, the clusters are found to

† **Authored by a contractor of the US Government under contract number W-31-109-ENG-38.**

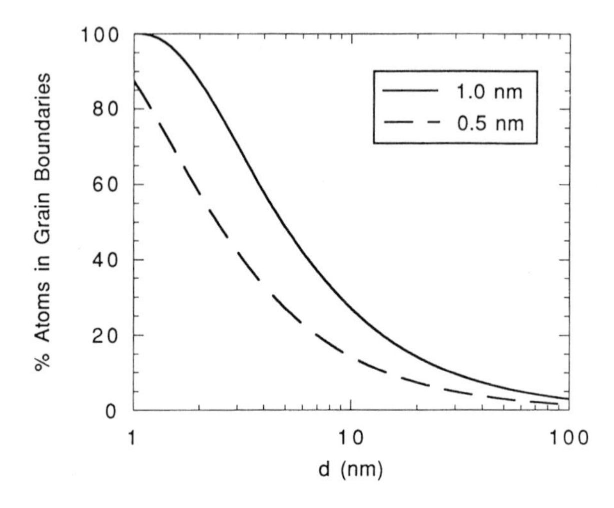

Figure 9.1. Percentage of atoms in grain boundaries (including grain boundary junctions) of a nanophase material as a function of grain diameter [5], assuming that the average grain boundary thickness ranges from 0.5 to 1.0 nm (about two to four atomic planes wide).

be aggregated. The degree and nature of aggregation can vary from essentially none in the case of chemically capped clusters in solution to rather open fractal arrays in the gas condensation method to a large amount of hard agglomeration in chemical solution routes when insufficient special additives (surfactants) have been used. The difficulties encountered in consolidating the hard equiaxed agglomerates of fine powders with surface contamination that can result from conventional wet chemistry synthesis routes are mostly avoided in the gas condensation method for synthesizing nanophase materials from clusters. In the case of wet chemical routes, hard agglomerates with strongly bonded arch-like structures are often formed in the drying process driven by the surface tension of the suspending medium. In contrast, however, the weak fractal agglomerates formed in the gas condensation process are easily broken down in either subsequent dispersion or consolidation processing. Further description of these effects can be found in chapter 8. The sample densities resulting from gas condensed cluster consolidation at room temperature have ranged up to about 97% of theoretical for nanophase metals and up to about 75–85% of theoretical for nanophase oxide ceramics. However, even the remaining porosity appears to be capable of being removed, when desired, by means of cluster consolidation at elevated temperatures and pressures without the occurrence of significant attendant grain growth.

9.2 STRUCTURES OF NANOPHASE MATERIALS

9.2.1 Grains

Our present knowledge of the grain structures of nanophase materials, whether formed by cluster consolidation, intense mechanical deformation, or crystallization from amorphous precursors, has resulted primarily from direct observations using transmission electron microscopy (TEM) [6–10]. A typical high-resolution image of a nanophase palladium sample formed by the consolidation of gas condensed clusters is shown in figure 9.2. TEM has shown that the grains in cluster consolidated nanophase compacts are essentially equiaxed, similar to the atom clusters from which they were formed, although departures from spherical structures are expected simply from the efficient packing of the clusters during consolidation. The grains also appear to retain the narrow (approximately $\pm 25\%$ FWHM) log-normal size distributions typical of the clusters formed in the gas-condensation method, since measurements of these distributions before or after cluster consolidation by dark-field electron microscopy yield similar results. Such a typical size distribution is shown in figure 9.3. The grain size distributions in deformation or crystallization produced nanostructructures tend to be somewhat broader than these, but similarly equiaxed.

The observations that the densities of nanophase materials consolidated from equiaxed clusters extend well beyond the theoretical limit (74%) for close packing of identical spheres indicate that an extrusion-like deformation of the clusters must occur during the consolidation process, filling (at least partially) the pores among the grains. A number of these observations indicate that cluster extrusion in forming nanophase grains may result from a combination of deformation and diffusion processes. Such processes are also evident from recent scanning tunneling microscopy (STM) and atomic force microscopy (AFM) observations [11] on nanophase palladium and silver. These investigations have clearly shown that the cluster morphologies accommodate to one another to help fill the volume. Further evidence that local extrusion is important in the cluster-consolidation process has recently been obtained in molecular dynamics simulations [12]. Palladium clusters before and after consolidation at 10 kbar for 4 ps are shown in figure 9.4, again indicating that inter-cluster voids are at least partially filled via extrusion processes.

Observations by electron and x-ray scattering indicate, however, that no apparent preferred orientation or 'texture' of the grains results from their uniaxial consolidation and that the grains in nanophase compacts are essentially randomly oriented with respect to one another. This is interesting with respect to the fact that deformation textures (such as those that result from uniaxial compression) in conventional grain size materials are the result of dislocation motion, which seems to be suppressed in nanophase materials. Indeed, only very few dislocations are observed within the grains in these ultrafine-grained

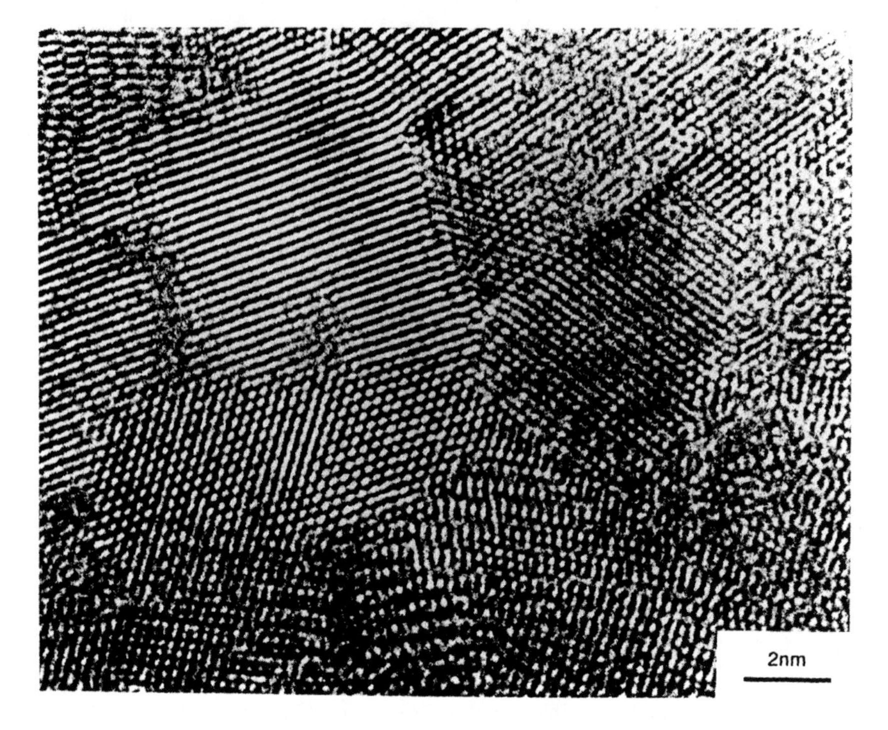

Figure 9.2. High-resolution transmission electron micrograph of a typical area in cluster consolidated nanophase palladium [8].

materials, and these are normally in locked (i.e. sessile) configurations. The cluster-consolidation process thus appears to resemble an Ashby–Verrall creep process [13] in which grains slide over one another accommodated by diffusional processes in the interfaces, but grain shape and orientation are retained while overall dimensional changes take place.

9.2.2 Atomic defects and dislocations

When an elemental precursor is evaporated in an inert gas atmosphere, as in the gas condensation method, the atom clusters formed and collected are the same material, only in a reconstituted form. In this case, the clusters and the subsequently consolidated nanophase material are expected to have only equilibrium concentrations of intrinsic atomic defects (e.g. vacancies or interstitials) present. This expectation is based on the fact that the mobility of these defects [14, 15] and the respective efficiencies of the surfaces and grain boundaries as sinks for them [16] are sufficiently high to maintain atomic defect equilibrium during synthesis and processing. No experimental evidence to the contrary exists at present.

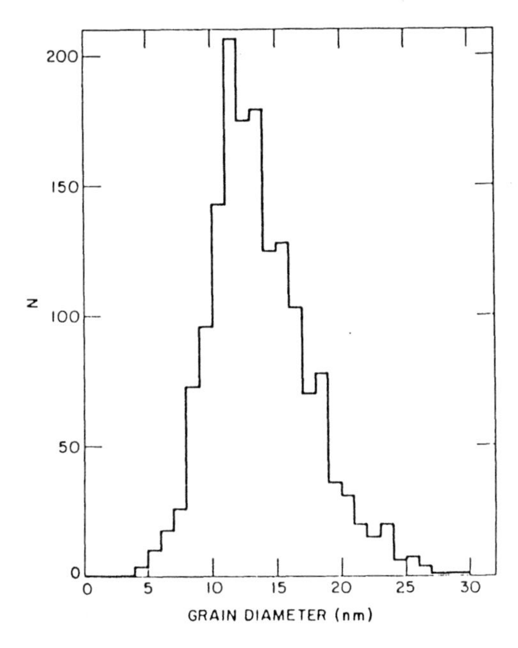

Figure 9.3. Grain size distribution for a nanophase TiO_2 sample determined by dark-field TEM [5].

However, if clusters of a metal oxide compound are to be made in order to synthesize a nanophase ceramic, the process is somewhat more complex and constitutional atomic defects can be present in significant concentrations. For example, to produce nanophase TiO_2 with a rutile structure, titanium metal clusters condensed in He are first collected and subsequently oxidized by the introduction of oxygen into the synthesis chamber [6]. A similar method has been used to produce α-Al_2O_3 [17] and Cr_2O_3 [18]. If the vapor pressure of a compound is sufficiently large, as in the cases of MgO and ZnO, it is possible to sublime the material directly from the oxide precursor in a He atmosphere containing a partial pressure of O_2 to attempt to maintain oxygen stoichiometry during cluster synthesis. Such a method has been used [17] to produce such nanophase oxides with average grain sizes down to about 5 nm. Frequently, however, oxygen stoichiometry is not maintained.

In the case of nanophase TiO_2 cited above, the oxygen deficiency is rather small and easily remedied as a result of the small grain sizes and short diffusion distances involved. Raman spectroscopy has been a useful tool in studying the oxidation state of nanophase titanium dioxide owing to the intense and well studied Raman bands in both the anatase and rutile forms of this oxide and the observation that these bands were broadened and shifted in nanophase samples [19]. The band broadening and shifting in both the anatase and rutile phases were confirmed [20] to be the result of an oxygen deficiency which

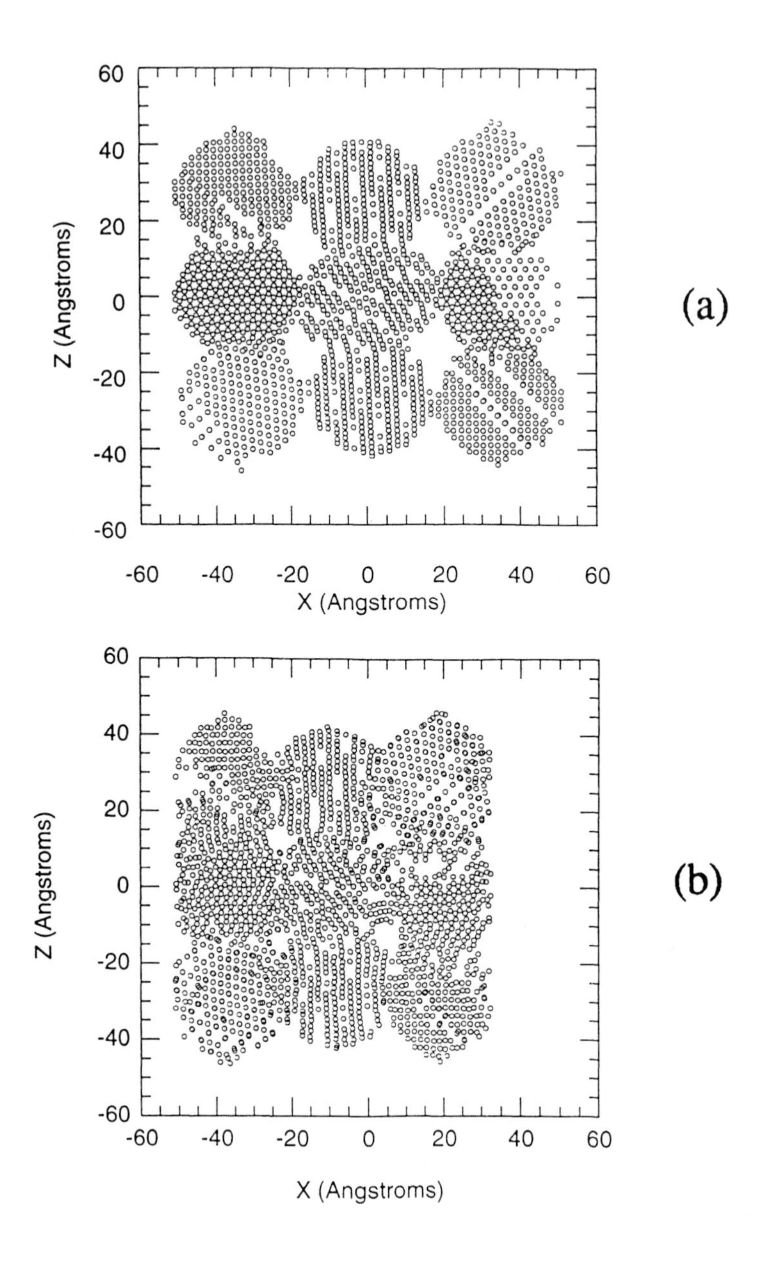

Figure 9.4. Sections 0.4 nm thick through palladium clusters before (*a*) and after (*b*) consolidation for 4 ps under a uniaxial pressure of 10 kbar in the *X* direction. The consolidation simulation was carried out using molecular dynamics and an embedded atom potential for Pd [12].

could be subsequently removed by annealing these samples in air. A subsequent calibration of this deviation from stoichiometry [21] indicated that $TiO_{1.89}$ was the actual material produced by the gas condensation method used, as shown in figure 9.5, but that it could be readily oxidized to fully stoichiometric TiO_2 without sacrificing its small grain size (12 nm). Also, if intermediate deviations from stoichiometry were sought, in order to select particular material properties sensitive to the presence of oxygen vacancy defects, they could be readily achieved as well.

The currently available experimental evidence suggests that dislocations are seldom present in nanophase materials [8, 22, 23]. When dislocations are observed, it is primarily in either materials at the upper end of the grain size range [24] or in immobile or locked configurations [8, 9]. The reason for this substantial lack of dislocations is that image forces exist in finite atomic ensembles that tend to pull mobile dislocations out of the grains, especially when they are small, in analogy with the forces on a point electrical charge near a free surface of a conducting body. This paucity of mobile dislocations can have a significant effect upon the mechanical behavior of nanophase materials [25, 26]; see also chapter 13. Since mobile dislocations are not initially available in sufficient numbers to effect plasticity in ultrafine-grained nanophase materials, new mobile dislocations must be created or other deformation mechanisms, such as grain boundary sliding, must come into play.

Dislocations can be created or can multiply from a variety of sources. A simple but representative example is the Frank–Read dislocation source in which a dislocation line, pinned between two pinning points that prevent its forward motion on a slip plane, can bow out between these pinning points to form a new dislocation, if the stress acting on the pinned dislocation is sufficient. The critical stress to operate such a Frank–Read source is inversely proportional to the distance between the pinning points and hence will also be limited by the grain size, which limits the maximum distance between such pinning points. This suggests that dislocation multiplication in nanophase metals will become increasingly difficult as the grain size decreases. The critical stress will eventually, at sufficiently small grain sizes, become larger than the yield stress in the conventional material and could even approach the theoretical yield strength of a perfect, dislocation-free single crystal.

9.2.3 Pores

Nanophase materials consolidated to date from clusters at room temperature have invariably possessed significant porosity ranging from about 25% to less than 5%, as measured by Archimedes densitometry, with a tendency for the larger values in ceramics and the smaller ones in metals. Porosity can also result from the deformation or crystallization synthesis of nanophase materials. Evidence for this porosity has been obtained by positron annihilation spectroscopy (PAS) [6, 27, 28] (discussed in detail in chapter 11 of this book), precise densitometry

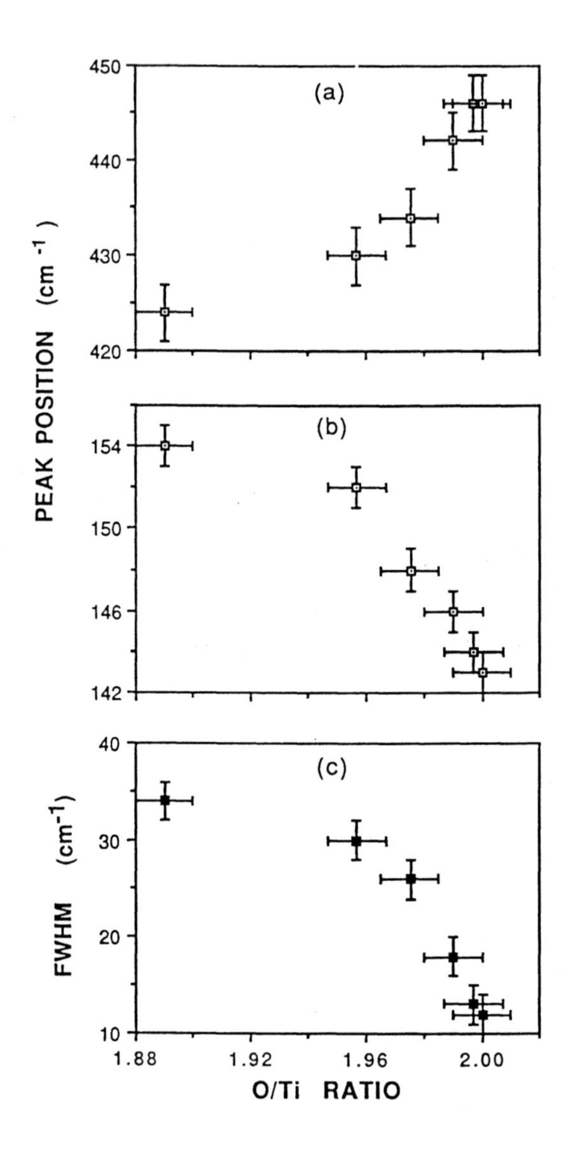

Figure 9.5. Variation with O/Ti ratio of the peak position of (*a*) the rutile '447 cm^{-1}' vibrational mode and (*b*) the anatase '143 cm^{-1}' vibrational mode, as well as (*c*) this anatase mode's FWHM [21].

[29] and porosimetry [30, 31] measurements, and small-angle neutron scattering (SANS) [31, 32].

PAS is primarily sensitive to small pores, ranging from single vacant lattice sites to larger voids, but can probe these structures enclosed in the bulk of the material. Porosimetry measurements using the BET (Brunauer–Emmett–Teller) N_2 adsorption method, on the other hand, probe only pore structures open to the free surface of the sample, but can yield pore size distributions (although with some questions regarding their validity at nanometer pore sizes), which are unavailable from PAS. Densitometry using an Archimedes method integrates over all densities in the sample, including grains, pores (open or closed), and density decrements at defect sites as well. SANS is quite sensitive to pores in the 1–100 nm size range, but deconvolution of the scattering data can be difficult when a broad spectrum of scattering centers is present, as is often the case in nanophase materials. However, even in such a case, it was possible [32] to analyze SANS data from cluster consolidated Pd in terms of a population of small (about 1 nm diameter) voids at grain boundary intersections and grain-sized voids.

A variety of such measurements have shown that the porosity in nanophase metals and ceramics is primarily smaller than or equal to the grain size of the material (although some larger porous flaws have been observed by scanning electron microscopy). The porosity is frequently associated with the grain boundary junctions (triple junctions) and, especially but not only in ceramics, it is interconnected and intersects with the specimen surfaces. Fortunately, consolidation at elevated temperatures should be able to uniformly remove this porosity without sacrificing the ultrafine grain sizes in these materials. Some evidence for this has already been obtained by means of sinter forging nanophase ceramics [33] and uniaxial pressing of nanophase metals at elevated temperatures [32, 34]. (Also see chapter 8.)

It should be noted that atomic diffusion in nanophase materials, which can have a significant bearing on their mechanical properties, such as creep and superplasticity, and other properties as well, has been found to be very rapid compared with conventional materials. Measurements of self- and solute diffusion [35–39] in as-consolidated nanophase metals (Cu, Pd) and ceramics (TiO_2) indicate that atomic transport can be orders of magnitude faster in these materials than in coarser-grained polycrystalline samples. However, this very rapid diffusion appears to be intrinsically coupled with the porous nature of the interfaces in these materials, since the diffusion can be suppressed back to conventional values by sintering samples to full density, as shown by measurements of Hf diffusion in TiO_2 before and after densification by sintering [35]. Thus, in fully densified nanophase materials it is most likely that normal grain boundary diffusion, enhanced by the large number of grain boundaries, will play a significant role in their mechanical behavior and in the ability to impurity dope these materials for a variety of applications.

9.2.4 Grain boundaries

Owing to their ultrafine grain sizes, nanophase materials have a significant fraction of their atoms in grain boundary environments, where they occupy positions relaxed from their normal lattice sites. For conventional high-angle grain boundaries, these relaxations generally extend over about two atom planes on either side of the boundary, with the greatest relaxation existing in the first plane [40]. For an average grain diameter range between 5 and 10 nm, where much of the research on nanophase materials has focused, grain boundary atom percentages range between about 15% and 50% (figure 9.1). Since such a large fraction of their atoms reside in grain boundaries, these interface structures may play a significant role in affecting the properties of nanophase materials.

A number of early investigations on nanocrystalline metals, including x-ray diffraction [41], Mössbauer spectroscopy [42], positron lifetime studies [27, 28], and extended x-ray absorption fine structure (EXAFS) measurements [43, 44], were interpreted in terms of grain boundary atomic structures that may be random, rather than possessing either the short-range or long-range order normally found in the grain boundaries of conventional coarser-grained polycrystalline materials. This randomness was variously associated [45] with either the local structure of individual boundaries (as seen by a local probe such as EXAFS or Mössbauer spectroscopy) or the structural coordination among boundaries (as might be seen by x-ray diffraction). However, direct observations by high-resolution electron microscopy (HREM) [7, 8] have indicated that their structures are rather similar to those of conventional high-angle grain boundaries, as has a very recent EXAFS study of nanophase Cu [46]. An earlier extensive review of many of these results has appeared elsewhere [47]. Very recent molecular dynamics modeling studies indicate that the constrained nature of nanophase grain boundaries may lead to significant detailed structural differences between them and corresponding bicrystalline grain boundaries, but that these differences lead to rather normal properties nonetheless [48].

The direct imaging of grain boundaries with HREM, supplemented by image simulations [49], can avoid the complications that may arise from porosity and other defects in the interpretation of data from less direct methods, such as x-ray scattering and Mössbauer spectroscopy. The only HREM study [7, 8] to date on a nanophase material that has included both experimental observations and complementary image simulations indicated no manifestations of grain boundary structures with random displacements of the type or extent suggested by earlier x-ray studies on nanophase Fe, Pd, and Cu [41, 43, 44]. As shown in figure 9.6, contrast features at the observed grain boundaries that might be associated with disorder did not appear wider than 0.4 nm, indicating that any significant structural disorder which may be present essentially extends no further than the planes immediately adjacent to the boundary plane. Such localized lattice relaxation features are typical of the conventional high-angle grain boundary structures found in coarse-grained metals. Indeed, the experimentally observed

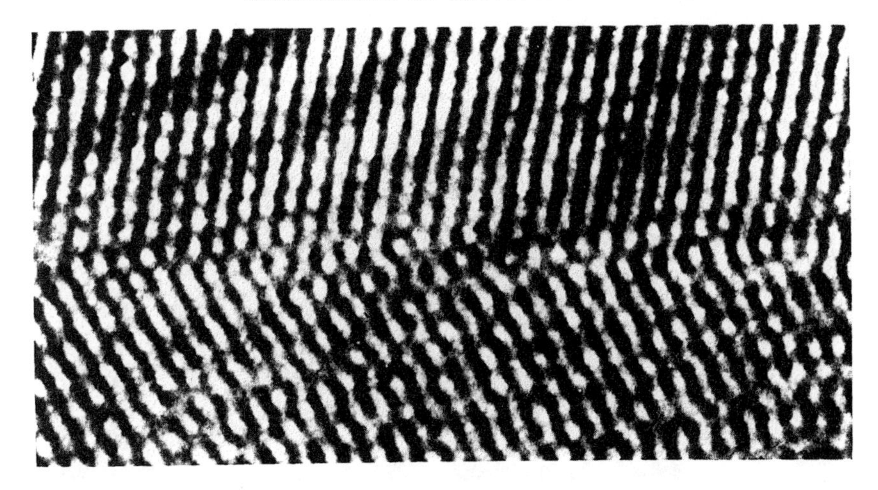

Figure 9.6. High-resolution transmission electron micrograph of a grain boundary in nanophase palladium from an area as shown in figure 9.2. The magnification is indicated by the lattice fringe spacings of 0.225 nm for (111) planes [7, 8].

grain boundary images for nanophase Pd were far more similar to those simulated for a 'perfect' boundary (figure 9.7) than to those for model boundaries possessing the degree of disorder suggested by the early interpretations [45] of less direct methods. Additional HREM observations of grain boundaries in nanophase Cu [10] and Fe alloys [50] produced by surface wear and high-energy ball milling, respectively, appear to support this view. However, there are indications [51, 52] that metastable grain boundary configurations do exist in some cases that can be transformed by low-temperature annealing to more stable states. Whether such behavior is associated with the intrinsic grain boundary structure itself or with extrinsic grain boundary dislocation configurations and/or strains remaining from synthesis or processing remains to be clarified.

Recent nuclear magnetic resonance (NMR) studies of cluster consolidated nanophase silver [53] have indicated that the grain boundaries in this material have an electronic structure that appears to be consistent with that expected for conventional high-angle boundaries. As shown in figure 9.8, it has been possible to approximately model the measured NMR spectrum from nanophase silver using a spin density decrement at the grain boundary of about 1% and an electronic width about twice that of the observed structural width. Since electronic healing distances for metals are about an interatomic spacing, such a picture seems quite natural.

The nanophase grain boundaries typically observed by HREM of cluster assembled nanophase materials appear to take up rather low-energy configurations exhibiting flat facets interspersed with steps, such as that seen in figure 9.6. Weak grain boundary grooving seen within islands of grains by STM and AFM [11] appears to support such a view. Such low-energy structures

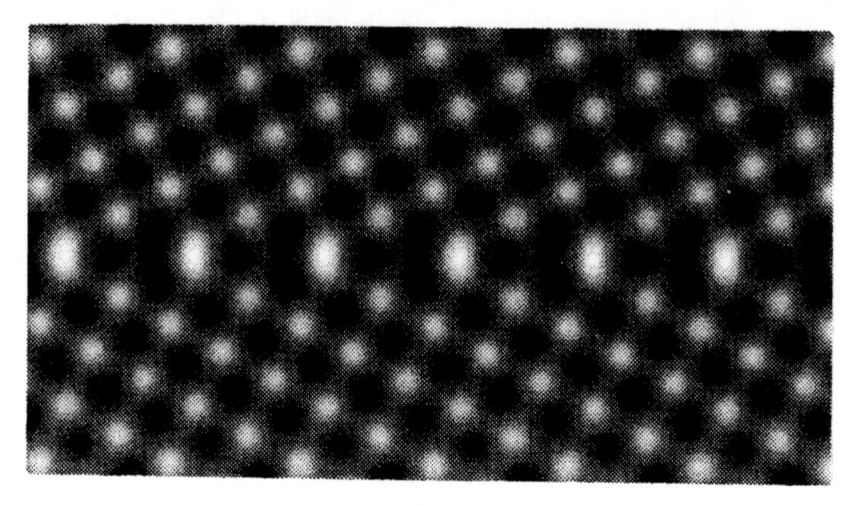

Figure 9.7. Image simulation for a 'perfect' $\Sigma 5$ symmetric $\langle 001 \rangle$ tilt boundary with no atomic displacements in 7.6 nm thick palladium using microscope parameters and imaging conditions consistent with those used during HREM experimental observations [7, 8], as shown in figure 9.6.

could only arise if sufficient local atomic motion occurred during the cluster-consolidation process to allow the system to reach a local energy minimum. These observations suggest at least two conclusions [54]: first, that the atoms that constitute the grain boundary volume in nanophase materials have sufficient mobility during cluster consolidation to accommodate themselves into relatively low-energy grain boundary configurations; and second, that the local driving forces for grain growth are relatively small, despite the large amount of energy stored in the many grain boundaries in these materials.

9.2.5 Stability

The narrow grain size distributions, essentially equiaxed grain morphologies, and low-energy grain boundary structures in nanophase materials suggest that the inherent resistance to grain growth observed for cluster assembled nanophase materials results primarily from a sort of frustration [55]. It appears that the narrow grain size distributions normally observed in these cluster assembled materials coupled with their relatively flat and faceted grain boundary configurations place these nanophase structures in a local minimum in energy from which they are not easily extricated. Such frustration would also likely be increased by the multiplicity of grain boundary junctions in these materials. There are normally no really large grains to grow at the expense of small ones through an Ostwald ripening process, and the grain boundaries, being essentially flat, have no local curvature to tell them in which direction to migrate. Their

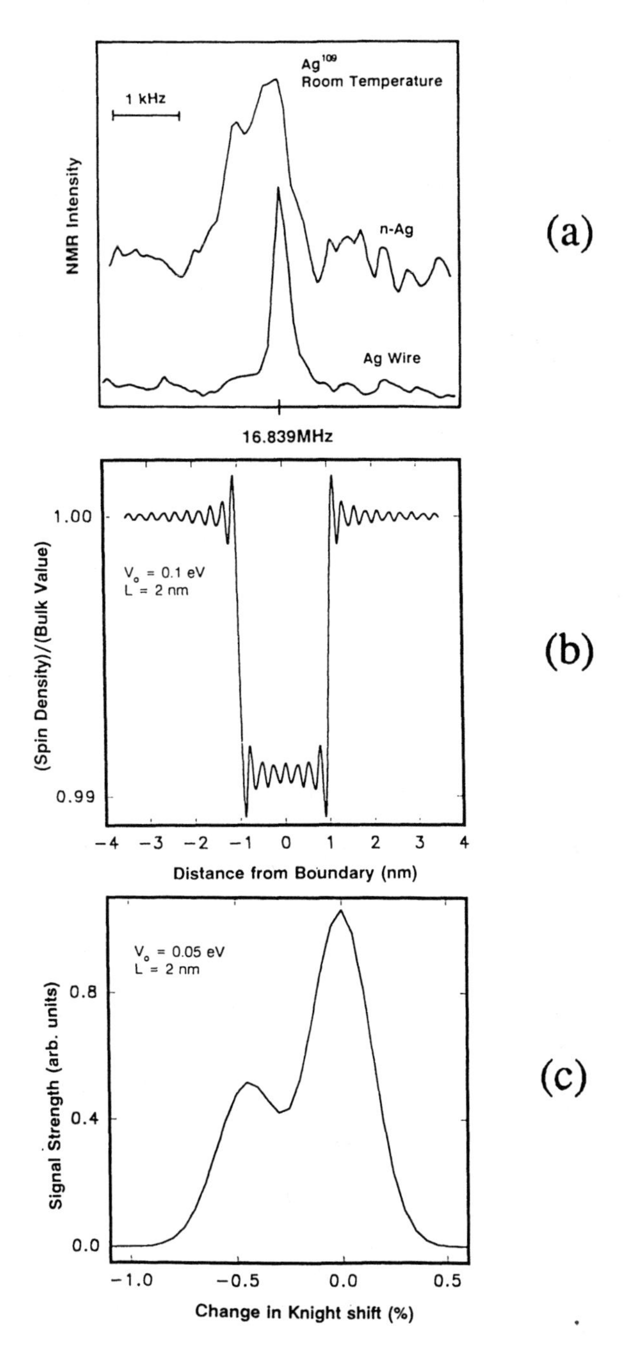

Figure 9.8. NMR spectra from nanophase silver and a coarse-grained silver wire (*a*), relative variations in the spin density in the vicinity of a model grain boundary (*b*), and the NMR spectrum (*c*) simulated using the simple model of (*b*) [53].

stability thus appears to be analogous to that of a variety of closed-cell foam structures with narrow cell size distributions, which are deeply metastable despite their large stored surface energy. Such a picture appears to have some theoretical support [56].

Exceptions to this frustrated grain growth behavior would be expected if considerably broader grain size distributions were accidentally present in a sample, which would allow a few larger grains to grow at the expense of smaller ones, or if significant grain boundary contamination were present, allowing enhanced stabilization of the small grain sizes to further elevated temperatures. Occasional observations of each of these types of behavior have been made. One could, of course, intentionally stabilize against grain growth by appropriate doping by insoluble elements or composite formation in the grain boundaries. For these cluster assembled materials, such stabilization should be especially easy, since the grain boundaries are available as readily accessible cluster surfaces prior to consolidation. The ability to retain the ultrafine grain sizes of nanophase materials is important when one considers the fact that it is their small grain size and large number of grain boundaries that determine to a large extent their special properties.

9.2.6 Strains

Strains are a natural component of nanophase materials. Simply owing to the large number of grain boundaries, and the concomitant short distances between them, the intrinsic strains associated with such interfaces [40, 57] are always present in these nanostructured materials. Beyond these intrinsic strains, however, there may also be present extrinsic strains associated with the particular synthesis method. For example, intense plastic deformation synthesis of nanophase materials may lead to additional residual strains [58] that can be subsequently relieved via low-temperature annealing, leaving only those intrinsic strains due to the presence of the high-angle grain boundaries. Evidence for the strains in nanophase materials is now becoming available. X-ray line broadening investigations of both cluster consolidated [29] and ball mill deformed metallic nanophase samples [59] have indicated residual strains of about 0.5–1%, which are consistent in magnitude with the strains expected from conventional high-angle grain boundaries [40] in the grain size ranges investigated. Recently, NMR measurements [60] on cluster consolidated nanophase copper have also yielded estimated strain values (about 0.7%) in this range. A very recent study [61] of the strains in nanophase Cu and Pd using x-ray line broadening has been able to separate the effects from strains and faults in these materials and to elucidate the grain-size dependence of the strains. Clearly, further careful work remains to elucidate the nature and magnitudes of the strains in nanophase materials and their contributions to the properties of these new materials.

9.3 CONCLUSIONS

After almost a decade of research, we are only really just beginning to understand the essential features of the structures of nanophase materials assembled from cluster building blocks created by a variety of methods or synthesized by other techniques. Much of the currently available information results from studies on nanophase materials made by the gas condensation method. These results have shown that clusters and resulting grains contain few if any mobile dislocations owing to image stresses acting on these finite atomic ensembles. However, sessile dislocations are observed along with frequent twin boundaries, presumably formed during the cluster synthesis. The mobile atomic defects in elemental nanophase solids are probably at equilibrium, but clusters of vacancies or voids may be present and deviations from stoichiometry in compounds are certainly rather common but easily controlled.

The grains formed by the consolidation of clusters are also essentially equiaxed and possess random crystallographic orientations, presumably a result of the grain boundary sliding mechanisms that are operative during consolidation. Yet the grains appear to be locally extruded, probably by diffusional mechanisms, in order to help fill in the open intergrain regions during consolidation. The interfaces formed during consolidation are similar in structure to those observed in coarse-grained polycrystals, but their detailed atomic structure needs to be further elucidated to see whether this similarity is exact or not. Owing to the high number density of grain boundaries, intrinsic strains are present in nanophase materials. The magnitude of these strains is consistent with the expectations from theoretical modeling studies, but better measurements are needed as are more realistic models. The nature and source of extrinsic strains from the various synthesis and processing routes for creating nanophase materials need to be more fully investigated.

Porosity exists on a variety of length scales from small vacancy clusters at the grain boundary junctions formed by consolidation to missing grains and larger porous flaws. However, by careful control of the synthesis and processing steps and consolidation under appropriate conditions, it appears that this porosity can be contolled to retain a desired degree of high surface area or removed without affecting the grain size of the nanophase material. Much work remains in all of these areas, however, if we are to fully understand the structures of nanophase materials and how their structures relate to their properties.

Finally, nanophase materials, in addition to having a variety of unique grain-size dependent properties, should also be valuable for studying the average properties of grain boundaries and interfaces in general. The high number density of these defects in nanophase materials enhances their influence on macroscopic properties, allowing their effects to be studied by a variety of experimental techniques. In order that the grain boundary properties can be accurately measured, however, specimen porosity will need to be removed via consolidation at elevated temperature and/or pressure so that its property contributions can be eliminated. Fortunately, this appears to be possible in cluster consolidated nanophase materials without sacrificing their ultrafine grain sizes. By varying the grain size in a set of experiments, the contributions from interfaces and their junctions in nanophase materials could be effectively separated. Our understanding of the atomic-scale nanostructures of nanophase materials and the properties of grain boundaries should thereby be permanently enriched.

REFERENCES

[1] Gleiter H 1989 *Prog. Mater. Sci.* **33** 223

[2] Hadjipanayis G C and Siegel R W (ed) 1994 *Nanophase Materials: Synthesis–Properties–Applications* (Dordrecht: Kluwer)

[3] Siegel R W 1994 *Encyclopedia of Applied Physics* vol 11, ed G L Trigg (Weinheim: VCH) p 173

[4] 1995 *Proc. 2nd Int. Conf. on Nanostruct. Mater; Nanostruct. Mater.* **6**

[5] Siegel R W 1991 *Annu. Rev. Mater. Sci.* **21** 149

[6] Siegel R W, Ramasamy S, Hahn H, Li Z, Lu T and Gronsky R 1988 *J. Mater. Res.* **3** 1367

[7] Thomas G J, Siegel R W and Eastman J A 1989 *Mater. Res. Soc. Symp. Proc.* **153** 13

[8] Thomas G J, Siegel R W and Eastman J A 1990 *Scr. Metall. Mater.* **24** 201

[9] Wunderlich W, Ishida Y and Maurer R 1990 *Scr. Metall. Mater.* **24** 403

[10] Ganapathi S K and Rigney D A 1990 *Scr. Metall. Mater.* **24** 1675

[11] Sattler K, Raina G, Ge M, Venkateswaran N, Xhie J, Liao Y X and Siegel R W 1994 *J. Appl. Phys.* **76** 546

[12] Liu C-L, Adams J B and Siegel R W 1994 *Nanostruct. Mater.* **4** 265

[13] Ashby M F and Verrall R A 1973 *Acta Metall.* **21** 149

[14] Balluffi R W 1978 *J. Nucl. Mater.* **69 & 70** 240

[15] Young F W Jr 1978 *J. Nucl. Mater.* **69 & 70** 310

[16] Balluffi R W 1980 *Grain Boundary Structure and Kinetics* ed R W Balluffi (Metals Park, OH: American Society of Metals) p 297

[17] Eastman J A, Liao Y X, Narayanasamy A and Siegel R W 1989 *Mater. Res. Soc. Symp. Proc.* **155** 255

[18] Balachandran U, Siegel R W and Askew T 1995 *Nanostruct. Mater.* **5** 505

[19] Melendres C A, Narayanasamy A, Maroni V A and Siegel R W 1989 *J. Mater. Res.* **4** 1246

[20] Parker J C and Siegel R W 1990 *J. Mater. Res.* **5** 1246

[21] Parker J C and Siegel R W 1990 *Appl. Phys. Lett.* **57** 943

[22] Gao P and Gleiter H 1987 *Acta Metall.* **35** 1571

[23] Milligan W W, Hackney S A, Ke M and Aifantis E C 1993 *Nanostruct. Mater.* **2** 267

[24] Morris D G and Morris M A 1991 *Acta Metall. Mater.* **39** 1763

[25] Siegel R W and Fougere G E 1994 *Nanophase Materials: Synthesis–Properties–Applications* ed G J Hadjipanayis and R W Siegel (Dordrecht: Kluwer) p 233

[26] Siegel R W and Fougere G E 1995 *Mater. Res. Soc. Symp. Proc.* **362** 219

[27] Schaefer H E, Würschum R, Scheytt M, Birringer R and Gleiter H 1987 *Mater. Sci. Forum* **15–18** 955
[28] Schaefer H E, Würschum R, Birringer R and Gleiter H 1988 *Phys. Rev.* B **38** 9545
[29] Nieman G W, Weertman J R and Siegel R W 1991 *J. Mater. Res.* **6** 1012
[30] Hahn H, Logas J and Averback R S 1990 *J. Mater. Res.* **5** 609
[31] Wagner W, Averback R S, Hahn H, Petry W and Wiedenmann A 1991 *J. Mater. Res.* **6** 2193
[32] Sanders P G, Weertman J R, Barker J G and Siegel R W 1993 *Scr. Metall. Mater.* **29** 91
[33] Owen D M and Chokshi A H 1993 *Nanostruct. Mater.* **2** 181
[34] Fougere G E, Weertman J R and Siegel R W 1995 *Nanostruct. Mater.* **5** 127
[35] Averback R S, Hahn H, Höfler H J, Logas J L and Chen T C 1989 *Mater. Res. Soc. Symp. Proc.* **153** 3
[36] Horváth J, Birringer R and Gleiter H 1987 *Solid State Commun.* **62** 319
[37] Horváth J 1989 *Defect Diffusion Forum* **66–69** 207
[38] Hahn H, Höfler H and Averback R S 1989 *Defect Diffusion Forum* **66–69** 549
[39] Schumacher S, Birringer R, Straub R and Gleiter H 1989 *Acta Metall.* **37** 2485
[40] Wolf D and Lutsko J F 1988 *Phys. Rev. Lett.* **60** 1170
Wolf D and Yip S (ed) 1992 *Materials Interfaces: Atomic-Level Structure and Properties* (London: Chapman and Hall)
[41] Zhu X, Birringer R, Herr U and Gleiter H 1987 *Phys. Rev.* B **35** 9085
[42] Herr U, Jing J, Birringer R, Gonser U and Gleiter H 1987 *Appl. Phys. Lett.* **50** 472
[43] Haubold T, Birringer R, Lengeler B and Gleiter H 1988 *J. Less-Common Met.* **145** 557
[44] Haubold T, Birringer R, Lengeler B and Gleiter H 1989 *Phys. Lett.* **135A** 461
[45] Birringer R and Gleiter H 1988 *Encyclopedia of Materials Science and Engineering* suppl. vol 1, ed R W Cahn (Oxford: Pergamon) p 339
[46] Stern E A, Siegel R W, Newville M, Sanders P G and Haskel D 1995 *Phys. Rev. Lett.* **75** 3874
[47] Siegel R W 1992 *Materials Interfaces: Atomic-Level Structure and Properties* ed D Wolf and S Yip (London: Chapman and Hall) p 431
[48] Phillpot S R, Wolf D and Gleiter H 1995 *J. Appl. Phys.* at press
[49] Balluffi R W 1984 *Ultramicroscopy* **14** 155
[50] Trudeau M L, Van Neste A and Schulz R 1991 *Mater. Res. Soc. Symp. Proc.* **206** 487
[51] Tschöpe A and Birringer R 1993 *Acta Metall. Mater.* **41** 2791
[52] Valiev R Z, Krasilnikov N A and Tsenev N K 1991 *Mater. Sci. Eng.* A **137** 35
[53] Suits B H, Siegel R W and Liao Y X 1993 *Nanostruct. Mater.* **2** 597
[54] Siegel R W and Thomas G J 1992 *Ultramicroscopy* **40** 376
[55] Siegel R W 1990 *Mater. Res. Soc. Symp. Proc.* **196** 59
[56] Rivier N 1992 *Physics and Chemistry of Finite Systems: from Clusters to Crystals* ed P Jena, S N Khanna and B K Rao (Dordrecht: Kluwer) p 189
[57] Cammarata R and Sieradzki K 1989 *Phys. Rev. Lett.* **62** 2005
[58] Valiev R Z 1993 *Mechanical Properties and Deformation Behavior of Materials Having Ultra-Fine Microstructures* ed M Nastasi, D M Parkin and H Gleiter (Dordrecht: Kluwer) p 303
[59] Eckert J, Holzer J C, Krill C E III and Johnson W L 1992 *J. Mater. Res.* **7** 1751; 1992 *Mater. Res. Soc. Symp. Proc.* **238** 745

[60] Suits B H, Meng M, Siegel R W and Liao Y X 1994 *J. Mater. Res.* **9** 336
[61] Sanders P G, Witney A B, Weertan J R, Valiev R Z and Siegel R W 1995 *Mater. Sci. Eng.* A **204** 7

Chapter 10

Characterization by scattering techniques and EXAFS

Jörg Weissmüller

10.1 INTRODUCTION

When new types of material are investigated, one of the major aims is the elucidation of the atomic structure, a knowledge of which is of central importance for the understanding of the properties of the material. Among the experimental techniques applied in structural investigations, scattering methods and extended x-ray absorption fine-structure spectroscopy (EXAFS) have often played key roles, largely because they can provide detailed information on the structure in real space. These techniques complement microscopic imaging methods, such as transmission electron microscopy (TEM). While microscopic techniques have the advantage of providing the most direct information on the real space structure, scattering techniques and EXAFS examine larger and sometimes more representative sample volumes. In addition, they can often be applied to the sample as it is, avoiding potential artifacts caused by the specimen preparation.

When the size of the crystallites in a polycrystalline material is progressively reduced from macroscopic dimensions down to the scale of a few nanometers, the consequences for the atomic structure are the following. First, interfaces are created, in which the local atomic arrangements (short-range order) are different from those of the crystal lattice. As the grain size is reduced, the number of interfaces increases, and the fraction of atoms on interface sites becomes comparable to the number of crystal lattice atoms. Hence, the overall properties of the solid will no longer be determined by the atomic interactions in the crystal lattice alone. Instead, the material can have novel properties reflecting the contribution from the interfaces. Second, the atomic structure in the interior of crystallites is modified through the introduction of defects, strain fields, or short-range correlated static or dynamic displacements of atoms from their ideal crystal lattice positions. In addition, nanometer-sized isolated crystallites and bulk nanocrystalline solids can differ from macroscopic crystals with respect to their crystallographic phase and their lattice constant. Many properties will reflect the combined effect of these structural changes, that is of the reduction of the length scale on which there is coherency in the atomic arrangement (long-range order), of the modified crystallographic structure, and of the introduction of interfaces with an atomic short-range order (SRO) different from the one in the crystal lattice. Understanding nanocrystalline solids requires a detailed characterization of all of these aspects of their atomic structure.

A number of early scattering and EXAFS studies of nanocrystalline materials, aimed at obtaining the sort of information addressed above, had led to contradictory results on some important issues, in particular on the question of the atomic SRO in the grain boundaries of pure nanocrystalline metals. Early experiments indicated a grain boundary phase with a 'gas like' atomic SRO, considerably less ordered and less dense than any other solid state of matter (figure 10.1(*a*)). Although the notion of a gas like grain boundary phase greatly stimulated the interest in nanocrystalline solids, it was not completely accepted, and evidence was brought forward which implied ordered, as opposed to disordered grain, boundaries (figure 10.1(*b*)). More recently, studies on samples of considerably improved quality have led to more convergent conclusions. Compared to earlier results these experiments reveal that reducing the scale of the microstructure of a material to the nanometer range induces more subtle, but still fascinating, changes in the atomic structure and in the physical properties. It is the purpose of this chapter to discuss the extent to which our present understanding and the available experimental data allow the characterization of the atomic-scale structure of these materials. The scope will be limited to materials where the nanometer-scale structure is the dominant structural property. We shall be concerned with those solids in which nanometer-scale particles are bounded by surfaces, high-angle grain boundaries, or incoherent interfaces, and for which the experimental or technological interest is focused on the modification of the material properties induced by the small particle size and by the presence of the interfaces. We shall distinguish *isolated particles*, that is particles bounded by free surfaces or imbedded in a medium which interacts only weakly with the matter in the particles (figure 10.1(*c*)), from *nanocrystalline solids*, that is polycrystals with a very fine grain size, in which the particles are bounded by high-angle grain boundaries or by solid–solid interfaces (figures 10.1(*a*),(*b*)).

Scattering studies are used routinely to characterize the crystallographic structure and microstructure of nanocrystalline solids, and there is consequently a large amount of experimental scattering data on a wide variety of systems in the literature. Work has been selected that is either exemplary for the way in which the experimental techniques can be used or else which supplies key evidence on those properties of nanostructured solids which are the most direct consequences of scaling and of the presence of grain boundaries or interfaces. Results from experimental methods other than scattering are reviewed elsewhere in this

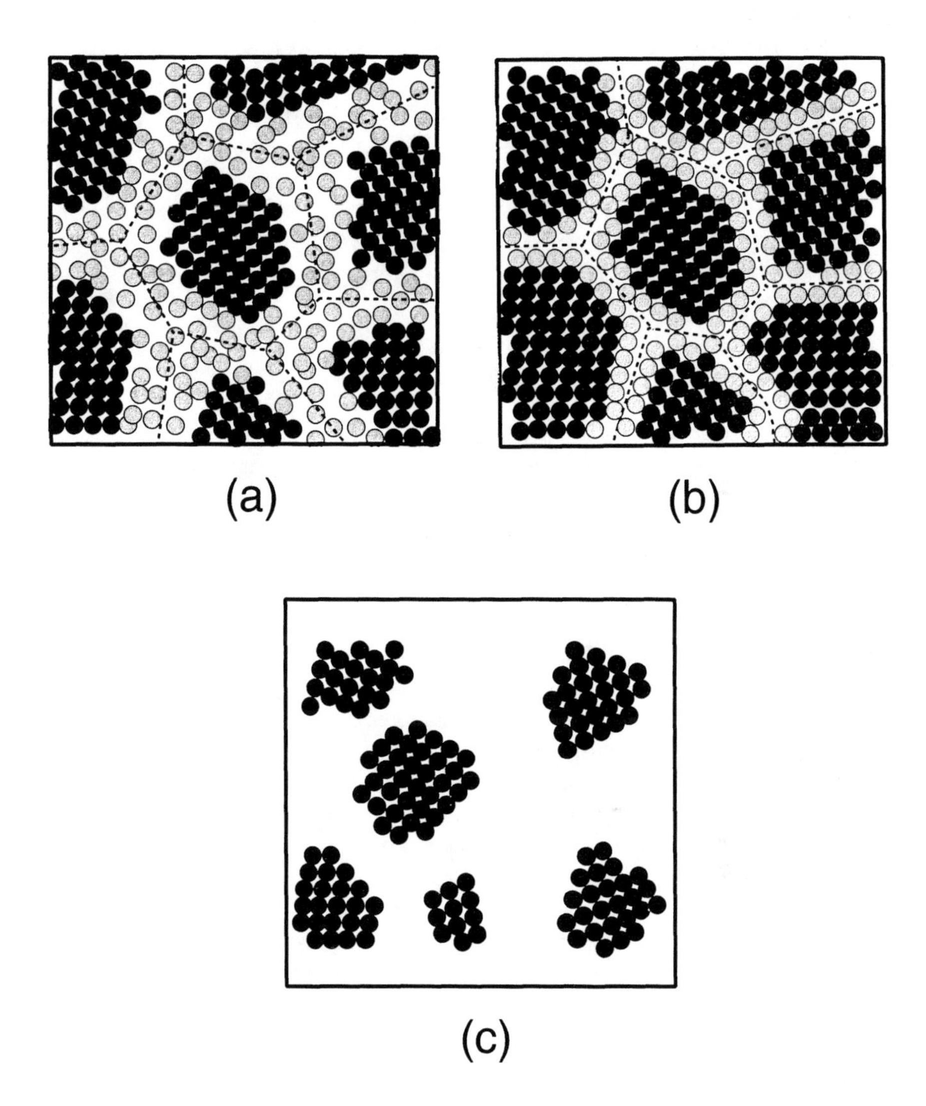

Figure 10.1. Schematic representation of nanocrystalline solids (a, b) and isolated nanometer particles (c). Circles symbolize the atoms in the interior of the crystallites (dark shaded) and at the interfaces (light shaded). Dashed lines denote the grain boundaries. (a) Nanocrystalline solid with a disordered grain boundary component (grain boundary atoms not on lattice sites); (b) nanocrystalline solid where all atoms, including those in the grain boundaries, are located on crystal lattice sites (compare the 'nonreconstructed nanocrystal' model in section 10.2.2).

book and in recent overview papers [1–6]. The most relevant consequences of the finite grain size and the modified atomic SRO in the interfaces of nanocrystalline solids for scattering and EXAFS studies are summarized in section 10.2 as a theoretical background for the discussion of experimental results in sections 10.3–10.8.

10.2 REAL-SPACE INFORMATION FROM EXPERIMENTAL RECIPROCAL-SPACE DATA: ATOMIC DISTRIBUTION FUNCTION

10.2.1 Real-space–reciprocal-space relations

The common feature of both scattering experiments and EXAFS is the interference between photon, electron, or neutron waves originating at scatterers centered at the atomic positions. For single-component systems, the most important difference between scattering experiments and EXAFS is that with the former, incoming and scattered waves are coherent on a length scale of the order of a micron, whereas in the latter the finite lifetime of the photoelectron limits the coherence to distances of the order of a few Å. Therefore, scattering data contain information on real-space structures extending from atomic dimensions over the dimension of the crystallites (lattice symmetry, extended defects such as strain fields, grain size) up to still larger dimensions characteristic of the microstructure (spatial arrangement of different phases, porosity), while EXAFS probes exclusively the atomic short-range order. For multi-component systems, the scattering intensity involves contributions from all components, whereas EXAFS probes only the atomic short-range order around the excited component.

Both scattering and EXAFS data allow the calculation of a function which is directly related to the real-space structure, the atomic distribution function $\rho(r)$. By definition, $\rho(r)$ is the spherical average of the density of atoms surrounding an average central atom at a distance r [7]. The effects of nanometer-scale structuring on the scattering intensity and EXAFS can be discussed in terms of this function. The atomic distribution function is generally determined in EXAFS studies, and to provide a background for the discussion of EXAFS results on nanostructures it is therefore sufficient to derive the atomic distribution function. Scattering data are, however, more frequently discussed in terms of scattering intensity or interference function. Hence, the effect of nanometer-scale structuring on these reciprocal-space functions will also be addressed in the following sections.

For isotropic materials, the kinematical scattering theory relates the interference function $P(k)$ to $\rho(r)$ by

$$P(k) = \int_{r=0}^{\infty} \left(4\pi r^2 (\rho(r) - \langle\rho\rangle) \frac{\sin(kr)}{kr} \right) \mathrm{d}r \tag{10.1}$$

and its back-transform

$$4\pi r(\rho(r) - \langle\rho\rangle) = \frac{2}{\pi}\int_0^\infty kP(k)\sin(kr)\,dk. \tag{10.2}$$

Here k is the wavevector, which is related to the scattering angle, 2θ, and to the wavelength λ, by $k = 4\pi\sin(\theta)/\lambda$, and $\langle\rho\rangle$ is the macroscopic atomic density of the sample. For the example of x-ray scattering, the coherently scattered intensity is

$$I(k) = V\langle\rho\rangle ff^*(P(k) + 1). \tag{10.3}$$

V is the sample volume; f and f^* denote the atomic form factor and its complex conjugate. Except for the thermal diffuse background (which is small at low k), $P(k) = -1$ and $I(k) = 0$ in coarse-grained polycrystalline, defect-free matter in the k-region between the Bragg reflections.

Experimental techniques and quantitative analysis of scattering data are described in [8]. Besides a finite angular resolution (instrumental line broadening), the most important experimental limitation is the accessible range in reciprocal space, the maximum experimental wavenumber being $k_{max} = 4\pi/\lambda$. From equation (10.2) and the convolution theorem of Fourier transforms, this is seen to imply a broadening of sharp peaks in the atomic distribution function to a width of the order of $\Delta r \approx \pi/k_{max}$. With laboratory x-ray sources, which are employed in most experimental studies, $\Delta r \approx 0.2$ Å for Mo Kα or Ag Kα radiation. Hence, part of the information on the detailed distribution of interatomic spacings, with a natural width of the order of 3% of the nearest-neighbor spacing for metals at room temperature (compare section 10.3.6), is missing from the experimental data. A better resolution can be achieved with smaller-wavelength radiation from synchrotron sources [9] and, with potentially less experimental effort, with energy filtered electron diffraction data [10]. These approaches have apparently not yet been employed in systematic structural studies of nanocrystalline solids.

10.2.2 Atomic distribution and interference functions for arrays of nanometer particles

The effect of a finite grain size on the atomic distribution and interference functions is reviewed in detail in a number of textbooks [7, 11–13]. Some of the particularly simple relations for $\rho(r)$ and $P(k)$ in isotropic matter are summarized below.

In an array of small particles or in a nanocrystalline solid, the atomic distribution function can be expressed as the sum of an 'intragrain' [14] part, consisting of contributions from pairs of atoms located in the same particle, and of an 'intergrain ' part, accounting for the remaining interatomic spacings. These two parts can be discussed in a relatively concise form if the arrangement of particles meets two conditions: first, the particles must be homogeneous, with all atoms on crystal lattice positions. This implies that the intragrain part of the atomic distribution function in the particulate system contains only those interatomic spacings which occur also in the extended lattice; it is only the number of the neighboring atoms which is reduced due to the finite size of the crystal lattice. The reduction is described by the intragrain correlation function $H(r)$, which is the probability that a vector of length r originating at a randomly chosen central atom terminates inside the particle containing the central atom (figure 10.2(*a*–*c*)) [7, 12]. The second condition requires the orientations of the crystal lattices in the particles to be random. This randomness implies that, for any given central atom, the *a priori* probability of finding an atomic neighbor in a small volume $\mathrm{d}V$ at a distance r in a neighboring particle has the same value, $\langle\rho\rangle^V\,\mathrm{d}V$, everywhere in that particle. The intergrain correlation function, $H'(r)$, is the probability that a vector of length r originating at a randomly chosen central atom terminates inside one of the other particles. For this simple 'non-reconstructed' nanocrystal model [15, 16], illustrated in figures 10.1(*b*) and 10.1(*c*), the atoms in the outer layers of each crystallite occupy lattice positions, in other words the grain boundary region is not reconstructed. The total atomic distribution function of the arrangement of particles, that is the sum of intragrain and intergrain parts, is therefore

$$\rho^{nn}(r) = \rho^V(r)H(r) + \langle\rho\rangle^V H'(r) \tag{10.4}$$

where $\rho^V(r)$ and $\langle\rho\rangle^V$ are the atomic distribution function and average atomic density, respectively, of an extended crystal lattice with the same structure as the interior of the particles (compare figure 10.2(*d*)).

With respect to nanocrystalline solids prepared by the inert gas condensation technique [17, 18], the two assumptions provide a not too unrealistic first approximation: with the possible exception of the grain boundary region, the materials are known to be essentially crystalline, and the specific preparation process, consolidation of randomly oriented nanometer-sized crystallites, is expected to result in a random distribution of crystal orientations and a random crystallographic orientation of the grain boundary planes.

From equation (10.1) and the convolution theorem of Fourier transforms, the interference function $P^{nn}(k)$ for the model nanocrystal is obtained as (convolution denoted by $\otimes$)

$$P^{nn}(k) = P^V(k) \otimes W(k) + S(k) \tag{10.5}$$

with

$$W(k) = \frac{1}{\pi}\int_{r=0}^{\infty} H(r)\cos(kr)\,\mathrm{d}r \tag{10.6}$$

and

$$S(k) = 4\pi\langle\rho\rangle^V \int_{r=0}^{\infty} r^2\left(H(r) + H'(r) - \frac{\langle\rho\rangle}{\langle\rho\rangle^V}\right)\frac{\sin(kr)}{kr}\,\mathrm{d}r. \tag{10.7}$$

The wide-angle scattering intensity is represented by the first term on the right-hand side of (10.5), which is the interference function of the extended lattice, $P^V(k)$, convoluted with the cosine transform of the intragrain correlation function, $W(k)$. Therefore, $W(k)$ is the broadening of the Bragg reflections due to the finite size of the particles. The second term is the sine transform of functions without atomic-scale structure, and does not contribute intensity in the wide-angle scattering region. Instead, this term represents the small-angle scattering of the nanocrystalline solid [12, 13].

An important property of the intragrain correlation function is the following: independent of the shape of the particles, H decreases from the value unity at $r = 0$ with an initial slope which is proportional to the surface area of the particle [11, 19]. In terms of the specific surface area, $\alpha_S = A/V_P$, with V_P the total volume occupied by all the particles in a sample and A the total free surface area of a system of isolated particles with the same size and shape, H at small r is given by [13, 16]

$$H(r) \approx 1 - \frac{\alpha_S}{4} r \qquad (r \ll \langle D \rangle). \tag{10.8}$$

In a dense nanocrystal, since two surfaces combine to form a grain boundary, the specific grain boundary area α_{GB} (total grain boundary area over sample volume) is half the specific surface area.

10.2.3 Comparison of a non-reconstructed nanocrystalline solid to isolated particles

Equation (10.5) has the important consequence that wide-angle scattering contains information on the intragrain part of the atomic distribution function only (it does not depend on the intergrain correlation function). Note that the total atomic distribution function of a nanocrystalline solid contains a distribution (the intergrain contribution $\langle \rho \rangle^V H'(r)$) of non-lattice interatomic spacings which, similar to that of a gas, has no atomic-scale structure. In contrast to what intuition might suggest, this broad distribution does not give rise to a diffuse background in the interference function. Instead, the considerations above demonstrate that the interference function of a non-reconstructed nanocrystal differs from that of a coarse-grained polycrystal only by the size induced peak broadening. Furthermore, the interference function in the wide-angle region of the scattering pattern is identical for the dense nanocrystal on the one hand and for isolated particles on the other (for a previous derivation of this result see [20]). Hence, the atomic short-range order in the plane of the grain boundary, involving interatomic distances between atoms located in different crystallites, cannot be determined in detail by wide-angle scattering studies. On the other hand, small-angle scattering does depend on the intergrain correlation function, and can in principle provide information on the geometry of the grain boundaries, e.g. the thickness or excess volume, in a bulk nanocrystalline solid. In the limiting case of a very dilute array of particles (isolated particle case), the intergrain correlation function vanishes, and equations (10.5) and (10.7) combine to the familiar equation for small-angle scattering of an array of isolated particles [13].

10.2.4 Nanocrystalline solid with a disordered grain boundary component

In a real poly- or nanocrystalline solid, the atomic structure in the grain boundary region may be reconstructed, that is, atoms are likely to be displaced from their lattice positions to new, non-lattice equilibrium positions. The distribution of the interatomic spacings involving these positions may display preferred distances, that is there may be SRO. On the other hand, it is conceivable that there are a large number of different local atomic structures, corresponding to different orientations of the neighboring crystal lattices relative to the grain boundary plane, and that on average over all those configurations there is little atomic SRO [17]. A somewhat simplistic description of this structure considers a grain boundary layer of finite thickness, in which the atomic positions are completely random, with an atomic density $\langle\rho\rangle^{GB}$ in the grain boundary (figure 10.1(*a*)). As a result of the randomness, the atoms in the layer do not interfere, in the wide-angle region of the scattering pattern, with each other or with the matter in the crystallites. Therefore, the wide-angle interference function of this 'disordered nanocrystal' model, $P^{dn}(k)$, can be computed from a two-phase expression for the atomic distribution function [15, 16]

$$\rho^{dn}(r) = x_L \rho^{nn}(r) + (1 - x_L)\langle\rho\rangle^{GB} \tag{10.9}$$

with x_L the fraction of atoms on crystal lattice sites. With equation (10.5), $P^{dn}(k)$ is

$$P^{dn}(k) = x_L P^V(k) \otimes W(k). \tag{10.10}$$

Due to the complete randomness of the atomic position in the grain boundary layer, the 'disordered nanocrystal model' involves interatomic distances which are considerably smaller than the atomic radius. Due to the strong repulsive forces between atoms at small interatomic distance, such configurations cannot occur in a real solid. The incorporation of these configurations in the model is justified *a priori* solely by the ease of the analysis of the scattering problem for the simple model. The good agreement of equation (10.9) with experimental results (see section 10.4.2) indicates that no serious error occurs because of the simplification.

These results suggest two alternative ways to obtain quantitative information on the atomic SRO in the grain boundaries. First, as the interference function of an extended ideal crystal lattice, $P^V(k)$ has the negative value -1 between the Bragg reflections, where the scattered intensity vanishes (compare equation (10.3)), equation (10.10) predicts a diffuse background of magnitude $V\langle\rho\rangle ff^*(1 - x_L)$ in the scattering pattern. In principle, x_L can therefore be determined from the scattering intensity. It is noted that the diffuse background may be hard to measure in experimental diffraction data; only when the crystals are very fine grained will there be a significant fraction of interface atoms (their number scales with the total grain boundary area). However, since the profile of the Bragg reflections widens when the grain size is reduced, there will also be appreciable overlap of the tails of the reflections in the fine-grained case. As the functional form of the reflection tails depends upon *a priori* unknown details of the distribution of the crystal size and shape, and of the lattice strain, it will be difficult to distinguish the reflection tail contribution from the diffuse background indicative of non-lattice atoms at the grain boundaries. The second alternative relies on an evaluation of the atomic distribution function. In contrast to the interference function, the atomic distribution function depends in a simple way on the average grain size, and it is little affected by lattice strain. Combining equations (10.4), (10.8), and (10.9), that part of the atomic distribution function which displays SRO is seen to be $x_L H(r) \rho^V(r)$. As $\rho^V(r)$ is known, the coefficient x_L can be determined from experimental atomic distribution functions. To do so, the relative reduction in coordination number, that is the ratio of the experimental number of atoms in the respective coordination shell in the nanocrystalline material, Z, over the coordination number of the ideal crystal lattice, Z^{ideal}, is plotted versus the interatomic distance, and the function $x_L H(r)$ is determined as the straight line of best fit to these data. As $H(0) = 1$ (compare (10.8)), the value of that function at $r = 0$ corresponds to the fraction of atoms on crystal lattice sites, x_L.

10.2.5 Particles with a distribution of sizes

In the context of scattering studies, a useful definition of the particle size D is the equivalent sphere diameter, that is the diameter of a sphere with the same volume V as the particle [21]: $D = (6V/\pi)^{1/3}$. Let the total volume occupied by all the particles in a sample be V_P. The size distribution $n(D)$ can be defined in such a way that in the sample there are $n(D) V_P \, dD$ particles with sizes between D and $D + dD$. The size distribution defined in this way has the units $(\text{length})^{-4}$; it is normalized so that $(\pi/6) \int_0^\infty n(D) D^3 \, dD = 1$.

Two important measures for the particle size can be expressed in terms of weighted averages of the distribution. The volume weighted average, $\langle D \rangle_{volume} = \int_0^\infty n(D) D^4 \, dD / \int_0^\infty n(D) D^3 \, dD$, determines the integral breadth of the Bragg reflections. The area weighted average $\langle D \rangle_{area} = \int_0^\infty n(D) D^3 \, dD / \int_0^\infty n(D) D^2 \, dD$, is related to the specific surface area α_S by $\langle D \rangle_{area} = 6 c_A / \alpha_S$. The constant c_A depends on the shape of the particles. For spherical particles $c_A = 1$.

The intragrain correlation function has been evaluated for a number of particle shapes [13, 22]; for the example of spherical particles of size D, it is [13]

$$H(r, D) = 1 - \frac{3}{2}\frac{r}{D} + \frac{1}{2}\left(\frac{r}{D}\right)^3 \qquad (0 \le r \le D). \tag{10.11}$$

The size broadening of a Bragg reflection for a spherical particle is obtained from equations (10.6) and (10.11) as

$$W^{sphere}(k, D) = \frac{1}{\pi} D w(kD) \qquad w(y) = \frac{3}{y^4}\left(1 - \cos(y) - y\sin(y) + \frac{y^2}{2}\right) \tag{10.12}$$

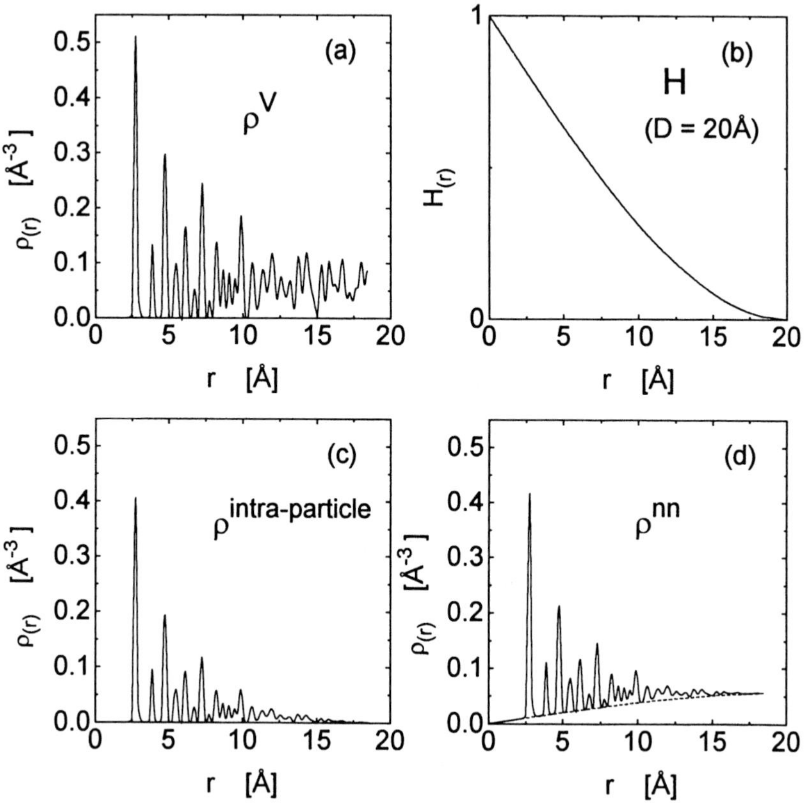

Figure 10.2. Schematic illustration of the atomic distribution function in small particles. (*a*) The atomic distribution function of the extended lattice, $\rho^V(r)$. (*b*) The intragrain correlation function, $H(r)$, for the example of spherical particles with diameter 20 Å. (*c*) The atomic distribution function of the isolated particles consists of the intragrain part alone $\rho^{intragrain}(r) = \rho^V(r)H(r)$. (*d*) The atomic distribution function of the nonreconstructed nanocrystal, $\rho^{nn}(r)$, is the sum of $\rho^{intragrain}(r)$ and of the intergrain part (dashed line) of the atomic distribution function, $\langle\rho\rangle^V(1 - H(r))$.

(see figure 10.3). In a similar way, equation (10.7) yields the small-angle scattering interference function for the isolated spherical particle

$$S^{sphere}(k, D) = 4\pi \langle\rho\rangle^V D^3 s(kD) \qquad s(y) = \frac{9}{y^6}(\sin(y) - y\cos(y))^2. \tag{10.13}$$

From the added-up scattering intensities of all the particles in the distribution (compare equations (10.5) and (10.3)), equation (10.3) yields the interference function [23]

$$P(k) = \int_0^\infty n(D)\frac{\pi}{6}D^3(P^V(k) \otimes W(k, D) + S(k, D))\,\mathrm{d}D. \tag{10.14}$$

The wide-angle part of this expression applies both to isolated particles and polycrystals, whereas the small-angle part applies only to the case of isolated particles (the intergrain correlation function is neglected—compare section 10.2.3). In other words, the profile of a Bragg reflection and the small-angle scattering intensity in the isolated particle case are simply the sums of intensity contributions from individual particles. For equiaxed particles these contributions can be approximated by the expressions in equations (10.12) and

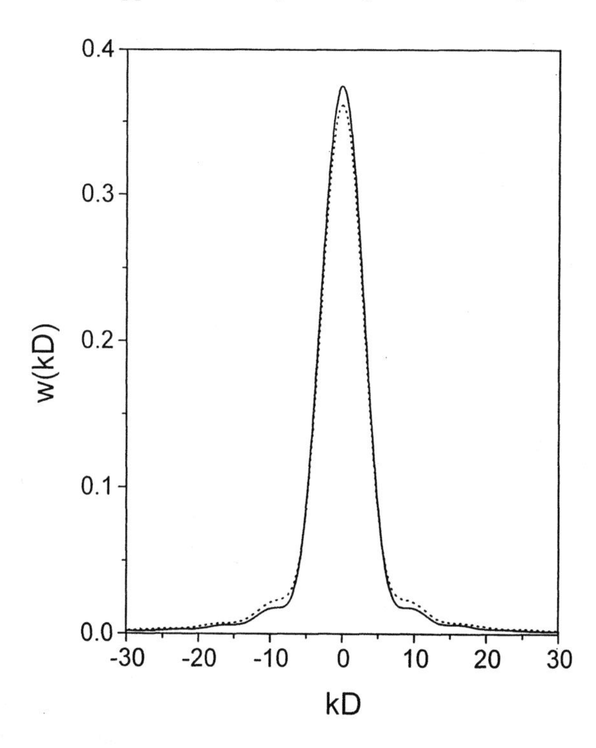

Figure 10.3. Bragg reflection profile $w(kD)$ for an arrangement of randomly oriented spherical (solid line, equation (10.12)) and cubic (dotted line) crystallites. From [23].

(10.13). Therefore, the size distribution $n(D)$ can be obtained from experimental wide-angle or small-angle scattering data by solving a linear set of equations (see sections 10.3.2 and 10.5).

On the basis of equation (10.14), it is readily shown that, if size is the only cause of reflection broadening, then the integral breadth of a Bragg reflection (the ratio of integrated intensity over intensity at the peak center) is proportional to the volume weighted average over the size distribution, $\langle D\rangle_{volume}$. In other words, the grain size determined from the integral breadth of the Bragg reflections is a volume weighted average grain size [12].

One of the consequences of equation (10.14) is instructive for the discussion of small-angle scattering by dense nanocrystalline solids. If $n(D)$ has the form of a power law, $n(D) = aD^{-b}$ $(3 < b < 7)$, then the small-angle part of the interference function, $P^{SAS}(k)$, can be determined analytically. Substituting $y = kD$ in the combined equations (10.14) and (10.13), one obtains

$$P^{SAS}(k) = \text{constant} \times k^{b-7} \qquad \text{constant} = \tfrac{2}{3}\pi^2 \langle\rho\rangle^V a \int_0^\infty y^{6-b} s(y)\,\mathrm{d}y \tag{10.15}$$

where the integral is independent of k. Thus, one of the possible origins of the power-law intensity variation, $I \sim k^{-n}$ $(n < 4)$ observed in experimental small-angle scattering data from dense nanocrystalline solids (compare section 10.5) is a power-law size distribution of scattering objects with the exponent $b = 7 - n$.

10.3 CHARACTERIZING THE CRYSTAL LATTICE

10.3.1 Crystallographic phase and lattice constant

Nanometer-sized isolated particles prepared by burning reactive metals or by condensation of elemental metal vapor in an inert gas atmosphere (see e.g. [24]) have been studied for several decades. A recent review [25] discusses crystallography and crystal habit. Well defined habits are generally found in nanometer-sized isolated metals [25] and ceramics [26]; however, transmission electron microscopy (TEM) studies of the nanocrystalline material consolidated from nanometer particles [27, 28] show irregularly shaped grains, indicating that the grain boundaries have essentially random crystallographic orientation, and that substantial atomic rearrangement takes place upon consolidation.

One of the most striking effects of the reduced crystal size is that in a number of materials the crystal lattice structure of fine-grained particles is found to differ from the thermodynamic equilibrium structure in the coarse-grained polycrystalline or single-crystal state. The crystallographic structure of as-prepared nanocrystalline Y_2O_3 (n-Y_2O_3) with an average grain size of 4 nm is that of the monoclinic high-pressure phase [29]. When the (consolidated) nanocrystalline material is annealed, it converts to the cubic ambient pressure equilibrium phase at the onset of grain growth. Isolated particles of n-ZrO_2 with average sizes of 6–12 nm are partly in the monoclinic equilibrium phase and partly in a tetragonal high-pressure phase [30]. The tetragonal phase transforms into the equilibrium phase upon consolidation. Fine particles of the bcc metals Cr [31,32], Mo [33,34] and W [33,35] can crystallize alternatively in the equilibrium bcc structure or in the A15 structure, which is not an equilibrium phase for the coarse-grained elements. In the case of Cr, the A15 phase forms under high-purity conditions, whereas oxygen impurities seem to favor the formation of the bcc phase [36]. Bulk nanocrystalline W can be prepared as pure A15 W or alternatively as pure bcc W, depending on the geometric details of the inert gas condensation setup [35]. In the case of Mo, self-arrangement of the individual particles into larger cubes made out of 3 by 3 by 3 smaller cubic particles has been reported [34].

The reason for the change in the crystallographic structure as a function of the particle size has apparently not been established conclusively. As the number of atoms in an isolated particle is reduced, the excess free energy associated with the surface constitutes an increasingly important fraction of the particle's total free energy. A small particle can reduce this excess free energy when changing its lattice structure to one that has a smaller surface free energy. This gain in excess free energy associated with the surface is achieved at the expense of excess free energy associated with the conversion of the lattice to an energetically less favorable crystallographic structure. The smaller the particle, the more important the surface term will be relative to the volume term. Hence, for a sufficiently small particle, it may be energetically favorable to form that crystal structure which has the lowest surface free energy. A different explanation has been proposed in the literature: for a spherical solid particle of radius R, the surface stress f (as opposed to the surface free energy) results in a hydrostatic pressure ΔP given, for isotropic f, by [37,38]

$$\Delta P = 2f/R. \tag{10.16}$$

It has been proposed that the pressure acting on the particle results in the transition of the crystal lattice structure towards that of a high-pressure phase [29,30]. A similar effect may arise from the stress associated with the grain boundaries in a nanocrystalline solid. The magnitude and the sign of the surface stress are known only for a small number of materials [39]. However, as the internal pressure induces a change in the particles' lattice constant, the surface stress can be determined from the experimental lattice constant and the relation [37,40]

$$f = -\frac{3}{2}K\frac{\Delta a}{a}R \tag{10.17}$$

where K is the bulk modulus, and $\Delta a/a$ is the relative change in the lattice constant. Apparently, the lattice contraction has not been determined for a system which undergoes a phase transition as a function of the particle size.

A number of investigations, by electron diffraction, of isolated particles supported on carbon foil have aimed at determining the interfacial stress in small noble metal particles. For Au and Ag, lattice contractions (relative to the lattice of the coarse-grained polycrystal) of the order of $\Delta a/a = -2 \times 10^{-3}$ and -4×10^{-3}, respectively, are found at a particle size of 5 nm [40–42]. In agreement with theoretical predictions, the surface stress is found to be positive and of a magnitude comparable to the surface free energy [39]. The small lattice contraction observed in 6 nm Au particles by x-ray diffraction [43] also agrees, within large error bars, with this observation. Considerably larger lattice contractions are obtained in room-temperature EXAFS studies on supported isolated metal particles in the size range from 10 nm down to clusters of a few atoms [44, 45]; however the potential influence of interactions between the particles and the supporting film (carbon or polymer) on the structure of the particles has been pointed out [45]. In fact, low-temperature EXAFS investigations on metal particles isolated in a solid argon matrix [46, 47] find no evidence for the larger contractions; instead, the observed lattice contraction is in at least qualitative agreement with electron diffraction data and with theory. The solid inert gas is expected to have a weaker interaction with the metal particles than carbon or polymer supports. The larger contractions observed in room-temperature EXAFS have also been attributed [48] to a problem with the standard EXAFS data analysis: at ambient temperature, anharmonic atomic displacements at the particle surface significantly affect the EXAFS spectrum. The simplifying assumption of symmetric distributions of interatomic spacings in the individual coordination shells, underlying standard EXAFS data analysis, will then result in considerable error in the experimental interatomic spacings and coordination numbers [49]. The accuracy of lattice constant analysis from Bragg reflection positions in diffraction data from very small particles has also been questioned [50]: the overlap between broad Bragg reflections induces error in the experimental peak positions. Interatomic spacings derived from diffraction based atomic distribution functions should be free of the limitations of both EXAFS and Bragg-reflection analysis and might therefore yield more accurate information.

In contrast to the case of isolated particles, where changes in the lattice constant as a function of particle size are well documented, similar experiments on nanocrystalline solids yield effects which are within or barely outside the error bars of the experiments. The interatomic spacings determined by EXAFS for n-Cu and n-Pd [51] and bcc n-W [52] agree, within large experimental errors, with those of the coarse-grained polycrystalline elements. In Pd with an average crystallite size of 8.3 nm, x-ray diffraction indicates a relative contraction of the lattice constant of less than 5×10^{-4} [53]. For the purpose of comparison with the lattice contraction observed in isolated particles with a size of 5 nm, the result for Pd can be extrapolated to this grain size: assuming a $1/D$ dependence of $\Delta a/a$, the extrapolated upper bound of 8×10^{-4} is obtained for the relative lattice contraction in n-Pd with $D = 5$ nm. This corresponds to a considerably smaller change in lattice constant than that observed in isolated Au or Ag particles [40, 42]. Comparative studies of isolated particles and nanocrystalline solids of the same material may yield information on the relative magnitude of surface and grain boundary stress. Apparently, such studies have not been reported. In addition, the experimental accuracy of presently available data appears inadequate for a quantitative characterization of the grain boundary stress and of the very small variation in the lattice parameter associated with it.

The thermal expansion of the crystal lattice in n-Pd in the temperature interval 16–300 K, as determined by x-ray diffraction, is identical to that in the coarse-grained material [53]. According to section 10.2.2, this result provides no information on the thermal expansion in the grain boundary layer. Indeed, dilatometric measurements, which probe a weighted average of boundary and lattice expansion, indicate an increased overall thermal expansion in n-Pd [3].

10.3.2 Grain size distribution

The interest in nanocrystalline materials arises from the fact that for very fine grain sizes a number of physical properties become grain-size dependent. Obviously, it is therefore essential to be able to accurately characterize the grain size or, in many cases, the distribution of sizes of the material under investigation.

As discussed in section 10.2.5, the refinement of the grain size has two basic consequences for the scattering pattern: first it results in a broadening of the Bragg reflections, and second it gives rise to small-angle scattering. This seems to suggest that the grain size or the size distribution can be analyzed from any of the two respective regions (wide- and small-angle) of the scattering pattern. In practice, a number of requirements have to be fulfilled for the analysis to be feasible. In the wide-angle scattering region, the Bragg reflection profile is the convolution of the profile of the reflections of the extended lattice, which is in general broadened by lattice defects (to be discussed below), with the size broadening function. Characterizing the grain size presupposes a separation of size and defect related broadening. Since only very simple assumptions can be made concerning the effects of defects on the broadening, an accurate separation requires that the defect induced broadening is small compared to the size induced broadening. Furthermore, neighboring Bragg reflections must be well separated in order to minimize overlap between the reflection tails. This requirement is generally satisfied in single-phase materials with a cubic lattice symmetry. The accuracy of the analysis may be limited in crystals with a lower symmetry or when reflections from more than one phase are present.

In contrast to wide-angle scattering, small-angle scattering depends only on the correlation functions, that is on the external shape and on the relative position of the crystallites, as opposed to their internal structure. Therefore, particle size distributions can be determined from small-angle scattering data even when the crystal lattice has a high defect density or when the particles are amorphous. However, the analysis requires that the experimental scattering data can be modeled as the sum of the intensities of the individual particles. This implies that the intergrain correlation function must be negligible compared to the intragrain correlation function, so that intergrain interference is small (compare equations (10.7) and (10.14)). This requirement is only satisfied in a dilute arrangement of isolated particles. This suggests that small-angle scattering is an adequate method for characterizing the size distribution of isolated particles, but that the grain size of bulk nanocrystalline solids cannot be characterized by this technique. The present section addresses Bragg reflection based techniques; small-angle scattering is discussed in section 10.5.

A crude estimate of the crystallite size can be obtained from the Scherrer equation [54], typically evaluated for the strongest reflection in the experimental diffraction pattern

$$\bar{D}_{Scherrer} = c\frac{\lambda}{\delta(2\theta)\cos(\theta)} = c\frac{2\pi}{\delta k} \tag{10.18}$$

where $\delta(2\theta)$ and δk are the reflections' breadth in units of the scattering angle 2θ and of the wavenumber k, respectively, and λ denotes the wavelength. The value of the constant c is close to unity. This method of determining the grain size

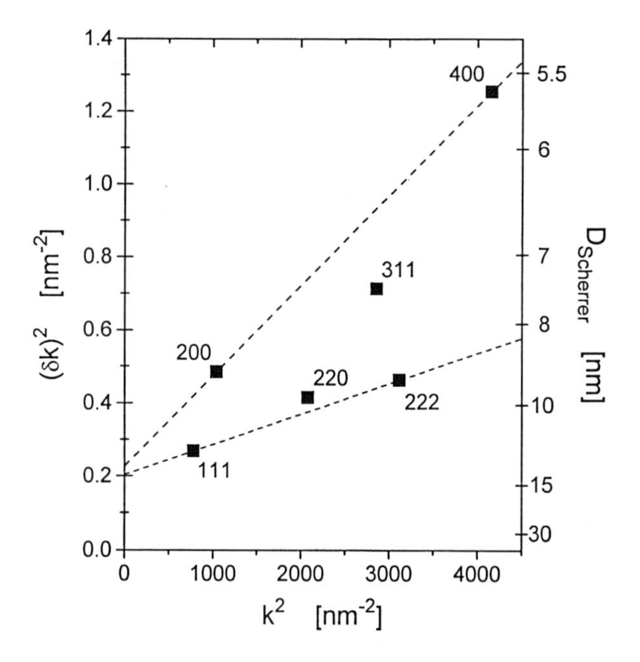

Figure 10.4. The square of the integral breadth of the first six Bragg reflections, $(\delta k)^2$, versus the square of the reflection wavevector k^2 for n-Pd prepared by inert gas condensation. The reflection indices are indicated in the figure. The right vertical axis shows the apparent grain size obtained from the Scherrer equation (10.18) for the respective peak breadth δk. The dashed lines are fits by equation (10.19) for the [111] and [100] lattice directions.

has the shortcoming of neglecting contributions other than size broadening to the breadth of the reflection. The importance of the issue is illustrated in figure 10.4, which shows the square of the integral peak breadth, $(\delta k)^2$, as a function of the square of the wavenumber of the reflection, k^2, for a nanocrystalline Pd sample produced by inert gas condensation. The reason for choosing this particular representation will be apparent below. The apparent grain size obtained by applying the Scherrer equation (10.18) to each reflection is indicated on the right vertical axis. It is seen that the grain size determined from the strongest reflection (111), 12 nm, is considerably larger than the one determined from the next strongest reflection (200), 9 nm, and that even smaller grain sizes would be obtained from reflections at higher wavenumbers. None of those values corresponds to the 'real' grain size. The problem arises from the neglect of the contribution to the reflection broadening from lattice strain. The respective effects can be separated, since the size broadening is independent of the order of a reflection, whereas the strain broadening is increased for the higher-order reflections. Under the assumption that both strain and size cause a Gaussian line-shape, the integral breadth due to the combined effects of strain and size is given by [55]

$$(\delta k)^2 = \frac{4\pi^2}{L^2} + 6.25\langle\varepsilon^2\rangle k^2 \qquad (10.19)$$

where $\langle\varepsilon^2\rangle$ denotes the mean square lattice strain (see also section 10.3.4). L is an average linear dimension of the crystals, which is related to the volume weighted average grain crystal size (section 10.2.5) by $\langle D\rangle_{volume} = 4L/3$ [12, 56, 57]. A fit to the experimental data in figure 10.4 by equation (10.19) is a straight line through the data points corresponding to the reflections of one particular set of planes. The grain size is determined from the intercept of the fit with the vertical axis at $k^2 = 0$, and the strain is obtained from the slope. In the example in figure 10.4, nearly identical grain sizes of 14 nm and 13 nm are obtained for the [111] and [100] directions. However, the values for $\langle\varepsilon^2\rangle^{1/2}$, 0.36% for [111] and 0.64% for [100], differ significantly. The difference in the strain values reflects the anisotropy of the elastic constants of Pd. A similar effect is observed in other metals (see e.g. [58]). In some studies, an average strain is determined by simultaneously fitting the breadths of all reflections, irrespective of the lattice direction, with a single pair of size/strain values (e.g. [59]).

For the example in figure 10.4, the grain size determined from equation (10.19) is only slightly larger than the one obtained by applying the Scherrer equation to the 111 reflection. However, considerably larger discrepancies between the two values can arise in pure metals with somewhat larger grain sizes and in alloys [58, 60, 61]. Correction for strain broadening is therefore indispensable for a reliable characterization of the grain size. Rather than assuming Gaussian broadening effects from both strain and size, the more realistic equivalent of equation (10.19) for Cauchy-type size broadening and Gaussian-type strain broadening should be employed [55]. That relation has a less illustrative form, but does also allow a separation of size and strain broadening by graphical means or, alternatively, by linear regression. This latter method has the advantage of providing an estimate for the uncertainty in the parameters from the errors in the fit [59]. It is seen that methods based on the evaluation of integral breadth data rely on more or less realistic assumptions on the character of the peak broadening. If scattering data of sufficient quality are available, then more reliable results are obtained by a method which does not require such assumptions. This method, the Warren–Averbach analysis [54, 62], is based on the Fourier analysis of the reflection profiles. Detailed descriptions of the individual techniques for analyzing grain size from diffraction data and comparative discussions of their relative merits can be found in textbooks [54, 55]; a few more recent aspects are discussed in [63–66].

While all the techniques used routinely for the characterization of the grain size supply a single value for an average crystallite size, it is known that in general there is actually a distribution of sizes. For as-prepared inert gas condensed metals [53, 67, 68] and ceramics [29, 30] TEM size histograms are well approximated by log-normal distributions—see figure 10.5 [68]. Contrary to this, anomalous grain growth at room temperature in metals has been reported to result in a bimodal size distribution [69]. It has been demonstrated repeatedly [3, 70] that the value of an average grain size depends strongly on which weight function is used when averaging over the size distribution. For the example of nanocrystalline Pd prepared by inert gas condensation, the value of the area weighted average grain size differs from that of the volume weighted average grain size by typically a factor of 2 [16]. The numerical value of the size depends also on which definition of size is adopted [21]. When relating physical properties to average grain sizes, it is therefore imperative to specify which definition of size is used, to identify the weight function relevant for the problem in question and to measure that particular weighted average. In this respect the volume weighted average size determined by integral breadth based methods is complemented by the area weighted average over the column length distribution determined by the Warren–Averbach analysis, which is readily converted to an area weighted average grain size [71].

Information on the distribution of the lateral dimensions of the coherently scattering units can be obtained by a number of more sophisticated scattering data evaluation techniques described in the literature. These are based on a relation originally deduced by Bertaut [19] and by Warren and Averbach [72], which expresses the intensity of the reflection of a given set of lattice planes as the sum of the intensities from all columns of lattice cells perpendicular to the set of planes. In other words, the line profile is expressed as a linear combination of intensity functions, each corresponding to a particular column length (see [73] for details). The column length distribution can be obtained from a double differentiation of the Fourier transform of the line profile [19, 74]. Alternative methods approximate the column-length distribution by a set of analytical functions [75] or solve the linear set of equations by an iterative

Gauss–Seidel least-square algorithm [76], yielding realistically smooth column-length distributions if combined with a stabilization procedure [77].

It has been pointed out that in the case which is generally encountered in practice, namely when there is a set of particles with a distribution of sizes, the column-length distribution does not provide an easy to interpret picture of the form of the crystal size distribution [78]. A method which does achieve a characteristic of the size distribution is described in [23]. The algorithm is based on the relation between the Bragg reflection line profile and the size distribution, equation (10.14). This relation implies that the experimental line profile can be expressed as a linear combination of the simulated experimental line profiles of particles with different sizes. These profiles are readily obtained by convoluting the experimental line profile of a suitable coarse-grained reference sample with the size broadening function $W(k)$ (equation (10.6)) for the respective size, and with a Gaussian accounting for the mean square lattice strain. The size distribution of the nanocrystalline sample is determined by a stabilized linear least-square fit to the data of the nanocrystalline sample, with lattice strain refined iteratively so as to obtain a simultaneous fit to two or more orders of the same reflection. Figure 10.6 is a schematic illustration of the deconvolution of the line profile. The number of size classes has been limited to five for the sake of readability of the figure. The distributions obtained by the indirect deconvolution technique are found in good agreement with TEM size histograms, and in particular the values for $\langle D \rangle_{area}$ and $\langle D \rangle_{volume}$ computed from weighted integrals of the distributions appear reliable [23].

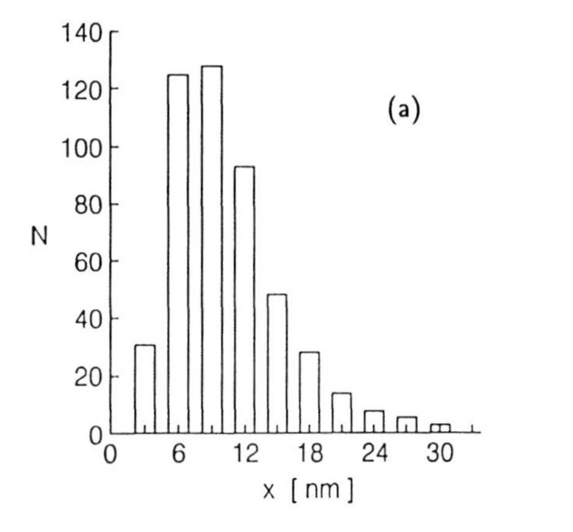

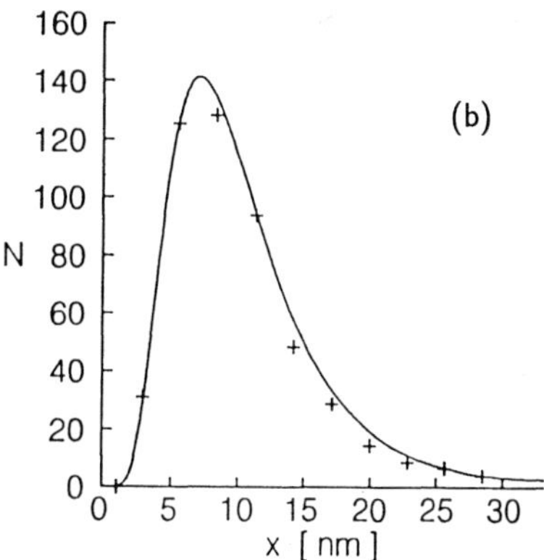

Figure 10.5. (*a*) TEM size histogram of inert gas condensed Pd particles. (*b*) Fit (solid line) to the same data (+) by a log-normal distribution of particle size. Reproduced by permission of Haas [68].

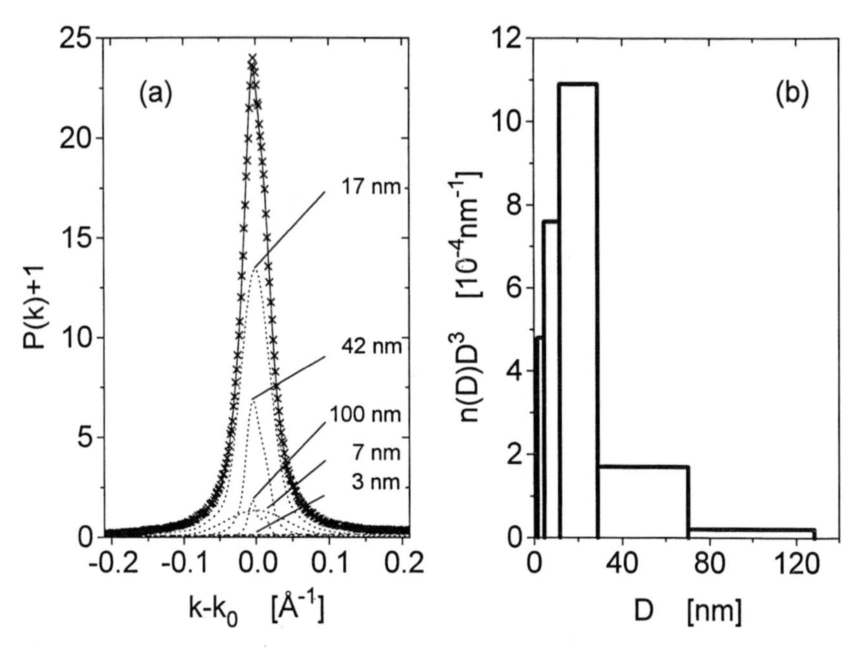

Figure 10.6. (*a*) Schematic illustration of the deconvolution of the experimental line profile of a 111 reflection in the interference function $P(k)$ of n-Pd into contributions by different particle size classes. k is the wavevector and k_0 the reflection center. Crosses: experimental data; dashed lines: contribution of the individual particle size classes as indicated in the figure. The line shape for each size class is obtained by convoluting the theoretical peak shape (equation (10.6)) with the instrumental and strain broadening functions. Solid line: sum of the individual contributions. (*b*) Particle size histogram deduced from the deconvolution. From [23, 79].

10.3.3 Grain growth and control of grain size

Bulk nanocrystalline metals prepared by inert gas condensation have been reported to exhibit grain growth at room temperature [15, 69, 79]. This observation indicates a high atomic mobility in the interfaces at room temperature. For n-Pd the grain size is seen, in figure 10.7, to increase from initially about 12 nm to up to 80 nm in a period of several months [15, 79]. Fits by a growth law $D(t) \sim (t - t_0)^n$ yield exponents in the range $n = 0.25$–0.46, which is inside the range of exponents observed in coarse-grained polycrystalline solids [80]. Hence, the growth mechanism in n-Pd appears to be similar to the one in coarse-grained polycrystals, where the growth rate is limited by the grain boundary mobility. This finding disagrees with recent results in vapor quenched thin-film Ag, where a linear growth law ($n = 1$) is argued to imply triple-line mobility limited kinetics [81]. Additional studies on grain growth in nanostructured materials are reviewed in [4, 6, 82].

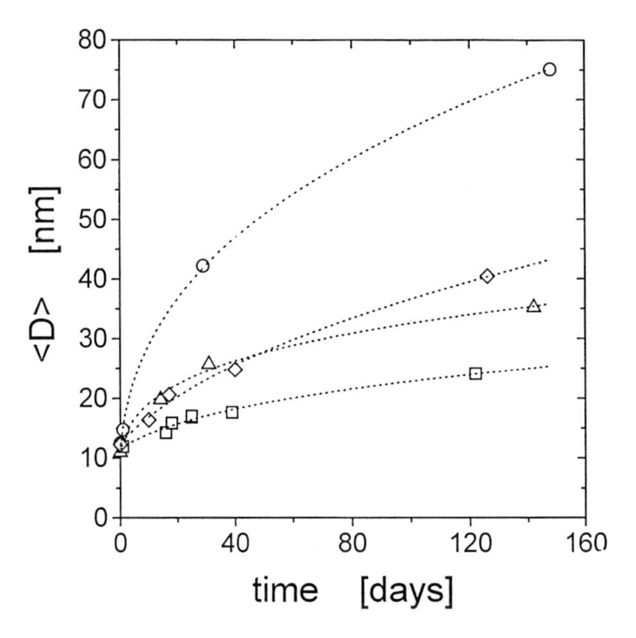

Figure 10.7. Variation of grain size $\langle D \rangle$ with sample age at room temperature for several Pd samples. The different symbols represent the individual samples. Dashed lines: fit for each sample as discussed in the text. From [15, 79].

Grain growth at room temperature has been observed only recently. It may be speculated that this is a result of the higher purity achieved with recent, improved preparation techniques. Impurities are known to slow grain growth by two mechanisms: first, segregated impurities induce a 'solute drag' on the movement of the grain boundaries [83–85]. Second, grain boundary segregation reduces the specific grain boundary energy, and thereby the driving force for grain growth [86]. Alternatively, the higher density of recent inert gas condensed samples may result in less efficient pinning of the grain boundaries by interaction with pores. These considerations imply that the 'stable' grain size (the size where grain growth at room temperature becomes imperceptibly slow) of an ideally pure and dense nanocrystalline pure metal may be comparatively large (of the order of 100 nm for Pd) and intrinsic to a given material, and therefore independent of the initial size achieved upon preparation. It is therefore important to evaluate concepts for controlling the grain size, or more generally the microstructure.

It has been predicted that grain-size refinement and possibly a thermodynamically metastable nanocrystalline state can be achieved in nanocrystalline solid solutions with a large enthalpy of grain boundary segregation [87, 96]. Several experimental results support this concept. Figure 10.8 shows the variation of the grain size with the alloy composition in nanocrystalline Y–Fe alloys prepared by inert gas condensation [88]. In the

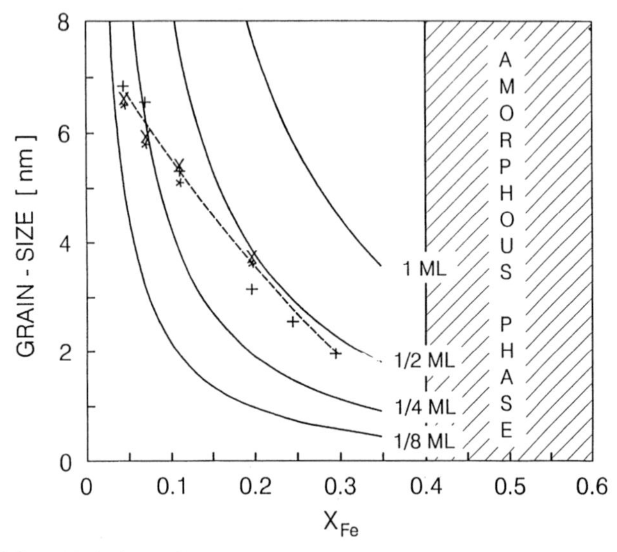

Figure 10.8. Variation of grain size with Fe molar fraction x_{Fe} and nanocrystal–amorphous metal transition for n-Y–Fe prepared by inert gas condensation. The dashed lines are theoretical predictions for different concentrations (in atomic monolayers) of Fe segregated to the grain boundaries, as indicated in the figure. From [88].

nanocrystalline alloy, the majority of Fe is segregated to the grain boundaries between hcp Y crystallites. Increasing the molar fraction of Fe, x_{Fe}, is seen to induce a progressive decrease in grain size until the nanocrystal–amorphous metal transition intervenes at a grain size of about 2 nm. At this very small grain size, it is difficult to distinguish between the nanocrystalline and the amorphous state on the basis of diffraction data alone. This is evident in figure 10.9, where the broad Bragg peaks of $Y_{70}Fe_{30}$ seem to indicate that the alloy is amorphous. However, EXAFS at the Y K-edge (figure 10.10) shows the SRO around Y for that alloy to be the one of the crystal lattice, thereby evidencing a nanocrystalline state of the alloy. It is only at the higher Fe content $Y_{65}Fe_{45}$ that a distinctive change in atomic SRO around Y is observed in the EXAFS spectra. This is indicative of the loss of long-range order of the Y crystal lattice at the nanocrystal–amorphous metal transition [88]. The combination of several experimental techniques may generally be required if detailed structural information on extremely fine-grained nanostructures is to be obtained. In the present case, the results agree qualitatively with the theoretical predictions. In alloys with a large heat of segregation, the preferred grain size is predicted to be the one where the total grain boundary area is just sufficient to accommodate all solute in the grain boundaries. Therefore, the grain size decreases with increasing solute concentration.

In an experimental study by diffraction and calorimetric methods, the grain

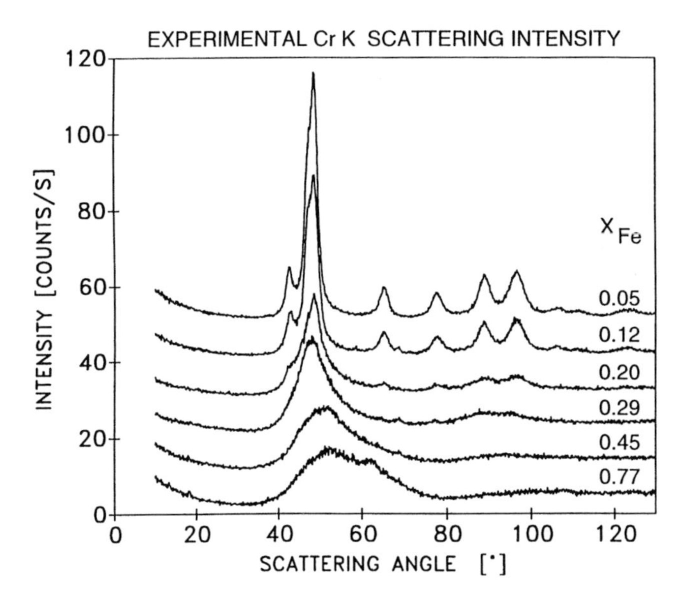

Figure 10.9. X-ray scattering intensity (Cr Kα radiation) of Y–Fe alloys with various compositions prepared by inert gas condensation. The different molar fractions of Fe, x_{Fe}, are indicated in the figure. From [88].

size of Pd–Zr solid solutions prepared by high-energy ball milling (HEBM) has also been found to decrease with increasing solute concentration, and the heat release upon annealing indicates that solute (Zr) segregates to the grain boundaries, thereby reducing the specific grain boundary energy and impeding grain growth [89]. The inhibition of grain growth by segregated solute has also been used to retain small grain sizes at the elevated sintering temperatures which are required in order to obtain dense ceramic bodies in coarse-grained polycrystalline [90] and in nanocrystalline [91] ceramics. Additional strategies for inhibiting grain growth are discussed in [6, 82, 92].

10.3.4 Lattice strain

In general, there is considerable lattice strain in nanocrystalline solids. Strain involves local deviations of the lattice constant from its mean value, which in itself may also be size dependent (see section 10.3.1). Diffraction data detect strain broadening in the Bragg reflections for atomic displacement fields which decay more slowly than $r^{-3/2}$, where r is the distance from the center [93]. Displacement fields with a faster decay give rise to an attenuation of the integrated intensity of the reflections, characterized by the Debye–Waller parameter (see section 10.3.6). Strain fields from dislocations and thermal atomic vibrations, respectively, are examples of the first and second type of defect [93].

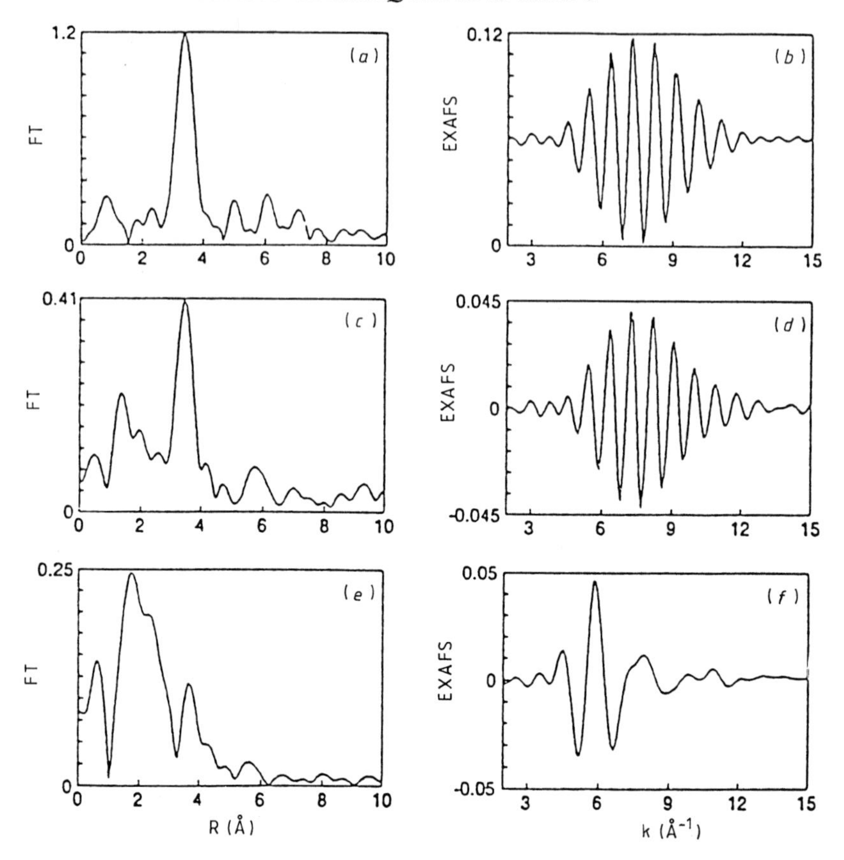

Figure 10.10. Y K-edge EXAFS Fourier transforms for coarse-grained polycrystalline Y (*a*) and for two bulk nanostructured alloys prepared by inert gas condensation, $Y_{60}Fe_{30}$ (*b*) and $Y_{55}Fe_{45}$ (*c*). From [88].

Strain data are reported in the literature for a number of nanocrystalline metals and alloys prepared by inert gas condensation and by HEBM. For pure fcc metals prepared by HEBM, the strain is found to increase with increasing melting temperature T_m and with decreasing grain size [59]. Figure 10.11 displays the variation, with grain size, of the root mean square (RMS) strain, as determined from the angular variation of the breadths of the Bragg reflection (assuming Cauchy size and Gaussian strain broadening [55]), of n-Pd prepared by inert gas condensation [15, 79]. A least-square fit by a power law $\langle \varepsilon^2 \rangle^{1/2} = 0.01(D/D_1)^{-n}$ (D_1 is the grain size for which the RMS strain is 1%) yields $D_1 = 5.4 \pm 0.4$ nm and $n = 1.23 \pm 0.06$ for samples consolidated at room temperature. In this experiment, the grain size was varied by changing the preparation conditions and by inducing grain growth through aging and annealing. For a given grain size, the strain was found to be independent of

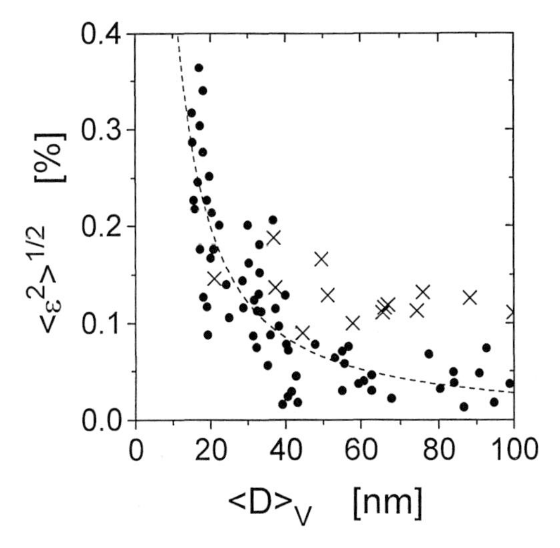

Figure 10.11. RMS strain $\langle \varepsilon^2 \rangle^{1/2}$ versus grain size $\langle D \rangle_{volume}$ for n-Pd. •: consolidated at room temperature and 1 GPa for 300 s; ×: consolidated at 100 °C and 3.2 GPa for 48 h. Dashed line: fit as described in the text. From [15].

whether the sample was as prepared or had undergone significant annealing or aging induced grain growth. This indicates that the observed strain may be an intrinsic property of the nanocrystalline state. In agreement with that idea, the amount of strain as a function of grain size appears also to be independent of the material and of the method of preparation. This is seen in figure 10.12, where the strain data from [59] for fcc metals prepared by HEBM are displayed together with the extrapolated fit to the inert gas condensed Pd from figure 10.11. With one exception (Rh), the data for samples prepared by HEBM follow the same strain–grain size relationship as those for inert gas condensed Pd. It is noted that a decrease of the mean square strain upon annealing and prior to the onset of grain growth has been reported for n-Ni [94]. In that study, Rietveld analysis was used to characterize size and strain, whereas the remaining data referred to above were obtained by the integral breadth method (see section 10.3.2).

The possible universal relation between strain and grain size may suggest that lattice strain is an intrinsic property of nanocrystalline solids. On the other hand, additional strain can be incorporated in the materials. This is evidenced by the data in figure 10.11 for inert gas condensed n-Pd which was subject to an additional, 48 h consolidation step at 3.2 GPa and 100 °C, subsequent to the conventional consolidation at 1 GPa and room temperature. This seems to indicate that the presumed intrinsic strain represents a lower limit for the strain level, and that additional, extrinsic strain may be induced in nanocrystalline solids. In other words, at given values for the thermodynamic parameters pressure, temperature, and amount of matter, the grain size (or the

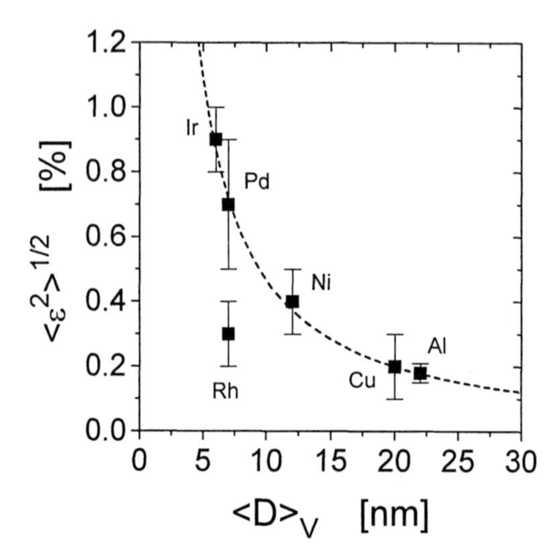

Figure 10.12. RMS lattice strain versus grain size for various fcc metals prepared by HEBM as indicated in the figure. The dashed line is the fit to the data for n-Pd prepared by inert gas condensation (see figure 10.11). HEBM data from [59].

total grain boundary area) is not the only additional thermodynamic variable required to describe the state of a nanocrystalline pure metal. This conclusion agrees with a recent experimental characterization of the energetics of the grain boundaries in n-Pt [95], and with results from a theoretical analysis of the alloy thermodynamics of nanocrystalline solid solutions [96]. It is also in agreement with studies of the grain boundary atomic SRO in nanocrystalline Pd (see section 10.4.2).

The total strain energy W_S stored in heavily deformed metals amounts to only a small fraction of the melting enthalpy [97]. A number of different relations between W_S and the x-ray mean square strain $\langle\varepsilon^2\rangle$ have been proposed, which all apparently underestimate the real value for W_S [97]. The closest agreement to calorimetric data for cold-worked metals is obtained from the relation [98]

$$W_S = \frac{15E}{4(1+\nu)}\langle\varepsilon^2\rangle \tag{10.20}$$

where E is Young's modulus and ν is Poisson's ratio. The values for W_S obtained from this relation and the n-Pd data in figure 10.11 are considerably smaller than the total energy stored in the grain boundaries. However, equation (10.20) and the experimental data predict the strain energy stored in the lattice increases with diminishing particle size faster than D^{-2}. Since the total grain boundary area A varies as D^{-1}, the conventional *ansatz* for the excess Gibbs free energy of a polycrystal, $\Delta G = \sigma A$ (see e.g. [87,96]), with the conjugate variables σ, the specific grain boundary energy, and A, the total grain boundary area, implies the slower increase $\Delta G \sim D^{-1}$. The different exponents in the size dependence have two implications: firstly, as the grain size is reduced, the strain energy stored in the lattice constitutes an increasing fraction of the total excess energy. Secondly, since the D^{-2} term is not accounted for by the *ansatz* $\Delta G = \sigma A$, it follows that if lattice strain is indeed intrinsic to the nanocrystalline state, an additional set of conjugate variables has to be considered in the thermodynamic functions of a nanocrystalline solid. Together with the additional degree of freedom arising from the apparent existence of non-equilibrium states for a nanocrystalline solid with a given grain size or size distribution, this suggests that the description of the thermodynamic state of a simple elemental nanocrystalline solid may turn out to be a complex undertaking.

The lattice strain in nanocrystalline solids prepared by HEBM has been proposed to originate from lattice dislocations [59, 99]. However, it has been questioned whether the lattice dislocation density is sufficient to account for the observed strain levels, and an alternative model, based on extrinsic grain boundary dislocations, has been suggested [100]. Quantitative studies of lattice and grain boundary dislocation density, which might distinguish between the two models, are needed. Yet another potential source of lattice strain is of a purely elastic origin. Part of the strong Hertzian contact stress, which is observed in loosely agglomerated nanometer particles [101], may persist in the bulk nanocrystalline material. Finally, anisotropic grain boundary stress and grain boundary torque, due to the dependence of grain boundary energy on crystallographic orientation [127], may also contribute to the observed lattice strain. It is seen that, in addition to the question of the microscopic nature of the lattice strain, there is a more fundamental question. Large RMS strains have not been reported for isolated particles. Hence, if strain is intrinsic to the dense nanocrystalline solids, then it must in some way be related to the presence of constraints from packing of the crystallites. In addition to the question of how the strain is accommodated (grain boundary or lattice dislocations, purely elastic distortion), there is the important question regarding the physical origin of the constraints themselves, and the factors which govern their magnitude. This issue deserves investigation in future studies (compare also [6]).

10.3.5 Stacking faults and twin boundaries

Defects in the stacking sequence of atomic planes, such as twin boundaries and stacking faults, are frequently observed by TEM in isolated small metal particles (e.g. [101–103]). The density of these defects in bulk nanocrystalline solids has attracted less attention. As described in [54, 58], their number is readily characterized using wide-angle scattering data. The detailed effects on the scattering pattern depend on the crystallographic structure; for the example of faulting on (111) planes in fcc, the effect is a change in the relative positions of the 111–200, 220–200, and 220–311 pairs of reflections. Since systematic experimental errors in the peak position, due, for example, to sample misalignment, cancel upon determination of the relative peak position, this quantity can be determined with considerable accuracy. In this way, nanocrystalline Cu–Co alloys prepared by HEBM have been shown to contain an appreciable density of stacking faults [104]. Figure 10.13 depicts the variation, with alloy concentration, of the stacking fault probability α in Cu–Co alloys. The average number of (111) planes between faults is α^{-1}. The stacking fault probability is seen to be highest for the material with the lowest stacking fault energy.

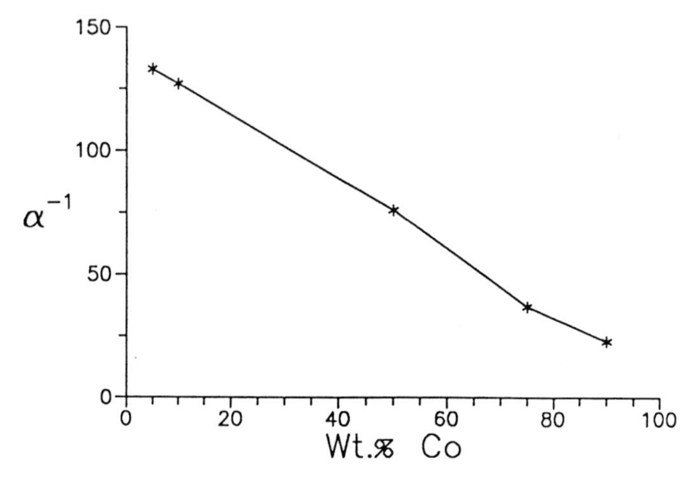

Figure 10.13. Inverse stacking fault probability, α^{-1}, versus alloy concentration for nanocrystalline Cu–Co prepared by HEBM. Reproduced by permission of Gayle [104].

In addition to the angular displacement of the Bragg reflections, faults and twin boundaries also contribute to the broadening of the reflections [54]. For the example of $Cu_{10}Co_{90}$, the apparent grain size determined when neglecting fault contributions is 12 nm. However, when stacking faults are taken into account, an average spacing between faults of 13 nm and a 'true' crystallite size of 129 nm, one order of magnitude larger than the apparent size, are found [104]. This illustrates that it is essential to characterize the fault probability when accurate values for the crystallite size are to be derived from diffraction data.

10.3.6 Short- and medium-range correlated displacements (dynamic/static)

As mentioned above, those atomic displacements from the ideal lattice sites for which the magnitude of the displacement diminishes faster than $r^{-3/2}$ result in a decrease in the integrated intensity in the Bragg reflections. Thermal vibrations of the atoms constitute examples for such displacement fields. Similar to the Bragg reflections in scattering data, the EXAFS oscillations are also attenuated when dynamic or static deviations from the ideal crystal lattice structure are present. Finally, the peaks in experimental atomic distribution functions are broadened due to the atomic displacements [16]. Experimental techniques based on the different effects yield different measures for the magnitude of the displacements. In contrast to the case of neutron scattering (compare section 10.8), x-ray and electron scattering and EXAFS data contain no information on the momentum of the atoms. Therefore, these methods supply no direct evidence on whether observed displacements are of a static (lattice defects) or dynamic (thermal vibrations) origin. However, by performing a series of experiments as a function of temperature one can separate temperature-dependent (presumably dynamic) from temperature-independent (presumably static) displacements.

Let the atoms be displaced at random from their crystalline lattice sites, with the probability $c(x, y, z)$ of finding an atom displaced by the vector (x, y, z) given by the normalized, three-dimensional Gaussian of variance σ_{disp}

$$c(x, y, z) = \frac{\exp(-x^2/\sigma_{disp}^2)\exp(-y^2/\sigma_{disp}^2)\exp(-z^2/\sigma_{disp}^2)}{(2\pi\sigma_{disp}^2)^{3/2}}. \quad (10.21)$$

Then the integrated intensity for a Bragg reflection centered at wavenumber k is reduced, relative to the one of the ideal crystal lattice, by the Debye–Waller factor $\exp(-2M)$, where $M = \langle u_x^2 \rangle k^2$ [105]. The mean square of the projection of the atomic displacements on the x axis, $\langle u_x^2 \rangle$, is the square of the variance of the Gaussian probability distribution: $\langle u_x^2 \rangle = \sigma_{disp}^2$. Similar to the probability distribution, the distribution of interatomic spacings in the coordination shells of the atomic distribution function is also a Gaussian. Since it involves the *relative* displacements of the central atom to its neighbors, the variance of the distribution of spacings in the atomic distribution function, σ_{ADD}, is larger than that relating to the displacement of the individual atoms: $\sigma_{ADD}^2 = 2\sigma_{disp}^2$ provided that the individual displacements are independent [106]. The EXAFS oscillations are attenuated by the EXAFS Debye–Waller factor $\exp(-2k^2\sigma_{ADD}^2)$ [107]. Hence, the measure for the displacement obtained from EXAFS and from analysis of the atomic distribution function, σ_{ADD}^2, refers to a different physical property than the one obtained from the scattering Debye–Waller factor, $\langle u_x^2 \rangle$. An additional difference between the EXAFS and scattering Debye–Waller factors is due to the fact that the EXAFS Debye–Waller factor for the jth coordination shell is a function of the mean square relative displacement of the central atom to its neighbors in that particular shell, and that this quantity depends on the nature of the correlations between the displacements. If the displacements are correlated so that neighboring atoms move in the same direction (figure 10.14), then the mean square relative displacement is reduced as compared to the case of random displacements [108, 109]. Long-wavelength phonons are an example of this situation.

The mean square atomic displacements in isolated Au particles and clusters in the size range 1.5–20 nm have been studied by x-ray [43] and electron diffraction [110] and by EXAFS [109]. In all studies, displacements are found

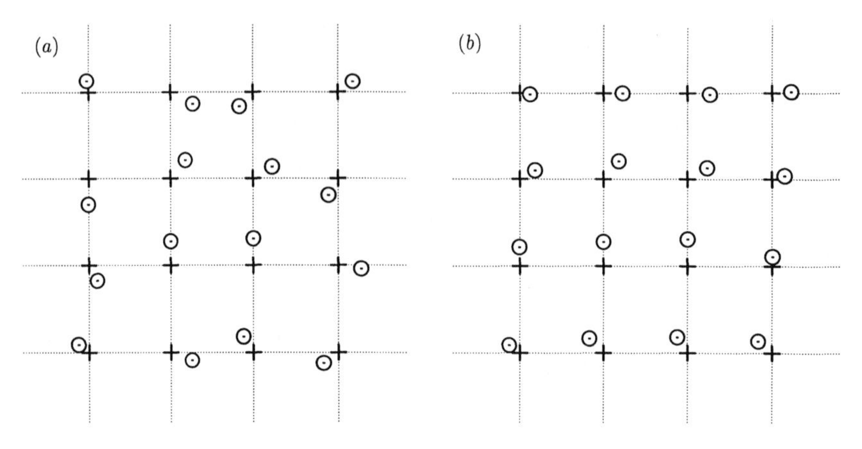

Figure 10.14. Schematic representation of atomic displacements from the ideal lattice sites (crosses). In (*a*), the displacements between neighboring atoms (circles) are uncorrelated; in (*b*) neighboring atoms are displaced preferentially in the same direction.

to increase with decreasing grain size. The small static contribution to the mean square displacement implies that enhanced thermal vibrations are the origin of the displacements [43]. All of the studies suggest a strong contribution by surface excitations to the vibrational spectra. This finding is in qualitative agreement with low-energy electron diffraction (LEED) data obtained on single-crystal surfaces [111].

The situation appears to be different for bulk nanocrystalline solids. Figure 10.15 displays the variation of the mean square displacements of atoms from their crystal lattice sites, as probed by the x-ray Debye–Waller parameter for n-Pd and for a coarse-grained reference sample [53]. Data for both samples show the same variation with temperature, indicating that the phonon spectrum in n-Pd is not significantly modified. Also in contrast to the findings on isolated particles, a significant temperature-independent additive contribution to the Debye–Waller parameter is found in n-Pd. This suggests that static displacements contribute to the mean square displacements in this material. At room temperature, the values for $\langle u_x^2 \rangle^{1/2}$ are 0.084 Å and 0.063 Å, respectively, in n-Pd and coarse-grained polycrystalline Pd [53].

Atomic distribution functions from a set of several coarse-grained reference and nanocrystalline Pd samples yield variances of the nearest-neighbor distance distribution of $\sigma_{ADD} = 0.074 \pm 0.005$ Å and 0.074 ± 0.01 Å for the reference and nanocrystalline samples, respectively [15, 16]. Assuming uncorrelated displacements, this implies mean square displacements from the lattice sites of $\langle u_x^2 \rangle^{1/2} = 0.052 \pm 0.004$ Å and 0.052 ± 0.007 Å. These values are significantly smaller than those obtained, for the same materials, from the x-ray Debye–Waller parameter, indicating that the assumption of uncorrelated displacements is incorrect. The data suggest that the atomic SRO in the crystal lattice is

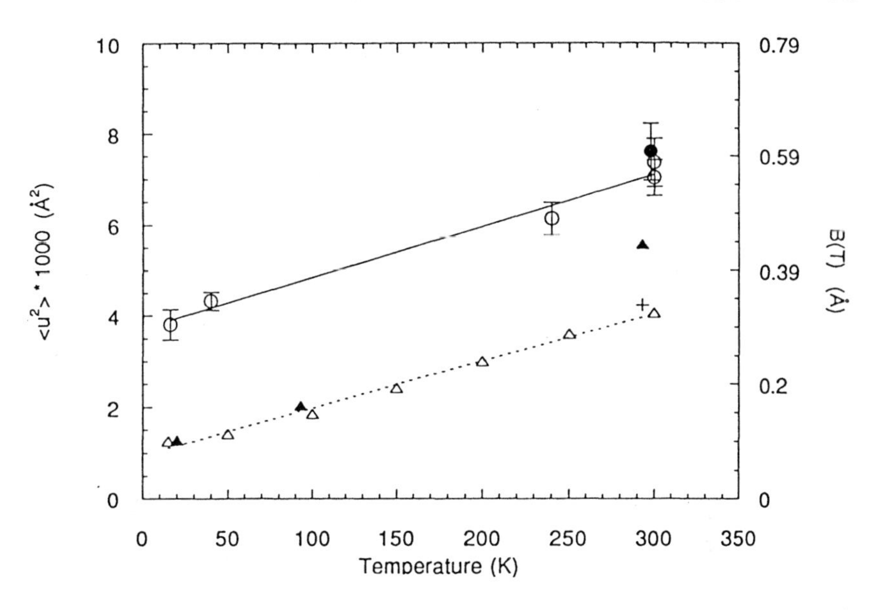

Figure 10.15. Mean square atomic displacement from the ideal lattice site, $\langle u_x^2 \rangle$, as determined from the x-ray Debye–Waller parameter, versus temperature for n-Pd (open and solid circles) and for coarse-grained polycrystalline Pd (triangles and cross). Reproduced by permission of Eastman [53].

not measurably affected by the reduction of the grain size to the nanometer scale. EXAFS mean square relative displacements also differ very little between coarse-grained Pd and n-Pd [112, 113]. Combining the information from the different techniques, it is concluded that the additional atomic displacements in the crystal lattice of n-Pd are static and involve correlated atomic displacements (figure 10.14(*b*)) with a range of several interatomic distances [15, 112, 113].

In a study of the x-ray Debye–Waller parameter in bulk nanocrystalline Cr, the static part of the mean square displacement is found to increase, with decreasing grain size, as $\langle u_x^2 \rangle \sim D^{-1}$ [114]. This is argued to suggest that the displacements occur in a layer of constant thickness at the grain boundaries. The volume of such a layer is approximately proportional to D^{-1}. It is interesting to note that the grain-size dependence of $\langle u_x^2 \rangle$ differs from that of the mean square lattice strain, which increases with diminishing grain size as $\langle \varepsilon^2 \rangle \sim D^{-2.4}$, that is considerably faster than the mean square displacement (compare section 10.3.4). The difference in grain-size dependence indicates that $\langle u_x^2 \rangle$ on the one hand and $\langle \varepsilon^2 \rangle$ on the other originate from different types of defect in the crystal lattice of the nanocrystalline solid.

10.4 CHARACTERIZING THE GRAIN BOUNDARIES

10.4.1 Relation between the grain boundary short-range order and the grain size

The atomic SRO in the grain boundaries is one of the central characteristics of nanocrystalline solids, and the one that has generated a great deal of controversy. There is conflicting evidence on the conclusion, reached in a number of experimental investigations (see references below), that the atomic SRO in the grain boundaries is modified dramatically as the grain size decreases. In these studies, the grain boundaries in nanocrystalline metals are argued to be considerably less ordered and of lower atomic density than in any other solid state of matter. There are a number of simple (but speculative) arguments which support the concept of a grain-size dependent grain boundary structure.

First, for any given crystallite the neighboring grains impose constraints on the rigid body relaxations both in the grain boundary plane and perpendicular to it. It is unknown to what extent these constraints can be accommodated through elastic or plastic deformation of the crystal lattice. Rigid body relaxations are of central importance for minimizing the energy of a grain boundary. If they are hindered, then a significant increase in the grain boundary free energy results.

A second, similar effect is that a given crystallite need not necessarily fit into the hole made up by the surrounding crystallites. Again, the misfit may be accommodated through deformation of the crystallite, or else it may induce a change in the local free volume, that is in the atomic density of the grain boundaries.

Thirdly, for very small grain sizes, the length of each boundary becomes comparable to the periodicity of the atomic arrangements in the grain boundary plane. It is unknown how this loss of long-range periodicity affects the atomic structure of the boundary. In agreement with these considerations, computer simulations of nanocrystalline materials indicate a widened grain boundary core at small grain sizes and an isotropic state of the grain boundaries (diminished variation of the energetics with the crystallographic parameters of the boundary) [115]. It has been pointed out [115] that in zeroth order these findings suggest a cement-like grain boundary phase reminiscent of early grain boundary models.

Finally, grain boundaries in nanocrystalline solids are generally created at considerably lower temperatures than their counterparts in coarse-grained polycrystals. The reduced atomic mobility at the low temperatures may prevent the boundaries from reaching the equilibrium structure during the finite time between preparation and experiment. Indeed, changes in the grain boundary structure at constant grain size have been inferred from calorimetric data. In n-Pt, grain growth is preceded by an exothermic process which is identified as a relaxation of the grain boundaries [116]. From the amount of heat released the grain boundary free energy of n-Pt in the as-prepared state is computed to be twice that of the relaxed state.

Because the grain boundary structure has a strong effect on many physical properties of nanocrystalline materials, the investigation of the dependence of their atomic structure on the grain size and on the preparation and annealing conditions is imperative for understanding their properties.

10.4.2 Grain boundary short-range order in nanocrystalline solids from scattering studies

Modeling the important diffuse background intensity found in an early x-ray diffraction study of n-Fe [117] requires that each atom in the three outer atomic layers of each crystallite be displaced randomly from its crystal lattice site, the maximum displacement being 50% of the nearest-neighbor distance in the outermost layer. This has been argued to indicate a 'gas like' state of the grain boundary component in nanocrystalline solids [117]. On the other hand, a more recent study of n-Pd [118] shows that in this material the diffuse intensity between the Bragg reflections can be explained by overlap of the reflection tails, if suitable assumptions on the functional form of the reflections are made. This finding has been discussed in terms of an ordered state of the grain boundaries in n-Pd [118].

The interpretation of the n-Pd x-ray data in terms of ordered grain boundaries is supported by high-resolution transmission electron microscopy (HRTEM) data [28] displaying narrow and facetted grain boundaries in n-Pd and n-Cu. However, such studies have remained inconclusive because of the difficulties in estimating the influence of the sample preparation on the grain boundary structure. Rigid body relaxations of the crystallites in and perpendicular to the grain boundary plane are essential for minimizing the grain boundary energy [127]. In nanocrystalline solids, these relaxations are subject to constraints imposed by the surrounding crystals, as discussed in section 10.4.1. For observation by HRTEM, the sample is thinned to a thickness of the order of one grain diameter. This process eliminates the constraints, and may thereby induce a relaxation towards a more ordered grain boundary structure with lower energy.

The analysis of the scattering problem (see section 10.2.4) indicates that the grain boundary SRO can be characterized most accurately on the basis of experimental atomic distribution functions, as opposed to interference functions discussed in earlier investigations by scattering methods. Two recent experiments apply this approach to the systems studied in the earlier work. In the first, the x-ray investigation of nanocrystalline Fe has been repeated on samples maintained under high-vacuum conditions [119]. The diffuse background in the interference function is found to be smaller than in the earlier investigation, which was carried out under atmospheric conditions. An evaluation on the basis of equation (10.9) (see figures 10.16 and 10.17) indicates that $95 \pm 6\%$ of the atoms are on crystal lattice sites, suggesting a considerably more ordered state of the grain boundaries in n-Fe than had been reported before.

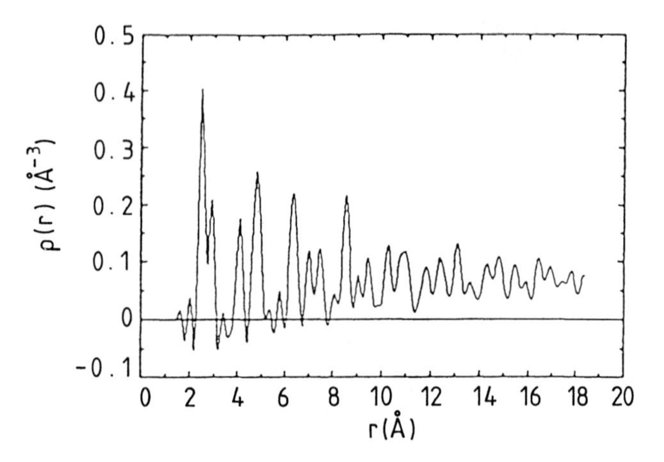

Figure 10.16. Atomic distribution function of nanocrystalline Fe. From [119].

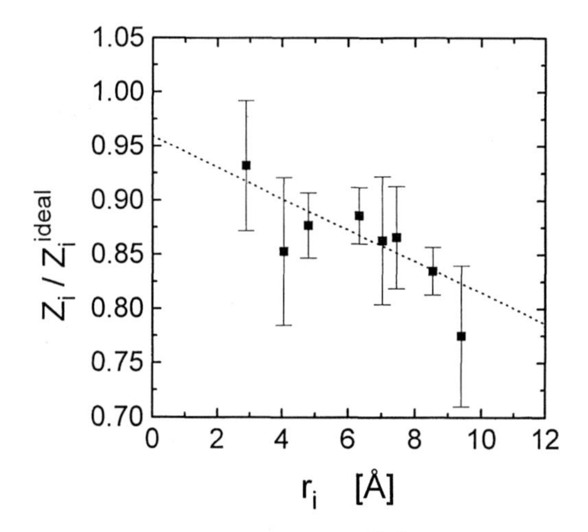

Figure 10.17. Relative coordination number Z/Z^{ideal} versus interatomic distance r for n-Fe. The dashed line is a fit by equation (10.9). From [119].

The second investigation re-examines n-Pd [16, 120]. A comparison of x-ray atomic distribution functions computed for coarse-grained polycrystalline Pd and for n-Pd produced by inert gas condensation is shown in figure 10.18. The coordination numbers in n-Pd are seen to be reduced as compared to those of the coarse-grained sample. The atomic distribution functions of a number of samples with different consolidation and annealing treatments are all in excellent agreement with equations (10.8) and (10.9). This is concluded from the fact that the experimental relative coordination numbers (compared to the ideal lattice) decrease linearly with the interatomic distance (figure 10.19). The fraction of atoms on crystal lattice sites, x_L, is determined from the intersect of the straight

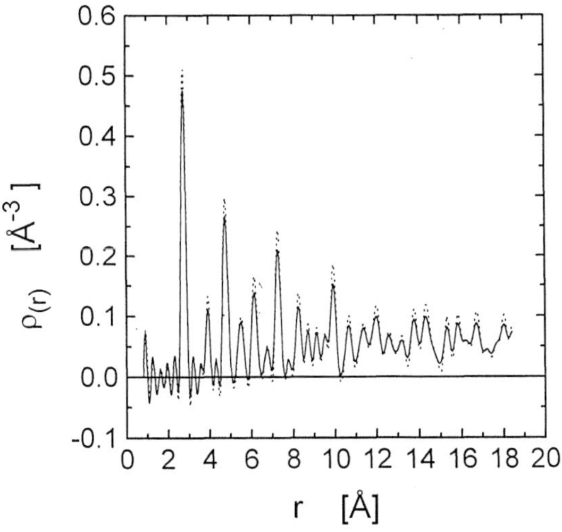

Figure 10.18. Atomic density distribution functions for nanocrystalline (solid line) and coarse-grained polycrystalline (dashed line) Pd. From [16].

line of best fit to the relative coordination numbers (compare section 10.2.4). Within error bars, x_L is unity for samples which are aged at room temperature for several months or annealed. For this group of samples, the data indicate an upper limit of 0.26 atomic monolayers of atoms on nonlattice sites in the grain boundary region. On the other hand, in samples investigated within less than ten days of preparation ('as-prepared' samples), about 10% of atoms are located on nonlattice sites (figure 10.19(*c*)) with no detectable atomic SRO. At the grain size of those samples, 12 nm, this corresponds to 1.8 atomic monolayers of nonlattice atoms in the disordered grain boundary layer. While as-prepared samples also tend to have the smallest grain size, the data (figure 10.20) indicate that the difference in grain boundary SRO is correlated to the sample age, rather than to the grain size. The results suggest that the grain boundaries in as-prepared n-Pd are in a nonequilibrium state with lower atomic SRO than conventional grain boundaries in polycrystalline metals.

The disordered state of the grain boundaries in as-prepared n-Pd is unstable at room temperature; it evolves, on a time-scale of several weeks, towards a more ordered state. This state is more ordered in the sense that a higher number of atoms occupy crystal lattice sites. However, as the experimental data supply no direct information on the SRO in the grain boundary plane (compare section 10.2.4), the distribution of interatomic spacings across the grain boundary may be random (disordered). The highly ordered state of the crystal lattice and the very small number of nonlattice atoms in the aged and annealed nanocrystalline solids imply that the topological defect associated with the grain boundary is localized in an essentially two-dimensional manner in the grain boundary plane (compare figure 10.1(*b*)). The available evidence

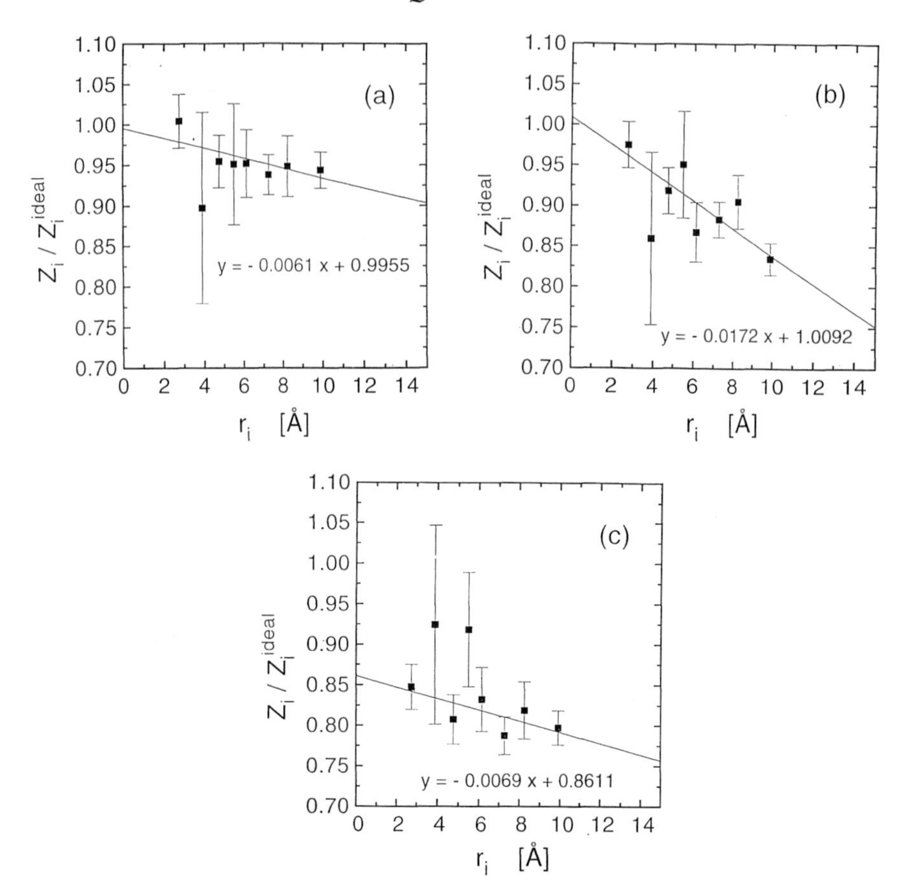

Figure 10.19. Relative coordination number Z/Z^{ideal} of Pd as a function of interatomic spacing r. (*a*) Coarse-grained reference obtained by annealing a n-Pd sample for 24 h at 700 °C. (*b*) Nanocrystalline Pd aged for several months at room temperature. (*c*) As-prepared nanocrystalline Pd. The dashed lines are fits to the experimental data (weighted according to the error bars) by equations (10.9) and (10.8). From [16, 120].

allows no conclusion on whether the grain boundary structure in this relaxed nanocrystalline state is identical to that in conventional polycrystals.

10.4.3 Grain boundary short-range order from EXAFS studies

EXAFS probes the atomic SRO around the atoms of the excited species. If the grain boundaries in an nanocrystalline metal are doped with an element which is not soluble in the crystal lattice, then the EXAFS signal at the absorption edge of that element arises exclusively from the grain boundary regions. In this way, a strongly enhanced signal characteristic of the grain boundary SRO is obtained.

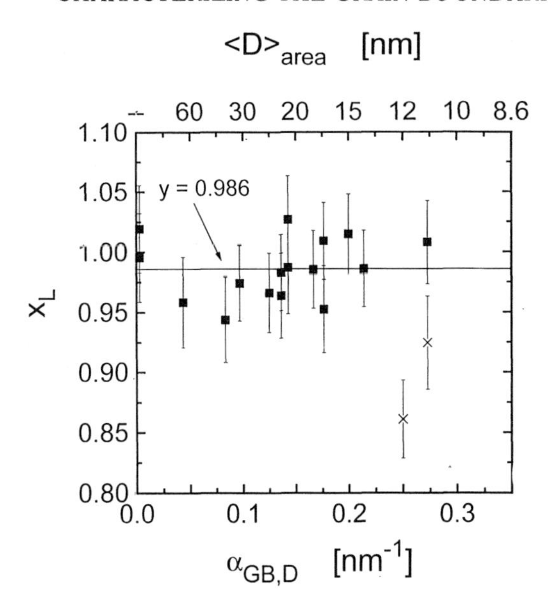

Figure 10.20. Fraction of atoms on lattice sites, x_L, versus specific grain boundary area α_{GB}, for inert gas condensed n-Pd. The grain size $\langle D \rangle_{area}$ is indicated on the upper horizontal axis. Solid squares: aged at room temperature for at least three months or annealed. Crosses: aged for less than ten days. The solid line is the average value for the aged and annealed data. From [16, 120].

The effect has been used in the study of n-Cu(Co) [121] and n-Cu(Bi) [122]. The results of those studies agree in indicating a strongly reduced coordination number for the grain boundary atoms. This indicates that the probe atoms in those studies, Co and Bi, may be located on those sites which are identified as 'nonlattice sites' in x-ray atomic distribution studies (compare section 10.4.2). The strongly reduced coordination number in EXAFS seems to suggest that there is, on average, very little atomic SRO around the probe atoms in the grain boundaries.

Most EXAFS investigations on pure elemental bulk nanocrystalline fcc metals Cu [51] and Pd [51, 123] and on bcc W [52] prepared by the inert gas condensation technique also report a large reduction in nearest-neighbor coordination number. This is interpreted in terms of a two-phase structure, comprising a crystalline component and a grain boundary component with random atomic arrangement [51]. A similar large reduction in the coordination number is reported for n-Fe prepared by HEBM [124]. The result is argued to suggest a large number of defects in the crystal lattice, as opposed to a disordered grain boundary component. Figure 10.21 shows experimental EXAFS k- and r-space data for n-Pd from [51].

For the purpose of comparison to the results from x-ray scattering,

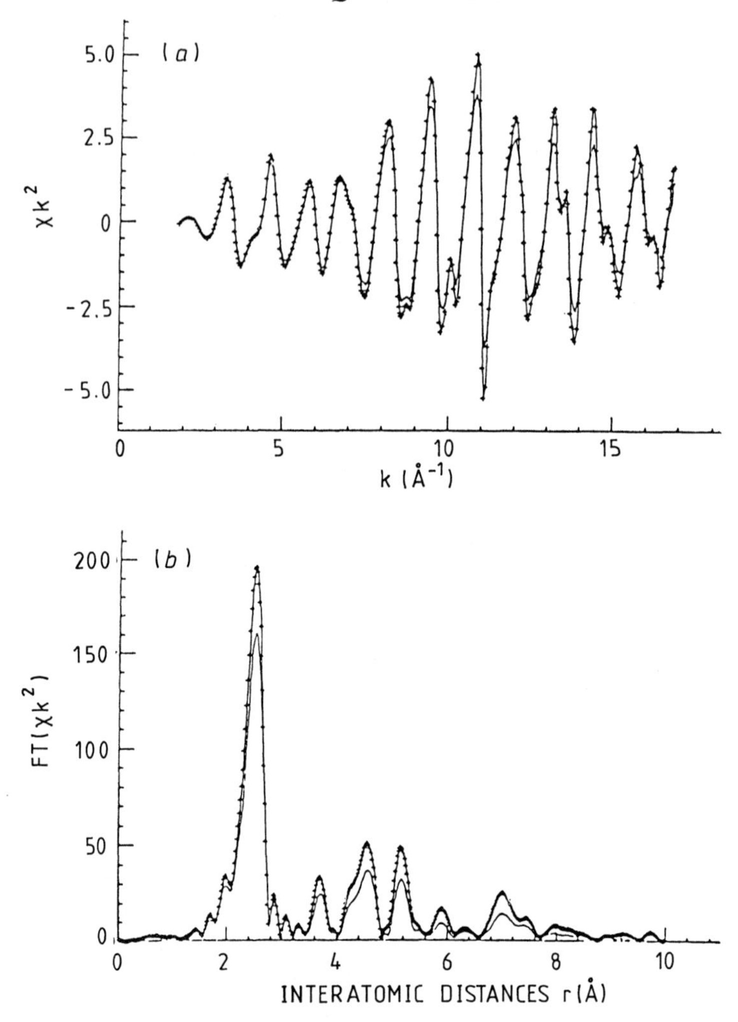

Figure 10.21. The weighted EXAFS χk^2 (*a*) and the Fourier transform FT(χk^2) (*b*) for n-Pd (crystallite diameter 16 nm; solid line) and for polycrystalline Pd (crossed line). Reproduced by permission of Haubold [51].

figure 10.22 displays the relative coordination numbers of n-Pd determined from EXAFS data in [51]. The results agree qualitatively with the recent x-ray data for aged and annealed n-Pd in that the fraction of atoms located on crystal lattice sites, as determined from the value of the straight line of best fit at zero interatomic distance, is close to unity. In other words, the data indicate that all atoms are located on crystal lattice sites, in contrast to the conclusion of a disordered grain boundary layer in the original study [51]. On the other hand, the EXAFS results in figure 10.22 disagree with the x-ray results in that the

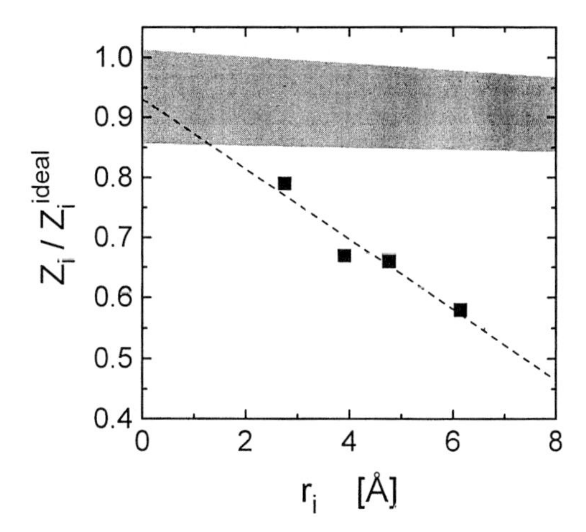

Figure 10.22. Relative coordination number Z/Z^{ideal} versus interatomic distance for n-Pd, determined from EXAFS data in [51]. Dashed line: self-correlation function determined from fit to the data. Shaded region: range of x-ray results for n-Pd at the same grain size.

grain size determined from the slope of the fit in the figure (assuming spherical particles with $\alpha_S = 6/D$; compare section 10.2.2), 2.6 nm, is considerably smaller than that obtained from x-ray line broadening analysis for the same sample, 16 nm [51]. The shaded region in figure 10.22 indicates the relative coordination numbers as obtained from x-ray atomic distribution functions at the grain size of the order of 16 nm. The coordination numbers determined from x-ray data are seen to be considerably larger than those obtained from EXAFS. Because of the good agreement of the x-ray data with theoretical predictions, the difference suggests that the apparent coordination numbers obtained by EXAFS can strongly underestimate the true coordination numbers.

A recent EXAFS investigation on inert gas condensed n-Pd [113] points out that a potential cause for erroneously low EXAFS coordination numbers resides in the nonuniform thickness of granular samples (see also [125, 126]). Compared to earlier results, the study finds a much smaller reduction in the coordination number. In agreement with the recent x-ray scattering results on aged samples the EXAFS data suggest non-reconstructed, incoherent grain boundaries.

10.5 DISTRIBUTION OF FREE VOLUME FROM SMALL-ANGLE SCATTERING

It has been shown above that wide-angle scattering experiments and EXAFS supply limited information on the atomic SRO in the grain boundary plane, that is on interatomic spacings between atoms located in different crystallites.

It is, however, precisely this intergrain part of the atomic distribution function that contains the information on the excess free volume of the grain boundary, $\{V\}$ [87]. This quantity determines important properties such as the grain boundary free energy [127]. In contrast to wide-angle scattering, small-angle scattering (SAS) does contain information on the intergrain part of the atomic distribution function (compare equation (10.7)) Therefore, SAS has the potential for determining $\{V\}$. Positron lifetime spectroscopy indicates that the distribution of free volume in the grain boundaries of nanocrystalline solids is indeed different from that in coarse-grained polycrystals (see chapter 11).

A complication in SAS data evaluation arises from the fact that the relation between the scattering pattern and real-space structure is generally not unique. More or less stringent assumptions on the nature and the arrangement of the scattering objects are required in order to extract real-space information from the data. If the arrangement of the scattering objects is dilute (e.g. in an array of well separated pores or particles), and if the shape of the objects is known, then their size distribution can be determined from the scattering data. Algorithms for the determination of the distribution are reviewed in [128]; they rely on the fact that in a dilute system interparticle interference is negligible. In other words, the total SAS intensity is the sum of the intensities of the individual scattering objects. Consequently the SAS intensity can be expressed as a linear combination of the known scattering intensities of individual objects belonging to different size classes (compare section 10.2.5). The requirement of negligible intergrain interference implies that this type of analysis yields exact results only for arrangements of well separated particles and well separated pores in nearly fully dense consolidated nanocrystalline matter.

When applied to SAS data from 60–70% dense n-Pd, the above analysis indicates scattering objects with a broad size distribution, with an average size of 2–3 nm [135]. For n-TiO_2 of similar relative density (by relative density we mean the ratio of the actual density over the value for a single crystal), the pore-size distribution determined in this way is found to be similar to that determined by BET nitrogen adsorption [129]. The SAS data indicate a considerable number of small pores, presumed closed and therefore not accessible to BET analysis. SAS pore-size analysis is argued to be able to monitor the evolution of the pore-size distribution on sintering of nanocrystalline ceramics [129]. The absence of intergrain interference in the far from dilute systems is conjectured to result from the wide distribution of shape and size of the pores.

More recent studies show significant increases in the uniformness of the particle size of ceramic nanocrystalline solids and in the green-body density achievable for nanocrystalline metals. For these more uniform systems, intergrain (or, more to the point, interpore) interference may be important even at high relative density, as the example of n-ZrO_2 [131] shows. The maximum in the SAS intensity observed in the material is argued to originate from interference between pores decorating the crystalline particles, that is separated by about one particle diameter. Therefore, the position of the maximum yields information on the size of the crystallites [130, 131]. This has allowed the simultaneous characterization of the evolution of the pore and of the crystallite structures in n-ZrO_2, suggesting that full density can be achieved at a crystallite size of 60 nm. In contrast to the case of ceramics, no maximum is observed in the interference function of nanocrystalline metals at high density. However, the increase in density coincides with a qualitative change in the scattering pattern as is discussed below. Interference peaks in the small-angle part of the scattering pattern are also observed in dilute arrangements of self-assembling, nanometer scale Mo particles [132]. SAS by the highly ordered cubic arrays of $2 \times 2 \times 2$, $3 \times 3 \times 3, \ldots$ particles (compare chapter 2) is described by a formalism inspired by scattering theory of crystalline structures, rather than the disordered structure approach adequate for the more commonly encountered nanocrystalline systems.

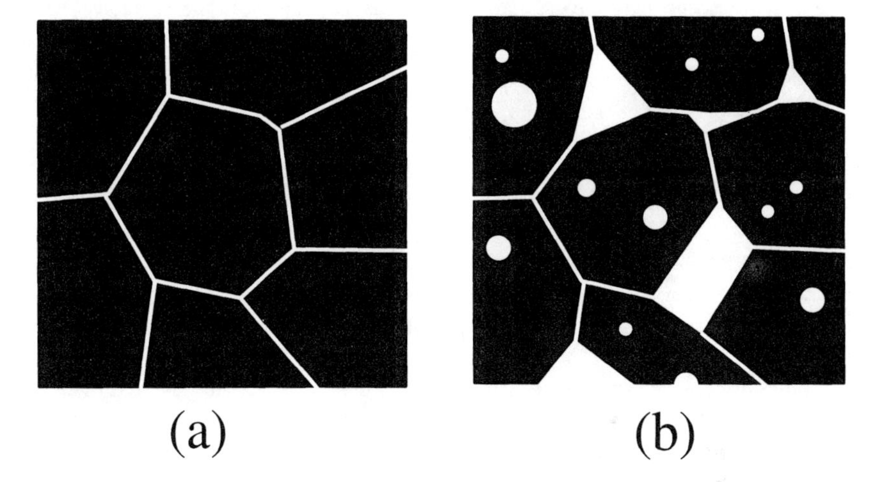

Figure 10.23. Schematic representation of an ideally dense nanocrystalline solid (*a*), where SAS arises exclusively from the grain boundaries with reduced density (white lines), and of a real nanocrystalline solid (*b*) where pores (white circles and polyhedra) dominate the scattering.

While the pore structure in highly dense nanocrystalline solids can be analyzed from SAS studies, no rigorous theoretical treatment of the scattering problem appears available for an ideally dense (pore-free) poly- or nanocrystalline solid, where intergrain interference is strong and where scattering contrast arises from the grain boundary free volume alone (see figure 10.23). However, the important effect of dense packing can be illustrated by considering a dense-packed structure for which the scattering problem can be solved analytically, the Poisson plate mosaic. The Poisson plate mosaic is a random arrangement of infinite plates with the average separation and the plate thickness as model parameters (see figure 10.24(*a*)); it resembles a nanocrystalline solid in that the scattering contrast arises from planar regions, similar to grain boundaries, separating the particles. An analytical solution for both the intragrain and the intergrain correlation functions is given in [133]. From this information, the SAS intensity can be computed both for the dense-packed arrangement in the mosaic and for a set of isolated particles with the same size and shape as those in the mosaic (compare section 10.2.3) [134]. The effect of dense packing of particles on the SAS intensity is illustrated in figure 10.24(*b*), for the example of a 300 nm sized spherical finite Poisson plate mosaic with the following parameters: average separation between planes (this is a measure for the 'grain size') 10 nm, and plane thickness of 0.1 nm. The intensity of the mosaic (dense-packed arrangement of particles) is seen to be considerably lower than the intensity of the isolated particles. The intensity at high wavenumbers follows a power law with an exponent −2 (characteristic of plates [13]) for the mosaic, as opposed to the exponent of −4 (characteristic of free surfaces [13]) for the isolated particles. While its similarity to a nanocrystalline solid makes the Poisson plate mosaic suitable for this qualitative illustration of the effect of dense packing on the SAS signal, the model's topology, featuring infinite planes as opposed to finite-size grain boundaries and four-grain junction lines as opposed to triple lines in a polycrystal, is not compatible with that of a poly- or nanocrystalline solid. The Poisson plate mosaic is therefore not suitable for a quantitative comparison to a real nanocrystal. At present, there appears to be no theoretical basis for the determination of particle size distributions from SAS data of nanocrystalline solids. In other words, the nature of an SAS pattern dominated by grain boundaries, as opposed to isolated pores or particles, is unknown.

Nevertheless, instructive results are obtained if scattering is assumed to arise from a two-phase structure, with constant scattering-length density in both phases. In a nanocrystalline single-component material, the first phase may be identified with the crystallites and the second phase may be a grain boundary layer of reduced density, or alternatively triple-line cores or porosity due to incomplete consolidation. For a single-component material, let v_P be the volume fraction occupied by the crystallites of atomic density $\langle\rho\rangle^V$ and scattering-length density $s\langle\rho\rangle^V$, and let the differences in atomic and scattering-length densities between the two phases be $\Delta\rho$ and $s\Delta\rho$, respectively. The macroscopic density of the sample, as determined e.g. by the Archimedes technique, depends in a linear way on v_P and $\Delta\rho$:

$$\langle\rho\rangle = v_P\langle\rho\rangle^V + (1 - v_P)(\langle\rho\rangle^V - \Delta\rho). \tag{10.22}$$

In contrast, the invariant of the SAS intensity, Q, depends on the squares of v_c and $\Delta\rho$ [13]

$$Q = \frac{1}{V}\int_0^\infty k^2 I(k)\,\mathrm{d}k = 2\pi^2 s^2 \Delta\rho^2 v_c(1 - v_c). \tag{10.23}$$

Therefore, the volume fraction and density of the second phase can be determined from a combination of the two independent measurements. In this way, the first

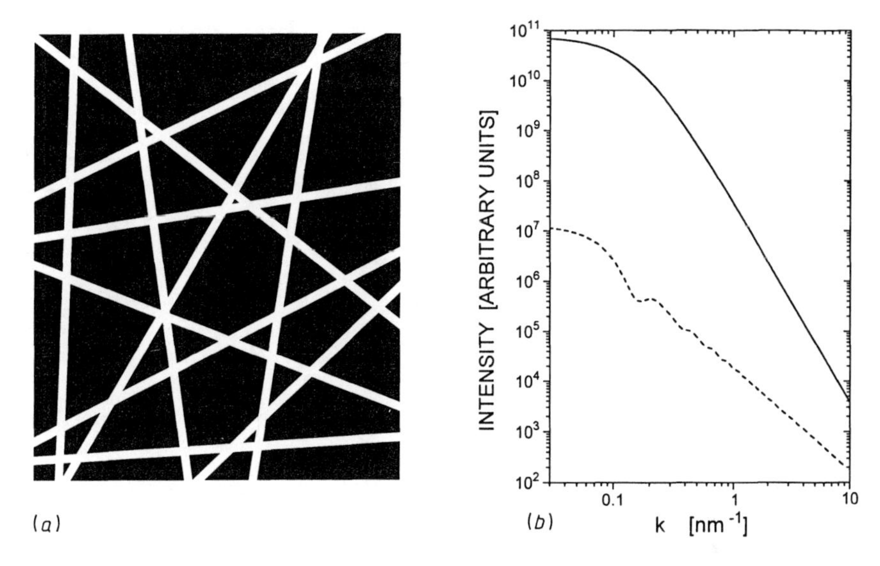

Figure 10.24. (*a*) Cross-section of a 3D Poisson plate mosaic. The black regions symbolize the particles, separated by randomly arranged, infinitely extended grain boundary planes of a constant thickness. (*b*) Dashed line: SAS intensity for the 3D Poisson plate mosaic; the asymptotic power law at high k is $I(k) \sim k^{-2}$. Solid line: SAS intensity for isolated, i.e. noninterfering particles with the same size and shape as those in the mosaic; the asymptotic power law at high k is $I(k) \sim k^{-4}$.

experimental study of n-Pd by small-angle neutron scattering (SANS) [135], neglecting porosity, derives volume fractions for crystallites and grain boundary components of 0.3 and 0.7, respectively, with a 50% reduction in atomic density for the grain boundary component. This result suggests that the state of the matter in the grain boundary component differs fundamentally from that in metallic glasses, where the atomic density is reduced by much less, of the order of 5%, in comparison to the crystalline state. On the other hand, more recent investigations [136] of n-Pd consolidated to higher macroscopic density find a 100% reduction in the scattering length density of the second phase. In other words, the scattering objects are identified as voids, as opposed to the grain boundary component as reported in [135]. While additional data point out that the high-k part of the SANS scattering pattern reported in [136], discussed in terms of triple-junction voids, is dominated by hydrogen incoherent scattering, the above conclusion on the density reduction is not affected [137].

Additional information on the nature of the scattering objects in n-Pd has been obtained by small-angle x-ray scattering (SAXS) [15]. In figure 10.25, the experimental values for the invariant Q are seen to agree, within error bars, with the value of the invariant determined from the relative density assuming zero atomic density in the scattering objects (equation (10.23)). This confirms the

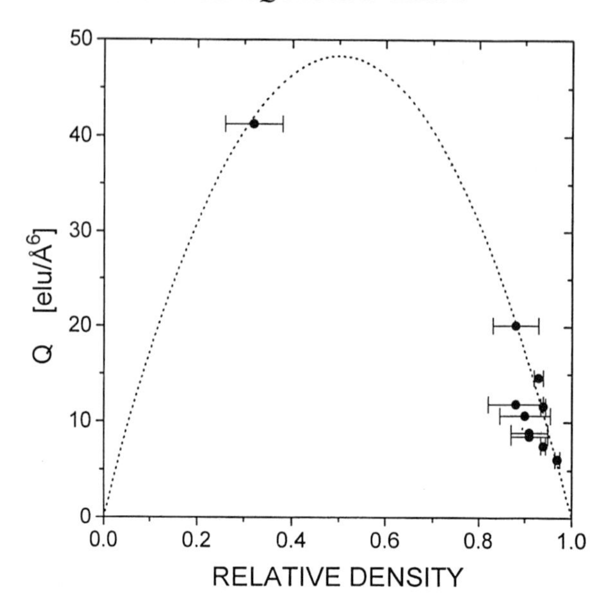

Figure 10.25. Invariant Q of the scattering intensity of inert gas condensed n-Pd versus sample density relative to the literature value for coarse-grained Pd. •: experimental data; dashed line: theory for a two-phase microstructure $\Delta\rho/\langle\rho\rangle_V = 1$. From [15].

conclusions in [136]. Figure 10.26 displays SAXS intensity data representative for n-Pd in different stages of consolidation. As $\langle\rho\rangle$ increases, the overall scattering intensity is seen to decrease, and the slope of the high-k part of the scattering curves is seen to decrease. The scattering intensity can be approximated by a power law $I(k) \sim k^{-n}$ at high k. Figure 10.27 displays the variation of power-law exponent, n, with macroscopic density $\langle\rho\rangle$. As $\langle\rho\rangle$ is increased, n is seen to decrease to values as small as 3.2. The deviation of n from four characteristic of free surfaces (Porod law of small-angle scattering [13]) indicates a fundamental change in the character of the objects dominating the scattering intensity. The nature of this change is understood qualitatively from equation (10.15), which relates the intensity power-law exponent to the size distribution of the scattering objects. For exponents less than four, the size distribution is also a power law, $n(D) \sim D^{-b}$, with $b > 3$. Consequently, the volume weighted distribution diverges for very small sizes.

The divergence of the volume weighted size distribution implies that the free volume and the SAS in the high-density nanocrystalline samples are dominated by very small objects with sizes down to below 1 nm. The nature of these objects is unknown. It may, however, be speculated that grain boundaries will give rise to a similar scattering pattern, since their thickness is of the order of 1 nm or below. The observed evolution of the power-law exponent to lower values as the density of the nanocrystalline system is increased is in qualitative

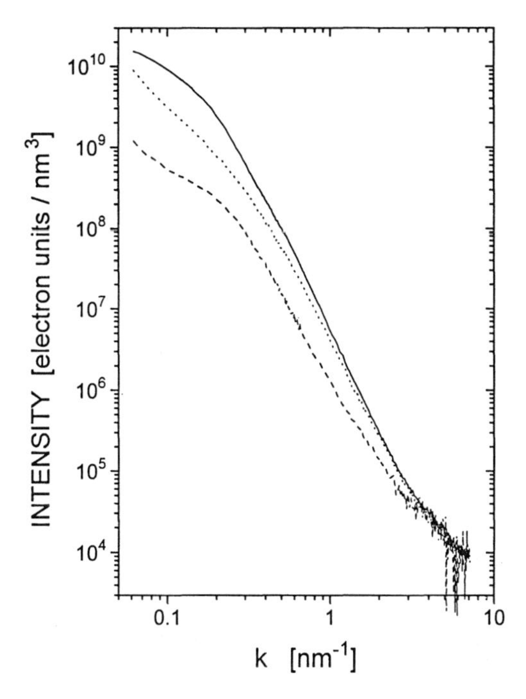

Figure 10.26. Experimental SAXS intensity versus wavevector for inert gas condensed n-Pd with 32% (solid line), 87% (dotted line), and 97% (dashed line) relative density. From [15].

agreement with the results obtained for the Poisson plate model. Hence, the change in power-law exponent may reflect a transition from pore dominated to grain boundary dominated scattering. This transition will necessarily occur if the porosity is progressively reduced until the grain boundaries are the only remaining scattering objects (compare figure 10.23). It is readily estimated that, if the grain boundaries in nanocrystalline metals had the same excess volume as conventional grain boundaries in metals, typically $\{V\}/a \leq 0.05$ [127] (a is the lattice constant), then the pore-free, 10 nm grain size nanocrystal would have a relative density $\geq$ 99.6%. In order for the scattering signal of those conventional grain boundaries to be observable, the nanometer-scale pore volume may have to be reduced to a value comparable with the total free volume of all the grain boundaries, that is a fractional porosity $\leq$ 0.4%. Experiments on nanocrystalline solids with this high density have not been reported. Future SAS studies may answer the question as to the existence of grain boundaries with highly increased free volume in nanocrystalline solids, or of other structures which may be intrinsic to the very fine-grained state of matter, such as dilute regions or voids at triple-junction lines or at higher-order grain junctions.

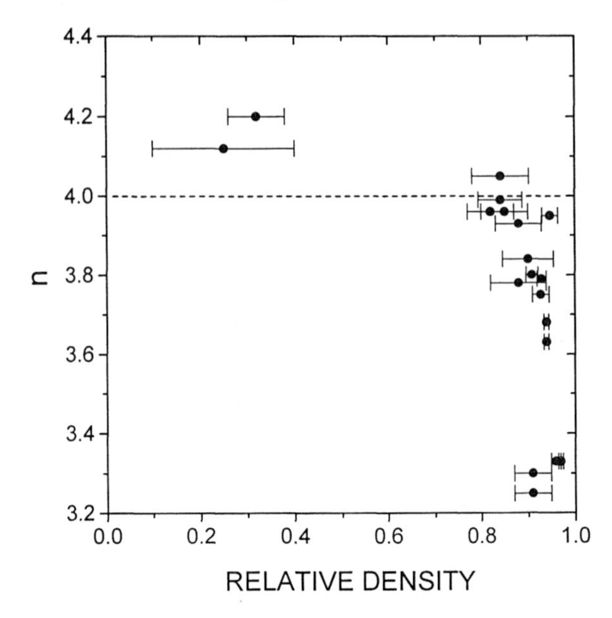

Figure 10.27. High-wavevector power-law exponent n versus relative density for n-Pd SAXS data. From [15].

10.6 NANOSTRUCTURED AMORPHOUS SOLIDS

Inert gas condensed nanometer particles of certain glass forming alloys are amorphous. These materials can be consolidated by the same process as used to prepare nanocrystalline metals to form bulk amorphous alloys. On the basis of early Mössbauer spectroscopy results [138] it has been proposed that these 'nanostructured amorphous solids' have a modified atomic SRO. The atomic SRO in a bulk amorphous metal (e.g. in a meltspun ribbon or in vapor quenched films with a thickness exceeding considerably the range of the coherent structural ordering) is generally very similar to that of the metallic glass [139]. Therefore, the atomic SRO of a bulk amorphous metal is determined by the thermodynamic equilibrium structure of the undercooled melt at the glass transition temperature. In contrast, the atomic SRO in the interfaces in nanostructured amorphous solids is proposed to be determined by the local atomic configurations which are generated when two originally free noncrystalline surfaces are contacted in random relative orientation. Hence, one expects an overall reduction in the atomic SRO.

Indeed, an x-ray diffraction study finds that the range of structural correlations, typically several interatomic distances in bulk amorphous metals and semiconductors, is reduced to one atomic nearest-neighbor distance in nanostructured amorphous Si–Au (figure 10.28) [140, 141]. The unusual reduced atomic SRO in the material was, however, found to be at least partially intrinsic

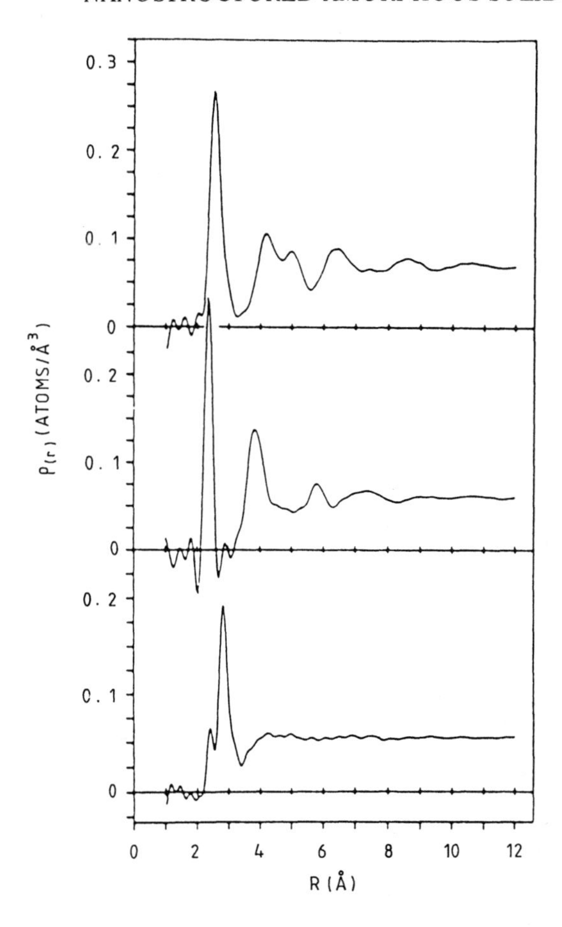

Figure 10.28. Comparative representation of the atomic distribution functions of a dense random packing of hard-sphere type metallic glass ($Fe_{78}B_{13}Si_9$, top), a continuous random network structure type amorphous semiconductor (Si–H, center), and nanostructured amorphous Si–Au (bottom). Note the reduced range of structural correlation in nanostructured Si–Au. From [141].

to the particular alloy system, with a pronounced effect of the large bond-length variations resulting from the mixed covalent and metallic character of the Si–Si bond on the atomic SRO. On the other hand, the strong SAS observed in the system evidences significant modulations of the structure on a nanometer scale, indicating that the particular mode of preparation will indeed result in a microstructure unlike that of a conventional metallic glass [142–144]. The non-Porod power laws observed in SAS from amorphous nanostructured Si–Au have been attributed to arrangements of amorphous particles with a fractal character [143, 144]. Contrary to this interpretation, gradients of the atomic density and/or composition in the interface regions have been argued to be responsible for the observed non-Porod SAS signal in nanostructured amorphous Ti–Pd [142]. At present, no independent experimental evidence in favor of one of the alternative interpretations appears to be available. For highly dense nanocrystalline Pd, which also displays a non-Porod power law, the notion of an atomic-scale fractal structure appears incompatible with the experimental finding that practically all atoms are located in a highly perfect crystal lattice (compare sections 10.3.6 and 10.4).

10.7 MAGNETIC STRUCTURE

Magnetic properties are among those which are most strongly affected by nanometer scaling. For ferromagnetic particles isolated by a nonmagnetic matrix, the coercive force increases as the particle size is reduced from micrometers to the nanometer scale. The coercive force is a maximum at the size where the particles become single-domain particles, typically 10–100 nm [145]. Nanocomposite materials with this particle size have large-scale commercial applications as hard magnets. Similar to conventional polycrystalline ferromagnets, the magnetization direction in the absence of an external magnetic field is aligned with the easy directions determined by the crystal lattice and shape anisotropy in each particle. If the particle size is reduced further, then the energy needed to rotate the magnetization vector out of the easy direction diminishes: at sufficiently small size, the thermal energy is sufficient to do this. The array of isolated particles as a whole will then lose its ferromagnetic properties and become superparamagnetic (see e.g. [146]). If small single-domain particles approach one another until their surfaces are within the magnetic exchange length, or if they are embedded in a ferromagnetic matrix, then the exchange interaction tends to align the magnetization vectors in neighboring particles. As a result, the magnetization vector is expected to be aligned on a scale considerably larger than the particle size. As the particle size can be comparable to or even smaller than the width of a domain wall in a coarse-grained polycrystal (typically several tens of nanometers in elemental ferromagnets), the character of the domain walls is expected to undergo a fundamental change as the particle size is reduced, to the extent that the concept of a domain wall may cease to apply.

In addition to these microstructure related issues, there may also be an effect from the variation of atomic SRO in the interfaces on the magnetic properties. In transition metals, the magnetic moment per atom depends on the interatomic spacing (Bethe–Slater curve [146]) and on the 3d and 4s electron densities (Slater–Pauling curve [146]). Due to the loss of atomic SRO and to the reduced atomic density in the grain boundary region, the values of both parameters differ from those of the crystal lattice. Therefore the magnetic moments of the grain boundary atoms may also be different from those in the crystal lattice. Similar effects are well known for free surfaces and isolated small particles (see reviews in [147, 148]). In certain magnetically ordered solids, the spin structure is modulated on a length scale which is much larger than the scale of the periodicity in the nonmagnetic state (see the case of Cr mentioned below [149]), that is much larger than the unit cell of the nuclear lattice. In this case, the size of the crystallites can become comparable to the characteristic length scale of the spin structure, and a pronounced effect of the fine particle size on the long-range magnetic ordering may be expected.

Because of the interaction of the neutron spin with the magnetic moment of the nucleus, the neutron scattering intensity is the sum of two contributions: the first, *nuclear* interference due to the atomic structure, has been discussed earlier in this work. In addition, *magnetic scattering* of a comparable magnitude arises from the magnetic structure, that is from the spatial distribution of magnitude and direction of the nuclear magnetic moments. In the wide-angle scattering region, this changes the relative intensities of the Bragg reflections, and may result in the appearance of reflections which are forbidden by the extinction rules of the nuclear lattice [150]. With respect to the magnetic microstructure, probed by SAS, the following argument can be made: conceptually, if all the local magnetization directions are aligned by a strong external magnetic field, then no nanometer-scale magnetic structure remains, and therefore the remaining scattering is exclusively due to the atomic structure. Hence, magnetic and atomic contributions can be separated when data are recorded both in the absence of an external field and in a field of sufficient strength to saturate the sample. In detail, there is still a magnetic contribution in the saturated sample, since the density of the magnetic moments is always proportional to the atomic density. For example, due to the change in magnetic scattering length at an internal surface, pores contribute strongly to the magnetic SAS signal. This contribution can be corrected for.

The last-mentioned type of characterization of the magnetic microstructure has been applied to n-Fe and n-Ni [151, 152]. The SAS intensity changes significantly when an external magnetic field is applied (figure 10.29). From the intensity difference, a distribution of magnetic domain sizes is determined. In the absence of a magnetic field, the magnetic correlations are found to extend across the interfaces and to result in the alignment of the magnetization on a scale of about 200 nm. The results further suggest a microstructural model involving ferromagnetic grains and nonmagnetic interfaces. This notion of nonmagnetic interfaces apparently contradicts measurements of the saturation magnetization for n-Ni, which indicate that the magnetic moment of the grain boundary atoms of this material is identical to that of the crystal lattice atoms [153, 154]. As for the case of SAS studies related to atomic microstructure, the relative magnitude of contributions from grain boundaries and from internal surfaces, due to porosity, appears to require further studies.

The high saturation magnetization of nanocrystalline elemental ferromagnets indicates that the reduction of the crystal size and the presence of the grain boundaries have little effect on the magnetic ordering in ferromagnets.

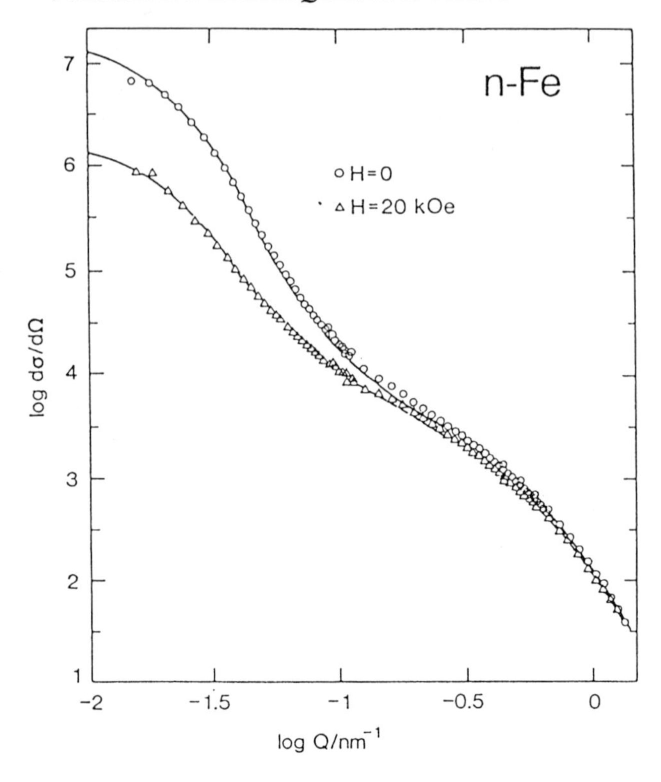

Figure 10.29. Experimental SANS differential cross-section $d\sigma/d\Omega$ as a function of wavevector Q for inert gas condensed n-Fe in zero magnetic field (circles), and in a magnetic field $H = 20$ kOe (triangles). The solid lines represent fits to the experimental data. Reproduced by permission of Wagner [151].

In contrast to this, a strong effect of the crystallite size on the antiferromagnetic ordering is observed in nanocrystalline Cr with crystal sizes of 16 nm and below [149]. From the absence of the (100) magnetic superstructure peak in neutron wide-angle scattering data, it is concluded that antiferromagnetic ordering in n-Cr is suppressed for temperatures below 20 K. In coarse-grained Cr, the superstructure peak is present below the Néel temperature of 311 K. In single-crystalline and coarse-grained polycrystalline Cr, the antiferromagnetic order of the spin system is modulated by a spin density wave with a wavelength of 6.6 nm, which is comparable to the particle size in n-Cr. It is pointed out in [149] that the suppression of antiferromagnetic ordering in n-Cr may be related to the confinement of the spin density wave in the small grains, possibly in combination with lattice strain.

10.8 DYNAMICS FROM INELASTIC NEUTRON SCATTERING

Diffraction studies require that the wavelength of the diffracted particles be comparable to the characteristic length scale in real space of the structure under investigation, that is of the order of 1 Å. As a result of its larger mass, the energy of a neutron with the required wavelength is small (1 eV) as compared to the energies of photons (10^4 eV) and electrons (10^5 eV) used in diffraction studies. Consequently, the energy transferred in inelastic interactions with thermally excited states of the crystal lattice (typically < 0.1 eV) is a very small fraction of the energy of the radiation used in the experiment for x-ray and electron diffraction studies, but constitutes a measurable fraction of the neutron's total energy. This enables studies of the dynamics of a solid by inelastic neutron diffraction. Various experimental techniques determine the scattering cross-section as a function of the two variables momentum and energy transfer (see eg. [155, 156]).

The modified lattice dynamics of small particles and their characteristic surface excitations are evidenced, in a more indirect way, by the studies of the Debye–Waller factor for elastic scattering described in section 10.3.6. The author is not aware of comparable studies of the dynamics of isolated small particles by inelastic scattering. However, the high specific surface area and the well defined habit planes of fine graphite and MgO particles have enabled studies of the dynamics and of phase transitions of thin layers of methane adsorbed on the particle surface [157, 158].

Two inelastic neutron scattering studies on bulk nanocrystalline samples have been reported. The vibrational excitations and the position of hydrogen in n-Pd were investigated in a combined inelastic neutron scattering and hydrogen solubility study [159, 160]. The study focussed on the hydrogen concentration regime where, at room temperature, no precipitation of H in the β-hydride phase is observed. In this concentration regime, solubility measurements show an enhanced overall solubility in n-Pd relative to coarse-grained Pd. The neutron scattering data indicate two distinctive vibrational modes for the dissolved hydrogen. The first can be attributed to hydrogen on interstitial sites in the crystal lattice. The second vibrational mode is similar to that found for H at free surfaces in coarse-grained polycrystalline and single-crystalline Pd; this mode is attributed to hydrogen at grain boundaries and free surfaces in n-Pd. This supports the conclusion, deduced from earlier measurements of the solubility as a function of the chemical potential [161, 162] that the additional hydrogen is incorporated at the internal interfaces in n-Pd. Results of an x-ray diffraction study of the phase transition between the Pd–H solid solution and the β-hydride phase in the nanocrystalline material have also been argued to agree with this conclusion [163]. However, contradictory explanations have been brought forward for the reduction of the amount of H which can be dissolved in the β-hydride phase [162, 163].

The second experiment on bulk nanocrystalline metal explores the excitations in pure n-Ni [164] by means of inelastic neutron scattering. As compared with coarse-grained polycrystalline Ni, nanocrystalline Ni has a higher phonon density of states (DOS) at energies below 15 meV. The additional component in the DOS is a linear function of the energy, which is argued to imply a two-dimensional nature of the excitations. Noncompacted nanocrystalline powder, investigated for comparison in the same experiment, shows less enhancement in the low-energy DOS. This supports the interpretation that the enhanced DOS is due to vibrational modes at grain boundaries, as opposed to free surfaces. The existence of vibrational modes specific to grain boundaries indicates that elastic moduli of grain boundaries may differ from those of the crystal lattice.

10.9 SUMMARY

In summary, scattering techniques and EXAFS provide insight into a number of aspects related to the structure of nanocrystalline solids. The characteristics of bulk nanocrystalline solids are found to differ in several respects from those of isolated or supported particles.

A contraction of the crystal lattice with decreasing particle size is well documented for isolated or supported metal particles. For bulk nanocrystalline solids, the little evidence that is available seems indicative of a considerably smaller change in lattice parameter. Increased short- and medium-range correlated displacements of the atoms from their ideal lattice sites are observed both in fine-grained isolated particles and in nanocrystalline solids. The displacements are of a dynamic origin in isolated particles, but of static character in nanocrystalline solids. As the grain size of nanocrystalline solids is decreased, long-range correlated displacements of the atoms from the ideal lattice sites (lattice strain) increase strongly. A similar effect has apparently not been reported for isolated particles. The presence of the static displacements suggests that the constraints imposed on each crystallite by the neighboring grains have an important effect on the atomic structure in nanocrystalline solids. However, the amplitudes of the displacements due to this effect are considerably smaller than the amplitudes of the thermally activated displacements (at room temperature) in single crystals. The atomic short- and long-range order in the crystal lattice in nanocrystalline solids are therefore remarkably perfect.

The effect of nanometer-scale structuring on the real-space information obtained from scattering data and from EXAFS, the atomic distribution function, and on the wide angle and small angle scattering interference functions can be formulated concisely. The results are well supported by experiment. Theory indicates how a quantitative measure for the atomic SRO in the grain boundaries, the fraction of atoms on crystal lattice sites, can be obtained from experimental data. The emerging picture is that the SRO in the grain boundaries in nanocrystalline solids is not unique, but depends on parameters such as the age or the thermal history of the sample. In nanocrystalline Pd a comparatively wide disordered grain boundary structure relaxes, upon aging for several weeks at room temperature, towards a structure where the topological defect is essentially localized in the two-dimensional grain boundary plane. The high atomic mobility required for the relaxations is also evidenced by the important grain growth at room temperature observed recently in nanocrystalline metals. This phenomenon, presumably connected with the higher purity achieved in recent samples, highlights the need for developing methods for stabilizing the grain size.

The free volume of the grain boundaries in nanocrystalline solids has been speculated to deviate significantly from that of grain boundaries in conventional coarse-grained polycrystals. These speculations have not been conclusively confirmed or disproved. The grain boundary excess free volume can in principle be determined by SAS. However, improved samples with significantly lower porosity and theoretical studies of the scattering problem are required to achieve this. On the other hand, SAS studies of the pore-size distribution have provided valuable insight into the microstructure and its evolution during sintering.

Small angle scattering studies conducted to characterize metals have been concerned with issues somewhat different from those involved in scattering studies of ceramics. Studies of metals have emphasized fundamental issues, whereas studies of ceramics have been motivated by the application oriented goal of achieving a dense microstructure. With refined experimental techniques and data analysis, diffraction experiments and EXAFS on improved samples will contribute further to promoting our understanding of nanostructured materials.

REFERENCES

[1] Gleiter H 1991 *Prog. Mater. Sci.* **33** 223

[2] Siegel R W 1991 *Annu. Rev. Mater. Sci.* **21** 559

[3] Birringer R *Proc. NATO ASI Nanophase Materials: Synthesis–Properties–Applications* ed G C Hadjipanayis and R W Siegel (Dordrecht: Kluwer) p 157

[4] Suryanarayana C 1995 *Int. Mater. Rev.* **40** 41

[5] Gleiter H 1995 *Nanostruct. Mater.* **6** 3

[6] Weissmüller J 1996 *Synthesis and Processing of Nanocrystalline Powder* ed D Bourell (Warrendale, PA: TMS)

[7] Azarroff L, Kaplow R, Kato N, Weiss R, Wilson A and Young R 1974 *X-Ray Diffraction* (New York: McGraw-Hill) section 2:II

[8] Wagner C N J 1980 *J. Non-Cryst. Solids* **31** 1

[9] Toby B H and Egami T 1992 *Acta Crystallogr.* A **48** 336

[10] Cockayne D J H and McKenzie D R 1988 *Acta Crystallogr.* A **44** 870

[11] Wilson A J C 1962 *X-ray Optics* 2nd edn (New York: Wiley) ch IV

[12] Guinier A 1963 *X-ray Diffraction in Crystals, Imperfect Crystals, and Amorphous Bodies* (San Francisco, CA: Freeman) ch 2, 5, 10 (reprint 1994 (New York: Dover))

[13] Porod G 1982 *Small-angle X-ray Scattering* ed O Glatter and O Kratky (London: Academic) p 17

[14] In an earlier publication [15], the intragrain part of the atomic distribution function has been referred to as the 'self' part.
[15] Weissmüller J, Löffler J and Kleber M 1995 *Proc. 2nd Int. Conf. on Nanostructured Materials, Nanostruct. Mater.* **6** 105
[16] Löffler J and Weissmüller J 1995 *Phys. Rev.* B **52** 7076
[17] Birringer R, Gleiter H, Klein H-P and Marquardt P 1984 *Phys. Lett.* **102A** 365
[18] Gleiter H and Marquardt P 1984 *Z. Metallkunde* **75** 263
[19] Bertaut F 1950 *Acta Crystallogr.* **3** 14
[20] Betts F and Bienenstock A J 1972 *J. Appl. Phys.* **43** 4591
[21] Matyi R J, Schwartz L H and Butt J B 1987 *Catal. Rev.-Sci. Eng.* **29** 41
[22] Glatter O 1982 *Small-angle X-ray Scattering* ed O Glatter and O Kratky (London: Academic) p 167
[23] Weissmüller J, Löffler J, Krill C, Birringer R and Gleiter H to be published
[24] Granqvist C G and Buhrman R A 1976 *J. Appl. Phys.* **47** 2200
[25] Uyeda R 1992 *Prog. Mater. Sci.* **35** 1
[26] Kaito C, Kazuo F, Shibahara H and Shiojiri M 1977 *Japan. J. Appl. Phys.* **16** 697
[27] Wunderlich W, Ishida Y and Maurer R 1990 *Scr. Metall.* **24** 403
[28] Thomas G, Siegel R and Eastman J 1990 *Scr. Metall.* **24** 201
[29] Hahn H 1993 *Nanostruct. Mater.* **2** 251
[30] Skandan G, Hahn H, Roddy M and Cannon W 1994 *J. Am. Ceram. Soc.* **77** 1706
[31] Kimoto K and Nishida I 1967 *Japan. J. Appl. Phys.* **6** 1047
[32] Ganqvist C G, Milanowski G J and Buhrman R A 1975 *Phys. Lett.* **54A** 245
[33] Saito Y, Mihama K and Uyeda R 1980 *Japan. J. Appl. Phys.* **16** 1603
[34] Edelstein A S, Chow G M, Altman E I, Colton R J and Hwang D M 1991 *Science* **251** 1590
[35] Krauss W 1995 *PhD Thesis* Universität des Saarlandes
[36] Kimoto K and Nishida I 1967 *J. Phys. Soc. Japan* **22** 744
[37] Shuttleworth R 1950 *Proc. Phys. Soc.* A **63** 444
[38] Cahn J W 1980 *Acta Metall.* **28** 1333
[39] Cammarata R and Sieradzki K 1994 *Annu. Rev. Mater. Sci.* **24** 215
[40] Mays C, Vermaak J and Kuhlmann-Wilsdorf D 1968 *Surf. Sci.* **12** 134
[41] Wassermann H and Vermaak J 1970 *Surf. Sci.* **22** 164
[42] Wassermann H and Vermaak J 1972 *Surf. Sci.* **32** 168
[43] Harada J and Ohshima K 1981 *Surf. Sci.* **106** 51
[44] Apai G, Hamilton J, Stöhr J and Thompson A 1979 *Phys. Rev. Lett.* **43** 165
[45] Balerna A, Bernieri E, Picozzi P, Reale A, Santucci S, Burattini E and Mobilio S 1985 *Phys. Rev.* B **31** 5058
[46] Montano P, Schulze W, Tesche B, Shenoy G and Morrison T 1984 *Phys. Rev.* B **30** 672
[47] Montano P, Shenoy G, Alp E, Schulze W and Urban J 1986 *Phys. Rev. Lett.* **56** 2076
[48] Hansen L, Stoltze P, Nørskov J, Clausen B and Niemann W 1990 *Phys. Rev. Lett.* **64** 3156
[49] Crozier E, Rehr J and Ingalls R 1988 *X-ray Absorption* ed D Koningsberger and R Prins (New York: Wiley) p 373
[50] Briant C L and Burton J J 1975 *Surf. Sci.* **51** 345
[51] Haubold T, Birringer R, Lengeler B and Gleiter H 1989 *Phys. Lett.* **135A** 461
[52] Haubold T, Krauss W and Gleiter H 1991 *Phil. Mag. Lett.* **63** 245

[53] Eastman J, Fitzsimmons M and Thompson L 1992 *Phil. Mag.* B **66** 667
[54] Warren B E 1990 *X-ray Diffraction* (New York: Dover)
[55] Klug H P and Alexander L E 1974 *X-ray Diffraction Procedures for Polycrystalline and Amorphous Materials* 2nd edn (New York: Wiley) p 661
[56] Patterson A L 1939 *Phys. Rev.* **56** 978
[57] Stokes A R and Wilson A J C 1942 *Proc. Camb. Phil. Soc.* **38** 313
[58] Wagner C N J and Aqua E N 1964 *Adv. X-ray Anal.* **7** 46
[59] Eckert J, Holzer J, Krill C and Johnson W 1992 *J. Mater. Res.* **7** 1751
[60] Hellstern E, Fecht H J, Fu Z and Johnson W L 1989 *J. Appl. Phys.* **65** 305
[61] Bennett L H, Takacs L, Swartzendruber L J, Weissmüller J, Bendersky L A and Shapiro A J 1995 *Scr. Metall.* **33** 1717
[62] Warren B and Averbach B 1950 *J. Appl. Phys.* **21** 595
[63] Langford J I 1992 *Int. Conf. Accuracy in Powder Diffraction II (NIST Special Publ 846)* p 110
[64] Balzar D 1992 *J. Appl. Crystallogr.* **25** 559
[65] van Berkum J G M, Vermeulen A C, Delhez R, de Keijser T H and Mittermeijer E J 1994 *J. Appl. Crystallogr.* **27** 345
[66] Wagner C N J 1992 *J. Non-Cryst. Solids* **150** 1
[67] Granqvist C and Buhrman R 1976 *J. Appl. Phys.* **47** 2200
[68] Haas V and Birringer R 1992 *Nanostruct. Mater.* **1** 491
[69] Günther B, Kumpmann A and Kunze H-D 1992 *Scr. Metall.* **27** 833
[70] Hilliard J E, Cohen J E and Paulson W M 1968 *Proc. 15th Sagamore Army Mater. Res. Conf.* (Syracuse, NY: Syracuse University Press) p 73
[71] Smith W L 1972 *J. Appl. Crystallogr.* **5** 127
[72] Warren B E and Averbach B L 1950 *J. Appl. Phys.* **21** 595
[73] Warren B E 1990 *X-ray Diffraction* (New York: Dover) ch 13.4
[74] Sashital S R, Cohen J B, Burwell R L and Butt J B 1977 *J. Catalysis* **50** 479
[75] Moraweck G, De Montgolfier Ph and Renouprez A J 1977 *J. Appl. Crystallogr.* **10** 191
[76] Hossfeld F and Oel H J 1966 *Z. Angew. Phys.* **20** 493
[77] Le Bail A and Louër D 1978 *J. Appl. Crystallogr.* **11** 50
[78] Krill C R, MacMahon G, Löffler J and Birringer R *Engineering Foundation Conf. Nanophase Materials (Davos, March 1994)* to be published
[79] Löffler J 1994 *Diploma Thesis* Fachbereich Physik, Universtät des Saarlandes
[80] Anderson M, Grest G and Srolovitz D 1989 *Phil. Mag.* B **59** 293
[81] Sursaeva V 1990 *Mater. Sci. Forum* **62–64** 807
[82] Malow T R and Koch C C 1996 *Synthesis and Processing of Nanocrystalline Powder* ed D Bourell (Warrendale, PA: TMS) p 33
[83] Cahn J W 1962 *Acta Metall.* **10** 789
[84] Lücke K and Stüwe H P 1971 *Acta Metall.* **19** 1087
[85] Srolovitz R E, Eykhold R, Barnett D M and Hirth J P 1987 *Phys. Rev.* B **35** 6107
[86] Hondros E and Seah M 1983 *Physical Metallurgy* ed R Cahn and P Haasen (Amsterdam: North-Holland) p 855
[87] Weissmüller J 1993 *Nanostruct. Mater.* **3** 261
[88] Weissmüller J, Krauss W, Haubold T, Birringer R and Gleiter H 1992 *Nanostruct. Mater.* **1** 439
[89] Krill C E, Klein R, Janes S and Birringer R *Proc. ISMANAM-94; Mater. Sci. Forum* **179–181** 443

[90] Hwang S L and Chen I W 1990 *J. Am. Ceram. Soc.* **73** 3269
[91] Terwillinger C D and Chiang Y M 1994 *Nanostruct. Mater.* **4** 651
[92] Bansal C, Gao Z Q and Fultz B 1995 *Nanosruct. Mater.* **5** 327
[93] Krivoglaz M *Theory of X-ray and Thermal Neutron Scattering by Real Crystals* (New York: Plenum) section 14
[94] Eastman J A, Beno M A, Knapp G S and Thompson L J 1995 *Nanosruct. Mater.* **6** 543
[95] Tschöpe A and Birringer R 1992 *J. Appl. Phys.* **71** 5391
[96] Weissmüller J 1994 *J. Mater. Res.* **9** 4
[97] Bever M B, Holt D L and Titchener A L 1973 *Prog. Mater. Sci.* **17** 1
[98] Faulkner E A 1960 *Phil. Mag.* **5** 519
[99] Hellstern E, Fecht H, Fu Z and Johnson W 1989 *J. Appl. Phys.* **65** 305
[100] Nazarov A, Romanov A and Valiev R 1994 *Nanostruct. Mater.* **4** 93
[101] Thölen A R 1990 *Phase Transitions* **24–26** 375
[102] Gao P 1990 *Z. Phys.* D **15** 175
[103] Hofmeister H and Junghans T 1993 *Nanostruct. Mater.* **3** 137
[104] Gayle F W and Biancaniello F S 1995 *Proc. 2nd Int. Conf. on Nanostructured Materials, Nanostruct. Mater.* **6** 429
[105] Krivoglaz M A 1969 *Theory of X-ray and Thermal Neutron Scattering by Real Crystals* (New York: Plenum) section 23
[106] Azarroff L, Kaplow R, Kato N, Weiss R, Wilson A and Young R 1974 *X-ray Diffraction* (New York: McGraw-Hill) section 2.II.F
[107] Stern E A 1988 *X-ray Absorption* ed D Koningsberger and R Prins (New York: Wiley) p 3
[108] Beni G and Platzmann P 1976 *Phys. Rev.* B **14** 9514
[109] Balerna A and Mobilio S 1986 *Phys. Rev.* B **34** 2293
[110] Solliard C, Flüeli M and Borel J-P 1988 *Helv. Phys. Acta* **61** 730
[111] Maradudin A 1980 *Handbook of Surfaces and Interfaces* vol III, ed L Dobryzynski (New York: Garland STPM)
[112] Boscherini F *Proc. 1993 CNR School on Nanostructured Materials* ed D Fiorani and G Sberveglieri (Singapore: World Scientific) p 88
[113] de Panfilis S, D'Acapito F, Haas V, Konrad H, Weissmüller J and Boscherini F 1995 *Phys. Lett.* **207A** 397
[114] Eastman J A and Fitzsimmons M R 1995 *J. Appl. Phys.* **77** 522
[115] Phillpot S R, Wolf D and Gleiter H J 1995 *J. Appl. Phys.* **78** 847
[116] Tschöpe A and Birringer R 1992 *J. Appl. Phys.* **71** 5391
[117] Zhu X, Birringer R, Herr U and Gleiter H 1987 *Phys. Rev.* B **35** 9085
[118] Fitzsimmons M, Eastman J, Müller-Stach M and Wallner G 1991 *Phys. Rev.* B **44** 2452
[119] Schlorke N, Weissmüller J, Dickenscheid W and Gleiter H 1995 *Proc. 2nd Int. Conf. Nanostructured Materials, Nanostruct. Mater.* **6** 593
[120] Löffler J, Weissmüller J and Gleiter H 1995 *Proc. 2nd Int. Conf. on Nanostructured Solids, Nanostruct. Mater.* **6** 567
[121] Haubold T, Boscherini F, Pascarelli S, Mobilio S and Gleiter H 1992 *Phil. Mag.* A **66** 591
[122] Haubold T 1993 *Acta Metall.* **41** 1769
[123] Eastman J, Fitzsimmons M, Müller-Stach M, Wallner G and Elam W 1992 *Nanostruct. Mater.* **1** 47

[124] Di Cicco A, Berrettoni M, Tizza S, Bonetti E and Cocco G 1994 *Phys. Rev.* B **50** 12 386

[125] Ottavio L, Filipponi A and Di Cicco A 1994 *Phys. Rev.* B **49** 11 749

[126] Stern E A and Kim K 1981 *Phys. Rev.* B **23** 3781

[127] Wolf D and Merkle K 1992 *Materials Interfaces* ed D Wolf and S Yip (London: Chapman and Hall) p 87

[128] Glatter O 1980 *J. Appl. Crystallogr.* **13** 7

[129] Wagner W, Averbach R, Hahn H, Petry W and Wiedenmann A 1991 *J. Mater. Res.* **6** 2193

[130] Allen A J, Long G G, Kerch H M, Krueger S, Skandan G, Hahn H and Parker J 1996 *Nanostruct. Mater.* **7** 113

[131] Allen A J, Krueger S, Skandan G, Long G G, Hahn H, Kerch H M and Parker J *J. Am. Ceram. Soc.* at press

[132] Martínez-Miranda L, Chow G and Edelstein A *Phys. Rev.* B submitted

[133] Hermann H 1991 *Stochastic Models for Heterogeneous Materials Mater. Sci. Forum* **78** (Zurich: Trans Tech) 1

[134] Due to the fact that the planes in the model are of infinite extension, the interference function, equation (10.1), diverges. The problem does not arise for a finite Poisson plate mosaic. The atomic distribution function of the finite mosaic is obtained by multiplying the atomic distribution function of the infinite mosaic with the correlation function of the finite volume containing the mosaic (equations (10.4) and (10.11)). For example, this volume may be a sphere with a size much larger than the average separation between the planes

[135] Jorra E, Franz H, Peisl J, Wallner G, Petry W, Birringer R, Gleiter H and Haubold T 1989 *Phil. Mag.* B **60** 159

[136] Sanders P, Weertman J, Barker J and Siegel R 1993 *Scr. Metall.* **29** 91

[137] Sanders P, Weertman J, Barker J and Siegel R 1996 *Mater. Sci. Eng* A **204** 7

[138] Jing J, Krämer A, Birringer R, Gleiter H and Gonser U 1989 *J. Non-Cryst. Solids* **113** 167

[139] Lee L, Chen J, Yuan M and Wagner C N J 1988 *J. Appl. Phys.* **64** 4772

[140] Weissmüller J, Birringer R and Gleiter H 1990 *Phys. Lett.* **145A** 130

[141] Weissmüller J 1992 *J. Non-Cryst. Solids* **142** 70

[142] Weissmüller J, Schubert P, Franz H, Birringer R and Gleiter H 1992 *The Physics of Non-crystalline Solids* ed W C LaCourse and H J Stevens (London: Taylor and Francis) p 26

[143] Wiedenmann A, Sturm A and Wollenberger H 1994 *Mater. Sci. Eng.* A **179/180** 458

[144] Sturm A, Wiedenmann A and Wollenberger H 1994 *Nanostruct. Mater.* **4** 417

[145] Luborsky F E 1961 *J. Appl. Phys.* **32** 171S

[146] Cullity B D 1972 *Introduction to Magnetic Materials* (Reading, MA: Addison-Wesley)

[147] Morrish A H, Haneda K and Zhou X Z 1994 *Nanophase Materials: Synthesis–Properties–Applications (NATO ASI Series E 260)* ed G C Hadjipanayis and R W Siegel (Dordrecht: Academic) p 515

[148] Falicov L M 1992 *Thin Solid Films* **216** 169

[149] Fitzsimmons M, Eastman J, Von Dreele R and Thompson J 1994 *Phys. Rev.* B **50** 5600

[150] Shull C G and Smart J S 1949 *Phys. Rev.* **76** 1256

[151] Wagner W, Wiedenmann A, Petry W, Geibel A and Gleiter H 1991 *J. Mater. Res.* **6** 2305

[152] Wagner W, Van Swygenhoven H, Höfler J and Wiedenmann A 1995 *Proc. 2nd Int. Conf. on Nanostructured Materials, Nanostruct. Mater.* **6** 929

[153] Aus M J, Szpunar B, El-Sherik A M, Erb U, Palumbo G and Aust K T 1992 *Scr. Metall.* **27** 1639

[154] Kisker H, Gessmann T, Würschum R, Kronmüller H and Schäfer H E 1995 *Proc. 2nd Int. Conf. on Nanostructured Materials, Nanostruct. Mater.* **6** 925

[155] Price D L and Sköld K 1986 *Neutron Diffraction* ed K Sköld and D L Price (New York: Academic) p 1

[156] Windsor C G 1986 *Neutron Diffraction* ed K Sköld and D L Price (New York: Academic) p 197

[157] Bienfait M, Gay J M and Blank H 1988 *Surf. Sci.* **204** 331

[158] Gay J M, Bienfait M, Coulomb J P, Suzanne J, Blank H and Convert P 1989 *Physica* B **156 & 157** 273

[159] Stuhr U, Wipf H, Udovic T J, Weissmüller J and Gleiter H 1995 *Proc. 2nd Int. Conf. on Nanostructured Materials, Nanostruct. Mater.* **6** 555

[160] Stuhr U, Wipf H, Udovic T J, Weissmüller J and Gleiter H 1995 *J. Phys.: Condens. Matter* **7** 219

[161] Mütschele T and Kirchheim R 1987 *Scr. Metall.* **21** 135

[162] Mütschele T and Kirchheim R 1987 *Scr. Metall.* **21** 1101

[163] Eastman J A, Thompson L J and Kestel B J 1993 *Phys. Rev.* B **48** 84

[164] Trampenau J, Bauszus K, Petry W and Herr U 1995 *Proc. 2nd Int. Conf. on Nanostructured Materials, Nanostruct. Mater* **6** 551

Chapter 11

Interfacial free volumes and atomic diffusion in nanostructured solids

Roland Würschum and Hans-Eckhardt Schaefer

11.1 INTRODUCTION

Interest is increasing in the physical properties of nanophase materials because of their potential applications [1–5]. The novel properties of nanophase materials are due to the small length scale of the modulation of the crystallite orientation or the chemical composition and due to the high number of atoms in interfaces associated with these modulations. In the case of nanocrystalline materials with a crystallite size in the range of 3 nm to 20 nm the crystallite interfaces comprise an atomic fraction of 10–50 at.% assuming an interface width of 1 nm.

Various modified properties like enhanced reactivities and solid solubilities or reduced elastic moduli of nanocrystalline metals (see references in [2]) as well as improved sintering characteristics [6, 7] and an enhanced plastic deformability [8] of nanocrystalline ceramics in comparison with the respective bulk crystalline materials are presumably a direct consequence of the low-density structure of the interfaces of nanophase materials. An understanding of these properties requires a detailed study of the atomic structure of the interfaces, particularly of their structural free volumes, and of the atomic transport behavior of the interfaces. After some introductory remarks on the preparation and characterization of nanocrystalline materials (section 11.2) a summary will be presented on the positron annihilation studies of the structural free volumes of nanophase materials (section 11.3) as well as on the diffusion data available at present for nanocrystalline materials (section 11.4). The results of a recent desorption study of He after implantation into nanocrystalline metals will be described in section 11.5 as another example of atomic transport phenomena in nanocrystalline metals.

11.2 PREPARATION AND CHARACTERIZATION OF NANOCRYSTALLINE MATERIALS

A variety of physical, chemical, and mechanical techniques are now available for the synthesis of nanophase materials, e.g. crystallization of amorphous alloys [9–11], severe torsional deformation [12], mechanical attrition [13, 14], or sol–gel processes [15–17]. The most detailed studies so far have been performed on nanophase materials synthesized by crystallite condensation and compaction [3, 8]. In this technique, coarse-grained ingots are evaporated from resistance heated boats, e.g. tungsten or boron nitride crucibles, into an He atmosphere (typically $p = 500$ Pa), where condensation and growth of nanometer-sized crystallites occurs. Assisted by a convective flow of the inert gas, the crystallites are collected at a cold finger from which they are finally stripped off under vacuum and compacted *in situ* under high pressure into disk-shaped specimens ($p > 1$ GPa). For the preparation of nanocrystalline metals of high-melting ingots or of nanophase alloys, ultrafine crystallites are prepared by direct current (dc) sputtering of targets in an Ar atmosphere ($p = 20$–200 Pa) [19, 20]. A controlled oxidation of the crystallites (e.g. Zr) prior to crystallite compaction enables, furthermore, the synthesis of nanocrystalline oxides by means of this technique [7, 21].

The as-prepared powder compacted materials have densities of 0.7–0.8 of the bulk density ρ_0 or lower values in the case of oxides as determined by pycnometric density measurements [20, 21]. Higher densifications [5] without substantial crystallite growth can be achieved by post-annealing (see section 11.3.2) or *in situ* compaction for hours at moderate temperatures.

Conventional transmission electron microscopy and x-ray diffraction are primarily used for microstructural characterization and phase identification of nanophase materials. A detailed profile analysis of the Bragg reflections observed by x-ray diffraction may provide information on internal stresses in addition to the determination of the crystallite size [22, 23]. In the case of compaction prepared nanocrystalline pure metals, the highest values of crystallite sizes occur for low-melting metals (e.g. Al, Cu, see section 11.3.2).

In addition to standard methods for the chemical analysis of nanophase materials such as an electron micro-beam in combination with an energy dispersive analysis of x-rays (EDX) for the detection of elements with atomic numbers $Z \geq 11$, highly sensitive and specific ion beam techniques are also available [24, 25]. By means of particle-induced emission of x-rays (PIXE) using 3.0 MeV $^4He^+$ ions as projectiles, total contents below 0.5 at.% of impurities with $Z \geq 11$, mainly tungsten and iron, were detected in the surface region of compaction prepared nanocrystalline Pd [26]. A quantitative analysis of light impurities appears to be needed to assess the physical properties of nanophase materials. The technique of nuclear reaction analysis (NRA) [24] enables the study of oxygen contamination by means of energy dispersive detection of protons (p) from the $^{16}O(d, p)^{17}O$ reaction initiated by a 0.9 MeV deuteron (d) beam [25]. In this way an oxygen content of 0.3 at.% distributed homogeneously within the measuring depth of 1 μm could be detected in the surface region of nanocrystalline Pd [26] or 0.5 at.% of oxygen in nanocrystalline Ni (n-Ni) that had been exclusively handled under vacuum conditions [27]. This impurity content, which is presumably caused by gas or moisture uptake after the preparation during the transfer of the specimens in ambient atmosphere, might be significantly lowered in future experiments by a more efficient *in situ* densification of the specimens and/or by *in situ* sealing of the specimens for subsequent studies of their physical properties.

11.3 STRUCTURAL FREE VOLUMES IN NANOPHASE MATERIALS AS PROBED BY POSITRONS

11.3.1 The technique of positron lifetime spectroscopy

The positron lifetime technique is a sensitive tool for the specific detection of free volumes on an atomic scale which can be used for the study of the interfacial structure of nanocrystalline solids. Positrons from a radioactive source are rapidly slowed down to thermal energies after implantation into solids [28]. The lifetime

$$\tau \propto \left\{ \int \Gamma(n_-(\boldsymbol{r}))|\Psi_+(\boldsymbol{r})|^2 \, \mathrm{d}\boldsymbol{r} \right\}^{-1} \tag{11.1}$$

of positrons in solids is determined by the overlap of the positron wavefunction $\Psi_+(\boldsymbol{r})$ with the local electron density $\Gamma(n_-(\boldsymbol{r}))$, taking into account the effect of the enhancement of the electron density due to electron–positron Coulomb correlation (see, e.g. [29–31]). In crystals without defects, positrons attain a 'free' extended Bloch state. This state is characterized by a high positron diffusivity which is mainly limited by scattering with acoustic phonons [32, 33]. On its diffusion path the positron may be trapped into a localized state at open volumes, for example, lattice vacancies or their agglomerates, which gives rise to a characteristic increase of the positron lifetime due to the reduced local electron density [29]. This provides a unique measure of the size of submicroscopic free volumes ranging from about one missing atom (vacancy) to vacancy agglomerates of about 0.9 nm in diameter applicable both to crystalline solids and to disordered condensed matter, such as amorphous alloys [34] or melts [35]. The characteristic positron lifetimes in single lattice vacancies are derived from detailed experimental studies in metals [36], intermetallic alloys [37], semiconductors [38], and metal oxides [39]. Likewise positron lifetimes are available for small vacancy agglomerates from theoretical [31, 40] and experimental studies [41]. In the field of ultrafine particles [42–46] positron annihilation has been used for the study of surface states, gas adsorption, and segregation phenomena.

A summary on the structural studies of nanocrystalline metals (section 11.3.2), of nanophase alloys (section 11.3.3) and of nanocrystalline Si (section 11.3.4) and ceramics (section 11.3.5) by means of positron lifetime spectroscopy will be given. For these studies the technique of $\gamma\gamma$ coincidence is used for measuring the positron lifetime [29]. Here, the time measurement is triggered by a prompt γ-signal provided by a suitable β^+-source such as ^{22}Na and stopped by one of the positron–electron annihilation γ-quanta. The coincidence spectrometer with a time resolution FWHM (full width at the half maximum) of 200 to 260 ps is equipped with either plastic or BaF_2 [47] scintillators. For the measurements on nanocrystalline materials at room temperature the positron source (typically 8×10^5 Bq of ^{22}NaCl) is carried by a thin Al foil (0.8 μm) and sandwiched between disk-shaped measuring specimens [48]. The positron lifetime spectra with a statistical accuracy of at least 2×10^6 coincidence counts are analyzed with standard numerical techniques (for details see [48]).

11.3.2 Nanocrystalline metals

11.3.2.1 Size of structural free volumes

According to the data compiled in table 11.1 [49] two dominant positron lifetime components τ_1, τ_2 as well as a weak third long-lived component $\tau_3 > 500$ ps can be resolved in nanocrystalline metals prepared by crystallite condensation and compaction. The positron lifetimes τ_1 are very similar to the lifetimes τ_{1V} in lattice vacancies in the corresponding coarse-grained metals (figure 11.1(*a*), table 11.1). The positron lifetimes τ_2 scale with calculated positron lifetimes in small agglomerates of about ten vacancies in crystals [31] (figure 11.1(*b*), table 11.1). Within the numerical uncertainties the same time constants τ_i are observed in specimens prepared either by evaporation or by sputtering of the crystallites (table 11.1). All positron lifetime components are higher than the free positron lifetime τ_f in crystals without defects, demonstrating that positrons are quantitatively trapped and annihilated at free volumes. The intensity ratio I_1/I_2 of the components τ_1 and τ_2, which in the case of saturation trapping is proportional to the ratio Σ_1/Σ_2 of the positron trapping rates of free volumes of two different sizes (see below, [48]), decreases with decreasing crystallite size (figure 11.2 , table 11.1). Similar positron lifetimes τ_1 and τ_2 to those in the compaction solidified nanocrystalline metals are observed in submicron-grained Cu and Ni (see table 11.1) [51] prepared by severe plastic deformation [12].

The components τ_1 and τ_2 are attributed to structural free volumes of the size of one to two vacancies and of the size of about ten missing atoms ('nanovoids'), respectively, in the interfaces between the crystallites. The latter open free volumes, the size of which corresponds to a diameter of 0.6 to 0.7 nm, are expected to be available in the intersections of interfaces, e.g. triple lines. Free volumes of this size cannot easily be detected by other techniques. The long-lived component (τ_3) which occurs with minor intensity is due to ortho-positronium formation presumably in larger voids of the size of, for example, a missing crystallite. The experimental evidence for this conclusion is as follows.

Table 11.1. Positron lifetimes τ_1, τ_2, τ_3 and relative intensities I_1, I_2 in nanocrystalline solids after compaction of crystallites prepared by evaporation (E) or sputtering (S) [49], in nanostructured Cu and Ni prepared by severe plastic deformation (SPD) and moderate annealing and in plasma deposited nanocrystalline Si layers (chemical vapor deposition (CVD)) [50] as well as the mean positron lifetimes $\overline{\tau}$ in nanostructured $Co_{33}Zr_{67}$, $Fe_{90}Zr_{10}$, $Fe_{73.5}Cu_1Nb_3Si_{13.5}B_9$ after crystallization of the amorphous alloys [51]. For n-Pd the decrease $\Delta\tau_i$ of τ_i ($i = 1, 2$) between the initial (or final) unloaded state and the state under the quasi-hydrostatic pressure $p_m = 4$ GPa is quoted. In addition, the 'free' lifetimes τ_f in undefected crystals [35], as well as positron lifetimes in monovacancies (τ_{1V}; see [36]), in agglomerates of i vacancies (τ_{iV}), and at dislocations (τ_d) are given; d: crystallite size.

Material		d (nm)	τ_1 (ps)	τ_2 (ps)	τ_3 (ns)	I_1	I_2	τ_f (ps)	τ_{1V} (ps)	τ_{iV} (ps)	τ_d (ps)
Al	E	40	253 ± 4	412 ± 7	1.97 ± 0.16	0.58	0.41	163	251	422 ($i = 13$)[a]	228[b]
Fe	E	10	161 ± 8	337 ± 6	0.9 ± 0.04	0.20	0.75	106	175	334 ($i = 10$)[a]	167
Ni	E, S	12	174 ± 1	363 ± 2	0.53 ± 0.07	0.28	0.72	94	180[c]	376 ($i = 13$)[a]	230[b]
Cu	E	20	175 ± 10	299 ± 40	0.47 ± 0.05	0.43	0.44	112	179		164[d]
Cu	S	14	182 ± 8	311 ± 13	0.6 ± 0.2	0.40	0.59				
Pd	E	12	182 ± 5	347 ± 5	1.08 ± 0.17	0.33	0.66	108[e]	Pt:168		
			$\Delta\tau_1 = 20(11)$	$\Delta\tau_2 = 31(23)$							
Mo	S	10	204 ± 9	345 ± 2		0.16	0.84	103	180	321 ($i = 9$)[a]	
Cu	SPD	100 to	171 ± 2	297 ± 9		0.83	0.17				
Ni	SPD	200	161 ± 1	330 ± 9		0.88	0.12				
Si	E	10	314 ± 35	422 ± 11	3.3 ± 0.3	0.2	0.79	219[f]	272[f]	399 ($i = 8$)[g,h]	
Si	CVD	10		400–450	6.8–12.5						
Pd_3Fe	S	9	170 ± 8	351 ± 3		0.16	0.84	Pd:108[e]			
ZrO_2	S	10	199 ± 2	378 ± 1	14.5 ± 0.6	0.31		175[i]			

	$\overline{\tau}$ (ps) nanocrystalline	$\overline{\tau}$ (ps) amorphous	$\overline{\tau}$ (ps) coarse-grained crystalline
$Co_{33}Zr_{67}$	191 ± 1	187 ± 1	148 ± 1
$Fe_{90}Zr_{10}$	158 ± 1	158 ± 1	145 ± 1
$Fe_{73.5}Cu_1Nb_3Si_{13.5}B_9$	145 ± 1	151 ± 1	Fe_3Si:114

[a] [31] [c] [53] [e] [55] [g] [40] [i] [56]
[b] [52] [d] [54] [f] [38] [h] [41]

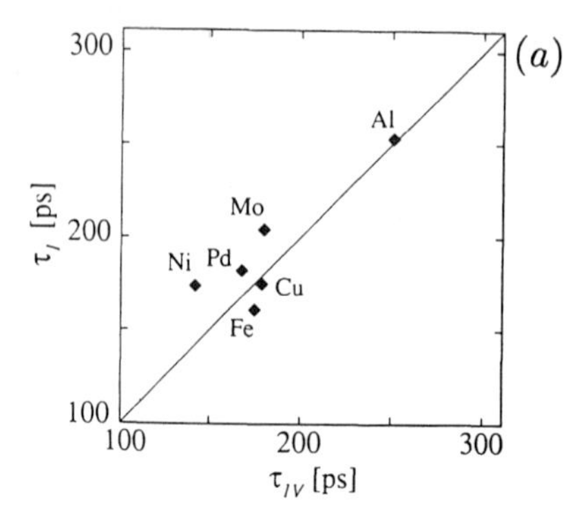

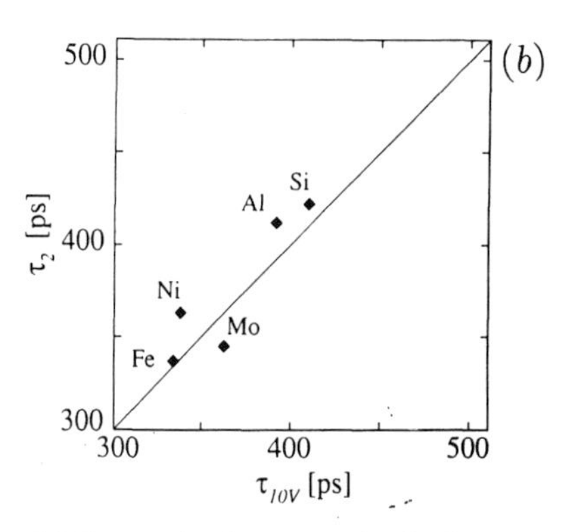

Figure 11.1. (*a*) Positron lifetime τ_1 in cluster synthesized nanocrystalline metals versus the lifetime τ_{1V} in lattice vacancies and (*b*) positron lifetime τ_2 in nanocrystalline solids versus the lifetime τ_{10V} in agglomerates of ten vacancies in crystals [20] (see table 11.1).

First of all, the quantitative trapping of positrons at structural free volumes in interfaces of nanocrystalline materials is a consequence of a crystallite size smaller than the positron diffusion length $L_{e^+} \approx 100$ nm in an undisturbed metal crystal. In diffusion-reaction theory for positron trapping at interfaces with a single type of vacancy sized positron traps (annihilation time constant τ_1), the mean positron lifetime [57]

$$\overline{\tau} \simeq \tau_1 - (\tau_1 - \tau_f)\left\{\frac{d}{6\alpha_{IF}\tau_f} - \left[\left(\frac{1}{3\alpha_{IF}\tau_f}\right)^2 - \frac{1}{15D\tau_f}\right]\frac{d^2}{4}\right\} \tag{11.2}$$

sigmoidally approaches τ_1 with decreasing crystallite diameter d. This yields $\overline{\tau} = \tau_1$, i.e. saturation trapping of positrons at the interfaces, for nanocrystalline materials with a specific positron trapping rate $\alpha_{IF} = 2 \times 10^3$ m s^{-1} of the interfaces (assumed to be similar to that of free surfaces [25, 58] or bubbles

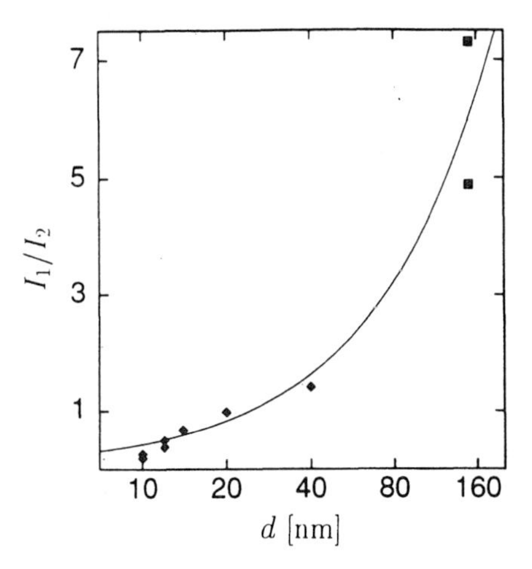

Figure 11.2. Intensity ratio I_1/I_2 of the positron lifetime components τ_1 and τ_2 in nanocrystalline metals and Si prepared by cluster-synthesis (♦) or plastic deformation (■) (see table 11.1). d denotes the crystallite diameter and the solid line depicts the ratio of the positron trapping rates $\Sigma_1 = 6\alpha_{IF}d^{-1}$ ($\alpha_{IF} = 2 \times 10^3$ m s^{-1}) of the interfaces and $\Sigma_2 = \sigma_{NV}C_{NV}$ ($\sigma_{NV} = 2 \times 10^{15}$ s^{-1}) of the nanovoids at the triple junctions of interfaces with a volume fraction $V_{TJ} = 3d^{-2}\delta^2 \simeq 0.01$ of triple junctions according to the polyhedron model of nanocrystalline materials [65] (see text). $C_{NV} = 0.1V_{TJ}$: concentration of nanovoids with the size of about ten missing atoms; $\delta = 1$ nm: thickness of crystallite interface.

[59]), with a positron diffusion coefficient $D = 10^{-4}$ m^2 s^{-1} [33], and with a crystallite diameter $d \simeq 20$ nm.

It appears unlikely that positrons are trapped in vacancy-like defects within the crystallites because the short lifetime τ_1 can be observed even after annealing of nanocrystalline (n-) Pd [60], n-Cu [50], or n-Fe [48] up to 600 K where lattice vacancies are annealed out in the coarse-grained solids [61]. Furthermore, the variation of τ_1 with pressure [20] (see below) clearly shows a compressibility of the positron trapping sites much higher than expected for lattice vacancies. Likewise, positron trapping and annihilation at dislocations in the crystallites with a characteristic lifetime of $\tau_d < \tau_1$ [62, 63] scarcely contribute to the intensity of the short-lived component τ_1 because, according to high-resolution transmission electron microscopy studies [64], on average only one crystallite (size e.g. 10 nm) out of ten contains a dislocation. It is unlikely as well that the component τ_2 is due to internal surfaces because, for example, τ_2 in n-Fe is significantly smaller than the positron lifetime of 440 ps characteristic for surface annihilation in nanocrystalline Fe powder [48]. However, the roles of dislocations and surfaces deserve additional studies.

Due to the positron diffusion length $L_{e^+} \propto \sqrt{D\tau_f} \simeq 100$ nm being much larger than the crystallite diameter of nanocrystalline materials, the trapping process of positrons at the interfaces is entirely controlled by the specific trapping rate α (see equation (11.2): $0.1d\alpha_{IF} \ll D$ for $d = 10$ nm). In this case the intensity ratio I_1/I_2 is given by the ratio of the trapping rates $\Sigma_1 = 6\alpha_{IF}d^{-1}$ of the interfaces with the characteristic positron time constant τ_1 and $\Sigma_2 = \sigma_{NV}C_{NV}$ of the nanovoids at the intersections of interfaces (time constant τ_2) which yields an atomic concentration $C_{NV} \simeq 10^{-3}$ of nanovoids with about ten missing atoms for $I_1/I_2 = 0.5$ ($d = 10$ nm, table 11.1, figure 11.2), assuming a specific trapping rate $\sigma_{NV} = 2 \times 10^{15}$ s^{-1} such as for larger vacancy agglomerates [66]. This concentration of nanovoids is in reasonable agreement with a volume fraction $V_{TJ} = 3d^{-2}\delta^2 \simeq 0.01$ ($d = 10$ nm, interface width $\delta = 1$ nm) of triple junctions between the crystallites according to the polyhedron model of nanocrystalline materials suggested by Palumbo *et al* [65]. Within the framework of this model the increase of I_1/I_2 with the crystallite size (table 11.1, figure 11.2) can be well understood by the increase of the volume ratio of crystallite boundaries (trap τ_1) and triple junctions (trap τ_2). Likewise, the reduced intensity I_2 in fine-grained Cu and Ni prepared by severe plastic deformation (table 11.1) is attributable to the reduced trapping fraction of positrons at nanovoids at the crystallite intersections due to the larger grain size, 100–200 nm (figure 11.2) [51].

The interpretation of the present lifetime data by positron trapping in interfacial free volumes is further supported by studies during crystallite growth. Here the positron lifetime is observed to decrease with the value of the 'free' lifetime in crystals without defects when the trapping of positrons at the interfaces becomes less important due to an increase of crystallite size beyond the free positron diffusion length $L_{e^+} \simeq 100$ nm and the onset of partial annihilation of positrons within the crystallites.

The long-lived positron lifetime τ_3 (see table 11.1), which mostly exceeds the low electron-density limit of positron lifetimes in solids ($\sim$500 ps), has to be attributed to ortho-positronium (o-Ps) formation and annihilation in larger free volumes, presumably associated with the residual porosity in compaction prepared nanocrystalline metals. Although the reduction of the mass density in the as-compacted state of nanocrystalline materials is considered to be mainly due to porosity, the trapping rate of positrons in pores, i.e. the intensity $I_3 = 1 - I_1 - I_2$, is small compared to the total positron trapping rate of the structural free volumes in the interfaces (components τ_1 and τ_2). This is due to the high total area of interfaces compared to the area of internal surfaces associated with larger voids or pores. Evidence for voids was also derived from studies of small-angle neutron scattering (SANS) [67, 68]. Ps formation at these larger voids may be favored when the internal surfaces are contaminated by gaseous impurities [69]. The effect of impurities on the interfacial free volumes of the size of vacancies (τ_1) or triple junctions (τ_2) is considered to be negligible because the total atomic impurity concentrations of about 1 at.%, e.g. in n-Pd ([26], see section 11.2), are much lower than the fraction of interfacial atoms. However, detailed studies of impurity effects should be conducted more thoroughly in the future.

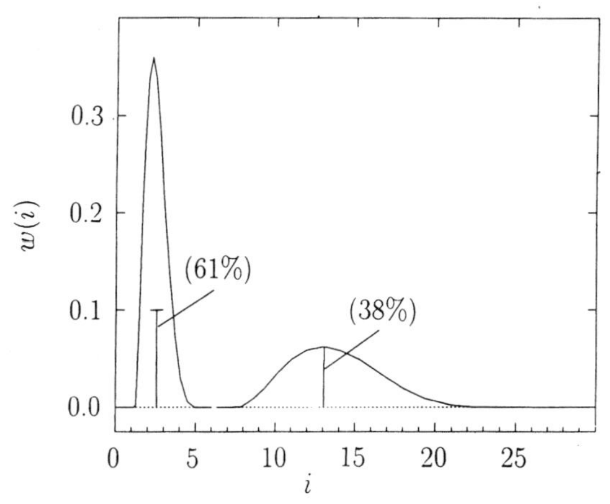

Figure 11.3. Size distribution of structural free volumes in nanocrystalline Mo as determined by Laplace transformation of a positron lifetime spectrum [71] making use of the relationship between the positron lifetime and the number i of vacancies per agglomerate [31]. The bars denote the free volume sizes with the intensities of the time constants according to the multicomponent analysis of the positron lifetime spectra.

At this point we should emphasize that the positron lifetimes τ_1, τ_2, and τ_3 discussed above were determined by standard numerical techniques for the decomposition of the positron lifetime spectra into discrete components (see [48]). However, these components and therefore the sizes of free volumes in nanocrystalline materials should be considered as the centers of mass of a size distribution. In addition, the size distribution is expected to extend to interfacial free volumes smaller than 'vacancies' which are, however, not detected due to insufficient positron trapping. Moreover, the positron lifetimes, and therefore the sizes of the interfacial free volumes, may depend slightly on the details of the interfacial structure and therefore of the specimen preparation as indicated by the slight variations of the results of various measurements (see [20, 48, 50, 60]). Information on the size distribution of the free volumes can be obtained from positron lifetime spectra with high statistical accuracy by means of numerical analysis of these spectra via Laplace transformation [70]. This is demonstrated for nanocrystalline Mo in figure 11.3 where relatively narrow distributions of free volume sizes around two maxima can be discerned. The positions of these maxima are in good agreement with those of the discrete time constants derived from a multicomponent analysis [71]. Recently it was shown that the interfacial structure of n-Pd can be modified by changing the preparation route as demonstrated by the changes in the size distribution of the submicroscopic free volumes [27].

The interfacial free volumes in cluster synthesized nanocrystalline metals with a mean size of one to two lattice vacancies (component τ_1) corresponds to the reduced atomic density in the interfaces [5]. At the interfaces there are a wide range of interatomic distances as deduced, for example, from studies of x-ray diffraction [72] and EXAFS (extended x-ray absorption fine structure) [73]. An enhanced specific free volume was also found for crystallite interfaces of deformation prepared nanocrystalline materials by means of dilatometric analysis [74].

The free volumes with a mean size of one to two vacancies are considered to represent structural elements of the metastable and highly disordered interfaces of nanocrystalline metals, which are unlikely to exist, for example, in symmetric tilt or twist boundaries of conventional bi- or polycrystalline metals. One may assume that the size of localized free volumes in these special equilibrium grain boundaries does not exceed the core volume of a grain boundary dislocation [75] or the total excess volume of a grain boundary per unit cell [76] which are both smaller than a lattice vacancy. One could hardly expect that structural free volumes in more general high-angle boundaries with shorter periodicity lengths of their structural units will exceed the size of a lattice monovacancy. Simulation studies have shown that, due to a strong atomic relaxation, a missing atom (i.e. a 'point defect') in a grain boundary gives rise to a local open volume which is smaller than that of a vacancy in a crystal lattice [77, 78].

11.3.2.2 Compressibility of interfacial free volumes

Since the positron lifetime is sensitive to the size of free volumes, this technique is useful for studying the compressibility of structural free volumes in nanocrystalline materials. In the interfacial structure with a wide interatomic distance distribution [72] and reduced elastic properties [79], a compressibility of free volumes higher than that of, for example, lattice vacancies is to be expected. Positron lifetime measurements performed on as-prepared n-Pd specimens under quasi-hydrostatic pressure in a piston-anvil device [80] show a change $\Delta\bar{\tau} = 30$ ps between 0 and 4 GPa (see figure 11.4). This effect is reduced to half this value after annealing at 463 K (see figure 11.4) indicating a stiffening of the interfacial free volumes with annealing concomitant with an elastic stiffening of the interfaces [79].

From comparative measurements on plastically deformed polycrystalline Pd with lattice dislocations as dominant positron traps, a single positron lifetime of 177 ps under a load of 4 GPa and a value of 183 ps after load release could be derived. This small pressure variation compared to n-Pd can be considered as further evidence that the component τ_1 in nanocrystalline metals is not to be attributed to lattice dislocations but to interfacial free volumes.

For an assessment of the pressure variation of the interfacial free volumes in n-Pd one may compare the decrease of the positron lifetime component τ_1 between $p_m = 0$ and 4 GPa (table 11.1) with the change expected for the positron lifetime in lattice vacancies. Assuming as an upper limit [81] that the lifetime of the vacancy trapped positron changes like the crystal volume ($\Delta V/V = -0.02$ for $p_m = 4$ GPa), one may estimate a much smaller change $\Delta\tau_{1V} = 4$ ps for

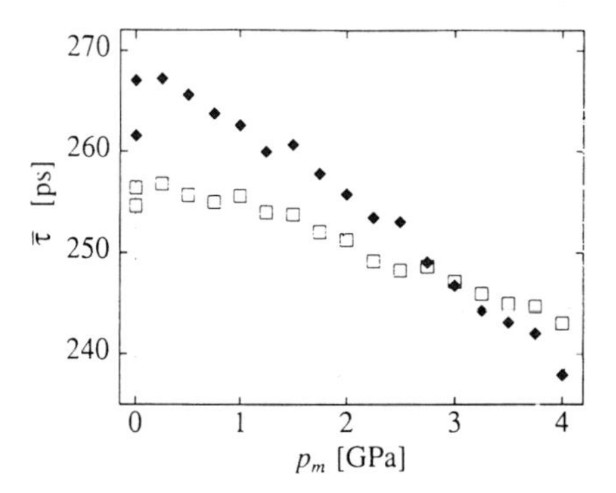

Figure 11.4. Pressure variation of the mean positron lifetime $\overline{\tau}$ measured on n-Pd before (♦) and after annealing (□; $T_a = 190\,^\circ\mathrm{C}$ for 0.5 h) [20]. The lower values for $p_m = 0$ refer to the state after unloading. The positron lifetimes are measured with an uncertainty of $\Delta\overline{\tau} = \pm 2$ ps.

a lattice vacancy in crystalline Pd. From the pressure variation of the vacancy formation volume an even smaller size change is expected for vacancies (see discussion in [49, 82]). The free volumes in nanocrystalline metals therefore appear to be more compressible than lattice vacancies and indicate a rather soft interfacial structure, giving rise to a softening of the elastic moduli [79], an increased specific grain boundary energy [83], and a strong pressure induced (1 GPa) isomer shift of the interfacial Mössbauer component in n-Fe [5] which greatly exceeds that in α-Fe crystals or in short-range ordered amorphous Fe alloys. The reduction of the pressure variation of the positron lifetime after annealing at 463 K (see figure 11.4) coincides with the structural relaxation of the interfaces as discussed below. A minor irreversible fraction of the pressure induced variation of $\overline{\tau}$ (see figure 11.4) is ascribed to a structural relaxation due to the applied pressure during the positron lifetime measurements ($\sim$ 10 h).

11.3.2.3 Structural relaxation of the interfaces

The small crystallite size of nanocrystalline materials together with the athermal process of crystallite compaction during preparation raises the question of thermal stability and structural relaxation of these materials. By means of positron lifetime spectroscopy on nanocrystalline Pd the structural relaxation of the disordered interfaces below the onset of substantial crystallite growth [84] could be observed (figure 11.5). The decrease of the intensity ratio I_1/I_2 at $150\,^\circ\mathrm{C}$ of the positron lifetime components τ_1 and τ_2 and the increases of τ_1 and τ_2 above $180\,^\circ\mathrm{C}$ (figure 11.5) are attributed to an atomic reordering and a subsequent agglomeration of interfacial free volumes, respectively [60]. This is accompanied by an increase in mass density (figure 11.6(*a*)) and a decrease in electrical resistivity (see figure 11.6(*b*)) [26]. Annealing of n-Pd at moderate temperatures (390 K) induces irreversible reordering in the interfaces as evidenced by an increase of the shear modulus [79] and a reduction of the specific grain boundary energy [83]. These annealing phenomena prior to grain growth show that the interfacial structure of compaction prepared nanocrystalline metals exhibits states far from equilibrium. States of lower energy are attained by atomic reordering as well as by growth and reduction of the number density of free volumes. This is consistent with theoretical considerations [85] of the energy of large-angle grain boundaries.

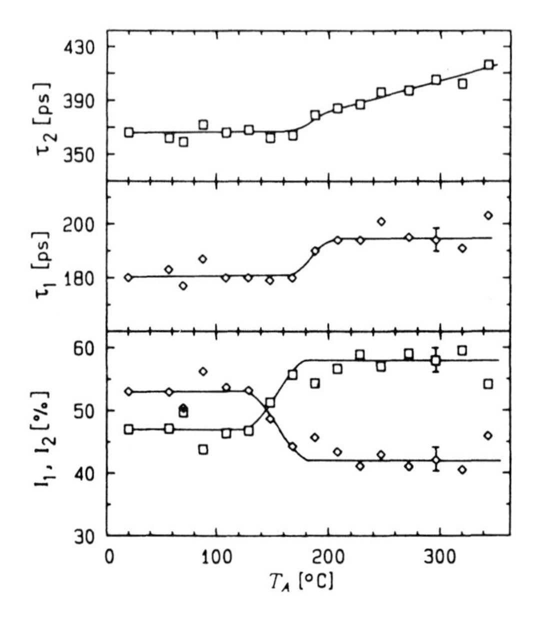

Figure 11.5. Positron lifetime components τ_1 and τ_2 with the corresponding relative intensities I_1 (◇) and I_2 (□) after isochronal annealing at T_a (t_a = 30 min) of n-Pd in high vacuum [60].

At higher annealing temperatures substantial crystallite growth occurs giving rise to a partial annihilation of positrons within the crystallites as detected by a decrease of the positron lifetime towards the 'free' lifetime in a crystal without defects [60] when the crystallite size exceeds the positron diffusion length. Crystallite growth mechanisms are not well understood as in the case of the formation of large crystallites of n-Cu even at ambient temperature [86–88].

11.3.3 Nanocrystalline alloys

In nanocrystalline Pd_3Fe prepared by compaction of Pd_3Fe crystallites [51], positron lifetimes τ_1 and τ_2 similar to those in nanocrystalline pure metals occur (table 11.1) with an intensity ratio I_1/I_2 which fits to the variation with

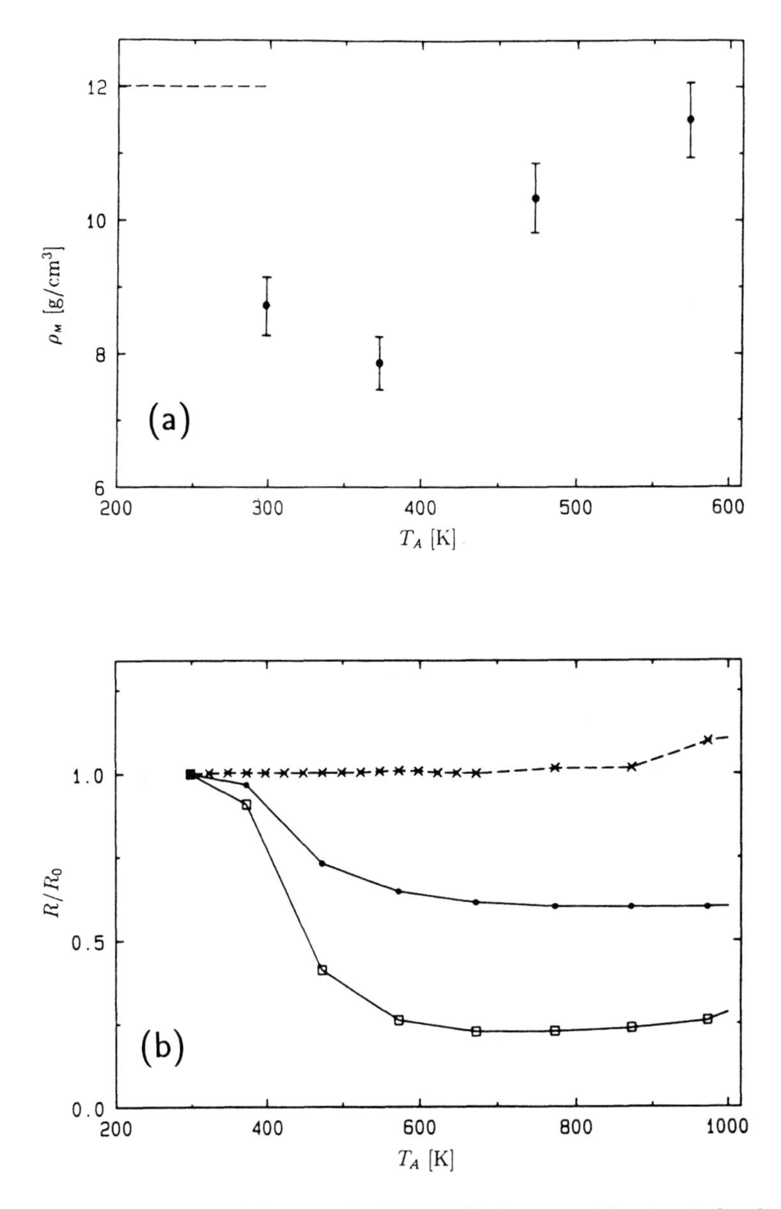

Figure 11.6. (*a*) Increase of the mass density and (*b*) decrease of the electrical resistance R at 5 K (□) and at 293 K (●) upon isochronal annealing ($T_a = 60$ min) of nanocrystalline Pd [26]. The bulk density ρ_0 (– – –) (*a*) and the resistance variation (×) (*b*) of coarse-grained Pd are depicted. R_0 denotes the resistance prior to annealing.

crystallite size according to the polyhedron model of nanocrystalline materials (see figure 11.2). This indicates a pattern of interfacial free volumes in Pd_3Fe similar to that in nanocrystalline pure metals.

Synthesis of nanophase alloys can be achieved by crystallization of melt spun amorphous alloys [9–11]. In the nanocrystalline alloys $Co_{33}Zr_{67}$, $Fe_{90}Zr_{10}$, $Fe_{73.5}Cu_1Nb_3Si_{13.5}B_9$ [51], or $Ni_{80}P_{20}$ [89] prepared by this technique, a single [51] or a predominant ($I = 0.95$) [89] component τ_1 with a lifetime similar to that in the amorphous state occurs characterizing free volumes slightly smaller than lattice vacancies [34]. In particular, lifetimes longer than $\tau_2 = 350$ ps signifying the presence of nanovoids could not be detected after crystallization of these alloys. The reduced size of interfacial free volumes and the absence of nanovoids reflect the high mass density of the materials close to the value of bulk crystals and demonstrate that the small crystallites are embedded in a dense multicomponent phase of amorphous material (see [90]) without the free volumes of nanovoids or triple lines detected in nanocrystalline pure metals after compaction.

11.3.4 Nanocrystalline silicon

Ultrafine-grained polycrystalline Si has attracted rising interest particularly due to the modified electrical, optical, and physicochemical properties associated with the high number of crystallite interfaces [90–93].

In nanocrystalline Si layers deposited by chemical transport in a hydrogen plasma [50] and in cluster synthesized nanocrystalline Si [49], a dominant positron lifetime $\tau_2 = 400$–450 ps and a weak long-lived component indicating ortho-positronium formation in pores are observed (table 11.1). The component τ_2 characterizes (in, for example, hydrogenated amorphous Si [94]) free volumes of the size of 8–15 vacancies [40, 41]. As in nanocrystalline metals these free volumes are considered as structural elements of the interfaces and are thought to be located at the intersections of interfaces. As demonstrated for plasma deposited n-Si, these free volumes still exist after annealing treatments above the recovery temperatures of vacancy agglomerates in crystalline Si [38, 50].

In nanocrystalline Si prepared by compaction of crystallites, an additional positron lifetime component $\tau_1 = 325$ ps with low intensity occurs (table 11.1 [49]). This component, although determined with considerable uncertainty, is typical for triple or tetravacancies [41] and may signify interfacial free volumes larger than one missing atom in the covalently bonded semiconductor. That these small free volumes with positron lifetime τ_1 are not found in plasma deposited nanocrystalline Si layers [50] may be due to the elevated preparation temperatures and a high degree of orientational correlation between the crystallites. As in the cases of nanocrystalline metals (section 11.3.2) and alloys (section 11.3.3), this demonstrates that the microstructure of nanocrystalline materials depends strongly on the preparation technique.

These measurements on nanocrystalline Si illustrate the utility of the positron annihilation technique for the study of the interaction of gaseous impurities with voids or bubbles [69, 95]. For example, long-term exposure to air or thermally induced evolution of hydrogen gives rise to a significant decrease or increase of the positron lifetime component τ_2 in plasma deposited Si [41]. There is a reduction of the positron lifetime in gas decorated voids. In addition, the characteristics of ortho-positronium formation and annihilation (component τ_3) change after hydrogen evolution, as in hydrogenated amorphous Si, which is ascribed to hydrogen dissociation from dangling bonds at void surfaces (see [50] for references).

11.3.5 Nanocrystalline ceramics

The recent evidence for low-temperature ductility [8] and improved sintering characteristics [6, 7] of cluster synthesized nanocrystalline ceramics initiated a number of studies in this field. Positron lifetime spectroscopy is well suited for the investigation of the structure and particularly of the sintering properties of nanocrystalline ceramics enabling the detection of ultrafine pores (nanovoids) and hence supplements other techniques commonly utilized for the study of porosity.

In the following a summary will be given on a recent positron lifetime and x-ray diffraction study [21] of nanocrystalline ZrO_2. This material is of considerable technological interest considering the importance of zirconia based solids as sensor materials or ion conductors [96] as well as for transformation toughening [97].

The cluster-synthesis of nanocrystalline ZrO_2 was performed by dc sputtering [19] of metallic Zr followed by slow oxidation at ambient temperature, and *in situ* compaction ($p = 1$ GPa) of the ZrO_2 crystallites [21]. In the n-ZrO_2 specimens prepared in this way a tetragonal crystallite structure instead of the monoclinic ambient-temperature structure of coarse-grained ZrO_2 is observed. This is in agreement with earlier observations on ultrafine ZrO_2 powders [98, 99] (figure 11.7(*a*)). Different specific interface and surface energies of tetragonal and monoclinic ZrO_2 were considered to give rise to this stabilization [99] which depends on the crystallite diameter (critical diameter $d_C = 10$ nm), on lattice distortions, and on the nucleation conditions at the interfaces for the initiation of the phase transformation. An additional contribution of cubic phase material due to an oxygen deficiency may be present initially.

The tetragonal-to-monoclinic transformation occurs in two stages: stage I (at 150 °C) and stage II (at 750 °C) upon isochronal annealing [21]. The partial transformation in stage I below d_C is accompanied by an increase of the positron lifetimes τ_1 and τ_2 and their intensity ratio I_1/I_2 (see table 11.2). In analogy to nanocrystalline metals (see section 11.3.2) the lifetimes τ_1 and τ_2 can be attributed to vacancy-sized free volumes in the crystallite interfaces and to nanovoids located in the triple lines between the crystallites, respectively. This model is supported by the increase of I_1/I_2 upon crystallite growth when the

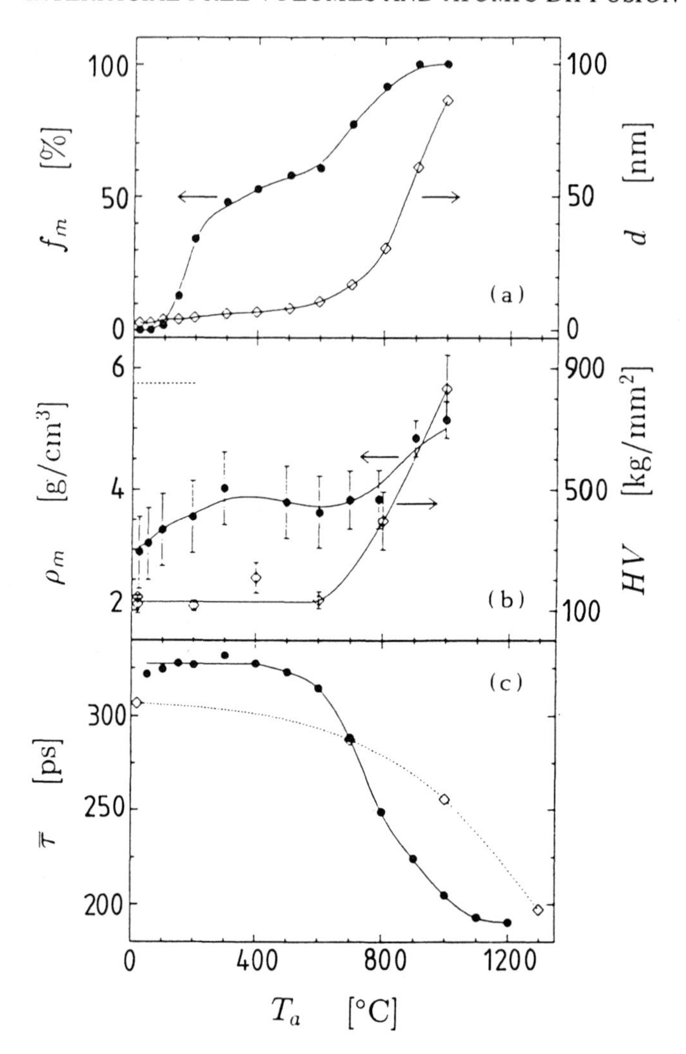

Figure 11.7. Annealing behavior (t_a = 60 min) of n-ZrO_2 [21]: (*a*) fraction of monoclinic phase and crystallite diameter d from x-ray diffraction; (*b*) mass density ρ_m (· · ·: density of monoclinic ZrO_2) and Vickers hardness HV; (*c*) mean positron lifetime $\overline{\tau}$ for gas phase prepared n-ZrO_2 and commercial ZrO_2 (◇, 7 nm) specimens.

relative contribution of triple lines decreases [65]. From a comparison with detailed positron lifetime studies in vacancy clusters in electron irradiated Al_2O_3 crystals [39], the τ_2-values in n-ZrO_2 (see table 11.2) can be attributed to nanovoids of the size of about 10 to 15 missing atoms. Partial annihilation of positrons at defects originating from the martensitic tetragonal-to-monoclinic (t–m) phase transformation with lifetimes near τ_1 may occur. A weak long-lived positron annihilation component (see table 11.2) indicates o-Ps formation in a few larger voids.

Table 11.2. Positron lifetimes τ_i and relative intensities I_i measured on nanocrystalline ZrO_2 after isochronal annealing at temperature T_a for t_a = 60 min [21].

T_a (°C)	τ_1 (ps)	τ_2 (ps)	I_1/I_2	τ_3 (ns)	I_3
50	199 ± 3	379 ± 2	0.44	14.5 ± 0.6	0.02
300	239 ± 7	404 ± 7	0.73	7.8 ± 0.4	0.022
600	242 ± 2	420 ± 4	1.36	5.2 ± 0.2	0.01
900	200 ± 1	460 ± 7	9.3		
1200	188 ± 1	763 ± 45	($I_1 = 0.997$)		

The transformation in stage I may be attributed to a structural relaxation of the interfaces which is evidenced by the agglomeration of interfacial free volumes (see τ_1, τ_2 increase, table 11.2), initial grain growth (figure 11.7(*a*)), and a release of strains within the crystallites. The incorporation of oxygen may contribute to this transformation. In stage II the tetragonal-to-monoclinic phase transformation is finally completed as the crystallite diameter exceeds the critical size of about 10 nm for stabilizing the tetragonal phase (figure 11.7(*a*)). Simultaneously with the crystallite growth, nanovoids and pores disappear in stage II as deduced from the increase of the positron lifetime component τ_2 (table 11.2), from the decrease of the mean positron lifetime $\overline{\tau}$ (figure 11.7(*c*)) and of the intensities I_2 and I_3 (table 11.2), as well as from the increase of the Vickers hardness and the mass density (figure 11.7(*b*)).

Above 1100 °C the mean positron lifetime $\overline{\tau}$ approaches the free bulk lifetime τ_f = 175 ps [21, 39] although the crystallite diameter is determined to be similar to the conventional diffusion length of positrons in crystals. This might be due to a strong relaxation or a positive charge state of interfacial free volumes hampering the trapping of positrons or due to competitive positron trapping in the crystallites, for example, at negatively charged substitutional impurities like in GaAs [100] with positron lifetimes similar to τ_f.

The present results of positron lifetime, mass density, and Vickers hardness measurements indicate that in n-ZrO_2 (as well as in condensation prepared TiO_2 [6, 7]) the sintering and densification temperatures are lowered compared to ultrafine ZrO_2 prepared by common techniques. Comparative positron lifetime measurements on a compacted conventional ZrO_2 powder (crystallite size $<$ 7 nm, Johnson Matthey Company), where the dissolution of pores is observed at higher temperatures (figure 11.7(*c*)), give further evidence of the advantageous sintering behavior of cluster synthesized ceramics.

11.4 DIFFUSION

The atomistic structure of interfaces, their structural relaxation, and, in particular, the role of free volumes may be further investigated by employing diffusion studies. These studies are of technological interest because of the initially observed strongly enhanced diffusivities in nanophase materials (see [101]). This enhanced diffusion can result in enhanced plastic deformability of ceramics and intermetallics, and may also have technological implications for sensors and solid state batteries.

Table 11.3. Apparent diffusion activation enthalpies (eV) for nanocrystalline solids and for volume diffusion (see [101–103]). The values for nanocrystalline solids were determined neglecting interfacial relaxation and crystallite growth during diffusion heating.

Diffusor Matrix	Cu Cu	Ag Cu	Ag Pd	Be Fe	F CaF_2	Hf TiO_2	O TiO_2
Nano-crystalline	0.64[a]	0.39–0.63[b]	0.44[c]	0.54[c]	0.33–1.0[d]	1.24[c]	1.29[c]
Volume	2.0[e]		2.67[f] (Pd in Pd)	0.82[g]	1.59[d]	2.45[h] (Ti in TiO_2)	2.6[i]

[a] [101]
[b] [105]
[c] [103]
[d] [108]
[e] [104]
[f] [106]
[g] [107]
[h] [109]
[i] [110]

A concise compilation of recent diffusion data in nanostructured solids is given in table 11.3. The apparent activation energies Q for self-diffusion (e.g. Cu in Cu or F in CaF_2) or for substitutionally dissolved foreign atoms (e.g. Ag in Cu or Pd) are drastically reduced to one quarter or one third of that of the volume diffusion activation energies. This reduction enhances the low-temperature diffusivities by many orders of magnitude [101].

Some information on the diffusion mechanism may be derived from the pressure variation of the diffusivity [103] yielding an apparent activation volume $\Delta V = 0.6\ \Omega$ for the Cu diffusivity in n-Pd, where Ω denotes the Pd atomic volume. Similar values of ΔV were observed for substitutional diffusion in the lattice or in a grain boundary [111]. The value of 0.6 Ω is therefore considered as evidence for substitutional Cu diffusion in n-Pd, for example via the vacancy-sized free volumes detected by positron lifetime studies (see section 11.3). In addition, the pressure sensitivity of the diffusivity makes an interpretation of the low activation enthalpies by diffusion processes along free internal surfaces unlikely.

It should be pointed out that the activation enthalpies listed in table 11.1 have to be considered as apparent or effective values in the sense that interfacial structural relaxation or crystallite growth during the diffusion heating treatment were not considered. Changes and densification of the interfacial structure with increasing temperature are demonstrated by the increase of the apparent Q-values and therefore a decrease of the diffusivities of Ag in n-Cu [105] or of F in n-CaF_2 [108] (see span of the Q-values in table 11.3). This is presumably due to a change in size and a reduction in number of interfacial free volumes as detected by positron lifetime studies.

The diffusion behavior of light interstitial atoms in nanocrystalline solids appears to differ from that of substitutional atoms. The diffusivities of boron in n-Fe [103] or of hydrogen in n-Pd at low H concentrations [112] are lower than the corresponding lattice diffusivities. This may originate from trapping of the light diffusors at interfacial free volumes. At higher H concentrations the hydrogen diffusivity is observed to increase [112] due to a successive filling of the traps.

Diffusion studies in nanocrystalline metal oxides are of particular interest because the low-temperature ductility [113] and an increased strain-rate sensitivity [114] observed in these otherwise brittle ceramics may be a result of enhanced atomic transport in the interfaces. In n-TiO_2 densified at 570 °C under pressure (1.5 GPa) to 95% of the bulk density, oxygen diffusivities about five orders of magnitude higher than in crystalline TiO_2 (rutile) are observed [103]. In contrast, Hf diffusion in n-TiO_2 is found to be relatively slow, presumably due to an interstitialcy process (involving interstitial sites) for cation self-diffusion evoked in bulk rutile (see [103]).

For a better understanding of the diffusion mechanisms in nanocrystalline solids and their interrelation with the atomistic interfacial structure, diffusion experiments on high-purity dense materials of small grain size without changing the interfacial structure or the crystallite size during the diffusion heating are needed. This means that the temperature variation of the diffusivity should be measured below the temperature of a preceding annealing treatment. Then the atomic diffusivities may be specifically used for studying the changes of the interfacial structure due to annealing treatments.

11.5 HELIUM DESORPTION AFTER IMPLANTATION

The effect of the interfacial structure of nanocrystalline metals on atomic transport phenomena may additionally be demonstrated by desorption studies of He after implantation with MeV energies [82, 115]. Helium with its closed electron shell of noble gases and its short-range repulsive potential in solids exhibits a low solubility, a high diffusivity, and a high binding energy ≥ 2.0 eV to vacancies so that it tends to form agglomerates and bubbles in crystalline solids (for a review see [116]) and cannot be removed even at high temperatures.

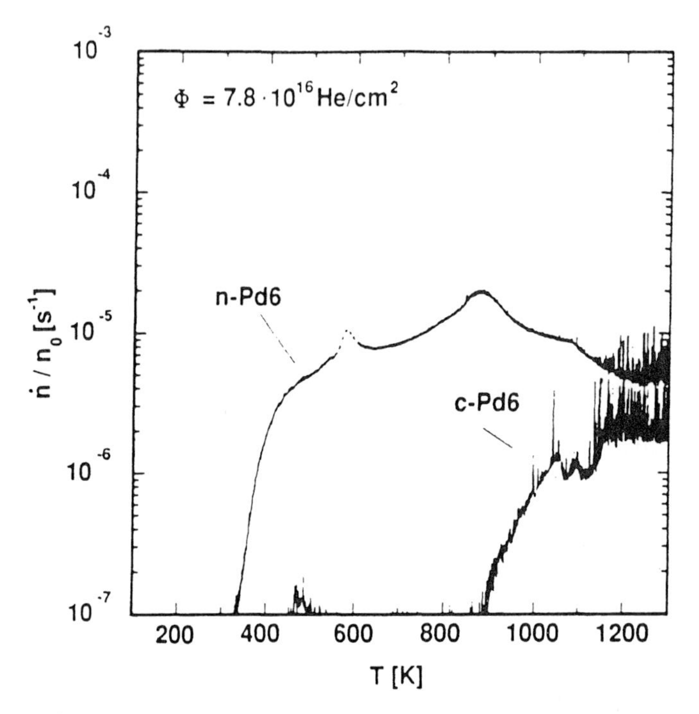

Figure 11.8. Temperature variation of the He desorption rates $\dot{n}/n_0$ (heating rate 100 K h^{-1}) from nanocrystalline Pd (n-Pd6) and coarse-grained Pd (c-Pd6) after low-dose 3 MeV implantation of He^+ ions (mean implantation depth 4.5 μm) at 100 K [82, 115].

In nanocrystalline metals the He desorption after implantation is facilitated as indicated by the following results:

(i) The He desorption from n-Pd after 3 MeV implantation of He ions (see figure 11.8) starts to occur at ambient temperature whereas the implanted He atoms are retained in the coarse-grained specimen (c-Pd) up to 900 K. At higher He implantation doses the He desorption was found to occur in n-Pd, at lower temperatures than in c-Pd [82, 115], and a desorption maximum in n-Pd at 650 K can be quantitatively described by a dissociation reaction with an activation energy of 2 eV and a pre-exponential factor of 8×10^{13} s^{-1} [82, 115].

(ii) After structural relaxation of the interfaces of n-Pd due to annealing at 495 K prior to He implantation, the He desorption is impeded and shifted to higher temperatures (see figure 11.9).

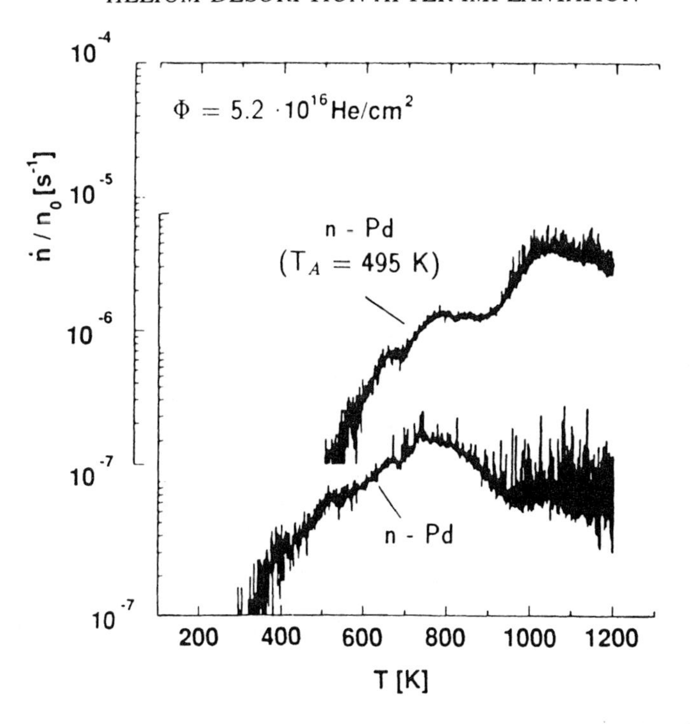

Figure 11.9. Comparison of the He desorption rates after 3 MeV implantation into as-prepared n-Pd and into an n-Pd specimen annealed at 495 K for 30 min [82, 115]. Note the shifted scales of the desorption rates.

(iii) The discontinuous He desorption observed for coarse-grained Pd due to He bubble desorption processes is strongly suppressed in n-Pd [82, 115].

From these experimental results it may be concluded that the initial He desorption from nanocrystalline Pd at relatively low temperatures is due to He dissociation from bubbles during bubble Ostwald ripening with subsequent fast diffusion of He via interfaces. This facilitates He diffusion in contrast to what happens in coarse-grained metals where the discontinuous desorption originates from He transport by bubble diffusion.

The shift of the He desorption process in n-Pd to higher temperatures after a pre-implantation annealing at $T_A = 495$ K, where no significant grain growth but densification (see figure 11.6(*a*)) and elastic stiffening [79] occur, may be ascribed to atomic relaxation in the interfaces with loss and agglomeration of free volumes (see section 11.3.2). This gives rise to a slowing down of the He transport in the interfaces. This transport may be directly affected by the jumping of metal atoms during structural relaxation of the interfaces. From the impeding of the He desorption by structural relaxation prior to substantial grain growth we may conclude that open porosity is unlikely to be involved in this process.

REFERENCES

[1] Siegel R W and Hahn H 1987 *Current Trends in the Physics of Materials* ed M Yussouff (Singapore: World Scientific) p 403

[2] Schaefer H-E, Wüschum R, Birringer R and Gleiter H 1988 *Physical Research* vol 8, ed K Hennig (Berlin: Akademie) p 580

[3] Gleiter H 1989 *Prog. Mater. Sci.* **33** 233

[4] Siegel R W 1991 *Annu. Rev. Mater. Sci.* **21** 559

[5] Gleiter H 1993 *Mechanical Properties and Deformation Behavior of Materials Having Ultra-Fine Microstructures (NATO ASI Series E: Applied Science 233)* ed M Nastasi, D M Parkin and H Gleiter (Dordrecht: Kluwer) p 3

[6] Siegel R W, Ramasamy S, Hahn H, Li Zongquan, Lu Ting and Gronsky R 1988 *J. Mater. Res.* **3** 1367

[7] Hahn H, Logas J, Höfler J, Kurath H J and Averback R S 1990 *Mater. Res. Soc. Symp. Proc.* **196** 71, 76

[8] Karch J and Birringer R 1990 *Ceram. Int.* **16** 291

[9] Yoshizawa Y, Oguma S and Yamauchi K 1988 *J. Appl. Phys.* **64** 6044

[10] Herzer G 1989 *IEEE Trans. Mag.* **MAG-25** 3327

[11] Buschow K H J 1982 *J. Less-Common Met.* **85** 221

[12] Valiev R Z, Kaibyshev O A, Kuznetsova R I, Musalimov R Sh and Tsenev N K 1988 *Dokl. Acad. Nauk SSSR* **301** 864 (in Russian)
Valiev R Z 1993 *Mechanical Properties and Deformation Behavior of Materials Having Ultra-Fine Microstructures (NATO ASI Series E: Applied Science 233)* ed M Nastasi, D M Parkin and H Gleiter (Dordrecht: Kluwer) p 303

[13] Shinghu P H, Nishitani S R and Nasu S 1988 *Suppl. Japan Inst. Met.* **29** 3

[14] Hellstern E, Fecht H-J, Fu Z and Johanson W L 1989 *J. Appl. Phys.* **65** 305

[15] Hausner H 1978 *Ber. Dtsch. Keram. Gesell.* **55** 194

[16] Spanhel L, Arpac E and Schmidt H 1992 *J. Non-Cryst. Solids* **147 & 148** 657

[17] Ponthieu E, Payen E and Grimblot J 1992 *J. Non-Cryst. Solids* **147 & 148** 598

[18] Gleiter H 1981 *Deformation of Polycrystals: Mechanisms and Microstructures* ed N Hansen *et al* (Roskilde: Risø National Laboratory) p 15

[19] Hahn H and Averback R S 1990 *J. Appl. Phys.* **67** 1113

[20] Würschum R, Greiner W and Schaefer H-E 1993 *Nanostruct. Mater.* **2** 55

[21] Würschum R, Soyez G and Schaefer H-E 1992 *Structure and Properties of Interfaces in Materials (MRS Symp. Proc.)* vol 238, ed W A T Clark, C L Briant and U Dahmen p 733; 1993 *Nanostruct. Mater.* **3** 225

[22] Nieman G W, Weertman J R and Siegel R W 1991 *J. Mater. Res.* **6** 1012

[23] Wagner C N J 1992 *J. Non-Cryst. Solids* **150** 1

[24] Mayer J W, Rimini E (ed) 1977 *Ion Beam Handbook for Material Analysis* (New York: Academic)

[25] Carstanjen, H-D, Decker W, Diehl J, Enders T, Emrick R M, Föhl A, Friedland E, Plachke D and Stoll H 1990 *Nucl. Instrum. Methods Phys. Res.* B **51** 152

[26] Weber H-J, Würschum R, Eckert W, Schaefer H-E, Carstanjen H-D, Enders T, Föhl A and Plachke D-W 1990 *Verhandl. DPG (VI)* **25** M22.2
Weber H-J 1989 *Diploma Thesis* Stuttgart University

[27] Schaefer H-E, Würschum R, Gessmann T, Stöckl G, Scharwaechter P, Frank W, Mulyukov R R, Fecht H-J and Moelle C 1995 *Nanostruct. Mater.* **6** 869

[28] Nieminen R M and Oliva J 1980 *Phys. Rev.* B **22** 2226

[29] Hautojärvi P (ed) 1979 *Positrons in Solids* (Berlin: Springer)
[30] Borónski E and Nieminen R M 1986 *Phys. Rev.* B **34** 3820
[31] Puska M J and Nieminen R M 1983 *J. Phys. F: Met. Phys.* **13** 333
[32] Seeger A 1972 *Phys. Lett.* **41A** 267
[33] Soininen E, Huomo H, Huttenen P A, Mäkinen J, Vehanen A and Hautojärvi P 1990 *Phys. Rev.* B **41** 6227
[34] Shiotani N 1982 *Positron Annihilation* ed P G Coleman, S C Sharma and L M Diana (Amsterdam: North-Holland) p 561
[35] Schaefer H-E, Eckert W, Briggmann J and Bauer W 1989 *J. Phys.: Condens. Matter* **1** SA97
[36] Schaefer H-E 1987 *Phys. Status Solidi* a **102** 47
[37] Schaefer H-E, Würschum R and Bub J 1992 *Mater. Sci. Forum* **105–110** 439
[38] Würschum R, Bauer W, Maier K, Seeger A and Schaefer H-E 1989 *J. Phys.: Condens. Matter* **1** SA33
[39] Forster M, Claudy W, Hermes H, Koch M, Maier K, Major J, Stoll H and Schaefer H-E 1992 *Mater. Sci. Forum* **105–110** 1005
Forster M 1991 *Dissertation* Stuttgart University
[40] Puska M J and Corbell C 1988 *Phys. Rev.* B **38** 9874
[41] Würschum R 1988 *Dissertation* Stuttgart University
[42] Paulin R and Ripon R 1974 *Appl. Phys.* **4** 343
[43] Tsuchiya Y, Noguchi S and Hasegawa M 1979 *Phil. Mag.* B **39** 181
[44] Matsuoka Y, Terakado S and Tanigawa S 1984 *Phys. Status Solidi* b **125** 823
[45] Eldrup M, Bentzon M D, Pedersen A S and Larson B 1993 *Mechanical Properties and Deformation Behavior of Materials Having Ultra-Fine Microstructures (NATO ASI Series E: Applied Science 233)* ed M Nastasi, D M Parkin and H Gleiter (Dordrecht: Kluwer) p 571
[46] Svirida S V, Linderoth S and Repin I A 1989 *Positron Annihilation* ed L Dorikens-Vanpraet, M Dorikens and D Segers (Singapore: World Scientific) p 521
[47] Bauer W, Major J, Weiler W, Maier K and Schaefer H-E 1985 *Positron Annihilation* ed P C Jain, R M Singru and K P Gopinathan (Singapore: World Scientific) p 804
[48] Schaefer H-E, Würschum R, Birringer R and Gleiter H 1988 *Phys. Rev.* B **38** 9545
[49] Würschum R, Greiner W, Soyez G and Schaefer H-E 1992 *Mater. Sci. Forum* **105–110** 1337
[50] Würschum R, Scheytt M and Schaefer H-E 1987 *Phys. Status Solidi* a **102** 119
[51] Würschum R, Greiner W, Valiev R Z, Rapp M, Sigle W, Schneeweiss O and Schaefer H-E 1991 *Scr. Metall. Mater.* **25** 2451
[52] Doyama M and Cotterill R M J 1979 *Positron Annihilation* ed R R Hasiguti and K Fujiwara (Sendai: Japan Institute of Metals) p 89
[53] Dlubek G, Brümmer O, Meyendorf N, Hautojärvi P, Vehanen A and Yli-Kauppila J 1979 *J. Phys. F: Met. Phys.* **9** 1961
[54] McKee B T A, Saimoto S, Stewart A T and Stott M J 1974 *Can. J. Phys.* **52** 759
[55] Alekseeva O K, Onishckuk V A, Shantarovich V P, Dekhtyar I Y and Shevchenko V I 1979 *Phys. Status Solidi* b **95** K135
[56] Schaefer H-E and Forster M 1989 *Mater. Sci. Eng.* A **109** 161
[57] Seeger A and Würschum R *Phil. Mag.* A at press
[58] Huttunen P A, Mäkinen J, Britton D T, Soininen E and Vehanen A 1990 *Phys. Rev.* B **42** 1560

[59] Jensen K O, Eldrup M, Linderoth S and Evans J H 1990 *J. Phys.: Condens. Matter* **2** 2081

[60] Schaefer H-E, Eckert W, Stritzke O, Würschum R and Templ W 1989 *Positron Annihilation* ed L Dorikens-Vanpraet, M Dorikens and D Segers (Singapore: World Scientific) p 79

[61] Schaefer H-E 1982 *Positron Annihilation* ed P G Coleman, S C Sharma and L M Diana (Amsterdam: North-Holland) p 369

[62] Häkkinen H, Mäkinen S and Manninen M 1989 *Europhys. Lett.* **9** 809

[63] Shirai Y, Matsumoto K, Kawaguchi G and Yamaguchi M 1992 *Mater. Sci. Forum* **105–110** 1225

[64] Wunderlich W, Maurer R and Ishida Y 1990 *Scr. Metall. Mater.* **24** 403

[65] Palumbo G, Thorpe S J and Aust K T 1990 *Scr. Metall. Mater.* **24** 1347

[66] Nieminen R M, Laakonen J, Hautojärvi P and Vehanen A 1979 *Phys. Rev.* B **19** 1397

[67] Jorra E, Franz H, Peisl J, Wallner G, Petry W, Birringer R, Gleiter H and Haubold T 1989 *Phil. Mag.* B **60** 159

[68] Sanders P G, Weerman J R, Barker J G and Siegel R W 1993 *Scr. Metall. Mater.* **29** 91

[69] Hasegawa M, Yoshinari O, Matsui H and Yamaguchi S 1989 *J. Phys.: Condens. Matter* **1** SA77

[70] Gregory R B 1991 *J. Appl. Phys.* **70** 4665

[71] Schaefer H-E, Würschum R, Hof P, Straub W and Gessmann T 1995 *Mater. Sci. Forum* **175–178** 505

[72] Zhu X, Birringer R, Herr U and Gleiter H 1987 *Phys. Rev.* **35** 9085

[73] Haubold T, Birringer R, Lengeler B and Gleiter H 1988 *J. Less-Common Met.* **145** 557

[74] Musalimov R Sh and Valiev R Z 1992 *Scr. Metall. Mater.* **27** 1685

[75] Aaron H B and Bolling G F 1976 *Grain Boundary Structure and Properties* ed G A Chadwick and D A Smith (London: Academic) p 107

[76] Wolf D 1990 *Acta Metall. Mater.* **38** 791

[77] Hahn H and Gleiter H 1981 *Acta Metall.* **29** 601

[78] Bristowe P D, Brokman A, Spaepen F and Balluffi R W 1980 *Scr. Metall.* **14** 943

[79] Weller M, Diehl J and Schaefer H-E 1991 *Phil. Mag.* A **63** 527

[80] Greiner W 1990 *Diploma Thesis* Stuttgart University

[81] Gupta R P and Siegel R W 1977 *Phys. Rev. Lett.* **39** 1212

[82] Schaefer H-E 1993 *Mechanical Properties and Deformation Behavior of Materials Having Ultra-Fine Microstructures (NATO ASI Series E: Applied Science 233)* ed M Nastasi, D M Parkin and H Gleiter (Dordrecht: Kluwer) p 81

[83] Tschöpe A and Birringer R 1993 *Acta Metall. Mater.* **41** 2791

[84] Landesberger C, Wallner G and Peisl J 1987 private communication

[85] Seeger A and Schottky G 1959 *Acta Metall.* **7** 495

[86] Kumpmann A, Günther B and Kunze H-D 1993 *Mechanical Properties and Deformation Behavior of Materials Having Ultra-Fine Microstructures (NATO ASI Series E: Applied Science 233)* ed M Nastasi, D M Parkin and H Gleiter (Dordrecht: Kluwer) p 309

Günther B, Kumpmann A and Kunze H-D 1992 *Scr. Metall. Mater.* **27** 833

[87] Siegel R W 1993 *Mechanical Properties and Deformation Behavior of Materials Having Ultra-Fine Microstructures (NATO ASI Series E: Applied Science 233)* ed M Nastasi, D M Parkin and H Gleiter (Dordrecht: Kluwer) p 509

[88] Hague D C and Mayo M J 1993 *Mechanical Properties and Deformation Behavior of Materials Having Ultra-Fine Microstructures (NATO ASI Series E: Applied Science 233)* ed M Nastasi, D M Parkin and H Gleiter (Dordrecht: Kluwer) p 539

[89] Sui K L, Lu K, Deng W, Xiong L Y, Patu S and He Y Z 1991 *Phys. Rev.* B **44** 6466

[90] Sinning H-R, Nicolaus M M and Haeßner F 1991 *Mater. Sci. Eng* A **133** 371

[91] Vepřek S, Iqbal Z, Oswald H R and Webb A P 1981 *J. Phys. C: Solid State Phys.* **14** 295

[92] Vepřek S, Iqbal Z, Kühne R O, Capezzuto P, Sarott F A and Gimzewski J K 1983 *J. Phys. C: Solid State Phys.* C **16** 6241

[93] Vepřek S 1984 *Proc. Mater. Res. Soc. Eur.* ed P Pinard and S Kalbitzer (Strasbourg: Les Ulis) p 425

[94] Schaefer H-E, Würschum R, Schwarz R, Slobodin D and Wagner S 1986 *Appl. Phys.* A **40** 145

[95] Jensen K O, Nieminen R M, Eldrup M, Singh B N and Evans J H 1989 *J. Phys.: Condens. Matter* **1** SA67

[96] Weppner W 1983 *Goldschmidt Inform.* **59** 6

[97] Lange F F 1982 *J. Mater. Sci.* **17** 225

[98] Blanchin M G, Nihoul G and Bernstein E 1992 *Phil. Mag.* A **65** 683

[99] Garvie J C 1978 *J. Phys. Chem. Solids* **82** 218

[100] Saarinen K, Hautojärvi P, Vehanen A, Krause R and Dlubek G 1989 *Phys. Rev.* **39** 5287

[101] Gleiter H 1992 *Phys. Status Solidi* b **172** 41

[102] Schaefer H-E 1993 *Defects in Insulating Materials* ed O Kanert and J-M Spaeth (Singapore: World Scientific) p 122

[103] Höfler H J, Hahn H and Averback R S 1991 *Defect Diff. Forum* **75** 195

[104] Maier K 1977 *Phys. Status Solidi* a **44** 567

[105] Schumacher S, Birringer R, Strauss R and Gleiter H 1989 *Acta Metall.* **37** 2485

[106] Peterson N L 1964 *Phys. Rev.* A **136** 568

[107] Golovin S A, Krishtal M A and Svobodov A N 1968 *Fiz. Kluin. Obrab. Mater.* **4** 119

[108] Puin W, Heitjans P, Dickenscheid W and Gleiter H 1993 *Defects in Insulating Materials* ed O Kanert and J-M Spaeth (Singapore: World Scientific) p 137

[109] Hoshino K, Peterson N L and Wiley C L 1985 *J. Phys. Chem. Solids* **46** 1397

[110] Lundy T S and Coghlan W A 1973 *J. Physique Coll.* **C9** 299

[111] Martin G, Blackburn D A and Adda Y 1967 *Phys. Status Solidi* **23** 223

[112] Kirchheim R, Mütschele T, Kieninger W, Gleiter H, Birringer R and Koblé T D 1988 *Mater. Sci. Eng.* **99** 457

[113] Karch J, Birringer R and Gleiter H 1987 *Nature* **330** 556

[114] Mayo M J, Siegel R W, Liao Y X and Nix W D 1992 *J. Mater. Res.* **7** 973

[115] Eckert W 1992 *Dissertation* Stuttgart University

[116] Schilling W 1982 *Point Defects and Defect Interaction in Metals* ed J I Takamura, M Doyama and Kiritani M (Tokyo: University of Tokyo Press) p 303

PART 6

PROPERTIES OF NANOSTRUCTURED MATERIALS

Chapter 12

Chemical properties

Debra R Rolison

12.1 INTRODUCTION

Chemistry plays an important role in creating nanoscale structures, where the modern usage of the term nanostructure implies that control, and often explicit design, of the chemistry (or physics) is required to generate the as-created nanostructure. Two recent review articles summarize the uses of chemistry in this small-scale (and sometimes Herculean) creative effort [1, 2] and use the term 'nanochemistry' to describe synthetic means to nanoscale ends.

Perspectives on nanostructures and nanoscale phenomena need to span the insights and limits of multiple scientific disciplines. The following paraphrases an exchange between a chemist and a physicist who mutually agreed with a 1991 editorial in *Science* accompanying the special section on nanoscale science [3] which stated 'This is an area currently getting a lot of hype' [4]. The physicist, who studies nanoscale materials, commented that the greater benefit of nanostructures would surely be their application in chemistry—particularly in catalysis—only to be countered by the chemist, who also studies nanoscale materials and felt certain that the quantum confinement effects exhibited by nanometer-sized materials would be best expressed in optical or electronic technology where the nanostructure could be isolated from encountering molecules and thereby avoid any risk of chemistry.

This exchange highlights another interpretation of the term 'nanochemistry' —an interpretation where the questions become: can chemistry be done with (or at) nanostructures; what remains of a nanostructure when it is exposed to molecules with which chemical reactions can occur; can the integrity of a nanostructure—its controlled composition and placement, often laboriously achieved—be assured when it acts as a catalyst to enhance the rate of a chemical reaction—or worse yet, when it acts as a (potentially unplanned) reactant in a chemical reaction; what are the issues when chemical reactions occur at nanostructures or within a nanostructured environment? It is the conceit of this author (a chemist) that these questions need to be acknowledged and thought about in greater detail than has been given to them to date; this discussion is an initial and undoubtedly biased and incomplete attempt to do so.

12.2 EXAMPLES OF NANOSTRUCTURES IN CHEMISTRY

One place to seek some of the answers to these questions is in the old chemical literature because the chemical preparation of a nanostructure, although bigger than the customary chemical yardstick (a molecule), is nonetheless nothing particularly new. A chart compiled by Whitesides *et al* [1] summarizes the different concepts of 'customary size' as defined by the practitioners of biology, chemistry, or physics, and highlights the relativity of size (what is large/what is small) within and between the three disciplines.

The research in such established small domains as colloids, micelles, nucleation phenomena, and supported metal or metal oxide catalysts developed much of the arsenal now being used to design nanochemistry (i.e. to synthesize nanomaterials by chemistry [1, 2]) and is a store of both useful information and lessons already learned about nanochemistry (i.e. chemical reactions at nanoscale structures). These general concepts include self-organization (e.g. self-assembly) and site specificity (e.g. templating). What has changed is primarily our current expectations of nanostructures and their properties. A brief summary of several of these long-established small domains provides a starting point for the interested.

12.2.1 Colloids

Just as preparations of nanomaterials via chemical processes are not new, neither are the use and characterization of nanomaterials in chemistry. Colloids exemplify both categories. Colloids are particles (of a vast range of chemical composition) formed in a condensed phase (solution or solid state) whose sizes range from 1–1000 nm. Colloids have either been considered as small pieces of material where the continuum properties of matter are exhibited [5] (i.e. nanoscale in size but not in effect) or as small pieces of material, particularly at the low end of the size range, that define the transition between molecular and materials properties [6] (i.e. nanoscale in size and effect). Colloids are age old and colloidal science dates from the turn of the century; Henglein [6] and Thomas [7], in their respective reviews of colloidal semiconductors and metals, take delight in pointing out that the title of the first major text on colloid science (by Ostwald in 1920) translates as *The World of Neglected Dimensions* [8].

Transitions in the physical and chemical properties of colloids as the size was diminished were observed (and then eventually expected) to show effects attributed to quantization. Indeed, as Henglein details in his review [6], the wavelength shift in the absorption of light by nanoscale semiconductors was first observed in the late 1960s for colloidal AgI [9] and AgBr [9, 10].

The practical demonstration of the effects of quantum confinement evolved from such observations, because the use of colloids readily permitted optical characterization of the effect of size on the position of the absorption maximum. Varying the known average sizes of polydisperse colloids (as established by various physical and reactive methods) [11–13] showed that the absorption maximum blue-shifted as the size of the colloid decreased. The optical properties of colloids as a function of size have been especially well studied for the cadmium chalcogenides (i.e. CdS and CdSe) when formed as free-standing colloids [13–15], surface-supported colloids [11, 16, 17], or as three-dimensionally encaged colloids within zeolites [18, 19], metal phosphonate lattices [20], or micelles [21, 22]. The optical characteristics of nanostructures are discussed more fully in chapters 4 and 16.

Another optical phenomenon to which colloids (and variations in their size and shape) have contributed is the study of the surface enhanced Raman effect. Ever since Fleischmann *et al* reported a millionfold increase in the Raman signal from pyridine adsorbed on an electrochemically roughened silver electrode [23], the relative contributions to the signal enhancement from electromagnetic factors or from chemical effects have been theoretically debated and experimentally probed. The ability to vary the particle size and morphology of a Raman-active surface by the deposition of silver and other metallic colloids, or by the study of the dispersed colloids [24], has provided an experimental approach to explore the influence of size and shape on the Raman intensity, without the randomness of size and shape characteristic of a metal surface after electrochemical roughening. These advantages can be lessened, however, once a neutral Raman probe has adsorbed to the surface of a colloid [24] and disrupted the electrostatic forces or steric interactions otherwise preventing aggregation: aggregation can lead to irreproducible spectroscopy [25]. Enhanced Raman intensity has been found for colloids sized at tens of nanometers and shaped as prolate spheroids, rather than for smaller particles shaped as perfect spheres, where the larger sizes and nonspherical shape can be correlated with the resonance wavelength of surface plasmon excitations in the metal [24].

12.2.2 Supported nanoscale catalysts

The use of sub-micrometer particles of metals or metal oxides on nonmetallic supports to perform heterogeneous catalytic reactions is much examined and the literature on this topic is vast—a minor subset of this literature will be discussed in this section. Had the workers in this area not entitled the 1–10 nm supported particles microcrystallites or microstructures, but christened them instead the nanocrystallites or nanostructures they actually are, nanoscale nomenclature might have been inspired decades earlier.

An important goal in a practical supported catalyst is to achieve a high degree of dispersion [26] since a high degree of dispersion maximizes the contact area of catalyst with reactant and support and minimizes the fraction of catalyst buried within large particles unable to participate directly in the catalyzed reaction. A sufficiently large catalyst particle does, however, buttress continuum electronic effects—which may be the desired effect in the catalytic reaction. But, just as for the size quantization effects observed in the optical properties of colloids, the possibility of quantum effects influencing catalytic/chemical reactions has long been postulated as the size of the catalyst particle is brought into the nanoscale regime [27].

Agglomeration of a supported catalyst into larger particles is one of the dreaded outcomes of long-duration reactions at supported catalysts [26, 28] (i.e. nanocrystallites). Minimization, although not outright prevention, of catalyst particle agglomeration has been one of the valid reasons to use microporous supports, especially the aluminosilicate zeolites where the nanometer scale of the pore/channel/cage architecture of zeolite crystals offers a means to template the formation of the catalyst particle and then corral it once formed [29]. The pore/channel/cage architecture that defines the structure of microporous zeolites can be used to further restrict the particle size and to limit the size of molecular sorbates with which the nanocrystallites can interact.

With the importance of zeolites for strong-acid catalyzed chemistry [30], the literature on forming metal, metal oxide, or semiconductor compounds supported on zeolites is also vast with entire books devoted to the preparation, characterization, and properties of such systems [31]. The zeolite community refers to such systems as microstructured materials because their emphasis is primarily on the zeolite host and secondarily on the nanometer-sized guest. The decades of work on this approach have provided to the nanoscale community both a very promising templating host and the synthetic schemes necessary to create the nanostructures more in vogue today, including synthesizing—as an encapsulated guest—nanoscale materials desirable for uses other than catalysis [2]. The principles leading to the design of such nanostructured composites are, however, the same as those pioneered in the study of zeolite supported catalysts.

12.2.3 Nucleation phenomena

The heart of either homogeneous or heterogeneous nucleation phenomena—whether on a surface, as in electrocrystallization [32], or from a solution or in the gas phase—lies in the production of a material arising from the level of single atom or molecule through nanoscale to microscale and possibly on to macroscopic scale. The growth of crystals has much about it that strikes one as a controlled assembly of structures, although the process is rarely quenched at the nanometer scale. The language currently being used to describe the growth of zeolite crystals [33], for instance, sounds reminiscent of the thinking behind self-assembly where entropic factors play guiding roles in the formation of the nanostructured material [1].

Electrochemically induced nucleation has long been favored in the study of nucleation as it allows experimental control of the rates of nucleation and growth by judicious control of the potential applied to the electrode. The recent use of ultramicroelectrodes [34, 35], which are electrodes with at least one dimension on the order of 0.1–10 μm, has permitted the creation, isolation, and characterization of a single electrocrystallization center; if excessive growth is not permitted, this nucleus can be quenched as a nanostructure isolated on the surface of the ultrasmall electrode [36]. Studies of single electrocrystallization centers of α-PbO_2 and β-PbO_2 (materials of technological importance whose electrocrystallization, growth, and morphology are of fundamental importance in the behavior of lead–acid batteries) on carbon or platinum disk ultramicroelectrodes have greatly simplified the ability to determine the kinetics of growth of the lead dioxide nucleus [37]. Limiting nucleation to the single center permits the use of Poisson statistics to describe explicitly the birth and death processes of the nucleus [38].

12.3 THE EFFECT OF NANOSCALE MATERIALS ON CHEMICAL REACTIVITY

Just as changes in the optical and electronic [39] properties of nanoscale materials are anticipated based on quantum confinement effects, as previously discussed, changes in the chemical reactivity of nanoscale materials have also been anticipated [6, 7, 27]. Although gas phase bare metal clusters are an ephemeral example of a nanostructured material, the research in this area offers another source of insight into size-dependent variations in chemical reactivity [40, 41]. As chemical reactions are governed by electrons, relative electron affinities (or ionization potentials), and electron orbital densities, a natural coupling exists between chemical reactivity and the electronic character of the reactants and any reaction catalysts. In his review of transition metal atom clusters, Morse [40] tabulates ionization thresholds as a function of the number of atoms and these data show that the ionization potential increases as the cluster size drops below the bulk limit. The increase in ionization potential does not, however, always vary monotonically with cluster size. A demarcation in the chemical reactivity of gas phase metal clusters, using the language developed to describe heterogeneous catalysis at supported catalysts, has been made between 'facile' reactions of high rate with minimal size dependence and 'demanding' reactions where the rate is dependent on cluster size and also exhibits crystal plane dependencies for the same reaction with the bulk metal [40, 41]. This latter reactivity path has been restated as one involving a correlation between ionization potential of the metal cluster for reactions involving electron transfer to the metal—particularly those involving reaction with hydrogen [42].

12.3.1 Chemical reactivity: metal nanocrystallites supported on oxides

Size quantization in the electronic properties of a catalyst, though much anticipated, may remain elusive, or, at the least, support or host dependent and possibly even reaction dependent. A recent summary by Sinfelt and Meitzner [43] of the characterization of oxide supported metal catalysts by studies of the x-ray absorption edge of the metal (which probes its electronic structure) led them to conclude that clusters of platinum, iridium, or osmium sized down to 1 nm and supported on alumina or silica exhibited electronic properties similar to those found for large crystallites of the metal (which show bulk electronic properties).

Recent results by Koningsberger and colleagues with platinum clusters (containing five to six atoms) fabricated within the channels of synthetic KL zeolite [44] are consistent with the above conclusions that the electronic structure of metal nanocrystallites supported on nonconducting oxides reflect bulk electronic characteristics. This study also offers significant insight into the importance of physical structure and chemical environment on the catalytic activity of supported nanocrystallites.

The Pt/KL catalyst has been used to produce benzene from *n*-hexane, but it is readily poisoned by sulfur [45] (creating the need for feedstocks with minimal sulfur content). Extended x-ray absorption fine structure (EXAFS) spectroscopy was used to show that Pt/KL exposed to H_2S (to a level that lowered the catalytic activity for benzene production by 70%) yielded larger platinum clusters (growing to 13 atoms of platinum), which, nonetheless still were metallic (i.e. exposure to sulfur had not formed a platinum sulfide). Hydrogen chemisorption measurements showed that not every surface platinum atom was associated with a sulfur atom as the available platinum surface had only diminished by 25%. The 70% decrease in chemical reactivity at the larger, still relatively sulfur-free platinum clusters was attributed to the loss of active metal surface as more platinum atoms became incorporated within the interior of the nanocrystallites and as growth of the larger particles apparently blocked the zeolite channels and restricted access of the *n*-hexane reactant to the Pt nanocrystallite catalyst [44].

12.3.2 Electrochemical reactivity

12.3.2.1 Metal nanocrystallites supported on conductors

A catalytic arena in which nanostructure size effects have been observed has been in electrochemical reactivity. The importance of the character of the support (or host) for the nanocrystallites can also be seen in such studies.

Platinum or palladium supported on conducting surfaces (chosen to be catalytically inert for the reaction of study) were studied by Bagotsky and Skundin to correlate electronic properties (as determined from electron photoemission) [46] with electrocatalytic activity [47]. They found that the photoemission current increased (relative to that for the bulk metal) for nanocrystallites of platinum supported on titanium or glassy carbon while the photoemission current for palladium nanocrystallites supported on glassy carbon decreased relative to that of bulk palladium (there was negligible difference between bulk palladium and palladium nanocrystallites supported on titanium); the photoemission currents also varied as a function of the crystallite size [46]. The trends in the photoemission current tracked the trends in the electrocatalytic activity of the nanoscale Pt or Pd for hydrogen evolution

$$2H^+ + 2e^- \rightarrow H_{2(g)} \tag{12.1}$$

and other reactions [46]. That is, as compared to the bulk metal, increased reactivity (or electrochemical current) for supported platinum nanocrystallites and decreased reactivity for glassy carbon supported palladium nanocrystallites [46, 47] was observed.

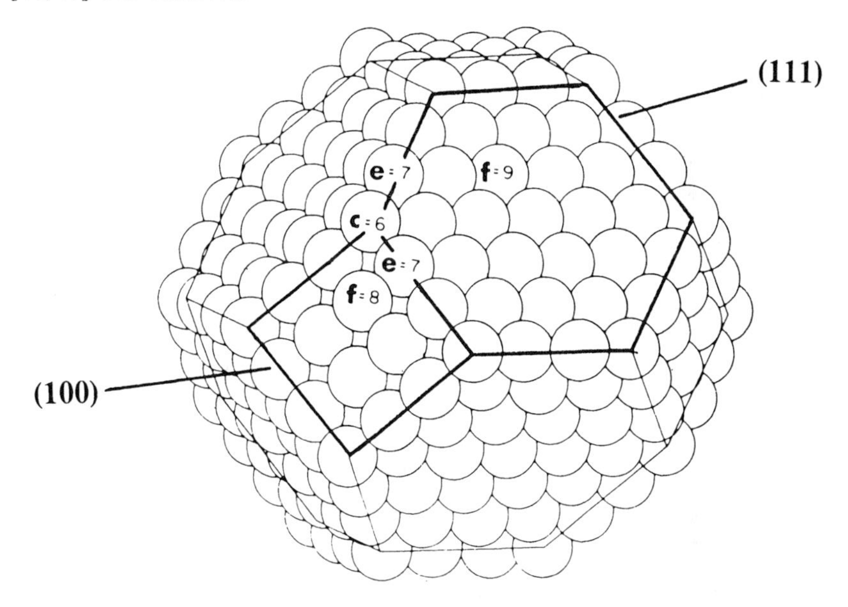

Figure 12.1. The surface atom arrangement for a face-centered cubic (fcc) crystallite in a cubo-octahedral configuration with the (100) and (111) faces outlined with a thick black line and specifying the number of nearest neighbors for edge (e), corner (c), and facial (f) atoms (after [73]).

Catalyst nanostructures supported on a conducting support have a long history of use under very practical technological conditions, e.g. in fuel cells where platinum or platinum group alloys (usually with Sn or Ru) supported on carbon blocks electrocatalyze both hydrogen oxidation at a Pt/C anode (the reverse of equation (12.1)) and oxygen reduction at a Pt/C cathode. Electrocatalytic efficacy correlates with particle size for these systems. Kinoshita has recently discussed the importance of the physical structure and size of carbon supported platinum nanocrystallites for oxygen reduction in acid electrolyte [48] (a common fuel-cell configuration). Kinoshita correlates his and other literature data with the relative fraction of platinum surface atoms on the faces of the particle (either a (100) or (111) face, as seen in figure 12.1 for fcc Pt in a surface-energy minimized cubo-octahedral configuration), and finds that the specific activity of the oxygen electroreduction reaction (defined as current per unit surface area of Pt) increases with increasing particle size and stabilizes above 10 nm [48]. There seems to be good agreement that if the efficacy of oxygen electrocatalysis at carbon supported Pt nanocrystallites is measured instead as a function of mass activity (defined as current per unit mass of Pt), the activity maximizes at Pt particle sizes of ~ 3 nm [48, 49].

12.3.2.2 Metal nanocrystallites supported on nonconductors

In work performed at the Naval Research Laboratory, Pt nanocrystallites supported on synthetic faujasite zeolite have been studied as ensembles of dispersed ultramicroelectrodes [50, 51]. In this approach, each zeolite supports Pt nanocrystallites on the external surface of the zeolite particle (extracrystalline Pt) and within the supercages of the faujasite structure (intracrystalline Pt); a direct electrically wired contact to each of these potential ultramicroelectrodes is not made. The ability to lower the ionic strength of the electrolyte is one of the advantages gained by the use of ultramicroelectrodes [34], which allowed these zeolite supported ultramicroelectrodes to be studied in salt-free liquids (pure water or acetonitrile or a benzene–water mixture) and electrified by applying large d.c. voltages (> 10 V). A schematic representation of the electrified dispersed zeolite supported Pt ultramicroelectrodes is seen in figure 12.2, which also shows the cage and pore structure of the synthetic faujasite zeolite type Y used for our studies with the siting of intracrystalline and extracrystalline Pt.

Using redox probes small enough to enter the pores and supercages of the zeolite, it was demonstrated that the intracrystalline Pt nanocrystallites were not contributing to the observed electrochemical reactions [51]. This was further verified by preparing two zeolite samples with the same weight loading of Pt but where the calcination temperature was varied during the preparation so that the intracrystalline Pt particles formed after calcination at 600 °C would be approximately twice as large as those formed after calcination at 300 °C [52]. Both preparations, however, had similarly sized extracrystalline Pt particles (as confirmed by x-ray photoelectron spectroscopy, XPS) [51]. As seen by the current–voltage (i–V) curves of figure 12.2, both materials behaved similarly, again indicating that the electrochemistry was not sustained at intracrystalline Pt, but rather occurred at extracrystalline Pt. The extracrystalline Pt nanocrystallites, whose size could be varied by changing the weight loading of Pt supported on the zeolite, were then shown to be more effective for the electrolysis of water (i.e. production of H_2 and O_2) when the extracrystalline Pt nanocrystallites were sized ≤ 2.5 nm relative to either a comparable surface area of bulk Pt or extracrystalline Pt particles sized at 5 nm or greater [51, 53].

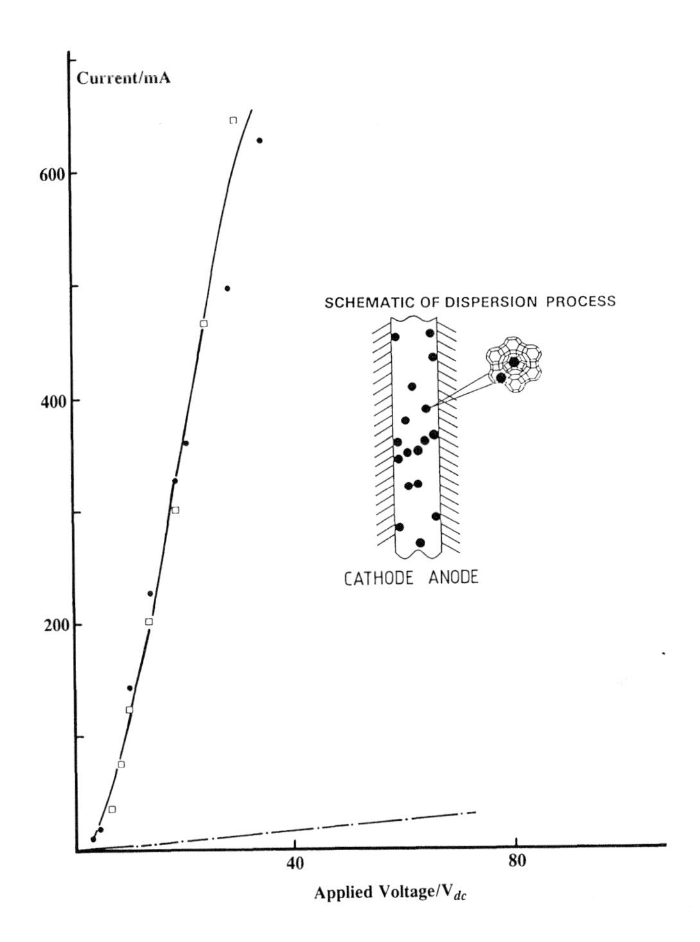

Figure 12.2. The current–applied voltage curves obtained at Pt feeder electrodes for the electrolysis of 18 MΩ cm water at 1 wt% Pt supported on synthetic faujasite zeolite Y (Pt–Y) where the preparation included calcination at: □ 600 °C; or ● 300 °C; suspension density = 0.1 g of Pt–Y in 20 ml water; — · —: the ionic background at the same Pt feeder electrodes for the filtrate from 20 mL of 18 MΩ cm H_2O exposed to 0.1 g of Pt–Y; the low linear current indicates minimal Faradaic reaction due to the electrolysis of water in the absence of Pt–Y. Inset: a schematic representation of the electrifed dispersed zeolite supported ultramicroelectrodes between two feeder electrodes; also shown is the siting of intracrystalline Pt within the faujasite supercage and extracrystalline Pt supported on the exterior of the faujasite unit cell.

12.3.3 Effect of nanostructures on mass transport

Another possible size related effect on chemical reactivity at nanostructures arises from the influence a nanometer-sized catalyst has on the mass transport of reactant from a homogeneous phase to it and of product away from it. The mass flux to nanometer-sized spheres (as for colloids) and planes or cylinders or hemispheres (as for ultramicroelectrodes [34, 35] or supported crystallites) rapidly converts from planar diffusion to nonplanar (i.e. hemispherical or radial) diffusion. This conversion produces enhanced, seemingly stirred or convective-like transport relative to that typical of a conventional surface in quiescent (unstirred) solution where diffusion remains planar. These differences in mass transport are illustrated schematically in figure 12.3, which also shows the characteristic current–potential (i–E) curves observed under these mass-transport conditions.

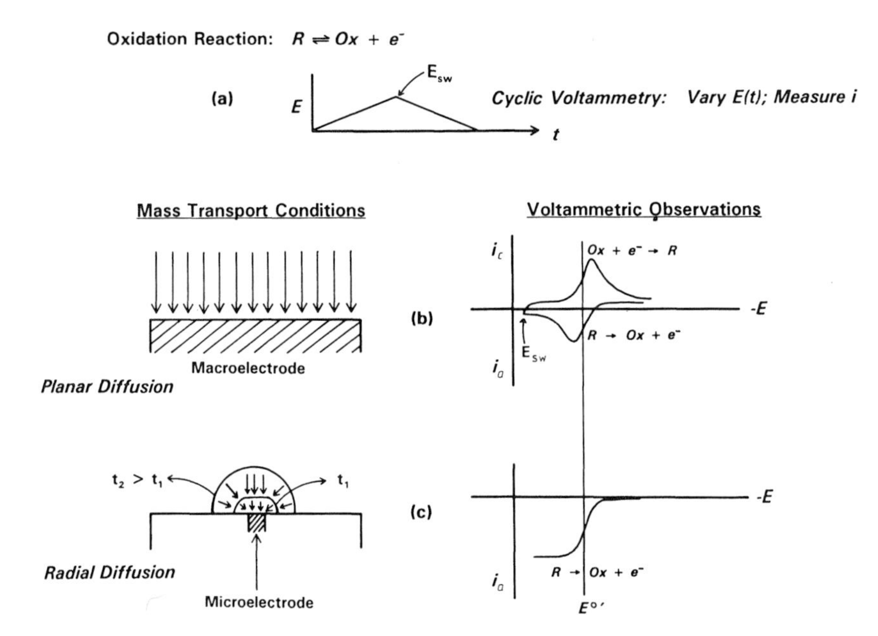

Figure 12.3. The effect of electrode size on mass transport as monitored by the voltammetric response of a reversible oxidation reaction. (*a*) The general half-cell reaction of a reversible oxidation and the variation of potential as a function of time for a cyclic voltammetric experiment; E_{sw} represents the switching potential where the linear potential ramp is reversed. (*b*) The voltammetric curve observed under conditions of planar diffusion at a macroelectrode; i_a and i_c represent anodic and cathodic current, respectively. (*c*) The voltammetric curve observed under conditions of radial diffusion at a microelectrode; the mass flux profile at short time (t_1) can be seen to approximate planar diffusion, but at $t_2 > t_1$ the conversion to hemispherical (radial) diffusion can be seen.

As discussed by Heinze [54, 55] unusually large current densities can be supported (up to 70 A cm^{-2}) at electrodes with dimensions below 20 μm because of the increase in the mass transport which arises from the transformation to nonplanar diffusion of a reactant undergoing electron transfer at an ultramicroelectrode. Of importance for chemical or electrochemical reactions at nanostructures is the rate of increase in the mass transport increases as the particle size decreases [55]. This allows reactions whose rates outrun diffusion, and are thereby limited, to be studied under conditions where the kinetics of the reaction may [56] control the rate. The increased mass transport also 'sweeps' product away from the reactive/catalytic surface; this feature diminishes the likelihood of unproductive (back or side) reactions at the catalyst surface.

Spiro and Freund [57] have studied the impact of mass transport on catalytic reactions at colloids in order to distinguish between cases of surface control (i.e. where the reaction at the surface of the colloid limits the rate), transport control, or mixed control using literature results obtained for the catalyzed photodecomposition of water at platinum [58] and gold [59] colloids in the presence of methylviologen (MV)

$$2MV + 2H_2O \rightarrow 2MV^{2+} + H_{2(g)} + 2OH^- \tag{12.2}$$

Plots relating the observed catalytic rates for this reaction to colloid size showed a linear increase in rate as the radius of the Pt colloid increased, indicating (according to the equations derived by Spiro and Freund) mixed control of the reaction rate, while the reaction rate at Au was essentially invariant with colloid size indicating that the rate was controlled by the kinetics of the reaction at the surface of the colloid [57].

To minimize the influence on mass transport that might result from evolution of a gaseous product at the colloid surface (the molecular hydrogen of equation (12.2)), Freund and Spiro chose a noble metal catalyzed reaction that did not lead to gaseous products

$$Fe(CN)_6^{3-} + S_2O_3^{2-} \rightarrow Fe(CN)_6^{4-} + \tfrac{1}{2}S_4O_6^{2-} \tag{12.3}$$

and studied the rate of reaction as a function of the size of Au colloids ($\sim$8–30 nm) [60]. This reaction was shown to remain surface controlled over the entire range of colloids sizes, i.e. the process was not mass-transport limited.

The interplay between the enhancement in mass transport at such small structures and the ability for rapid reaction kinetics to control rates can be seen by the degree to which a linear voltage scan can be increased at ultramicroelectrodes to obtain voltammetric features characteristic of a planar diffusion controlled reversible (or quasi-reversible) reaction. At standard voltammetric electrodes (with a radius on the order of millimeters) a voltage scan of the order of 10^{-1} V s^{-1} yields the peaked i–E curve seen in figure 12.3, but by shrinking the characteristic dimension of the electrode to sub-micrometer, voltage scan rates up to 10^6 V s^{-1} may be required to attain a similar i–E shape [54, 55, 61].

The long-postulated influence of the physical structure of the catalytic site and the morphology of small catalyst particles [27, 28, 62] on reactivity may, in light of the above information, be of greater relevance in the endeavor to correlate size effects on catalytic reactivity. As the examples above demonstrate, size effects on the electronic structure of supported metallic nanocrystallites remain controversial and predictively complicated.

12.4 THE EFFECT OF CHEMISTRY ON NANOSTRUCTURES

Consider now the question of whether elegant ensembles of nanocrystallites can sustain their manufactured physical structure, physical location, and chemical composition under serious chemical reaction conditions. The previous discussion on the effect of nanostructures on chemical reactivity would indicate that retention of a nanostructure's size, siting, or chemical composition in the face of reactive chemicals should not be assumed.

Nanoscale particles exhibit unusual behavior relative to bulk: their equilibrium vapor pressures, chemical potentials, and solubilities are greater than that for the same material expressed as large particles [63]. To minimize the high surface energy of such small particles, Ostwald ripening occurs [63], where larger particles grow at the expense of smaller particles. Anything that enhances the prospect of atomic/molecular motion also enhances particle growth and aggregation. Exposure to increased temperatures (sintering) or certain chemicals can increase the size of a nanostructure.

The deliberate design of mild chemical and/or physical interactions between the surface atoms/molecules of a nanocrystallite and its support has been the guiding rationale behind the use of high-surface-area supports in heterogeneous catalysis [26–29]. A recent high-resolution microscopy study [64] has visibly confirmed this rationale by following the rapid (on the order of milliseconds) coalescence in real time of nanoscale Au particles supported on silica, a support with which Au has minimal interaction. The rate of coalescence could be slowed a hundredfold for larger Au particles supported on silica and by many orders of magnitude for nanoscale Au supported on MgO, a support with which Au has stronger interactions [64]. Alternatively, designing a mild chemical interaction between the surface atoms/molecules of a nanostructure and an ion or molecule has led to the use of molecular caps to terminate and stabilize nanostructures (e.g. Nosaka and co-workers' use of thiols to terminate CdS nanocrystallites in solution [65]) and has long been recognized in colloidal science as a means to create electrostatic repulsion between colloids and thereby minimize agglomeration [63, 66].

An *overly* strong chemical interaction, however, between the surface atoms of a nanostructure and a molecule or ion is known as poisoning and is another longstanding and serious concern in heterogeneous catalysis with nanocrystallites. Design of an interaction to decrease particulate or atomic mobility is a successful strategy in many respects, but will begin to fail under rigorous chemical reaction conditions. Because of the importance of carbon supported noble metal catalysts to electrocatalytic reactions in fuel cells, the agglomeration of the catalyst nanocrystallites with use has been much studied [67], as has the sintering process in heterogeneous catalysis for the oxide supported catalysts [28]. Regeneration at high temperatures of supported heterogeneous catalysts can burn off poisons and help revive catalyst dispersion, but a procedure this drastic may have one effect for zeolite supported nanostructured catalysts and quite another if applied to nanostructures created in the absence of a templating microenvironment, such as those formed by x-ray lithography or deposited with the tip of a scanning tunneling microscope (STM).

Further evidence that it is hard to outrun thermodynamics can be seen by two recent electrochemically designed nanostructured surfaces. In one, Penner and co-workers [68] created a nanoscale battery when they used an STM tip to deposit silver and copper nanoscale pillars next to each other (separated by $\sim$40 nm) on highly oriented pyrolytic graphite. Immediately upon formation of the copper metal pillars next to the preformed silver structures, the STM monitored the partial destruction of the copper pillars and an increase in the size of the structures sited where the silver pillars were formed. Galvanic coupling between the nanostructured metals through the conductive graphite support created potential differences between the copper and the silver such that anodic dissolution of the copper occurred to form copper ions which then electroplated as copper metal onto the silver nanostructures [68]. After several monolayers of copper coated the silver structures, dissolution of the copper pillars ceased.

In the second example, Stuve and co-workers [69] documented the creation of a two-dimensional battery as they studied (by *ex situ* XPS) the surface state of electrodeposited adlayers of lead on Pt(111) after emersion (removal of the electrode from the plating electrolyte at a controlled and known potential) followed by re-immersion into pure water or lead- (or lead-ion-) free electrolyte. Although not the intent of their study, this experimental protocol mimics the natural sequence that would be followed after electrochemical fabrication of a nanostructure and its subsequent use in another liquid medium.

The electroformed surfaces emersed under potentials where lead should be oxidized were found upon re-immersion to contain Pb(0) as well as Pb(II), while those surfaces emersed under potentials where lead should be metallic were found to contain, after re-immersion, some degree of oxidized lead; the relevant reaction is:

$$\mathrm{Pb}^{2+}_{ads} + 2\mathrm{e}^- \rightarrow \mathrm{Pb}^{0}_{ads} \tag{12.4}$$

The fractional amounts of adsorbed Pb(0) and Pb(II), as determined by XPS, were found to match the amounts defined by the Nernst equation (as given in equation (12.5) below) for the specific emersion potential. As written for the surface confined electrochemical reaction, this equation relates the emersion potential, E, to the ratio of the number density of adsorbed Pb^0, n^0, and adsorbed

Pb^{2+}, n^{2+}:

$$E = E^0 - (RT/nF)\ln(n^0/n^{2+}) \tag{12.5}$$

where E^0 represents the standard equilibrium potential, R the gas constant, n the number of electrons transferred, and F the Faraday constant. In keeping with this interpretation and knowing that the standard equilibrium potential for the reduction of Pb^{2+} to Pb^0 occurs at 0.65 ± 0.05 V, re-immersion of an electrodeposited adlayer of Pb emersed at 0.68 V produced no current transient upon re-immersion and had, by XPS, equal amounts of oxidized and reduced lead [69].

Finally, a recent effort has been made to counter the mobility of supported nanoscale Pt particles by creating a conducting support which controls the size and dispersion of Pt particles. By coordinating a Pt(0) based organometallic molecule to diacetylene linkages (–C≡C–C≡C–) in poly(phenylenediacetylene), Callstrom and colleagues have been able to create atomically dispersed Pt centers along the carbon polymer backbone [70, 71]. Thermalizing this modified polymer at 600 °C produces a conducting glassy carbon incorporating nanoscale clusters of Pt, where now the Pt is contained *within* a carbon matrix rather than supported *on* the surface of graphite powder, as is usual in the preparation of carbon supported fuel-cell catalysts.

Micrometer-thick films of the catalyst can be prepared by casting the modified polymer as a thin film on a substrate and then thermalizing; the thermalized glassy carbon films exhibit electrochemistry indicative of metallic Pt [71]. These films also exhibit catalytic activity for both of the half-cell reactions characteristic of a fuel cell (i.e. hydrogen oxidation and oxygen reduction) [72]. It is anticipated that dispersing the Pt clusters through the rigid carbon matrix may offer the means to minimize particle coalescence during use at the demanding rates of electrochemical reaction required of fuel-cell catalysts [70–72].

12.5 CONCLUSIONS

Have some of the questions raised in the introduction been, at least partially, answered?

(i) *Can chemistry be 'done' at nanostructures?*
Absolutely. Not only can chemistry be done at nanostructures, it is hard to avoid it. Given the preponderance of oxygen and water on the planet and the high surface-to-volume ratio of nanoscale particles, exposure of a formed nanostructure to even ambient air offers an opportunity for chemical reaction. Ample research described in the chemical, chemical engineering, and materials science literatures contains decades' worth of information on chemical and catalytic reactions studied at nanoscale structures. One hopes these are lessons that can be built upon rather than repeated.

(ii) *What remains of a nanostructure when exposed to molecules with which chemical reactions can occur?*
The fate of the nanostructure depends on its nature (chemical identity, size, and morphology) and the character of the chemical reaction. Inertness should never be assumed.

(iii) *Can the integrity of a nanostructure—its controlled composition and placement—be assured when it acts as a catalyst for chemical reactions or as a reactant in a chemical reaction?*
No.

(iv) *What are the issues when chemical reactions occur at nanostructures or within a nanostructured environment?*
It again depends on the nanostructure and the character of the chemical reaction. Strategies to bolster the stability of nanostructured metals and semiconductors exist, as discussed above, and may be the first steps in the right direction to create designs of ever greater complexity of the physical and chemical environment surrounding the nanostructure. Just as biological systems have complex nanostructured and microstructured environments for specific chemical reactions or classes of reactions, so, too, may environmental complexity be necessary for durable chemical catalysis at nanostructures.

REFERENCES

[1] Whitesides G M, Mathias J P and Seto C T 1991 *Science* **254** 1312
[2] Ozin G A 1992 *Adv. Mater.* **4** 612
[3] 1991 *Science* **254** 1300
[4] Brauman J I 1991 *Science* **254** 1277
[5] Prost J and Rondelez F 1991 *Nature* **350** 11
[6] Henglein A 1988 *Top. Curr. Chem. (Electrochem. 2)* **143** 113
[7] Thomas J M 1988 *Pure Appl. Chem.* **60** 1517
[8] Ostwald W 1920 *Die Welt der vernachläßigten Dimensionen* (Dresden: Steinkopff)
[9] Berry C R 1967 *Phys. Rev.* **161** 848
[10] Meehan E J and Miller J K 1968 *J. Phys. Chem.* **72** 1523
[11] Henglein A 1982 *Ber. Bunsenges. Phys. Chem.* **86** 301
[12] Rossetti R, Ellison J L, Gibson J M and Brus L E 1984 *J. Chem. Phys.* **80** 4464
[13] Weller H, Schmidt H M, Koch U, Fojtik A, Baral S, Henglein A, Kunath W, Weiss K and Dieman E 1986 *Chem. Phys. Lett.* **124** 557
[14] Ramsden J J and Grätzel M 1984 *J. Chem. Soc. Faraday Trans.* I **80** 919
[15] O'Neil M, Marohn J and McLendon G 1990 *J. Phys. Chem.* **94** 4356
[16] Smotkin E S, Lee C, Bard A J, Campion A, Fox M A, Mallouk T E, Webber S E and White J M 1988 *Chem. Phys. Lett.* **152** 265
[17] Colvin V L, Alivasatos A P and Tobin J G 1991 *Phys. Rev. Lett.* **66** 2786
[18] Wang Y and Herron N 1987 *J. Phys. Chem.* **91** 257
[19] Herron N, Wang Y, Eddy M M, Stucky G D, Cox D E, Moller K and Bein T 1989 *J. Am. Chem. Soc.* **11** 530

[20] Cao G, Rabenberg L K, Nunn C M and Mallouk T E 1991 *Chem. Mater.* **3** 149
[21] Steigerwald M L and Brus L E 1990 *Accounts Chem. Res.* **23** 183
[22] Meyer M, Wallberg C, Kurihara K and Fendler J H 1984 *J. Chem. Soc. Chem. Commun.* 90
[23] Fleischmann M, Hendra P J and McQuillan A J 1974 *Chem. Phys. Lett.* **26** 163
[24] Creighton J A 1982 *Surface Enhanced Raman Scattering* ed R K Chang and T E Furtak (New York: Plenum) p 315
[25] Laserna J J, Cabalín L M and Montes R 1992 *Anal. Chem.* **64** 2006
[26] Romanowski W 1987 *Highly Dispersed Metals* (Chichester: Ellis Horwood)
[27] Wynblatt P and Gjostein N A 1975 *Progress in Solid State Chemistry* vol 9, ed J O McCaldin and G Somorjai (Oxford: Pergamon) p 21
[28] Stevenson S A, Dumesic J A, Baker R T K and Ruckenstein E (ed) 1987 *Metal–Support Interactions in Catalysis, Sintering and Redispersion* (New York: Van Nostrand Reinhold)
[29] Sachtler W M H 1993 *Accounts Chem. Res.* **26** 383
[30] Venuto P B and Habib E T Jr 1979 *Fluid Catalytic Cracking with Zeolite Catalysts* (New York: Dekker)
[31] Jacobs P A, Jaeger N I, Jírü P and Schulz-Ekloff G (ed) 1982 *Stud. Surf. Sci. Catal.* **12**
[32] Gunawardena G, Hills G, Montenegro I and Scharifker B 1982 *J. Electroanal. Chem.* **138** 225
[33] Dutta P K 1993 *Proc. 9th Int. Zeolite Conf.* ed R von Ballmoos, J B Higgins and M M J Treacy (Stoneham, MA: Butterworth–Heinemann) p 181
[34] Fleischmann M, Pons S, Rolison D R and Schmidt P P (ed) 1987 *Ultramicroelectrodes* (Morganton, NC: Datatech Systems)
[35] Wightman R M and Wipf D O 1989 *Electroanalytical Chemistry* vol 15, ed A J Bard (New York: Dekker) p 267
[36] Fleischmann M, Pons S, Sousa J and Ghoroghchian J 1994 *J. Electroanal. Chem.* **366** 171
[37] Li L J, Fleischmann M and Peter L M 1989 *Electrochim. Acta* **34** 459
[38] Fleischmann M, Li L J and Peter L M 1989 *Electrochim. Acta* **34** 475
[39] Sundaram M, Chalmers S A, Hopkins P F and Gossard A C 1991 *Science* **254** 1326
[40] Morse M D 1986 *Chem. Rev.* **86** 1049
[41] Kappes M M 1988 *Chem. Rev.* **88** 369
[42] Cox D M, Kaldor A, Fayet P, Eberhardt W, Brickman R, Sherwood R, Fu Z and Sondericher D 1990 *Novel Materials in Heterogeneous Catalysis (ACS Symposium Series 437)* ed R T K Baker and L L Murrell (Washington, DC: American Chemical Society) p 172
[43] Sinfelt J H and Meitzner G D 1993 *Accounts Chem. Res.* **26** 1
[44] Vaarkamp M, Miller J T, Modica F S, Lane G S and Koningsberger D C 1992 *J. Catal.* **138** 675
[45] Hughes T R, Buss W C, Tamm P W and Jacobson R L 1986 *Proc. 7th Int. Zeolite Conf.* ed Y Murakami, A Iijima and J W Ward (Tokyo: Kodansha) p 725
[46] Bagotzky V S and Skundin A M 1985 *Electrochim. Acta* **30** 899
[47] Bagotzky V S and Skundin A M 1984 *Electrochim. Acta* **29** 757
[48] Kinoshita K 1990 *J. Electrochem. Soc.* **137** 845
[49] Peuckert M, Yoneda T, Dalla Betta R A and Boudart M 1986 *J. Electrochem. Soc.* **133** 944

[50] Rolison D R, Nowak R J, Pons S, Ghoroghchian J and Fleischmann M 1988 *Molecular Electronic Devices III* ed F L Carter, R E Siatkowski and H Wohltjen (Amsterdam: Elsevier) p 401
[51] Rolison D R, Hayes E A and Rudzinski W E 1989 *J. Phys. Chem.* **93** 5524
[52] Gallezot P, Alarcon-Diaz A, Dalmon J-A, Renouprez A J and Imelik B 1975 *J. Catal.* **39** 334
[53] Rolison D R 1990 *Chem. Rev.* **90** 867
[54] Heinze J 1991 *Angew. Chem. Int. Edn. Engl.* **30** 170
[55] Heinze J 1993 *Angew. Chem. Int. Edn. Engl.* **32** 1268
[56] Smith C P and White H S 1993 *Anal. Chem.* **65** 3343
[57] Spiro M and Freund P L 1983 *J. Chem. Soc. Faraday Trans.* I **79** 1649
[58] Kiwi J and Grätzel M 1979 *J. Am. Chem. Soc.* **101** 7214
[59] Miller D S, Bard A J, McLendon G and Ferguson J 1981 *J. Am. Chem. Soc.* **103** 5336
[60] Freund P L and Spiro M 1985 *J. Phys. Chem.* **89** 1074
[61] Wightman R M 1991 *Microelectrodes: Theory and Applications* ed M I Montenegro, M A Queirós and J L Daschbach (Amsterdam: Kluwer) p 177
[62] Somorjai G 1991 *Langmuir* **7** 3176
[63] Hunter R J 1991 *Foundations of Colloid Science* vol 1 (Oxford: Oxford University Press)
[64] Iijima S and Ajayan P M 1991 *J. Appl. Phys.* **70** 5138
[65] Nosaka Y, Yamaguchi K, Miyama H and Hayashi H 1988 *Chem. Lett.* 605
[66] Henglein A 1993 *J. Phys. Chem.* **97** 5457
[67] Ehrburger P 1984 *Adv. Coll. Interface Sci.* **21** 275
[68] Li W, Virtanen J A and Penner R M 1992 *J. Phys. Chem.* **96** 6529
[69] Borup R L, Sauer D E and Stuve E M 1993 *Surf. Sci.* **293** 10
[70] Pocard N L, Alsmeyer D C, McCreery R L, Neenan T X and Callstrom M R 1992 *J. Am. Chem. Soc.* **114** 769
[71] Schueller O J A, Pocard N L, Huston M E, Spontak R J, Neenan T X and Callstrom M R 1993 *Chem. Mater.* **5** 11
[72] McCreery R L, Pocard N, Alsmeyer D, Huston M, Huang W and Callstrom M 1993 *Abstract No. 602 Extended Abstracts* vol 93–2 (Pennington, NJ: The Electrochemical Society) p 925
[73] Gates B C, Katzer J R and Schuit G C A 1978 *Chemistry of Catalytic Processes* (New York: McGraw-Hill) p 245

Chapter 13

Mechanical properties

J R Weertman and R S Averback

13.1 LOW-TEMPERATURE PROPERTIES: YIELD STRENGTH

13.1.1 Nanocrystalline metals and alloys

Metallurgists have known for many years that grain refinement often leads to an improvement in the properties of metals and alloys. For example, reducing grain size lowers the ductile-to-brittle transition temperature in steel. Of interest in this chapter is the influence of grain size reduction on the mechanical behavior of materials. The well known empirical Hall–Petch equation [1, 2] relates yield stress σ_y to average grain size d

$$\sigma_y = \sigma_0 + \frac{k}{\sqrt{d}} \tag{13.1}$$

where σ_0 is a friction stress and k is a constant. A similar relationship exists between hardness and grain size. Nanocrystalline metals represent the ultimate in grain refinement. If equation (13.1) were valid down to grain sizes in the nanometer range with the same value of k found at conventional grain sizes, remarkable increases in strength would be realized. Reducing d from 10 μm to 10 nm would increase the strength by a factor of about 30. Although very significant improvement is seen, for all nanocrystalline metals measured to date, the amount has fallen short of this rosy prediction. Reasons for failure to live up to the promise of equation (13.1) will be discussed later. In the part of this chapter that treats nanocrystalline metals, we first examine the deformation mechanisms that are expected to be important in the nanometer grain size range, then survey the experimental data available on mechanical properties, and evaluate the results. Discussion of the outstanding questions and challenges will appear at the end of the chapter.

13.1.1.1 Expected deformation mechanisms for nanocrystalline metals

Derivations of the Hall–Petch equation are based on the interactions of dislocations with grain boundaries. In the well known model of Cottrell [3], yielding occurs when the concentrated stress ahead of a pile-up of dislocations stopped by a grain boundary is sufficient to activate a Frank–Read source in the neighboring grain. The length of the pile-up is identified with the grain size. A simple calculation shows that it is meaningless to speak of a pile-up if the grain size is in the 5–10 nm range. Thus Cottrell's theory cannot be extrapolated to predict the dependence of σ_y on d in nanostructured materials. A model developed by Meyers and Ashworth [4], based on initial localized yielding and hardening in the grain boundary region before general yielding takes place, predicts a decrease in the constant k in equation (13.1) as d decreases to sub-micrometer values. Such a lowering of the Hall–Petch slope at grain sizes below about 1 μm was observed in the experiments of Thompson [5, 6]. In other investigations the turnover, if seen at all, is found at much lower grain sizes (e.g. [7, 8]). Like the Cottrell model, the Meyers–Ashworth theory cannot be extrapolated to nanocrystalline grain sizes. Nonetheless it is reasonable to expect that the additional hindering of dislocation multiplication and movement with continued decrease in grain size eventually will lead to less and less improvement in strength since the dislocations already are largely immobilized. The yield strength will fall below the value predicted by an extrapolation of equation (13.1) with constants determined from data taken on large-grain-size material.

While the suppression of dislocation motion with grain refinement is expected to cause nanocrystalline metals to be very strong, at the finest grain sizes a softening mechanism may take over as deformation by diffusional accommodation becomes significant. Coble [9] has shown that under conditions in which grain boundary diffusion predominates over bulk diffusion, the diffusion creep rate $\dot{\varepsilon}$ is given by

$$\dot{\varepsilon} = \frac{B\sigma\Omega\delta D_b}{d^3kT} \tag{13.2}$$

where B is a constant, σ the applied stress, Ω the atomic volume, δ is the effective grain boundary thickness, D_b is the grain boundary diffusion coefficient, and k and T have their usual meaning. A calculation of the magnitude of $\dot{\varepsilon}$ predicted by equation (13.2) for nanocrystalline metals crept at room temperature shows that, despite the large contribution of the d^{-3} factor, diffusional creep will be significant only if D_b is enhanced over the value measured for materials with conventional grain size. Such an enhancement has been reported [10]. Superplastic behavior [11] also might be expected, especially at elevated temperatures if grain sizes can be kept stable, as discussed in section 13.2.

13.1.1.2 Survey of experimental results

Because the amount of material available often is limited in size, most of the studies of the strength of nanocrystalline metals have been limited to microhardness measurements. Less information is obtained with this technique than is furnished by stress–strain tests, and the light loads typically used in microhardness measurements can lead to erroneous results [12]. Nonetheless, hardness determination is a simple and fast method to survey strength that can be performed on very small pieces of material or even on unconsolidated powders from ball-milling experiments. More extensive information can be obtained with the use of nanoindentation techniques [13].

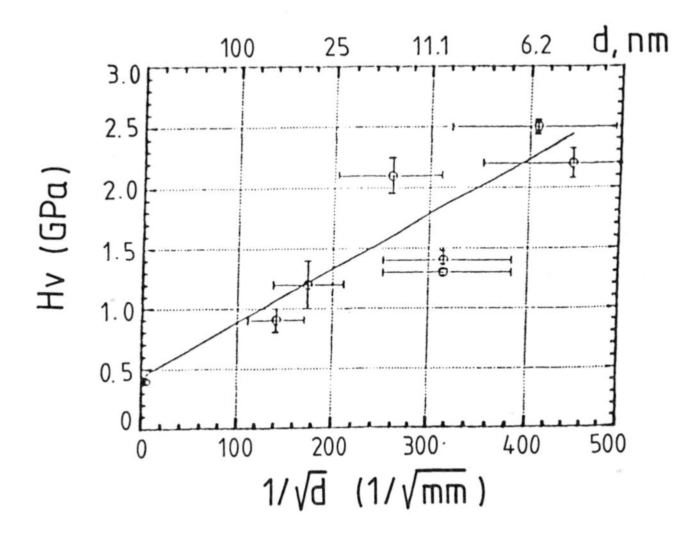

Figure 13.1. Hall–Petch plot of Vickers microhardness H_v for nanocrystalline Cu. (From [14].)

Hardness measurements of nanocrystalline metals typically are carried out as a function of grain size and, following tradition, the results are presented in the form of Hall–Petch plots. The scatter in the data usually is too great to allow the actual form of the dependence of the hardness H_v on grain size to be determined, i.e. whether H_v is proportional to $d^{-1/2}$ or d^{-1}, etc. A number of investigations have shown increased hardening with continuing grain refinement down to the finest grain sizes measured [7, 8, 14, 15]. In most cases the rate of hardening with increasing values of $1/\sqrt{d}$ is considerably less in the nanometer range than occurs at conventional grain sizes, and may be approaching zero [8, 14]. Figures 13.1–13.4 show examples of continual hardening with decreasing grain size. Figure 13.3 indicates a sharp drop in slope whereas figure 13.2 does not. Both curves refer to electroplated Ni and the results actually are quite similar except for the points corresponding to the finest grain sizes in each plot. Many

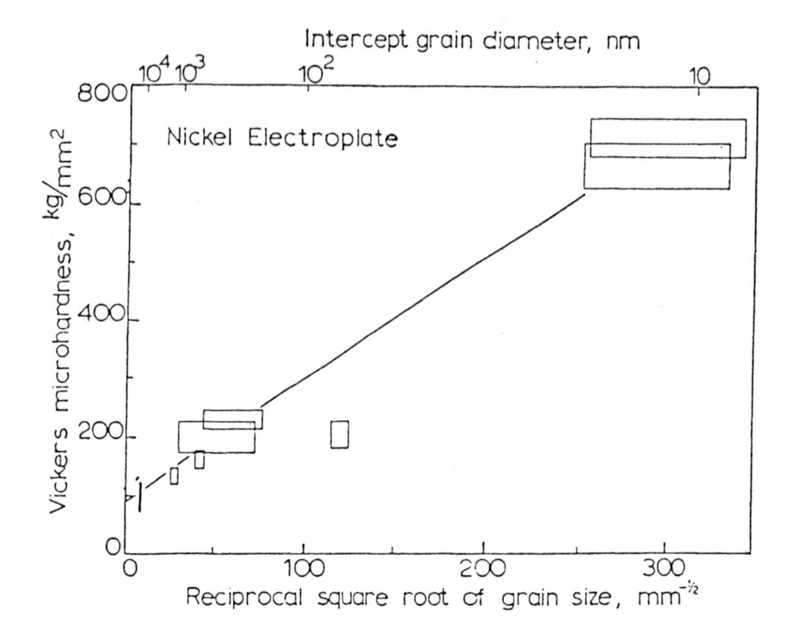

Figure 13.2. Hall–Petch plot for electrodeposited Ni. Rectangles define 95% confidence limits. (From [7].)

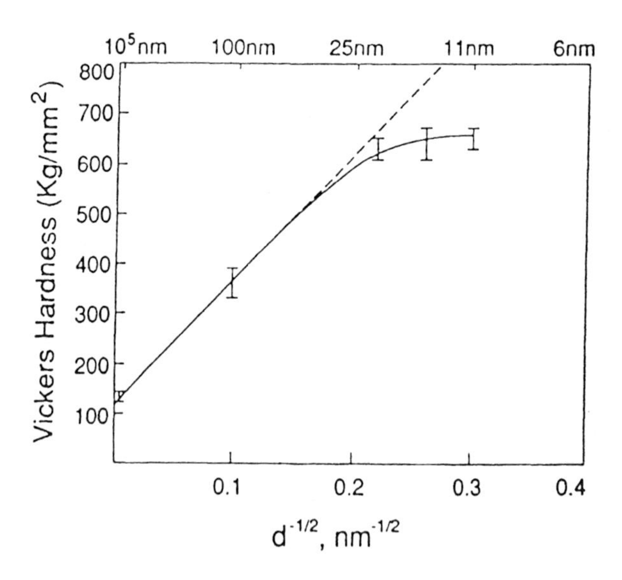

Figure 13.3. Hardness as a function of (grain size)$^{-1/2}$ for as-prepared nanocrystalline Ni electrodeposits. (From [8].)

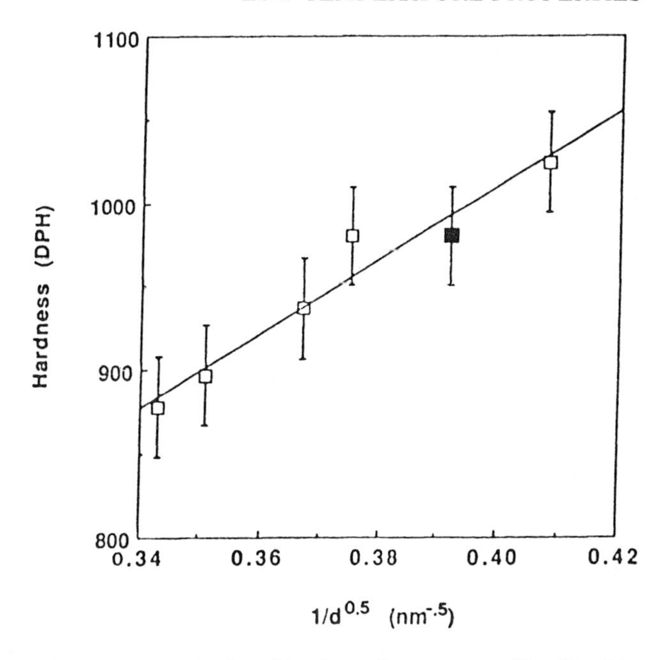

Figure 13.4. Hall–Petch plot of hardness for nanocrystalline Fe. (From [15].)

of the samples used in the Hughes *et al* [7] study were the same as those used by Thompson in his earlier study [5]. The latter investigation extended down in grain size only to 0.12 μm and indicated a slope approaching zero in the Hall–Petch plot. The continuation of the measurements by Hughes *et al* to the nanometer-sized range appears to show that the 0.12 μm sample is not consistent with the general trend in strength versus grain size of the other samples. These examples point out the difficulty in ascertaining mechanical behavior when the grain size range and number of samples is seriously limited.

In contrast to the results of hardness measurements described in the preceding paragraph, a number of studies (e.g. [16–21]) have found a drop in strength with increasing grain refinement at the finest grain sizes, i.e. a negative Hall-Petch slope. Figure 13.5, from the work of Chokshi *et al* [16], clearly shows a softening in the nanometer-grain-size regime for Cu and Pd material made by inert gas condensation and compaction. Similar softening has been seen in intermetallics [19–21] and Ni–P alloys [17, 18].

A relatively small number of tensile tests have been carried out on nanocrystalline metals (e.g. [14, 22–25]). The limited quantities of material available when synthesis is by inert gas condensation and compaction require specialized test equipment [14] for the tiny tensile specimens. A Hall–Petch plot of the yield stress σ_y for Cu [22] appears similar to that found for H_v versus $1/\sqrt{d}$ (figure 13.1). However the slope k (equation (13.1)) is only about $\frac{1}{10}$ that of figure 13.1, rather than the usual factor of about $\frac{1}{3}$ [26].

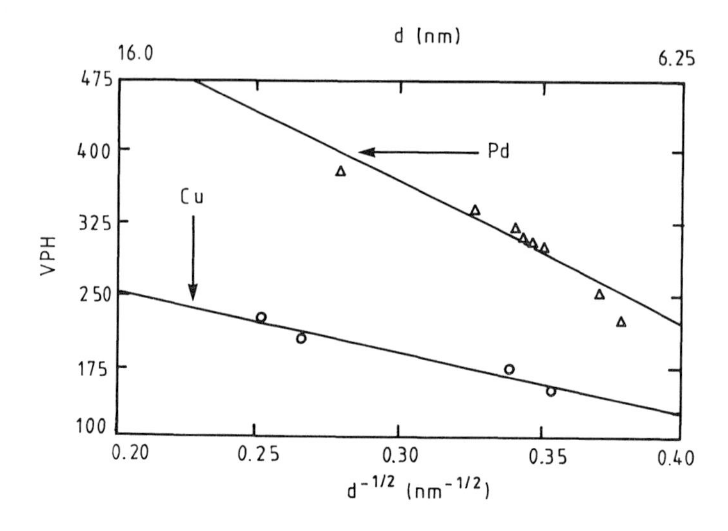

Figure 13.5. Hall–Petch plots for nanocrystalline Pd and Cu. (From [16].)

In general, although the increase in hardness H_v or yield stress σ_y upon going from conventional to nanometer-grain-size material does not live up to the predictions of the Hall–Petch equation, nonetheless a substantial improvement in strength is found. Ratios ranging from about four to six have been measured between the values of H_v for nanocrystalline and coarse-grain Cu [14, 16], while the ratio is about three for σ_y values [22]; H_v increases about fourfold in going from coarse-grain to nanocrystalline Pd [14]; for Ni, the factor is seven to eight [7, 8]. Chang *et al* [19] note a doubling of the strength of nanocrystalline TiAl compared to the cast alloy. Especially promising results were reported by Morris and Morris [24] for a Cu–Zr alloy strengthened by about 7% volume fraction of ≈ 5 nm size oxides and carbides of zirconium that form during the ball-milling operation. Yield stresses in excess of 1300 MPa were measured. The low ductility of 1% can be improved by heat treatment to 3–4%, though with some sacrifice of strength. The authors attribute the high strength of the alloy to Orowan bypassing [27, 28] for grain sizes of the Cu matrix in excess of about 100 nm, with grain boundary strengthening becoming dominant below this value.

13.1.1.3 Discussion

It is likely that the flaws and porosity present in many, if not most, samples of nanocrystalline material seriously affect the results of mechanical tests. Figure 13.6, taken from [14], shows that the extent of the polishing given to tensile test specimens greatly influences their (apparent) tensile strength. Removing both the defects and the residual stresses left on the surface by the high-stress compaction step in the synthesis process drastically increases the fracture stress. Polishing also has been shown to lessen the spatial variability in measurements of microhardness [29]. The inherent strength of nanocrystalline metals probably is considerably greater than indicated by most measurements.

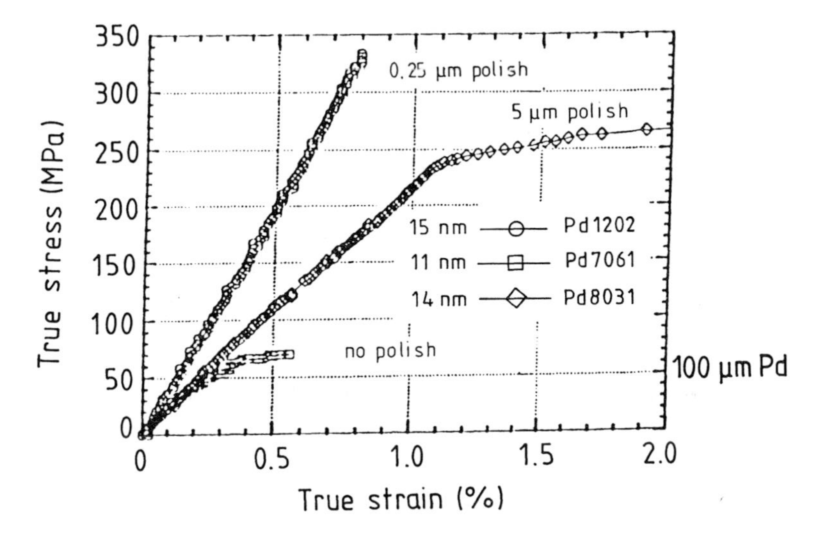

Figure 13.6. Effect of polishing on the stress–strain behavior of nanocrystalline Pd. (From [14].)

As noted in the previous section, there have been numerous observations of a negative Hall–Petch slope at ultrafine grain sizes. The first explanation for this anomalous behavior was given by Chokshi *et al* [16], who invoked the softening effect of Coble creep (equation (13.2)). Using an enhanced value for the grain boundary diffusivity measured by Horvath *et al* [10], the authors calculated a room-temperature strain rate from equation (13.2) of about 6×10^{-3} s^{-1} for 5 nm grain size Cu under a stress of 100 MPa. It was believed that this creep rate is fast enough to account for the observed softening. However, direct measurement by Nieman *et al* [23] of the room-temperature creep rate of nanocrystalline Pd and Cu under stresses greater than 100 MPa showed creep rates of only 10^{-8}–10^{-9} s^{-1}.

Nieh and Wadsworth [30] explained the negative Hall–Petch slopes on the breakdown of the Hall–Petch equation when a grain becomes too small to support a pile-up of even the smallest number of dislocations possible, namely, two. The critical grain sizes for Cu and Pd were calculated to be greater than the grain sizes of the nanocrystalline samples used in the experiments of Chokshi *et al*, whereas for Ni [7] and Fe [15] the critical grain sizes are much smaller and lie below the values of the grain sizes of any of the Ni and Fe samples. Hence softening at the smallest grain sizes would be predicted for the first two metals, in accordance with observation [16], whereas Ni and Fe should continue to harden with decreasing grain size, again in agreement with experimental results [7, 15]. However, the concept of a dislocation pile-up consisting of only two dislocations is far removed from the original theory of Cottrell, and the Nieh–Wadsworth model does not explain the conflicting results of Nieman *et al* [14, 22] in which nanocrystalline Cu and Pd are seen to harden to the smallest grain sizes tested.

Scattergood and Koch [31] have used the decrease in line tension of a dislocation that results from a drop in its long-range stress field cut-off distance to explain observations of softening at the finest grain sizes. (It is assumed that this cut-off distance scales with grain size.) The authors postulate that the process of cutting through dislocation forests is the mechanism that controls yielding at large grain sizes. While the cutting stress is relatively insensitive to grain size, the Orowan stress for looping the obstacles depends on the long-range stress field cut-off distance (and thus the grain size) through the dislocation line tension [28]. Calculations show that unless the material is especially fine grained, the stress to loop the dislocation obstacles in the network exceeds the cutting stress. However, below a critical grain size the Orowan looping becomes the favored deformation mechanism. The authors, by incorporating into their theory the idea of Li [32] that grain boundary ledges provide the source for dislocations, show that below the critical grain size the Hall–Petch slope can become negative. They are able to fit negative-slope hardness data in the literature with their equations using reasonable values for the relevant parameters. The ideas of Scattergood and Koch are especially noteworthy in that they take into account changes in dislocation self-energy produced by small grain sizes that even begin to approach the extent of dislocation cores. Their model is based on specific dislocation mechanisms. It should be noted that few dislocations have been observed to date in nanocrystalline material (e.g. [24]).

Palumbo *et al* [33] have pointed out the importance of triple junctions to the mechanical behavior of nanocrystalline materials. Calculations [34] show that as grain sizes dip below 10 nm the volume fraction associated with triple junctions becomes appreciable and eventually overtakes the value of the volume fraction occupied by grain boundaries. In a material with random grain orientation a substantial fraction of the triple junctions have a disclination character [33]. These triple junctions are expected to contribute to the ductility of the nanocrystalline sample, and because the volume fraction of triple junctions increases with decreasing grain size a softening is expected at the finest grain sizes.

King and Zhu [35] have advanced another argument for the presence of disclinations at the triple junctions of nanocrystalline materials. The misorientation across a very short grain boundary can only partially be accomplished by an array of dislocations. The balance of the misorientation must be made up by a disclination dipole, the two components situated at the ends of the boundary.

A survey of the literature [36] has shown that, in many cases, observation

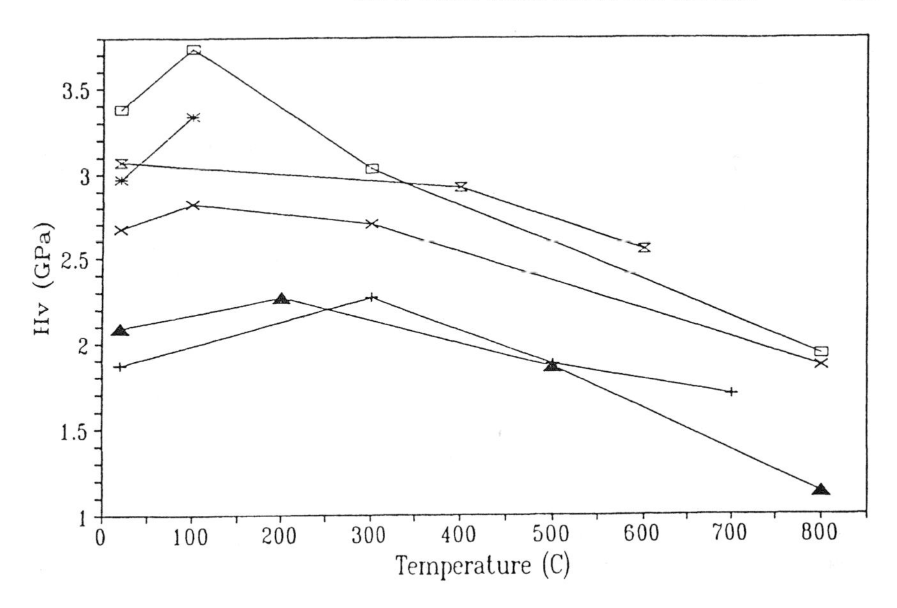

Figure 13.7. Effect of heat treatment on hardness of nanocrystalline Pd samples. Straight lines have been drawn between adjacent points. (From [37].)

of a positive Hall–Petch slope down to the finest grain size tested occurs when all samples in the experiment are in the as-prepared state, whereas a negative Hall–Petch slope at the smallest grain sizes is obtained if a single sample is repeatedly annealed to change the grain size. It appears that a low-temperature anneal strengthens nanocrystalline materials. Figure 13.7 shows the influence of 100 minute anneals on the microhardness of nanocrystalline Pd [37]. In order to ensure uniformity among the samples in a given series, a sample disk was cut into several pieces. One was left in the as-compacted state and the others were subjected to annealing at various temperatures. It can be seen from the figure that, despite modest grain growth, annealing in the range 373–573 K produces a substantial increase in microhardness. Strengthening with heating also has been observed in alloys with a submicrometer grain size produced by intensive straining [38].

Several factors may account for the increase in strength with heating. These include: decrease in free volume at the interfaces; drop in internal strains, believed to be high in nanocrystalline metals [14]; decrease in porosity; and possible healing of microcracks. Thermal experiments by Tschope and Birringer [39] on nanocrystalline Pt have shown a decrease in enthalpy with heating that they associate with atomic rearrangements at the grain boundaries. Such rearrangements lower the energy by bringing the interfacial structure into a metastable configuration. This view is supported by positron lifetime experiments by Schaefer [40] on nanocrystalline Pd. A large population of vacancy-size voids was observed which was identified with excess free volume at the grain boundaries.

Small-angle neutron scattering (SANS) experiments [41,42] were carried out on several series of nanocrystalline samples to investigate their internal structure and its dependence on annealing temperature. The results were interpreted on the basis of the scattering arising primarily from pores. The scatterers in the size range amenable to SANS characterization were found to be about the size of the grains and thus are designated 'missing grain' voids [40]. Comparison of the size and density of the voids in an as-compacted sample with the corresponding values in annealed samples shows that these quantities scale with the grain size as predicted by geometrical considerations. Additional scattering came from the presence of hydrogen in those samples for which precautions had not been taken during processing to drive off adsorbed gases [42], and from features too large to be quantified by SANS. Even after correction for the hydrogen scattering, the total volume fraction of the voids as determined by the SANS measurements agrees well with the results of density measurements. This agreement lends support to the assumption that the scattering arises primarily from porosity. The observed decrease in porosity with annealing may contribute to strengthening.

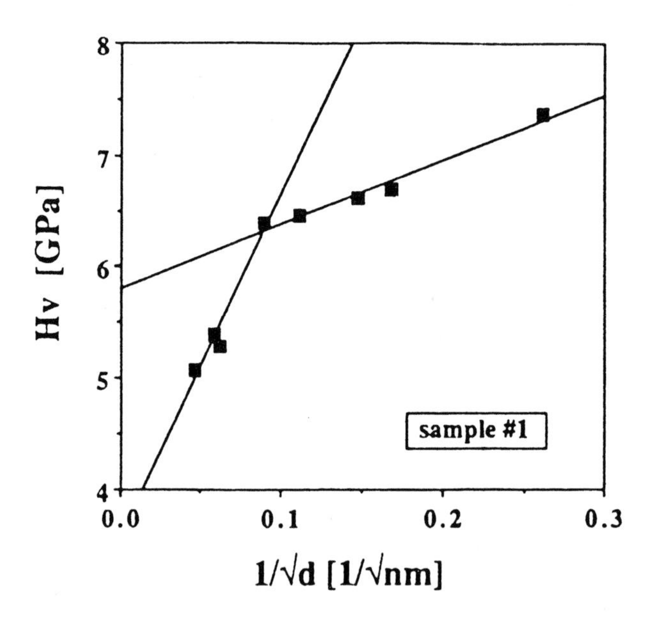

Figure 13.8. Hall–Petch plot for nanocrystalline TiO_2. (From [40].)

13.1.2 Nanocrystalline ceramics

Although work reported on the mechanical properties of nanocrystalline ceramics is scarce, it is mentioned here for completeness. In figure 13.8, a Hall–Petch plot for nanocrystalline TiO_2 is shown [43]. The grain size in this study was increased by thermal annealing treatments. An inverse Hall–Petch behavior is not observed in this case, although the slope in this plot decreases when the grain sizes fall below 200 nm. A clue to this different behavior appears in the work on nanocrystalline TiAl [44]. In that study, the peak in the hardness shifted to smaller 'critical' grain sizes as the indentation temperature was decreased. This might suggest that the reduction in hardness at small grain sizes is mediated by a mechanism involving grain boundary diffusion. Since 300 K is a lower homologous temperature in TiO_2 than in TiAl, it is possible that diffusion is so limited in TiO_2 at room temperature that the critical grain size in n-TiO_2 would be fractions of a nanometer, and therefore non-existent. Consistent with this picture is the result that the hardness of TiO_2 begins to decrease at temperatures above about 400 °C whereas it decreases at much lower temperatures in most metals, including n-TiAl [44, 45]. Karch and Birringer reported similar values of hardness in dense TiO_2 nanocrystals as those shown in figure 13.8 [46]. They also reported exceptionally high values of fracture toughness in this material, ≈ 14 MPa $m^{-1/2}$, as determined by indentation methods. It is noteworthy that this high value is about the same as that in tungsten carbide; a much smaller value, however, was found by Höfler *et al* [43] using similar methods. No models have been proposed to explain the hardness of nanocrystalline ceramics.

13.2 HIGH-TEMPERATURE PROPERTIES: SUPERPLASTICITY

While the Hall–Petch relation has stimulated research on nanocrystalline materials for strengthening of ductile materials, the theory of Coble creep, as expressed in equation (13.2), provides a similarly fascinating possibility that traditionally brittle materials can be ductilized by reducing their grain sizes. It was noted above that the softening behavior of metals at very small grain sizes may be due to diffusional creep, even at room temperature. If this softening can be exploited for processing, nanocrystalline materials could gain enormous technological significance. Brittle materials like ceramics and intermetallics, which are needed for high-temperature applications, could be superplastically deformed at modest temperatures in nanocrystalline form and then heat-treated for grain growth to recapture their high-temperature strength. Examination of equation (13.2) illustrates that by reducing the grain size from microns, which is typical of rapidly solidified metals, to nanometers, creep rates could be enhanced by six to eight orders of magnitude. In the rest of section 13.2, we first examine why nanocrystalline materials are expected to be superplastic, we then relate what experimental evidence there is for superplastic nanocrystalline materials, and finally we discuss the obstacles that are presently impeding a more rapid development of their technological potential.

13.2.1 Background of superplasticity

Superplasticity refers to the capability of some polycrystalline materials to undergo extensive tensile deformation without necking or fracture. Although superplastic forming is now an established industrial process for many metals [47], it is only recently becoming a possibility for the more brittle classes of materials. Many of these brittle materials, however, are important for their excellent high-temperature properties: creep and oxidation resistance, and low densities. A few of these materials have been shown to be superplastic at high temperatures. Polycrystalline tetragonal zirconia is the most celebrated example of a superplastic ceramic [48] and TiAl has been demonstrated to be a superplastic intermetallic [49], but these are clearly the exceptions. The attractive feature of nanocrystalline materials is that they are all believed to be intrinsically superplastic, i.e. nanocrystalline materials share all of the common microstructural features of superplastic materials. Currently, however, no nanocrystalline material has been clearly demonstrated to be truly superplastic. The common characteristices of superplastic materials are [50] as follows.

(i) *Small grain sizes.* Typically, grain sizes are less than approximately 5 μm. The small grain size reflects the central role of grain boundary diffusion in the deformation process. Although many theories of superplasticity have been proposed, they all require grain boundary diffusion.

(ii) *Equiaxed grains.* Since grain boundary sliding is an important mechanism in superplasticity, it is necessary to establish a large shear stress across the grain boundary. For grains shaped as platelets, this will not occur.

(iii) *High-energy grain boundaries.* Grain boundary diffusion and sliding are fastest along high-energy boundaries. In nanocrystalline materials, diffusion studies suggest that the grain boundary diffusion is faster, and that grain boundary energies are higher, than in more conventional polycrystalline materials [51].

(iv) *Presence of a second phase.* At temperatures where grain boundary diffusion and sliding are significant, grain growth is generally rapid. Additions of second-phase particles are necessary to restrict grain growth. Grain growth in nanocrystalline materials has been a serious problem. The ability to synthesize nanocomposites [52] however should alleviate this problem.

Experimental evidence motivating the development of ultrafine-grain materials for superplastic applications is illustrated in figure 13.9 where flow stress and elongation are plotted as a function of strain rate for tetragonal zirconia and metal alloys of varying grain size. Again the trend of easier flow and ductility with small grain sizes is clear, but as yet no data are available for grains sizes below some tenths of microns.

Various models of superplasticity have been proposed over the past 20 years, and several review papers are available [50, 53, 54]. For most diffusional creep

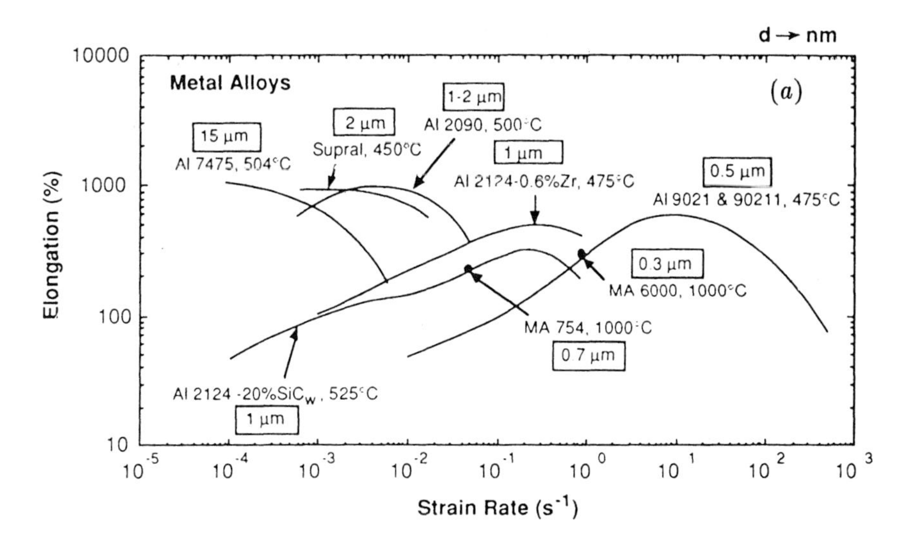

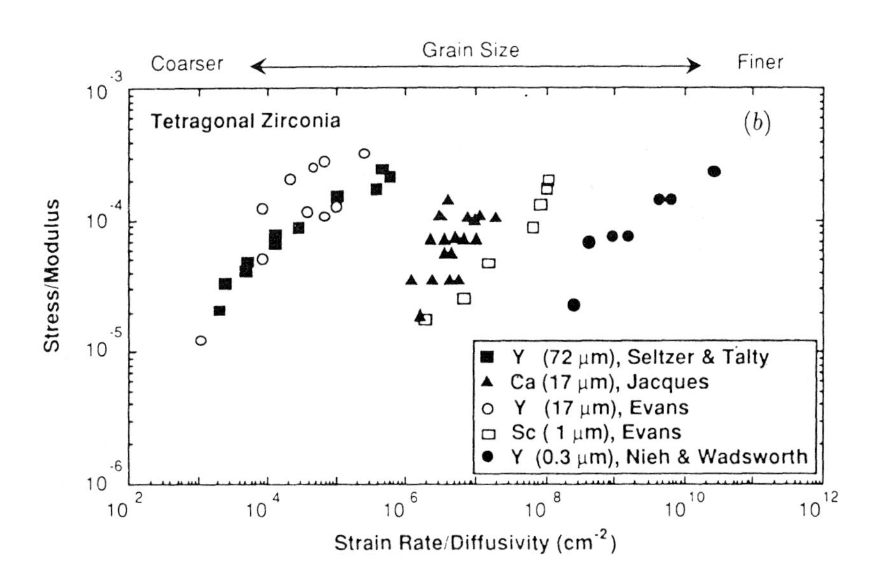

Figure 13.9. (*a*) Modulus compensated flow stress as a function of diffusion compensated strain rate of several ceramics. (From [65].) (*b*) Overview of superplastic behavior in metal alloys illustrating the effect of grain size. (From [65].)

mechanisms, it is convenient to generalize equation (13.2) to

$$\dot{\varepsilon} = \frac{AD_bGb}{kT}\left(\frac{b}{d}\right)^p\left(\frac{\sigma}{G}\right)^n \tag{13.3}$$

where A is a constant, G is the shear modulus, b is the Burgers vector, n is the stress exponent, and p is the grain size exponent. Other symbols have the same meaning as in equation (13.2). A condition for superplasticity is that the strain rate sensitivity, m ($m = \partial \ln \sigma / \partial \ln \varepsilon$), be greater than $\approx \frac{1}{3}$ [50]. For Coble creep, $n = 1$ ($m = 1$) and $p = 3$; however, most data reported on materials with grain sizes $\lesssim 10\ \mu$m show $n \geqslant 2$ ($m \leqslant 0.5$) and $p \approx 2$ [50, 53].

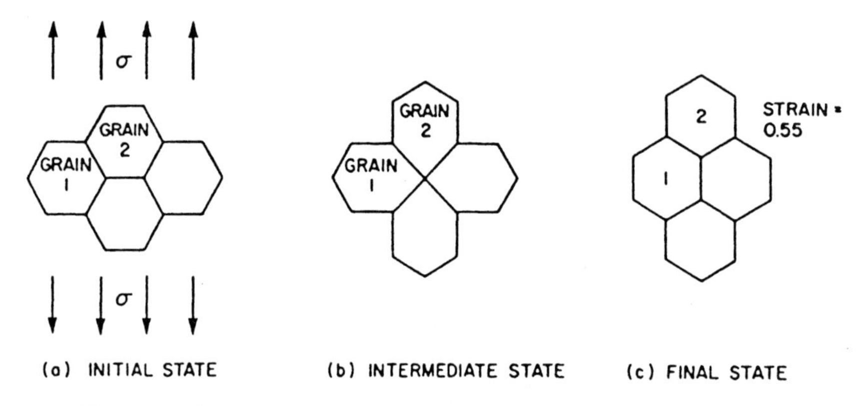

Figure 13.10. Grain switching model of Ashby and Verrall. (From [55].)

Grain boundary sliding is now recognized to be an important mechanism in high-temperature creep of fine-grained materials [53]. A fundamental model of high-temperature deformation apropos of nanocrystalline materials is the 'grain switching' model of Ashby and Verrall [55]. This model is illustrated in figure 13.10. Grains are assumed to slide past one another as deformation proceeds. However, to avoid void formation at the triple junctions, the grains must distort while they slide. This is achieved by grain boundary diffusion. The distortion, however, increases the grain boundary area and, with it, the free energy of the system. Work must be perfomed, therefore, by the applied stress, thereby establishing a threshold stress for deformation. This effect requires equation (13.3) to be rewritten as [55]

$$\dot{\varepsilon} = \frac{A}{kTd^2}\left(\frac{\sigma - \sigma_0}{E}\right) D_v \left(1 + \frac{3.3}{d}\frac{\delta D_b}{D_v}\right). \tag{13.4}$$

The magnitude of the threshold stress in this model is

$$\sigma_0 = \frac{0.72\Gamma}{d} \tag{13.5}$$

where Γ is the grain boundary energy, ≈ 1 J m^{-2}, and D_v is the lattice diffusion coefficient. For materials with grain sizes of ≈ 1 μm, the threshold stress is ≈ 1 MPa and for the most part, negligible. For nanocrystals, however, the grain sizes are $\approx$ 5–10 nm, and the threshold stress increases to 100–200 MPa. Ashby and Verrall further point out that when the grain size is very small, grain boundary diffusion becomes limited by the sources and sinks for point defects, i.e. grain boundary dislocations. They assume that the number of grain boundary dislocations is proportional to the applied stress so that equation (13.4) takes the form

$$\dot{\varepsilon} = A\left\{\frac{\sigma}{d}\right\}\left(\sigma - \frac{0.72\Gamma}{d}\right)D_b. \tag{13.6}$$

Although the Ashby–Verrall model has not been useful in predicting diffusional creep behavior in larger-grained materials, it may nevertheless be important for nanograined materials. The model shows moreover that the dominant mechanisms of creep might be very different than those in larger-grained materials.

13.2.2 Survey of experimental results

As previously noted in this article, a clear demonstration of superplastic behavior in nanocrystalline materials is still lacking. Nevertheless, a considerable literature is being established on superplasticity in 'near'-nanocrystalline materials, on nanocrystalline composites where one phase has nano-dimensions, and on encouraging signs of superplasticity in fully nanocrystalline materials. This literature is now discussed.

13.2.2.1 Ceramics

Superplasticity in ceramic materials became of technical interest around 1986 when Wakei and co-workers demonstrated that yttria stabilized tetragonal zirconia polycrystals (Y–TZP) could be elongated over 100% in tension [48]. The grain size in these samples and those used later in similar tests by Nieh *et al* [56] were $\approx$ 300 nm, i.e. nearly nanocrystalline. Composite specimens of Y–TZP containing 20 wt% Al_2O_3 were also shown to be superplastic by Wakai and Kato [57] and Nieh *et al* [58]. The grain sizes are stabilized in these materials by the presence of second phases (the Y–TZP is 90% tetragonal and 10% cubic). The constitutive behaviors of these materials have been examined. Wakai reported that in Y–TZP, the grain size exponent, $p = 1.8$, and the stress exponent $n = 2$ (see equation (13.3)); no indication of a threshold stress was reported (note that the term $0.72G/d$ in the Ashby–Verrall model is only $\approx$ 2 MPa for their samples). Nieh *et al* [58] reported $n = 3$ for the same material. For the composite material, Wakai *et al* [57] obtained $n = 2.1$ which is in good agreement with Nieh *et al* [58] who found $n = 2$, and $p = 1.5$. These pioneering studies established the capability to synthesize superplastic ceramic materials. Note that with a grain size exponent, $p = 1.7$, reducing the grain size from 0.3 μm (a grain size now commercially available) to 10 nm results in an enhanced deformation rate of two orders of magnitude, down from the six noted above but, nevertheless, impressive.

The first study of creep in nanocrystalline ceramics with $d < 20$ nm was reported by Karch *et al* who illustrated that CaF_2 and TiO_2 could be deformed in bending at 80 °C and 180 °C respectively [59]. For TiO_2 they further showed an apparent brittle-to-ductile transition at room temperature in a microhardness indentation test by varying the rate at which the load was applied. It is quite likely, however, that the materials in these studies were not fully dense. Similarly, Mayo *et al* [60] reported values of the strain sensitivity, m, in nanocrystalline ZnO and TiO_2 at room temperature. They noted that the value decreased from 0.04 to 0.015 during sintering, and suggested the decrease was primarily due to grain growth. They noted that the larger value of the strain-rate sensitivity in the smaller-grained samples might indicate some improvement toward superplasticity in nanograined ceramics, but the highest value reported, 0.040, is still far below the value of $m = \frac{1}{3}$ needed for superplasticity. Hot hardness tests on fully dense nanocrystalline TiO_2 showed, in contrast, no evidence of softening until 400 °C [45].

Figure 13.11. Deformation of nanocrystalline TiO_2 at $\approx$800 °C. (From [58].)

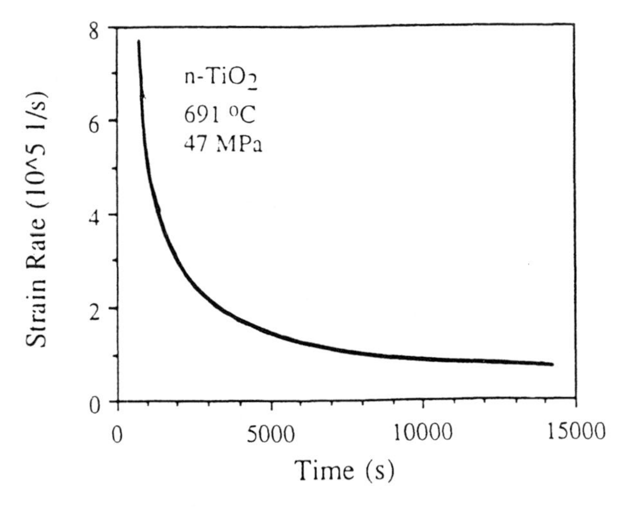

Figure 13.12. Strain rate plotted as a function of time during the deformation of nanocrystalline TiO_2 at 691 °C, which is less than half the melting temperature of TiO_2. (From [58].)

Direct creep measurements on nanocrystalline TiO_2 were performed in compression by Hahn and Averback [61]. Figure 13.11 shows the deformation of a fully dense cylinder to a strain of 0.6 under a load of 38 MPa for 15 h at 800 °C. The initial and final grain sizes were 40 nm and 1 μm respectively. Figure 13.12 illustrates the strain versus time in such a sample. The rapid reduction in the strain rate is a consequence of concomitant grain growth in the sample. Analyses of these data yield $n \approx 3.5$ and $p = 1.5$. The rather large value of the stress exponent is not consistent with most models of superplasticity, but it does seem to fit with the empirical result that the stress exponent increases with decreasing grain size [53], moreover $n = 3.5$ is close to that ($n = 3$) in Y–TZP reported by Nieh *et al* [56].

13.2.2.2 Metals

Even less work for superplasticity in nanocrystalline metals has been performed than in ceramics. The slow progress in developing superplastic nanocrystalline metals has been largely due to the difficulty in controlling grain growth during both the fabrication of bulk materials and the deformation testing. Ball milling has been a principal means of producing nanograined materials. Recently, Ameyama *et al* [49] produced a series of TiAl alloys, TiAl(γ), $Ti_3Al(\alpha_2)$, and an $(\alpha_2+\gamma)$ two-phase structure. Although these alloys were demonstrated to be superplastic, the heat treatment used for consolidating the powders caused the grain size to increase to 0.3–1.5 μm prior to the creep measurements. In related work, Higashi [62] prepared a series of nanocrystalline Al alloys by mechanical alloying. Of direct interest here are the alloys prepared from amorphous Al–14 wt% NiMm–1 wt% Zr powders, where Mm is so-called 'mischmetal'. After hot extrusion, the mean grain size of the matrix was 50–80 nm, while Al_3Mm and Al_3Ni particles of grain sizes 10–50 nm and metastable Al_3Zr precipitates of grain size ≈ 15 nm were present at a volume fraction of 30%. The fine extruded structures, however, were unstable above ≈ 773 K, and after superplastic forming at 873 K, the grain sizes had increased to ≈ 1.5 μm. Table 13.1 lists the properties of some of these advanced Al alloys.

Table 13.1. Deformation properties of advanced fine-grained Al alloys [49].

Material	Temperature (K)	Strain rate (s^{-1})	Stress (MPa)	m	Elongation (%)	Grain size (nm)
IN9052	863	10	15	0.6	330	500
IN905XL	848	20	12	0.6	190	400
IN9021	823	50	18	0.5	1250	500
SiCp/IN9021	823	5	5	0.5	600	500
Al–Ni–Mm	885	1	15	0.5	650	1000
Al–Ni–Mm–Zr	873	1	15	0.5	650	800

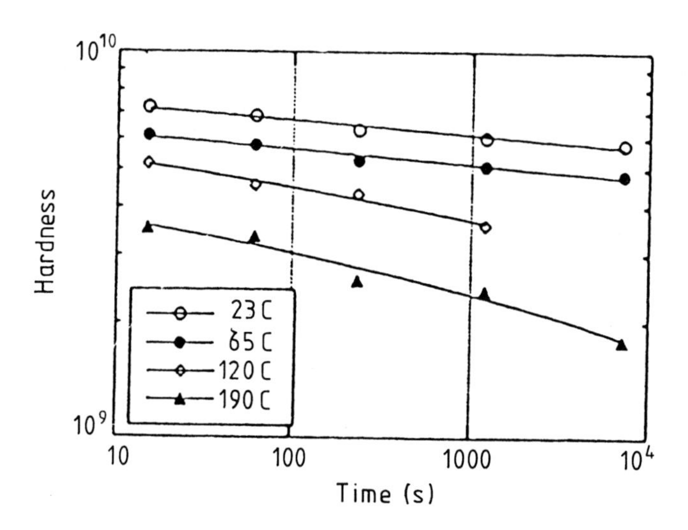

Figure 13.13. Vickers hardness as a function of time and temperature in nanocrystalline Fe. (From [60].)

The only reported studies on high-temperature deformation properties of truly nanocrystalline metals have employed hot-hardness measurements. An example is illustrated in figure 13.13, where the time-dependent hardness of nanocrystalline Fe, with a grain size of ≈ 10 nm, is plotted for several temperatures [63]. The data show that the hardness decreases approximately logarithmically with time and exponentially with increasing temperature, These data are consistent with the idea that deformation involves diffusional creep. Data for nanocrystalline Cu [63] and TiAl [44] showed similar behavior. Kim *et al* [63] have subsequently interpreted these data according to a modified model of Coble creep in which the grain boundary diffusion coefficient is an exponentially decreasing function of strain. The basis for this model is that nanocrystalline materials have high grain boundary energies in their 'as-prepared' state. However, as grain boundaries slide and migrate during deformation, they continuously lower their average energy. This picture is entirely consistent with the ideas of grain boundary relaxation deduced from calorimetry [36] and diffusion [51] measurements, which were discussed above.

13.2.3 Sinter forging

Grain growth during consolidation of nanocrystalline materials into fully dense components has been one of the primary impediments to their technological development. A full discussion of this problem is considered in chapters 5 and 8. One method of consolidation, which has significance for near-net shaping, is sinter forging, i.e. sintering under an applied uniaxial stress. Since this is a high-temperature creep process, it is briefly discussed here.

Three mechanisms of densification are associated with sinter forging: (i) improved packing; (ii) hot isostatic pressing, i.e. 'HIPing'; (iii) shear deformation. The first mechanism concerns initial packing of the powder. Usually the green compact contains large flaws or pores due to particle agglomeration, and these pores are difficult to sinter. Such flaws are the cause of mechanical weakness if not removed from the final product. During sinter forging at low densities, the grains readily slide past one another and fill large empty spaces. Even in the absence of sliding, three-dimensional voids will collapse into two-dimensional platelets by viscous flow of matter under the shear stress of sinter forging [64]. During HIPing, in contrast, shear component of stress is absent, and large pores can remain stable. The second mechanism derives from the isostatic component of stress that develops on application of a uniaxial stress, σ_a, i.e. $P = \frac{1}{3}\,\mathrm{Tr}\,\sigma = \sigma_a/3$. Thus the 'sintering pressure' σ is increased to [64]

$$\sigma = \sigma_s - \sigma_a/3 \tag{13.7}$$

where σ_s is the internal sintering pressure in the absence of an applied stress. Since $\sigma_s \approx 2\gamma/r$, where γ is the surface energy (≈ 1 J m^{-2}) and r is the pore size, it is clear that for nanocrystalline materials applied stresses must be greater than ≈ 25 MPa for this mechanism to be of much significance. The third mechanism, which has the most significance for the present context, is the shear deformation under uniaxial load, since it provides information about creep mechanisms. For an isotropic medium, a macroscopic constitutive creep law can be expressed in the form [65]

$$\frac{\dot{\varepsilon}_{zz}}{\dot{\varepsilon}_0} = \frac{9\pi}{16\sqrt{3}}\left(\frac{\rho-\rho_0}{1-\rho}\right)^{1/2}\left(\frac{(1-\rho_0)}{3\rho^2(\rho-\rho_0)}\right)^n (1+2^{(n+1)/n})^n \left(\frac{\sigma}{3\sigma_0}\right)^n \tag{13.8}$$

and

$$\frac{\dot{\varepsilon}_{xx}}{\dot{\varepsilon}_0} = \frac{\dot{\varepsilon}_{yy}}{\dot{\varepsilon}_0} = \frac{9\pi}{16\sqrt{3}}\left(\frac{\rho-\rho_0}{1-\rho}\right)^{1/2}\left(\frac{(1-\rho_0)}{3\rho^2(\rho-\rho_0)}\right)^n \times (1+2^{(n+1)/n})^{n-1}(1-2^{1/n})^n \left(\frac{\sigma}{3\sigma_0}\right)^n. \tag{13.9}$$

In this model, the ratio of radial to axial strain rates, therefore, is

$$\lambda = \frac{(1-2^{1/n})}{(1+2^{(n+1)/n})}. \tag{13.10}$$

Here μ is the bulk shear modulus, K is the bulk viscosity, ρ is the density and ρ_0 is the density of a randomly, close-packed powder ($\rho_0 = 0.64$). The constitutive behavior of the particles is contained in the term $\dot{\varepsilon}_0$ and the stress exponent, n.

Several experiments have now been performed on sinter forging on nanocrystalline ceramics. Panda *et al* [66] first showed that, unlike sinter forging in larger-grained ceramics, the stress exponent, n, in nanograined Y–TZP at 1200 °C has a value of three. Note, this is the same value as that found by Nieh *et al* [58] during superplastic deformation of dense Y–TZP (see above). More complete investigations were performed on nanocrystalline TiO_2 for which a constitutive law of densification was obtained of the form

$$\dot{\varepsilon} = A\frac{\sigma^n}{d^q}\exp\left(\frac{-Q}{RT}\right)\exp(\beta\rho) \tag{13.11}$$

with $n = 2.3$, $q = 1.7$, $\beta = 30$. The activation enthalpy for creep deformation increased from 1.6 eV at densities below ≈80%, to 2.2 eV for densities over 80% [67]. Except for the lower activation enthalpy at low densities, the different parameters in the constitutive law are quite similar to those obtained in dense nanocrystalline TiO_2, again suggesting that similar mechanisms of deformation may be operating in porous and dense materials. The increase in activation enthalpy with increasing density, or strain, fits well with the idea that grain boundary energies increase as grain boundaries slide and grains grow, as suggested by Kim *et al* [63]. A threshold stress of ≈50 MPa was reported in these experiments at densities $\gtrsim$ 89% [68]; however, its origin has not yet been elucidated. These sinter-forging experiments also revealed that the radial strain rate was nearly zero until the relative density exceeded ≈93%; this is in agreement with equation (13.10) and an important consideration for near-net shaping of components. It is noteworthy that complete densification and shaping of nanocrystalline TiO_2 could be achieved at 700 °C, whereas conventional TiO_2 powder is typically sintered to full density at ≈1200 °C.

13.3 FUTURE DIRECTIONS

A number of important questions regarding the mechanical behavior of nanocrystalline metals remain unanswered; fundamentally the underlying mechanisms of deformation at both low and high temperatures are unknown. The major research challenge in this area currently lies in the areas of synthesis and processing. Until nanocrystalline materials can be produced which are dense, free of flaws, of high purity, and of sufficient quantity to carry out a variety of mechanical tests, measured values will be influenced by extrinsic factors rather than intrinsic properties. For high-temperature applications, a great amount of work needs to be performed in controlling grain sizes so that the correct balance of ductility, formability, and strength are obtained. Despite these challenging problems, nanocrystalline materials offer a wide variety of opportunities for fabricating new materials with precisely engineered microstructures where strength and ductility can be finely tuned. Moreover, the ability to structure materials on the nanoscale level makes it possible to test fundamental ideas about grain boundaries, dislocations, and deformation.

REFERENCES

[1] Hall E O 1951 *Proc. Phys. Soc. London* B **64** 747
[2] Petch N J 1953 *J. Iron Steel Inst.* **17** 25
[3] Cottrell A H 1958 *Trans. TMS-AIME* **212** 192
[4] Meyers A and Ashworth E 1982 *Phil. Mag.* A **46** 737
[5] Thompson A W 1975 *Acta Metall.* **23** 1337
[6] Thompson A W 1977 *Acta Metall.* **25** 83
[7] Hughes G D, Smith S D, Pande C S, Johnson H R and Armstrong R W 1986 *Scr. Metall.* **20** 93
[8] El-Sherik A M, Erb U, Palumbo G and Aust K T 1992 *Scr. Metall. Mater.* **27** 1185
[9] Coble R L 1963 *J. Appl. Phys.* **34** 1679
[10] Horvath J, Birringer R and Gleiter H 1987 *Solid State Commun.* **62** 319
[11] Meyers M A and Chawla K K 1984 *Mechanical Metallurgy* (Englewood Cliffs, NJ: Prentice-Hall) p 673
[12] Li H, Ghosh A, Han Y H and Bradt R C 1993 *J. Mater. Res.* **8** 1028
[13] Oliver W C, Lucas B N and Pharr G M 1993 *Mechanical Properties and Deformation Behavior of Materials Having Ultra-Fine Microstructures* ed M Nastasi, D M Parkin and H Gleiter (Dordrecht: Kluwer) p 417
[14] Nieman G W, Weertman J R and Siegel R W 1991 *J. Mater. Res.* **6** 1012
[15] Jang J S C and Koch C C 1990 *Scr. Metall. Mater.* **24** 1599
[16] Chokshi A H, Rosen A, Karch J and Gleiter H 1989 *Scr. Metall.* **23** 1679
[17] Lu K, Wei W D and Wang J T 1990 *Scr. Metall. Mater.* **24** 2319
[18] MacMahon G and Erb U 1989 *Microstruct. Sci.* **17** 447
[19] Chang H, Hofler H J, Alstetter C J and Averback R S 1991 *Scr. Metall. Mater.* **25** 1161
[20] Christman T and Jain M 1991 *Scr. Metall. Mater.* **25** 767
[21] Bohn R, Haubold T, Birringer R and Gleiter H 1991 *Scr. Metall. Mater.* **25** 811

[22] Nieman G W, Weertman J R and Siegel R W 1991 *Mater. Res. Soc. Symp. Proc.* **206** 581
[23] Nieman G W, Weertman J R and Siegel R W 1990 *Scr. Metall. Mater.* **24** 145
[24] Morris D G and Morris M A 1991 *Acta Metall. Mater.* **39** 1763
[25] Gunther B, Baalmann A and Weiss H 1990 *Mater. Res. Soc. Symp. Proc.* **195** 611
[26] Meyers M A and Chawla K K 1984 *Mechanical Metallurgy* (Englewood Cliffs, NJ: Prentice-Hall) p 494
[27] Dieter G E 1986 *Mechanical Metallurgy* 3rd edn (New York: McGraw-Hill) p 218
[28] Meyers M A and Chawla K K 1984 *Mechanical Metallurgy* (Englewood Cliffs, NJ: Prentice-Hall) p 409
[29] Nieman G W, Weertman J R and Siegel R W 1989 *Scr. Metall.* **23** 2013
[30] Nieh T G and Wadsworth J 1991 *Scr. Metall. Mater.* **25** 955
[31] Scattergood R O and Koch C C 1992 *Scr. Metall. Mater.* **27** 1195
[32] Li J C M 1963 *Trans. TMS-AIME* **227** 239
[33] Palumbo G, Erb U and Aust K T 1990 *Scr. Metall. Mater.* **2** 2347
[34] Palumbo G, Thorpe S J and Aust K T 1990 *Scr. Metall. Mater.* **24** 1347
[35] King A H and Zhu Y 1993 *Phil. Mag.* A **67** 1037
[36] Fougere G E, Weertman J R and Siegel R W 1993 *Nanostruct. Mater.* **3** 379
[37] Weertman J R and Sanders P G 1994 *Solid State Phenom.* **35** 249
[38] Valiev R Z 1993 *Mechanical Properties and Deformation Behavior of Materials Having Ultra-Fine Microstructures* ed M Nastasi, D M Parkin and H Gleiter (Dordrecht: Kluwer) p 303
[39] Tschope A and Birringer R 1993 *Acta Metall. Mater.* **41** 2791
[40] Schaefer H-E 1993 *Mechanical Properties and Deformation Behavior of Materials Having Ultra-Fine Microstructures* ed M Nastasi, D M Parkin and H Gleiter (Dordrecht: Kluwer) p 81
[41] Sanders P G, Weertman J R, Barker J G and Siegel R W 1993 *Scr. Metall. Mater.* **29** 91
[42] Sanders P G, Wertman J R, Barker J G and Siegal R W 1994 *Mat. Res. Soc. Proc.* **351** 319
[43] Höfler H J and Averback R S 1990 *Scr. Metall. Mater.* **24** 2401
[44] Chang H, Alstetter C J and Averback R S 1992 *J. Mater. Res.* **7** 2962
[45] Guermazi M, Höfler H J, Hahn R and Averback R S 1991 *J. Am. Ceram. Soc.* **74** 2672
[46] Karch J and Birringer R 1990 *Ceram. Int.* **16** 291
[47] See, for example, Hamilton C H and Paton N E (ed) 1988 *Superplasticity and Superplastic Forming* (Warrendale, PA: Metallurgical Society)
[48] Wakei F, Sakaguchi S and Matsuno Y 1986 *Adv. Ceram. Mater.* **1** 259
[49] Ameyama K, Miyazaki A and Tokizane M 1991 *Superplasticity in Advanced Materials* ed S Hori, M Tokizane and N Furushiro (Japanese Society of Research on Superplasticity) p 317
[50] See, for example, Sherby O D and Wadsworth J 1989 *Prog. Mater. Sci.* **33** 169
[51] Höfler H J, Averback R S, Hahn H and Gleiter H 1993 *J. Appl. Phys.* **74** 3832
[52] Herr U, Jing J, Gonser U and Gleiter H 1990 *Solid State Commun.* **76** 197
[53] Langdon T G 1993 *Mater. Sci. Eng.* **166** 67
[54] See, for example, Kaibyshev O A 1992 *Superplasticity of Alloys, Intermetallides and Ceramics* (Berlin: Springer)
[55] Ashby M F and Verrall R A 1973 *Acta Metall.* **21** 149

[56] Nieh T G and Wadsworth J 1990 *Acta Metall. Mater.* **38** 1121
[57] Wakai F and Kato H 1988 *Adv. Ceram. Mater.* **3** 71
[58] Nieh T G, McNally C M and Wadsworth J 1989 *Scr. Metall.* **23** 457
[59] Karch J, Birringer R and Gleiter H 1987 *Nature* **330** 556
[60] Mayo M J, Siegel R W, Liao Y X and Nix W D 1992 *J. Mater. Res.* **7** 973
[61] Hahn H and Averback R S 1991 *J. Am. Ceram. Soc.* **74** 2918
[62] Higashi K 1992 *Mater. Sci. Eng.* A **166** 109
[63] Kim L, Höfler H J, Daykin A, Averback R S and Altstetter C 1995 *Interfaces II (Mat. Sci. Forum 189–190)* ed B C Muddle p 367
Kim L, Höfler H J, Averback R S and Altstetter C unpublished
[64] Venkatachari K R and Raj R 1987 *J. Am. Ceram. Soc.* **70** 514
[65] Kuhn L T and McMeeking R M 1992 *Int. J. Mech. Sci.* **34** 563
[66] Panda P C, Wang J and Raj R 1988 *J. Am. Ceram. Soc.* **71** C-507
[67] Höfler H J and Averback R S 1993 *Mater. Res. Soc. Symp. Proc.* **286** 9
[68] Uchic M, Höfler H J, Flick W J, Tao R, Kurath P and Averback R S 1991 *Scr. Metall. Mater.* **26** 791

Chapter 14

Magnetic and electron transport properties of granular films

K M Unruh and C L Chien

14.1 INTRODUCTION

Granular metals are phase separated nanostructured composites that consist of a metallic component in combination with an immiscible insulator, semiconductor, or metal. Although these unique materials have been studied since the early 1960s, it was not until the late 1960s and early 1970s that the first systematic studies were carried out [1]. The majority of this work was devoted to the study of structural, transport, and magnetic properties. More recently, granular metals have again become the focus of considerable experimental and theoretical interest. As a result of these efforts, it is now known that granular metals exhibit many interesting, and in some cases unexpected, physical properties. While considerable progress has been made in the understanding and application of these properties, many opportunities for further study and development remain.

The physical properties of granular metals are determined by the chemical composition and microstructure of each constituent phase, and by the consequences of the compositional heterogeneity of the microstructure as a whole. The compositional heterogeneity of granular metals often results in unusual combinations of properties normally not associated with homogeneous materials. For example, granular Ag–Al_2O_3 and Ni–Al_2O_3 films exhibit the high optical reflectivity and magnetic properties of Ag and Ni respectively, in combination with the mechanical hardness of Al_2O_3 [2]. In addition, the relative concentrations of the constituent materials are not limited by structural or chemical constraints, and the properties of these materials can therefore be continuously tuned by simply changing the relative concentrations of the constituents. From an alternative standpoint, the nanometer length scale which characterizes the microstructure of granular solids itself gives rise to unusual size-dependent physical properties which differ from those of the bulk components. These properties arise from the large fraction of surface or interface atoms whose atomic environments differ from those of the bulk material, and from the fact that the length scales of many fundamental physical processes are comparable to that of the microstructure.

The primary purpose of this chapter is to present a brief overview of the salient magnetic and transport properties of sputter deposited metal–insulator and metal–metal based granular films. Our discussion will begin by describing the microstructure of these materials in terms of the respective volume fractions of each phase, and the size of the particles or grains of the metallic phase(s). Several characteristic magnetic properties of granular metal–insulator films consisting of isolated single-domain ferromagnetic particles will then be introduced. More detailed treatments of the magnetic properties of these materials have recently appeared [3]. The next topic of discussion will be the transport behavior of metal–insulator granular films. This section will emphasize the evolution in the electrical resistivity as a function of composition and temperature. A recent review of this subject has provided an in-depth treatment of many theoretical issues related to electronic transport in granular films [4]. Finally, the newly discovered 'giant magnetoresistance' behavior observed in a number of granular ferromagnetic metal–nonmagnetic metal based films will be discussed. More detailed discussions of this subject have also appeared [5]. In each instance, we will first present an overview of the current experimental situation, and then discuss the underlying physical mechanisms that give rise to the observed behavior. While space restrictions have made it necessary to omit several important topics (granular superconductivity, for example), the references provided at the end of this chapter should provide an adequate starting point for a detailed study of the existing literature.

14.2 MICROSTRUCTURE

The purpose of this section is to introduce two microstructural variables that are very useful in parameterizing the magnetic and transport properties of vapor deposited granular metal films. The first variable is the volume fraction of each component (in the remainder of this work all compositions will be given in terms of volume fractions). The volume fraction is a well defined quantity which only depends on the molecular weight, density, and relative composition of each component, and in particular is independent of the microstructure. The second microstructural variable is a measure of the length scale(s) which characterizes the structure and extent of each phase. Although this length scale is less well defined than the volume fraction, it is generally taken to be the mean particle size r of the metallic particles or grains in metal–insulator and metal–semiconductor based granular films, and the mean particle size of the minority component when both components are metallic. Unlike the volume fraction, the mean particle size generally depends on the type of granular system (e.g. metal–insulator, metal–semiconductor, or metal–metal) and on the details of the preparation process (e.g. the substrate temperature), as well as on the relative concentrations of each component (e.g. metal rich or metal poor). Each of these cases will be discussed in the following paragraphs.

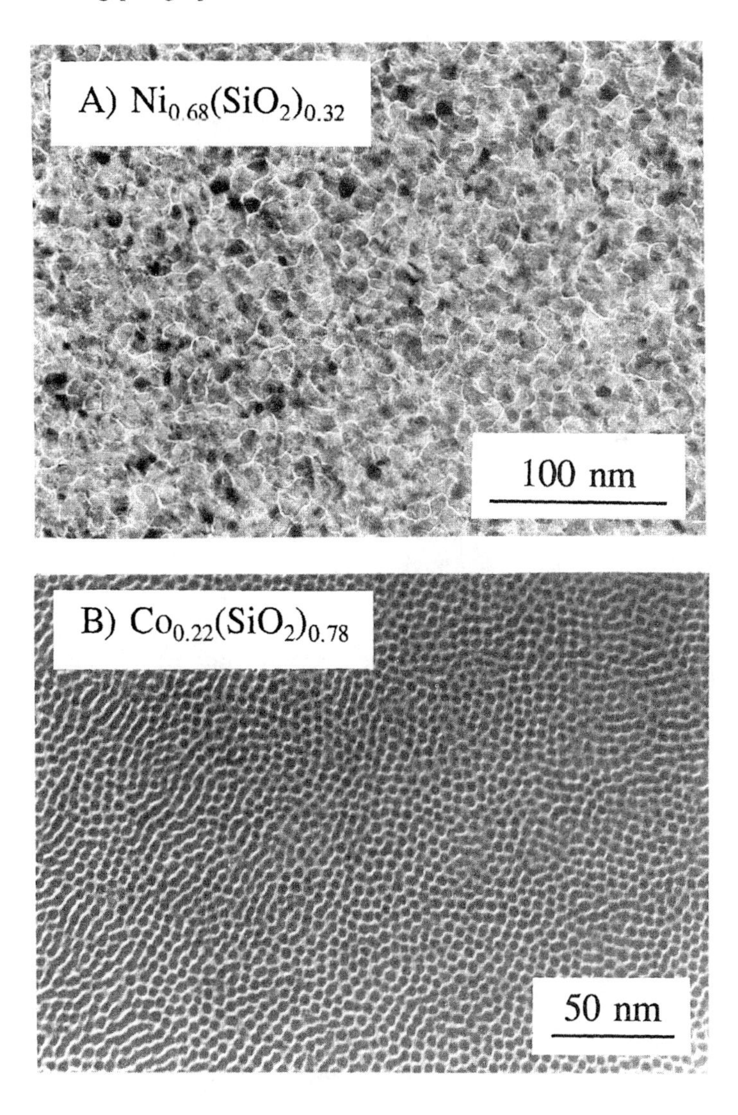

Figure 14.1. Bright-field TEM micrographs of (*a*) a sputter deposited granular $Ni_{0.68}(SiO_2)_{0.32}$ film and (*b*) a sputter deposited granular $Co_{0.22}(SiO_2)_{0.78}$ film. The dark regions correspond to Ni or Co, and the light regions are amorphous SiO_2.

Two distinct microstructural regimes occur in granular metal–semiconductor and metal–insulator films depending on the volume fraction of the metallic phase [1]. As shown in the bright-field transmission electron micrograph of figure 14.1(*a*), metal-rich films consist of a connected network of small irregularly shaped particles or grains of the metallic phase and isolated inclusions of the semiconductor or insulator phase. The metallic phase is usually crystalline and the semiconductor or insulator phase is usually amorphous when the granular films are deposited onto cooled or room-temperature substrates. In this regime the complex network of interconnected particles cannot be simply characterized in terms of a few structural parameters. Nevertheless, a mean particle size or structural coherence length can still be assigned based on the results of x-ray line broadening measurements, or by the direct imaging of the micro-structure in a transmission electron microscope (TEM). As the metal volume fraction is reduced the metallic network becomes less connected until, at a critical volume fraction x_c, no connected network of metallic particles spans the sample. Below x_c, the metallic phase consists of isolated, nearly spherical particles embedded in a continuous amorphous matrix as shown in figure 14.1(*b*). In this microstructural regime the isolated metal particles can usually be easily distinguished from the amorphous matrix, and the mean particle size and particle size distribution provide a relatively complete description of the microstructure.

In comparison with metal–insulator and metal–semiconductor granular films, both phases of metal–metal based films are generally crystalline. As a result, x-ray and electron diffraction studies are often complicated by the presence of broad and poorly resolved diffraction lines as can be seen in figure 14.2(*a*) for the case of a granular $Co_{0.20}Ag_{0.80}$ film. Low-resolution TEM images of the microstructure are also more difficult to interpret due to the presence of randomly aligned nanometer-sized crystallites of both phases and, in a number of interesting materials such as Co–Cu, a lack of phase contrast between metallic components of similar molecular weight. In these cases, high-resolution TEM studies can often provide useful information [6]. A high-resolution TEM micrograph of a granular $Co_{0.14}Ag_{0.86}$ film is shown in figure 14.2(*b*).

When a series of different compositions is prepared under nominally identical deposition conditions, it has been found that the mean particle size decreases with decreasing volume fraction as can be seen in figure 14.3. When several samples of the same composition are prepared under different deposition conditions, the mean particle size can be varied at constant volume fraction. From a practical standpoint this is most often accomplished by changing the substrate temperature. Lower substrate temperatures result in smaller particles while elevated substrate temperatures result in larger particles. By varying the composition and deposition conditions, granular metal films with particle sizes between a few nanometers and several tens of nanometers can easily be obtained.

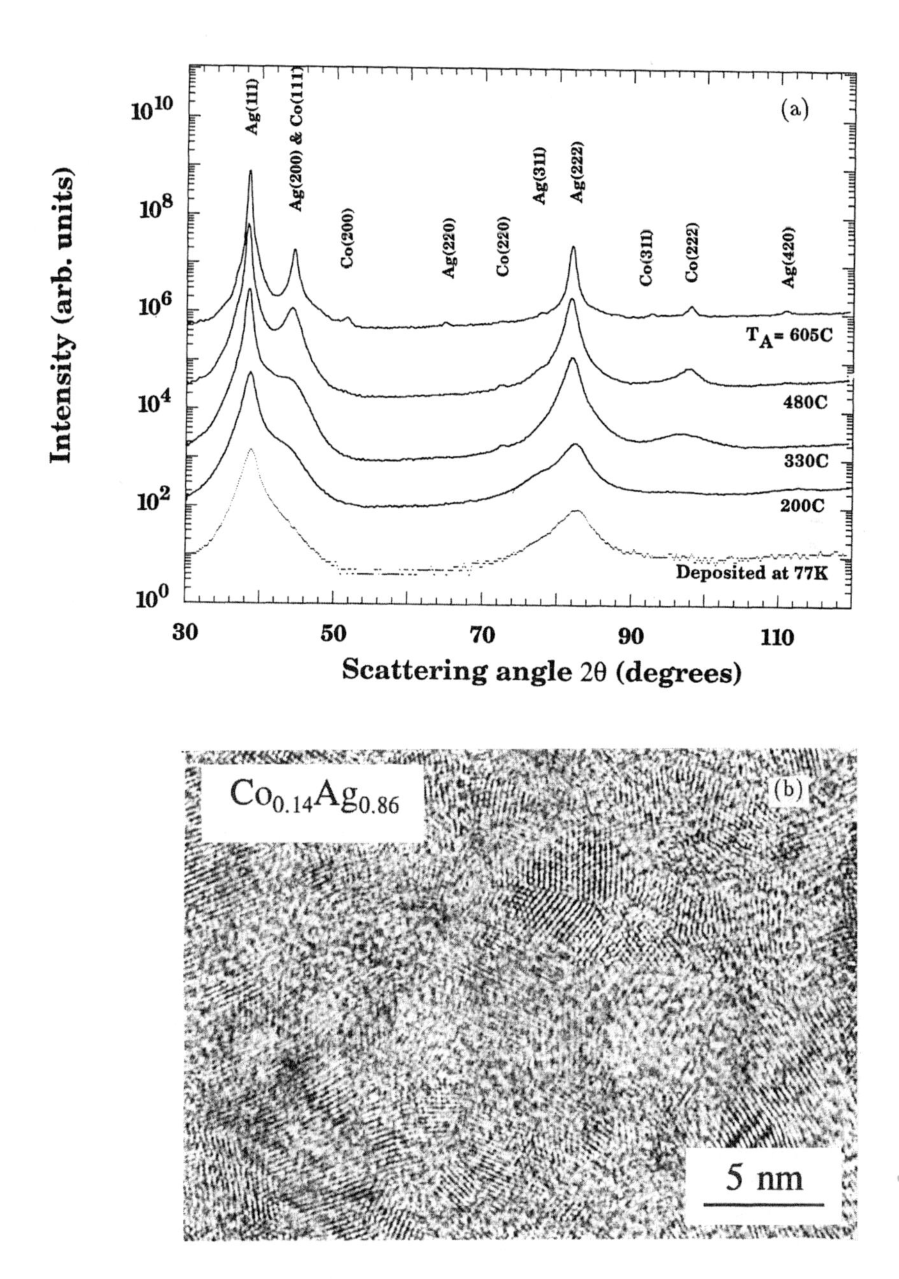

Figure 14.2. (*a*) X-ray diffraction patterns of a sputter deposited granular $Co_{0.20}Ag_{0.80}$ film and (*b*) high-resolution TEM micrograph of a sputter deposited granular $Co_{0.14}Ag_{0.86}$ film. The lattice fringes in the TEM micrograph correspond to the (200) reflection of fcc Co.

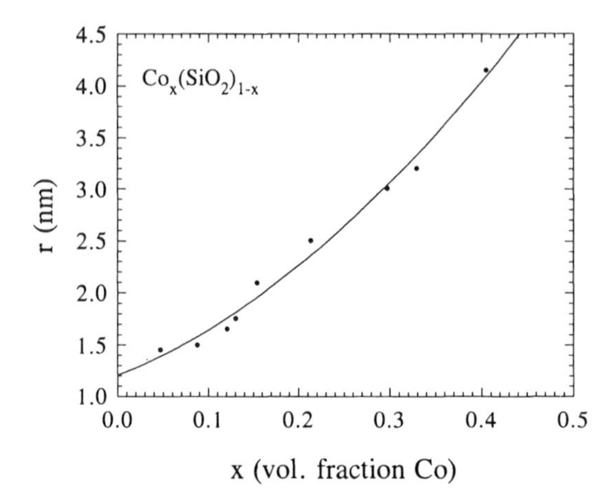

Figure 14.3. Mean Co particle radius as a function of the Co volume fraction x for a series of granular $Co_x(SiO_2)_{1-x}$ films. The mean particle sizes were obtained from bright- and dark-field transmission electron micrographs of each sample.

14.3 MAGNETIC PROPERTIES

It is often convenient to classify the properties of bulk matter as either intrinsic or extrinsic. Intrinsic properties arise from interactions on an atomic length scale, and are material specific and independent of sample size, shape, and microstructure. Extrinsic properties arise from longer-range interactions, and even for the same material vary with the size, shape, and microstructure of the sample. In the case of bulk ferromagnetic materials, the saturation magnetization per unit volume M_s, the magnetocrystalline anisotropy constant per unit volume K, and the Curie temperature T_C are intrinsic properties, while the coercivity H_c and the remanent magnetization per unit volume M_r are extrinsic.

The magnetic properties of nanometer-sized particles differ in several important respects from the properties of their bulk counterparts [7]. The large fraction of atoms located on surfaces or interfaces for example, whose local environments differ greatly from those of the interior atoms, leads to a blurring of the distinction between intrinsic and extrinsic properties. In small ferromagnetic particles, one might then expect M_s, K, and T_C, as well as H_c and M_r, to differ from their bulk values in a size-dependent way. A second phenomenon unique to nanometer-sized ferromagnetic particles must also be considered. Unlike bulk ferromagnetic materials, which usually form multiple magnetic domains, sufficiently small ferromagnetic particles consist of only a single magnetic domain. Although the critical particle size for single-domain formation depends in general on a number of different factors, critical sizes for the ferromagnetic 3d elements Co, Ni, and Fe are typically several tens of nanometers [7]. The realization that the effects of large surface-to-volume ratios and single-domain behavior could lead to new and unusual magnetic properties has motivated many studies of granular ferromagnetic films.

The following three sections will be devoted to a discussion of the magnetic properties of granular metal–insulator films consisting exclusively of isolated nanometer-sized ferromagnetic particles embedded in an insulating matrix (essentially bulk magnetic properties are observed in metal-rich compositions when the metallic particles become connected). In the first section an overview of the magnetic properties of these materials will be presented. It will be shown that the observed magnetic properties are qualitatively consistent with the behavior expected for an assembly of uniaxial single-domain particles. This description gives rise to a 'blocking' temperature T_B, below which the system exhibits ferromagnetic properties and above which paramagnetic properties are observed. The second section focuses on the relaxation properties of uniaxial single-domain particles, and will emphasize the importance of the relationship between the relaxation time and the characteristic time over which a particular magnetic measurement is carried out. Finally, the ferromagnetic properties exhibited by systems of single-domain particles at temperatures $T < T_B$ will be described with particular emphasis placed on a discussion of the greatly enhanced values of the coercivity observed in these materials.

14.3.1 Overview

The existence of single magnetic domains in small ferromagnetic particles was first predicted in 1930 [8], although detailed theoretical studies of these systems did not appear until more than a decade later [9]. As a result of this work it was shown that in an assembly of non-interacting ellipsoidal single-domain particles, each with a magnetocrystalline anisotropy constant K and saturation magnetization M_s, the coercivity H_c could be simply expressed as $H_c = 2K/M_s$ if the particle moments rotated coherently. Based on these considerations, significantly enhanced coercivities were expected in single-domain particles (about 600 Oe in single-domain Fe as compared with about 10 Oe in bulk Fe). Experimental studies on single-domain particles prepared in the form of precipitates, amalgams, supported catalysts, and suspensions soon followed [7], and enhanced values of H_c were indeed observed in qualitative agreement with the predicted result. The interpretation of these studies was often hampered, however, by problems associated with wide particle size distributions, environmental contamination, and small sample masses. Granular ferromagnetic metal–insulator films provide an important new class of materials for studies of this kind, and the study of their single-domain magnetic properties has been a fruitful field of research for many years.

Well below the Curie temperature T_C, all of the ferromagnetically coupled atomic moments in a single-domain particle are aligned in essentially the same direction, and give rise to a total magnetic moment μ that can be much larger than that of a single atom (a 10 nm Fe particle, for example, contains about

5×10^4 atomic moments). The equilibrium magnetic properties of a large assembly of noninteracting uniaxial single-domain particles, each of volume V_p, are then largely determined by the relative magnitudes of three characteristic energies: the thermal energy $E_T = k_B T$ where k_B is Boltzmann's constant, the anisotropy energy $E_A = CV_p$ where C is the total anisotropy energy per unit particle volume, and the magnetostatic energy $E_M = \boldsymbol{\mu} \cdot \boldsymbol{H}$ where $\boldsymbol{H}$ is the applied magnetic field. At temperatures where $E_T \gg E_A$, and in zero applied magnetic field, the direction of $\boldsymbol{\mu}$ rapidly fluctuates in time, and the system exhibits no net or global magnetization (as distinguished from the nonzero magnetization of each individual particle). In the presence of an applied field, a global magnetization is observed whose field dependence exhibits no magnetic hysteresis, and which resembles that of a classical paramagnet if the atomic moment is replaced by the particle moment μ. With decreasing temperature however, a 'blocking' temperature T_B, which depends on the particle size, applied field, and measuring instrument, is eventually reached below which ferromagnetic and history-dependent magnetic properties first appear. The ferromagnetic properties observed below T_B are not the result of a ferromagnetic coupling between the moments of the individual particles, but rather arise because the relaxation time for the particle moments becomes greater than the instrumental measuring time. Such systems are referred to as superparamagnetic.

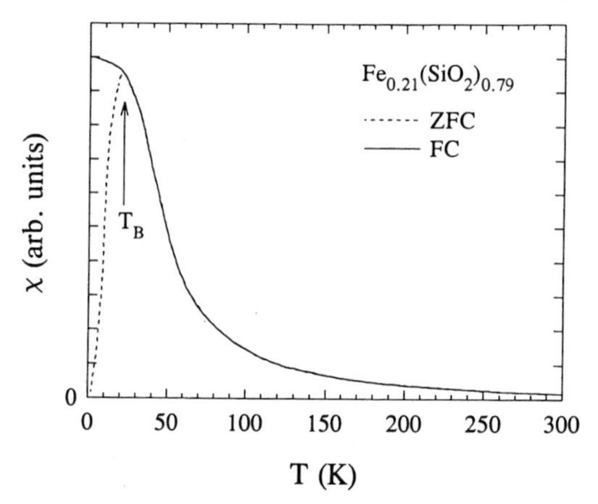

Figure 14.4. Zero-field cooled (ZFC) and field cooled (FC) susceptibility χ of a sputter deposited granular $Fe_{0.21}(SiO_2)_{0.79}$ film as a function of the temperature T. Both measurements were carried out in an applied field of 50 Oe. The peak in $\chi(T)$ locates the blocking temperature T_B. The two curves superimpose at temperatures greater than T_B.

Several characteristic magnetic properties of granular ferromagnetic metal–insulator films are illustrated in figures 14.4 and 14.5. In figure 14.4, for example, the temperature dependence of the low-field susceptibility $\chi(T)$ is shown for a granular $Fe_{0.21}(SiO_2)_{0.79}$ film. The curve displaying a clear peak was obtained

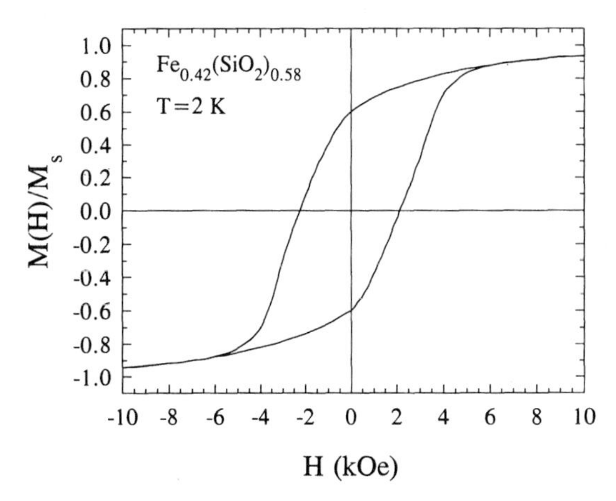

Figure 14.5. Magnetic hysteresis loop of a sputter deposited granular $Fe_{0.42}(SiO_2)_{0.58}$ film at 2 K.

by first cooling an as-prepared sample in zero applied field, and then measuring the sample's magnetization on warming in an applied field of 50 Oe. The second curve was obtained by recooling the same sample, also in a 50 Oe field. The peak in the zero-field cooled curve marks the blocking temperature for this particular sample and measuring instrument (a SQUID magnetometer in this case). Above T_B, the two magnetization curves superimpose, and measurements of the magnetization as a function of the applied magnetic field show no hysteresis. Below T_B, these materials exhibit a characteristic magnetic hysteresis loop with the associated ferromagnetic properties of nonzero coercivity and remanent magnetization. Figure 14.5 shows a typical hysteresis loop for a granular $Fe_{0.42}(SiO_2)_{0.58}$ film measured at a temperature of $T = 2$ K (well below T_B for this sample).

Enhanced values of the coercivity H_c in comparison to the corresponding bulk material are an additional magnetic characteristic of many granular films which contain a ferromagnetic component [10–12] and in particular greatly enhanced values of H_c have been observed in ferromagnetic metal–insulating oxide films [11, 12]. A typical example of this behavior is shown in figure 14.6 where it can be seen that the coercivity of a series of granular $Fe_x(SiO_2)_{1-x}$ films systematically evolves with Fe concentration, and near an Fe volume fraction of $x \approx 0.5$ reaches a maximum value of $H_c \approx 2500$ Oe at 2 K (more than two orders of magnitude greater than for bulk Fe). Even at room temperature, coercivities as large as $H_c \approx 500$ Oe were observed. Similar studies carried out on granular Co–SiO_2 films have also revealed enhanced coercivities, although not of the same magnitude as found in the case of the Fe based films [13]. When the temperature dependence of the coercivity is studied at fixed composition, $H_c(T)$ has been found to vary approximately as the square root of T [11], a result predicted as early as 1963 [14].

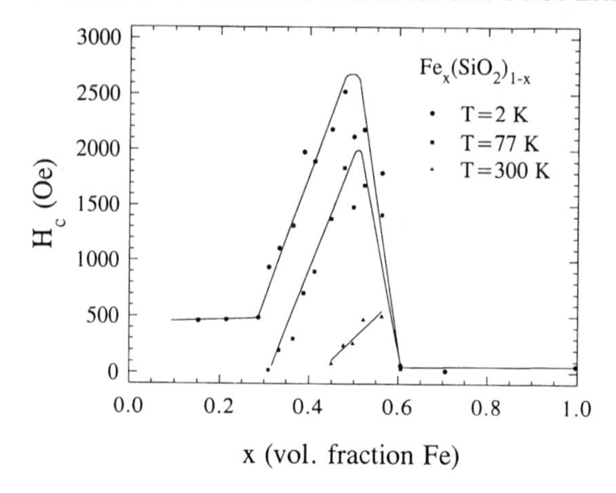

Figure 14.6. Evolution in the coercivity H_c as a function of the Fe volume fraction x in a series of sputter deposited granular $Fe_x(SiO_2)_{1-x}$ films at temperatures of $T = 2$, 77, and 300 K. Note that the mean Fe particle size is approximately proportional to x (see figure 14.3) and, as a result, H_c also depends on the particle size.

14.3.2 Superparamagnetic properties

At finite temperatures and in zero applied field, the ferromagnetically aligned magnetic moments within a uniaxial single-domain particle fluctuate between their two energetically degenerate ground states on a time scale, assuming simple Arrhenius behavior, given by

$$\tau = \tau_0 \exp\left(\frac{CV_p}{k_B T}\right) \tag{14.1}$$

where τ is the relaxation time, τ_0 is a constant estimated to be between about 10^{-13} and 10^{-9} s, and CV_p is the total anisotropy energy of the particle [15]. An expression similar to equation (14.1) can also be obtained when an applied magnetic field is present, but its exact form depends on the details of the field direction in comparison to the direction of the particle's anisotropy axis (in order to avoid this problem it is advantageous to carry out magnetization measurements in as small applied fields as possible). The strong temperature and particle-size dependence of τ plays an important role in determining the characteristic magnetic properties observed in systems of noninteracting single-domain particles.

The results of a magnetic measurement on an assembly of superparamagnetic particles depends not only on the relaxation time defined in equation (14.1), but also on the characteristic time of the measuring instrument. For a measurement carried out in zero applied field with an instrument whose characteristic measuring time is τ_i, the relaxation time and the instrumental time become equal at a temperature T_B given by

$$T_B = \frac{CV_p}{k_B \ln(\tau_i/\tau_0)}. \tag{14.2}$$

The temperature T_B defined in equation (14.2) is the same blocking temperature as described earlier, and defines the temperature below which the particle moments fail to equilibrate during the time scale of the measurement. The measuring instrument detects no net magnetization above T_B, because the particle moments fluctuate many times over the time interval of the measurement. On the other hand, history-dependent ferromagnetic properties are observed below T_B. Similar considerations also apply in the presence of a magnetic field.

Because each instrument has its own specific τ_i, different instruments will determine different values of T_B for the same sample. Thus, by using two or more instruments of very different measuring times, one can separately determine CV_p and τ_0 from equation (14.2). These measurements have been carried out [11], and it has been found that in many different granular ferromagnetic metal–insulator systems τ_0 has a value of about 10^{-13} s, and that C is much larger than the corresponding magnetocrystalline anisotropy energy of the bulk material. For example, while $K_1 \approx 10^5$ erg cm^{-3} for bulk Fe [16], $C \approx 10^7$ erg cm^{-3} has been observed in granular Fe–SiO_2 films. Evidently, the effective anisotropy energy must contain contributions from additional mechanisms such as shape, strain, and exchange anisotropies, as well as the magnetocrystalline anisotropy.

At temperatures above T_B, the temperature and field dependence of the measured global magnetization $M(T, H)$ and susceptibilty $\chi(T, H)$ are similar to those of a classical paramagnet if the atomic moment is replaced by the particle moment μ (unlike the atomic moment, however, $\mu = \mu(T, H)$ is itself a function of the temperature and applied field). The global magnetization can therefore be expressed as

$$M(T, H) = xM_p L\left(\frac{\mu H}{k_B T}\right) = xM_p\left[\coth\left(\frac{\mu H}{k_B T}\right) - \frac{k_B T}{\mu H}\right] \tag{14.3}$$

where L is the Langevin function and M_p is the magnetization of each particle per unit particle volume. As is the case for ordinary paramagnets, plots of $M(T, H)$ versus H/T for noninteracting assemblies of single-domain particles at temperatures above T_B must approximately superimpose when $k_B T \gg \mu H$ The magnetic susceptibility can be obtained in the usual way by differentiating equation (14.3) with respect to the applied field H. Assuming that $k_B T \gg \mu H$, one obtains for the temperature dependence of the susceptibility

$$\chi(T) = \frac{xV_p M_p^2}{3k_B T}. \tag{14.4}$$

Equation (14.4) can be extended to the case of interparticle interactions by rewriting the susceptibility in the Curie–Weiss form $\chi(T) \propto M_p^2/(T - T_0)$ where

T_0 is the analog of the Curie–Weiss temperature, and provides a measure of the correlations among the superparamagnetic particle moments [15].

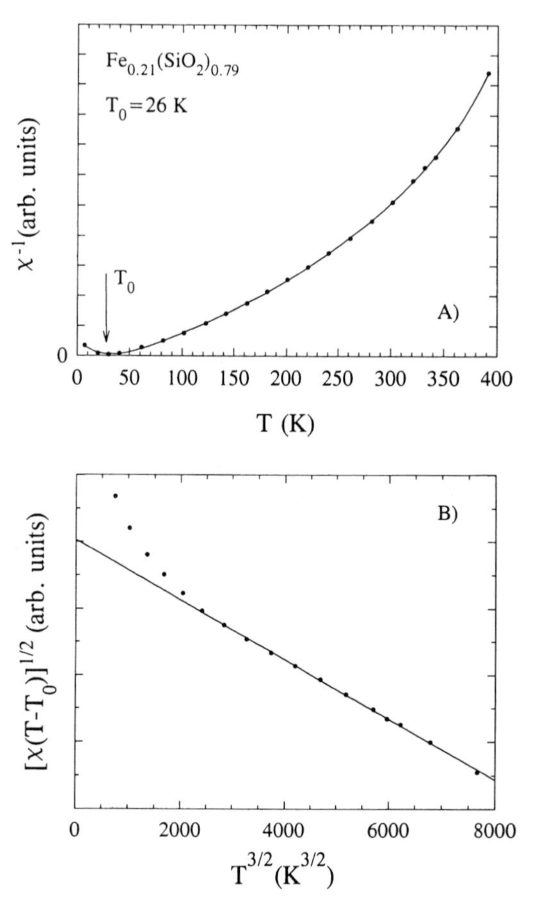

Figure 14.7. (*a*) Inverse susceptibility χ^{-1} as a function of the temperature T for a granular $Fe_{0.21}(SiO_2)_{0.79}$ film and (*b*) the quantity $[\chi(T - T_0)]^{1/2}$ as a function of $T^{3/2}$ for the same sample.

Due to the temperature dependence of M_p in equation (14.4), one would not necessarily expect a plot of χ^{-1} versus T to be linear in T for a granular sample. Figure 14.7(*a*) verifies this expectation and, furthermore, shows that $\chi^{-1} = 0$ at a finite temperature, indicating the presence of interparticle interactions. By explicitly introducing the temperature dependence of M_p in the Bloch form $M_p(T) = M_s(1 - BT^{3/2})$ where M_s is the saturation magnetization and B is the spin-wave stiffness constant, and then plotting $[\chi(T - T_0)]^{1/2}$ versus $T^{3/2}$ as shown in figure 14.7(*b*), one obtains a relatively straight line for temperatures greater than about 150 K. From the measured slope of this line, B has been determined and found to have a value approximately an order of magnitude greater than for bulk Fe [17]. The enhanced B values may arise because of the larger fluctuations in the surface moments as has been previously suggested in the case of magnetic thin films [18], or from a maximum allowed spin-wave wavelength due to finite particle size.

Equations (14.3) and (14.4) can be modified to incorporate a distribution of particle sizes as found in real materials. If $P(r)$ is the probability of finding a particle of size r, equation (14.3) can be rewritten as

$$M(H,T) = xM_p \int_0^{\infty} L\left(\frac{\mu H}{k_B T}\right) P(r)\,\mathrm{d}r. \tag{14.5}$$

The inversion of equation (14.5) could, in principle, directly yield the particle size distribution $P(r)$. The inversion of integral equations, however, poses formidable numerical challenges, and a more common approach has been to assume a particular distribution function. By varying the parameters of the probability distribution to obtain an optimal fit to the measured data, particle size information can be obtained. Log-normal distributions (i.e. the logarithm of the particle size is normally distributed) have been used in analyses of this kind, and have yielded values for the mean and standard deviation of the particle size distribution in relatively good agreement with the results of direct TEM imaging experiments [17].

14.3.3 Ferromagnetic properties

At very low temperatures, the total magnetic moment μ of a single-domain particle is essentially the sum of the individual atomic moments, and can be written as $\mu = M_s V_p$ where M_s is the saturation magnetization per unit particle volume V_p. In the absence of an applied magnetic field, the moment μ of a uniaxial particle is aligned either parallel or antiparallel to a single magnetic axis determined by the total magnetic anisotropy of the particle (which, as noted earlier, may include contributions from shape, strain, and exchange anisotropies, as well as the magnetocrystalline anisotropy). The parallel and antiparallel orientations of μ relative to the magnetic axis are energetically equivalent, but separated by an anisotropy energy barrier $E_A = CV_p$ where C is the total magnetic anisotropy energy per unit volume. In a large assembly of equivalent particles the magnetic axis of each particle is randomly oriented, and the measured global magnetization of the system in the absence of an applied magnetic field is zero. If such a system of single-domain particles is placed in an applied magnetic field H, the low-temperature magnetization per unit volume becomes $M(H, T \approx 0) = xM_s\langle\cos\theta\rangle$ where $\langle\cos\theta\rangle$ is the average angle between the individual particle moments and the applied field. With increasing applied field $M(H, T \approx 0)$ will increase, and eventually approach xM_s as all of the magnetic moments become aligned in the direction of H. If the applied field is then decreased, the global magnetization will decrease to a remanent magnetization $M_r = xM_s/2$ because all of the moments are now randomly oriented in only one hemisphere due to the anisotropy energy barrier giving $\langle\cos\theta\rangle = \frac{1}{2}$. Values of $M_r = xM_s/2$ following saturation at low temperatures are commonly taken as direct evidence for assemblies of uniaxial single-domain particles.

Unlike the low-temperature remanent magnetization, the coercivity of granular ferromagnetic metal–insulator films has been found to vary greatly from system to system, and depends, for example, on the particle size and the insulating matrix, as well as on the details of the fabrication process. The largest coercivities have been observed in Fe particles embedded in insulating oxide matrices in general, and SiO_2 matrices in particular. It has been reported that if an Fe–Si alloy forms during the deposition process significantly reduced coercivities will result [19]. While several qualitative explanations for the enhanced coercivities of these materials will be described in the following paragraph, a detailed theoretical understanding of these phenomena remains elusive.

One class of explanations for the enhanced values of the coercivity observed in granular materials has been based on the effects of the surface or interface that necessarily separates the ferromagnetic particles from the surrounding matrix. The presence of this interface must result in an effective surface anisotropy arising from surface magnetocrystalline, magnetoelastic, and dipolar shape anisotropies [20]. If the effective surface anisotropy is sufficiently large, it can then pin the magnetic moment of the particle in a preferred direction and enhance the coercivity. A second approach focuses on the assumed presence of a thin magnetic shell (usually a magnetic oxide if the matrix contains oxygen) surrounding a ferromagnetic core of the elemental metal. It can then be shown that if the shell orders antiferromagnetically or ferrimagnetically and exchange couples to the core, a significant additional energy beyond the magnetocrystalline anisotropy energy may be required to rotate the magnetization [21]. Direct experimental evidence for the presence of oxide shells has come from the results of Mössbauer effect and x-ray photoelectron spectroscopy measurements [22]. In the case of Fe, Co, and Ni particles embedded in an SiO_2 matrix, for example, these oxides have been identified as Fe_3O_4, Fe_2O_3, FeO, CoO, Ni_2O_3, and NiO respectively. In each case, at least one of these oxides is either antiferromagnetic or ferrimagnetic. It should be noted, however, that enhanced coercivities cannot exclusively arise from interactions of this kind because enhanced coercivities (although not to the extent observed in the case of oxide matrices) have also been observed in several granular systems that do not contain oxygen [10, 23].

14.4 ELECTRICAL TRANSPORT PROPERTIES

In addition to their unique magnetic properties, granular metal–insulator based films offer many opportunities for the study of metallic conduction in the presence of strong disorder, nonmetallic conduction due to thermally activated tunneling processes, the metal–insulator transition, superconductivity, and dimensionality effects. Although none of these phenomena is unique to granular metal films, the ability to continuously vary the size and volume fraction of the metallic component in these materials allows an unusually wide range of electrical transport properties to be systematically studied.

The discussion of the electrical transport properties of granular metal–insulator films has been divided into three sections. The first section presents an overview of the general types of conduction observed in granular films, and in particular the evolution in the temperature dependence of the electrical resistivity as a function of the metal volume fraction. The next two sections will then focus on the two distinct conduction regimes associated with metal-rich and metal-poor films. In the metal-rich or metallic regime, the emphasis will be on the logarithmic temperature dependence of the resistivity observed at low temperatures and on the film-thickness dependence of the conduction. In the metal-poor or insulating (in the sense that the resistivity becomes infinite at $T = 0$) regime, the low-field conduction will be described and discussed in terms of an electron hopping picture.

14.4.1 Overview

The temperature dependence of the sheet resistance $R_{\square}(T)$ in a series of sputter deposited $Ni_x(SiO_2)_{1-x}$ films about 100 nm in thickness is shown as a function of the Ni volume fraction x in figure 14.8. Several features of these data are typical of all granular metal films. Although the resistance increases by many orders of magnitude with decreasing values of x, two distinct conduction regimes can be identified [1]. The metallic regime is characterized by large values of x, relatively small values of the resistivity, and small positive temperature coefficients of the resistivity (TCR) near room temperature. In this conduction regime, the film resistivity slowly increases and the room-temperature TCR slowly decreases with decreasing x. The insulating regime, on the other hand, is characterized by smaller values of x, large values of the resistivity, and negative TCRs. In this regime, the film resistivity rapidly increases and the room-temperature TCR becomes increasingly negative with decreasing x. The transition from the metallic to the insulating regime occurs at a critical metal volume fraction x_c that experimentally is found to be between approximately 0.4 and 0.6 in as-prepared samples, and near 0.5 in well annealed samples [1, 24]. An example of the very sharp transition from metallic to nonmetallic transport behavior found in granular systems with particularly narrow particle size distributions is shown in figure 14.9.

The evolution in the conduction behavior of granular metal films as a function of the metal volume fraction is intimately connected to the corresponding evolution in the film microstructure. The relatively small values of the resistivity and positive TCRs that characterize the metallic conduction regime arise because the microstructure consists of connected metal particles which form

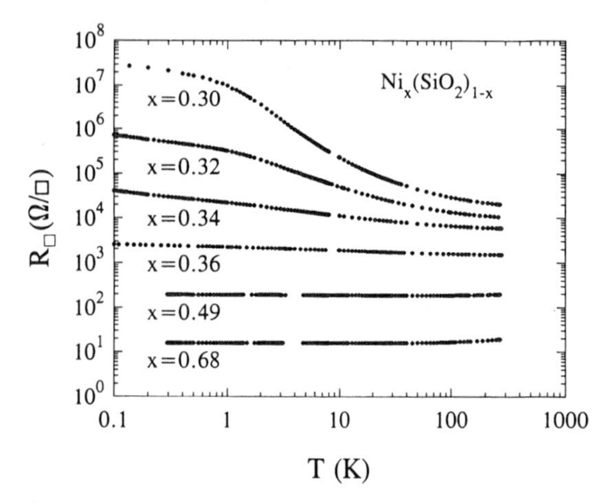

Figure 14.8. The temperature dependence of the sheet resistance $R_\square(T)$ in a series of sputter deposited granular $Ni_x(SiO_2)_{1-x}$ films as a function of the temperature T and the Ni volume fraction x.

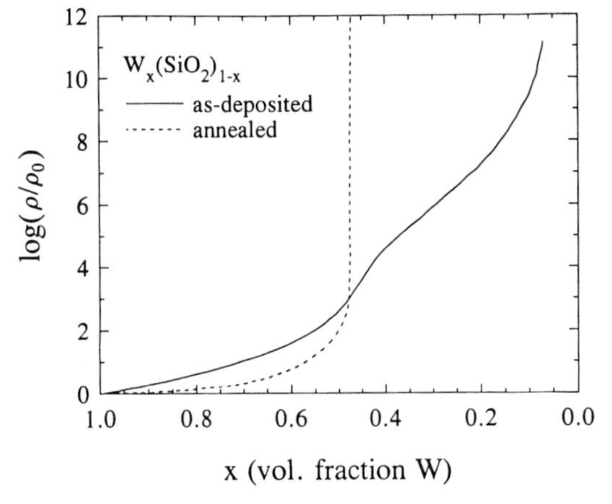

Figure 14.9. Evolution of the room-temperature resistivity $\rho(x)$ in a series of as-deposited and annealed sputter deposited granular $W_x(SiO_2)_{1-x}$ films as a function of the W volume fraction x. (Adapted from [24].)

a percolating network. With decreasing metal volume fraction the film resistivity increases and TCR becomes less positive due to the increasing tortuosity of the percolating conduction paths and a corresponding increase in the strength of the disorder. As will be discussed in greater detail in the following section, resistivity minima have also been observed as a function of the temperature in a number of different granular systems, even in relatively thick films [25]. Below the minima the resistivity increases in proportion to the logarithm of the temperature. In the insulating regime the film microstructure consists of isolated metal particles separated by insulating barriers. No percolating conduction paths exist, and the primary mechanism of electron transport is by thermally assisted hopping or tunneling between isolated metal particles. In this conduction regime the resistivity can be well described by a relation of the form $\rho(T) \propto \exp[(T_0/T)^{1/2}]$ where $T_0 = T_0(x)$ is a monotonically increasing function of x over many orders of magnitude in the resistivity, and from temperatures as low as several K to well above room temperature [1,26]. In several different granular metal–insulator systems, however, in which the resistivity has been studied at temperatures below a few K, it has recently been observed that T_0, and consequently $d\rho/dT$, becomes essentially independent of x [27].

14.4.2 Metallic regime

It has already been noted that the resistivities and TCRs of granular metal–insulator films are similar to those found in other disordered metallic materials. It is therefore interesting to ask whether other conduction properties associated with disordered metallic systems might also be found in the metallic regime of granular metals. This indeed turns out to be the case, and the purpose of this section is to describe several of these phenomena, focusing in particular on the temperature and film-thickness dependence of the resistivity.

The transport behavior of two of the metallic $Ni_x(SiO_2)_{1-x}$ films shown in figure 14.8 has been replotted in figure 14.10 in the form (see below) $\Delta R_\square/R_\square^2(T_0)$ versus $\ln(T)$ where $\Delta R_\square = R_\square(T) - R_\square(T_0)$ and $T_0 = 0.3$ K. Two obvious features of these data are worth noting. In the first instance, each sample (and in fact all of the metallic films shown in figure 14.8) exhibits a composition-dependent resistivity minimum which shifts to higher temperatures with decreasing x. Below the minimum, the resistivity increases in proportion to the logarithm of the temperature to the lowest measured temperatures, although the rate of increase varies from sample to sample.

Several different mechanisms, including the scattering of conduction electrons from certain isolated magnetic moments (Kondo effect) [28], as well as other quantum mechanical effects such as weak localization and electron–electron interactions in 2D systems, can give rise to a logarithmic temperature dependence of the resistivity at low temperature [29]. As previously noted, however, metal-rich granular metal–insulator films are ferromagnetic when the metallic component itself is ferromagnetic, so one would not expect a Kondo effect in this class of material. In the following paragraphs we will discuss the extent to which the observed low-temperature resistivity is consistent with the general predictions of weak localization and electron–electron interaction theory. Numerical studies of the localization properties of granular systems have recently addressed these issues in depth [4].

In disordered materials at low temperatures it is possible for an electron to undergo many elastic scattering events before an inelastic scattering event takes place. Generally speaking, such a material is said to be effectively

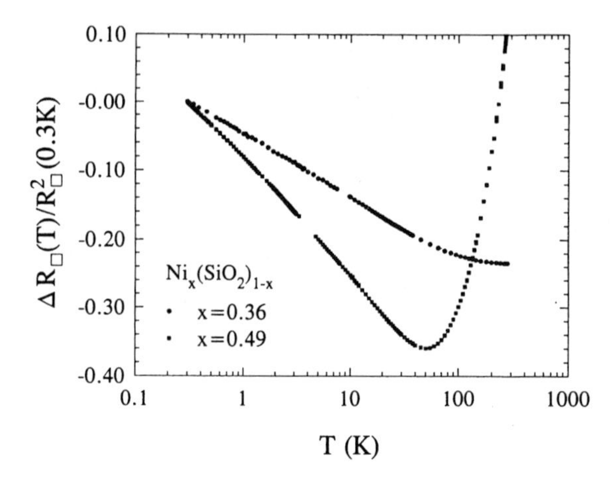

Figure 14.10. Conductance change $\Delta R_{\square}(T)/R_{\square}^{2}(0.3\ \mathrm{K})$ of two metallic sputter-deposited granular $Ni_x(SiO_2)_{1-x}$ films with Ni volume fractions of $x = 0.36$ and $x = 0.49$ as a function of $\ln(T)$.

two dimensional if its thickness is less than the diffusion distance or phase breaking length over which the phase shift of the inelastically scattered electron approaches 2π (that is the electron loses its phase memory). The theory then shows that, in the absence of an applied magnetic field and neglecting the effects of spin–orbit and spin–spin scattering, the change in the conductance of the sample can be expressed in the form

$$\frac{\Delta R_{\square}(T)}{R_{\square}^{2}(T_0)} = -\frac{e^2}{2\pi^2\hbar} K \ln\left[\frac{T}{T_0}\right] \tag{14.6}$$

where e and $\hbar$ have their usual meaning, and T_0 is a reference temperature [29, 30]. K is a temperature-independent factor arising from weak electron localization and/or electron–electron correlation effects. In the former case, $K = \alpha p$ where α is a constant of order unity, and the temperature dependence of the inelastic scattering time t_i is taken to be $t_i \propto T^{-p}$. If the dominant inelastic scattering mechanism is due to electron–phonon scattering, then p is expected to be between two and four for 3D phonons (even in thin films the phonons should behave three dimensionally if the film is supported on a thick substrate), while electron–electron scattering results in a value of $p = 1$. In the case of electron–electron correlation effects, $K = 1 - 0.75F$ where F is a screening factor which can take on values between 0 and 1. Weak localization and electron–electron effects are additive to first order [29], and one would therefore expect K to roughly fall between one and five. When the low-temperature data of figures 14.8 and 14.10 are fit to equation (14.6), one finds values of K between about two and six.

The apparent agreement between the measured K values described in the

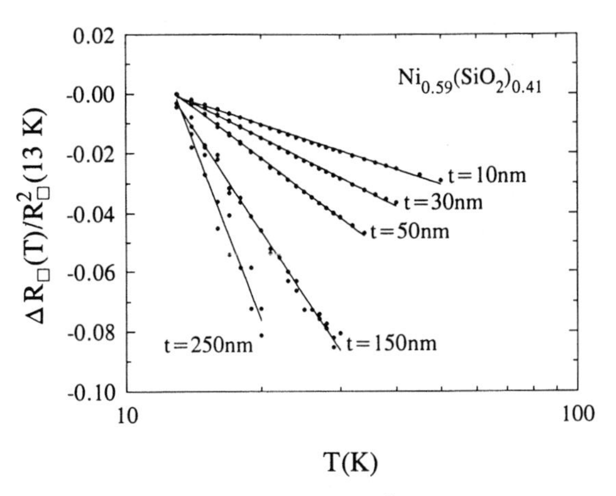

Figure 14.11. Conductance change $\Delta R_{\square}(T)/R_{\square}^2(13\ \mathrm{K})$ of several metallic sputter deposited granular $Ni_{0.59}(SiO_2)_{0.41}$ films as a function of the temperature T and film thickness.

preceding paragraph and the prediction of equation (14.6) is somewhat surprising due to the fact that the film thickness of about 100 nm is considerably larger than would normally be expected for 2D behavior (based on the characteristic scattering lengths for weak localization and/or electron–electron interaction effects). In an attempt to clarify this problem, the results of a series of resistivity measurements on films of varying thickness and the same nominal composition of $x \approx 0.59$ are shown in figure 14.11. Even the thickest film studied exhibited a logarithmic increase in the resistance at low temperature, while the rate of increase clearly becomes smaller with decreasing film thickness. No evidence for a thickness-independent slope was found, even in the thinnest samples. Clearly more work in this area is needed before a complete understanding of the transport properties of these materials can be obtained.

14.4.3 Insulating regime

Below the critical metal volume fraction x_c, the microstructure of granular metal–insulator films consists of isolated metal particles or grains embedded within a continuous insulating matrix. No long-range connected metallic conduction paths exist, and at finite temperatures electrons are transported by thermally assisted hopping or tunneling between isolated metal particles. In this conduction regime many different experiments have shown that the temperature dependence of the resistivity $\rho(T)$ can be well described by an expression of the form

$$\rho(T) = \rho_0 \exp\left(\frac{T_0}{T}\right)^{1/2} \tag{14.7}$$

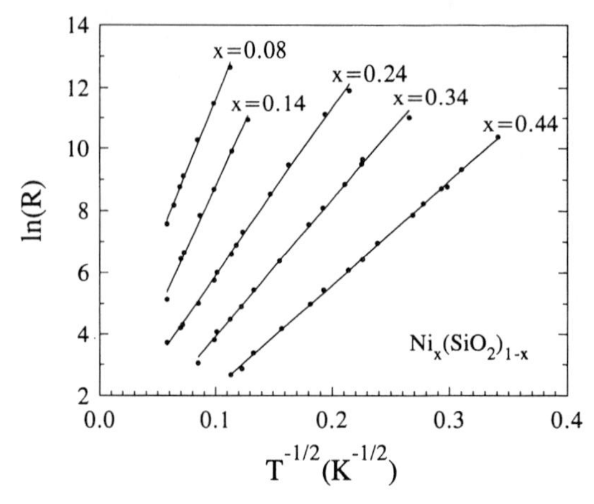

Figure 14.12. Logarithm of the resistance R as a function of $T^{-1/2}$ for a series of sputter deposited granular $Ni_x(SiO_2)_{1-x}$ films in the insulating conduction regime. (Adapted from [1].)

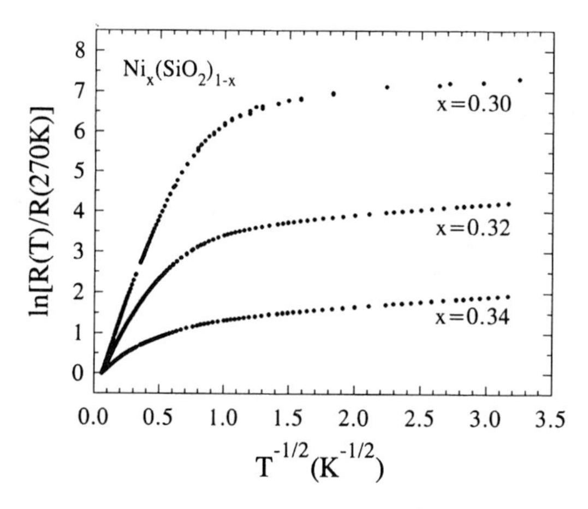

Figure 14.13. Logarithm of the reduced resistance as a function of $T^{-1/2}$ for a series of sputter deposited $Ni_x(SiO_2)_{1-x}$ films in the insulating conduction regime.

where $T_0 = T_0(x)$ is composition dependent [1]. Figure 14.12 shows an example of the transport behavior of a series of granular $Ni_x(SiO_2)_{1-x}$ films in the insulating conduction regime.

While equation (14.7) has been very successful in describing the temperature dependence of the resistivity over a wide range of temperatures in many granular metal–insulator films with metal volume fractions $x < x_c$ [1, 26], deviations from equation (14.7) have been observed in measurements at temperatures below several K [27]. This behavior is illustrated in figure 14.13 for three sputter deposited $Ni_x(SiO_2)_{1-x}$ granular films with Ni volume fractions of $x = 0.34, 0.32$, and 0.30. The resistivity of each film can be seen to fall into one of two regimes. At temperatures greater than several K, the data can be well described by equation (14.7). At lower temperatures, however, a weaker temperature dependence is observed, and $\mathrm{d}\rho/\mathrm{d}T$ appears to become essentially independent of the film composition.

Two different approaches have generally been used to justify the experimentally observed equation (14.7) for the temperature dependence of the resistivity in the insulating conduction regime. The first approach focuses on the non-negligible charging energy E_c (proportional to $1/r$) associated with the transfer of an electron from one metallic particle to a neighboring particle [31] By assuming that the charge carriers are thermally activated with a number density proportional to $\exp(-E_c/2k_BT)$, and that the carrier mobility is given by the tunneling probability $\exp(-2\alpha s)$ where α is the tunneling exponent and s is the distance between metallic particles, one then obtains for the conductivity $\sigma \propto \exp(-2\alpha s - E_c/2k_BT)$. In the original version of this argument it is further assumed that the particle size r and interparticle separation distance are such that their product is a constant, minimizing the argument of the exponent in the conductivity to an expression consistent with equation (14.7) [1,32]. The assumption that the product rs is a constant is open to some criticism [33], but in fact is not actually needed to obtain the necessary $T^{-1/2}$ temperature dependence. A second approach is based on including the Coulomb interaction between electrons in a variable-range hopping model, which results in a parabolic gap in the density of states at the Fermi level [34]. By assuming that all relevant energies are greater than k_BT, the resistivity can then be obtained in the form of equation (14.7) where the constant T_0 can be expressed in terms of the optimal hopping energy and distance.

14.5 GIANT MAGNETORESISTANCE

While granular metal–insulator based films have been studied since the early 1960s, granular metal–metal films have only been systematically fabricated and studied for the last several years [10]. Nevertheless, a host of novel transport properties have already been discovered, beginning with the observation of dramatically enhanced values of the magnetoresistance $\Delta\rho/\rho$ (conventionally defined as $\Delta\rho/\rho = [\rho(H) - \rho(0)]/\rho(0)$ or $\Delta\rho/\rho = [\rho(H) - \rho(H_s)]/\rho(H_s)$), in the granular Co–Cu system [35]. This effect, which is now commonly referred to as 'giant magnetoresistance' or GMR, will be the focus of this short section.

Giant magnetoresistance was first discovered in 1988 [36] in Fe/Cr multilayers (see chapter 6), and was subsequently observed in many other multilayer systems consisting of alternating layers of ferromagnetic and nonmagnetic metals [37]. For an appropriate choice of the nonmagnetic metal layer thickness the magnetic layers could be made to antiferromagnetically couple, resulting in a zero net magnetization in the absence of an applied magnetic field. In the presence of a sufficiently large applied magnetic field, the magnetization of each magnetic layer could be aligned. The large difference in resistance in the two spin states is then responsible for the observed GMR. Numerous experimental studies, including the oscillatory behavior of GMR as a function of the nonmagnetic layer thickness and its dependence on the crystalline orientation and multilayer quality, have generally confirmed this basic qualitative mechanism for the GMR.

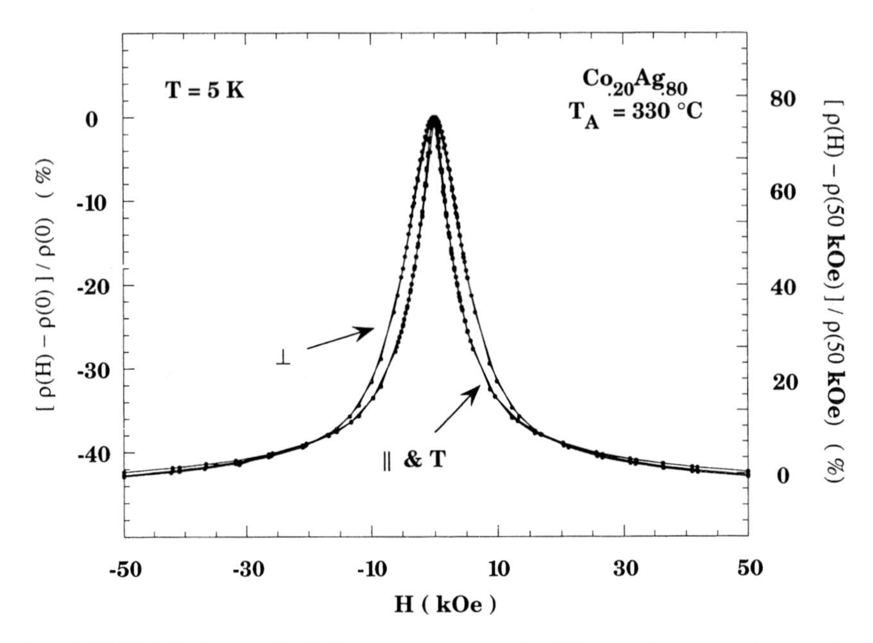

Figure 14.14. The longitudinal (||) and transverse (⊥: *H* perpendicular to the sample plane, *T*: *H* in the sample plane) magnetoresistance ρ of a granular $Co_{0.20}Ag_{0.80}$ film annealed at 603 K.

In addition to the GMR originally observed in the granular Co–Cu system, similar behavior has now also been observed in a much wider range of granular metal–metal solids, demonstrating that a multilayer structure is not required for the GMR phenomena [38]. Very large values of the magnetoresistance have been obtained in these materials as shown in figure 14.14 for a granular $Co_{0.20}Ag_{0.80}$ sample after annealing at 500 K.

Although magnetoresistance data have historically been described in terms of a relative shift in the resistivity (or resistance), it has recently been shown that the absolute change in the resistivity is of greater physical significance [5]. In particular the quantity $\rho_m = [1 - F(M/M_s)]$, where M is the global magnetization and M_s is the saturation magnetization, is the relevant parameter in describing the GMR. $F(M/M_s)$ must be an even function of M/M_s with the limiting values of $F(M/M_s) \to 0$ as $M \to 0$ (when $H = 0$ or H_c) and $F(M/M_s) \to 1$ as $M \to M_s$ (when $H \geq H_s$). The field dependence of the

GMR is contained in $F(M/M_s)$ which, in most cases, can be well described by $F(M/M_s) \approx (M/M_s)^2$, and further improved by retaining a term in $(M/M_s)^4$ and writing $F(M/M_s) = \alpha(M/M_s)^2 + (1-\alpha)(M/M_s)^4$ where α is a parameter with a value less than 1. These expressions account for the experimental fact that the global magnetization always saturates faster than that of the GMR.

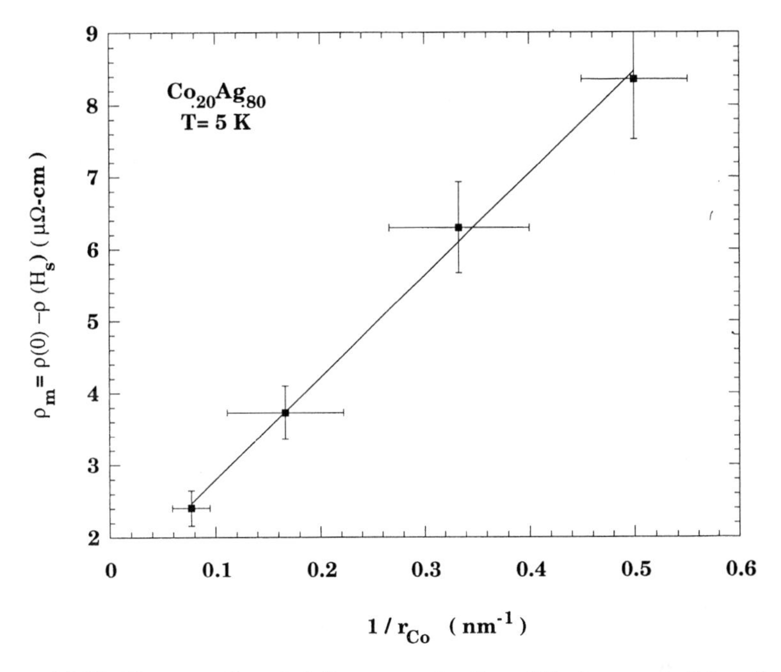

Figure 14.15. The magnetic resistivity ρ_m as a function of the inverse particle radius for a series of granular $Co_{0.20}Ag_{0.80}$ films annealed at various temperatures.

The quantity ρ_m, which depends on the sizes and the density of the magnetic particles, dictates the magnitude of GMR. The dependence of ρ_m on particle size can be revealed in samples with a fixed volume fraction of the magnetic species, but with different particle sizes as shown in figure 14.15. In this case it has been found that $\rho_m \propto 1/r$. Since the interface area/volume in a granular system is $S \approx 3x/r$, evidently $\rho_m \propto S$. Thus GMR is due to magnetic scattering at the interfaces.

The dependence of ρ_m on x is shown in figure 14.16, where the values of ρ, ρ_m, and GMR of Co–Ag with $0 \leq x_v \leq 1$ are displayed. The value of ρ initially rises when Co is first introduced, reaching a plateau for a broad range of Co contents, and decreases towards the pure Co limit. The value of ρ_m increases from 0, reaching a maximum at about $x_v \approx 25\%$, before decreasing to very small values at $x_v \approx 55\%$, which is just the percolation threshold, beyond which a connecting network of Co is formed, and there is no GMR.

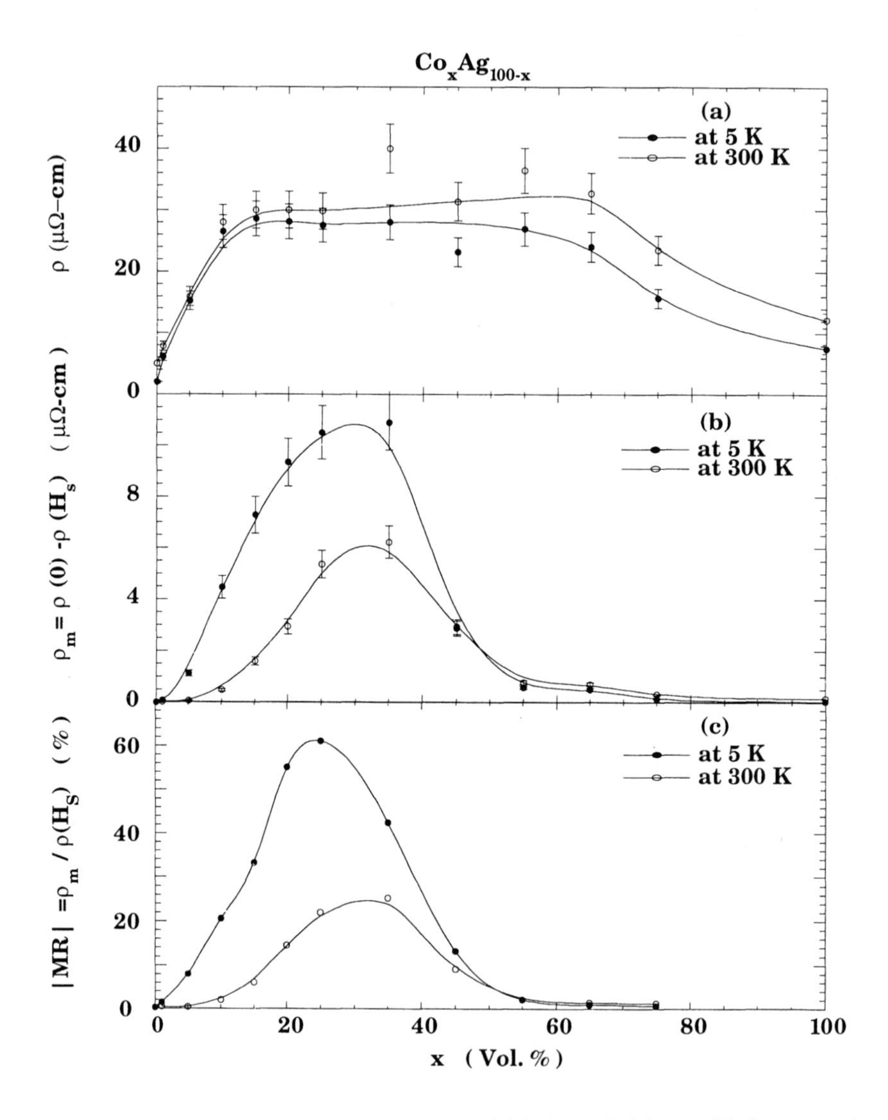

Figure 14.16. The concentration dependence of (*a*) the resistivity ρ, (*b*) the magnetic resistivity ρ_m, and (*c*) the magnetoresistance. Solid and open circles correspond to temperatures of 5 and 300 K respectively.

14.6 SUMMARY

The purpose of this chapter has been to provide a brief overview of several characteristic magnetic and transport properties of sputter deposited granular films consisting of nanometer-sized metallic particles embedded in either an insulating or a metallic matrix. In both cases, the properties of these materials have been shown to differ significantly from the corresponding properties of their bulk counterparts, and in some cases, to exhibit completely unexpected properties. These unusual properties arise from the combination of often disparate properties of the individual constituents from which these materials are formed, and from the nanometer length scale that characterizes the microstructure of these materials.

REFERENCES

[1] For the first general reviews describing the fabrication and physical properties of granular metal films, as well as an extensive list of references to the pre-1975 literature, see Abeles B, Sheng P, Coutts M D and Arie Y 1975 *Adv. Phys.* **24** 407 and Abeles B 1976 *Applied Solid State Science* ed R Wolfe (New York: Academic) pp 1–117

[2] Schlesinger T E, Cammarata R C, Gavin A, Xiao J Q, Chien C L, Ferber M K and Hayzelden C 1991 *J. Appl. Phys.* **70** 3275

[3] See, for example, Hadjipanayis G C and Prinz G A (ed) 1991 *Science and Technology of Nanostructured Magnetic Materials* (New York: Plenum)
Hadjipanayis G C and Siegel R W (ed) 1994 *Nanophase Materials: Synthesis–Properties–Applications* (Dordrecht: Kluwer)

[4] Sheng P 1992 *Phil. Mag.* B **65** 357

[5] Chien C L, Xiao J Q and Jiang J S 1993 *J. Appl. Phys.* **73** 5309. For a basic introduction to the related problem of multilayer GMR, see, for example, White R L 1992 *IEEE Trans. Magn.* **MAG-28** 2482 and the four subsequent papers in that journal issue.

[6] Tsoukatos A, Wan H, Hadjipanayis G C and Li Z G 1992 *Appl. Phys. Lett.* **61** 3059
Wan H, Tsoukatos A, Hadjipanayis G C, Li Z G and Liu J 1994 *Phys. Rev.* B **49** 1524

[7] See, for example, Jacobs I S and Bean C P 1963 *Magnetism III* ed G T Rado and H Suhl (New York: Academic) pp 271–350

[8] Frenkel J and Dorfman J 1930 *Nature* **126** 274

[9] Kittel C and Galt J K 1946 *Phys. Rev.* **70** 965
Stoner E C and Wohlfarth E P 1948 *Phil. Trans. R. Soc.* A **240** 599
Kittel C 1949 *Rev. Mod. Phys.* **21** 541
Kittel C and Galt J K 1959 *Solid State Phys.* **3** 437

[10] Childress J R and Chien C L 1991 *Phys. Rev.* B **43** 8089; 1991 *J. Appl. Phys.* **70** 5885

[11] Xiao G, Liou S H, Levy A, Taylor J N and Chien C L 1986 *Phys. Rev.* B **34** 7573
Chien C L, Xiao G, Liou S H, Taylor J N and Levy A 1987 *J. Appl. Phys.* **1** 3311
Xiao G and Chien C L 1987 *Appl. Phys. Lett.* **51** 1280

Liou S H and Chien C L 1988 *Appl. Phys. Lett.* **52** 512; *J. Appl. Phys.* **63** 4240
[12] Kanai Y and Charap S H 1991 *J. Appl. Phys.* **69** 4478
[13] Tsoukatos A, Wan H and Hadjipanayis G C 1993 *J. Appl. Phys.* **73** 6967
Tsoukatos A 1994 *PhD Thesis* University of Delaware
[14] Kneller E F and Luborsky F E 1963 *J. Appl. Phys.* **34** 5
[15] For a more sophisticated treatment of the relaxation time see Chantrell R W and Wohlfarth E P 1983 *J. Magn. Magn. Mater.* **40** 1 and references therein
[16] Morup S, Topsoe H and Clausen B S 1982 *Phys. Scr.* **25** 713
[17] Xiao G and Chien C L 1988 *J. Appl. Phys.* **61** 3308
Chien C L 1991 *J. Appl. Phys.* **69** 5267
[18] Mills D L 1971 *Comments Solid State Phys.* **4** 28; 1972 *Comments Solid State Phys.* **4** 95
Pierce D T, Celotta R J, Unguris J and Siegmann H C 1982 *Phys. Rev.* B **26** 2566
[19] Holtz R L, Edelstein A S, Lubitz P and Gossett C R 1988 *J. Appl. Phys.* **64** 4251
[20] Bruno P and Renard J-P 1989 *Appl. Phys.* A **49** 499
[21] Meiklejohn W H and Bean C P 1956 *Phys. Rev.* **102** 1413; 1957 *Phys. Rev.* **105** 904; 1962 *J. Appl. Phys.* **33** 1328
Malozemoff A P 1987 *Phys. Rev.* B **35** 3679; 1988 *J. Appl. Phys.* **63** 3874
[22] Dormann J-L, Gibart P and Renaudin P 1976 *J. Physique Coll.* **37** C6 281
Dormann J-L, Sella C, Renaudin P, Kaba A and Gibart P 1979 *Thin Solid Films* **58** 25
Papaefthymiou V, Tsoukatos A, Hadjipanayis G C, Simopoulos A and Kostikas A 1995 *J. Magn. Magn. Mater.* **140–144** 397
Paparazzo E, Dormann J L and Fiorani D 1983 *Phys. Rev.* B **28** 1154
Shah S I and Unruh K M 1991 *Appl. Phys. Lett.* **59** 3485
[23] Edelstein A S, Das B N, Holtz R L, Koon N C, Rubinstein M, Wolf S A and Kihlstrom K E 1987 *J. Appl. Phys.* **61** 3320
[24] Abeles B, Pinch H L and Gittleman J L 1975 *Phys. Rev. Lett.* **35** 247
[25] Carl A, Dumpich G and Hallfarth D 1987 *Phys. Rev.* B **39** 915; 1989 *Phys. Rev.* B **39** 3015
Lee S J, Ketterson J B and Trivedi N 1992 *Phys. Rev.* B **46** 12 695
[26] Gittleman J I, Goldstein Y and Bozowski S 1972 *Phys. Rev.* B **5** 3609
McAlister S P, Inglis A D and Kayll P M 1985 *Phys. Rev.* B **31** 5113
[27] Gershenfeld N, VanCleve J E, Web W W, Fischer H E, Fortune N A, Brooks J S and Graf M J 1988 *J. Appl. Phys.* **64** 4760
Unruh K M, Patterson B M, Beamish J R, Mulders N and Shah S I 1990 *J. Appl. Phys.* **68** 3015
Chui S-T 1991 *Phys. Rev.* B **43** 14 274
[28] Kondo J 1969 *Solid State Physics* vol 23, ed F Seitz and D Turnbull (New York: Academic) pp 183–281
[29] See, for example, Bergmann G 1984 *Phys. Rep.* **107** 1 and Lee P A and Ramakrishnan R V 1985 *Rev. Mod. Phys.* **57** 287 and references therein
[30] Lin J J and Giordano N 1987 *Phys. Rev.* B **35** 545
[31] Gorter C J 1951 *Physica* **17** 777
Neugebauer C A and Webb M B 1962 *J. Appl. Phys.* **33** 74
[32] Sheng P, Abeles B and Arie Y 1973 *Phys. Rev. Lett.* **31** 44
Sheng P and Klafter J 1983 *Phys. Rev.* B **27** 2583
Klafter J and Sheng P 1984 *J. Phys. C.: Solid State Phys.* **17** L93

[33] Adkins C J 1987 *J. Phys. C: Solid State Phys.* **20** 235

[34] Afros A L and Shlovskii B I 1975 *J. Phys. C: Solid State Phys.* **8** L49
Shlovskii B I and Efros A L 1984 *Electronic Properties of Doped Semiconductors* (Berlin: Springer) pp 82–9

[35] Berkowitz A, Young A P, Mitchell J R, Zhang S, Carey M J, Spada F E, Parker F T, Hutten A and Thomas G 1992 *Phys. Rev. Lett.* **68** 3745
Xiao J Q, Jiang J S and Chien C L 1992 *Phys. Rev. Lett.* **68** 3749

[36] Baibich M N, Broto J M, Fert A, Nguyen Van Dau F, Petroff F, Etienne P, Creuzet G, Friederich A and Chazeles J 1988 *Phys. Rev. Lett.* **61** 2472

[37] Parkin S S P, Bhadra R and Roche K P 1991 *Phys. Rev. Lett.* **66** 2152
Pratt W P Jr, Lee S F, Slaughter J M, Loloee R, Schroeder P A and Bass J 1991 *Phys. Rev. Lett.* **60** 3060
Krebs J J, Lubitz P, Chaiken A and Prinz G A 1989 *Phys. Rev. Lett.* **63** 1645

[38] See, for example, the series of papers in the *Proc. 37th Ann. Conf. on Magnetism and Magnetic Materials, Part A* (published in 1993 *J. Appl. Phys.* **73**)

Chapter 15

Magnetic and structural properties of nanoparticles

G C Hadjipanayis, K J Klabunde, and C M Sorensen

15.1 INTRODUCTION

This chapter focuses primarily on the magnetic properties of nanoparticles. As elsewhere in the book, the motivations for studying this area are both scientific and technological. Among the many applications [1–6] of nanoparticles, such as catalysts, medical diagnostics, color imaging, drug delivery systems, and pigments in paints and ceramics are those based primarily on magnetic properties such as magnetic tapes, ferrofluids, and magnetic refrigerants. Because the particle size is often less than the magnetic domain size, nanoparticles can have the special characteristic of exhibiting single-domain magnetism.

The techniques for the preparation of fine particles, discussed in part 2, including chemical reduction [7–12], thermal decomposition [13], spark erosion [14, 15], mechanical alloying [16], aerosolization [17], vapor deposition [18–24], and sputtering [25–29] can be used to produce magnetic nanoparticles. For example, Luborsky [7] prepared elongated Fe and Fe–Co fine particles with coercivity (H_c) about 1000 Oe by electrodeposition of the metal in mercury. A higher coercivity (2140 Oe) has been obtained [30] in elongated but coarser Fe particles produced by reducing geothite (α-FeOOH) particles with hydrogen gas. Reduction of transition metal ions by $NaBH_4$ or KBH_4 has been used [8–12, 30] to prepare fine Fe, Fe–Co, Fe–Co–B, and Fe–Ni–B particles with sizes in the range of 100–1000 Å, with H_c = 200–1000 Oe, and saturation magnetization σ_s = 100–140 emu g^{-1}. The technique of Co doping of the particles has been widely used [31–34] to obtain much better recording properties. Various kinds of metallic and nonmetallic additive have been used [35, 36], especially as sintering agents, in order to enhance the coercivity. Spark erosion was also used to prepare amorphous Fe(Co)–Si–B particles [14] in the range of 5 nm–25 μm, and Ni particles [15] with a wide range of sizes. Recently mechanical alloying has been employed to prepare large quantities of alloy particles [16]. Aerosol pyrolysis has also been used to prepare Fe oxide [17] and barium ferrite particles (see below). The latter particles have also been produced in glassy matrices [20]. Co and Ti doping has been used to obtain higher coercivities in barium ferrite particles [35]. Gas evaporation is the most widely used technique [18, 19, 21–24] for the production of uniform fine particles. This has led to particles with very high magnetization (200 emu g^{-1}) and coercivity (1580 Oe) [23, 24]. A similar behavior has been observed in granular solids where crystalline Fe [25, 28] and amorphous Fe–Si [37] fine particles have been embedded in insulating SiO_2 and BN matrices by sputtering.

Characterization of the surface and microstructure of the particles has included x-ray diffraction [38, 39], x-ray photoelectron spectroscopy (XPS) [39], conventional transmission electron microscopy (TEM) and high-resolution transmission electron microscopy (HRTEM) [40], and Mössbauer spectroscopy [24, 41–46]. The surfaces of the metallic particles are normally passivated for their environmental stability. HRTEM [40] and Mössbauer spectroscopy [45] have shown beyond doubt that the surface coating in such passivated Fe particles is not continuous but consists of microcrystals of Fe oxides (γ-Fe_2O_3 and/or Fe_3O_4). It has also been reported [38–40] that the 30 Å thick surface coating is actually a double layer, consisting of a 20 Å thick layer of Fe_3O_4 which is further covered with 10 Å of γ-Fe_2O_3. In acicular Co doped γ-Fe_2O_3 particles, the surface shell of $CoFe_2O_4$ shows an epitaxial growth [40]. For the spherical passivated metallic particles, the crystal orientation of the surface shell microcrystals is still not known.

The magnetic properties of fine particles have been found to be very different from the bulk. The reports are still controversial on the size dependence of magnetization. Luborsky [47] claimed that the σ_s of ultrafine Fe particles prepared by electrodeposition was invariant with size even for particles 15 Å in size, indicating a complete ferromagnetic coupling. However, most of the metallic particles are protected by a thin oxide layer, and exhibit much lower values of magnetization than the bulk [23, 24]. The existence of magnetic dead layers as a source for the lower magnetization in these particles has now been ruled out. Morrish *et al* [48] and Coey and Khalafella [49] were the first to point out that the lower values of magnetization observed in passivated metallic particles were due to surface spin canting. In a recent study, Pankhurst and Pollard [50] have argued this to be due to the nonsaturation effects because of the random distribution of anisotropy axes in ferrimagnetic particles. Parker and Berkowitz [51] have suggested that the canting occurs in the whole particle because of quantum size effects. On the other hand, the microscopic surface magnetic moment has been found to be higher. Tamura and Hayashi [41] found the hyperfine field coming from the interface layer of Fe passivated particles to be 8% larger than the value of bulk Fe, indicating a larger moment for the interface atoms. Similar results have also been found [52] in thin films of Fe, Ni, and Cr where the surface moments were found to be 30%, 20%, and 150%

higher than their bulk values respectively.

Finite-size effects have been shown to alter the Curie temperature (T_c) of the material [53]. A distinct increase in T_c was found in an epitaxial Fe layer upon adsorption of monolayers of Fe, Pd, Ag, or O_2; similar results have been obtained in Co and Ni thin films and fine powders [54–57]. Recently Tang *et al* [58] have observed a considerable increase in the T_C of ferrimagnetic $MnFe_2O_4$ particles. Finite-size effects also appear in the temperature dependence of magnetization.

Magnetic hysteresis studies on fine particles [17, 23] show values of effective anisotropy and coercivity orders of magnitude higher than in the bulk. Such large anisotropy could arise from magnetic surface anisotropy, unidirectional anisotropy caused by the exchange interaction at the core–shell interface, and/or the magnetoelastic energies caused by stress/pinning induced by the lattice mismatch at the coated surface. Values of K as large as 10^8 erg cm^{-3} have been reported [48] for the surface layer of γ-Fe_2O_3 particles. The magnetic state of the surface oxide plays a very important role in the temperature dependence of coercivity $H_c(T)$ of the whole particle [23, 59]. However, the hysteresis behavior of ultrafine particles is not explained consistently by any of the existing models [60].

In this chapter we will discuss the structural and magnetic properties of free-standing particles made by various techniques. These particles do not exhibit any of the matrix induced effects found in granular solids such as alloying and strain effects. In section 15.2 we describe the magnetic properties of metallic smoke particles prepared by the inert gas evaporation technique. In section 15.3 we describe metallic particles prepared by the solvated metal atom dispersion technique and compare them to the smoke particles. Section 15.4 describes mixed metal oxide particles prepared by aerosol spray pyrolysis.

15.2 SMOKE PARTICLES

15.2.1 Particle size variation and distribution

The inert gas evaporation technique [21] has been used to prepare Fe and Co particles with three different surface chemistries as described below.

(i) Passivated metal powders: the metallic powders were passivated with a constant amount of argon/air mixture to be protected from further oxidation.

(ii) Metal/Ag samples: the metallic particles were sandwiched between two Ag films as shown in figure 15.1. The metal/Ag samples were prepared in an attempt to reduce the amount of surface oxide and study its effect on the magnetic properties.

(iii) Metal/Mg samples: Fe(Co)-oxide-free metal/Mg samples were prepared (Mg was used since it is an oxygen scavenger and does not form any alloy with Fe). The Fe(Co) particles were produced in an Mg + MgO matrix after annealing at a higher temperature. The particle size obtained in this study could be varied from 50 to 400 Å, by controlling: (a) the height of the substrate,

(b) the source temperature, (c) the inert gas pressure and its atomic weight, and (d) the substrate temperature. Figure 15.2 shows the effect of gas pressure on particle size in Fe particles. As the argon pressure increases from 1 to 5 Torr, the particle size remains constant within experimental error, and above 5 Torr there is a clear increase in the particle size with argon pressure. As the inert gas pressure increases, the velocity of the particles decreases and so the radius increases. Also, when the gas pressure is low, the metal vapor covers a greater region of the substrate compared with the evaporation at higher pressure, where the smoke is confined to a much smaller surface area on the substrate (just above the source region).

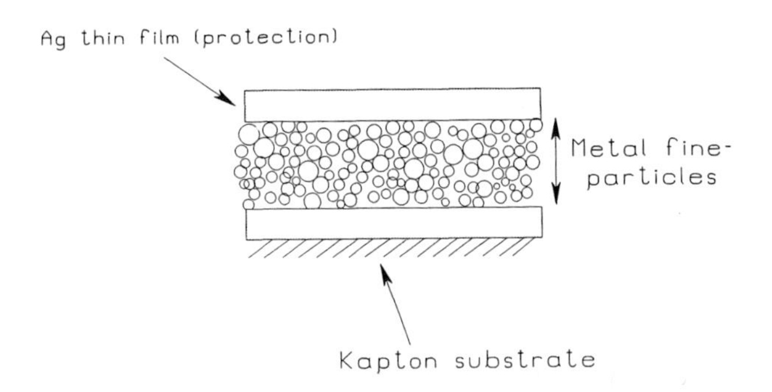

Figure 15.1. Schematic diagram showing metal particles sandwiched between two thin films of Ag.

An increase in particle size from 50 to 180 Å was observed as the substrate temperature increased from -196 to $200\,^{\circ}\mathrm{C}$, keeping other parameters constant, followed by a smaller increase with any further increase in the substrate temperature. This shows the effect of temperature on the coalescence mechanism. The particle size distribution which results from a particular preparation process depends on the growth mechanism. The mechanism of growth is by coalescence for which the logarithm of the particle volume can be approximated to be Gaussian [21]. For spherical particles the particle size distribution function is a log-normal distribution (see equation (2.16) in chapter 2). The log-normal size distribution function is a function of d the particle diameter, $\bar{d}$ the most probable diameter, and σ the geometrical standard deviation. Figure 15.3 shows the experimental distribution for an Fe sample. Smaller particles had narrower size distributions than the larger ones (figure 15.4).

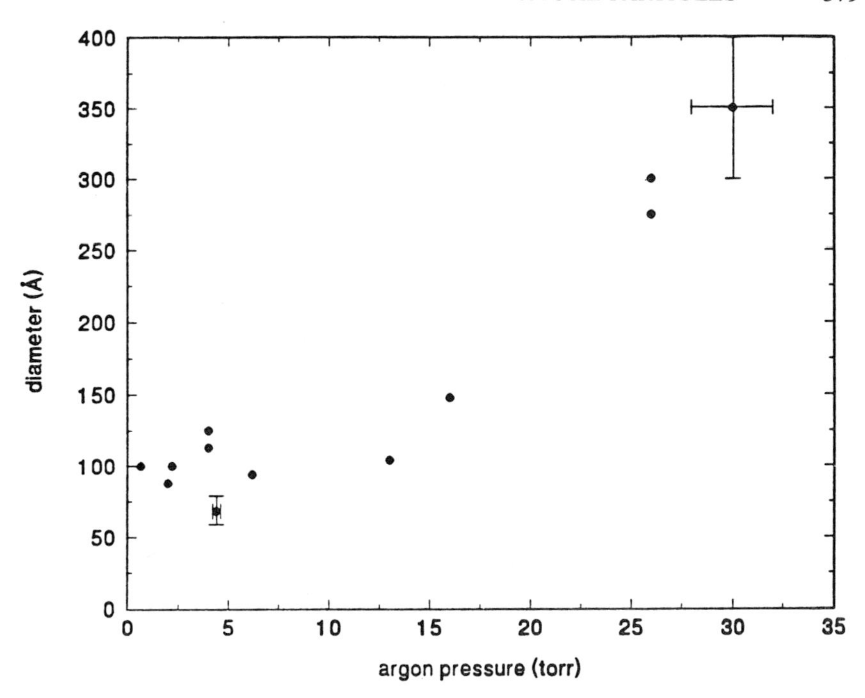

Figure 15.2. Particle size as a function of argon pressure during evaporation for passivated Fe particles.

15.2.2 Crystal structures and particle morphology

Particles produced by vapor deposition are chemically very pure and are also free of pores and other morphological irregularities. Figure 15.4 shows the typical morphology of passivated Fe particles of median diameter 140 Å. The particle shape is nearly spherical. Deviation from sphericity is higher as the particles grow larger. The maximum aspect ratio is always less than two. Similar morphologies are obtained in the case of Co particles. The particle sizes obtained from dark-field micrographs of passivated particles are smaller than those from the corresponding bright-field pictures. This indicates that the phase of the center of the particle is different from that on the surface. The tendency to form particle chains increased as the particle size increased, because of the higher magnetization of the particles and therefore higher magnetostatic interactions.

The electron diffraction patterns of passivated Fe particles show a mixture of bcc α-Fe with γ-Fe_2O_3/Fe_3O_4 oxides. No evidence of FeO and α-Fe_2O_3 has been found. Passivated Co shows fcc-Co with CoO/Co_3O_4 oxides. The diffraction lines due to the metal become broader as the particle decreases; whereas the lines due to the oxide phase are always broad and diffuse. Metal/Ag samples have a structure similar to the passivated particles, except that no oxides could be detected from the electron or x-ray diffraction studies.

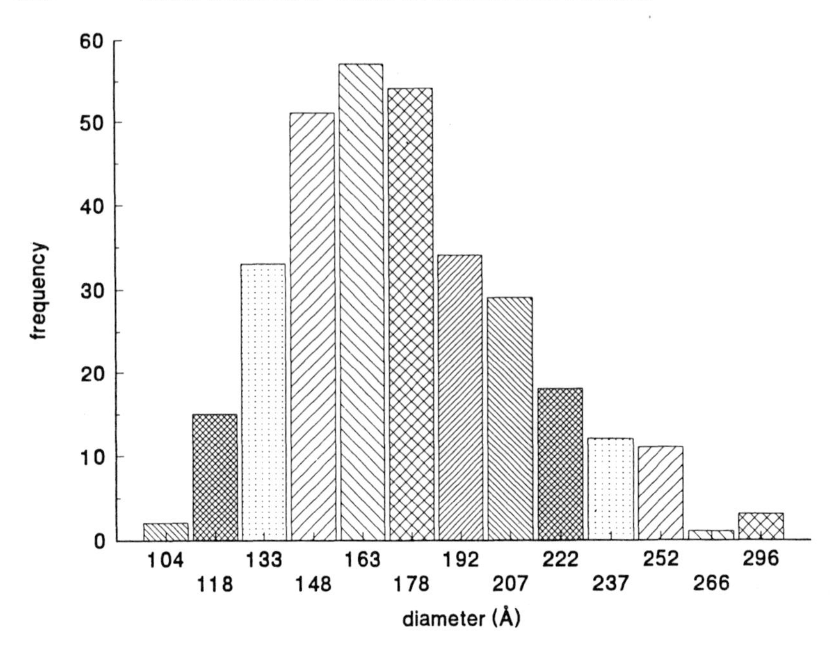

Figure 15.3. Particle size distribution in a typical Fe sample showing a log-normal distribution. The particles have a most probable diameter of 163 Å and geometrical standard deviation of 1.23.

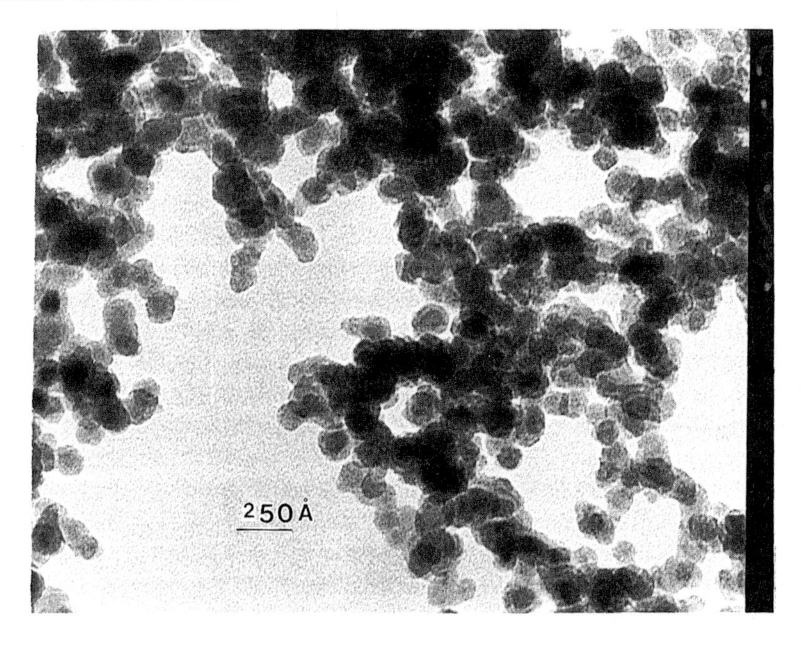

Figure 15.4. Bright-field TEM of Fe particles with a most probable diameter of 140 Å.

Fe–Mg samples were heat-treated at temperatures between 200 and 400 °C to obtain particles in the size range of 30–400 Å. As-made Fe–Mg samples showed sharp Mg peaks in addition to a very broad peak near the α-Fe (bcc) [100] peak. The initial sample was characterized by Mössbauer spectroscopy as amorphous Fe_xMg_{1-x}, where x is less than 0.3. As the sample was heat-treated, Fe segregated out and its diffraction lines became stronger with increasing annealing temperature. Also some lines due to MgO were observed.

15.2.3 Effects of particle size and surface chemistry on magnetic properties

15.2.3.1 Magnetization behavior

Figure 15.5 shows the change in σ_s with particle size in passivated samples. The maximum value of σ_s obtained was about 90% of the corresponding bulk value in Fe and 78% in Co, for particles with a size near 300 Å [61]. For particle sizes below 100 Å, σ_s drops very sharply. The reduction in σ_s with decreasing size is expected because of the presence of surface oxides due to passivation: the volume ratio of the oxides to metal becomes larger as the particle size decreases. However, much lower σ_s values are observed than those corresponding to γ-Fe_2O_3 and/or Fe_3O_4 (average σ_s of γ-Fe_2O_3 and Fe_3O_4 is about 90 emu g^{-1}). As the samples age, some initial deterioration of the magnetization happens during the first couple of weeks and then it stabilizes to about 70–80% of its initial value [23].

Figure 15.6 shows the Mössbauer spectra of a 113 Å passivated Fe particle sample. The 300 K sextet is due to α-Fe and the broad component which splits at lower temperatures is due to γ-Fe_2O_3 and/or Fe_3O_4. The presence of the broad doublet in the oxide spectra (instead of a sharp one) indicates interaction between the metallic Fe and the Fe oxides [41–45]. This feature confirms the presence of oxides around the particles instead of isolated oxide particles. The atomic fraction of the Fe and Fe oxides obtained from Mössbauer spectra was used to calculate the expected magnetization of the sample and compare it to that obtained from superconducting quantum interference device (SQUID) magnetization measurements (table 15.1). As the particle size decreases below 100 Å, the difference between the experimental and expected σ_s values grows larger, indicating a loss of moment. Using the linear weight ratios of the shell and the core [10], the estimated effective shell thickness was found to be about 25 Å for Fe and 12 Å for Co. These numbers are consistent with recent high-resolution TEM studies shown in figure 15.7. The core lattice corresponds to Fe (Co) and the polycrystalline shell is of thickness similar to that expected from the magnetic data.

Our magnetic, structural, and Mössbauer results on small spherical metallic oxide coated particles (< 100 Å) support the presence of spin canting. The most probable configuration seems to be one in which the surface/interface spins are canted resulting in an overall decrease of the magnetization. Our preliminary

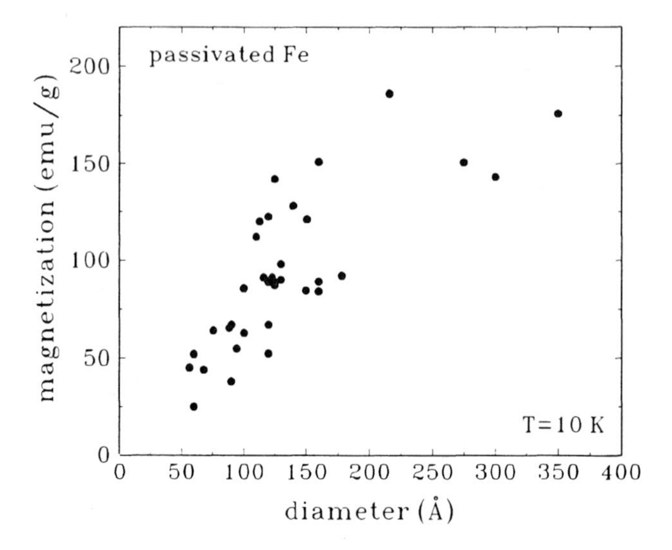

Figure 15.5. Variation of saturation magnetization with particle size in passivated Fe particles.

Table 15.1. Comparison of saturation magnetization obtained from Mössbauer spectroscopy and SQUID magnetometer.

Particle diameter (Å)	Fe (wt%)	Fe oxide (wt%)	Magnetization Mössbauer (emu g^{-1})	Magnetization SQUID (emu g^{-1})	Error (%)
275	41.6	58.4	144.6	150	4
214	27.5	72.2	126.6	135	6
113	30.0	70.0	129.5	120	−8
100	16.5	83.5	112.0	91	−23
88	7.3	92.7	100.2	65	−54

data on Mg coated Fe particles show that σ_s is about 200 emu g^{-1}, and is independent of particle diameter (25–400 Å). This may be a very profound result indicating that σ_s does not show finite-size effects in pure metallic particles. The reduction in σ_s as observed in various studies has been attributed to the ferrimagnetic species present in the samples. A very recent scanning tunneling microscopy study [24] on a magnetite single crystal has shown that the large surface anisotropy caused by Fe^{2+} ions is responsible for the surface pinning of moments and hence the reduction in magnetization.

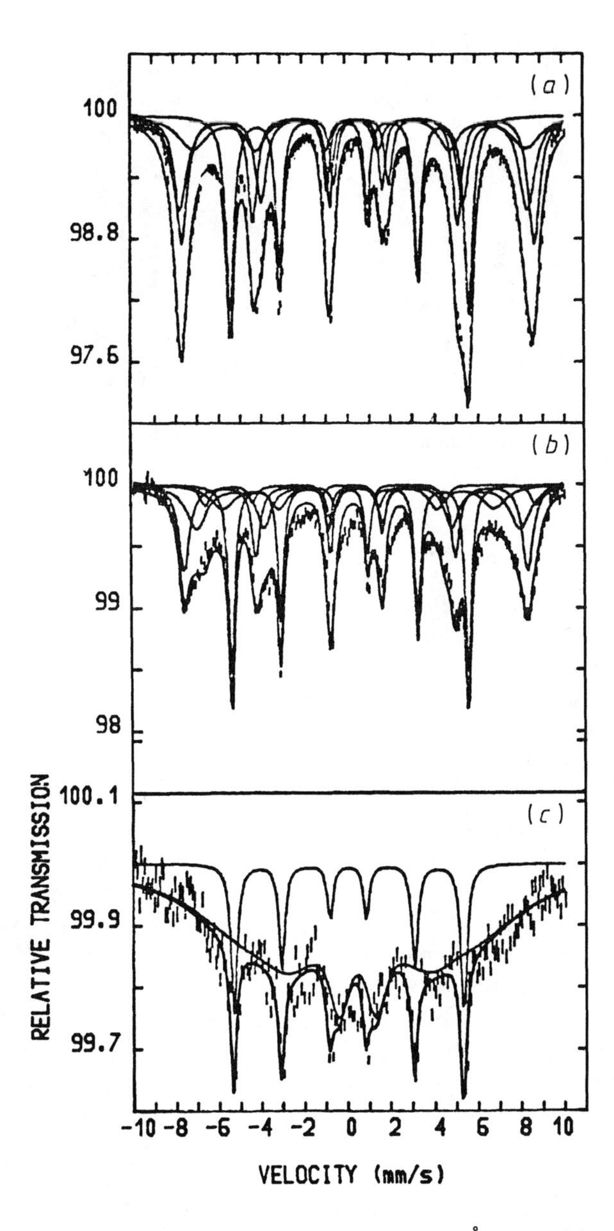

Figure 15.6. Mössbauer spectra and model fits of a 113 Å Fe particle sample at (*a*) 4.2 K, (*b*) 85 K, and (*c*) 300 K.

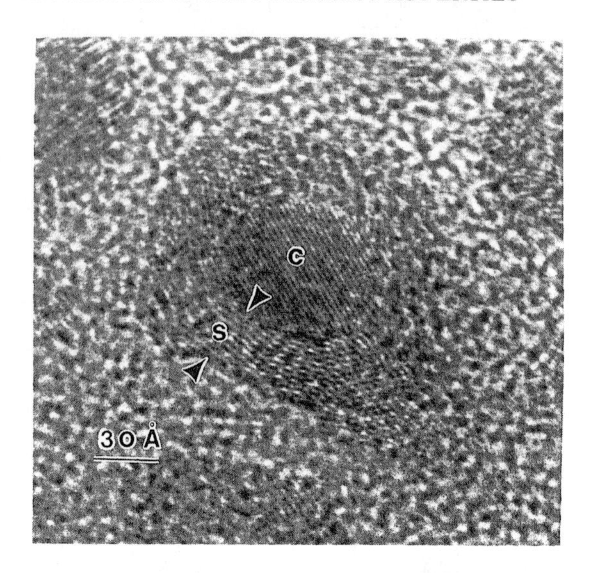

Figure 15.7. High-resolution TEM of an Fe sample showing a clear 'core–shell' morphology, where 'c' or core is of α-Fe, and 's' or shell is of γ-Fe_2O_3/Fe_3O_4.

15.2.3.2 Coercivity behavior

Passivated particles. Passivated metallic particles show a strong size dependence of H_c. This was first demonstrated by Kneller and Luborsky [62] in Fe–Co alloy particles. Figure 15.8 shows the size and temperature dependence of coercivity in Fe particles. The maximum room-temperature coercivity obtained is 1150 Oe (275 Å) in Fe, and 1500 Oe (350 Å) in Co [63]. Previously, H_c values of around 1000 Oe have been reported in passivated Fe particles [27]. Particles with a size less than 70 Å diameter are superparamagnetic below room temperature. The size dependence of H_c is reversed at cryogenic temperatures with the smallest particles having higher values of H_c. This can be related to the fact that in very small particles the surface areas, and therefore the oxide fraction, are large. Similar results were also obtained in passivated Co particles. In Co the temperature dependence is more involved at cryogenic temperatures and details have been reported elsewhere [64].

M/Ag particles. In the M/Ag series the amount of oxide was minimized (to less than 10 at.%) and this was confirmed using Mössbauer spectroscopy and XPS. Figure 15.9 shows the temperature dependence of H_c for various sizes of Fe/Ag samples. The anomalous behavior observed in the passivated particles (figure 15.2) has now disappeared and at 10 and 300 K the coercivity increases with particle size. The maximum room-temperature coercivity in the case of Fe is now 1400 Oe for 172 Å particles and for Co 1100 Oe for 130 Å particles. The multidomain behavior in M/Ag samples begins at a smaller total diameter than the passivated counterparts, indicating the importance of the core size [64].

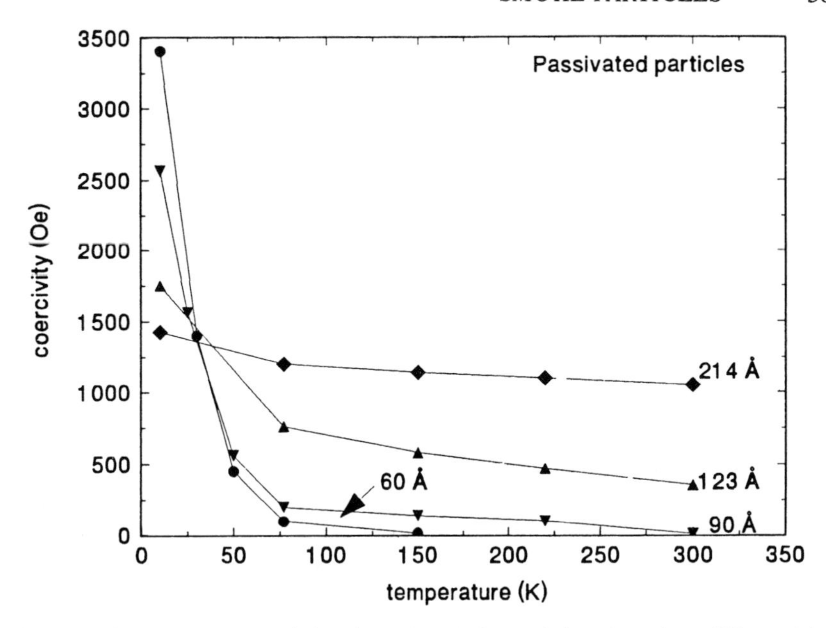

Figure 15.8. Temperature and size dependence of coercivity of passivated Fe particles.

Fe–Mg particles. Our preliminary data on co-evaporated Fe–Mg samples did not show any Fe oxides but only α-Fe, Mg, and MgO. The particle size was varied from 25 to 400 Å (measured by x-ray line broadening and TEM) by annealing the Fe–Mg samples at different temperatures (200–450 °C). The coercivity does not change much with temperature [64]. The particles with a diameter greater than 200 Å have much lower coercivities (200–300 Oe) at room temperature than their passivated counterparts. These results are remarkable as they illustrate the crucial role played by the surface chemistry in dictating the entire magnetic hysteresis behavior of metallic particles. The coercivities of the particles with the three different surface chemistries are compared in figure 15.10.

15.2.4 'Core–shell' particle model

The coercivity values obtained in passivated Fe and Co particles are much higher than those expected from classical magnetization reversal models [65–67]. As mentioned earlier, the coercivity can be influenced by the particle morphology. In this study, all the experimental data including magnetization, Mössbauer spectroscopy, and HRTEM show that the passivated particles have a 'core–shell' type of morphology (figures 15.7 and 15.11). The core is metallic and the shell is composed of metal oxides. The magnetic interaction between the core and the shell and its strong temperature dependence affect the hysteresis behavior drastically. When the oxide fraction is completely removed, both the large values

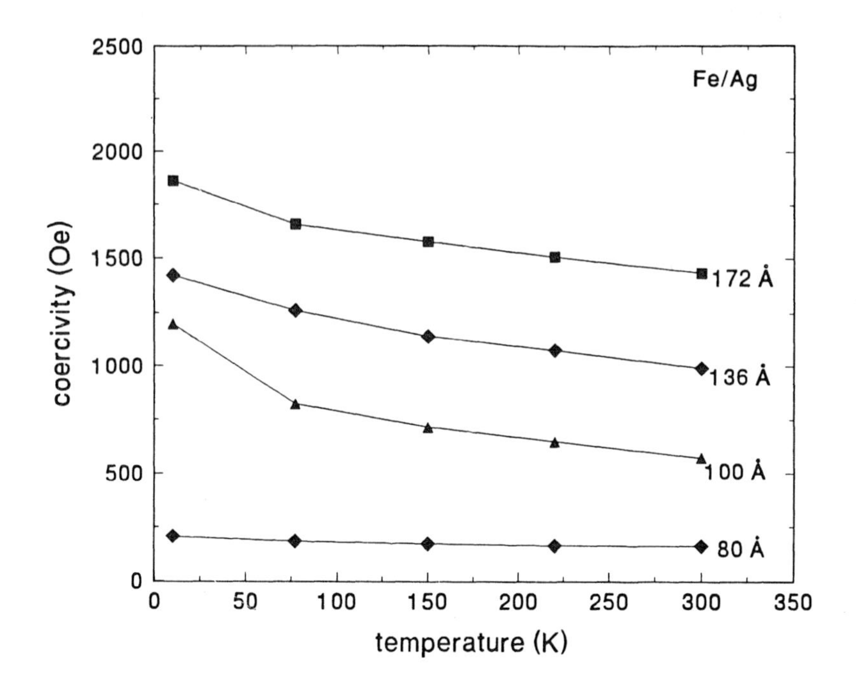

Figure 15.9. Temperature and size dependence of coercivity of Fe/Ag particles.

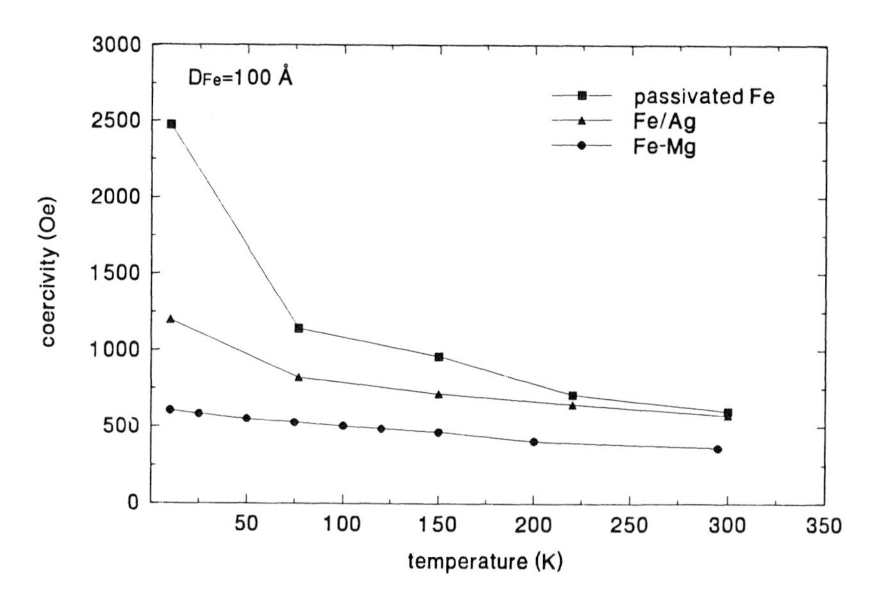

Figure 15.10. Comparison of hysteresis among the different surface morphologies.

of coercivity and its strong temperature dependence disappear (figures 15.9 and 15.10). Thus magnetic interactions among the particles and the exchange term

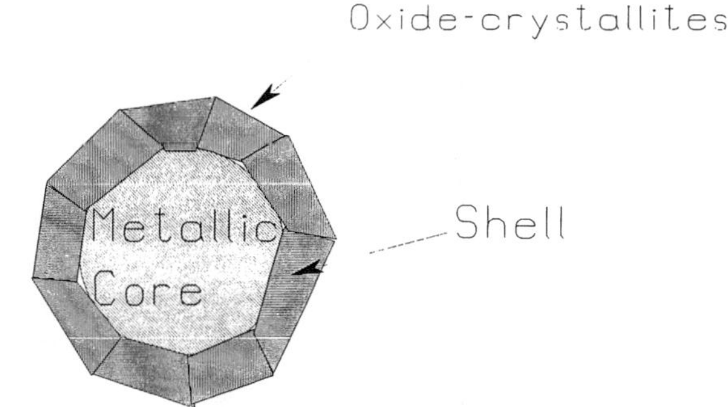

Figure 15.11. 'Core–shell' particle morphology.

which couples the shell crystallites to the core must be considered in addition to the other energy terms when modeling the magnetization reversal of the coated small magnetic particles. The oxide shell does not grow epitaxially with respect to the core as seen in figure 15.7. One could minimize the total energy to find the equilibrium angles between the core moment with the various shell moments. It may be that the easy axes of the individual oxide crystallites are oriented in a certain crystallographic direction with respect to the core, because the oxide shell grows in the presence of the demagnetization field of the core. The exchange coupling between the two magnetic species (core and shell) is also a function of a length ratio, namely the total diameter to the core diameter. The temperature dependence of the exchange coupling constant (or unidirectional anisotropy) and the superparamagnetism of the oxide shell determine the temperature dependence of the coercivity of the whole particle.

15.3 SOLVATED METAL ATOM DISPERSION TECHNIQUE (SMAD)

15.3.1 The synthetic method

In the solvated metal atom dispersion (SMAD) technique [68, 69], metal atoms are codeposited with a large excess of organic substrate at a low temperature, thereby promoting reaction between the metal atoms and the substrate and suppressing recombination to the bulk metal. This technique has led to fundamentally new materials, such as non-aqueous colloidal metal solutions [70], solvated metal atom dispersed catalysts [71], bimetallic clusters [72], and magnetic cluster/particles [73–76]. The principle is that the reactive metal atom is moved to a zone where a diluent (solvent) can be added at low temperatures. The metal atom is usually in its ground state, but is very reactive since it possesses no steric restrictions and its orbits are readily available. However, since the interaction of the metal atom with the substrate usually occurs at low temperatures (10–150 K), only low-energy processes proceed. If such a low-energy reaction pathway does not exist, the metal atoms simply repolymerize back to the bulk solid material. The agglomeration of the metal atoms competes with the reaction between metal atoms and the solvent. Interestingly, intermediate cases can be found where metal atom clustering is so profoundly affected that particle growth is stopped by solvation. Thus competing pathways with the solvent are possible, and lead to small metal particles containing some organic material which changes the chemical and magnetic properties of the particles.

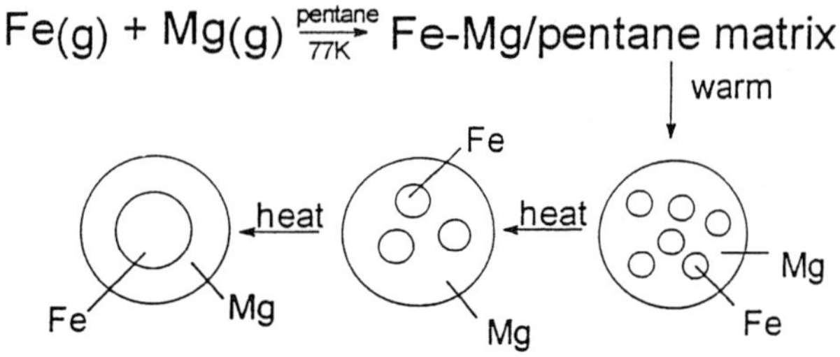

$$\mathrm{Fe(g)} + \mathrm{Mg(g)} \xrightarrow[77K]{pentane} \text{Fe-Mg/pentane matrix}$$

Figure 15.12. Reaction scheme for SMAD production of magnesium coated iron particles.

The Fe–Li/pentane and Fe–Mg/pentane bimetallic systems [75, 76] have proven to be very interesting. In addition to the metal–substrate interactions mentioned above, metal–metal interaction is also interesting since Fe is immiscible with Li and Mg in the bulk. At room temperature, very small α-Fe crystallites embedded in a matrix of nanocrystalline Li or Mg were obtained. Particle sizes averaged 210 Å with α-Fe crystallite regions of about 23–38 Å. Heat-treatment of the fresh powders caused further phase separation of Fe and Li(Mg), with an α-Fe crystallite growth. Air passification formed a protective outer layer. By choosing proper temperature and time of heating, the crystallite size of α-Fe could be controlled (figure 15.12). These materials had a very high surface area, usually over 100 $m^2\ g^{-1}$. Growth of the Fe–Li and Fe–Mg particles in this way (out of a cold pentane matrix) allowed free-flowing powders to be formed. Furthermore, α-Fe crystallite size could be controlled by heat treatment, and, after passification, air stable powders were obtained [75, 76]. The Fe–Mg system proved to be the most interesting [77].

15.3.2 Magnetic properties

Thus, for the Fe–Mg system two methods have been used for their synthesis. In section 15.2 we described the 'smoke' method and in this section the 'SMAD' method. How do the magnetic properties compare?

15.3.2.1 Magnetization behavior

Magnetization data were collected for a series of powder Fe–Mg samples where α-Fe crystallite sizes ranged from 30 to 200 Å, and measurement temperatures were 5, 150, and 300 K. Interestingly, σ_s rose from 50 emu g^{-1} Fe to 150 emu g^{-1} as particle size ranged from 30 to 70 Å. Then a rather sharp increase to 170 emu g^{-1} was found for 100 Å α-Fe crystallites, and then a slow rise from 170 to 200 emu g^{-1} for the α-Fe crystallites of size range 130–200 Å.

These results are very similar to those of passivated Fe particles (see figure 15.5) but distinctly different from the Fe–Mg particles obtained by the 'smoke' method. These results show how important the initial microstructure is on the magnetic properties of the particles formed.

15.3.2.2 Coercivity behavior

The SMAD generated particles were protected from oxidation much in the same way as the smoke generated Fe–Mg particles. In the absence of iron oxides, the same remarkable behavior was observed. When the diameter of the α-Fe crystallites was > 70 Å, the samples showed superparamagnetic behavior as identified by σ versus H/T curves after a small ferromagnetic fraction was subtracted. As α-Fe size increased, eventually the sample became wholly ferromagnetic. Coercivity measurements as a function of size indicated very small H_c values (50–150 Oe) independent of size, and did not change much with T [77].

Overall, these results are very satisfying since both the SMAD and the 'smoke' Fe–Mg core/shell particles behaved essentially the same. Furthermore, they reinforce the idea that surface chemistry can dominate the various parameters that can affect σ_s and especially H_c in nanoscale particles.

15.4 AEROSOL SPRAY PYROLYSIS

15.4.1 The synthetic method

In the aerosol spray pyrolysis technique, aqueous metal salts are sprayed as a fine mist, dried, and then passed into a hot flow tube where pyrolysis converts the salts to the final products. This technique has two main advantages in synthesizing fine particles. First, materials are mixed in solution, hence they are homogeneously mixed on the atomic level at the start. Second, only subsintering temperatures are necessary to form crystallized particles. This method of synthesis creates particle sizes typically in the range of 50 to 5000 Å.

In our work using spray pyrolysis [17, 78–81], we have created a number of small-particle ferrites and garnets. We discovered that a novel phase may appear, and that completion of chemical reactions to desired products occurs quickly. Both of these results are apparently due to the small particle sizes obtained compared with more conventional methods.

15.4.2 $BaFe_{12}O_{19}$

To make barium ferrite particles [81], $Fe(NO_3)_3 \cdot 9H_2O$ and $Ba(NO_3)_2$ were dissolved in distilled deionized water with an Fe/Ba atomic ratio of 12. This solution was spray pyrolyzed. The as-received particles were spherical with an average diameter of 800 Å and amorphous. Annealing under N_2 crystallized these amorphous particles to $BaO \cdot 6Fe_2O_3$. The morphology changed from spherical to hexagonal platelets with average thickness of 350 Å and a 3.5 to 1 aspect ratio. The saturation magnetization was 70.6 emu g^{-1} at 300 K which is essentially the known bulk value (72 emu g^{-1}). Some variation of σ_s and the coercivity was seen with annealing temperature. The highest coercivity obtained was 5360 Oe after a 1000 °C anneal.

15.4.3 $Gd_3Fe_5O_{12}$

Gadolinium iron garnet is one of a general class of rare earth garnets and is a magneto-optical material. We created it by spray pyrolysis of an aqueous $Gd(NO_3)_3$ and $Fe(NO_3)_3$ solution [79]. Once again the as-prepared particles were spherical and amorphous, and post-aerosol annealing was necessary to obtain the garnet. During annealing, intermediate phases of $GdFeO_3$ and Fe_2O_3 appeared but with sufficient time or temperature the more stable garnet phase resulted. Once again bulk magnetic saturation values were obtained (96 emu g^{-1} at 0 K). These submicron particles also displayed the bulk compensation temperature at 290 K and Curie temperature at 560 K.

15.4.4 $MnFe_2O_4$

Manganese ferrite was prepared by spray pyrolysis of an aqueous solution of $MnCl_2$ or $Mn(NO_3)_2$ with $FeCl_3$ or $Fe(NO_3)_3$ [80]. Once again intermediates such as Mn_2O_3, Mn_3O_4, and Fe_2O_3 occurred but less so with decreasing NO_3^-

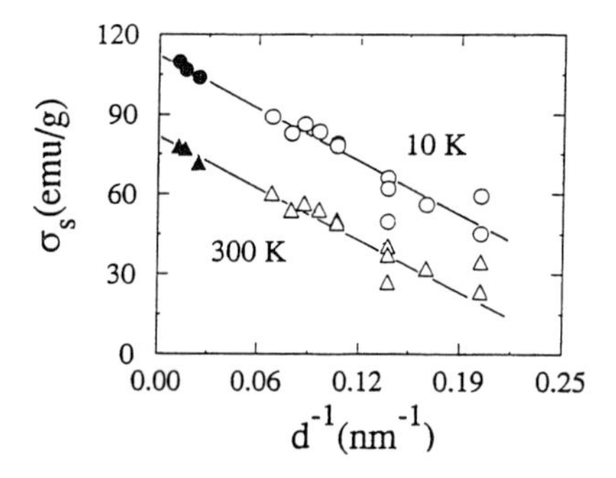

Figure 15.13. Saturation magnetization versus inverse mean diameter for $MnFe_2O_4$ particles prepared by aerosol spray pyrolysis (solid symbols) and aqueous phase precipitation (open symbols).

concentration. We found that with sufficient temperature (> 900 °C) the ferrites could be created in one aerosol step. The morphology was solid, spherical to polygonal particles with mean size that we could vary from 400 to 800 Å. Once again bulklike magnetic properties were obtained with size dependent coercivities in the range 20 to 360 Oe. One interesting small-particle property observed was that the saturation magnetization was slightly smaller than in the bulk and consistent with particles prepared by aqueous phase precipitation [82] as shown in figure 15.13. This dependence of $\sigma_s \propto d^{-1}$ is consistent with a constant-thickness dead layer on the particles.

15.5 SUMMARY

Metal and metal oxide nanoscale particles have been prepared by a variety of techniques. A great deal has been learned about synthesis, controlling particle size and the effect of particle size, and particle coating on the magnetic properties. Several important discoveries and lessons have come out of this work.

(i) Surface protection of metallic particles is absolutely necessary if one is to observe unaltered magnetic properties of the metal. We have developed a novel approach to this by vapor deposition of two immiscible metals in low-temperature matrices (with or without hydrocarbon diluent). A metastable alloy is formed such as Fe–Li, Fe–Mg, or Fe–Ag. Upon heat treatment, phase separation to form an Fe core coated with Li (or Mg) takes place; the Fe core size is controlled by temperature. It is perfectly protected from environmental oxidation by the sacrificial Li or Mg.

(ii) For core–shell structures, surface physics can dominate magnetic properties. We have developed procedures for the preparation of nonmagnetic coatings (e.g. Li, Mg, and Ag, as mentioned above), antiferromagnetic coatings (Fe coated with FeS by heat-treatment with organic thiols), and ferrimagnetic coatings (see below).

We have extensively studied vapor deposited Fe and Co particles in the size range 50–350 Å. Particles consisted of a single-phase metallic core surrounded by a polycrystalline oxide shell (10–20 Å thick). The magnetic properties showed strong size effects due to the magnetic shell. The magnetization was low and this may be due to spin canting in the particles. Coercivities were large (up to 7 kOe), dependent on shell thickness and strongly dependent on temperature, a result of a strong unidirectional exchange coupling between core and shell. This exchange vanished at 150 K in Co and 40 K in Fe particles. Interparticle interactions were found to affect hysteresis behavior as shown by packing experiments. Remanence curves demonstrated negative interactions regardless of particle size.

(iii) Stoichiometric complex metal oxide and ferrite, and garnet submicron particles of high purity, theoretical density, and bulk magnetic properties can be produced by aerosol pyrolysis. High coercivities were found to be due to single-domain particle effects. We found that the best precursors are those easily decomposed. A large variety of intermediate phases appears during pyrolysis, but with high temperature or long heat duration, single-phased dense particles are formed.

REFERENCES

[1] Matijevic E 1989 *Mater. Res. Bull.* **18–20**

[2] Matijevic E 1985 *Annu. Rev. Mater. Sci.* **15** 483

[3] Ozaki M 1989 *MRS Bull.* **35–40**

[4] White R M 1985 *Science* **229** 11

[5] Halperin W P 1986 *Rev. Mod. Phys.* **58** 533

[6] Shull R D, McMichael R D, Swartzendruber L J and Bennett L H 1992 *Studies of Magnetic Properties of Fine Particles and their Relevance to Materials Science* ed J L Dormann and D Fiorani (Amsterdam: Elsevier) p 161

[7] Luborsky F E 1961 *J. Appl. Phys.* **32** 1715

[8] Dragieva I, Gavrilov G, Buchkov D and Slavcheva M 1979 *J. Less-Common. Met.* **67** 375

[9] Watanabe A, Uehori T, Saitoh S and Iamaoka Y 1981 *IEEE Trans. Magn.* **MAG-17** 1455

[10] van Wonterghem J, Morup S, Koch C J W, Charles S W and Wells S 1986 *Nature* **322** 622; 1989 *J. Physique Coll.* C8 1369

[11] Nafis S, Hadjipanayis G C, Sorensen C M and Klabunde K J 1989 *IEEE Trans. Magn.* **MAG-25** 3641

[12] Yiping L, Hadjipanayis G C, Sorensen C M and Klabunde K J 1989 *J. Magn. Magn. Mater.* **79** 321

[13] van Wonterghem J, Morup S, Charles S W, Wells S and Villadsen J 1985 *Phys. Rev. Lett.* **55** 410

[14] Berkowitz A E, Walter J L and Wall K F 1981 *Phys. Rev. Lett.* **46** 1484

[15] Slade S B, Berkowitz A E and Parker F T 1991 *J. Appl. Phys.* **69** 5127

[16] Gilman P S and Benjamin J S 1983 *Annu. Rev. Mater. Sci.* **13** 279

[17] Tang Z X, Nafis S, Sorensen C M, Hadjipanayis G C and Klabunde K J 1989 *J. Magn. Magn. Mater.* **80** 285

[18] Tasaki A, Tomiyana S and Iida S 1965 *Japan. J. Appl. Phys.* **4** 707

[19] Hayashi C 1987 *J. Vac. Sci. Technol.* A **5** 1375

[20] Nanna L C, Arajs S and Anderson E E 1989 *APS Bull.* **34** 976

[21] Granqvist C G and Buhrman R A 1976 *J. Appl. Phys.* **47** 2200

[22] Tasaki A, Oda M, Kashu S and Hayashi C 1979 *IEEE Trans. Magn.* **MAG-15** 1540

[23] Gangopadhyay S, Hadjipanayis G C, Dale B, Sorensen C M, Klabunde K J, Papaefthymiou V and Kostikas A 1992 *Phys. Rev.* B **45** 9778

[24] Du Y W, Wu J, Lu H, Wang T, Hiu Z Q, Tang H and Walker J C 1987 *J. Appl. Phys.* **61** 3314

[25] Xiao G and Chien C L 1987 *J. Appl. Phys.* **61** 3308

[26] Xiao G and Chien C L 1987 *J. Appl. Phys.* **51** 1280

[27] Xiao G, Liou S H, Levy A, Taylor J N and Chien C L 1986 *Phys. Rev.* B **34** 7573

[28] Edelstein A S, Das B N, Holtz R L, Koon N C, Rubinstein M, Wolt S A and Kihlsran K E 1987 *J. Appl. Phys.* **61** 3320

[29] Chow G W, Pattnaik A, Schlesinger T E, Cammarata R C, Twigg M E and Edelstein A S 1991 *J. Mater. Res.* **6** 737

[30] Oppergard A L, Darnell F J and Miller H C 1961 *J. Appl. Phys.* **32** 1845

[31] Hallerand W D and Collins R M 1971 *US Patent* 3 573 980

[32] Umeki S, Saitoh S and Imaoka Y 1974 *IEEE Trans. Magn.* **MAG-10** 655

[33] Tokuoka Y, Umeki S and Imaoka Y 1977 *J. Physique Coll.* **38** C1 377

[34] Imaoka Y, Umeki S, Kubota Y and Tokuoka Y 1978 *IEEE Trans. Magn.* **MAG-14** 649

[35] Kubo O, Ido T and Yokoyama H 1982 *IEEE Trans. Magn.* **MAG-18** 1122

[36] Templeton T L, Arrott A S, Curzon A E, Frindt R F, Gee M A and Li X Z 1992 *IEEE Trans. Magn.* **MAG-28** 2698

[37] Holtz R L, Edelstein A S, Lubitz P and Gossett C R 1988 *J. Appl. Phys.* **64** 4251

[38] van Diepen A M, Vledder H J and Langeris C 1977 *Appl. Phys.* **15** 163

[39] Xiukui S, Wenxiu C, Jian X, Xueshu F, Wenduo W and Wenhao W 1993 *Nanostruct. Mater.*

[40] Kishimoto M, Kitahata S and Amemiya M 1986 *IEEE Trans. Magn.* **MAG-22** 732

[41] Tamura I and Hayashi M 1984 *Surf. Sci.* **146** 501

[42] Shinjo T, Shigematsu T, Hosaito N, Iwasaki T and Takai T 1982 *Japan. J. Appl. Phys.* **21** L220

[43] Haneda K and Morrish A H 1979 *Nature* **282** 186

[44] Papaefthymiou V, Kostikas A, Simopoulos A, Niarchos D, Gangopadhyay S, Hadjipanayis G C, Sorensen C M and Klabunde K J 1990 *J. Appl. Phys.* **67** 4487

[45] Haneda K and Morrish A H 1978 *Surf. Sci.* **77** 584

[46] Morup S and Topsoe S 1976 *Appl. Phys.* **11** 63

[47] Luborsky F E 1958 *J. Appl. Phys.* **29** 309

[48] Morrish A H, Haneda K and Schiurer P J 1976 *J. Physique Coll.* C6 301

[49] Coey J M D and Khalafella D 1972 *Phys. Status Solidi* **11** 229

[50] Pankhurst Q A and Pollard R J 1991 *Phys. Rev. Lett.* **67** 248

[51] Parker F T and Berkowitz A E 1991 *Phys. Rev.* B **44** 7437

[52] Hasegawa H and Herman F 1988 *J. Physique Coll.* C8 1677

[53] Berkowitz A E and Walter J L 1982 *Mater. Sci. Eng.* **55** 275

[54] Weber W, Kerkmann D, Pescia D, Wesner D A and Guntherodt G 1990 *Phys. Rev. Lett.* **65** 2058

[55] Schneider C M, Bressler P, Schuster P, Kirschner J, de Miguel J J and Miranda R 1990 *Phys. Rev. Lett.* **64** 1059

[56] Rodmacq B and dos Santos Carlos A 1992 *J. Magn. Magn. Mater.* **109** 298

[57] Du Y W, Ku M, Wu J, Shi Y and Lu H 1991 *J. Appl. Phys.* **70** 5903

[58] Tang Z X, Sorensen C M, Klabunde K J and Hadjipanayis G C 1991 *Phys. Rev. Lett.* **67** 3602

[59] Gangopadhyay S, Hadjipanayis G C, Sorensen C M and Klabunde K J 1992 *IEEE Trans. Magn.* **MAG-28** 3174

[60] Aharoni A 1987 *IEEE Trans. Magn.* **Mag-22** 478

[61] Coey J M D, Shvets I V, Wiesendanger R and Guntherodt H-J 1993 *J. Appl. Phys.* **73** 6742

[62] Kneller E F and Luborsky F E 1963 *J. Appl. Phys.* **34** 656

[63] Gangopadhyay S, Hadjipanayis G C, Shah S I, Sorensen C M, Klabunde K J, Papaefthymiou V and Kostikas A 1991 *J. Appl. Phys.* **70** 5888

[64] Gangopadhyay S, Hadjipanayis G C, Sorensen C M and Klabunde K J *Nanophase Materials: Synthesis—Processes—Applications (series E, Appl. Sciences 260)* ed G C Hadjipanayis and R W Siegel (Dordrecht: Kluwer) p 573

[65] Schabes M E 1991 *J. Magn. Magn. Mater.* **95** 249

[66] Paul D and Cresswell A *Phys. Rev.* B submitted

[67] Victora H 1987 *Phys. Rev. Lett.* **58** 1788

[68] Klabunde K J, Timms P L, Skell P S and Ittel S 1979 *Inorg. Synth.* **19** 59

[69] Groshens T J and Klabunde K J 1987 *Experimental Organometallic Chemistry (ACS Symposia Series 357)* ed A L Wayda and M Y Darensbourg (Washington, DC: American Chemistry Society) p 190

[70] Lin S T, Franklin M T and Klabunde K J 1986 *Langmuir* **2** 259

[71] Klabunde K J and Imizu Y 1983 *J. Mol. Catal.* **21** 57

[72] Klabunde K J and Imizu Y 1984 *J. Am. Chem. Soc.* **106** 2721

[73] Kernizan C F, Klabunde K J, Sorensen C M and Hadjipanayis G C 1990 *Chem. Mater.* **2** 70

[74] Kernizan C F, Klabunde K J, Sorensen C M and Hadjipanayis G C 1990 *J. Appl. Phys.* **672** 5897

[75] Glavee G N, Kernizan C F, Klabunde K J, Sorensen C M and Hadjipanayis G C 1991 *Chem. Mater.* **3** 967

[76] Glavee G N, Easom K, Klabunde K J, Sorensen C M and Hadjipanayis G C 1992 *Chem. Mater.* **4** 1360

[77] Klabunde K J, Zhang D, Glavee G N, Sorensen C M and Hadjipanayis G C 1994 *Chem. Mater.* **6** 784

[78] Nafis S, Tang Z X, Dale E B, Sorensen C M, Hadjipanayis G C and Klabunde K J 1988 *J. Appl. Phys.* **64** 5835

[79] Xu H K, Sorensen C M, Hadjipanayis G C and Klabunde K J 1992 *J. Mater. Res.* **7** 712

[80] Li Q, Sorensen C M, Klabunde K J and Hadjipanayis G C 1993 *Aerosol Sci. Technol.* **19** 453

[81] Tang Z X, Nafis S, Sorensen C M, Hadjipanayis G C and Klabunde K J 1989 *IEEE Trans. Magn.* **MAG-25** 4236

[82] Tang Z X, Sorensen C M, Hadjipanayis G C and Klabunde K J 1991 *J. Colloid Interface Sci.* **146** 38

Chapter 16

Optical characterization and applications of semiconductor quantum dots

N Peyghambarian, E Hanamura, S W Koch, Y Masumoto, and E M Wright

16.1 INTRODUCTION

The possibility of manipulating the optical properties of semiconductors through various degrees of dimensional or quantum confinement has attracted considerable attention during the last decade. Even though the best known examples are still quantum-well structures which confine one motional degree of freedom of the excited electron–hole pair, quantum wires and quantum dots are becoming increasingly important since they yield two-dimensional and three-dimensional quantum confinement, respectively.

Three-dimensional quantum-confinement effects in semiconductor microcrystallites occur when the particle size approaches the bulk exciton Bohr radius. This confinement gives rise to interesting new effects, leading to novel optical properties. These properties of semiconductor microcrystallites make them potentially attractive for applications in optoelectronic devices, such as optical data storage and high-speed optical communication.

There has been a growing interest in searching for systems that exhibit such three-dimensional confinement effects and in understanding their behavior. A number of laboratories have attempted to fabricate quasi-zero-dimensional structures using a variety of techniques, including colloidal suspension of semiconductor particles [1, 2], electron-beam lithography [3–5], and semiconductor microcrystallites in glass matrices [6–13].

It has been shown that special glasses doped with CdS, CdSe, CuCl, or CuBr crystallites [6] can be fabricated that clearly exhibit quantum confinement. The microcrystallites in these glasses form out of the supersaturated solid solution of the basic constituents originally brought into the glass melt. The crystallites are more or less randomly distributed in the glass matrix. Ekimov *et al* [6] report crystallite growth following the growth law $R \propto t^{1/3}$, where R is the crystallite size and t is the duration of the heat treatment during which the crystallites grow. Average crystallite sizes from around 10 Å up to several 100 Å have been obtained. The experimental results discussed in this chapter are all related to such semiconductor confined particles in glass. Data are included for CdS, CdSe, and CdTe quantum dots in glass.

The theoretical analysis is based on the fact that the spatial dimensions of the quantum dots are usually large compared with the lattice constant of the semiconductor material. Therefore, it is a reasonable first assumption to consider the band structure of these mesoscopic systems as only weakly changed in comparison to the corresponding bulk material. This means that the Bloch part of the one-particle wavefunctions is the same as that in bulk, but the exponential part of the bulk electron wavefunction is replaced by a function which satisfies the quantum mechanical boundary conditions of the microstructure. In practice, one often uses the effective mass approximation for the envelope function describing the states in the vicinity of the fundamental absorption edge. This means that the envelope function is simply the wavefunction of a particle in the appropriate three-dimensional potential well.

Based on this envelope function model, we discuss a numerical method which allows the computation of the linear and third-order nonlinear optical response of quantum dots. The results of this theory are compared to experimental observations allowing us to identify the linear and nonlinear optical contributions of the one- and two-electron–hole-pair states. For large quantum dots we discuss model calculations predicting superradiance from microcrystallites and interacting Frenkel excitons.

We also describe the application of quantum dots as new lasers, switching elements, and optical storage devices. We explore the merits of using quantum dots for all-optical waveguide switching applications. Basically, most nonlinear all-optical switching devices require the accumulation of a nonlinear phase difference between two waveguides. However, absorption clearly limits the useful device size, and there is a basic trade-off between nonlinear refractive index and absorption. Here we explore this trade-off for quantum dots by introducing a material figure of merit [14]. We then describe the potential for using quantum dots for optical data storage applications. Spectral hole burning is a mechanism for this application. We show experimental results demonstrating room-temperature spectral hole burning.

16.2 OPTICAL CHARACTERIZATION

16.2.1 Linear optical properties

To understand the optical absorption in quasi-zero-dimensional semiconductors, we consider spherical semiconductor microcrystallites with radii R and background dielectric constant ϵ_2 embedded in another material with background dielectric constant ϵ_1. The radius of the quantum dot is usually on the order of a few tens of ångströms (Å), comparable to the bulk exciton Bohr radius. This closely models the system of semiconductor crystallites in glass which has been studied extensively.

For microcrystals with radius R in the range $a \ll R \simeq a_B$, where a is the lattice constant of the semiconductor and a_B is the exciton Bohr radius, the single-electron properties are determined by the periodic lattice. Hence, the quantum dot has a macroscopic size in comparison to the unit cell, but it is small compared with all other macroscopic scales. In such microcrystallites, which are usually categorized as mesoscopic structures, the effective-mass approximation is valid. The electrons and holes are, thus, assumed to have the effective masses m_e and m_h, respectively, of the bulk material. Optically excited electron–hole pairs are influenced by the small size of the microcrystals, leading to quantum-confinement effects.

For the experimentally relevant case of spherical quantum dots, the single-particle Schrödinger equations for the electron and the hole in the absence of Coulomb interaction can be written as

$$-\frac{\hbar^2}{2m_i}\nabla^2\zeta_i(\boldsymbol{r}) = \mathcal{E}_i\zeta_i(\boldsymbol{r}) \tag{16.1}$$

where $i = e$ or h. The boundary condition of ideal quantum confinement dictates that

$$\zeta_i(\boldsymbol{r}) = 0 \qquad \text{for } r = R. \tag{16.2}$$

The solution of the Schrödinger equation (16.1) with the spherical boundary condition, equation (16.2), is

$$\zeta_i(\boldsymbol{r}) = \sqrt{\frac{1}{4\pi R^3}}\frac{j_\ell\left(\alpha_{n\ell}\frac{r}{R}\right)}{j_{\ell+1}(\alpha_{n\ell})}Y_\ell^m(\theta,\phi) \tag{16.3}$$

where j_ℓ is the ℓth-order spherical Bessel function, $Y_\ell^m(\theta,\phi)$ are the spherical harmonics, $\alpha_{n\ell}$ is the nth root of the ℓth-order Bessel function. The quantum numbers for the particle are n, ℓ, and m. The boundary condition equation (16.2) is satisfied if

$$j_\ell\left(\alpha_{n\ell}\frac{r}{R}\right)\bigg|_{r=R} = 0 \tag{16.4}$$

or

$$j_\ell(\alpha_{n\ell}) = 0. \tag{16.5}$$

Equation (16.5) is satisfied for

$$\begin{array}{llll} \alpha_{10} = \pi & \alpha_{11} = 4.4934 & \alpha_{12} = 5.7635 & \alpha_{20} = 6.2832 \\ a_{21} = 7.7253 & \alpha_{22} = 9.0950 & \alpha_{30} = 9.4248 \text{ etc.} & \end{array} \tag{16.6}$$

Inserting equation (16.3) into equation (16.1) gives the discrete energy eigenvalues as

$$\mathcal{E}_i = \frac{\hbar^2}{2m_i}\left(\frac{\alpha_{n\ell}}{R}\right)^2. \tag{16.7}$$

It is customary to refer to the $n\ell$ eigenstates as 1s, 1p, 1d, etc, where s, p, d, etc, correspond to $\ell = 0, 1, 2, \ldots$ respectively ($\alpha_{10} = \alpha_{1s}, \alpha_{11} = \alpha_{1p}$, etc). Note the somewhat unusual notation for atomic spectroscopists, for whom a 1p state would not be possible. The difference arises from the fact that we are not dealing with a Coulomb potential, but a spherical confinement potential.

The lowest-energy states are those with the lowest $\alpha_{n\ell}$ values. Examination of equation (16.6) shows that these lowest energies are those with α_{1s}, α_{1p}, α_{1d}, etc. Taking the zero of energy at the top of the valence band, the electron and hole energy levels are then given by using equation (16.7)

$$\mathcal{E}^e = E_g + \frac{\hbar^2}{2m_e}\left(\frac{\alpha_{n_e\ell_e}}{R}\right)^2 \tag{16.8}$$

and

$$\mathcal{E}^h = -\frac{\hbar^2}{2m_h}\left(\frac{\alpha_{n_h\ell_h}}{R}\right)^2. \tag{16.9}$$

The lowest two energy levels are plotted schematically in figure 16.1. We see from this figure that the usual three-dimensional band structure is drastically modified and has become a series of quantized single-particle states.

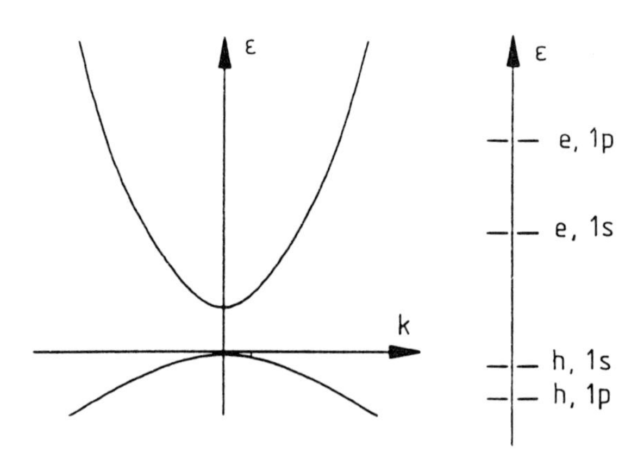

Figure 16.1. Schematic plot of the single-particle energy spectrum in bulk semiconductors (left) and in small quantum dots (right).

As we mentioned before, the single-particle spectrum does not correspond to the optical absorption spectrum, since the electron–hole Coulomb effects are excluded. The Schrödinger equation for one electron–hole pair is

$$\left(-\frac{\hbar^2}{2m_e}\nabla_e^2 - \frac{\hbar^2}{2m_h}\nabla_h^2 + V_c\right)\phi(\boldsymbol{r}) = \mathcal{E}\phi(\boldsymbol{r}) \tag{16.10}$$

with the usual spherical boundary condition $\phi(r = R) = 0$. Here V_c is the Coulomb potential. The Schrödinger equation (16.10), along with the corresponding boundary condition, can be solved analytically in the absence of the Coulomb interaction, resulting in

$$\mathcal{E} = \mathcal{E}^e + \mathcal{E}^h = E_g + \frac{\hbar^2}{2m_e}\left(\frac{\alpha_{n_e\ell_e}}{R}\right)^2 + \frac{\hbar^2}{2m_h}\left(\frac{\alpha_{n_h\ell_h}}{R}\right)^2 \tag{16.11}$$

and

$$\phi(\boldsymbol{r}_e, \boldsymbol{r}_h) = \zeta(\boldsymbol{r}_e)\zeta(\boldsymbol{r}_h) \tag{16.12}$$

where

$$\zeta(\boldsymbol{r}) = \sqrt{\frac{1}{4\pi R^3}}\frac{j_\ell\left(\alpha_{n\ell}\frac{r}{R}\right)}{j_{\ell+1}(\alpha_{n\ell})}Y_\ell^m(\theta, \phi). \tag{16.13}$$

Equation (16.11) shows that the absorption is blue-shifted with respect to the bulk bandgap E_g. The shift varies with crystal size R, like $1/R^2$, being larger for smaller sizes. Furthermore, this equation states that the energy spectrum consists of a series of lines corresponding to the electron–hole transitions. Figure 16.2 exhibits the schematic representation of the one-electron–hole-pair states. The selection rules for the dipole allowed interband transitions are $\Delta\ell = 0$ in the absence of Coulomb interaction. For example, the $\mathcal{E}_{1s-1s}$ transition, where electron and hole are both of 1s type, is allowed.

When the Coulomb interaction is included, the problem can no longer be solved analytically and a numerical approach is needed. The absolute value of the one- and two-pair states is only weakly shifted by Coulomb effects, since the kinetic energy terms dominate for dots with $R \simeq a_B$. The electron–hole pair wavefunctions are, however, modified enough to strongly influence the nonlinear effects. The selection rules stated earlier are no longer valid, and transitions with $\Delta\ell \neq 0$ become weakly allowed (see section 16.2.3 for more details).

The linear absorption spectra of two samples of CdS quantum dots in glass at room temperature are shown in figure 16.3. The spectrum labeled 'bulk' refers to a glass with semiconductor microcrystallites large enough to retain the three-dimensional bulk properties. This spectrum is typical of bulk CdS absorption spectrum at room temperature with a sharp band edge and structureless Coulomb enhanced continuum absorption at higher photon energies. The average crystallite size becomes smaller for lower heat-treatment temperatures. The spectrum labeled 'QD' refers to a sample with small crystal sizes. The quantum-confinement effects are clearly observable in this sample. The absorption has shifted to higher energies as expected for the confinement

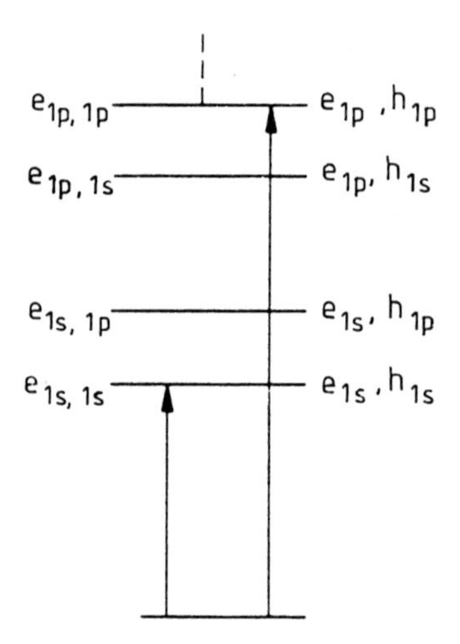

Figure 16.2. Schematic representation of the one-electron–hole-pair transitions in a semiconductor quantum dot. The notation e_{1s}, h_{1p}, etc refers to the electron being in the 1s state, the hole being in the 1p state, etc.

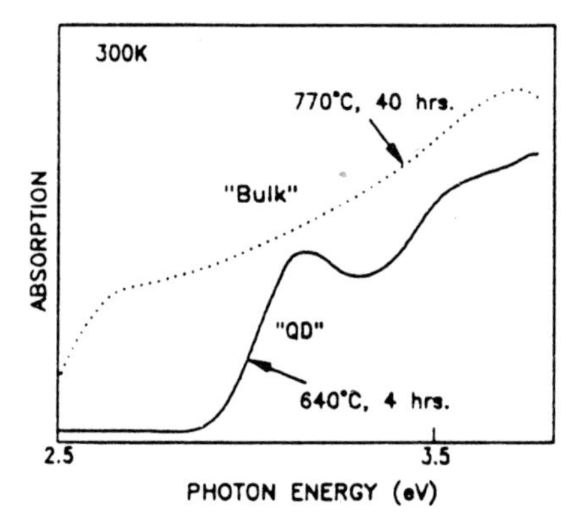

Figure 16.3. The linear absorption spectra at room temperature of two samples of CdS microstructures in glass. The crystal sizes in the sample labeled 'bulk' are large enough to have bulk behavior. In the 'quantum-dot' sample the crystals have very small sizes (the samples were grown by S Risbud and co-workers).

effect. Furthermore, discrete quantum-confined electron–hole pair states appear in the 'quantum-dot' sample.

Figure 16.4 displays the absorption spectra for similar samples at a temperature of $T = 10$ K. The lower temperature reduces the phonon broadening and, consequently, the transitions become narrower. The quantum-confined transitions are more clearly observed. In this figure, three samples with different quantum-dot sizes are displayed.

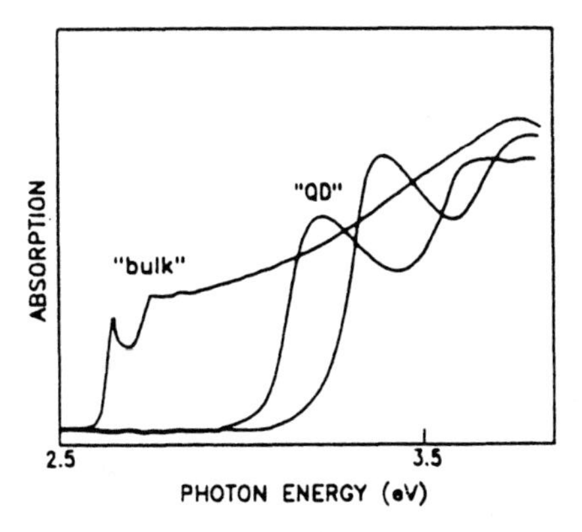

Figure 16.4. The linear absorption spectra at $T = 10$ K of CdS crystallites in glass. The two spectra labeled 'quantum dot' refer to two samples with small crystal sizes in the quantum-confinement regime. The 'bulk' sample has large crystal sizes, exhibiting bulk CdS properties (the samples were grown by S Risbud and co-workers).

The transition lines in figure 16.4 are much broader than transitions in bulk materials. The width of the transitions is a result of homogeneous and inhomogeneous broadening mechanisms. The homogeneous component is due to phonon and other scattering mechanisms. The inhomogeneous broadening comes from the fact that the crystallites do not have the same radii. Each radius has its own transition frequency ($\mathcal{E} \sim 1/R^2$), making the effective linewidth broad. This broadening can be taken into account theoretically. Using a density-matrix approach, the optical susceptibility and, consequently, optical absorption can be computed. The absorption coefficient of a single quantum dot as a function of photon frequency in such a calculation is given by

$$\alpha(\omega) = \frac{4\pi\omega}{\hbar c\sqrt{\epsilon_2}} \sum_i |d_{oi}|^2 \frac{\gamma_i}{\gamma_i^2 + (\omega_i - \omega)^2} \tag{16.14}$$

where ϵ_2 is the background dielectric constant of the semiconductor and d_{oi} is the transition dipole matrix element for the transition $o \to i$. The index o refers to the ground state (a state without any electron–hole pairs), while the index i refers to the one-electron–hole-pair state. The energy eigenvalue of the quantum-confined electron–hole transition is $\hbar\omega_i$, as given by equation (16.7), $\mathcal{E}_i = \hbar\omega_i$. Thus, $\hbar\omega_i$ explicitly depends on the size of the crystallites R. The homogeneous linewidth of the transition is γ_i. Equation (16.14) shows that the absorption spectrum of a single quantum dot consists of a series of Lorentzian peaks centered around the one-electron–hole-pair energies $\hbar\omega_i$. The inhomogeneous broadening may be taken into account by assuming that the particles have a size distribution given by $f(R)$ around a mean value $\bar{R}$. Since $\alpha(\omega)|_R$ in equation (16.14) is the absorption coefficient for a given radius R, the average absorption is then

$$\alpha(\omega)|_{average} = \int_0^\infty \mathrm{d}R\, f(R)\alpha(\omega)|_R. \tag{16.15}$$

Using a Gaussian distribution around the mean radius, $\bar{R} = 20$ Å, for $f(R)$ and different Gaussian distribution widths, we calculate the results shown in figure 16.5 for CdS quantum dots. It is clear that the quantum-confined transitions broaden and merge to a continuous structure with increasing width of the size distribution. The spectrum shown by the full line closely resembles the observed spectrum in figure 16.4. This suggests that the samples in figure 16.4 have a spread in their size distribution of 15% to 20%.

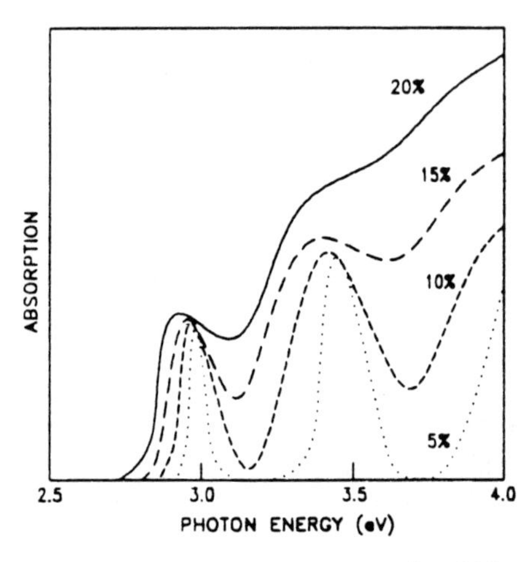

Figure 16.5. The calculated linear absorption spectra for CdS quantum dots with a Gaussian size distribution around a mean radius of 20 Å. The different curves correspond to different widths of the Gaussian distribution. For example, a 20% width corresponds to a 4 Å width for the Gaussian size distribution around a mean radius of 20 Å.

16.2.2 Electro-optical properties

The electroabsorptive effect in multiple-quantum-well structures has been the subject of intensive research, leading to quantum-confined Franz–Keldysh and quantum-confined Stark effects and the well known self electro-optic effect devices (SEED) [15]. The effect of an electric field on a quantum-dot system has also drawn considerable interest [16–19].

Here we review some of the electroabsorption effects observed in CdTe quantum dots [18]. The samples were prepared by the usual heat-treatment technique. The measured linear absorption spectrum of one of the samples is shown in figure 16.6(a). The peaks are the result of transitions to various quantum-confined levels. To provide sufficient interaction length, the sample was configured in a waveguide, and the field was applied transverse to the light propagation direction. The sample ends were polished flat and reasonably square, and then the samples, originally 1–2 mm thick, were ground and polished to the desired 100–200 μm thickness. Approximately 10 nm thick electrodes were applied to the sample center using a mask, keeping about 1 mm separation from the sample edges. Silver paint was used as an electrode. The applied voltage of up to 2 kV was modulated at 50 Hz so that drift and instability could be averaged out during data taking. Gated photon counting, with voltage on and off, was done in synchronization with the applied voltage.

The electroabsorption measured at 8 K in the sample is shown in figure 16.6(*c*). The effect of the electric field strength was examined at several points on the curve, and the required quadratic behavior was verified. These measurements were compared with calculations of the quantum-confined Franz–Keldysh effect [17]. The theory followed procedures adopted by Miller *et al* [16]. Infinitely high potential barriers were assumed and the Coulombic interaction neglected. The main result of the electric field, as shown in figure 16.6(*b*), is the red-shift of the 1s–1s transition and the opening of previously disallowed transitions.

16.2.3 Nonlinear optical properties

As discussed in previous sections, the linear absorption spectrum of quantum dots consists of a series of lines corresponding to transitions between quantum-confined electron–hole states. The origin of optical nonlinearity [20–24] in small intrinsic quantum dots is mainly the state filling effect, resulting in bleaching of the quantum-confined transitions. Screening of the Coulomb interaction by excited carriers is not important in quantum dots. This may be understood by noting that screening occurs in bulk semiconductors by carrier excitation because the excited electrons and holes can easily undergo intraband transitions, since a large number of states are available for such transitions. Such transitions lead to a rearrangement of carriers in real space, resulting in the weakening of the Coulomb potential by screening. In quantum dots, however, the energy states that electrons or holes may occupy are quantized with a relatively large energy separation. Therefore, transitions between those states require a significant amount of energy transfer and are very unlikely to occur as a consequence of Coulomb effects alone. Consequently, plasma screening in small quantum dots is basically absent. However, the Coulomb interactions between the charged particles is still important and leads to some other interesting nonlinear effects that will be discussed shortly. Therefore, since the Pauli exclusion principle prevents occupation of an electronic state by more than one identical particle, the quantum-confined transitions undergo absorption saturation if we try to excite additional charged particles into the quantum-confined states. This process of state filling is mainly responsible for intrinsic quantum-dot optical nonlinearities.

An example of the experimentally measured absorption change for resonant

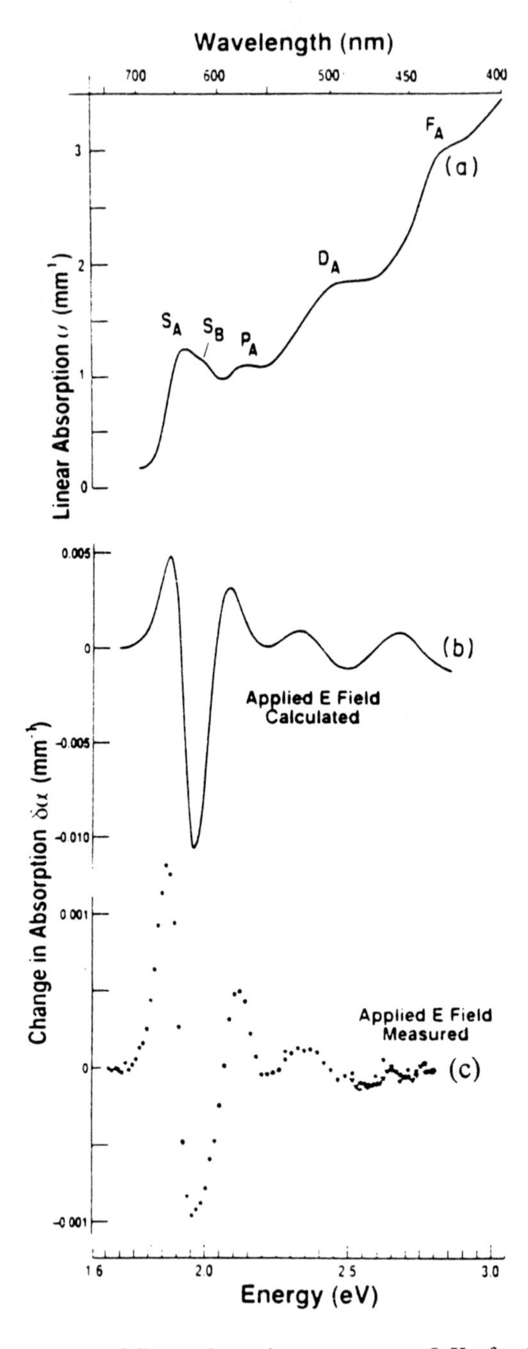

Figure 16.6. (*a*) Measured linear absorption spectrum at 8 K of a CdTe quantum-dot sample approximately 1.5 mm thick. The surface reflections ($\alpha L \simeq 0.08$) and scattering have not been subtracted. (*b*) and (*c*) Changes in the absorption spectra: (*b*) calculated for a uniform external field of 50 kV cm^{-1}; (*c*) observed for a uniform external field of 58 kV cm^{-1} at 8 K (sample was grown by S Risbud and co-workers [75]).

excitation into the energetically lowest quantum-confined transition is shown in figure 16.7 [20, 22]. The top spectrum in figure 16.7 displays the linear absorption of a CdS doped glass quantum-dot sample at $T = 10$ K. This spectrum has two peaks at energies of approximately 2.95 eV and 3.26 eV. The low-energy peak originates from the transition between the 1s-hole and 1s-electron states. The high-energy peak is due to the transition between the 1p-hole and 1p-electron states.

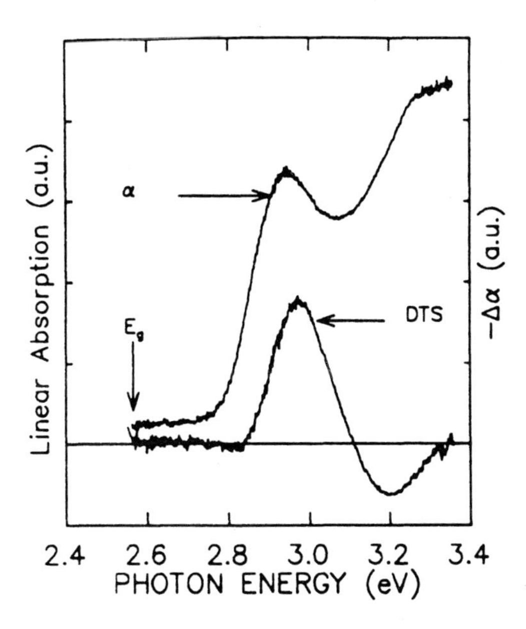

Figure 16.7. The top curve is the linear absorption spectrum of a sample of CdS quantum dots in glass. The lower trace is the change in the linear absorption spectrum as a result of excitation by the pump pulse.

To investigate the optical nonlinearity of these transitions, a pump-probe technique was employed. The photon energy of the pump pulse was tuned inside the energetically lowest transition. A photon from the pump excites the first electron–hole pair in the dot. A broadband probe pulse detects the changes made in the absorption spectra as a result of excitation by the pump pulse. Probe absorption without previous excitation by the pump is equivalent to the generation of one electron–hole pair in the unexcited quantum dot. If, however, a pump photon has been absorbed previously, the probe beam generates a second electron–hole pair in the presence of the pump generated pair. These situations are schematically shown in figure 16.8 for the two cases. When only the pump or only the probe is present, one pair exists in the dot (figure 16.8(*a*)), but when both pump and probe pulses are present, two electron–hole pairs are inside the dot (figure 16.8(*b*)).

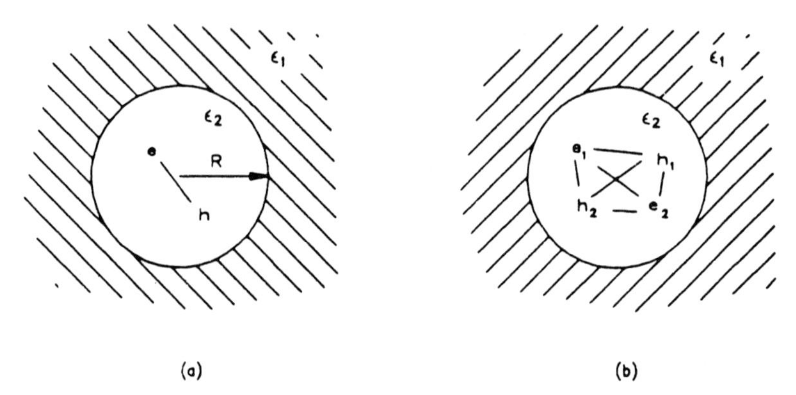

Figure 16.8. (*a*) One electron–hole pair inside a semiconductor quantum dot of dielectric constant ϵ_2 embedded in a glass matrix of dielectric constant ϵ_1. (*b*) Two electron–hole pairs in the quantum dot. The Coulomb interaction is attractive between the electrons and holes; it is repulsive between electrons and electrons and between holes and holes.

The lower spectrum in figure 16.7 shows the measured changes in the absorption, $-\Delta\alpha d = d[\alpha_{probe}(\text{without pump}) - \alpha_{probe}(\text{with pump})]$. In such a spectrum, a positive peak corresponds to bleaching and a negative peak indicates an induced absorption, i.e. an increase in the probe absorption as a consequence of the presence of the pump generated electron–hole pair. The nonlinear spectrum in figure 16.7 shows a positive peak around the $1s_h$–$1s_e$ transition, centered around the pump, and a negative peak on the high-energy side, centered around 3.178 eV. The bleaching of the $1s_h$–$1s_e$ transition is the result of state filling. The generation of one electron–hole pair by the pump causes saturation of the one-pair transition (bleaching of the linear absorption peak).

The origin of the induced absorption feature (the negative peak in the lower spectrum of figure 16.7) is assigned to the generation of two-pair states (or biexciton states) in the quantum dot. The pairs are created by absorption of one pump and one probe photon. The theoretical analyses of the linear and third-order nonlinear optical properties are discussed later in this chapter. These investigations consistently lead to the conclusion that Coulomb effects are important even for the smallest quantum dots. Using numerical matrix diagonalization techniques, the energies and wavefunctions were obtained for the one- and two-pair ground states and for all the excited pair states. With these wavefunctions, the various dipole matrix elements were evaluated for transitions between the ground state and the one- and two-electron–hole-pair states. The changes in the absorption, $-\Delta\alpha d$, were then calculated, as shown in figure 16.9. Like the experiment, the theory also exhibits a decreasing absorption feature (bleaching) around the lowest quantum-confined transition and an increasing absorption feature on the high-energy side. The details of the calculations are given in the theory section.

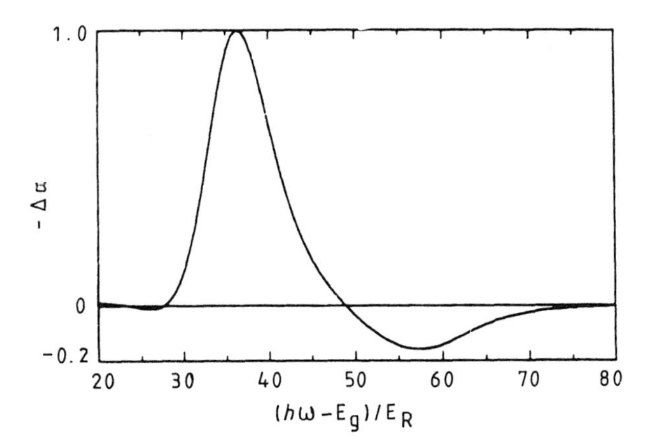

Figure 16.9. Computed absorption changes ($-\Delta\alpha$) for a semiconductor quantum dot of radius $R = 6.5\ a_B$ (where a_B is the exciton Bohr radius), assuming a pump and probe geometry and pumping into the energetically lowest one-pair state.

Under strong excitation conditions or for extended irradiation of the quantum-dot samples in glass, one observes that the spectral changes persist for a long time, longer than 1 μs. The spectral shape of these persistent spectra appear to be different from that of the transient spectra. It is concluded that the long-lived nonlinear optical effects occur because one or both of the carriers is trapped at the surface or outside the quantum dot under the long-time exposure condition, reducing the electron–hole overlap and decay rate and maintaining a $\delta\alpha$ through the Coulomb interaction of these charges. Here 'trap' is a catch-all word used to refer to the reduction of electron–hole overlap needed to explain such a long recovery time. These effects of surface polarization have been calculated [25]. In figure 16.10 we show the comparison of two calculated linear absorption spectra. The solid curve is the linear absorption spectrum of one electron–hole pair inside the dot. The dashed curve is the computed absorption for the case when two electron–hole pairs are present, but one of the holes is trapped at the surface and the other three carriers are free to move inside the dot. Figure 16.10(b) shows that the difference between the two linear absorption spectra of figure 16.10(a), α(1 pair) $-$ α(3 charges + a surface charge), consists of a negative peak for low energies (red-shift of the lowest transition) and a positive peak on the high-energy side (blue-shift of the second-lowest transition). This spectrum deviates strongly from the absorption changes shown in figures 16.7 and 16.9.

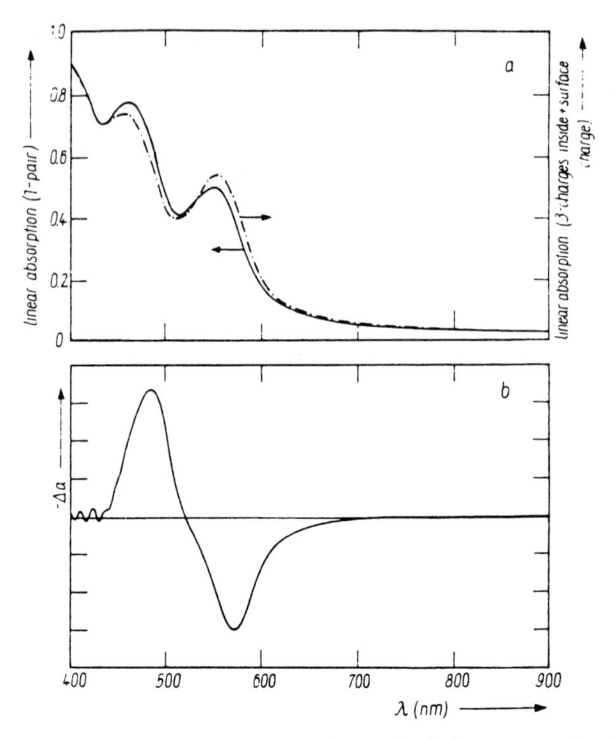

Figure 16.10. (*a*) Linear absorption spectra for an intrinsic quantum dot (solid curve) and for a quantum dot with a second electron–hole pair whose hole is trapped at the surface (dashed curve). (*b*) The difference between the two linear absorption spectra, i.e. α(1 pair) $-$ α(3 charges inside + a surface charge).

16.2.4 Two-photon absorption and confinement induced mixing of valence bands

Even though simple parabolic valence- and conduction-band models have been successful in describing the optical nonlinearities of quantum dots measured in pump-probe experiments, recent theoretical publications [26, 27] indicate that this model needs to be modified to account for mixing of the valence bands caused by the spherical confining potential of small quantum dots. To obtain the experimental verification of such confinement induced valence-band mixing, we measured and compared one- and two-photon absorption spectra (see figure 16.11(*a*)) of CdS quantum dots in glass. Because of the large number of poorly known system parameters, one-photon absorption measurements alone are not sufficient to unambiguously determine the energies between the quantum-confined states. However, the combination of one- and two-photon absorption spectra allows a direct comparison of electron–hole pair states with total angular momentum of zero and one, thus eliminating many of the experimental

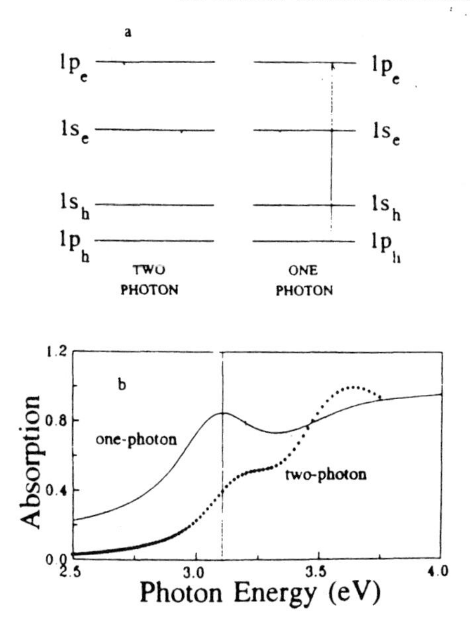

Figure 16.11. (*a*) One- and two-photon allowed transitions between the discrete states in the spherical potential well. (*b*) Theoretical calculation of one- and two-photon absorption based on a parabolic band model.

uncertainties. The results show significant deviations from calculations based on the parabolic valence-band approximation.

To demonstrate the significance of the confinement induced valence-band mixing effects, we first calculated the one- and two-photon absorption spectra in the simple parabolic band model. Figure 16.11(*b*) shows the results of such a calculation, where effective masses appropriate for the A band in bulk wurzite CdS, $M_e \simeq 0.2m_0$ and $m_h \simeq 1.35m_0$ ($m_\perp = 0.7m_0$, $m_\parallel = 5m_0$, and thus $m_h = \sqrt[3]{(m_\perp^2 m_\parallel)} = 1.35m_0$) have been used. A phenomenological broadening parameter has been introduced to account for the homogeneous linewidth and the inhomogeneous contribution from the distribution of dot sizes. The first two-photon resonances are expected to be situated between the first two one-photon transitions as seen in figure 16.11(*b*). The splitting between the one- and two-photon absorption peaks is a result of the nondegeneracy of the valence-band states. By experimentally measuring both the one- and two-photon absorption spectra, we can check the predictions of this model, which assume a single uncoupled parabolic valence band for these quantum dots.

For our experiments, we used samples containing CdS microcrystallites in a glass matrix. The sample with the smallest microcrystallites had average radii of 0.9 ± 0.7 nm, while the sample with the largest microcrystallites had average radii of 8.0 ± 0.7 nm.

Optical characterizations consisted of both one- and two-photon absorption measurements for each sample. One-photon spectra were obtained through direct transmission measurements or by excitation spectroscopy. Two-photon spectra were obtained through two-photon excitation spectroscopy, where the photoluminescence (PL) intensity is detected as the excitation frequency is varied at constant input intensity excitation. Most of the PL from our samples was emitted in a broad band, peaking near 800 nm. These PL spectra were observed to be independent of the excitation photon energy, allowing us to monitor a constant band of PL while tuning the excitation source to obtain the two-photon absorption spectrum. The validity of this technique was checked by obtaining one-photon absorption spectra through PL excitation, comparing it to absorption spectra obtained from transmission measurements, and verifying that the two techniques give identical results (see figure 16.12).

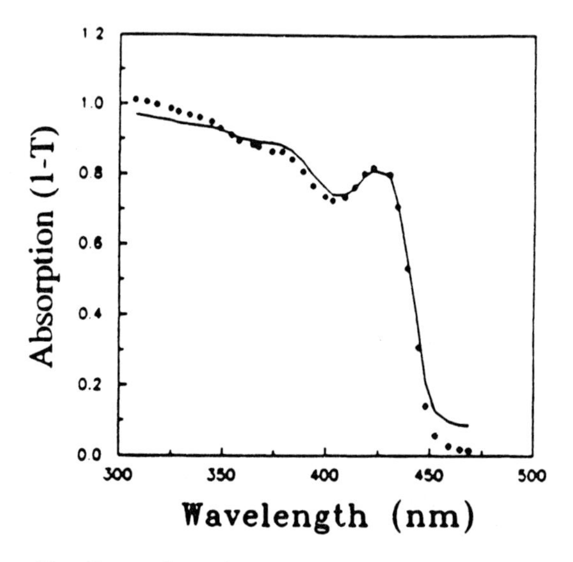

Figure 16.12. The linear absorption spectrum of the CdS quantum-dot sample measured using transmission (solid curve) and luminescence excitation (dotted curve) spectroscopies.

Examples of one- and two-photon spectra for two samples with different average quantum-dot radii between 1 and 2 nm are shown in figure 16.13(*a*). It is apparent that the one- and two-photon absorption peaks occur at the same energies. Similar spectra were obtained for other quantum-dot samples with different average sizes. The poor agreement between the experimental spectra in figures 16.13(*a*) and (*b*) and the theoretical spectra in figure 16.11(*b*) shows that the independent parabolic valence-band approximation is insufficient to explain the optical transitions of CdS quantum dots.

To improve the model, we performed a theoretical analysis, which included

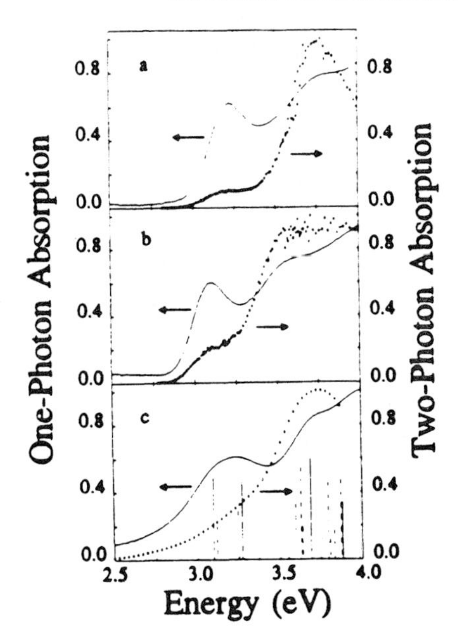

Figure 16.13. (*a*) Experimental results of one-photon (solid line) and two-photon (dots) absorption spectra for a quantum-dot sample heat-treated at 640 °C for 1 h. No observable differences are seen in the transition energies between the one- and two-photon spectra. (*b*) Similar experimental results for a quantum-dot sample heat-treated at 640 °C for 3 h, thus having a larger radius. (*c*) Calculations of one-photon (solid line) and two-photon (dots) absorption spectra based on the Luttinger Hamiltonian. The vertical lines are calculated with no broadening, while the dashed and solid curves represent spectra for a broadening of $\gamma = 8E_R$, where E_R is the bulk exciton binding energy of 27 meV. The magnitude of a vertical line represents the absorption strength. The other parameters used in the calculations are $R/a_B = 0.5$, with $\gamma_1 = 2.97$ and $\mu = 0.75$. The experimental data of (*a*) and (*b*) should be compared with (*c*).

mixing of the heavy- and light-hole valence bands induced by the spherical confining potential in a Luttinger Hamiltonian approach. The details of this calculation are given in the next section. Here we only describe the results. The numerically computed one- and two-photon absorption spectra from this model, using parameters appropriate for cubic CdS ($\mu = 0.75$), are shown in figure 16.13(*c*). The vertical lines in this figure represent the energetic positions of the transitions and their oscillator strengths. There are several predicted new resonances that satisfy the selection rules of total angular momentum, including quantum confinement and Coulomb interaction. The solid and dashed curves show the one- and two-photon absorption spectra calculated with a broadening of $\gamma = 8E_R$, where $E_R = 27$ meV is the bulk exciton binding energy. We see that the resonances merge and form an absorption curve that agrees well with the experimental data.

The near degeneracy of the one- and two-photon transition energies observed here results from the fact that the two lowest-energy hole states are roughly degenerate for the case of CdS quantum dots. This can be easily seen in figure 16.14, where the calculated energy of the two lowest hole states, labeled $s_{3/2}$ and $p_{3/2}$ to indicate their s- and p-like characteristics, are plotted as a function of the coupling constant μ for a quantum-dot radius equal to half of the bulk exciton radius. In the vicinity of $\mu = 0.75$, which is appropriate for CdS, we see that the two curves nearly touch, suggesting near degeneracy caused by the confinement induced modification of the original heavy- and light-hole states. Figure 16.14 indicates that for materials with smaller coupling constant μ values, these states are nondegenerate and the one- and two-photon transition energies are not expected to coincide. It is noted that the results of figure 16.14 are not strongly size dependent for the range of quantum-dot sizes analyzed in this study. For very large dot sizes the assumptions made in the calculations are no longer valid and the treatment does not apply.

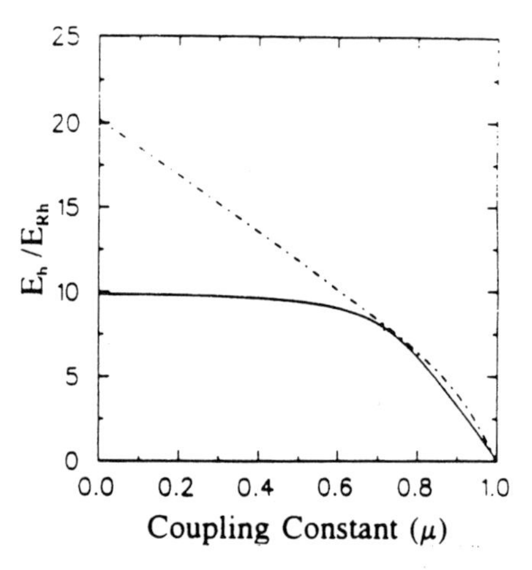

Figure 16.14. Calculated normalized energies of the two lowest valence-band states, $s_{3/2}$ and $p_{3/2}$, in the quantum dot with crystallite size of $R = a_{hB}$. Here E_{Rh} and a_{hB} are the Rydberg energy and the Bohr radius computed for the hole. The solid line (dashed line) represents the $s_{3/2}(p_{3/2})$ states.

16.3 THEORY OF NONLINEAR OPTICAL RESPONSE

16.3.1 Nonlinear optical response

For simplicity we discuss our theoretical analysis [28] of quantum-dot resonances assuming the simple parabolic band approximation. This theory has been applied to include more elaborate band structures, but since these lead to a cumbersome complication of the notation, we do not stress these aspects here.

We describe the motion of the electrons and holes inside the dot using the many-body electron–hole Hamiltonian in position representation. Electrons and holes interact via the Coulomb interaction, which is modified in comparison to bulk semiconductors because of the presence of induced dielectric surface charges [1]. The basic Hamiltonian is given by

$$\hat{H} = \hat{H}_e + \hat{H}_h + \hat{V}_{ee} + \hat{V}_{hh} + \hat{V}_{eh} + \hat{W}$$

where

$$\begin{aligned}
\hat{H}_e &= \sum_s \int \mathrm{d}r\, \psi_e^\dagger(r,s)\left(E_g - \frac{\hbar^2}{2m_e}\nabla^2\right)\psi_e(r,s) \\
\hat{H}_h &= \sum_s \int \mathrm{d}r \psi_h^\dagger(r,s)\left(-\frac{\hbar^2}{2m_h}\right)\nabla^2\psi_h(r,s) \\
\hat{V}_{ee} &= \frac{1}{2}\sum_{ss'} \int\int \mathrm{d}r\, \mathrm{d}r' \psi_e^\dagger(r,s)\psi_e^\dagger(r',s')V(r-r')\psi_e(r',s')\psi_e(r,s) \qquad (16.16) \\
\hat{V}_{hh} &= \frac{1}{2}\sum_{ss'} \int\int \mathrm{d}r\, \mathrm{d}r' \psi_h^\dagger(r,s)\psi_h^\dagger(r',s')V(r-r')\psi_h(r',s')\psi_h(r,s) \\
\hat{V}_{eh} &= -\sum_{ss'} \int\int \mathrm{d}r\, \mathrm{d}r' \psi_e^\dagger(r,s)\psi_h^\dagger(r',s')V(r-r')\psi_h(r',s')\psi_e(r,s)
\end{aligned}$$

and $\psi_e(r,s)$ and $\psi_h(r,s)$ are the annihilation operator of, respectively, an electron and a hole with spin s at position r. The field operators obey the usual Fermi anti-commutation rules. $\hat{W}$ is the correction to the Coulomb interaction due to the surface polarization of the quantum dots [1]. The boundary conditions at the surface of the quantum dot require that

$$\psi_e(r,s) = \psi_h(r,s) = 0 \qquad \text{when } |r| \geq R \qquad (16.17)$$

where R is the radius of the dot.

The state structure of our system is determined by the solutions of the eigenvalue problem given by the Hamiltonian (16.16). To discuss the optical response we couple the system to a light field using the interaction Hamiltonian

$$\begin{aligned}
\hat{H}_{int} &= -E(t)P^+ + \mathrm{HC} \\
&= -p_{cv}E(t)\sum_s \int \mathrm{d}r\, \psi_e^\dagger(r,s)\psi_h^\dagger(r,s) + \mathrm{HC} \qquad (16.18)
\end{aligned}$$

where HC is the Hermitian conjugate of the first term and P^+ is the polarization operator. To obtain the optical properties of the quantum dot we must calculate the expectation value of the polarization operator from which we can extract the optical susceptibility.

As an example of numerical results we show in figure 16.15 the computed changes in optical transmission, $-\Delta\alpha$, of a quantum-dot sample, which is proportional to the imaginary part of the susceptibility. The parameters are $R/a_0 = 1$, $\epsilon_2/\epsilon_1 = 1$, and $m_e/m_h = 0.24$. In figures 16.15(*a*)–(*c*) we present the results for the frequency regime around the lowest exciton resonance E_1 for different damping constants $\hbar\Gamma_{ij} = \gamma$. Figure 16.15 displays a positive peak around the pump frequency which indicates the saturation (bleaching) of the one-pair transition. Additionally, we observe negative structures on the low- and high-energy side of the positive peak. These negative peaks show increasing probe absorption due to the generation of two-electron–hole-pair states via absorption of one pump and one probe photon. The resonance on the low-energy side of the positive peak is caused by the ground-state biexciton. This resonance is visible in quantum dots only for relatively small broadening γ and it is suppressed by the saturating one-pair resonance for increasing γ.

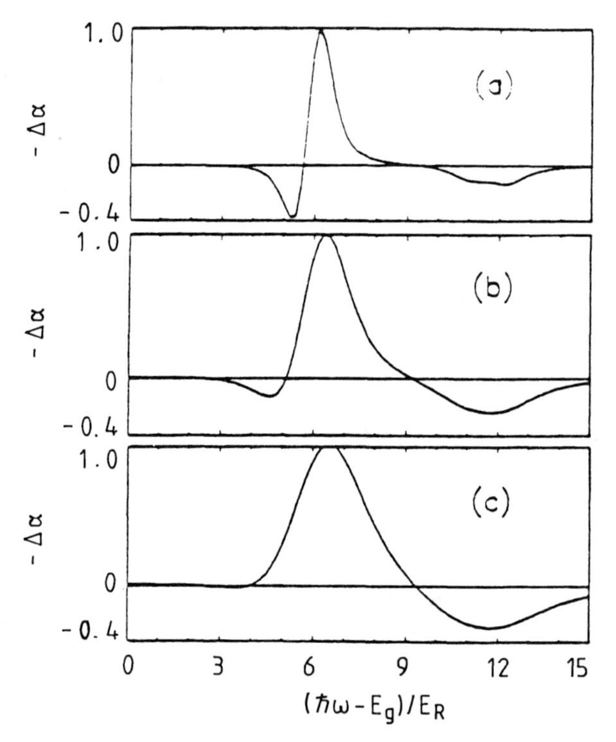

Figure 16.15. Change in optical transmission, $-\Delta\alpha$, computed from the imaginary part of χ_3, equation (16.22), as a function of probe-energy detuning from the bulk semiconductor bandgap E_g (in units of the bulk-exciton Rydberg energy E_R). The shown results are for $R/a_0 = 1$, $\epsilon_2/\epsilon_1 = 1$, $m_e/m_h = 0.24$ and all damping constants $\hbar\Gamma_{ij} = \gamma = E_R$ (*a*), $\gamma = 2E_R$ (*b*), and $\gamma = 3E_R$ (*c*) respectively.

The induced absorption on the high-energy side of the saturating one-pair resonance is caused by transitions to excited-state biexcitons. These transitions are possible since the Coulomb interaction changes the selection rules for dipole transitions. Taking a closer look at the possible excited two-pair states shows that the energetically lowest of these states are actually those where one or two of the heavier holes are not in their ground state. The relevant examples are where the main quantum numbers of the state (e_1, e_2, h_1, h_2) are $n = (1, 1, 1, 1)$, $\ell = (0, 0, 1, 1)$ and $n = (1, 1, 1, 2)$, $\ell = (0, 0, 0, 0)$ respectively. These pure product states could not make a dipole transition to the one-pair state. In reality, however, such a transition becomes possible since the Coulomb interaction causes a mixing of the independent particle states. In simple terms, one can explain this induced absorption as a consequence of the symmetry breaking through the presence of the pump generated electron–hole pair. In this situation, the probe photon generates an electron–hole pair in the presence of the pump generated pair. Since the Coulomb interaction changes the dipole selection rules, the possibilities for dipole transitions involving the probe photon are different from those of the pump photon. In this sense, the induced absorption resembles the 'excited state absorption' in atomic physics.

The theoretical results in figures 16.6 and 16.9–16.11 are obtained using the theory outlined in this section. For the calculations in figure 16.13(*c*) the hole Hamiltonian $\hat{H}_h$ in equation (16.16) was replaced by the Luttinger Hamiltonian.

16.3.2 Excitonic superradiance from microcrystallites

In contrast to the analysis of small quantum dots in the previous section, we discuss in this section now the case of weak confinement of excitons within semiconductor microcrystallites. Such a situation is realized in CuCl microcrystallites with the radius R substantially larger than an exciton Bohr radius $a_B = 6.7$ Å. The exciton energy is 3.2 eV in CuCl, while the matrix NaCl has the bandgap energy 10 eV. As a result, the center-of-mass energy of the exciton is quantized within CuCl microcrystallites as follows

$$E_n = E_g - E_R + \frac{1}{2M}\left(\frac{\hbar\pi n}{R}\right)^2 \qquad (n = 1, 2, \ldots). \tag{16.19}$$

Here E_g is the energy gap between the valence and conduction bands, E_R is the exciton binding energy in the lowest electron–hole relative motion 1s, and n is the principal quantum number for the center-of-mass motion with the mass M. The transition dipole moment

$$P_n = \frac{2\sqrt{2}}{\pi}\left(\frac{R}{a_B}\right)^{3/2}\frac{1}{n}\mu_{cv} \qquad (n = 1, 2, \ldots) \tag{16.20}$$

has a mesoscopic enhancement $(2\sqrt{2}/\pi)(R/a_B)^{3/2}$ in comparison to the band-to-band transition μ_{cv}. It is noted that the oscillator strength is almost concentrated on the lowest state, $n = 1$.

The dominant effect of a mesoscopic transition dipole moment is superradiative decay of the lowest exciton (1s, $n = 1$). We consider such a microcrystallite to have radius R larger than the microscopic size of the exciton Bohr radius a_B, but much smaller than the wavelength of the radiation field λ which can excite this exciton. Therefore, this system is called mesoscopic. Then the radiative decay rate is calculated [29] in terms of the Fermi golden rule as

$$2\gamma \equiv \frac{1}{T_1} 64\pi \left(\frac{R}{a_B}\right)^3 \frac{4\mu_{cv}^2}{3\hbar\lambda^3}. \tag{16.21}$$

Note that it has a mesoscopic enhancement, $64\pi(R/a_B)^3$, in comparison to the band-to-band transition rate $4\mu_{cv}^2/3\hbar\lambda^3$. This exciton has the maximum cooperation number in the sense that this exciton state is made by a coherent superposition of all atomic excitations composing the microcrystallite. For example, the exciton in microcrystallites with radius 80 Å can superradiatively decay in 100 picoseconds. Note that the exciton in the larger microcrystallites can radiate more rapidly. However, the energy separation of equation (16.19) decreases as R^{-2} so that scattering by defects and phonons smears out the quantized levels in larger microcrystallites and, consequently, the superrradiant phenomenon.

Itoh *et al* [30] succeeded in making microcrystallites in an NaCl matrix, the radius of which ranged from 17 Å to more than 100 Å. The absolute values as well as the size dependence of the radiative decay time were observed in good agreement with the theoretical predictions, without any adjustable parameters. On the other hand, the size dependence of 2γ was observed to be $R^{2.1}$, deviating from the R^3 dependence for CuCl microcrystallites embedded in glasses [31]. These two effects may be caused by the fact that excitations penetrate into the glasses because the glasses have smaller energy barriers than those of NaCl.

16.3.3 Superradiance from interacting Frenkel excitons

In the previous section we discussed the superradiance of a single Wannier exciton in a microcrystallite. It is hard to describe the superradiant phenomena from many Wannier excitons in a single microcrystallite. Therefore, we study the superradiance master equation for many Frenkel excitons in a linear chain, which can be solved exactly [32–34]. As the emitted radiation field adiabatically follows the electronic system, the superradiant process can be represented by the master equation for the density operator of the electronic system $\rho(t)$ [35, 36]. The superradiance is characterized by the pulse profile and the time resolved emission spectrum, both of which can be evaluated in terms of $\rho(t)$. We can evaluate these quantities exactly using numerical methods. From this we can conclude that the superradiance from multi-Frenkel excitons has two characteristic features which are missing in Dicke's superradiance for atomic or molecular systems.

First, when the transition dipole moment d is equal to the static one μ induced in the excited state, we have the same superradiant response as for Dicke's model. For simplicity, let us consider the case in which both the dipole moments are parallel to the chain axis. Secondly, for the general case $d \neq \mu$, the slow-emission component is inevitable because the excitation can transfer to other molecules before it radiatively decays so that the cooperation number decreases in the radiative process. Thirdly, for the case $d > \mu$, the time resolved emission spectrum shows the blue-shift and red-shift, respectively, in the first and second halves of the superradiant pulse. This frequency shifting will be used for pulse compression by a grating pair at the output. For the case $\mu > d$, the sign of the frequency shift is reversed.

These predictions may be observed from a strongly pumped system of bound excitons in the case where the trapping centers such as isoelectronic traps or donor (acceptor) centers are equally spaced. Excitons in anthracene-PMDA and J-aggregates of dyes correspond to the present model. It is, however, difficult to realize the complete population inversion, and the partially pumped state is brought about by the femtosecond laser pulse. The differential transmission spectrum (DTS) of this system is interesting in showing the existence of excitonic n-string states [37, 38]. The DTS is evaluated for these systems with several parameters for the ratio of dipole moments μ/d [39]. For the case $\mu > d$, the absorption saturation is expected at the exciton frequency and the induced absorption on the lower-energy side due to the transition from a single exciton into a two-string state. The transition from n-string to $(n+1)$-string states gives the induced absorption at the lower-frequency side, under the stronger pumping or just after the strong pumping. It is interesting to confirm such a bound state of multiexcitons, more than two, which is impossible for a system of Wannier excitons.

16.3.4 Enhancement of optical nonlinearity

The mesoscopic transition dipole moment of a Wannier exciton results in an excitonic enhancement of the third-order optical polarization

$$\langle P^{(3)}(\omega)\rangle = \chi^{(3)}(\omega; \omega, -\omega, \omega)|E|^2 E \mathrm{e}^{-\mathrm{i}\omega t}. \tag{16.22}$$

Here we use a single beam with E being the effective electric field in the crystal and with the local field effects induced by the external field taken into account. An ideal boson or a harmonic oscillator cannot show any nonlinearity. A few factors work to make the excitons deviate from harmonic oscillators. First, the exciton–exciton interaction works repulsively for two singlet excitons with same spins and increases inversely proportionally to the volume of the microcrystallite. A bound state of two excitons, which is called an excitonic molecule, is formed for two singlet excitons of different spins. When the exciton is resonantly pumped with a detuning much smaller than the molecular binding energy ω_b or the exciton–exciton interaction energy ω_{int}, the electronic system can be treated as a two-level system with a mesoscopic dipole moment. Second, the decay and relaxation of the exciton system, which depend on the number of excitations, make the excitons deviate from ideal bosons. As a result, $\chi^{(3)}(\Omega; \Omega, -\Omega, \Omega)$ is expected to be mesoscopically enhanced under resonant pumping of excitons in the microcrystallites (MCs) as follows [40, 41].

(*a*) $\omega_{int}, \omega_b > |\Omega - \omega_1| > \Gamma$:

$$\chi^{(3)} = \frac{2N_c|P_1|^4}{\hbar(\Omega - \omega_1)^3}\left(1 + \frac{\gamma'}{\gamma}\right) \propto R^3 \tag{16.23}$$

where $N_c \equiv 3r/(4\pi R^3)$ is the number density of the MCs with r being a constant volume fraction, and $\hbar\omega_1 = E_1$.

(*b*) $\omega_{int}, \omega_b > \Gamma > |\Omega - \omega_1|$:

$$\chi^{(3)} = -\mathrm{i}\frac{2N_c|P_1|^4}{\hbar^3(\gamma + \gamma')^2\gamma}. \tag{16.24}$$

For case (*a*), $\chi^{(3)}$ is almost real and increases as R^3 because the fourth power of P_1 gives an R^6 dependence and overcomes the R^{-3} dependence of N_c. For case (*b*), $\chi^{(3)}$ becomes almost imaginary and increases as R^3 if the longitudinal decay rate 2γ is determined not by the superradiance but by size-independent processes. On the other hand, when the pure dephasing γ' is much larger than γ, and 2γ is determined by the superradiance, $\chi^{(3)}$ is almost independent of R. For example, a system of CuCl MCs with $R = 80$ Å and the volume ratio $r = 10^{-3}$ will show $\mathrm{Im}\,\chi^{(3)} = -10^{-3}$ esu when we assume $\hbar\Gamma = 0.1$ meV and $2\hbar\gamma = 0.03$ meV.

Masumoto *et al* [42] observed the absorption saturation effects by the pump-probe method, which depend on the size of the CuCl MCs. The observed results, $-\mathrm{Im}\,\chi^{(3)} \propto R^{2.6}$ and $\mathrm{Im}\,\chi^{(3)} = -10^{-3}$ esu for 0.12 CuCl MCs with radius of 100 Å, are nearly in agreement with theory. Nakamura *et al* [43] also observed a large $\chi^{(3)}$ value under nearly resonant pumping of the exciton in CuCl doped glasses and obtained an optimum size for the largest $\chi^{(3)}$ value, which depends on the lattice temperature. These systems give the enhancement of the figure of merit $|\chi^{(3)}|/\alpha\tau$ when the switching is determined by the superradiant decay.

The existence of excitonic trapped states results in interesting nonlinear optical phenomena [41]. First, the ground-state electron is missing, while the excited electron is in the trapped state, so that the third-order susceptibility, whose imaginary part describes the absorption saturation, is enhanced by the factor $\Gamma_{n\to t}/\Gamma_{t\to g}$ times the dominant term for the case without any trapped states. Here $\Gamma_{n\to t}$ is the trapping rate of a free exciton and $\Gamma_{t\to g}$ is a decay rate of the trapped electron into the ground state. The enhancement factor $\Gamma_{n\to t}/\Gamma_{t\to g}$ is on the order of 10^3–10^6, but the switching time is determined by the slowest channel, $1/\Gamma_{t\to g} \simeq 10^{-6}$ s. Therefore, the figure of merit is conserved.

Secondly, the four-wave mixing spectrum is evaluated as a function of detuning between the pump Ω_1 and probe Ω_2 frequencies under nearly resonant pumping of the exciton with the trapped state. Then the spectrum consists of the hierarchical structure: (1) the very sharp Lorentzian structure (like a spike) with the half-width $\Gamma_{t\to g} \simeq 10^6$ s^{-1} around $\Omega_1 - \Omega_2 = 0$, (2) the rather sharp Lorentzian with the half-width $\Gamma_n \simeq 10^9$ s^{-1} around $\Omega_1 - \Omega_2 = 0$, and (3) the broad structure with the width $\Gamma_{ng} \simeq 10^{12}$ s^{-1} with the center at $\Omega_1 \simeq \omega_1$ or $\Omega_2 \simeq \omega_1$. Here $h\omega_1$ is the lowest exciton energy, and Γ_n and Γ_{ng} are the longitudinal and transverse relaxation rates of the exciton. Thirdly, the differential transmission spectrum is also calculated as a function of detuning, $\Omega_1 - \Omega_2$, under nearly resonant pumping of the exciton. Under resonant exciton pumping, the absorption saturation signal shows a peak at $\Omega_2 - \Omega_1 = 0$ with spectral width Γ_n while the induced absorption overcomes the exciton saturation under a slight off-resonant exciton pumping. This induced absorption is observed as a sharp dip at $\Omega_2 - \Omega_1 = 0$ with the smallest width $\Gamma_{t\to g}$. Defects at the interface between MCs and the matrix play important roles in nonlinear optical and dynamical processes, but we have no microscopic understanding of these defects yet. The pump-probe spectroscopy introduced here is helpful in clarifying their microscopic origins.

16.4 APPLICATIONS

16.4.1 Quantum-dot lasers

Semiconductor low-dimensional quantum structures are expected to be promising laser devices [44–46]. As the dimension is lowered, the modified density of states concentrates carriers in a certain energy range. This concentration is expected to give the system more gain for lasing. Zero-dimensional quantum confinement of carriers converts the density of states to a set of discrete quantum levels. This is desirable for the semiconductor laser because the discrete spectral gain region is compressed. In addition, spatial carrier confinement is also favorable for high optical gain.

The CuCl quantum dots analyzed in the previous section provide unique opportunities for laser applications. The large binding energy of biexcitons in a CuCl crystal, 32 meV, makes the biexciton stable and allows us to observe biexciton absorption and luminescence at low temperatures [47]. Biexciton luminescence is also observed in CuCl quantum dots, even at 77 K [30]. In photo-pumped bulk CuCl crystals, optical gain due to the transition from biexciton to longitudinal Z_3 exciton is high [48, 49] and lasing takes place at this transition [50]. The laser action takes place in CuCl as a result of the population inversion between excitons and biexcitons. In this sense, it is a three-level laser, where the three levels are ground, exciton, and biexciton states. This section describes the observation of lasing in CuCl quantum dots. The optical gain of the CuCl microcrystals is examined and compared with that of CuCl bulk crystals.

The samples used in this study were CuCl microcrystals embedded in NaCl crystals. They were grown by the transverse Bridgman method from the melt of a mixture of NaCl powder and CuCl powder, and were annealed in order to control the size distribution of the CuCl microcrystals. The mean size of the microcrystals was determined by small-angle x-ray scattering measurements (see chapter 10). The mean size of the microcrystals was determined to be 5.0 nm using this method. The molar fraction was 0.3 mol%, as measured by inductively coupled plasma emission spectroscopy. Three pieces of samples were made by cleaving a grown and heat-treated crystal. The thinnest piece, 0.12 mm thick, was used for the optical absorption measurement. The second piece was used for the study of luminescence. The third piece was a rectangular parallelepiped and its size was $3.2 \times 5.6 \times 0.58$ mm^3. The laser device was the third piece placed in a cavity composed of two parallel dielectric mirrors with 90% reflectivities (see inset of figure 16.16). The shortest side, 0.58 mm, was placed perpendicular to the mirror face. The cavity length was 0.62 mm.

The excitation sources used were a nitrogen laser (345 nm), an XeF excimer laser (359 nm), or the third harmonics of a Nd^{3+}:YAG laser (363 nm). Ultraviolet laser pulses were used to excite the sample placed in liquid nitrogen or in a temperature-variable cryostat. The pulse widths of the nitrogen laser, the excimer laser, and the third harmonics of the Nd^{3+}:YAG laser were 10 ns, 20 ns, and 5 ns respectively. For the lasing experiments, both longitudinal and transverse pumping geometries were used. Lasing was observed in both geometries.

The optical gain measurement was done by observing the intensity of the stimulated emission as a function of the excitation length using a cylindrical lens to focus the output of the excimer laser on a line. A part of the line was cut by a slit and was refocused on the sample surface using another lens. The excitation length was varied by the slit. The stimulated emission propagating along the sample surface was observed from the extension of the excited line. The minimum excitation length was measured to be 30 μm, which gives the spatial resolution of the experiment. The optical gain of two samples was observed by this method. One sample was a CuCl microcrystal in an NaCl crystal with mirrors for the lasing experiment, and the other sample was a CuCl bulk crystal.

The absorption and luminescence spectra of the sample at 77 K are shown in figure 16.16. The Z_3 exciton line shows a blue-shift of 6 meV compared with bulk CuCl. The blue-shift is ascribed to the quantum confinement of excitons. The shift and the mean radius of CuCl microcrystals deduced from the small-angle x-ray scattering experiment are consistent with the previous measurement which gave the relation between the blue-shift of the Z_3 exciton energy and the mean size of the CuCl microcrystals [51]. The luminescence spectra were measured under the excitation of the nitrogen laser. With the increase in the excitation intensity, a lower-energy band appears around 391 nm and grows. The lower-energy band overwhelms the exciton band around the excitation density of 3 MW cm^{-2}. The luminescence spectra under the low-density excitation

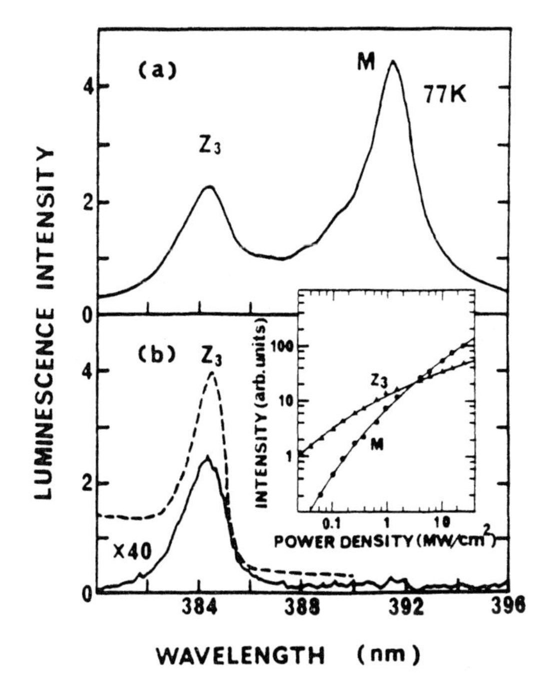

Figure 16.16. The luminescence spectra of CuCl microcrystals embedded in an NaCl crystal at 77 K under the nitrogen laser excitation of (*a*) 24 MW cm^{-2} and (*b*) 39 kW cm^{-2}. With the increase of the excitation intensity, the M band appears and grows. The absorption spectrum of CuCl microcrystals embedded in an NaCl crystal at 77 K is shown by the dashed line. The peak of the Z_3 exciton shows a blue-shift of 6 meV compared with the energy position of the Z_3 exciton in a bulk CuCl crystal. In the inset, the excitation density dependence of the Z_3 exciton luminescence and that of the M biexciton luminescence are shown. The excitation density dependence of the Z_3 exciton luminescence is fitted by the expression $\log_{10} I_{ex} = 0.0144(\log_{10} I)^3 - 0.0872(\log_{10} I)^2 + 0.504(\log_{10} I) + 1.11$, while that of the M biexciton luminescence is fitted to the expression $\log_{10} I_{be} = 2\log_{10} I_{ex} - 1.339$, where I_{ex} is the Z_3 exciton luminescence intensity, I_{be} the M biexciton luminescence intensity, and I is the excitation density in units of MW cm^{-2}.

and the highest-density excitation are also shown in figure 16.16. The 391 nm band is ascribed to the M band, which corresponds to biexciton recombination leaving an exciton in a crystal [30, 46]. The M band is broad and does not seem to be composed of two bands which are usually observed in bulk CuCl crystals at liquid helium temperature. The excitation-dependent luminescence of the Z_3 exciton and the M biexciton bands is shown in the inset of figure 16.16. Although the biexciton luminescence intensity is not simply proportional to the excitation intensity or the square of this intensity, it is proportional to the square of the

Z_3 exciton luminescence. This is reasonable because biexcitons are formed as a result of the binary attractive interaction of excitons in CuCl microcrystals.

Lasing occurs as soon as the sample, which is placed in a cavity, is excited by a nitrogen laser and the excitation intensity exceeds a threshold. Figure 16.17 shows the emission intensity as a function of the excitation power density. At 77 K the emission intensity grows critically at the threshold power density of 2.1 MW cm^{-2} under the transverse pumping condition. The emission spectrum around the lasing threshold is shown in figure 16.18. The broad M band is observed below the threshold. Above the threshold the sharp emission spectrum with a maximum peak at 391.4 nm, is composed of a few longitudinal modes of the laser cavity, which are separated from each other by 0.07 nm. The separation almost agrees with the calculated longitudinal mode interval, 0.08 nm. It is obtained by using the refractive index of NaCl, 1.567, the spacing between two mirrors, 0.62 mm, and the thickness of an NaCl crystal, 0.58 mm. The emission is directional with an emission solid angle of about 0.03 steradian. The photograph of the lasing device under the excitation shows halation when taken from the lasing direction. These observations clearly indicate that the device shows lasing for temperatures up to 108 K.

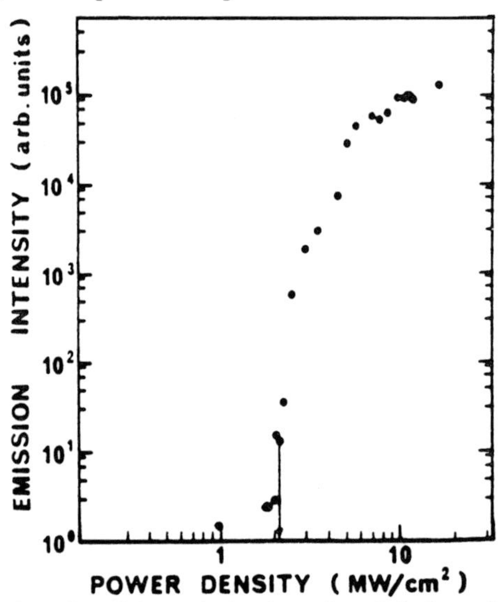

Figure 16.17. The log–log plot of the emission intensity of the laser device as a function of the excitation power density for the transverse pumping configuration at 77 K. The arrow shows the threshold for lasing.

In this system, the laser action takes place in three levels, ground, exciton, and biexciton states. Ultraviolet laser light (337 nm) corresponding to the band-to-band transition generates electron–hole pairs in CuCl microcrystals. They quickly form excitons and biexcitons. If the excitation density is high enough to generate a biexciton rather than an exciton, the population inversion takes

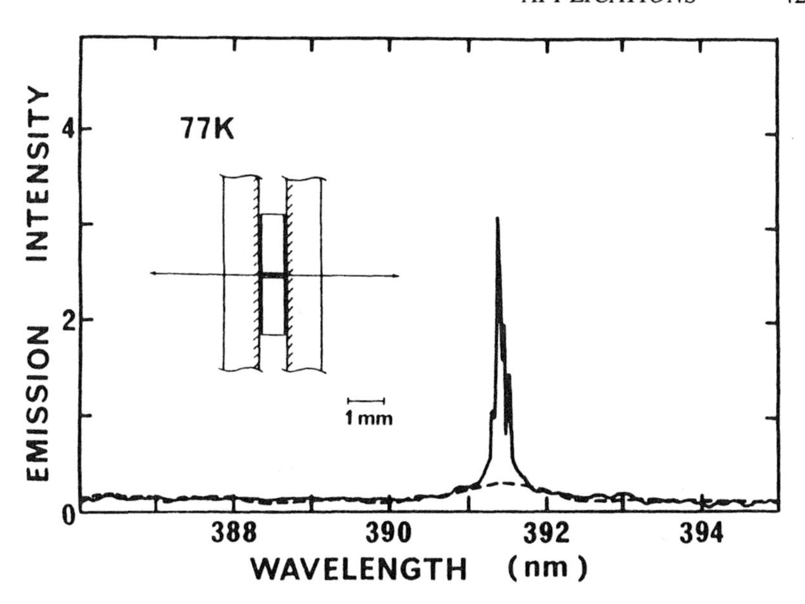

Figure 16.18. Emission spectra of the laser device at 77 K below and above the lasing threshold. The threshold I_{th} is about 2.1 MW cm^{-2}. The solid line shows the spectrum under the excitation of $1.08I_{th}$. The dashed line shows the spectrum under the excitation of $0.86I_{th}$. The inset shows a sketch of the laser device.

place. The optical gain is formed as a result of the population inversion. The threshold excitation density for lasing and the excitation density where the M luminescence exceeds the Z_3 exciton luminescence almost coincide with each other.

The optical gain spectrum was observed by means of the pump-and-probe method. The pump was the third harmonic of the Nd^{3+}:YAG laser, and the probe was the amplified spontaneous emission of an LD390 dye solution pumped by the third harmonics of the Nd^{3+}:YAG laser. Figure 16.19 is the pump-and-probe spectrum of the CuCl microcrystals with a mean radius of 3.8 nm. The absorption spectrum with pump shows the bleaching and the blue-shift of the Z_3 exciton, and optical gain at 392 nm. The peak of the optical gain is slightly red-shifted from the peak of the M luminescence band. The concentration and size dependence of the optical gain were also measured. The optical gain seems to be proportional to the CuCl concentration and to decrease with the decrease of the size of CuCl microcrystals.

It is an interesting test to examine the optical gain of CuCl microcrystals in an NaCl crystal (quantum-dot sample) in comparison with that of a CuCl bulk crystal (B sample). The test may clarify the merits of the semiconductor quantum dots. The optical gain of the samples, quantum-dot and B, was measured by the Shaklee method [48]. The optical gain of the quantum-dot

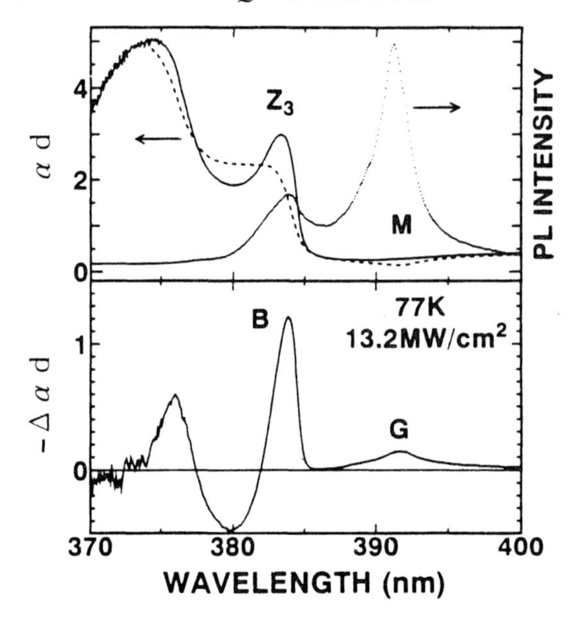

Figure 16.19. Pump-and-probe spectrum of CuCl microcrystals at 77 K. The mean radius of microcrystals is 3.8 nm. The solid line shows the absorption spectrum without pump. The dashed line shows the spectrum with pump. The lower figure is the differential absorbance. The dotted line shows the luminescence spectrum.

sample was compared with that of the B sample. The mean radius of CuCl microcrystals in the quantum-dot sample is 2.8 nm. An XeF excimer laser was used because the output of the laser is intense and has good spatial uniformity. The laser light (351 nm) also corresponds to the band-to-band excitation. Similar luminescence spectra and excitation density dependences were observed, as shown in figure 16.16. The optical gain was measured under the excitation density of 5 MW cm^{-2}. Although the absorption coefficient of the B sample is much larger than the quantum-dot sample, both quantum-dot and B samples are thick enough to absorb all of the laser fluence except the reflection loss. The experimental configuration and the result of the gain measurement are shown in figure 16.20. The data show the optical gain because the emission intensity is proportional to $[\exp(g\ell) - 1]/g$, where g is the small-signal optical gain and ℓ is the excitation length. The small-signal optical gains of two samples are comparable to each other, although the molar fraction of CuCl is only 0.14 mol% in the quantum-dot sample. This seems to indicate that the optical gain of microcrystals per unit volume of active medium for lasing, CuCl, is 700 times larger than that of bulk crystals. It is well known that the optical gain increases in proportion to the concentration of the active ions in the ion doped solid state laser materials below a certain concentration where the ion–ion interaction starts broadening the linewidth of the transition [52]. Therefore, we can expect further enhancement of the optical gain in CuCl microcrystals embedded in an NaCl crystal by increasing the CuCl concentration.

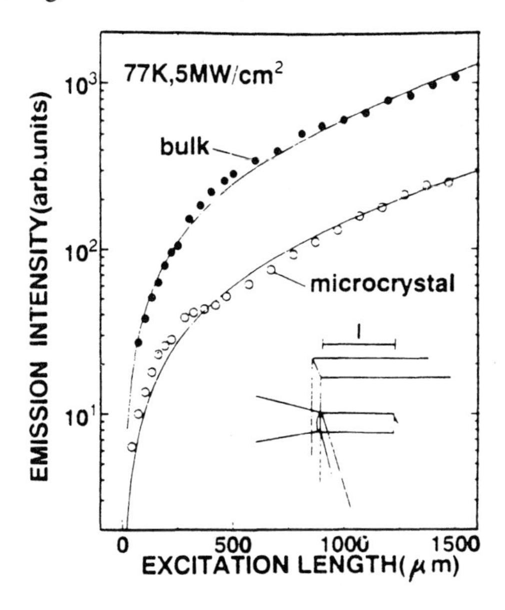

Figure 16.20. Stimulated emission intensity of M luminescence as a function of excitation length. The excitation wavelength is 351 nm and its density is 5 MW cm^{-2}. Open circles show the data of CuCl microcrystals in an NaCl crystal (quantum-dot sample), while solid circles show that of a CuCl bulk crystal (B sample). The small-signal optical gains for the quantum-dot sample and the B sample are 7 cm^{-1} and 8 cm^{-1}, respectively. In the inset, excitation and observation geometries are shown.

The optical gain due to the population inversion is expressed by $g = 2\pi \Delta N |\mu_{be}|^2 \omega_{be}/\hbar c \Delta\omega_{be}$, where $\Delta N = N_{be} - N_{ex}$ is the population inversion between biexcitons and excitons per unit volume of samples, μ_{be} is the dipole moment of the transition between biexciton and exciton states, $\hbar\omega_{be}$ is the transition energy, and $\hbar\Delta\omega_{be}$ is the spectral linewidth [53]. We can compare the values of μ_{be} and $\Delta\omega_{be}$ for CuCl microcrystals and CuCl bulk crystals. Calculations show that the value of μ_{be} for the same volume of CuCl is almost the same for CuCl bulk crystals and CuCl microcrystals if the radius of microcrystals is larger than 3 nm [54]. The value of $\hbar\Delta\omega_{be}$ corresponds to the observed broadening of the M band, 1.5 meV, due to the thermal distribution on the dispersion of biexcitons in a bulk crystal. On the other hand, it corresponds to the inhomogeneous broadening of 1.5 meV due to the size distribution of CuCl microcrystals. They are almost equal to each other. The experimental result shows that the optical gains of CuCl microcrystals in an NaCl crystal and a CuCl bulk crystal are almost equal to each other. This means that the optical gain of microcrystals per unit volume of active medium for lasing, CuCl, is 700 times larger than that of bulk crystals. Therefore, the population inversion ΔN for the quantum-dot sample is 700 times larger than that for the B sample.

The large population inversion for CuCl microcrystals is probably due to the spatial carrier confinement effect. The luminescence quantum efficiency of CuCl microcrystals is much higher than that of CuCl bulk crystals [55]. The impurity isolation effect observed in AgBr quantum dots [56, 57] and the absence of polariton propagation or exciton diffusion may explain the high quantum efficiency of luminescence in CuCl microcrystals. They can explain the high concentration of excitons leading to the large population inversion between biexcitons and excitons. Further study is necessary to clarify why the population inversion for CuCl microcrystals is so large. Because the homogeneous width of the transition in CuCl microcrystals is expected to be much narrower than the inhomogeneous broadening, similar to the exciton linewidth [58], we can expect the further enhancement of the optical gain if we can prepare CuCl microcrystals whose size distribution is much narrower than that used in this experiment.

16.4.2 All-optical switching using quantum dots

For all-optical switching applications using semiconductor microcrystallites, a large nonlinear refractive-index effect is sought while at the same time minimizing the detrimental effects of absorption. In material terms this calls for a system whose resonance conditions can be varied so that the trade-off between refractive index and absorption can be adjusted for a given laser frequency. This degree of flexibility is realized in quantum dots, since the quantum confinement leads to a tunable blue-shift in the lowest exciton resonance which increases with decreasing dot radius. In addition, the operating device length can be scaled by varying the quantum-dot concentration. Here we theoretically examine the trade-offs presented by quantum dots from the perspective of all-optical switching. In particular, we consider the prototypical all-optical switch, namely the nonlinear directional coupler (NLDC) [59, 60].

The linear directional coupler consists of two identical optical waveguides which are in sufficiently close proximity that their evanescent fields overlap and, hence, there is periodic power transfer between the two waveguides with propagation distance [61]. For initial excitation of only one waveguide and a beat length device $L = L_b$, the output appears only in the input or bar waveguide, whereas for a half-beat length device, $L = L_c = L_b/2$, the output appears in the adjacent or cross waveguide, where L_c is the coupling length. In linear integrated optics the half-beat length directional coupler is operated as an optical switch using the electro-optic effect [62, 63]: applying a d.c. field to the bar waveguide causes a wavevector mismatch, $\Delta\beta = k_0 \Delta n$, between the waveguides which inhibits the coupling, and the output exits through the bar state with the d.c. field applied. The key idea of all-optical switching using an NLDC is that the input field itself generates a nonlinear change in propagation wavevector, $\Delta\beta^{NL} = k_0 \Delta n^{NL}$, where Δn^{NL} is the nonlinear change in refractive index, so that the input intensity determines the output state [59, 60]. The basic requirement for efficient all-optical switching to occur is that the light induced nonlinear phase shift for an isolated waveguide obeys [14, 64, 65]

$$k_0 \Delta n^{NL} L = \Delta\beta^{NL} L > 4\pi \tag{16.25}$$

where L is the device length. Equation (16.25) will form the basis of our discussion of all-optical switching of an NLDC using quantum dots.

The basic operation of an NLDC can be modeled in the framework of coupled-mode theory [59, 60]. Specifically, we expand the field in the coupler as a coherent superposition of the transverse modes of the individual waveguides

$$\boldsymbol{E}(\boldsymbol{r}, t) - \boldsymbol{x}\tfrac{1}{2}[\mathcal{E}_1(z)u_1(x, y) + \mathcal{E}_2(z)u_2(x, y)]\mathrm{e}^{\mathrm{i}(\beta z - \omega t)} + \mathrm{CC} \tag{16.26}$$

where CC is the complex conjugate, z is the propagation direction, $\boldsymbol{x}$ is the unit polarization vector, $u_j(x, y)$ is the transverse mode of waveguide j, $\mathcal{E}_j(z)$ is the amplitude of the jth mode, ω is the central carrier frequency of the field, and β is the propagation wavevector of the identical waveguides. Here the propagation wavevector β is assumed to account for the optical properties of the host glass, which is treated as a transparent dielectric. The linear and nonlinear optical properties of the quantum dots are incorporated through the following polarization source term for each individual waveguide (nonlinear coupling between the waveguides is neglected)

$$\begin{aligned} P_j(z) &= P_j^{(1)}(z) + P_j^{(3)}(z) \\ &= \chi^{(1)}(\omega)\mathcal{E}_j(z) + \chi^{(3)}(\omega)|\mathcal{E}_j(z)|^2\mathcal{E}_j(z) \end{aligned} \tag{16.27}$$

where $\chi^{(1)}(\omega)$ is the linear susceptibility of the quantum-dot system and $\chi^{(3)}(\omega) \equiv \chi^{(3)}(-\omega, \omega, -\omega, \omega)$ is the third-order nonlinear susceptibility. Substituting equations (16.26) and (16.27) into Maxwell's equations yields the following coupled-mode equations in the usual slowly varying envelope approximation [64, 66]

$$\frac{\mathrm{d}\mathcal{E}_1}{\mathrm{d}z} = \frac{\mathrm{i}\pi}{2L_c}\mathcal{E}_2 - \tfrac{1}{2}(\alpha_0 - \alpha_3|\mathcal{E}_1|^2)\mathcal{E}_1 + \Delta\beta_1^{NL}\mathcal{E}_1 \tag{16.28a}$$

and

$$\frac{\mathrm{d}\mathcal{E}_2}{\mathrm{d}z} = \frac{\mathrm{i}\pi}{2L_c}\mathcal{E}_1 - \tfrac{1}{2}(\alpha_0 - \alpha_3|\mathcal{E}_2|^2)\mathcal{E}_2 + \Delta\beta_2^{NL}\mathcal{E}_2. \tag{16.28b}$$

The first term on the right-hand side of equations (16.28) describes the linear mode coupling due to the evanescent field overlap. In the absence of the quantum dots these terms correctly predict the periodic energy exchange between the waveguides. The next two terms involving α_0 and α_3, defined as

$$\alpha_0 = \frac{4\pi\omega^2\eta}{\beta c^2}\,\mathrm{Im}(\chi^{(1)}(\omega)) \qquad \alpha_3 = -\frac{4\pi\omega^2\eta}{\beta c^2}\,\mathrm{Im}(\chi^{(3)}(\omega)) \tag{16.29}$$

describe linear absorption and absorption saturation ($\alpha_3 > 0$) respectively. Finally, $\Delta\beta_j^{NL}$ is the nonlinear change in the model wavevector of the jth waveguide

$$\Delta\beta_j^{NL} = \frac{2\pi\omega^2\eta}{\beta c^2}\,\mathrm{Re}(\chi^{(3)}(\omega))|\mathcal{E}_j|^2. \tag{16.30}$$

The parameter η in equations (16.29) and (16.30) is the volume fill fraction of quantum dots: $\eta = 0$ corresponds to the absence of quantum dots.

A material figure of merit for the NLDC can now be obtained using equation (16.25) and equations (16.28)–(16.30) as follows. It is clear that linear absorption places a limit on the useful device length, so we require $\alpha_0 L_c \simeq 1$ for a half-beat length coupler. Also, it follows from equations (16.28) that the nonlinearity will saturate when the field strength approaches $|\mathcal{E}_s| \simeq \alpha_0/\alpha_3$. Substituting this in equation (16.30) and using equation (16.29), we obtain the peak phase shift over the device length as $\Delta\beta^{NL} L_c \simeq \mathrm{Re}(\chi^{(3)}(\omega))/2\,\mathrm{Im}(\chi^{(3)}(\omega))$, where we used $\alpha_0 L_c = 1$. According to equation (16.25), we require this phase shift to exceed 4π for efficient all-optical switching, so we define the material figure of merit [14]

$$W = \frac{1}{8\pi}\frac{\mathrm{Re}(\chi^{(3)}(\omega))}{\mathrm{Im}(\chi^{(3)}(\omega))} > 1. \tag{16.31}$$

This figure of merit is identical in form to that obtained for two-photon absorption as a limitation to all-optical switching [67]. Note that W is geometry independent and independent of the volume fill fraction η. The dependence on the fill fraction η is contained in the device length which is derived using $\alpha_0 L_c = 1$ as

$$L_c = \frac{\beta c^2}{2\pi\omega^2\eta\,\mathrm{Im}(\chi^{(1)}(\omega))} \tag{16.32}$$

that is, the device length is inversely proportional to η.

From the perspective of all-optical switching in NLDCs, one wants to make the material figure of merit as large as possible, $W \geq 1$ (switching may still occur for $W < 1$ but at the expense of poor transmission). For a material to be suitable for fabricating an NLDC, it is desirable that $\mathrm{Re}(\chi^{(3)})$ is 25 or more times greater than $\mathrm{Im}(\chi^{(3)})$. Whether or not this is possible for a given material depends on its resonance structure. We have evaluated the material figure of merit W for the case of quantum dots embedded in glass using the theory of Hu *et al* [22] which provides both the linear susceptibiliity $\chi^{(1)}$ and the third-order susceptibility $\chi^{(3)}$. Figure 16.21 shows the frequency variation of (*a*) $\mathrm{Im}(\chi^{(1)})$, and (*b*) $\mathrm{Re}(\chi^{(3)})$ and $\mathrm{Im}(\chi^{(3)})$ for the semiconductor CdS and a quantum-dot radius $R = a_0$, with a_0 being the bulk exciton radius. Here the frequency is represented by the dimensionless detuning $\Delta = (\hbar\omega - E_G)/E_R$, where E_G is the bandgap of the bulk material and E_R is the exciton Rydberg. For CdS, $E_G = 2520$ meV, $E_R = 27$ meV, and a homogeneous line broadening $\gamma = 2E_R$ was included in the calculations. The $\chi^{(3)}$ values in figure 16.21(*b*) can be converted to units of m^2 V^{-2} by multiplying by $10^{-16}\eta$. Then, for example, for a detuning of $-20E_R$ we have $\text{Re}(\chi^{(3)}) \simeq 10^{-20}$ m^2 V^{-2} and $\text{Im}(\chi^{(3)}) \simeq 2 \times 10^{-21}$ m^2 V^{-2} for $\eta = 10^{-2}$, in reasonable agreement with the measured values reported by [68].

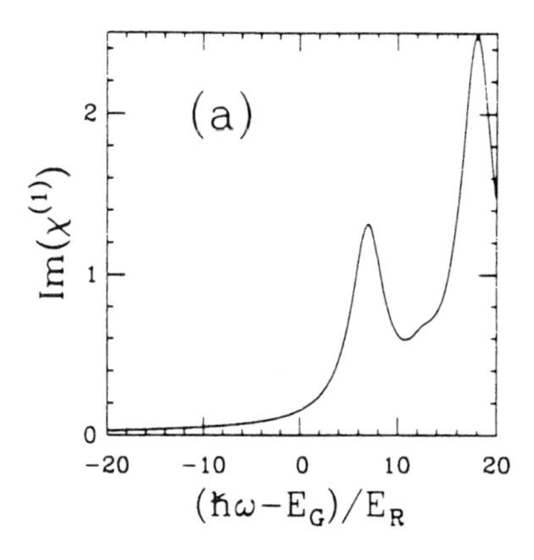

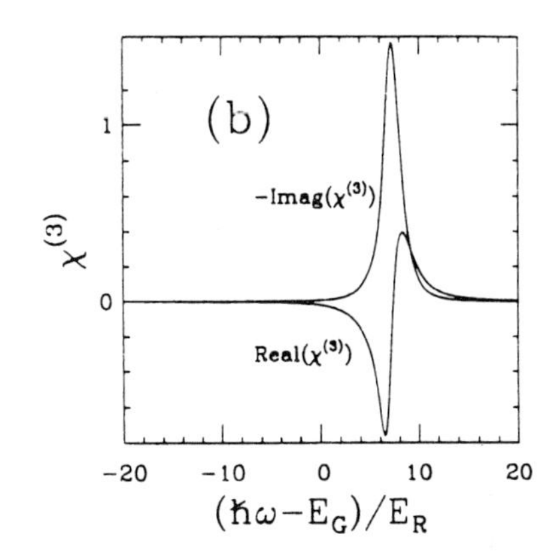

Figure 16.21. Frequency variation of (*a*) $\text{Im}(\chi^{(1)})$ and (*b*) $\text{Re}(\chi^{(3)})$ and $\text{Im}(\chi^{(3)})$, for the semiconductor CdS, $R = a_0$, $E_G = 2520$ meV, $E_R = 27$ meV, and a homogeneous line broadening $\gamma = 2E_R$.

Figure 16.22 shows the material figure of merit W versus detuning Δ for CdS quantum dots and two values of the dot radius. For $R = a_0$, W remains less than one all the way out to $-20E_R$. In contrast, for $R = 0.5a_0$, W is greater than one for detunings below $-16E_R$. The variation of W with detuning Δ may be understood by realizing that the lowest exciton resonance saturates like a two-level system in which the effective detuning is that between the field and the 1s exciton, $\delta = (\hbar\omega - E_{1s})/E_R$ [22, 69]. In this case, $\text{Re}(\chi^{(3)})$ varies as $\delta/(1+\delta^2)$, and $\text{Im}(\chi^{(3)})$ varies as $-1/(1+\delta^2)$, so that according to equation (16.31), W varies as

$$W \simeq \frac{1}{8\pi}\left[-\Delta + \frac{(E_{1s} - E_G)}{E_R}\right]. \tag{16.33}$$

On the basis of this scaling argument we expect that the slope of the W versus Δ curve should be independent of the dot radius, and this is clearly satisfied in figure 16.22. Furthermore, since the blue-shift of the 1s exciton energy E_{1s} varies with dot radius as roughly R^{-2} in the quantum-confined regime, it follows from equation (16.31) that for a given Δ the figure of merit W is larger for smaller dots. This property is satisfied in figure 16.22.

The results shown in figure 16.22 are for fixed values of the dot radius. These results show that if the dot radius can be controlled, then it is possible to engineer the material so that it is useful for all-optical switching applications. In quantum-dot samples there is a statistical distribution of dot radii, which adds inhomogeneous broadening to the problem. This inhomogeneous broadening is

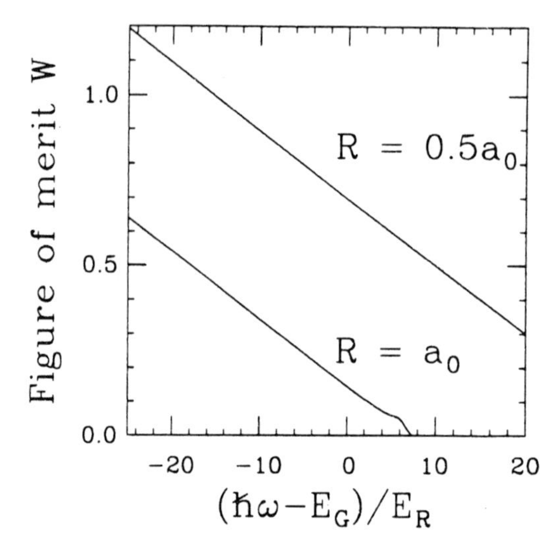

Figure 16.22. Figure of merit W versus detuning for CdS and $R = 0.5a_0$ and $R = a_0$.

typically of the same order as the homogeneous broadening, and tends to wash out any strong dependence on the dot radius. However, there is no fundamental reason why the inhomogeneous broadening cannot be reduced, and the figure of merit of semiconductor quantum dots enhanced by producing small quantum dots with well controlled radii.

Figure 16.23 shows an example of the predicted all-optical switching for CdS quantum dots and $R = 0.5a_0$, and a detuning of $-20E_R$. These results were obtained by the numerical solution of equations (16.28), incorporating the theoretical results for the linear and nonlinear susceptibilities in figure 16.21. In this case, $W > 1$ and all-optical switching is expected. The transmission of both the bar and cross waveguides with respect to the total output intensity is shown versus the dimensionless input intensity $|\mathcal{E}_1(0)|^2$ in figure 16.23(a), and all-optical switching is evident. (The fields are converted to V m^{-1} by multiplying $\mathcal{E}$ by 10^8.) For an input intensity of 1000, the output exits through the cross waveguide, whereas for an input intensity of 2600 the output exits the bar state. This is also seen in figure 16.23(b) which shows the output intensity versus the input intensity. Here we see that the actual transmission of the device increases at high input intensity due to the saturation of the linear absorption. Assuming a value of $\eta = 10^{-2}$ for the fill fraction, the predicted coupling length is $L_c = 1/\alpha_0 \simeq 5$ mm. In contrast, for a dot size $R = a_0$ and a detuning of $-20E_R$, no all-optical switching is observed in the device simulations, in agreement with the observation that $W < 1$ (see figure 16.22).

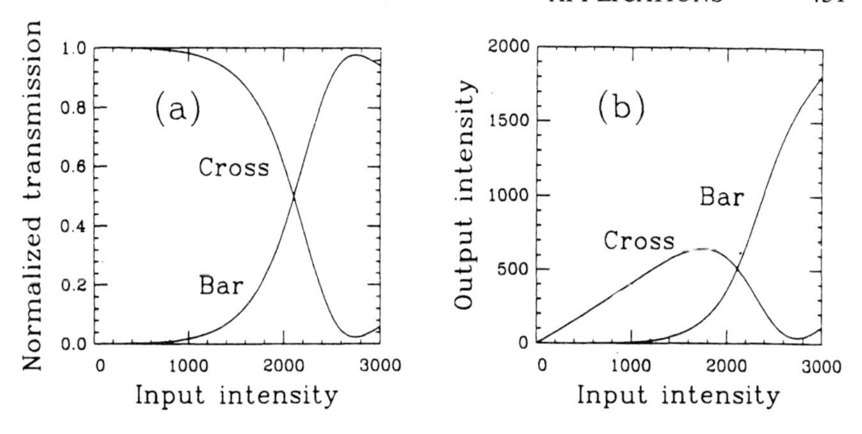

Figure 16.23. Switching characteristics of a nonlinear directional coupler in CdS quantum dots with $R = 0.5a_0$: (*a*) normalized transmission for the bar and cross waveguides versus dimensionless intensity, and (*b*) output intensity for the bar and cross waveguides versus input intensity.

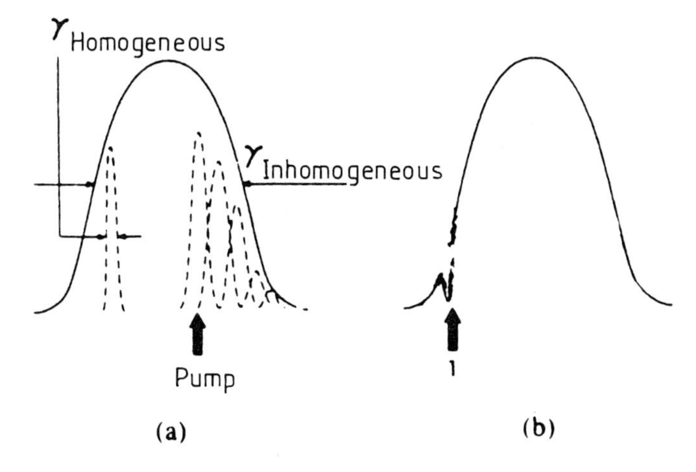

Figure 16.24. (*a*) Schematic absorption spectrum of an inhomogeneously broadened absorption line consisting of homogeneously broadened features. (*b*) The application of a laser beam at the spectral position shown by the arrow results in the creation of a hole, corresponding to binary '1'.

16.4.3 Quantum dots for optical data storage

Spectral hole burning has been employed in the past to investigate frequency-domain optical data storage [70, 71]. In such an application, a recording material is required that may undergo spectral hole burning with certain well defined characteristics. Most of the materials that have been used so far have required liquid helium refrigeration. This requirement limits the practical applications of

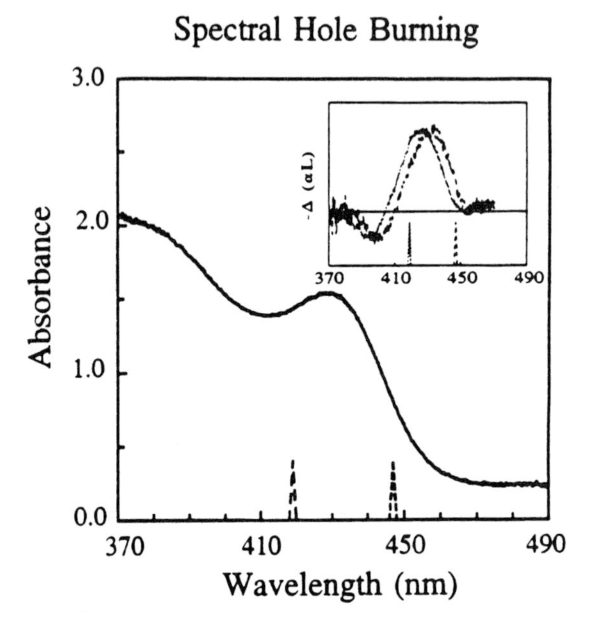

Figure 16.25. The linear absorption spectrum of a CdS quantum-dot sample at room temperature. The sample was pumped at two spectral positions inside the first quantum-confined transition. The positions of the pumps are shown by the small peaks at wavelengths of 419 nm and 447 nm. The inset shows the change in the absorption ($-\Delta\alpha L$) obtained as a result of this pumping. Absorption bleaches and spectral holes are generated, which shift as the pump wavelength is changed. This figure clearly demonstrates transient spectral hole burning at room temperature in a quantum-dot sample (sample was grown by S Risbud and co-workers).

such materials for optical data storage systems.

Figure 16.24 schematically shows the concept of the spectral hole burning technique. The absorption spectrum of the recording material has to be inhomogeneously broadened with a total width of γ_{inh}, consisting of homogeneously broadened components with linewidths of γ_{hom}, as shown in figure 16.24(a). When a laser beam with a well defined wavelength irradiates the sample, only the homogeneous line that is in resonance with the laser can be excited, and consequently gets bleached. Figure 16.24(b) shows such a bleaching of the homogeneously broadened line, which is referred to as spectral hole burning. For various tuning of the laser frequency inside the inhomogeneously broadened absorption spectrum, holes at different spectral positions can be burned. Each burned hole may be considered as a logic '1', while a no-hole is a logic '0'. The number of holes that may be burned (which is related to the storage density) is on the order of the ratio of the inhomogeneous width to the homogeneous width, $R = \gamma_{inh}/\gamma_{hom}$. The larger γ_{inh}, or the smaller γ_{hom}, the

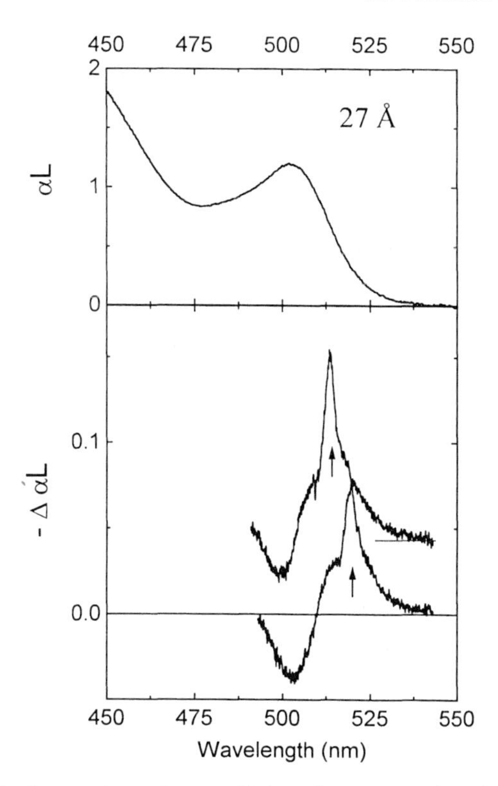

Figure 16.26. The linear absorption coefficient (upper curve) and the change in the absorption ($-\Delta\alpha L$) for pumping at 515 nm and 520 nm (sample was prepared by S Gaponenko and co-workers).

larger is the number of burned holes. The laser may be focused to a spot size of $\simeq 1\ \mu$m, and within this small spot a number of spectral holes may be burned.

Quantum dots may be appropriate for data storage. This application stems from the fact that quantum dots in various media have different sizes, leading to an inhomogeneously broadened absorption spectra. The larger the particle size distribution, the broader the total absorption linewidth is. Also, the persistence of quantum-confined absorption peaks at room temperature in quantum-confined structures makes it possible to realize this inhomogeneous broadened lineshape at room temperature.

Spectral hole burning experiments in CdS and CdSe quantum dots in glass at room temperature were performed. Figure 16.25 shows a typical result for the CdS dots in glass. The linear transmission spectrum of the sample, the spectral positions of the pump pulses, and the differential transmission of the sample (inset of figure 16.25) as a result of the presence of the pump are plotted. The spectral holes burned by the pumps are clearly observable. The spectral hole moves with the pump wavelength, as expected for an inhomogeneously broadened transition. The width of the spectral hole increases as the pump intensity increases. These measurements show that spectral hole burning is possible in quantum-dot samples. However, the width of the burned hole is broad in most of the samples studied so far [20, 72], indicating that not many holes can be burned in the excitation spot by changing the wavelength.

This situation can be improved for samples with better surface quality and for specially prepared samples. For example, spectrally narrow holes have been seen in quantum dots in organic matrices [73]. Recently, narrow spectral holes were observed in quantum dots in glass after the sample was strongly illuminated [74]. An example of such a narrow spectral hole burning measured in our laboratories is shown in figure 16.26. Spectral holes with widths as low as a few meV have been achieved. The data in figure 16.26 were obtained after the sample was irradiated with ≈ 30 MW cm^{-2} for two hours. The origin of the narrow holes has been attributed to the growth process, i.e. they have been observed only in the samples prepared in the nucleation and normal growth stages, whereas the broad holes are observed in the sample prepared in the coalescence stage.

REFERENCES

[1] Brus L E 1984 *J. Chem. Phys.* **80** 4403

[2] Brus L E 1986 *IEEE J. Quantum Electron.* **QE-22** 1909 and references to the author's earlier work

Brus L E 1983 *J. Chem. Phys.* **79** 5566

Bawendi M G, Carroll P J, Wilson W L and Brus L E 1992 *J. Chem. Phys.* **92** 946

[3] Kash K, Scherer A, Worlock J M, Craighead H G and Tamargo M C 1986 *Appl. Phys. Lett.* **49** 1043

Kang K I, McGinnis B P, Sandalphon Hu Y Z, Koch S W, Peyghambarian N, Mysyrowicz A, Liu L C and Risbud S H 1992 *Phys. Rev.* B **45** 3465

[4] Reed M A, Bate R T, Bradshaw K, Duncan W M, Frensley W R, Lee J W and Shih H D 1986 *J. Vac. Sci. Technol.* **4** 358

Rama Krishna M V and Friesner R A 1991 *Phys. Rev. Lett.* **67** 629

Roussignol P, Ricard D, Flytzanis C and Neuroth N 1989 *Phys. Rev. Lett.* **62** 312

[5] Cibert J, Petroff P M, Dolan G J, Pearton S J, Gossard A C and English J H 1986 *Appl. Phys. Lett.* **49** 1275

Efros Al L and Efros A L 1982 *Sov. Phys.–Semicond.* **16** 772 (in Russian)

[6] Ekimov A I, Efros Al L and Onushchenko A A 1985 *Solid State Commun.* **56** 921

Ekimov Al and Onushchenko A A 1982 *Sov. Phys.–Semicond.* **16** 775 (in Russian)

Bawendi M G, Wilson W L, Rothberg P L, Carroll P J, Jedjin T M, Steigerwald M L and Brus L E 1990 *Phys. Rev. Lett.* **65** 1623

Brus L E 1991 *Appl. Phys.* A **53** 465

Bladereschi A and Lipari N O 1973 *Phys. Rev.* B **8** 2697

[7] Borrelli N F, Hall D W, Holland H J and Smith D W 1987 *J. Appl. Phys.* **61** 5399

[8] Peyghambarian N and Koch S.W 1987 *Rev. Phys. Appl.* **22** 1711

[9] Jain R K and Lind R C 1983 *J. Opt. Soc. Am.* **73**

[10] Yao S S, Karaguleff C, Gabel A, Fortenbery R, Seaton C T and Stegeman G 1985 *Appl. Phys. Lett.* **46** 801
[11] Roussignol P, Ricard D, Rustagi K C and Flytzanis C 1985 *Opt. Commun.* **55** 1431
[12] Nuss M C, Zinth W and Kaiser W 1987 *Appl. Phys. Lett.* **49** 1717
[13] Olbright G R, Peyghambarian N, Koch S W and Banyai L 1987 *Opt. Lett.* **12** 413
[14] Stegeman G I and Wright E M 1990 *Opt. Quantum Electron.* **22** 95
[15] Miller D A B, Chemla D S, Damen T C, Gossard A C, Wiegmann W, Wood T H and Burrus C A 1984 *Appl. Phys. Lett.* **45** 13
[16] Miller D A B, Chemla D S and Schmitt-Rink S 1988 *Appl. Phys. Lett.* **52** 2154
[17] Esch V, Fluegel B, Khitrova G, Gibbs H M, Jiagin X, Kang K, Koch S W, Liu L C, Risbud S H and Peyghambarian N 1990 *Phys. Rev.* B **42** 7450
[18] Henneberger F, Puls J, Spiegelberg Ch, Jungnickel V and Ekimov A J 1991 *Semicond. Sci. Technol.* **6** 441
[19] Buot F A 1987 *Int. J. Comp. Math. Electron. Elect. Eng. COMPEL* **6** 45
[20] Peyghambarian N, Fluegel B, Hulin D, Migus A, Joffre M, Antonette A, Koch S W and Lindberg M 1989 *IEEE J. Quantum Electron.* **QE-25** 2516
[21] Peyghambarian N, Koch S W and Mysyrowicz A 1993 *Introduction to Semiconductor Optics* (Englewood Cliffs, NJ: Prentice-Hall)
[22] Hu Y Z, Koch S W, Lindberg M, Peyghambarian N, Pollock E L and Abraham F F 1990 *Phys. Rev. Lett.* **64** 1805
[23] Yumoto J, Shinojima H, Uesugi N, Tsunetomo K, Nasu H and Osaka Y 1990 *Appl. Phys. Lett.* **57** 2393
[24] Shinojima H, Yumoto J and Uesugi N 1992 *Appl. Phys. Lett.* **60** 298
[25] Park S H, Morgan R A, Hu Y Z, Lindberg M, Koch S W and Peyghambarian N 1990 *J. Opt. Soc. Am.* B **7** 2097
[26] Xia J-B 1989 *Phys. Rev.* B **40** 8500
[27] Vahala K J and Sercel P C 1990 *Phys. Rev. Lett.* **65** 239
[28] Hu Y Z, Lindberg M and Koch S W 1990 *Phys. Rev.* B **42** 1713
[29] Hanamura E 1988 *Phys. Rev.* B **38** 1228
[30] Itoh T, Jin F, Iwabuchi Y and Ikehara T 1989 *Nonlinear Optics of Organics and Semiconductors* ed T Kobayashi (Berlin: Springer) p 76
[31] Nakamura A, Yamada H and Tokizaki T 1989 *Phys. Rev.* B **40** 8585
[32] Hanamura E, Tokohiro T and Manabe Y 1992 *Proc. Int. Workshop on Science and Technology of Mesoscopic Systems (Nara, Japan, 1992)*
[33] Tokohiro T, Manabe Y and Hanamura E 1993 *Phys. Rev.* B **47**
[34] Manabe Y, Tokihiro T and Hanamura E 1993 *Phys. Rev.* submitted
[35] Lehmberg R H 1970 *Phys. Rev.* A **2** 883 and 889
Luttinger J M 1956 *Phys. Rev.* **102** 1030
[36] Gross M and Haroche S 1992 *Phys. Rep.* **93** 301
[37] Gonokami M and Peyghambarian N 1992 private communication
[38] Minoshima K and Kobayashi T 1992 private communication
[39] Ezaki H, Tokihiro T and Hanamura E 1993 *Phys. Soc. Meeting of Japan*
[40] Hanamura E 1988 *Phys. Rev.* B **37** 1273
[41] Hanamura E 1992 *Phys. Rev.* B **46** 4718
[42] Masumoto T, Yamazaki M and Sugawara H 1988 *Appl. Phys. Lett.* **53** 1527i
[43] Nakamura A and Tokizaki T 1992 *Nonlinear Optics* ed S Miyata (Amsterdam: Elsevier) p 139

[44] Weisbuch C and Nagle J 1990 *Science and Engineering of One- and Zero-Dimensional Semiconductors* (New York: Plenum) p 309
[45] Arakawa Y and Sakaki H 1982 *Appl. Phys. Lett.* **40** 939
[46] Asada M, Miyamoto Y and Suematsu Y 1986 *IEEE J. Quantum Electron* **QE-22** 1915
[47] Ueta M, Kanzaki H, Kobayashi M, Toyozawa Y and Hanamura E 1986 *Excitonic Processes in Solids* (Berlin: Springer) ch 3
[48] Shaklee K L, Leheny R F and Nahory R E 1971 *Phys. Rev. Lett.* **26** 888
[49] Ojima M, Oka Y, Kushida T and Shionoya S 1977 *Solid State Commun.* **24** 845
[50] Weinberger D A, Peyghambarian N, Rushford M C and Gibbs H M 1984 *Proc. 1984 Annu. Meeting Opt. Soc. Am. (San Diego, 1994)* (Optical Society of America) p 31
[51] Itoh T, Iwabuchi Y and Kataoka M 1988 *Phys. Status Solidi* b **145** 567
[52] Kaminskii A A 1981 *Laser Crystals* (Berlin: Springer) ch 6
[53] Shimoda K 1986 *Introduction to Laser Physics* (Berlin: Springer)
[54] Takagahara T 1989 *Phys. Rev.* B **39** 10206
[55] Itoh T, Furumiya M, Ikehara T and Gourdon C 1990 *Solid State Commun.* **73** 271
[56] Kanzaki H and Tadakuma Y 1991 *Solid State Commun.* **80** 33
[57] Masumoto Y, Kawamura T, Ohzeki T and Urabe S 1992 *Phys. Rev.* B **46** 1827
[58] Wamura T, Masumoto Y and Kawamura T 1991 *Appl. Phys. Lett.* **59** 1758
[59] Jensen S M 1982 *IEEE J. Quantum Electron.* **QE-18** 1580
[60] Maier A 1982 *Sov. J. Quantum Electron.* **12** 1490
[61] Marcatili E A J 1969 *Bell. Syst. Tech. J.* **48** 2071
[62] Kurazono S, Iwasaki K and Kumagai N 1972 *Electron. Commun. Japan* **55** 103
[63] Tada K and Hirose K 1974 *Appl. Phys. Lett.* **25** 561
[64] Stegeman G I, Caglioti E, Trillo S and Wabnitz S 1987 *Opt. Commun.* **63** 281
[65] Wright E M, Koch S W, Ehrlich J E, Seaton C T and Stegeman G I 1988 *Appl. Phys. Lett.* **52** 2127
[66] Caglioti E, Trillo S, Wabnitz S and Stegeman G I 1988 *J. Opt. Soc. Am.* B **5** 472
[67] Mizrahi V, DeLong K W, Stegeman G I, Saifi M A and Andrejco M J 1989 *Opt. Lett.* **14** 1140
[68] Cotter D, Burt M G and Manning R J 1992 *Phys. Rev. Lett.* **68** 1200
[69] Schmitt-Rink S, Miller D A B and Chemla D S 1987 *Phys. Rev.* B **35** 8113
[70] Castro G D, Haarer D, Macfarlane R M and Trommsdorff H P 1978 *US Patent* 4 101 976
[71] Ortiz C, Macfarlane R M, Shalby R M, Lenth W and Bjorklund G C 1981 *Appl. Phys. Lett.* **25** 87
[72] Spiegelberg C, Henneberger F and Puls J 1991 *SPIE Proc.* **1362** 935
[73] Alivisatos A P, Harris A C, Levinos N J, Steigerwald M L and Brus L E 1988 *J. Chem. Phys.* **89** 4001
[74] Woggon U, Gaponenko S, Langbin W, Uhrig A and Klingshirn C 1993 *Phys. Rev.* B
[75] Liu L C and Risbud S H 1990 *J. Appl. Phys.* **68** 28

PART 7

SPECIAL NANOMATERIALS

Chapter 17

Porous silicon nanostructures

S M Prokes

17.1 INTRODUCTION

Porous silicon was first produced by Uhlir [1] and Turner [2] in their studies of electropolishing of silicon in dilute hydrofluoric (HF) solution. It was found that a material consisting of a network of pores could be formed in the dilute hydrofluoric acid in which the current densities were below those for electropolishing. Until recently, the major application of porous silicon was for Si-on-insulator (SOI) technology, where the active devices could be dielectrically isolated by the oxidation of the underlying porous silicon. The formation of this oxide layer was relatively easy in porous silicon, since it contained a large number of internal surfaces, which were highly reactive at elevated temperatures [3].

Recently, significant attention has been directed toward this material due to its visible photoluminescence (PL) [4]. The first visible PL in porous silicon was reported by Pickering *et al* [5] in 1984, but it was not until 1990 that Canham pointed out the importance of this phenomenon [4]. It is well known that improvements in electronics require increased speed, decreased dimensions, and increased functionality. As dimensions decrease, the device interconnects become two or three times larger than the actual device. This of course becomes a problem since the properties and failure modes in the interconnects become the overriding feature. Therefore, opto-electronics may be the next generation in device structures, where optical interconnects may replace metal interconnects. Most advances in this area have been in the III–V and II–VI systems, but Si is the material of choice in the electronics industry. It is very abundant, a factor of ten or more cheaper than other materials, has good passivating characteristics, good thermal and mechanical properties, and a vast majority of current technology and processing is based on it. The difficulty with silicon, from the perspective of opto-electronics, is that it is a poor opto-electronic material. Silicon is an indirect gap semiconductor in which interband transitions need phonons. This means that transition probabilities in bulk silicon are a 100 times smaller than in direct gap materials, resulting in a quantum efficiency in the range of $10^{-4}\%$, with the radiative recombination producing light in the infrared (1.1 eV). Thus, the reports of visible light emission (1.6–2.1 eV) [6] of high quantum efficiency (1–10%) [6] from porous silicon are quite surprising, and have generated intense interest.

Since the porous silicon light emitting structures can be fabricated using a relatively simple and inexpensive etching technique, and since they can easily be integrated with existing silicon technology, a significant worldwide research effort has emerged. The research has been aimed at understanding the mechanism of the light emission process, which will impact possible technological applications, such as optical interconnects or light emitting diodes.

17.2 POROUS SILICON FORMATION

The simplest way to produce porous silicon is to etch silicon in an electrochemical cell containing dilute HF, at constant current [1, 2]. Canham [5] reported a two-step etching process, consisting of anodization in dilute HF at low current densities, followed by an open-circuit pore widening treatment in dilute HF. The anodization generates an array of holes or pores perpendicular to the surface of the substrate. The open-circuit etching of the porous silicon structures in HF has been used to increase the pore size and produce a blue-shift in the visible luminescence [4, 7]. However, it has also been shown that the pore widening treatments are not necessary to observe visible luminescence [8, 9]. For a more detailed discussion of the electrochemical formation of porous silicon, see Zhang *et al* [10], Searson and Zhang [11], and Foell [12].

One of the reasons that light emission in this system is so controversial is that porous silicon can differ significantly in structure and surface properties from sample to sample, depending on the doping, and to some extent on the etching conditions [13]. Initially, the structure of porous silicon was assumed to consist of quantum wires [4], which were formed by open-circuit etching to create a material with a porosity above 78%, as shown in figure 17.1(*a*). Alternatively, Cullis *et al* [14] suggested that the structure resembled undulating wires, as shown in figure 17.1(*b*), and other results from high-porosity samples suggest a structure comprising of crystalline silicon particles embedded in amorphous silicon or oxide [15, 16], as shown in figure 17.1(*c*).

In general, the morphology can be grouped according to substrate doping and etching conditions. A summary of the various morphologies exhibited by porous silicon has been compiled by Searson *et al* [13].

(i) 1 Ω cm p-type porous silicon: exhibits a 'sponge'-like structure, with particle sizes on the nanometer scale [8, 15, 17].

(ii) Low-doped p-type silicon (>1 Ω cm): interconnected network of nanometer-sized silicon particles, with porosities on the order of 40–60%

I. "Quantum Wires"

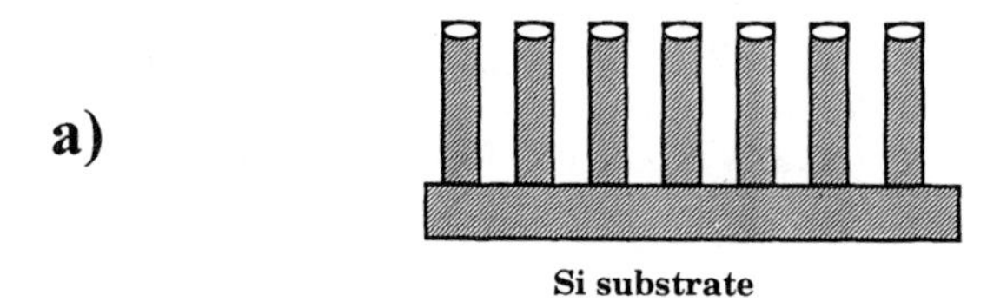

II. "Undulating" Wires or Embedded Particles

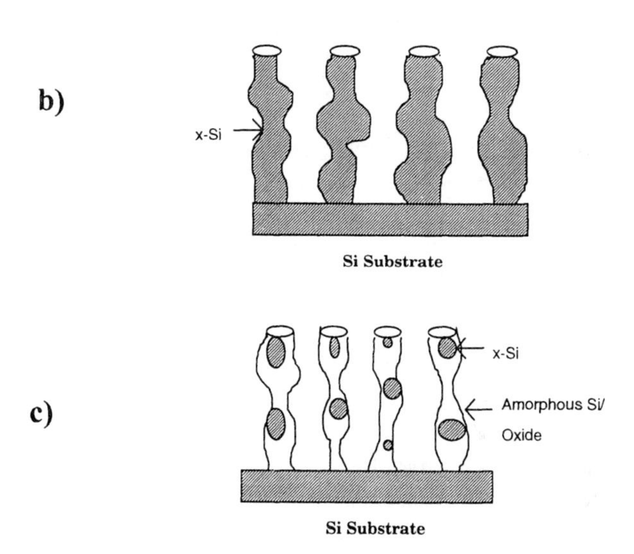

Figure 17.1. Suggested microstructure of high-porosity porous silicon: (*a*) quantum wires, (*b*) 'undulating' wires, and (*c*) embedded crystallites, (x-Si).

[19, 20, 22, 23].

(iii) p-type silicon (0.1 to 0.01 Ω cm): pores along ⟨100⟩, extensive branching noted. Pore dimensions are 10 nm or less and the porosity is in the range of 20–80% [8, 18–23].

(iv) n-type silicon (0.1–0.01): pores have square cross-sections, 100 nm or less, and are separated by micron-scale regions of bulk silicon [18].

(v) Low-doped n-type silicon (>1 Ω cm): morphology similar to n-type material; pore dimensions can be as large as one micron [18]. Porous structures formed from 10 Ω cm n-type silicon can exhibit one micron square pores, growing in the ⟨100⟩ direction [18].

Reports indicate some crystalline character of the porous silicon [24–29], surrounded by various amounts of amorphous silicon or oxide, the fraction of each varying with etching conditions and extent of porosity [30–34]. Yet, despite the variety of morphologies, particle sizes, and surface chemistry, porous silicon has been shown to exhibit very similar visible PL [8].

17.3 SURFACE CHEMISTRY

In addition to crystalline silicon, porous silicon also consists of surface species created during the etching process, or adsorbed from the atmosphere. The etching process has been examined, and the formation of amorphous silicon on the surface of Si has been reported [6, 35]. In addition, the surface of porous silicon after various etching treatments also contains hydrides and/or oxyhydrides and oxide [16]. In some cases, it has been reported that exposing porous silicon samples to the atmosphere resulted in a decrease in the PL intensity [36, 37]. In other experiments, prolonged exposure to atmosphere led to a significant PL intensity increase [38–40], and oxidation of the porous silicon also resulted in PL intensity increases [39, 41]. In view of these results and the large amount of surface area per unit volume of material, it is clear that the chemistry at the surface must also be considered.

17.4 RELATIONSHIP BETWEEN STRUCTURE, SURFACES, AND OPTICAL PROPERTIES

Since bulk silicon has an indirect bandgap at 1.1 eV at room temperature, only very inefficient emission in the infrared spectrum is expected. In fact, both optical absorption and emission are very weak in comparison to direct gap materials. The weak emission in indirect gap semiconductors can be overcome if the requirement for momentum conservation was relaxed. This would be possible if periodicity were removed locally or globally within the semiconductor. This occurs naturally at grain boundaries, line defects, and other structural defects, as well as in amorphous semiconductors, where long-range order is not present. Periodicity can also be artificially destroyed by deliberately growing alternate layers of different materials, such as in multiple quantum wells and superlattices. In addition, material in the form of lines or dots also removes periodicity. This leads to a spatial confinement of electrons and holes in the wells, wires, or dots, which in turn enhances the emission intensities. In addition, a series of localized states are formed whose energies scale inversely as the square of the size of the confinement region. In the case of Si, particle sizes with dimensions below 5 nm would enhance the oscillator strength, as well as produce a sizeable blue-shift of the optical gap from 1.1 eV into the range of 1.5 to 1.9 eV [34]. The subject of quantum dots is discussed in chapter 16.

17.4.1 Quantum confinement mechanism

Canham [4] attributed the visible PL in porous silicon to quantum confinement. Open-circuit etching was believed to enlarge the pores, thereby reducing the size of the silicon walls. For these types of sample, luminescence (800 nm to 700 nm) was noted, blue-shifting toward the orange with increased pore widening treatments. Assuming quantum confinement, Canham [4] predicted that porous Si would produce visible luminescence, provided that the porosity was high enough (greater than 78%).

An important property of all of the PL spectra in porous Si was that the line widths were very large, in the range of 300 to 400 meV [44]. It has been suggested that this rather broad PL emission line in the visible range could be attributed to a distribution of particle sizes in porous Si [4, 7, 42]. This particle size distribution would then imply a distribution of PL energies, as predicted by the quantum-confinement model.

Before quantum confinement can be determined as the mechanism of visible PL in porous Si, the PL and particle sizes must be correlated in a systematic fashion. One attempt to do this was to monitor the PL as a function of open-circuit etching. This type of experiment [4, 7] indicated a PL blue-shift with increased etching time in a dilute HF solution, with some samples having particles sizes as small as 3 nm [15, 28] (in certain p-type porous silicon). Furthermore, anodic oxidation was also used [7] to reduce the silicon wall thicknesses and to stabilize the porous silicon in the atmosphere. In this case, the PL was also seen to blue-shift with increasing anodic oxidation. However, no direct experimental data showing a consistent PL blue-shift with particle size reduction were presented. Dry oxidation for 1 min at high temperatures (up to 1100 °C) has also been reported [41, 43], in which the PL line increased in intensity and blue-shifted with increasing oxidation temperature, which was interpreted within the quantum-confinement model. This interpretation was given because it was assumed that no hydrides remained on the surface at these high temperatures [41]. In another experiment, porous silicon was dry oxidized for up to two hours at various temperatures above 700 °C, and no PL shifts were noted after an initially small blue-shift [44]. Raman spectroscopy and transmission electron microscopy (TEM) indicated significant material loss, and in some cases, the formation of pure oxide was noted. These results were not consistent with the suggested quantum-confinement model, and will be discussed later. A direct correlation has been reported between the PL peak energy and average particle size, obtained from Raman spectroscopy [28, 29]. However, other experiments show no correlation between particle size and PL peak energy [45, 46] obtained by TEM, optical absorption, and Raman spectroscopy [46].

It is clear that a relationship between particle size and PL emission energy is a necessary requirement for the quantum-confinement model. If quantum confinement is the mechanism of the PL, then we know that the luminescence energy is inversely proportional to the square of the particle size. Also, PL from smaller particles would have faster decay rates, as suggested by the work of Hybersten [42]. This would predict a multiexponential PL time decay, as has been reported [47], with measured lifetimes in the 10 μs–1 ms range and faster decay at the higher-energy end. It should be pointed out that this type of behavior is not unique to quantized particles. In fact, this would be expected of any localized radiative center, such as those in α-Si:H [48], various chalcogenide systems, defect states, and siloxene derivatives, which will be discussed later. Furthermore, Calcott *et al* [49] have reported the observation of momentum conserving phonons of the PL in porous silicon at $T = 2$ K, suggesting crystalline Si forms the luminescing material.

17.4.2 Difficulties with quantum confinement

Clearly, some data exist which are consistent with the quantum-confinement model. Unfortunately, the key result, which would show convincingly a direct relationship between the particle size and PL energy, does not exist. This is critical in the case of porous silicon, which can have very large variations in structure and particle sizes, yet still luminesce in a similar visible energy range [8]. To examine particle size and the PL energy relationship, four very different samples, of different initial doping, etched under various conditions were analyzed using in-line Raman spectroscopy and PL [45]. The resultant particle sizes were obtained from the Raman spectra, as in previous works [28, 29]. The experimentally obtained PL peak energies of these samples and the expected PL peak energies which would result from the different particle sizes in the case of 3D confinement are shown in figure 17.2 and table 17.1. As can be seen, the experimentally determined PL peak energy has little correlation to the energies expected for these samples if based on particle sizes. In fact, the PL peak energies of the various samples are relatively constant, in the 1.75 eV range. The work of Kanemitsu *et al* [46] also supports this result. In fact, Raman, optical absorption, and transmission electron microscopy (TEM) experiments were performed on various porous silicon samples. Results showed that a significant change in the particle size occurred from sample to sample, yet no shift in the PL peak energy was ever noted.

There are other problems with the quantum-confinement picture. Low-temperature desorption experiments of 0.1 Ω cm porous silicon resistively heated in ultrahigh vacuum (UHV) [50] resulted in large PL intensity reductions and PL peak energy red-shift at temperatures below 350 °C (shown in figure 17.3). This was accompanied by significant hydrogen evolution and a drop in the concentration of SiH_x species [50]. No structural changes were noted in the experiment at these low temperatures. This is consistent with previous results [21, 23] on the structural stability of porous silicon at low temperatures. Collins *et al* [51] have also reported a significant PL drop and red-shift in some samples, as a function of ultraviolet irradiation at room temperature, which was accompanied by hydrogen loss. Since no heating was performed in this

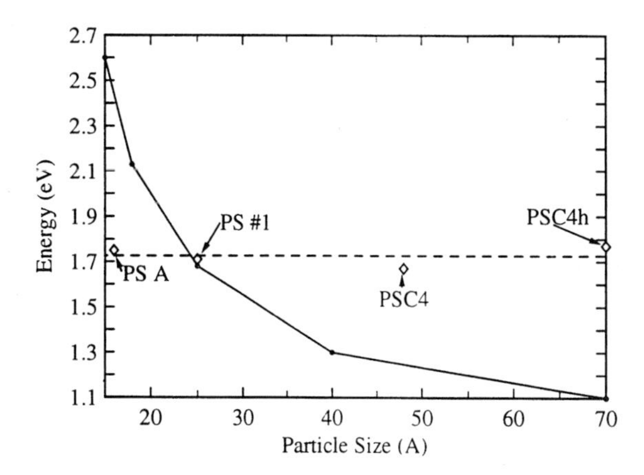

Figure 17.2. Experimentally determined PL emission energies for four various porous silicon samples prepared as listed in table 17.1. Particle size was obtained using Raman spectroscopy. The solid line is the expected PL energy for the various samples according to 3D quantum confinement.

Table 17.1. Porous silicon particle size obtained from Raman spectroscopy as a function of preparation conditions.

Sample	Sample preparation	Raman shift (cm^{-1})	$\Delta\omega$	QC particle size (nm)
C4	0.1 Ω cm, p 25 mA cm^{-2} 25% HF/ethanol	517	3	7
C4 (60 min) HF	C4 60 min 25% HF/ethanol	515	5	4.8
PS No 1	7 Ω cm, p 49 mA cm^{-2} 25% HF/ethanol	507	13	2.5
PS A	1 Ω cm, p 30 mA cm^{-2} 25% HF/ethanol	501	19	~1.6

case, it is impossible to attribute the loss in the PL intensity and the red-shift in the peak energy to structural collapse. Further evidence of the PL red-shift and intensity loss phenomenon was reported for a 1 Ω cm p-type porous silicon sample annealed in an argon atmosphere, up to temperatures of 690 °C [52]. A cooled Ge detector was used, which was more sensitive in the near infrared. In this case, the PL red-shift noted was as large as 0.3 eV, with extremely weak PL intensity. A similar red-shift was later also noted in dry oxidation experiments for temperatures below 700 °C [42].

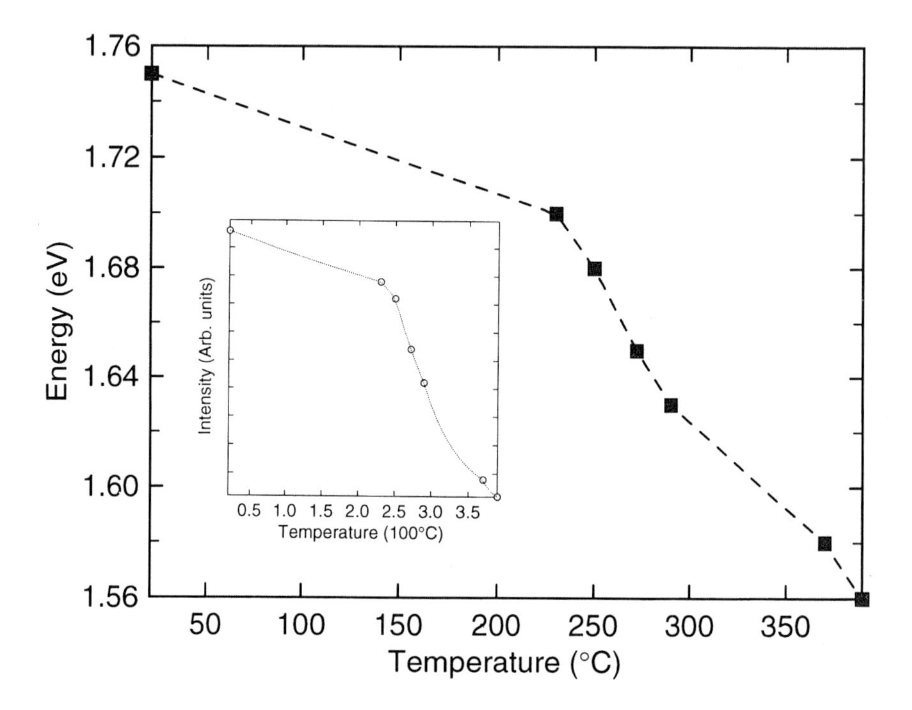

Figure 17.3. Red-shift of the PL peak energy and intensity drop, as a function of annealing 0.1 Ω cm p-type porous silicon sample under UHV conditions. Temperatures in insert are in units of 100 °C.

Although a PL blue-shift upon initial oxidation between 700 °C and 1000 °C was reported [42, 44], it is not at all clear that this blue-shift is the result of PL from quantum-confined particles. In fact, very short oxidation of porous silicon at 830 °C can lead to either PL peak energy blue-shifts or red-shifts, depending on the initial PL peak energy, as shown in figure 17.4. Also, extended oxidations at 700 °C and 900 °C [44] have led to no further shifts in the PL peak energy (shown in figure 17.5), although a significant PL intensity drop was noted for extended oxidation. In some of the thin oxidized samples, the PL became almost undetectable, and complete oxide formation was noted. Since oxide formation is a continuous process well characterized in porous silicon [4], one would expect to obtain a continuous PL blue-shift as the particle size continuously decreased, if the PL were the result of quantum confinement. Furthermore, the particle size distribution must become narrower as more and more particles oxidize, which should also lead to a significant PL peak narrowing in the case of quantum confinement. In fact, no PL peak energy shift was noted [48], even for extended oxidations, and the PL peak width broadened with longer oxidation times.

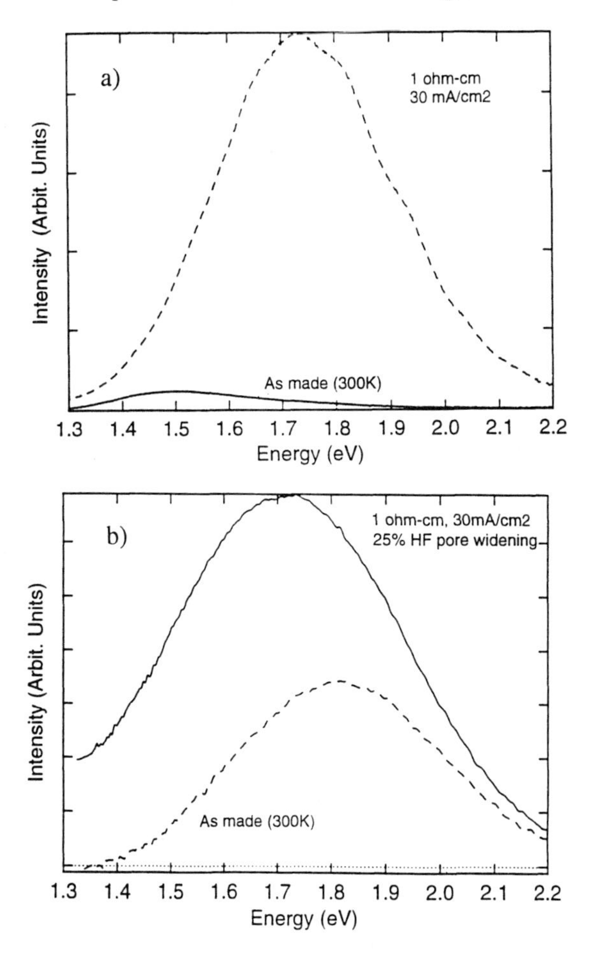

Figure 17.4. PL peak energy shift after laser annealing at 830 °C of (*a*) 1 Ω cm p-type porous silicon etched at 30 mA cm^{-2}, and (*b*) same sample open-circuit etched for 30 min in 25% HF/ethanol prior to short laser heating.

A different approach to significantly increase the PL intensity and change the particle size is to heat the existing porous silicon particles using an intense laser line [39]. As already discussed, the Si microcrystallites in porous silicon

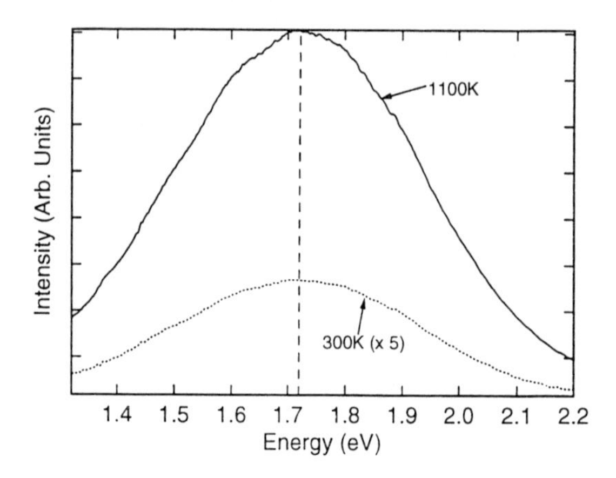

Figure 17.5. PL spectrum of a porous silicon sample, during laser annealing at 1100 K for 2 min, and at room temperature (300 K).

are thought to be embedded in an amorphous silicon/oxide shell, as shown in figure 17.1(*c*). Due to poor thermal contact, the silicon crystallites cannot shed heat efficiently, and thus can be heated to temperatures as high as 1100 K by laser annealing in the atmosphere. The extent of the Raman red-shift while heating was used to monitor the temperature, which was obtained from the Stokes–anti-Stokes shift [39]. The PL line of the porous silicon sample while at 1100 K, and while at room temperature (300 K) is shown in figure 17.5. Clearly, the PL lines are identical in shape and PL peak energy. As we know, the gap of bulk silicon is temperature dependent, and in fact shrinks by 0.3 eV at 1100 K [52]! Thus, this PL cannot be a property of silicon, be it caused by quantum confinement or not. If the PL originated within quantum-confined silicon particles, it would have to exhibit a shift, just as silicon must.

Further evidence for surface related red PL in silicon structures has been reported by Fisher *et al* [53]. In this case, very uniform Si posts of 20 nm diameter were produced by reactive ion etching (RIE), followed by wet etching in HF to remove the damage from the dry etching step. Interestingly, the silicon posts resulted in a 720 nm red PL, although the posts were directly connected to the silicon substrate, and were of a uniform 20 nm diameter. Since no tiny microcrystallites were present, and 20 nm sizes are essentially bulk when considering quantum confinement, this result also suggests a surface-type luminescence mechanism.

It is also known that solvents which do not etch silicon have been reported to cause shifts in the PL peak energy as well as in the PL intensity [54, 55]. These solvents can not affect the particle size; they can only affect the surface properties of porous silicon. With the above results in mind, alternative models, involving species other than silicon nanoparticles, must be considered in order to determine the origin of the PL. Since different species have been reported on the surface of these nanocrystals, let us examine their luminescence behavior.

17.4.3 Photoluminescence from hydrides/polysilanes

It has been shown that the surface of as-prepared porous silicon is saturated by SiH_x structures [50, 56–58] and the desorption of these structures occurs at low temperatures [50, 56]. Due to certain similarities between the PL from porous silicon and that expected from hydrides/polysilanes, models for the PL based on these structures have been suggested [48, 56]. Let us now examine the luminescence behavior observed in α-Si:H [48, 59] (which consists of SiH_x structures) and polysilanes [60].

The low-temperature luminescence features of α-Si:H [48] consist of lower-energy peaks associated with dangling bonds (around 0.8–0.9 eV), along with a higher-energy peak (above the Si bandgap) due to radiative transition between band tail states. Wolford *et al* [59] reported room-temperature PL resulting from α-Si:H deposited by homogeneous chemical vapor deposition (HOMOCVD) at low temperatures. Their results indicate a 5 K luminescence in the range of 1.3 to 2.05 eV, blue-shifting with increasing hydrogen content. The samples with the highest hydrogen content (PL at 1.8–2.0 eV) exhibited room-temperature luminescence. Hydrogen complexes (SiH, SiH_2, SiH_3, or $(SiH_2)_n$) were viewed as the source of this room-temperature visible luminescence. The polysilane chains $(SiH_2)_n$ result in a material with a larger bandgap than in Si [60], changing with the length of the chain. Tight-binding descriptions of Si–H bonding [61, 62] show that the presence of monohydrides leads to new bonding states deep within the silicon valence bands, which increase the gap. These calculations have shown that a large gap increase occurs as a function of increasing hydrogen content in the complex, and for 30% hydrogen, the bandgap increases to 1.7 eV [62]. This result has also been confirmed experimentally [63], where a 500 °C anneal of a-Si:H resulted in a significant hydrogen loss, and in a bandgap reduction of 0.2 eV. Similarly, polysilane chains can also lead to a depression of the valence bandedge. Photoemission studies have shown a 0.4 eV valence band recession for amorphous silicon of low hydrogen content, and up to a 0.8 eV recession for polysilanes [64]. As noted earlier, the PL time decay in a-Si:H would also be expected to be multiexponential, as in porous silicon.

Thus, there are some similarities between the PL in porous silicon and that attributed to hydrides/polysilanes. This not only includes the fact that porous silicon contains hydrides/polysilanes on the surface, which have been reported to luminesce in the 1.7–2.0 eV range at room temperature, but the PL energy in a-Si:H red-shifts with hydrogen loss, which is very similar to the behavior of porous silicon [50]. One of the major problems, however, is associated with the oxidation results. As mentioned earlier, a more intense red PL is observed after high-temperature short oxidations. Since most of the hydrides desorb by 500 °C or so, it is unlikely that they can be playing a role in the material oxidized at 800 °C or higher.

17.4.4 Silicon bandtail state model

The model suggested by Koch *et al* [66] to explain the red PL in porous silicon was twofold. It was proposed that carriers were created within the quantum confined silicon nanoparticles, but that recombination, leading to light emission, occurred at the surface. The recombination at the surface, resulting in the red PL, was thought to originate from silicon bandtail states created at the surface by strain and disorder. One significant advantage of this model compared to the quantum-confinement model was that the red PL no longer required identical particle size distribution for various samples, since the PL energy was not as dependent on the particle size. Furthermore, this model would predict a shift in the PL with loss of hydrogen, since the presence of hydrogen can affect the position of the bandtail states. This model may also be consistent to some extent with the oxidation results [42, 44], in which the high PL intensity would be the result of a change in the surface distortions and/or the reduction of particle size.

One of the serious problems with this model is the fact that the red PL in the oxidized porous silicon does not shift in energy at high temperatures (shown in figure 17.6). Since this model assumes that recombination occurs within the silicon related surface bandtail states, they must be temperature sensitive, as is amorphous silicon [67, 68], and thus the PL would also be expected to shift! In addition, since it is assumed in this model that the PL originates from the silicon surface states, one would expect significant changes in the elastic strain, e, experienced by the particles as a function of radius, r, since $e \propto 1/r$ [69]. Thus, a large change in particle size should lead to a large change in the distortion within the particle, as well as the surface. If the PL is a result of these distortions, a significant change in this distortion should lead to a significant change in the PL, which does not appear to be the case [46].

17.4.5 Siloxene luminescence mechanism

A mechanism suggested by Brandt *et al* [70] for PL in porous silicon involves luminescence from Si–O–H compounds of siloxene ($Si_6O_3H_6$). In general, the structures of these materials are believed to consist of linear Si chains interconnected by oxygen or Si layers with alternating OH or H terminations. The important property of these materials is that the luminescence energies can be tuned over a large spectral range, including the visible regions, by the substitution of hydrogen by various ligands, such as OH groups, alcohols, or halogens. In addition, PL tuning can also be achieved by annealing of siloxene in air [71].

Thus, the resulting PL can appear very similar to that seen in porous silicon, and, in addition, the infrared and Raman spectra of siloxene also exhibit very similar behavior to porous silicon [72]. The authors of reference [70] discuss several main observations made in porous silicon within the framework of a siloxene model. It is suggested that the PL peak energy blue-shift with extended open-circuit etching of porous Si, reported by Canham [5] and Bsiesy *et al* [7], could be explained by the substitution of hydrogen by OH or other ligands in the Si ring of siloxene. Also, the electroluminescence reported in porous silicon [73] appears similar to the chemiluminescence reported during oxidation of siloxene in $KMnO_4$ [70]. The PL decay after pulsed excitation of both porous silicon and siloxene is highly nonexponential [47, 74], and depends on the PL peak energy, resulting in a large distribution of carrier lifetimes.

The essentials of the siloxene model suggested by Brandt *et al* [70] are as follows. The sixfold silicon rings with various ligands are what determines the PL of porous silicon, and the PL peak energy can shift, depending on the ligand that has been substituted in the rings. The rate at which radiative recombination occurs is governed by specific lifetimes of the excited states in the ring, as well as the rate of carrier transfer into these rings, which could account for the large lifetime distributions.

Siloxene appears to exhibit many similarities to porous silicon. However, as with all the models discussed so far, there are certain drawbacks. First of all, as in the case of the hydrides, the H and OH ligands would be expected to desorb by 500 °C, leaving an SiO_2-type structure. In fact, it has been reported that annealed siloxene, which appears amorphous, best resembles the optical properties of porous silicon [72]. This is clearly no longer an ordered molecular structure such as siloxene, and thus it should not be referred to as siloxene, but a variation of amorphous silicon/O/H. Furthermore, since siloxene itself does not require the presence of crystalline silicon particles, amorphous porous silicon should also luminesce if siloxene is responsible for the red PL, yet it does not [74]. In addition, the Raman spectra of porous silicon can shift quite noticeably from sample to sample, whereas the Raman spectrum of siloxene does not usually exhibit such a range [70]. These drawbacks suggest that the molecular structure known as siloxene cannot explain the emission properties of porous silicon.

17.4.6 Oxide related interfacial states

Let us now examine PL from a surface related localized defect, similar to the non-bridging oxygen hole center (NBOHC) [75]. These centers have been reported in silica optical fibers and they luminesce in the red (600 nm to 670 nm) at room temperature, with a peak width (FWHM) in the range of 0.35 eV [76–78]. An example of the PL from these centers, compared to that of porous silicon, is shown in figure 17.6.

At this point, three different NBOHC types have been identified [78], which vary in PL energy and quantum efficiencies. The first type of NBOHC is the $Si–O^-$, which exhibits a lower-energy PL (more red) with lower quantum efficiencies, and does not appreciatively shift with heat treatments [78]. The second type of NBOHC is stabilized with a hydrogen bond, such as $Si–O^- \cdots H–Si$. This type of defect is seen in silica containing a high concentration of hydroxyls or hydrides, and exists only at temperatures

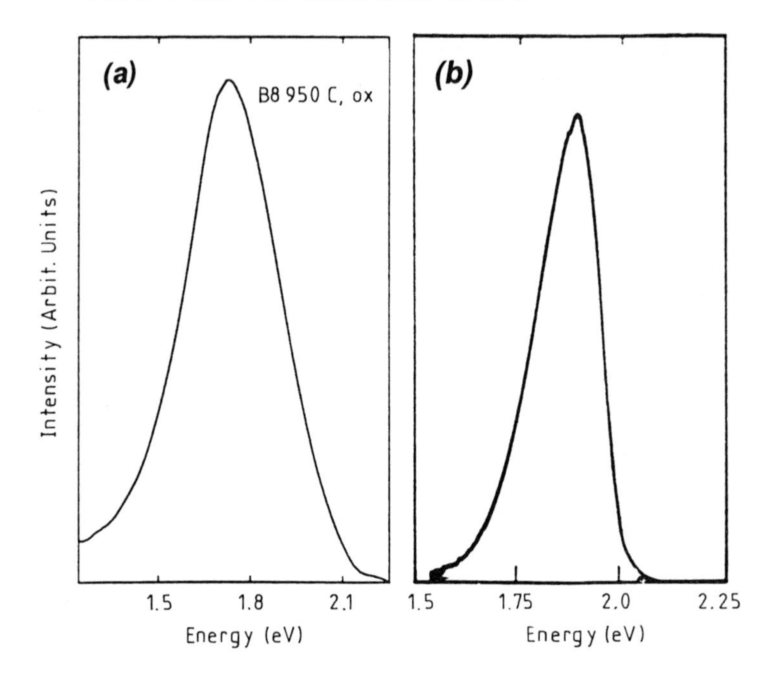

Figure 17.6. PL spectra of as-made 4 Ω cm p-type porous silicon (*a*), and that of the NBOHC (*b*).

below 350 °C. It has been shown to blue-shift with increasing hydrogen content, along with an increase in quantum efficiency [77–79]. The reverse is also true, in that the PL red-shifts and drops in intensity with hydrogen loss [78]. The least understood NBOHC is likely caused by the strain of bonding at an interface between two materials of different bond lengths, density or structure. It has been reported to occur after drawing of fibers which contain glass cladding and after high-temperature annealing [79]. Finally, the NBOHCs which are stable at higher temperatures (as high as 800 °C) do not show any PL shift or shape change during heating to high temperature as compared to PL at room temperature [78].

Now let us examine some results obtained from porous silicon. Room-temperature PL has been reported, which generally occurs in the red [4], with a PL peak width in the 0.3 eV range [4, 6, 7]. The PL peak energy has been reported to blue-shift with open-circuit etching in an HF solution, which can also increase the hydride content [30]. Along with a blue-shift in the PL peak energy, a PL intensity increase has also been noted [4, 7]. This result appears consistent with the second type of NBOHC, which behaves in a similar fashion. The PL peak energy in porous silicon has been reported to red-shift, which appears to be associated with the loss of hydrogen, achieved by heating [50] or by room-temperature UV irradiation [51]. This behavior has also been reported for NBOHCs. Furthermore, since the NBOHCs are localized defects, the PL time decay is non-exponential [76], as has been reported in the case of porous silicon [47]. The most convincing evidence, however, can be obtained from the temperature behavior of the PL. As shown in figure 17.5, the PL spectrum of porous silicon taken while at 1100 K and cooled to room temperature exhibits no PL peak energy shift or line-shape change, exactly the same behavior as reported for the high-temperature NBOHCs [78]! As already pointed out, this behavior cannot be explained if the PL in porous silicon originates in the silicon nanostructures. In addition, it has been noted from our experiments that porous silicon samples which remain in the atmosphere for six months or more always exhibit a much more intense PL, and generally blue-shift to some extent. Identical behavior has also been reported in the case of NBOHCs, which pick up OH groups from the atmosphere, leading to a PL increase and blue-shift, due to the stabilizing quality of hydrogen on the NBOHC [79]. Significant similarities have also been noted in the excitation spectra of the NBOHC and that of porous silicon [80].

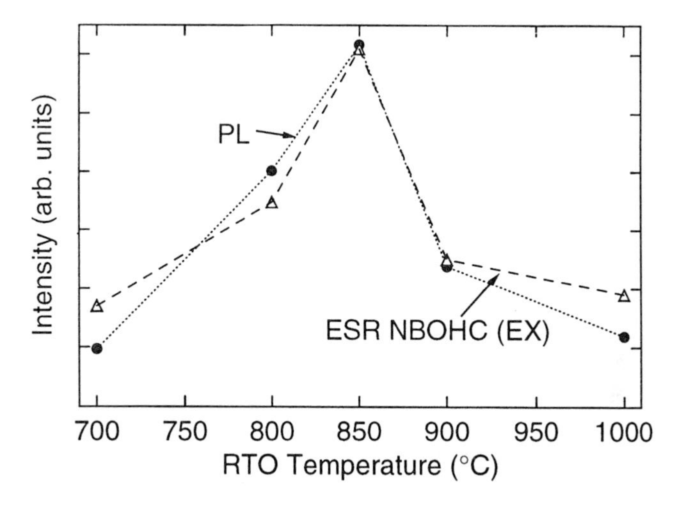

Figure 17.7. Electron spin resonance signal (NBOHC cluster) (broken line) and red PL integrated intensities (dotted line) for porous silicon samples oxidized for 30 s at temperatures between 700 °C and 1000 °C, using rapid thermal oxidation (RTO).

Thus, a model based on the interfacial NBOHCs has been suggested [44] to explain the red PL in porous silicon, and is as follows. The porous silicon structure has been modeled [39] to consist of a crystalline silicon core, surrounded by some amount of amorphous silicon, which may contain hydrides and oxygen. An SiO_2 layer surrounds this structure [15,74], especially in samples which have had contact with the atmosphere or which were heated in the atmosphere or oxidized. It has been suggested [39] that the PL intensity should scale with the total number of NBOHCs created at the crystalline Si/SiO_2 interface. For a sharp interface, the number of these defects would be larger than in a more gradual interface containing amorphous silicon. This is due to the fact that the interfacial strain would be distributed over a larger volume in a gradual interface, leading to lower PL intensities. Also, the relative ratios of the three types of NBOHC could affect the intensity in various samples. The NBOHCs have already been reported to luminesce in the red, and the extent of interface curvature and oxygen and hydride content would be expected to affect the PL energy. In fact, recent results have shown the presence of NBOHC clusters in both as-made and oxidized porous silicon, which track extremely well with the resultant red PL [81, 82]. This result can be seen in figure 17.7, in which the electron spin resonance signal for the NBOHC clusters and the red PL intensity are plotted as a function of short high-temperature oxidations.

As discussed above, this model appears to best describe the red photoluminescence exhibited by porous silicon. However, it is still unclear how absorption in porous silicon occurs and how these defects are activated or created.

17.5 CONCLUSION

The emission of visible light in silicon is quite exciting scientifically, and could prove to be very important technologically for optoelectronic applications. The proposed mechanism of quantum confinement has been analyzed with respect to the available data. It has been found that a simple particle-in-a-box model can not fit all the available data, especially in view of the fact that the red PL can not be a property of the Si particles. Several surface related luminescence models have been discussed in terms of their agreement with current data and their shortcomings. What appears to be the most successful model involves luminescence from interfacial non-bridging oxygen hole centers (NBOHCs). This model predicts a broad (0.3 eV FWHM) PL in the red at room temperature, should drop in intensity and redshift with hydrogen loss, will exhibit nonlinear PL time decay, no shifts in the PL during high-temperature anneals should occur, and photoluminescence excitation (PLE) absorption onsets expected would be in the 2 eV, 3eV, and 4.8 eV regions.

In terms of the technological applications, the important parameters are the stability of the PL under atmospheric conditions, as well as with prolonged irradiation, and producing efficient electroluminescence from this system. As already discussed, exposure to the atmosphere can cause large changes in the PL intensity, which is not an attractive quality for device applications. Also, a significant decrease in the PL intensity upon prolonged irradiation has been reported [52], which needs to be addressed. Finally, although some electroluminescent devices using porous silicon have been obtained [6], the efficiency is quite low, due to the fact that the particles are isolated and the electrons must tunnel through oxide, which requires high current densities.A more promising structure at this point may be lithographically fabricated silicon poles, which also exhibit red photoluminescence [53], but do not require electron tunneling through significant oxides.

REFERENCES

[1] Uhlir A 1956 *Bell System Tech. J.* **35** 333

[2] Turner D R 1958 *J. Electrochem. Soc.* **105** 402

[3] Unagami T 1980 *Japan. J. Appl. Phys.* **19** 231

[4] Canham L T 1990 *Appl. Phys. Lett.* **57** 1046

[5] Pickering C, Beale M I J, Robbins D J, Pearson P J and Greef R 1984 *J. Phys. C: Solid State Phys.* **17** 6535

[6] Fauchet P M, Tsai C C, Canham L T, Shimizu I and Aoyahi Y (ed) 1993 *Microcrystalline Semiconductors: Materials Science and Devices* vol 283 (Pittsburgh, PA: Materials Research Society)

[7] Bsiesy A, Vial J C, Gaspard F, Herino R, Ligeon M, Muller F, Romestein R, Wasiela A, Halimaoui A and Bomchil G 1991 *Surf. Sci* **254** 195

[8] Macauley J M, Ross F M, Searson P C, Sputz S K, People R and Friedersdorf L E *Mater. Res. Soc. Symp. Proc.* vol 256, ed S S Iyer, R T Collins and L T Canham (Pittsburgh, PA: Materials Research Society) p 47

[9] Friedersdorf L E, Searson P C, Prokes S M, Glembocki O J and Macauley J M 1992 *Appl. Phys. Lett.* **60** 2285

[10] Zhang X G, Collins S D and Smith R L 1989 *J. Electrochem. Soc.* **136** 1561

[11] Searson P C and Zhang X G 1990 *J. Electrochem. Soc.* **137** 2539

[12] Foell H 1991 *Appl. Phys.* A **53** 8

[13] Searson P C, Macauley J M and Prokes S M 1992 *J. Electrochem. Soc.* **139** 3373

[14] Cullis A G, Canham L T and Dosser O D 1992 *Mater. Res. Soc. Symp. Proc.* vol 256 (Pittsburgh, PA: Materials Research Society) p 7

[15] Cullis A G and Canham L T 1991 *Nature* **353** 335

[16] Fauchet P M, Tsai C C, Canham L T, Shimizu I and Aoyagi Y (ed) 1993 *Mater. Res. Soc. Symp. Proc.* **283**

[17] Lehmann V and Gosele U 1992 *Adv. Mater.* **4** 114

[18] Searson P C, Macauley J M and Ross F A 1992 *J. Appl. Phys.* **72** 253

[19] Beale M I J, Chew N G, Uren M J, Cullis A G and Benjamin J D 1985 *Appl. Phys. Lett.* **46** 86

[20] Beale M I J, Benjamin J D, Uren M J, Chew N G and Cullis A G 1985 *J. Cryst. Growth* **73** 622

[21] Sugiyama H and Nittono O 1989 *Japan. J. Appl. Phys.* **28** L2013

[22] Bomchil G, Herino R, Barla K and Pfister J C 1983 *J. Electrochem. Soc.* **130** 1611

[23] Bomchil G, Herino R and Barla K 1985 *Proc. 1985 MRS Meeting (Strasbourg, 1985)* vol 4, ed V T Nguyen and A G Cullis (Pittsburgh, PA: Materials Research Society) p 463

[24] Barla K, Bomchil G, Herino R, Pfister J C and Baruchel J 1984 *J. Cryst. Growth* **68** 721

[25] Phillip F, Urban K and Wilkens M 1984 *Ultramicroscopy* **13** 379

[26] Barla K, Herino R, Bomchil G, Pfister J C and Freund A 1984 *J. Cryst. Growth* **68** 727

[27] Young I M, Beale M I J and Benjamin J D 1985 *Appl. Phys. Lett.* **46** 1133
[28] Sui Shifeng, Leong P P, Herman I P, Higashi G S and Temkin H 1992 *Appl. Phys. Lett.* **60** 2086
[29] Tsu R, Shen H and Dutta M 1922 *Appl. Phys. Lett.* **60** 112
[30] Perez J M, Villalobos J, McNeill P, Prasad J, Cheek R, Kelber J, Estrera J P, Stevens P D and Glosser R 1992 *Appl. Phys. Lett.* **61** 563
[31] Prokes S M, Carlos W E and Bermudez V M 1992 *Appl. Phys. Lett.* **61** 1447
[32] Yao T, Konishi T, Daito H and Nishiyama F 1992 *Bull. Am. Phys. Soc.* **37** 564
[33] Vasquez R P, Fathauer R W, George T, Ksendzov A and Lin T L 1992 *Appl. Phys. Lett.* **60** 1004
[34] Lehmann V and Gosele U 1991 *Appl. Phys. Lett.* **58** 856
[35] Peter L A, Blackwood D J and Pons S 1989 *Phys. Rev. Lett.* **62** 308
[36] Canham L T, Houlton M R, Leong W Y, Pickering C and Keen J M 1991 *J. Appl. Phys.* **70** 422
[37] Tischler M A, Collins R T, Stathis J H and Tsang J C 1992 *Appl. Phys. Lett.* **60** 639
[38] Guilinger T R, Kelly M J, Tallant D R, Redman D A and Follstaed D M 1993 *Mater. Res. Soc. Proc.* **283** 115
[39] Prokes S M and Glembocki O J 1994 *Phys. Rev.* B **49** 2238
[40] Iyer S S, Collins R T and Canham L T (ed) 1992 *Light Emission from Silicon (Mater. Res. Soc. Proc.* **256**) (Pittsburgh, PA: Materials Research Society)
[41] Petrova-Koch V, Muschik T, Kux A, Meyer B K, Koch F and Lehmann V 1992 *Appl. Phys. Lett.* **61** 943
[42] Hybertsen M S 1992 *Mater. Res. Soc. Symp. Proc.* vol 256 (Pittsburgh, PA: Materials Research Society) p 179
[43] Ito T, Yasumatsu T, Watabi H and Hiraki A 1990 *Japan. J. Appl. Phys.* **29** L201
[44] Prokes S M 1993 *Appl. Phys. Lett.* **62** 3244
[45] Prokes S M 1993 *Bull. Am. Phys. Soc.* **38** 157
[46] Kanemitsu Y, Uto H, Matsumoto Y, Futagi T and Mimura H 1993 *Phys. Rev.* B **48** 2827
[47] Xie Y H, Wilson W L, Ross F M, Muha J A, Fitzgerald E A, Macauley J M and Harris T D 1992 *J. Appl. Phys.* **71** 2403
[48] Street R A 1976 *Adv. Phys.* **25** 397 and references therein
[49] Calcott P D J, Nash K J, Canham L T, Kane M J and Brumhead D 1993 *Mater. Res. Symp. Proc.* **283** 143
[50] Prokes S M, Glembocki O J, Bermudez V M, Kaplan R, Friedersdorf L E and Searson P C 1992 *Phys. Rev.* B **45** 13 788
[51] Collins R T, Tischler M A and Stathis J H 1992 *Appl. Phys. Lett.* **61** 1649
[52] Prokes S M, Freitas J A Jr and Searson P C 1992 *Appl. Phys. Lett.* **60** 3295
[53] Fischer P B, Dai K, Chen E and Chou S 1993 *J. Vac. Sci. Technol.* B **11** 2524
[54] Lauerhaas J M, Credo G M, Heinrich J L and Sailor M J 1992 *Mater. Res. Soc. Symp. Proc.* **256** 137
[55] Coffer J L, Liley S C, Martin R A, Files-Sesler L A 1993 *Mater. Res. Soc. Proc.* **283** 305
[56] Tsai C, Li K H, Sarathy J, Shih S, Campbell J C, Hance B K and White J M 1991 *Appl. Phys. Lett.* **59** 2814
[57] Venkateswara Rao A, Ozanam F and Chazalviel J N 1991 *J. Electrochem. Soc.* **138** 153

[58] Gupta P, Colvin V L and George S M 1988 *Phys. Rev.* B **37** 8234

[59] Wolford D J, Scott B A, Reimer J A and Bradley J A 1983 *Physica* B **117 & 118** 920

[60] Pitt C G, Bursey M M and Rogerson P F 1970 *J. Am. Chem. Soc.* **92** 519

[61] Ching W Y, Lam D J and Lin C C 1980 *Phys. Rev.* B **21** 2378

[62] Papaconstantopoulos D A and Economu E N 1981 *Phys. Rev.* B **24** 7233

[63] Yamasaki S, Hata N, Yoshida T, Oheda H, Matsuda A, Okushi H and Tanaka K 1981 *J. Physique Coll.* **42** C4 297

[64] von Roedern B, Ley L and Cardona M 1977 *Phys. Rev. Lett.* **39** 1576

[65] Zhou Weimin, Shen H, Harvey J F, Lux R A, Dutta M, Lu F, Perry C H, Tsu R, Kalkhoran N M and Namavar F 1992 *Appl. Phys. Lett.* **61** 1435

[66] Koch F, Petrova-Koch V, Muschik T, Nikolov A and Gavrilenko V 1993 *Mater. Res. Soc. Proc.* **283** 197

[67] Fischer R 1979 *Topics in Applied Physics: Amorphous Semiconductors* ed M H Brodsky (Berlin: Springer) p 159

[68] Knights J C 1980 *CRC Crit. Rev. Solid State Mater. Sci.* **210** and references therein

[69] Cammarata R C and Eby R K 1991 *J. Mater. Res.* **6** 888

[70] Brandt M S, Fuchs H D, Stutzmann M, Weber J and Cardona M 1992 *Solid State Commun.* **81** 302

[71] Hirabayashi I and Morigaki K 1983 *J. Non-Cryst: Solids* **59/60** 645

[72] Stutzman M, Weber J, Brandt M S, Fuchs H D, Rosenbauer M, Deak P, Hopner A and Breitschwerdt A 1992 *Adv. Solid State Phys.* **42**
Stutzmann M, Brandt M S, Rosenbauer M, Fuchs H D, Finkbeiner S, Weber J and Deak P 1993 *J. Lumin.* **57** 321

[73] Halimaoui A, Oules C, Bomchil G, Bsiesy A, Gaspard F, Herino R, Ligeon M and Muller F 1991 *Appl. Phys. Lett.* **59** 304

[74] Jung K H, Shih S, Hsieh T Y, Campbell J C, Kwong D L, George T, Lin T L, Liu H Y, Zavada J and Novak S 1992 *Mater. Res. Soc. Symp.* vol 256 p 31

[75] Griscom D L 1991 *J. Ceram. Soc. Japan.* **99** 923

[76] Skuja L N and Silin A R 1979 *Phys. Status Solidi* a **56** K11

[77] Nagasawa K, Ohki Y and Hama Y 1987 *Japan. J. Appl. Phys.* **26** L1009

[78] Munekuni S, Yamanaka T, Shimogaichi Y, Tohmon R, Ohki Y, Nagasawa K and Hama Y 1990 *J. Appl. Phys.* **68** 1212

[79] Nagasawa K, Hoshi Y, Ohki Y and Yahagi K 1986 *Japan. J. Appl. Phys.* **25** 464

[80] Prokes S M and Glembocki O J 1995 *Phys. Rev.* B **51** 11 183

[81] Prokes S M, Carlos W E and Glembocki O J 1994 *Phys. Rev.* B **50** 17 093

[82] Prokes S M and Carlos W E 1995 *J. Appl. Phys.* **78** 2671

Chapter 18

Biological nanomaterials

Dominic P E Dickson

18.1 INTRODUCTION

Biological nanomaterials differ from the other materials discussed in this book in that their properties have been refined by evolutionary processes over a very long timescale. This leads to a high level of optimization compared with many synthetic materials. Our understanding of the biological nanomaterials can be used as a guide to the production of synthetic materials with similar characteristics. This forms part of the approach known as biomimicking, which is a very widely applicable concept.

Biological systems contain many nanophase materials. For example, in many situations living systems produce mineral materials (e.g. bone) with particle sizes and microscopic structures in the nanometer size range. The process of biomineralization involves the operation of delicate biological control mechanisms that produce materials with very well defined characteristics. Indeed, in view of the importance of nanoscale structures in living organisms, life itself could be regarded as a nanophase system!

The nanomaterials found in living organisms could be used as a direct source of novel materials. These materials can be modified by *in vivo* procedures, i.e. by changing the biological situation in which they are produced. The biological material can also be subjected to *in vitro* manipulation following extraction. Indeed biological materials could be used as the starting material for many of the standard procedures for the synthesis and processing of nanomaterials, such as vapor techniques, mechanical attrition, etc. So far work in this area has been limited but a number of examples [1, 2] exist to suggest that there may be considerable potential for future developments.

Another aspect of biological nanomaterials is that they may provide excellent models which can assist in our understanding of the behavior of nanomaterials generally. A corollary of this is that the general understanding of nanomaterials can provide insight for the interpretation of nanomaterial behavior in biological systems.

In this chapter a number of examples of biological nanomaterials will be explored. The coverage is not intended to be exhaustive but rather to indicate different aspects of this field. The examples are drawn from the work of the author and collaborators and co-workers and include current developments and work in progress. The three areas to be considered are (i) ferritins and related iron-storage proteins in their native forms and as the basis for synthesizing novel small-particle magnetic materials, (ii) the highly optimized magnetic small-particle systems found in magnetotactic bacteria, and (iii) the small-iron-oxide-particle mechanical reinforcement system used in molluscan teeth.

18.2 FERRITINS AND RELATED PROTEINS

18.2.1 Background

A class of proteins, known as ferritins, provides a system whereby living organisms can synthesize and deal with nanometer-sized particles of iron oxyhydroxides and oxyphosphates. These proteins have been found in many types of living organism, from bacteria to man. The biological functions of these proteins relate to the storage, transport and detoxification of iron in the organism.

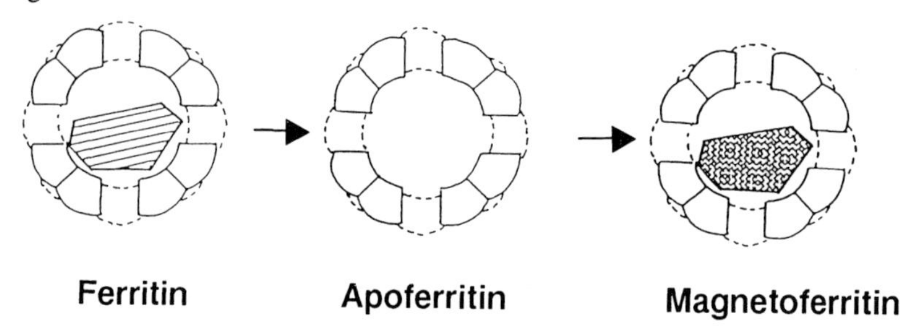

Figure 18.1. Schematic diagram of ferritin and the production of magnetoferritin.

The ferritin molecule consists of a spherical protein shell with an outer diameter of 12 nm and an inner diameter of 8 nm [3], as shown in figure 18.1. This spherical shell is made up of 24 protein subunits that self-assemble to form a structure with channels which allow iron ions to pass in and out. Within the spherical cavity mammalian ferritin molecules contain a particle of the mineral ferrihydrite (iron(III) oxyhydroxide, with some phosphate) having a diameter of typically 6 nm. In bacterial ferritins the composition of the core material may be much closer to iron(III) phosphate [4]. Even more diversity in the mineral form of the iron-containing cores is found in hemosiderins, a related form of iron protein associated with certain pathological conditions [5].

Ferritins provide a range of possibilities in the context of the synthesis and study of nanomaterials. Native ferritin consists of well separated nanometer-sized magnetic particles, produced by biologically controlled mineralization within the cavity in the protein shell. It should be noted that native ferritin from a number of different organs and species is commercially available. This mineralization can be modified to some extent *in vivo* by varying the growth conditions, and there is also a considerable species dependence. In addition, as the ferritin molecules are soluble, the separation between the magnetic particles can be varied. Thus, the native material can provide a range of magnetic small-particle model systems for the investigation of nanomaterial behavior by a variety of techniques. The interpretation of this behavior can of course assist in our understanding of the extracted material and, in turn, the system from which it has come. The ferritin protein shell can be regarded as a reaction vessel, within which small particles of new materials can be created. This can and has been done in a number of ways. One of these is to remove the ferritin from the shell and by using appropriate and controlled wet chemical conditions to synthesize different materials within the shell [1]. These can vary from small modifications of the ferrihydrite particle structure produced by subtle changes in the reconstitution conditions to the synthesis of completely different materials, not necessarily containing iron. Alternatively the native ferritin core can be modified within the shell by means of suitable chemical treatments [6].

18.2.2 Ferritin as a superparamagnetic model system

In a small particle ($\sim$10 nm) of a magnetically ordered material, the magnetization vector undergoes reversals below the magnetic ordering temperature as a result of thermal excitations over the anisotropy energy barrier. This process is usually referred to as superparamagnetic relaxation and leads to the disappearance of magnetic remanence and coercivity of the particle, as observed by a particular technique, at temperatures above a temperature known as the blocking temperature T_B. The value of the blocking temperature, as will be seen below, depends on the measuring technique used to determine it.

This process is often quantified in terms of the equation originally due to Néel [7], in which the timescale for magnetization reversal, τ, is a function of a pre-exponential factor, τ_0, and the anisotropy energy, E_A, which is approximately proportional to the particle volume.

$$\tau = \tau_0 \exp(E_A/kT). \tag{18.1}$$

This equation is strictly only valid in cases of uniaxial anisotropy, but it provides a useful phenomenological approach in many situations where the details of the anisotropy are unknown. In a sample of many fine particles, E_A must be replaced by a distribution $p(E_A)$ and this leads to a distribution of superparamagnetic relaxation times $p(\tau)$ at any particular temperature.

There is considerable interest in whether the above equation gives a reasonable representation of the behavior and the values of τ_0 and E_A in real systems. In order to explore this a material with well separated, single-phase, magnetically ordered small particles is required and one for which the superparamagnetic behavior lies in the readily accessible temperature range between 4 and 300 K. Mammalian ferritin provides just such a material. It should be noted that in a ferritin core the ordering is thought to be antiferromagnetic, but with a net magnetic moment per particle as a result of uncompensated spins.

In order to investigate the values of E_A and τ_0 in a sample, two measurement techniques with very different measurement times are required. In Mössbauer spectroscopy the magnetic hyperfine field (which is directly related to the particle magnetization) at the iron nucleus is measured on a timescale of 2.5×10^{-9} s [8], with a magnetically split sextet spectrum being observed if the superparamagnetic relaxation is slower than this. The thermal decay of remanence technique involves the measurement of the magnetic moment of a sample at a fixed time (e.g. 100 s) after the removal of a saturating magnetic field, at various temperatures [9]. Both techniques monitor the excitation of the particle magnetic moments over the anisotropy energy barrier as a function of temperature, but on very different timescales. The temperature at which the behavior of half of the sample corresponds to fast or slow superparamagnetic relaxation is the median blocking temperature and it is this quantity which is obtained by observing the variation in the slow-relaxing fraction of the sample as a function of temperature.

Figure 18.2 shows the data obtained from both thermal decay of remanence and variable-temperature Mössbauer spectroscopy measurements on ferritin, which lead to median blocking temperatures of 9 and 36 K respectively [9]. Note the large difference in the temperature dependence of the magnetic signal as measured by these two techniques. By putting these blocking temperatures, together with the measurement times for the two techniques, into the Néel equation, and making certain assumptions about the distribution of anisotropy energies, iron atoms, and magnetic moments among the particles, it is possible to obtain values for τ_0 and the median anisotropy energy. The value of τ_0 obtained in this way for horse spleen ferritin is approximately 10^{-12} s [9]. This value is considerably different from the values of around 10^{-9}–10^{-10} s which are frequently used for this parameter. The value of τ_0 obtained for the ferritin samples is much closer to the values of approximately 10^{-13} s obtained for iron metal particles in Fe–SiO_2 granular films [10] or 2×10^{-12} s recently obtained for amorphous Fe–C particles [11]. This work confirms the validity of the techniques used to measure τ_0 and the appropriateness of ferritin as a nanoparticle magnetic model system. In addition to τ_0 a value of the median anisotropy energy per ferritin particle can also be calculated from the values of blocking temperature. This gives a value of 3.9×10^{-21} J for the median anisotropy energy of the ferritin particles in the sample examined. Using a value of 6 nm for the mean particle diameter this corresponds to an anisotropy energy constant of 3.45×10^4 J m^{-3}.

An alternative approach to the anisotropy behavior can be obtained by using

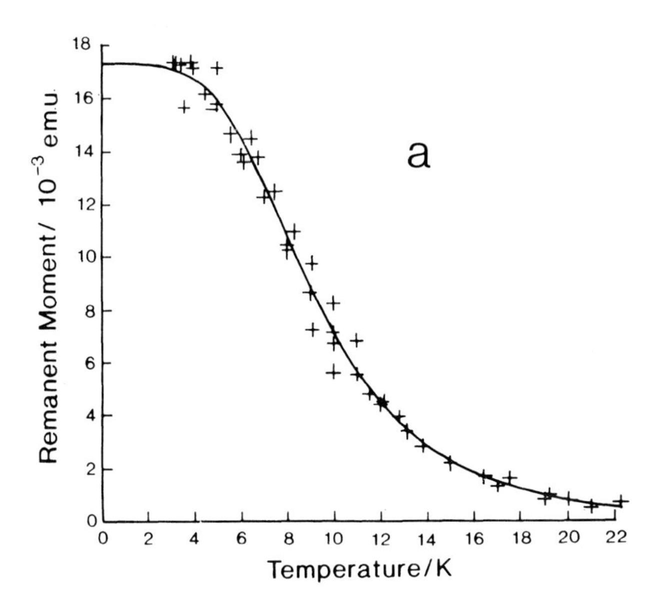

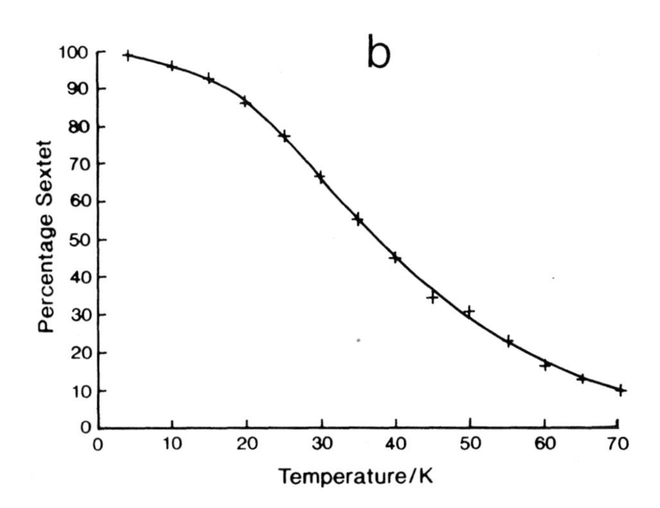

Figure 18.2. Decrease in the blocked behavior of a horse spleen ferritin sample as a function of increasing temperature as measured by (*a*) the thermal decay of remanence and (*b*) the sextet fraction in the Mössbauer spectrum.

Mössbauer spectroscopy to examine the response of a polycrystalline sample of ferritin to a large applied magnetic field. For ferritin, this method yields a value of 1.72 T for the anisotropy field experienced by each individual atomic magnetic moment [12]. This value gives an anisotropy energy per ferritin particle of 9×10^{-20} J, and a corresponding Mössbauer blocking temperature of around 900 K. The discrepancy between this value and the experimentally observed value of 36 K indicates an underlying problem in relating the anisotropy behavior of individual magnetic moments with that of the whole particle. It appears that either the superparamagnetic relaxation process involves a smaller 'domain' than that of the whole particle or that the magnetization reversal mechanism corresponds to an 'effective' anisotropy energy barrier which is considerably lower than that obtained by summing the anisotropy energies of the individual atomic magnetic moments. These ideas may relate to the concept of an open fractal structure for the ferritin cores leading to smaller domains within the particle [13] or to the concept of an activation volume for magnetization reversal [14].

Thus, the use of the biological nanomaterial ferritin as a superparamagnetic model system has led to experimental results which may have important consequences for our understanding of magnetic behavior on the nanoscale size range.

18.2.3 Different native forms of ferritin and hemosiderin

The magnetic properties of magnetic nanomaterials encompass both superparamagnetic relaxation behavior and magnetic ordering behavior. Superparamagnetic behavior can only be observed for a magnetically ordered material, and thus can only arise when the blocking temperature is lower than the ordering temperature. The ordering temperature is a function of the separation of the magnetic atoms and the magnitude of the exchange interaction between them, and is essentially a bulk property. The blocking temperature, determined by a particular measurement technique, is a function of the anisotropy energy distribution for the particles making up the sample as discussed in the previous section. What behavior is observed using a particular technique depends on the relative magnitudes of the blocking (T_B) and ordering temperatures (T_{ord}). These parameters are essentially independent of each other and therefore any one of three possibilities (i.e. $T_B < T_{ord}$, $T_B > T_{ord}$, $T_B \sim T_{ord}$) may occur in a particular small-particle system.

These cases have enabled three different forms of the iron storage protein ferritin to be identified on the basis of the character of their temperature-dependent Mössbauer spectra. The three forms are: mammalian ferritin, which exhibits a blocking temperature of around 35 K (i.e. $T_B < T_{ord}$); bacterial ferritin, which exhibits an ordering temperature of around 3 K and no blocking behavior, together with evidence that the ordering temperature is considerably higher (i.e. $T_B > T_{ord}$); and molluscan ferritin which exhibits a blocking temperature of around 20 K, with evidence that the ordering temperature is only slightly higher (i.e. $T_B \sim T_{ord}$) [15]. The differences among these three forms of ferritin, has led to the magnetic behavior described above, have been confirmed by evidence from structural techniques, such as electron diffraction and EXAFS, and biochemical and chemical analysis, which clearly show that the cores have a different mineral composition and different degree of crystallinity in the three forms.

In certain diseases there is a build up of iron in the body. These diseases fall into two main groups. In primary hemochromatosis, the iron overload is a consequence of a breakdown of a 'switch' in the gut which controls the uptake of iron. In secondary hemochromatosis the excess iron results from multiple blood transfusions administered because of a genetic blood disease. Certain animals also exhibit iron overload. Under normal conditions any excess iron is contained within the iron-storage protein, ferritin, which holds it in a nontoxic form (unlike free iron) and makes it available for the synthesis of iron-containing proteins and enzymes. When large quantities of iron are present, as in cases of iron overload, the iron is predominantly in the form of insoluble granules, of a similar size to that of ferritin cores, with associated protein, known as hemosiderin [16]. Determining the form of the iron in the hemosiderin should considerably advance our understanding of the nature of these diseases and help in devising appropriate treatment strategies.

Striking differences in the magnetic behavior of the hemosiderins associated with different iron overload conditions have been observed by means of Mössbauer spectroscopy [17]. The hemosiderins extracted from iron overloaded animals, normal humans, and untreated primary hemochromatosis patients show superparamagnetic behavior, very similar to that of ferritin, but with a slightly lower blocking temperature of around 25 K. The hemosiderin extracted from treated secondary hemochromatosis patients also shows superparamagnetic behavior, but with a much higher blocking temperature of around 70 K, indicating a material with much greater magnetic anisotropy. The hemosiderin extracted from treated primary hemochromatosis patients has very different temperature-dependent Mössbauer spectra. In this case, there is no evidence for superparamagnetic behavior, with the characteristic coexistence of doublet and sextet spectral components, but instead there is the gradual collapse of the magnetically split sextet spectrum as the temperature is raised, indicating a magnetic ordering transition at around 3 K. The ordering temperature of around 3 K is indicative of a material with much smaller magnetic exchange.

The results described above show that the small-particle magnetic behavior of these hemosiderins provides a valuable diagnostic tool for identifying and characterizing them. This information has important consequences for understanding the nature of the iron overload in these relatively widespread diseases and for improving their treatment.

18.2.4 Magnetoferritin

As discussed above, ferritin provides a valuable model system for studying the properties of magnetic small particles and is an important focus for investigations into iron metabolism in living organisms. However, it can also be used to produce novel magnetic fine-particle materials, with larger magnetic moments than are obtained with the native protein (in which the core consists of antiferromagnetic material with some uncompensated spins).

It has recently been shown that, following removal of the ferrihydrite core from native horse spleen ferritin, it is possible to reconstitute the empty protein shell (apoferritin) under controlled oxidative conditions, tailored to the synthesis of magnetite rather than ferrihydrite [1] as indicated in figure 18.1. The material produced, which has been named magnetoferritin, was characterized by electron diffraction, which indicated that the mineral form was either magnetite or maghemite. The black colour and the restricted oxidation conditions used in the preparation suggested that magnetite was the more likely possibility.

Figure 18.3 shows electron micrographs of a recent sample of magnetoferritin, together with those of a control sample, corresponding to the same chemical preparation conditions, but in the absence of the apoferritin. The images in the micrograph are determined primarily by the electron density. As can be clearly seen in the figure, the magnetoferritin sample consists of well defined, essentially spherical, small particles of approximately 4 nm mean diameter, contained within the protein envelope. This micrograph is essentially identical to those obtained from native ferritin, the mineral cores in both cases appearing essentially the same. On the other hand the electron micrograph of the control sample shows agglomerates of much larger crystals, indicative of a completely different structural form. Thus, figure 18.3 gives an excellent representation of magnetoferritin as a novel material, and the importance of the protein in its synthesis.

Room-temperature magnetization measurements have been made on these magnetoferritin samples [18]. They show Langevin function behavior, characteristic of ferrimagnetically ordered small particles. There is no evidence for any hysteresis, and the magnetic moment per particle is consistent with either magnetite or maghemite particles, with sizes typical of those of ferritin cores. Mössbauer spectra have also been obtained from these magnetoferritin samples [18] at various temperatures and in large applied magnetic fields. The applied magnetic field measurements, in particular, indicate that the cores in these magnetoferritin samples correspond to maghemite rather than magnetite, which has implications for understanding the oxidation processes involved in the reconstitution.

Magnetoferritin provides an excellent example of the potential for the production of novel nanophase materials derived from biological materials. It may have considerable importance as a biocompatible ferrofluid, with many possible biomedical applications.

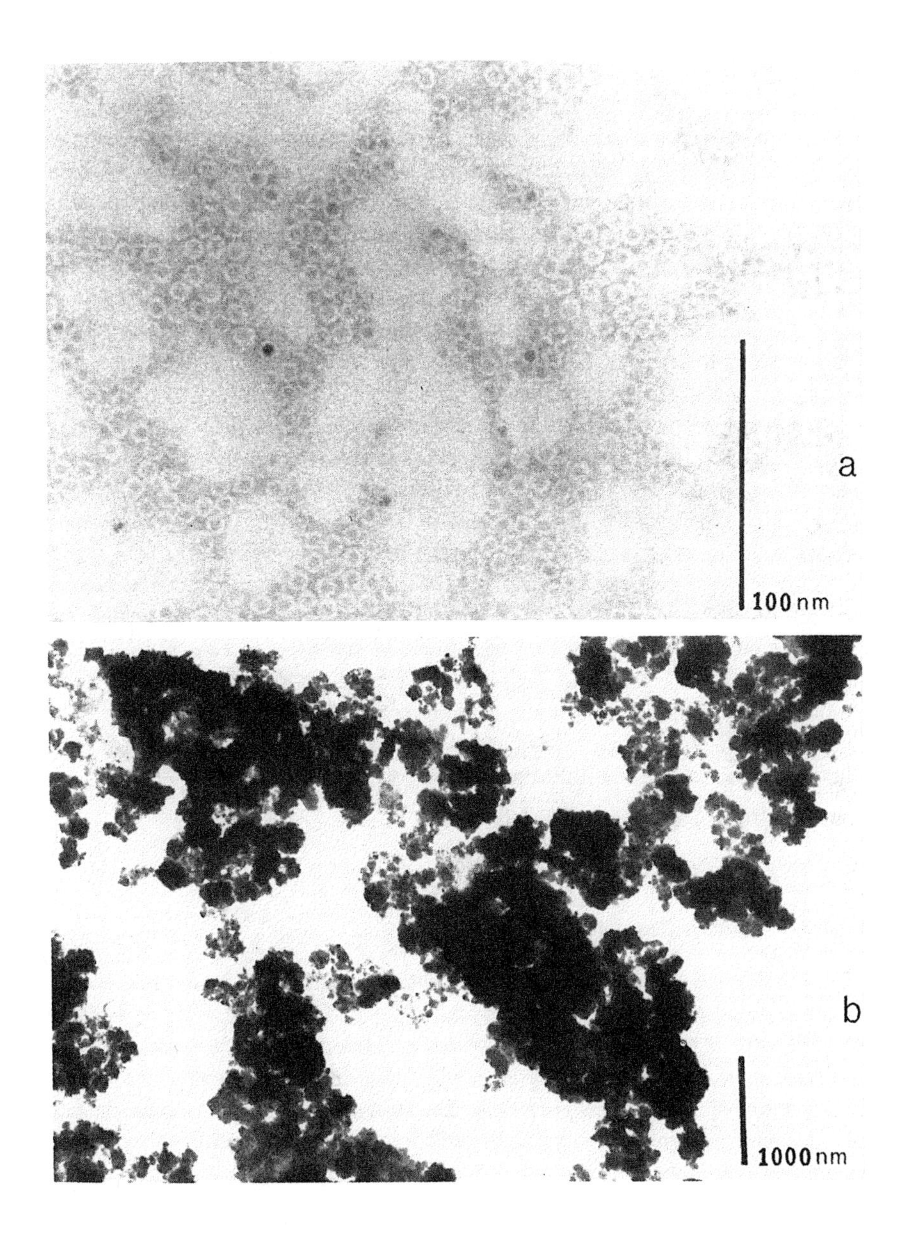

Figure 18.3. Electron micrographs of: (*a*) magnetoferritin and (*b*) a control sample produced under the same chemical preparation conditions, but in the absence of apoferritin. (Reproduced by permission of Dr T Douglas.)

18.2.5 Other ferritin derivatives and future possibilities

Magnetoferritin, described above, is just one of a series of possible ferritin derivatives with potential as novel nanophase materials.

The ferritin cores are isolated from each other by the protein shell and any material in which the protein is dissolved. Thus ferritin provides a macroscopic system containing a series of separated nanometer-sized particles. These particles may be randomly arranged or may be in a 'superlattice' if the ferritin molecules are in a ferritin single crystal. A highly desirable system would be one in which these isolated nanometer-scale, three-dimensional particles contain semiconductor material. Such nanometer-sized semiconducting particles are known as 'quantum dots'. In order to achieve this it is necessary to replace the iron oxyhydroxide in native ferritin with a material such as cadmium sulfide. (Nanophase crystallites of cadmium sulfide have been obtained using a biological route in another series of experiments [2].)

A first stage in the route to cadmium sulfide (or related semiconductor compound) cores in ferritin is to change the native iron oxyhydroxide cores to iron sulfide. This may be achieved by passing hydrogen sulfide gas through a solution of ferritin. The first attempt to accomplish this was only partially successful as only the surface of the ferritin cores was converted to iron sulfide [6]. These experiments did however produce a novel three-phase nanostructured material, with an inner core of ferrihydrite $\sim$3 nm diameter, coated with a thin layer of iron sulfide $\sim$0.5 nm thick, contained within a protein shell with an overall diameter of $\sim$12 nm. More recent experiments [19] have produced a complete conversion of the ferritin core to iron sulfide. These results represent an important first step in opening up a whole range of possibilities for novel materials based on the ferritin protein shell and core structure.

18.3 MAGNETOTACTIC BACTERIA

18.3.1 Magnetic direction finding

There is evidence for a magnetic direction finding ability in many species and this suggests some interesting biomimicking possibilities in relation to both the magnetic sensor and the transducer systems. However, this is very much for the future, as, currently, it is only in bacteria that the mechanism is understood. A large group, known as magnetotactic bacteria, use the Earth's magnetic field lines to enable themselves to orientate and move in the direction of nutritional or chemical gradients.

The process of magnetotaxis was originally found by accident [20]. Bacteria collected from marine and freshwater muds were observed to accumulate at the north side of drops of water and sediment on a microscope slide. Experiments with bar magnets and Helmholtz coils showed that the bacteria from the northern hemisphere swim along the field lines in the direction of the field. These bacteria are therefore referred to as north-seeking. Reversals in the field direction lead the bacteria to do a U-turn. Killed bacterial cells orient in the direction of the field but do not move along the field lines. Thus, motile magnetotactic bacteria behave like self-propelled permanent magnetic dipoles [21]. The bacteria propel themselves forwards by means of a propeller-like appendage, the flagellum. Magnetotactic bacteria can have one of two magnetic polarities, depending on the orientation of a magnetic dipole contained within the cell.

The situation described above is essentially unidirectional. The magnetotaxis enables the bacterium to follow the vertical component of the geomagnetic field and move downwards, towards the sediment (i.e. nutrient) and away from the toxic oxygen-rich conditions higher up in the water. Thus, in the northern hemisphere such bacteria are north-seeking, while in the southern hemisphere they are south-seeking. Moving away from the magnetic equator it has been found that an angle of inclination of only 6 to 8 degrees is sufficient to select one predominant polarity [22]. Some species of magnetotactic bacteria are bidirectional in that they have a flagellum at either end and can move in either direction relative to their internal magnetic moment. In these species, the magnetotaxis provides them with a vertical axis, so that they can move either up or down in order to optimize their position relative to a chemical or nutrient gradient. Thus, the magnetotaxis enables the bacterium to simplify a three-dimensional searching problem into a one-dimensional problem.

18.3.2 The magnetosomes

The permanent magnetic dipole moment of each magnetotactic cell is due to intracellular membrane bounded single-domain inorganic particles, known as magnetosomes, which are usually arranged in chains [21]. The magnetosomes of most of the magnetotactic bacteria that have been studied to date contain particles of magnetite, Fe_3O_4, in the 40 to 100 nm size range. However, magnetotactic bacteria from high-sulfide marine habitats have been found to contain ferrimagnetic greigite, Fe_3S_4, which is isostructural with magnetite [23]. An example of a chain of magnetosomes within a bacterial cell is shown in figure 18.4.

The nanometer-scale hierarchical structure of the magnetosome chain is crucial to its magnetic properties [24]. Firstly, consider the size of the individual particles. Large particles of any magnetic material, including magnetite, can lower their magnetostatic energy by forming magnetic domains. When the particle dimensions become comparable with the domain wall width, domains cannot form and the particle is forced to remain a single magnetic domain, with a magnetization which is uniform and has its maximum value. Calculations give 76 nm as the upper limit for the single-magnetic-domain size range for equidimensional particles of magnetite [25]. Because of shape anisotropy, the single-magnetic-domain volume increases with the aspect ratio for non-equidimensional particles and thus magnetite particles with long dimensions

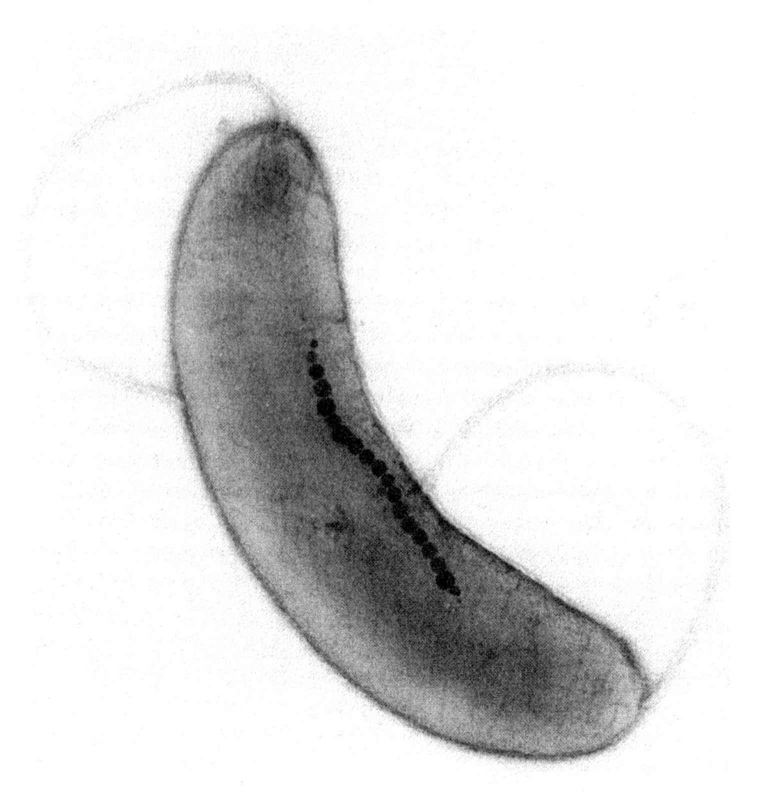

Figure 18.4. Electron micrograph of a magnetotactic bacterium showing the chain of magnetosomes. (Reproduced by permission of Dr R P Blakemore.)

of the order of 120 nm or less are still single magnetic domains. As explained in section 18.2.2 the thermal stability of the magnetization in small particles is related to the anisotropy energy E_A and hence to the particle volume. Only for single-magnetic-domain particles above a certain volume will the superparamagnetic relaxation rate be negligible such that the particles retain a permanent magnetization. For magnetite at 300 K, particles with dimensions greater than or equal to about 35 nm will be permanently magnetized. Thus magnetite particles with long dimensions between about 35 and 120 nm are permanent single magnetic domains at ambient temperature. The magnetite particles produced by magnetotactic bacteria are typically within this size range. Thus, these bacteria are producing magnetic mineral particles in a nanometer-sized range giving the maximum possible magnetic moment per particle.

When the particles are arranged in chains, the magnetic interactions between them cause their magnetic dipole moments to orient parallel to each other along the chain direction. The total magnetic dipole moment of the chain is thus the sum of the moments of the individual particles. By organizing the particles into chains, a bacterium is essentially constructing a permanent magnet with a sufficiently large magnetic moment to orient the cell in the geomagnetic field as it swims at ambient temperature in water. The speed of migration in the field direction v_B is then determined by the average projection of the magnetic dipole moment along the field direction: $v_B = v_0 L(\mu B/kT)$, where v_0 is the forward speed of the cell, $L(\mu B/kT)$ is the Langevin function, μ is the permanent magnetic dipole moment of the cell, and B is the local magnetic field. For $\mu B/kT$ greater than 10, the speed of migration in the field direction v_B is greater than 0.9 times the forward speed of the cell v_0. The number of magnetic particles in the magnetosome chain is typically 10–20 which corresponds to a value of $\mu B/kT$ of approximately 10. Because $L(\mu B/kT)$ approaches 1 asymptotically as $\mu B/kT$ increases, increasing μ beyond the value actually found would only produce a marginal increase in the migration velocity in the geomagnetic field direction [24]. The bacteria are therefore producing an adequate magnetic moment for the purpose, but no more than is necessary.

For a particular type of bacterium, the crystalline magnetosome cores of magnetite have uniform size, shape, crystal morphology, and arrangement within the cell. The biomineralization process, by which the production of the magnetically optimized magnetite single crystal is under biological control, is of considerable interest and has been extensively studied [26]. The bulk magnetic properties of bacterial magnetite have also been extensively measured and these data show that the biologically controlled mineralization produces magnetic particles with well defined magnetic properties which optimize their magnetic effectiveness [27].

The magnetotactic bacteria are producing a nanostructured composite magnetic system with optimized properties. In principle, these bacteria could be used as a biological source of magnetic small particles of higher quality than those that could be made synthetically. For certain applications requiring only relatively small quantities this could provide a feasible route for the production of technologically useful materials. It is, however, more likely that improved magnetic materials may result from the improved understanding of the controlled mineralization processes involved in the magnetosome synthesis. Again, as in the case of ferritin, it is the production of a mineral within the cage provided by a biological macromolecular structure that appears to be the crucial factor.

18.4 MOLLUSCAN TEETH

18.4.1 Introduction

Certain marine mollusks such as the limpet, *Patella vulgata*, and the chiton, *Acanthopleura hirtosa*, incorporate iron minerals into composite materials that make up the hard structural components of the teeth on their radula. The radula is a long thin organ used for scraping algae from the rocks and is continuously worn away in the process. There is therefore a necessity for the continual and rapid production of iron biominerals to replace those lost with the discarded teeth. Thus the radula provides a sequence of the biomineralization process [28]. The limpets and chitons are frequently found living side by side. It is interesting to note that the mineral found in limpet teeth is goethite (α-FeOOH) while that found in chiton teeth is magnetite (Fe_3O_4). As magnetite is harder than goethite, this may provide an advantage for one species over the other, in the ecological environments where they coexist.

The stages of mineralization that can be readily recognized in the teeth of the limpet are associated with the appearance and development of brown goethite deposits. The progressive rows of teeth along the radula provide examples of the various stages of maturation from immature teeth in the first 20 or so rows to fully mature teeth from rows 60 or so to the total of around 200 rows. In the mature teeth the predominant mineral form is that of goethite. In the immature teeth a ferrihydrite-like mineral form, similar to that in the iron-storage protein, ferritin, which is the source of iron for biomineralization, is found in addition to goethite. The relative proportions of these mineral forms are a function of the stage of maturation of the teeth [29].

18.4.2 Biomineralization of structural components

Goethite deposition in the tooth cusps takes place initially parallel to the tooth wall in the form of thin fibrous particles [30]. High-resolution transmission electron microscopy shows that these particles contain single-domain crystals, extensively elongated along the [001] direction. The width of these fibrous crystals is around 15 to 20 nm, but there are substantial irregularities in the crystal thickness [29]. The growth of these crystals from solution must be sufficiently slow to maintain their single-crystal nature and must also be significantly influenced by the local biochemical environment to account for the marked crystal growth anisotropy and the orientation of the long axis parallel to the tooth wall. Maturation of these goethite crystals occurs primarily by increases in thickness resulting in well formed acicular crystals [30] as shown in figure 18.5. However, it is particularly interesting to note that many crystals show marked morphological distortions, which are not associated with structural imperfections in the crystal lattice structure or domain boundaries (see figure 18.6). A possible explanation for these crystal forms is that they represent the presence of spatial constrains in the crystallization environment, such that the crystal grows to fill the space made available to it within the biological tissue. Since the goethite crystals are closely associated with organic filaments throughout the cusp [30], it is possible that spatial deviations in the organic matrix are ultimately reflected in the morphology of the growing goethite crystals. Thus the organization of matrix components within the teeth may play a crucial role in the control and organization of the biomineralization.

In addition to the iron mineral components in the limpet teeth, there are

Figure 18.5. Anterior region of the tooth cusp of the limpet *Patella vulgata* showing the aligned acicular crystals of goethite. Bar is 0.6 μm. (Reproduced with permission from [30].)

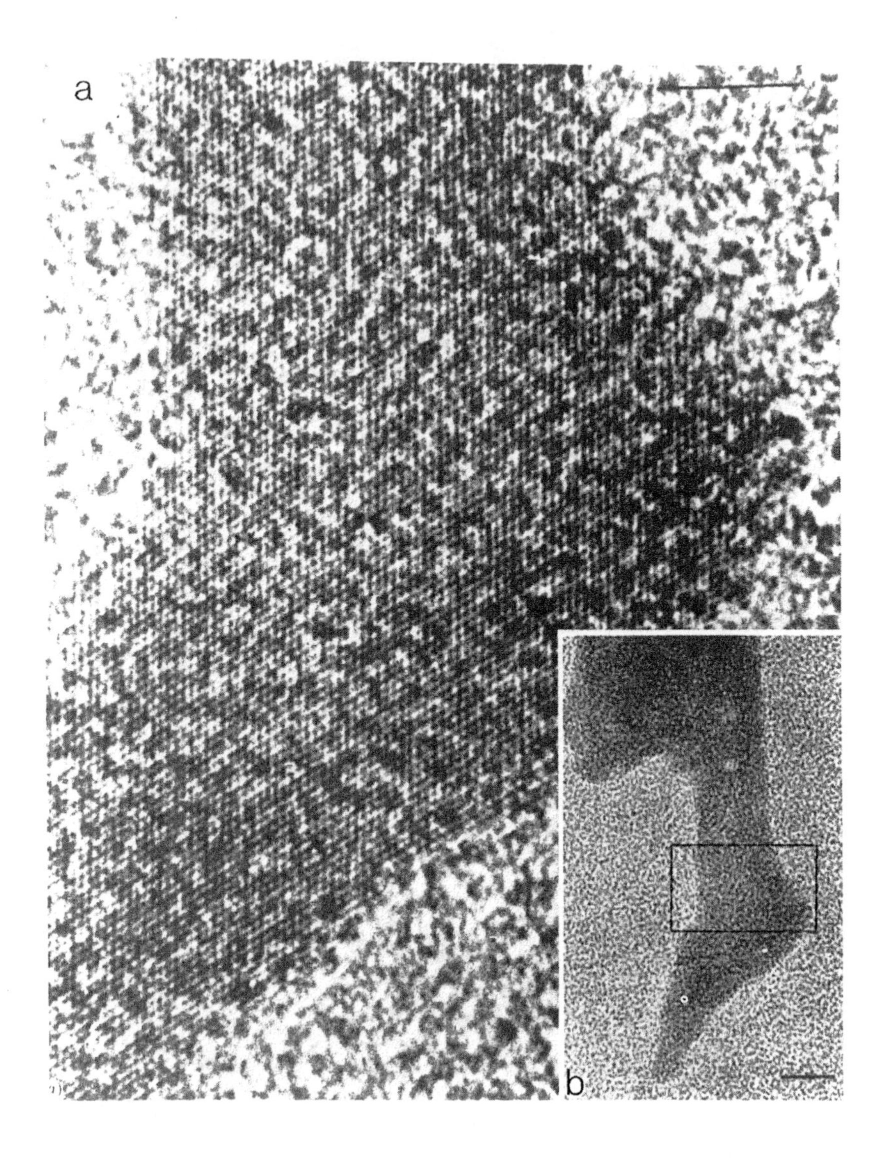

Figure 18.6. (*a*) High-resolution electron micrograph of a mature goethite crystal showing continuous lattice fringes throughout the crystal area. The box indicates the region shown in (*a*). Bar is 5 nm. (*b*) Lattice image of the end of a mature goethite crystal. Bar is 10 nm. (Reproduced with permission from [29].)

also silica components in the mineralized tissue. Treatment of the teeth with acid leads to a removal of the iron-containing components and allows the silica components to be investigated. Electron microscopy of acid treated teeth reveals a range of structural motifs constructed from aggregates of 5–15 nm amorphous silica particles. These include globular structures, fibers, and hollow tubes. Significantly, many of these silicified hollow tubes have diameters similar to the width of the goethite crystals, suggesting that the iron oxide mineralization occurs within the organic tubular structures which are subsequently impregnated with amorphous silica [30]. It should be noted that there are many nanophase silica materials found in a very wide range of biological systems [31].

18.5 CONCLUSIONS

The examples discussed in this chapter attempt to provide an indication of the possible links between the nanophase materials found in biological systems and their relevance to the synthesis, properties, and uses of nanostructured materials generally. This field is very much in its infancy and there are many exciting possibilities for the future, both in the use of biological materials as a source of inspiration for new synthetic materials and as the starting point for new synthesis pathways.

REFERENCES

[1] Meldrum F C, Heywood B R and Mann S 1992 *Science* **257** 522

[2] Dameron C T, Reese R N, Mehra R K, Kortan A R, Carroll P J, Steigerwald M L, Brus L E and Winge D R 1989 *Nature* **338** 596

[3] Ford G C, Harrison P M, Rice D N, Smith J M A, Treffry A, White J L and Yariv J 1984 *Phil. Trans. R. Soc.* B **304** 551

[4] Moore G R, Mann S and Bannister J V 1986 *J. Inorg. Biochem.* **28** 329

[5] Ward R J, O'Connell M J, Dickson D P E, Reid N M K, Wade V J, Mann S, Bomford A and Peters T J 1989 *Biochem. Biophys. Acta* **993** 131

[6] St Pierre T G, Chua-Anusorn W, Sipos P, Krou L and Webb J 1992 *Inorg. Chem.* **32** 4480

[7] Néel L 1949 *Ann. Geophys.* **5** 99

[8] Wickman H H 1966 *Mössbauer Effect Methodology* vol 2, ed I J Gruverman (New York: Plenum) p 39

[9] Dickson D P E, Reid N M K, Hunt C, Williams H D, El-Hilo M and O'Grady K 1993 *J. Magn. Magn. Mater.* **125** 345

[10] Gang Xiao, Liou S H, Levy A, Taylor J N and Chien C L 1986 *Phys. Rev.* B **34** 7573

[11] Linderoth S, Balcells L, Labarta A, Tejada J, Hendriksen P V and Sethi S A 1993 *J. Magn. Magn. Mater.* **124** 269

[12] Hunt C, Pankhurst Q A and Dickson D P E 1994 *Hyperfine Interact.* **91** 821

[13] Frankel R B, Papaefthymiou G C and Watt G D 1987 *Hyperfine Interact.* **33** 233

[14] Lyberatos A, Chantrell R W and O'Grady K 1996 *Nanophase Materials: Synthesis, Properties and Applications* ed G C Hadjipanayis and R W Siegel (Dordrecht: Kluwer) p 653

[15] St Pierre T G, Bell S H, Dickson D P E, Mann S, Webb J, Moore G R and Williams R J P 1986 *Biochem. Biophys. Acta* **870** 127

[16] Selden C, Owen M, Hopkins J M P and Peters T J 1980 *Br. J. Haematol.* **44** 593

[17] Dickson D P E, Reid N M K, Mann S, Wade V J, Ward R J and Peters T J 1989 *Hyperfine Interact.* **45** 225

[18] Pankhurst Q A, Betteridge S, Dickson D P E, Douglas T, Mann S and Frankel R B 1994 *Hyperfine Interact.* **91** 847

[19] Douglas T, Dickson D P E, Charnock J, Garner C D and Mann S 1995 *Science* **269** 54

[20] Blakemore R P 1975 *Science* **190** 337

[21] Frankel R B and Blakemore R P 1989 *Bioelectromagnetics* **10** 223

[22] Torres de Araujo F F, Germano N A, Goncalves L L, Pires M A and Frankel R B 1990 *Biophys. J.* **58** 549

[23] Mann S, Sparks N H C, Frankel R B, Bazylinski D A and Jannasch H W 1990 *Nature* **343** 258

[24] Frankel R B 1984 *Annu. Rev. Biophys. Bioeng.* **13** 85

[25] Butler R F and Banerjee S K 1975 *J. Geophys. Res.* **80** 4049

[26] Mann S and Frankel R B 1989 *Biomineralisation: Chemical and Biochemical Perspectives* ed S Mann, J Webb and R J P Williams (Weinheim: VCH) p 388

[27] Moskowitz B M, Frankel R B, Flanders P J, Blakemore R P and Schwartz B B 1988 *J. Magn. Magn. Mater.* **73** 273

[28] Webb J, Macey D J and Mann S 1989 *Biomineralisation: Chemical and Biochemical Perspectives* ed S Mann, J Webb and R J P Williams (Weinheim: VCH) p 345

[29] St Pierre T G, Mann S, Webb J, Dickson D P E, Runham N W and Williams R J P 1986 *Proc. R. Soc.* B **228** 31

[30] Mann S, Perry C C, Webb J, Luke B and Williams R J P 1986 *Proc. R. Soc.* B **227** 179

[31] Perry C C 1989 *Biomineralisation: Chemical and Biochemical Perspectives* ed S Mann, J Webb and R J P Williams (Weinheim: VCH) p 223

Chapter 19

Synthesis, structure, and properties of fullerenes

Donald R Huffman

19.1 INTRODUCTION

In 1985 the team of Harry Kroto from the University of Sussex and Richard Smalley from Rice University, along with their colleagues, found clear evidence [1] in the mass spectra of laser ablated graphite that clusters of 60 carbon atoms were unusually stable. A second and smaller peak occurred corresponding to 70 carbon atoms. In searching for the explanation of the C_{60} cluster, the team proposed an elegant soccerball-shaped molecule having icosahedral symmetry. The proposed molecule consisted of threefold-coordinated carbon atoms linked into a spherical structure consisting of 12 pentagons and 20 hexagons. The pentagons serve to curve the otherwise graphitic plane into a closed surface with no dangling bonds. Because of its similarity to geodesic domes, it was named buckminsterfullerene, after the American architect Buckminster Fuller, who studied such structures. Buckyball became the familiar name. The structure of the C_{70} cluster was proposed to be a slight modification of the buckyball having an extra band of 10 carbon atoms around its waist, giving it a slightly elongated shape. Many other even-numbered carbon clusters were found in mass spectra, and the terminology fullerenes was coined for the group of cage-shaped carbon clusters.

Although often discussed in the literature after 1985, until 1990 buckminsterfullerene and the other members of its family could only be studied in molecular beam experiments. Then the team of Krätschmer, Lamb, Fostiropoulos, and Huffman [2] announced the discovery of a route to the macroscopic synthesis of buckminsterfullerene and the third crystalline form of carbon consisting of crystallized buckyballs. The production technique was so simple that many laboratories worldwide quickly reproduced the results and began making and experimenting with the new nanoscale cluster and the materials that could be made from it.

Indeed the buckyball proved to be an almost perfectly formed prototype of a nanoparticle. Its effective diameter for packing into a crystal is almost exactly one nanometer (within a fraction of one per cent), and it is perhaps the most spherical molecule known. C_{60} has already proven itself to be an extremely versatile building block, as it forms readily into single crystals and thin-film solids. The molecule itself has proven to be much more reactive than originally thought so that numerous chemical appendages have been attached to it. Its large, hollow interior has allowed various atoms to be incorporated into it, and molecules have been fused together to form connected buckyballs and polymeric material.

It has been realized, since 1990, that the carbonaceous soot produced in the first step of the Krätschmer–Huffman process has much more in the way of nanoparticles than just C_{60}—buckminsterfullerene. There is also an abundant supply of the slightly elongated C_{70}, along with many other cage-shaped molecules of varying symmetry. Some of the 'magic numbers' that have thus far been abundant enough to isolate [3] are C_{76} (including a chiral isomer), C_{84}, C_{92}, and C_{96}. So called 'giant fullerenes' have been detected in the fullerene soot with sizes ranging up to hundreds of carbon atoms [4].

The wonders have not stopped here, however. Embedded in the residue from the carbon arc electrodes was found a vast collection of 'buckytubes' [5]—highly elongated and nested tubes of graphitic sheets with curved caps on the ends. These are similar to the tubes discussed earlier by Iijima [6], but now the arc technique succeeds in easily producing gram-sized quantities of these potentially useful materials. Then there are the nested fullerene structures consisting of concentric, closed graphitic layers, sometimes occurring with apparently hollow interiors [7].

This chapter gives brief details of the production technique which produces these submicroscopic entities, describes the structure of the C_{60} molecule and the crystalline solid built from it, and discusses some of the known properties of C_{60}. The field of fullerene research has exploded so rapidly that a chapter such as this can only touch on a few points. Overviews of the field are found in survey articles by Curl and Smalley [8] and by Huffman [9], and special issues in *Accounts of Chemical Research* [10] and *Carbon* [11]. Although some mention is given to fullerenes of larger size than C_{60} including tubes and 'onions', the main emphasis is on C_{60}, in both its molecular and solid forms.

19.2 SYNTHESIS

The breakthrough in macroscopic production of fullerenes came as a result of many years of studying the ultraviolet spectra of small particles of graphitic carbon. Because of long-term interest in the optical properties of small particles [12] and an interest in interstellar dust as examples of such finely dispersed particles, our Tucson laboratory had been producing many types of nanoparticle by vaporizing solids using electric arcs and other forms of heating since the early 1970s [13]. In 1983, collaborative work with Wolfgang Krätschmer varying the parameters in small-particle carbon production led to an unusual ultraviolet absorption peak with three sub-structures. No obvious explanation for such optical behavior was apparent until the work of Kroto, Smalley, and co-workers [1] proposed the new magic-number cluster of carbon—C_{60}. Although buckminsterfullerene immediately was considered as the source of our unusual smoke spectra, conclusive experimental evidence came slowly. The ultraviolet spectrum in comparison to calculations by Larsson, Volosov, and Rosén [14] suggested C_{60}, but the observation of four distinct peaks in the infrared spectrum, as predicted for the icosahedral molecule, provided more convincing evidence [15]. When the corresponding four peaks from material prepared with 99% pure ^{13}C showed the expected shift for a pure carbon molecule [16], the identification became clearer. The big breakthrough came in 1990, when the method for extracting the C_{60} from the soot was discovered, leading to the growth of crystals of C_{60}—the third crystalline form of carbon. Various experiments could then be done to prove with certainty that the buckminsterfullerene molecule as envisaged by the Kroto–Smalley group was indeed being made in macroscopic quantities. The simple technique has permitted the rapid expansion of fullerene work worldwide.

19.2.1 Synthesis of C_{60} and C_{70}

Figure 19.1 shows schematically our method for producing fullerene smoke. The important first step in producing and isolating C_{60} is the vaporization of graphite rods in a pure atmosphere of helium or other inert gases. While the 1990 paper described electrical heating of carbon rods with reduced tip diameters, more efficient heating using a small arc gap between electrodes (as was used in the earlier carbon smoke production) [13] has been adopted frequently as a means of producing larger quantities of soot [17]. Vaporization of the graphite feed material by means of an electron beam and by sputtering [18] has also produced fullerene smoke. The common process in these various modifications of the technique is the vaporization by heating of graphitic material in an inert gas atmosphere.

The presence of the inert gas serves to thermalize the carbon atoms that are removed from the graphite feed material and allows them to aggregate into larger clusters and small particles of carbon. Amazingly, an appreciable amount of the vaporized carbon clusters curl to form the C_{60} molecule, whose 12 pentagons give rise to the curvature. Small clusters of carbon atoms tend to be very reactive because of the unsatisfied bonds at the edges. If the clusters grow into graphitic sheets, the edges of these will also be highly reactive. However, if a totally closed C_{60} (or other cage-shaped fullerene molecule) forms, it will have no dangling bonds. It becomes a survivor in the rapidly changing soup of growing, reactive carbon clusters that ultimately ends up as carbon soot on the walls of

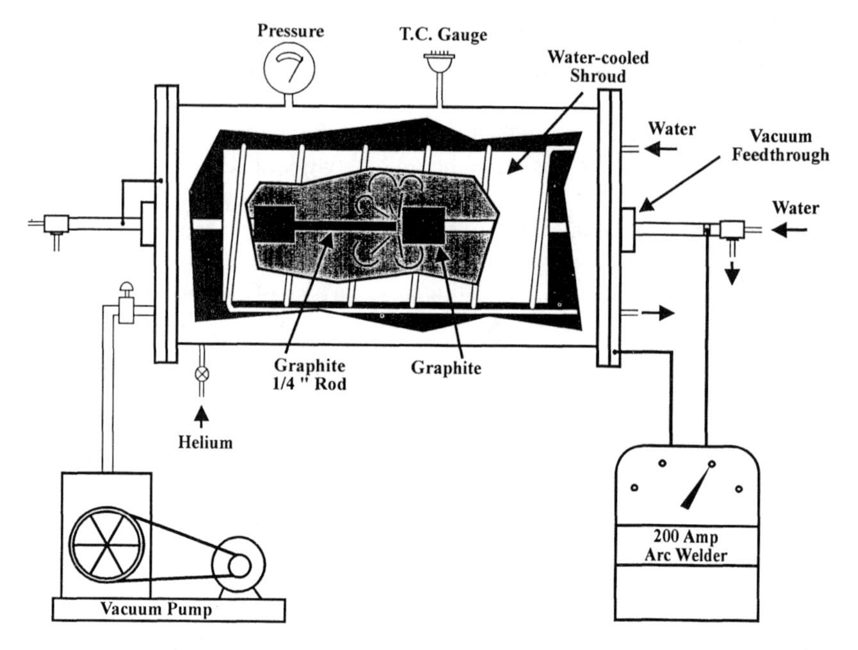

Figure 19.1. Fullerene soot production chamber.

the production chamber.

The second step in the production process for fullerenes is the extraction from the carbon soot. This can be accomplished in two ways. The fullerenes can be sublimed from the soot or dissolved and separated. C_{60} and other small fullerenes can be dissolved by dispersing the fullerene soot in a nonpolar organic solvent such as benzene or toluene, and filtering out the residual soot from the reddish brown resulting liquid. Ultrasonic dispersal of the carbon promotes the process, and soxhlet extractors are frequently used. The solution of fullerenes is then dried (using a rotary evaporator, for example) to leave a microcrystalline residue of solid material consisting of typically 85% C_{60}, 13% C_{70}, and the remainder of higher fullerenes such as C_{76}, C_{78}, and C_{84}. Chromatographic separation of the fullerene material can be accomplished by passing it through a high-performance liquid chromatography (HPLC) column containing neutral alumina and eluting with hexane or hexane–toluene mixtures [19, 20]. Such solutions of C_{60} and C_{70} can be 99.9% pure. More recent work [21] has produced milligram quantities of C_{76}, C_{78}, C_{82}, C_{84}, C_{90}, and C_{96}.

19.2.2 Synthesis of giant fullerenes

The use of solvents with higher boiling points allows fullerenes with much higher masses to be extracted from the soot [4]. Similarly, it is possible to use toluene at high pressure and high temperature to dissolve many higher fullerenes [22]. Fullerene soot confined in a pressure bomb with toluene and heated to several hundred degrees dissolves some of the larger fullerenes. The fullerenes that are dissolved remain in solution as the pressure bomb contents are returned to ambient conditions. Mass spectra show that there is a large range of fullerene molecules now present in the solution, although there is the possibility that some of these were produced either in the pressure cooking or in the mass spectrometric analysis. Scanning tunneling microscope (STM) images of the high-pressure extract clearly show rounded fullerene shapes of various sizes, with no evidence for highly elongated tubes. When all the evidence on extraction is taken together, it has been suggested [4] that almost half of the mass of the fullerene soot may be composed of fullerene molecules. Figure 19.2 shows a spectrum of a soot sample using laser desorption time-of-flight mass spectrometry.

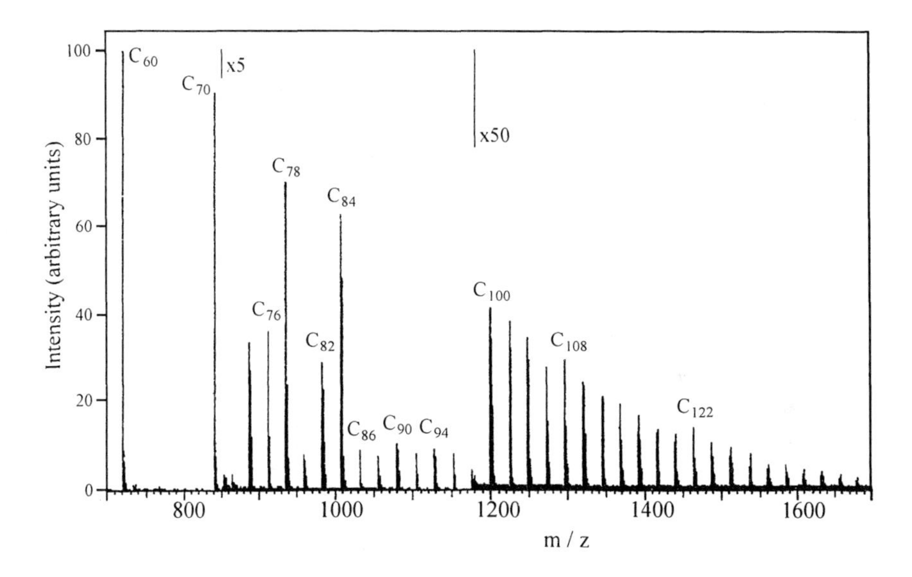

Figure 19.2. Negative ion chemical ionization mass spectrum of fullerene soot taken using thermal desorption. (Courtesy of Steve McElvaney, Naval Research Laboratory.)

An alternative method for extracting fullerenes [2] is by simply subliming it from the soot and collecting it on surfaces suitably placed. It is also possible to achieve some separation of C_{60} and C_{70} by subliming these molecules from the soot in a tube having a decreasing temperature gradient away from the sublimation zone (i.e. a tube hanging out of the end of a tube furnace). Condensation of the vapor occurs at slightly different positions for each of the two molecular species.

19.3 MOLECULAR AND CRYSTALLINE STRUCTURE

Following the discovery of the unusual abundance of C_{60} clusters in ablated graphite molecular beams, it was proposed [1] that this unusually stable molecule had the structure of a truncated icosahedron. Neighboring carbon atoms were considered to be connected in a threefold coordination employing predominantly sp^3 bonding as in graphite. The relationship of the proposed structure to the geometry of the icosahedron is shown in figure 19.3. Truncation of the 12 points of the icosahedron at an appropriate depth gives rise to 12 pentagons connected by 20 hexagons. The depth of the truncation affects the edge lengths of the hexagons, which become bond lengths in the molecule. Thus the edge lengths of the hexagons may be different from the edge lengths of the pentagons, while the structure still maintains the symmetry of the icosahedron usually denoted by I_h. As pointed out by Weeks and Harter [23], Hammermesch in one of the early works on group theory [24] explained that the icosahedral point group is the highest symmetry allowed in three-dimensional Euclidean space, but nothing in nature had this symmetry. Later, several natural cases of icosahedral symmetry were discovered [25].

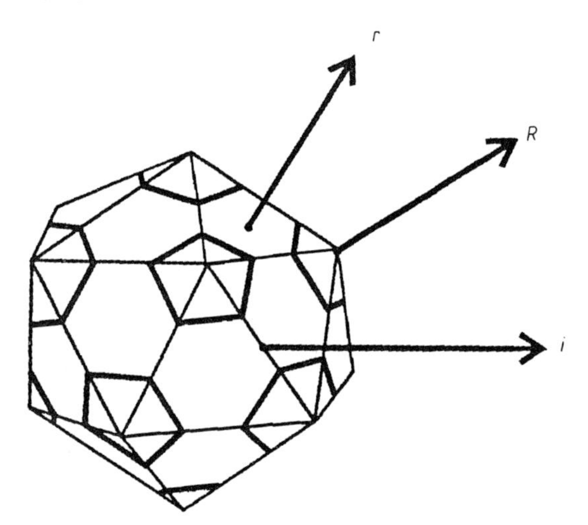

Figure 19.3. Truncation of the icosohedron.

19.3.1 The C_{60} molecule—buckminsterfullerene

Although much theoretical work was done on the proposed C_{60} structure before 1985, experimental verification of the structure had to await the production of macroscopic quantities in 1990, and the flood of experimental work that followed. One of the first and most exciting observations of macroscopic C_{60} was the viewing in a low-power microscope of hexagonal platelets of single-crystal fullerenes. These were displayed on the cover of the 27 September, 1990, issue of *Nature* [2] over the caption 'A new form of carbon'. Electron diffraction, x-ray diffraction, and scanning tunneling microscopy quickly confirmed that the crystals were composed of spherical molecules with effective diameters of about 1 nm, packed in a close-packed crystalline form. Infrared spectroscopy revealed only four vibrational absorption bands, as predicted by several groups of theorists for the highly symmetrical icosahedral molecule [26]. The ready supply of C_{60} molecules made available by the macroscopic production technique allowed confirmation of the molecular structure proposed in 1985 as well as the structure of the new crystalline form of carbon.

19.3.2 The solid—fullerite

The structure of crystalline C_{60}, dubbed fullerite by the Krätschmer–Huffman group, is shown in figure 19.4 in comparison to the other two crystalline forms of carbon—diamond and graphite. Each buckyball is composed of the 60 tightly bound carbon atoms of buckminsterfullerene, with the balls arranged in a close-packed array on a face-centered-cubic (fcc) crystalline lattice. The van der Waals forces between C_{60} molecules are much weaker than the covalent interatomic bonds. In contrast, each carbon atom in diamond is bound to its neighbors by four covalent bonds of sp^3 hybridization to make up a cubic crystal with great strength and hardness. Thus diamond is the strongest of crystals. Graphite, on the other hand, is a highly anisotropic solid which crystallizes into layers of hexagonal planes in which each carbon atom is bound to its three nearest neighbors by covalent bonds of sp^2 hybridization. Crystalline C_{60} can be compared with graphite by considering the planes of graphite to be curled into 60-atom balls. This requires the introduction of pentagons into the hexagonal graphite planes. In this comparison, the buckyballs in the crystal are analogous to graphite planes; the intermolecular interaction in crystalline C_{60} is analogous to the interplanar interaction in graphite. Both are comparatively weak. In graphite, the planes easily shear from one another, and in fullerite the balls detach from one another rather easily. Within the C_{60} molecular cage, the bonding is very strong, as is the bonding within the graphitic planes. The distance between adjacent carbon cages in solid C_{60} is about 0.29 nm compared with 0.335 nm—the distance between atomic planes of graphite. The two interatomic distances within a C_{60} molecule are 0.140 nm between the two carbon atoms shared by adjacent hexagons and 0.145 nm between the two carbon atoms shared by a hexagon and a pentagon. These values compare with 0.142 nm for the separation of nearest-neighbor atoms in graphite.

Solid-state nuclear magnetic resonance (NMR) studies [27] have shown that the molecular balls in solid C_{60} are randomly rotating at room temperature. This is a result of their being extremely smooth spherical objects with rather weak forces connecting them. The molecules spin so readily that they actually rival gas phase C_{60} molecules in their rotational characteristics [28]. Upon cooling

Figure 19.4. The three forms of solid carbon. The corner cut from the C_{60} crystal cube reveals a (111) crystal plane where molecules stack in hexagonal arrays. The structures of diamond and graphite are shown at the middle and bottom left. (Drawing ©1991, Henry Hill Jr.)

to about 255 K, a phase transition takes place [29] in which the fcc structure changes to a simple cubic structure. This is due to a partial rotational ordering of the molecules. At still lower temperatures there is a gradual freezing out of the rotational motion into a rotationally disordered glass [30].

Because of the large sizes of the C_{60} molecules, the interstices between molecular balls in the solid are correspondingly large. Tetrahedral sites have a diameter of 0.224 nm and octahedral sites have a diameter of 0.414 nm [31]. This gives rise to the possibility of incorporating rather large atoms and molecules into the interstices, producing both desirable and undesirable consequences. One of the most exciting of the properties of solid C_{60} is superconductivity [32], discovered in March 1991, produced by doping the solid material with alkali metal atoms such as potassium, rubidium, and cesium, which populate the interstices. An often less desirable effect is produced by the incorporation into the lattice of solvent molecules such as benzene or toluene, which have been used in the extraction process. Although these solvent molecules are harmless for some purposes, they cause difficulty in other experiments and uses.

19.3.3 Larger fullerenes

The C_{70} molecule, which is next most abundant to C_{60}, can be thought of as a C_{60} with an extra band of ten carbons around its waist. In the series of higher fullerenes above C_{70}, peaks in the mass spectrum show certain other fullerenes to be unusually abundant compared to those of neighboring mass, but still an order of magnitude or more less abundant than C_{60} and C_{70}. See figure 19.2. These include C_{76}, C_{78}, C_{84}, and C_{90}, which have been isolated in milligram quantities [3].

Although fullerenes above C_{100} have not yet been separated in macroscopic amounts, coalescence of isolated, pure fullerenes may lead the way to relatively pure higher fullerenes. Coalescence of C_{60} to form fullerenes in the vicinity of C_{120} and C_{180} has been reported [33] along with similar results for the coalescence of C_{70}. It has also been reported that, during the destruction of C_{60} by ozone, some C_{119} has been formed [34], presumably as a result of chemically removing one carbon atom from one buckminsterfullerene and somehow attaching the molecule to another C_{60}. This is the first example of odd-numbered fullerene molecules. Since trigonal sp^2 bonding only gives rise to even-numbered fullerene cages, this result suggests the additional presence of sp^3 bonds. The ability to combine two or more molecules that can be made in abundance in the pure form, such as C_{60} and C_{70}, seems to offer new hope for possible production and extraction of *some* of the higher fullerenes in macroscopic quantities.

In the study of solid fullerenes (fullerites) only C_{60} and C_{70} have been available for any detailed investigation. Since every new fullerene that can be produced in abundance should also crystallize into a new and unique solid, there is a whole new field of solid materials that should be forthcoming.

19.3.4 Tubes and onions

In 1991 Iijima obtained transmission electron micrographs of elongated carbon nanoparticles [35] that appeared to consist of cylindrical graphitic layers capped on the ends with fullerene-like domes. Several theoretical papers [36] reported on the possible physical properties of such buckytubes or carbon nanotubes. Not long thereafter, Ebbesen and Ajayan [5] succeeded in making macroscopic quantities of carbon nanotubes. The method was essentially that of figure 19.1, with the exception that the tubes were harvested from the core of a residue built up on one of the carbon arc electrodes. Such tubes are expected to be very strong, leading to possible improvements in carbon composite materials where carbon fibers are embedded in a matrix of either carbon or of another material. As in the original discovery of macroscopic fullerene production [2], the availability of copious quantities of the tube material and the simple method for making them has led to a small explosion of interest and study of carbon nanotubes. It has also been possible to create carbon nanotubes containing various metals or metallic compounds [37] as crystallites on the inside of carbon tubes. There has been speculation [38] on possible nanowires as a product of these developments.

Tubes made with the Ajayan–Ebbesen method are not the theoretically studied single-walled buckytubes. They characteristically contain from three or four to 20 or more cylindrical layers with concentric caps formed on the ends. More recently, success in producing the single-walled buckytubes has been reported by two different groups [39], who combined either iron or cobalt with carbon in the arc production process.

Even before the advent of macroscopic production of fullerenes, Iijima [40] had shown transmission electron micrographs of carbon particles showing concentric layers. This precipitated a discussion [41] as to whether buckyballs might be nuclei for carbon soot particles in general, which frequently grow as almost spherical particles. Ugarte [42] has shown how to make carbon particles with very perfectly concentric spherical layers described as bucky onions. These remarkably beautiful structures were produced from particles made in a conventional fullerene smoke chamber, that were subjected to electron bombardment from the beam of a transmission electron microscope.

19.4 PROPERTIES

Crystalline C_{60} is a molecular solid. That is, the crystals are formed from an ordered array of the C_{60} molecules, which retain their identity and many of their molecular properties. This is in contrast to the other crystalline forms of carbon—diamond and graphite, where there is no identifiable molecular sub-unit. Therefore many of the properties of solid C_{60} follow directly from those of the molecule. Table 19.1 lists separately some of the properties of the molecule and of the solid which are discussed in this section.

Table 19.1. Properties of C_{60}.

Property	Value	References
The molecule—C_{60}		
Size: cage diameter	0.7 nm	[2]
van der Waals diameter	1.0 nm	
Symmetry	icosahedral	[1]
Bond distances: five–six bonds	0.1404 nm	[47]
six–six bonds	0.1448 nm	
Electron affinity	2.65 eV	[48]
First ionization potential	7.58 eV	[49]
Cohesive energy	7.4 eV/atom	[50]
The solid		
Density	1.68 g cm^{-3}	[2]
Crystal structure	fcc	[51]
Lattice constant	1.417 nm	[52]
Nearest-neighbor distance	1.004 nm	
Index of refraction at 630 nm	2.2	[53, 54]
Cohesive energy per C_{60} molecule	1.5 eV	[50]
Bulk modulus	18 GPa	[62]
Electron bandgap	1.85 eV	[55]
Ionization potential	7.6 eV	[56]
Thermal conductivity (300 K)	0.4 W mK^{-1}	[57]
Thermal expansion coefficient	6.2×10^{-5}	[52]
Structural phase transition	255 K	[52, 51]

19.4.1 Properties of the C_{60} molecule

Because of the high symmetry of the C_{60} molecule (icosahedral) every carbon atom sits in an identical environment at the point of intersection of two hexagons and one pentagon. Thus the C_{60} molecule displays a C^{13} NMR spectrum consisting of a single line [43], which provided a striking confirmation of the buckyball model. The molecule, because of its especially stable structure, is highly resistant to photofragmentation [44]. It is a very resilient survivor of collisions against surfaces, as shown by both experiments [45] and theory [46].

The molecule C_{60} has a large electron affinity. Initially, two- and three-electron reduction was shown in cyclic voltammetry [58], and more recently [59] up to six electrons have been added reversibly. C_{60} has a closed-shell electronic structure with a gap energy of about 1.9 eV between the highest occupied molecular orbital and the lowest unoccupied molecular orbital (the HOMO–LUMO gap). The cohesive energy is 7.4 eV per carbon atom.

During the years from 1985 to 1990, when there was much discussion of the buckyball but no experiments outside of molecular beams, it was widely believed that the C_{60} molecule would be very unreactive due to the complete absence of dangling bonds. This would also make it very abundant in the universe, since it would be stable against chemical destruction. When the molecule became available in abundance, it was found that many chemical species react easily with it. Reaction has been shown to occur with oxygen, hydrogen, fluorine, chlorine, and bromine. A common feature of such reactions is that a complex mixture of products is produced. For example, when the first halogenation of C_{60} was carried out [60] a mixture of products was obtained, although it was dominated by $C_{60}H_{36}$, which had also been found in the hydrogenation experiments of Haufler *et al* [58]. Similar effects have been found for $C_{60}Br_{36}$. The growing list of molecules based on C_{60} has been categorized in a review by Taylor and Walton [61].

19.4.2 Properties of the solid

Since the solid is made up of tightly bound buckyballs which are weakly bound together by van der Waals forces, the mechanical properties of the solid can be characterized as light, weak, and soft with respect to the other carbon forms. The density is 1.7 g cm^{-3} compared to that of graphite (2.3 g cm^{-3}) and diamond (3.5 cm^3). Mechanical properties vary widely in the three forms of carbon because of the different bonding illustrated in figure 19.4. Diamond is a very strong isotropic solid whereas graphite is a highly anisotropic solid which yields easily to shear forces separating the planes, while maintaining high strength within the planes. Under the action of hydrostatic pressure, solid C_{60} compresses easily with a bulk modulus of 18 GPa [62], which is similar to the interplanar compressibility of graphite. As the hydrostatic pressure on solid C_{60} is increased to 20 GPa, the solid becomes less compressible. Initially the electron clouds separating molecules yield easily under pressure. However, after the rigid cages begin to interact, the solid becomes much stiffer.

The electronic properties of the C_{60} molecule are largely preserved as the solid is formed. The minimum bandgap in the solid of about 1.5 eV is similar to the HOMO–LUMO gap in the molecule with a slight broadening of the electronic energy bands in the solid. Thus solid C_{60} is a large-bandgap semiconductor. Its electrical conductivity is very low. It is transparent in the near infrared and into the visible region where transitions near the bandgap energy give it an orange color in thin films. The powder is usually gray to black in color depending on the size of the crystallites. The index of refraction in the visible is about 2.2 with some absorption, compared with diamond which has an index of refraction of about 3.2.

19.4.3 Clusters of clusters

In common with many nanoparticles, C_{60} is formed when vaporized carbon atoms are slowed down in collisions with inert gas molecules to conditions of temperature and density where aggregation into atomic clusters takes place. This subject is discussed in more detail in chapter 2 of this volume. If solid C_{60} is now vaporized in an inert gas atmosphere, aggregation of the C_{60} molecules into aggregates of C_{60} occurs. This 'smoke' of C_{60} was reported in the original paper on production of solid C_{60}. One of the first published STM images of solid C_{60} [63] showed a 'smoke' particle about 30 nm in diameter composed of the 1 nm C_{60} molecules. It looks something like a raspberry. Because of the large size of the molecule, images of individual molecules could be seen on the same picture as the entire particle. Recently Martin and colleagues [64] have been able to mass-analyze aggregates of C_{60} clusters. Magic numbers found in the mass spectra are interpreted in terms of icosahedral clusters of C_{60}. Thus the icosahedral symmetry which was once rare in the laboratory has shown up as icosahedral clusters of icosahedral clusters.

19.4.4 Various forms of fullerene materials

Adding to the potential value of C_{60} and other fullerene materials is the relative ease with which they can be produced in other material forms discussed in this volume. Some of these forms are listed in table 19.2.

Table 19.2. Various forms of fullerene materials.

Microcrystalline powder
Sub-micron smoke
Solutions, for example in toluene, benzene, etc
Spray coatings
Evaporated films on substrates
Multilayer films forming superlattices
Unsupported thin films
Single crystals grown from solution 5 mm in size with trapped solvent
Single crystals grown from vapor phase—1 mm size
Pressed-powder solids
Heated vapor
Molecular beams

Two properties of solid C_{60} that enable it to be rather easily transformed into other useful forms are its solubility and its comparatively low (about 450 °C) sublimation temperature. Solid C_{60} is soluble in a number of nonpolar organic solvents such as benzene, toluene, hexanes, carbon disulfide, etc. Some examples of solubilities are given in table 19.3, taken from the work of Ruoff *et al* [65].

Table 19.3. Solubilities of C_{60} in some organic solids (in mg ml^{-1}).

n-hexane	0.043
carbon tetrachloride	0.32
benzene	1.7
trichloroethylene	1.4
xylene	5.2
1,2-dichlorobenzene	33
ethanol	0.0001
carbon disulfide	7.9

The fullerenes are the only forms of pure carbon that are soluble in common solvents. This may give rise to ways of manipulating carbon that have not been previously available. Solutions may be used to carry the carbon into voids and cavities with the solvent subsequently removed. The solubility makes it possible to readily spray-coat C_{60} onto a variety of surfaces. Once the C_{60} is in place, it should be possible to convert it into graphitic carbon by heat or bombardment with electrons or other energetic particles.

C_{60} sublimes readily in a vacuum starting at a temperature of about 450 °C, and forms high-quality coatings on many types of substrate. In addition to the thin films and coatings produced in this way, one can also produce multilayer superlattices with C_{60} as one set of layers. We have found that unsupported films of C_{60} can be floated off substrates using techniques that have been developed to produce graphitic carbon foils for beam foil spectroscopy. Etching techniques have also succeeded in producing unsupported thin films of C_{60} [66].

Single crystals have been grown both from solution and by vapor transport. The solution grown crystals are typically larger, but invariably incorporate some of the solvent. Vapor transport has been used to grow solvent-free single crystals. To date, the reported size has been about a millimeter or less in linear dimensions. A special issue of *Applied Physics* A [67] has been devoted to the preparation and properties of single-crystal fullerenes.

Microcrystalline powders of C_{60} can be compressed into self-supporting forms. Pressed solids of C_{60} can be formed which approach the density of single-crystal material. The material holds together quite well.

19.4.5 Applications of the fullerenes

At the time of this writing there are no large-scale industrial applications of fullerenes, but few doubt these will come. Some of the many suggestions that have been made concerning possible applications are lubricants (molecular ball bearings), selective adsorption of gases, sensors, hydrogen storage, photochromic goggles, ion rocket propulsion, diamond seeding, diamond production, xerographic materials, new semiconductors, patterned diamond films, patterned superconductors, carbon–carbon composites, and anti-viral agents.

Perhaps the biggest potential use for fullerenes is for use as a diverse set of building blocks for creating new organic chemicals and solid materials. Already many derivatives and adducts of C_{60} have been reported [61]. Interstitial doping has led to the new family of superconductors. But only a few possibilities have thus far been explored, and the number of possibilities is enormous. Every atom of the periodic table could conceivably fit inside a C_{60} cage, creating a whole periodic chart of new molecules. On the outside of the cage, the three-dimensional lattice of pentagons and hexagons with intermediate hybridization presents the possibility of huge numbers of new chemical adducts. Each of the hundreds of higher fullerenes as well as the endohedral compounds formed from them hold the promise of a new solid material. And each such new solid can be doped interstitially in a variety of ways. Clearly the number of new molecules and new materials is increasing at a rapid pace because of the versatility of fullerene building blocks.

Before specific fullerene-based commercial products become available, however, it will be necessary for new generations of research to investigate the properties of these many new materials and chemicals.

REFERENCES

[1] Kroto H W, Heath J R, O'Brien S C, Curl R F and Smalley R E 1985 *Nature* **318** 162

[2] Krätschmer W, Lamb L D, Fostiropoulos K and Huffman D R 1990 *Nature* **347** 354

[3] Diederich F, Ettl R, Rubin Y, Whetten R L, Beck R, Alvarez M, Anz S, Sensharma D, Wudl F, Khemani K C and Koch A 1991 *Science* **252** 548

Kikuchi K, Nakahara N, Honda M, Suzuki S, Saito K, Shirommaru H, Yamauchi K, Ikemoto I, Kuramoti T, Hino S and Achiba Y 1991 *Chem. Lett.* **9** 1607

[4] Parker D H, Chatterjee K, Wurz P, Lykke K R, Pellin M J, Stock L M and Hemminger J C 1992 *Carbon* **30** 1167

Smart C, Eldridge B, Reuter W, Zimmerman J A, Creasy W R, Rivera N and Ruoff R S 1992 *Chem. Phys. Lett.* **188** 171

[5] Ebbesen T W and Ajayan P M 1992 *Nature* **358** 220

[6] Iijima S 1987 *J. Phys. Chem.* **91** 3466

[7] Ugarte D 1992 *Nature* **359** 707

[8] Curl R F and Smalley R E 1991 *Sci. Am.* **265** 32

[9] Huffman D R 1991 *Phys. Today* **44** 22

[10] 1992 *Accounts Chem. Res.* **25**

[11] 1992 *Carbon* **39** 1139

[12] Huffman D R 1977 *Adv. Phys.* **26** 129

[13] Day K L and Huffman D R 1973 *Nat. Phys. Sci.* **243** 50

[14] Larsson S, Volosov A and Rosén A 1987 *Chem. Phys. Lett.* **137** 501

[15] Krätschmer W, Fostiropoulos K and Huffman D R 1990 *Dusty Objects in the Universe* ed E Bussoletti and A A Vittone (Dordrecht: Kluwer) p 89
[16] Krätschmer W, Fostiropoulos K and Huffman D R 1990 *Chem. Phys. Lett.* **170** 167
[17] Haufler R , Conceicao J, Chibante L P F, Chai Y, Byrne N E, Flanagan S, Haley M M, O'Brien S C, Pan C, Xiao Z, Billups W E, Ciufolini M A, Huage R H, Margrave J L, Wilson L J, Curl R F and Smalley R E 1990 *J. Phys. Chem.* **94** 8634
[18] Bunshah R F, Jou S K, Prakash G K S, Doerr H J, Issacs L, Wehrsig A, Yeretzian C, Cynn H and Diederich F 1992 *J. Phys. Chem.* **96** 6866
[19] Hare J P, Kroto H W and Taylor R 1990 *Chem Phys. Lett.* **177** 394
[20] Ajie H, Alvarez M M, Anz S J, Beck R D, Diederich F, Fostiropoulos K, Huffman D R, Krätschmer W, Rubin Y, Schriver K E, Sensharma D and Whetten R L 1990 *J. Phys. Chem.* **94** 8630
[21] Kikuchi K, Nakahara N, Wakabayashi T, Honda M, Matsumiya H, Moriwaki T, Suzuki S, Shiromaru H, Saito K, Yamauchi Y, Ikemoto I and Achiba Y 1992 *Chem. Phys. Lett.* **188** 177
[22] Lamb L D, Huffman D R, Workman R K, Howell S, Chen T, Sarid D and Ziolo R F 1992 *Science* **255** 1413
[23] Weeks D E and Harter W G 1989 *J. Chem. Phys.* **90** 4744
[24] Hammermesch M 1972 *Group Theory and its Applications to Physical Problems* (Reading, MA: Addison-Wesley) p 51
[25] Mackay A L 1990 *Nature* **347** 336
[26] Wu Z C, Jelski D A and George T F 1987 *Chem. Phys. Lett.* **137** 291
Stanton R E and Newton M D 1988 *J. Phys. Chem.* **92** 2141
Weeks D E and Harter W G 1988 *Chem. Phys. Lett.* **144** 366
Cyvin S J, Brendsdal E, Cyvin B N and Brunvoll J 1988 *Chem. Phys. Lett.* **143** 377
Negri F, Orlandi G and Zerbetto F 1988 *Chem. Phys. Lett.* **144** 31
[27] Yanoni C S, Johnson R D, Meijer G, Bethune D S and Salem J R 1991 *J. Phys. Chem.* **95** 9
Tycko R, Haddon R C, Dabbagh G, Glarum S H, Douglass D C and Mujsce A M 1991 *J. Phys. Chem.* **95** 518
[28] Johnson R D, Yanoni C S, Dorn H C, Salem J R and Bethune D S 1992 *Science* **255** 1235
[29] Heiney P A, Fischer J E, McGhie A R, Romanow W J, Denenstein A M, McCauley J P Jr and Smith A B III 1991 *Phys. Rev. Lett.* **66** 2911
[30] Prassides K, Kroto H W, Taylor R, Walton D R M, David W I F, Tomkinson J, Haddon R C, Rosseinsky M J and Murphy D W 1992 *Carbon* **30** 1277
[31] Stevens P W, Mihaly L, Lee P L, Whetten R L, Huang S M, Kaner R B, Diederich F and Holczer K 1991 *Nature* **351** 632
[32] Hebard A F, Rosseinsky M J, Haddon R C, Murphy D W, Glarum S H, Palstra T T M, Ramirez A P and Kortan A R 1991 *Nature* **350** 600
Hebard A F 1992 *Phys. Today* **45** 26
[33] Yeretzian C, Hansen K, Diederich F N and Whetten R L 1992 *Nature* **359** 44
[34] McElvaney S W, Callahan J H, Ross M M, Lamb L D and Huffman D R 1992 *Science* **260** 1632
[35] Iijima S 1991 *Nature* **354** 56
[36] Mintmire J W, Dunlap B I and White C T 1992 *Phys. Rev. Lett.* **68** 631
Dresselhaus M S, Dresselhaus G and Saito R 1992 *Phys. Rev.* B **45** 6234

Tanaka K, Okahara K, Okada M and Yanabe T 1992 *Chem. Phys. Lett.* **191** 469

[37] Seraphin S, Zhou D, Jiao J, Withers J C and Loutfy R 1993 *Nature* **362** 503

Ajayan P M and Ijima S 1993 *Nature* **361** 333

[38] Jin C, Guo T, Chai Y, Lee A and Smalley R E 1992 *Proc. 1st Italian Workshop on Fullerenes: Status and Perspectives (Bologna, 1992)* ed C Taliani *et al* (Singapore: World Scientific)

[39] Iijima S and Ichihashi T 1993 *Nature* **363** 603

Bethune D S, Kiang C H, DeVries M S, Gorman G, Savoy R, Vazquez J and Beyers R 1993 *Nature* **363** 605

[40] Iijima S 1980 *J. Cryst. Growth* **50** 675

[41] Zhang Q L, O'Brien S C, Heath J R, Liu Y, Curl R F, Kroto H W and Smalley R E 1986 *J. Phys. Chem.* **90** 525

Ebert L B 1990 *Science* **247** 1468

Kroto H 1988 *Science* **242** 1139

[42] Ugarte D 1992 *Chem. Phys. Lett.* **198** 596

[43] Taylor R, Hare J P, Abdul-Sada A K and Kroto H W 1990 *J. Chem. Soc. Chem. Commun.* **20** 1423

[44] O'Brien S C, Heath J R, Curl R F and Smalley R E 1988 *J. Chem. Phys.* **88** 220

Curl R F 1992 *Carbon* **30** 1152

[45] Beck R D, St John P, Alvarez M M, Diederich F and Whetten R L 1991 *J. Phys. Chem.* **95** 8402

[46] Mowrey R C, Brenner D W, Dunlap B I, Mintmire J W and White C T 1991 *J. Phys. Chem.* **95** 7138

[47] Hedberg K, Hedberg L, Bethune D S, Brown C A, Dorn H C, Johnson R D and de Vries M 1991 *Science* **254** 410

[48] Wang L-S, Conceicao J J, Jin C M and Smalley R E 1991 *Chem. Phys. Lett.* **182** 5

[49] de Vries J, Steger H, Kamke B, Menzel C, Weisser B, Kamke W and Hertel I V 1992 *Chem. Phys. Lett.* **188** 159

[50] Saito S and Oshiyama A 1991 *Phys. Rev. Lett.* **66** 2637

[51] Fleming R M, Siegrist T, Marsh P M, Hessen B, Kortan A R, Murphy D W, Haddon R C, Tycko R, Dabbagh G, Mujsce A M, Kaplan M L and Zahurak S M 1991 *Mater. Res. Soc. Symp. Proc.* **206** 691

[52] Heiney P A, Vaughan G B M, Fischer J E, Coustel N, Cox D E, Copley J R D, Neuman D A, Kamitakahara W A, Creegan K M, Cox D M, McCauley J P Jr and Smith A B III 1992 *Phys. Rev.* B **45** 4544

[53] Ren S L, Wang Y, Rao A M, McRae E, Holden J M, Hager T, Wang K, Lee W-T, Ni H F, Selegue J and Eklund P C 1991 *Appl. Phys. Lett.* **59** 2678

[54] Saeta P N, Greene B I, Kortan A R, Kopylov N and Thiel F A 1992 *Chem. Phys. Lett.* **190** 184

[55] Kremer R S, Rabenau T, Maser W K, Kaiser M, Simon A, Haluiska M and Kuzmany H 1993 *Appl. Phys.* A **56** 211

[56] Lichtenberger D L, Jatcko M E, Nebesney K W, Ray C D, Huffman D R and Lamb L D 1991 *Mat. Res. Soc. Proc.* **206** 673

[57] Tea N H, Yu R-C, Salamon M B, Lorents D C, Malhotra R and Ruoff R S 1993 *Appl. Phys.* A **56** 219

[58] Haufler R E, Conceicao J, Chibante L P F, Chai Y, Byrne N E, Flanagan S, Haley M M, O'Brien S C, Pan C, Xiao Z, Billups W E, Ciufolini M A, Huage R H, Margrave J L, Wilson L J, Curl R F and Smalley R E 1990 *J. Phys. Chem.* **94** 8634

Allemand P M, Koch A, Wudl F, Rubin Y, Diederich F, Alvarez M M, Anz S J and Whetten R L 1991 *J. Am. Chem. Soc.* **113** 1050

[59] Dubois D, Kadish K, Flanagan S and Wilson L J 1991 *J. Am. Chem. Soc.* **113** 7773

[60] Selig H, Lifshitz C, Peres T, Fisher J E, McGhie A R, Romanow W J, McCauley J P Jr and Smith A B III 1991 *J. Am. Chem. Soc.* **113** 5475

[61] Taylor R and Walton D R M 1993 *Nature* **363** 685

[62] Duclos S J, Brister K, Haddon R C, Kortan A R and Thiel F A 1991 *Nature* **351** 380

[63] Wragg J L, Chamberlain J E, White H W, Krätschmer W and Huffman D R 1990 *Nature* **348** 623

[64] Martin T P, Naher U, Schaber H and Zimmermann U 1993 *Phys. Rev. Lett.* **20** 3079

[65] Ruoff R S, Tse D S, Malhotra R and Lorents D C 1993 *J. Phys. Chem.* **97** 3379

[66] Eom C B, Hebard A F, Trimble L E, Xeller G K and Haddon R C 1993 *Science* **259** 1887

[67] 1993 *Appl. Phys.* A **56** 159

PART 8

NANOFABRICATION AND NANOELECTRONICS

Chapter 20

Nanofabrication and nanoelectronics[†]

Elizabeth A Dobisz, Felixberto A Buot, and Christie R K Marrian

20.1 INTRODUCTION

The microelectronics industry is driven, as with other industries, by the need to remain financially viable. In the area of microelectronics circuitry, profitability has been connected to the reduction in size of the elements that make up these circuits. There are two reasons for this. First, the future success of these industries depends on their continuing to be at the cutting edge of the technology. For example, if the trend in miniaturization halts, the fabrication technology will stabilize and it would then become possible to fabricate microelectronics circuits without the need for a highly skilled and educated work force. Second, lower unit costs (e.g. the cost per byte of memory) are seen to be possible through increased miniaturization and integration. These trends have the added benefits of leading to faster speeds and lower energy (i.e. lower cost) per unit operation. Since the invention of the integrated circuit (IC), the minimum feature size of individual structures in the circuits has halved every five years or so. This trend was noted in 1975 by Moore [1] and the 'Moore curve' (as it is often called), shown in figure 20.1, has proved to be a remarkably accurate prediction of progress over the intervening two decades.

Today, the 64 MB (megabyte) DRAM (dynamic random access memory) chip is in full-scale production and contains features down to 350 nm in size. Manufacturing prototyping is in the 250 nm range (256 MB DRAM). By the beginning of the next decade, production manufacturing will be approaching 100 nm design rules. Indeed, in research and development (R&D) laboratories, conventional devices are currently being fabricated with feature sizes close to 100 nm. However, industry is reaching a critical point in its continued device down-scaling. On one hand, as the miniaturization continues, the

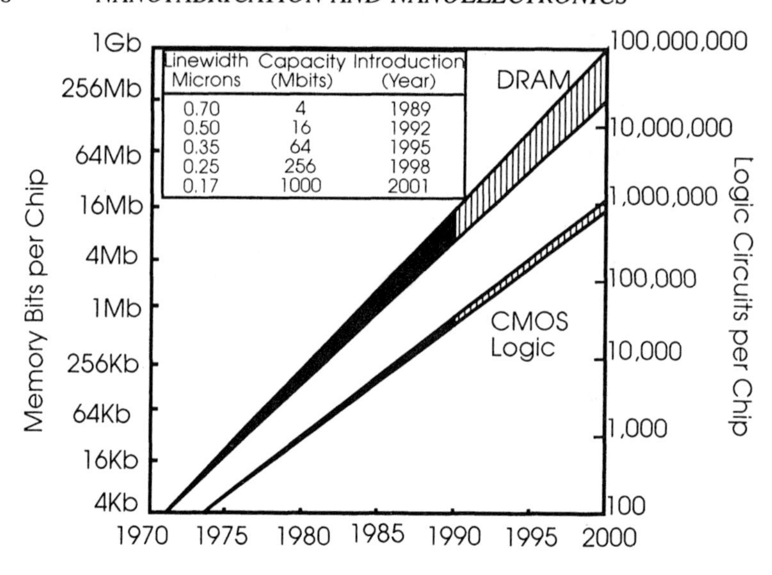

Figure 20.1. The miniaturization trend in microelectronics (from Moore [1]).

cost of fabrication and the necessary R&D for tools and processes increases dramatically. Second, there are many basic physics and reliability related issues which indicate the need for other directions for the continued down-scaling of the size of electronics devices and ICs. Lithographic miniaturization is approaching a limit beyond which dimensional down-scaling of conventional devices will not be possible. The large capacitances and dielectric breakdown in the large electric fields imposed by the small lateral dimensions make conventional Si MOS (metal oxide semiconductor) and bipolar devices impractical. A new field of electronics (and optoelectronics) has evolved to explore a wide range of novel devices with new logic functionality. This presents a fresh set of challenges both to the fabrication and device specialist. Here we discuss future directions of the methodologies of lithography and fabrication and the kinds of electronic device which will exploit the physics of the nanometer regime. Nanoelectronics, as this field is now called, is usually considered to be descriptive of devices and structures with dimensions less than 100 nm [2].

When a device size becomes comparable to the charge-carrier inelastic coherence length, and the charge-carrier confinement dimension approaches the Fermi wavelength, then classical device physics based on the motion of particles and ensemble averaging begins to breakdown. The carrier coherence length is the distance over which a carrier maintains coherency of the phase of its wavefunction and interferes as a wave. When the confinement dimension approaches the Fermi wavelength, then the carrier 'sees' the boundaries and the conduction band (or valence band) energy in the confining well become quantized, much like the particle-in-the-box in school physics. At these dimensions, macroscopic properties, such as electron transport, reflect the microscopic structure of the device. This boundary between the macro- and microscopic led to the birth of mesoscopic physics in the late 1970s. Stimulated by intriguing theories predicting novel physical phenomena that can be observed in the mesoscopic regime [3–7], the electronics community has identified novel device concepts. Furthermore, the field of nanoelectronics provides a direction for future IC technology.

The distinction between mesoscopic physics and nanoelectronics is used here in the same sense as the distinction between solid-state physics and solid-state electronics. A rapid growth of mesoscopic physics and nanoelectronics started around 1985, with mesoscopic physics driven by the IBM related community [8], nanoelectronics IC research by Texas Instruments [9], and discrete 'mesa' nanodevice design and circuit demonstrations by AT&T Laboratories [10] and Fujitsu [11].

Modern electronic materials growth techniques such as molecular beam epitaxy (MBE) provide interfaces with close to atomic-scale precision in the growth direction. Lateral patterning of these structures with fabrication techniques utilizing beams of energetic particles or radiation with etching are evolving well into the nanometer scale. Thus the length scales required for nanoelectronic devices (< 100 nm) can be achieved in the research environment. The advent of the scanning tunneling microscope (STM) [12] and atomic-scale fabrication using the principle of scanning tunneling microscopy have added a new dimension to fabrication techniques leading toward atomic resolution [13–15]. Analogous to the term nanoelectronics, the term nanofabrication is appropriately applied to modern epitaxial growth and lateral patterning techniques that are capable of making artificial structure with sub-100 nm feature sizes.

Section 20.2 describes nanofabrication tools and processes. The ability to fabricate small devices is the chief limitation to the minimum size for a nanoelectronic device. The lithographic pattern definition step is described by both extensions of existing techniques as well as new developments such as proximal probe lithography. The section begins with a discussion of resists, the radiation sensitive materials into which a pattern is defined with a masked or focused radiation source. The advantage of a resist based fabrication process is that a significant degree of flexibility is introduced in the materials and processes that can be used for nanofabrication. Although, at first glance, it may appear simpler to develop a 'resistless' fabrication technology, it is then necessary to develop specialized lithographic and pattern transfer techniques for each different material system. It is probable that electron-beam, ion-beam, and x-ray lithography can be successfully refined for applications at least down to 10 nm. Some key issues are discussed but there is every expectation that they will be overcome in a timely fashion for the development of nanoelectronic device concepts. A new lithographic technology that is particularly suited to the 'nano' regime (sub-10 nm) is that based on proximal probes such as the scanning tunneling microscope (STM). STM provides a means to manipulate material on the molecular scale. The advances in this area are very exciting. Section 20.2 concludes with a discussion of pattern transfer, the transfer of the lithographic pattern in a layer of resist into an electronic material. Metal lift-off and the variety of etch processes are described together with their limitations. A key problem is that of surface and sub-surface electronic damage that is introduced by energetic etch processes and the large surface to volume ratio of nanostructures. Innovative approaches that are at the forefront of materials science research are outlined.

Section 20.3 is devoted to nanoelectronic devices. At this stage several device concepts based on the physics of the nanometer-scale regime are being pursued. Broadly speaking they can be divided into those that depend on quantum effects (i.e. quantum confinement, interference, or tunneling) and those based on single-electron effects (i.e. Coulomb blockade). Of the quantum effect based devices, resonant tunneling transistors are the most 'advanced' in terms of being closest to implementation in commercial applications. The device action depends on the classically forbidden tunneling of electrons through an electronic barrier. With an appropriate bias the electron energy levels line up and resonant tunneling occurs. As the bias is changed in either direction, the resonance condition is lost, resulting in an increase in resistance of the structure. Various implementations and refinements to this class of device are described in some detail, particularly those suited to manufacturing production. Single-electron devices are intriguing as they suggest the ultimate degree of miniaturization of an electronic device: switches or memory elements controlled by a single electron. This class of device is based on the finite amount of energy which is required to move an electron onto a small capacitor. With suitable ingenuity the device can be designed so that the passage of one electron will impede the passage of further electrons until an external bias is changed. However, this type of structure imposes a considerable set of challenges in moving beyond the proof of principle demonstrations. Their first applications will probably be found in specialist high-precision circuits where cryogenic operation is acceptable. Room-temperature operation will require device dimensions below 10 nm. The section concludes with a description of device simulation. This is becoming an increasingly viable and attractive method of pursuing nanoelectronic device concepts which cannot be fabricated with any practical yield. It is of course necessary to include the requisite physics into these simulation programs. The combination of this approach with carrier based Monte Carlo techniques presents a promising approach to the development of a tractable simulation technique which will include the relevant properties (such as quantum effects, imperfect surfaces, and interfaces) of a real nanostructure or nanodevice. The chapter concludes with a short summary where the authors' perspectives on the key issues facing the development of nanoelectronics are discussed.

20.2 NANOFABRICATION

Nanofabrication involves the lithographic formation of a pattern on an electronic material and the transfer of that lithographic pattern to form an electronic or optical device. A schematic description of two fabrication processes is shown in figure 20.2. First, a pattern is formed in an imaging layer (resist) on top of semiconductor substrate by selective area irradiation. The radiation can be in the form of electrons, ions, ultraviolct photons, or x-rays. Patterned regions of the resists are then selectively dissolved in a chosen solvent. This process is known as development. Resists are either positive tone, where the exposed regions are dissolved, or negative tone, where the irradiated regions are insoluble. The resist pattern can either be directly etched into the underlying substrate (or device layer) or transferred to a material subsequently deposited in the developed resist pattern, as shown on the left. The deposited material pattern can either be used directly in a device or as an etch mask. This section is divided into three parts: resists, lithography, and pattern transfer. Several approaches for each step are discussed and compared. The physical processes that determine the limitations of nanofabrication are outlined in detail, below. The discussion concentrates on processes and approaches for the fabrication of electronic/optical structures of 100 nm or smaller.

20.2.1 Resists

The resist is the imaging layer material on the surface of the substrate in which the pattern for a device is defined. The lithographic process involves creating a chemical change, through selective irradiation of a radiation sensitive material. The key requirements for a resist in nanolithographic applications are high resolution, high contrast, and compatibility with the transfer of the pattern to the device. Commercial resists are polymeric materials, due to their facile deposition, high sensitivity, and flexibility. Radiation changes the molecular weight of the polymer by chain scission (positive resist) or cross-linking (negative resist) which results in a change in solubility of the resist in the developer. In the case of a positive resist, the irradiated lower-molecular-weight material dissolves in the developer. In a negative resist, the irradiated material is insoluble and remains on the substrate during development. In polymethylmethacrylate (PMMA), a high-resolution positive resist, linewidths of 5 nm have been reported [16]. Generally cross-linked polymers are more robust to solvents and other chemical environments than the linear polymeric chains used in most positive resists. However, negative resists generally exhibit lower contrast and resolution than the positive resists. The recently developed chemically amplified resists offer the advantage of a cross-linked material with high contrast and state-of-the-art resolution [17]. In these resists, the irradiation activates an acid catalyst (negative resist) or an acid inhibitor (positive resist). Following exposure, the acid, in the acid activated regions, catalyzes a cross-

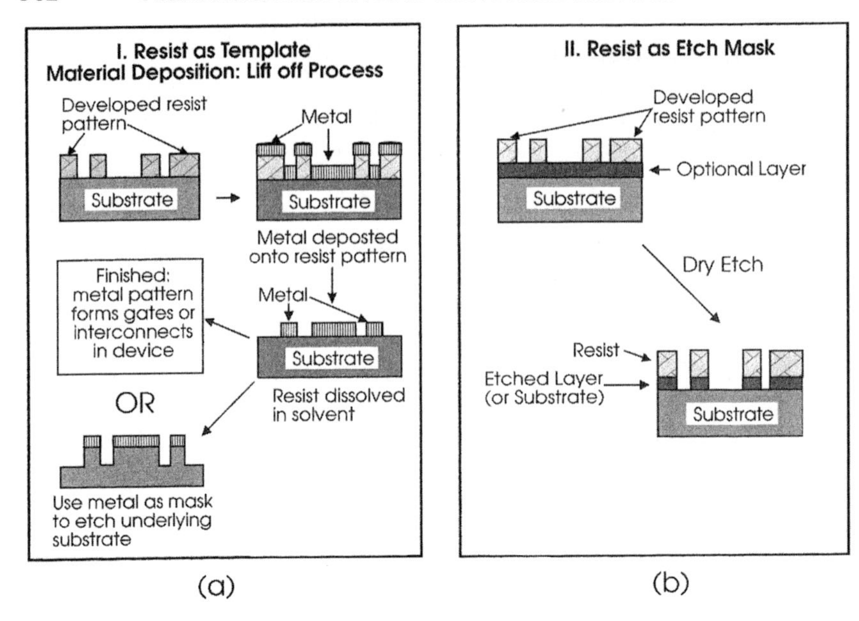

Figure 20.2. A schematic flow chart of two widely used nanofabrication processes. A pattern is first defined in a radiation sensitive material (resist). The post-exposure flow chart on the left illustrates the processing sequence for a lift-off process. The one on the right illustrates the use of the resist as an etch mask. (If the resist is not a sufficiently robust etch mask (as most frequently the case) the pattern must first be replicated in a suitable etch mask.)

linking reaction, during post-exposure annealing. The cross-linked regions of the positive or negative tone resist are insoluble in the developer, as in the case of conventional negative resists [18]. Linewidths under 25 nm have been reported in a chemically amplified resist [19].

Several inorganic resists, such as the metal halide salts [20], metal oxides [21], and semiconductor oxides [22, 23], undergo a radiolysis reaction under electron irradiation. Feature sizes of 1–5 nm have been produced in NaCl, AlF_3, LiF, and MgF_2, on Si_3N_4 membranes [20]. The metal halogen salts can be self-developing or developed by selective dissolution in water. However, the sensitivities of the resists are low (10^{-3}–20 C cm^{-2} as compared to 10^{-5}–10^{-6} C cm^{-2} for commercial resists) and the transfer of the pattern to the semiconductor has not been demonstrated. More recently, semiconductor oxide resists have been under investigation in several research laboratories. Semiconductor oxide resists are attractive because they can function as both resist and etch mask and are compatible with subsequent semiconductor processing. The oxide can be patterned through electron-beam exposure, STM lithography, or milling with a focused ion beam. (These will be discussed below in detail.) The exposed resist can be self-developing or developed through preferential etching [24]. Linewidths of 10 nm and below have been electron-beam patterned in oxide and etched into the underlying Si wafer [23]. Surface oxides on GaAs have been used as a mask for etching the substrate with thermal Cl_2, *in situ* in a MBE system, prior to regrowth [22, 25]. STM lithography has produced dimer wide lines of oxide on Si but these have not as yet been transferred [26].

An exciting area in resist research is surface layer imaging with monomolecular layer resists. These materials represent an entirely new class of materials made by molecular engineering [27]. The molecules can be engineered to possess the functionality at one end to chemically bond with the substrate surface. The other end of the molecule is designed so that its functionality is modified by the radiation in the lithographic tool and subsequent layers can be selectively attached. Since the molecules (and subsequent layers) bond to the surface, they are thermodynamically driven to coat the surface. The monomolecular layers can be easily deposited by spinning or dipping the substrate into a beaker. The use of self-assembling materials is attractive for very high-resolution lithography because the sizes of the molecules ($\sim$ 1 nm in length) are much smaller than commercial polymeric resists and they offer the possibility of much higher resolution. In addition, the self-assembling monolayers (SAMs) produce an ultrathin, homogeneous, densely packed layer [28]. Such schemes are particularly well suited to proximal probe lithography which is envisioned as a tool for sub-10 nm lithography on bulk substrates. The resist can be self-developing or the irradiation can destroy the functionality of chemical groups on the surface or of a thin monolayer on the surface. In the former case, 50 nm lines have been transferred to a GaAs substrate from an electron-beam patterned self-assembled monolayer resist [29]. In the latter case, the non-irradiated regions of the surface can be selectively plated with a metal [27, 30, 31]. The deposited metal layer is of sufficiently high quality to be used as an interconnect or electrode or as a very robust mask to subsequent etching of the underlying thick resist or semiconductor. Lines in the metal pattern of width 15 nm have been etched into the underlying Si substrate [58].

20.2.2 Lithography

Lithographic patterns are formed by the exposure of a resist material to a type of radiation. Sub-100 nm features have been patterned with electrons, ions, and x-rays. The first two methods involve the serial writing of a pattern with a small focused probe. The small probe is formed through an electrostatic/magnetic lens system or by proximity to an STM tip. In x-ray lithography, an entire mask pattern is printed at one time by a flood exposure of x-rays through a mask or reflected from a mask (in the case of projection x-ray lithography).

Electron-beam lithography has been the most widely used method of fabricating nanostructures. This has largely been due to a mature electron optics technology base and the physics of electron-beam–resist interactions permitting very small features to be formed. Ion optics has not been as well developed, but the wide variety of ion processes in materials make it a versatile tool for nanofabrication, as discussed below. X-ray lithography has the advantages of speed due to the parallel nature of the process. Although a nanofabrication tool, the technique has not been demonstrated to show as high resolution as electron-beam lithography. Parallel flood exposure lithography with electrons [32] or ions [33] has been demonstrated, but the lithographic systems are under development. The newest lithographic tool is the STM, which at present is a research tool. It is a very rapidly growing field and offers a range of potential lithographic applications. These techniques are discussed below.

20.2.2.1 Electron-beam lithography

Electron-beam nanolithography employs a highly focused Gaussian-shaped beam of electrons of energy 2 keV to 300 keV, which writes a pattern in a resist layer on a semiconductor substrate. Until recent STM work [13], the smallest lithographic patterns (1–5 nm in fluoride resist [20], $\sim$5 nm in oxide films [21] and 5 nm in polymeric resist [16]) had been written with electron-beam lithography. Electron-beam nanowriters can be made by modifying a scanning electron microscope (SEM) or a scanning transmission electron microscope (STEM), or can be purchased commercially.

The major limitation in electron-beam lithography is that the exposure profile is larger than the electron probe and determined by electron scattering. The beam broadens in the resist by small-angle elastic scattering. The resist is also exposed by electrons elastically backscattered from the substrate. In addition electrons from inelastic processes play an important role. The forward scattering results in an enlarged feature size. The electrons from the substrate cause the feature size and point to point (or line to line) resolution to be a function of pattern density. These are known as proximity effects. Small-angle forward scattering in the resist can be minimized by using thin resists and high electron-beam voltages. For example, the beam broadening of a 50 kV Gaussian beam in 100 nm of resist is < 1 nm [34]. Although many nanostructures and nanodevices have been defined in this manner, the thin resist limits the pattern transfer options and has too many pinholes for manufacturing applications. The number or area density of electrons elastically scattered from the substrate can be minimized by the use of very high-voltage beams and membrane substrates.

In most practical applications, one is interested in writing patterns on bulk substrates rather than on a membrane. Resist exposure due to electrons scattered from the substrate is a difficult problem to minimize. The higher the atomic weight of the substrate material, the more severe the problem. There are two approaches to the problem. One is to use very high-energy electron beams (≥ 50 keV), which penetrate deep into the substrate. In this case, the resulting backscattered electrons are scattered over a very large area relative to the incident probe. The diffuse fog dose (per unit area) of backscattered electrons is lower than that of the primary electron-beam exposure and does not expose a high-

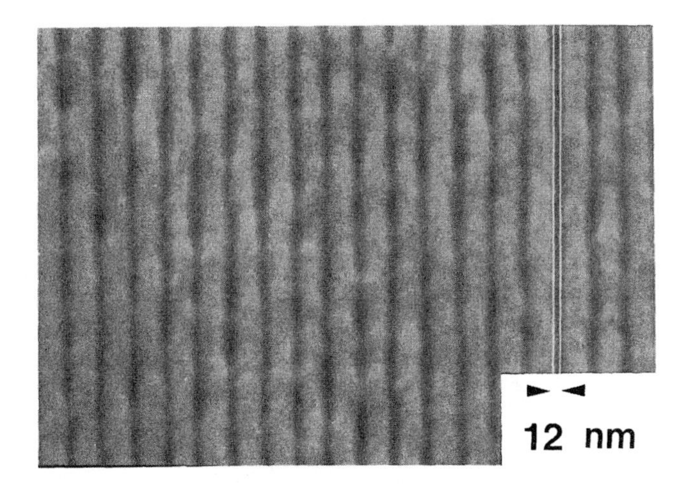

Figure 20.3. 50 kV electron-beam nanolithography. 12 nm lines in a 60 nm period grating in 60 nm thick PMMA on W [35].

contrast resist. With a high-contrast resist, it is possible to write 12 nm lines in a 60 nm period grating on a high-atomic-number material, such as W with a 50 kV electron beam, shown in figure 20.3 [35]. The other approach is to use very low-energy electrons that do not scatter over large distances. The reduction in proximity effects with a 2 kV electron-beam exposure of PMMA on Si has been illustrated by the report of a constant gate length both between large (> 10 μm) source and drain pads [36]. A similar absence of proximity effects was found with even lower-energy electrons (< 50 eV) by STM lithography as discussed below. The key issues in the high-voltage approach are scattered electrons, electron irradiation damage to the semiconductor, and the complexity and cost of high-voltage columns. The key issues with the low-voltage approach are the electron optics, the sensitivity of the electron optical column to noise, and the need for ultrathin resists.

Our understanding of resist exposure is based on the energy deposited in the resist by the electron beam. Monte Carlo simulations calculate the energy deposited into the resist by scattered primary electrons, backscattered electrons, and fast secondary electrons. The exposure of the resist is expected to be proportional to the amount of energy deposited in the resist. Our present understanding of resolution in a resist breaks down with dimensions below 100 nm and in applications to resists other than PMMA. Although in the case of PMMA, linewidths of 8–10 nm can be fabricated, the line-to-line spacing is three to five times larger. The resolution of other polymeric commercial resists is not as good as illustrated by the resolution of Microposit SAL-601, a chemically amplified negative polymeric resist available from the Shipley Corporation. With a focused 50 kV electron-beam that could write 12 nm lines on a 60 nm period grating in PMMA on a W substrate, the smallest linewidths in Microposit SAL-601 were 50 nm in a 100 nm period grating [35]. Our results suggest that Microposit SAL-601 may be more sensitive to low-energy secondary electrons than PMMA. Shown in figure 20.4 are the results from a Monte Carlo simulation of resist exposure by fast secondary electrons generated by interaction of the primary electron beam with the substrate. It can be seen that most of the fast (2–8 keV) secondary electrons enter the resist within a 100 nm radius of the incident beam. A thin (50–300 nm) layer of a dielectric of silicon nitride or silicon dioxide substantially reduces the secondary electrons entering the resist and dramatically improves the resolution of Microposit SAL-601 [35]. Finally, the effects of photoelectrons generated both within the resist and at the substrate surface on a resolution of dimensions of about 10 nm has not been adequately addressed.

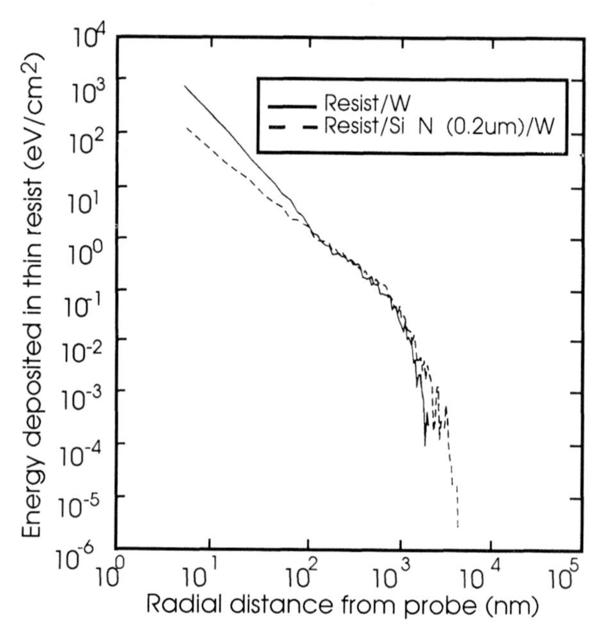

Figure 20.4. Energy deposited in resist per unit area by fast secondary electrons as a function of radial distance from a primary electron beam consisting of a point source of 50 kV electrons. The effect of incorporating an intermediate SiN_x layer is shown [35].

20.2.2.2 *Ion-beam lithography*

Focused ion-beam (FIB) lithography is performed by writing with a focused Gaussian ion beam. The main advantage of FIB patterning is the versatility of the technique. Lithographic patterns can be written by one of several processes including the exposure of a resist [37], ion milling [38], ion induced etching [39], ion stimulated deposition of material [40], or selective area ion implantation [41]. True nanolithography has been demonstrated by 8–10 lines written in 30 nm of PMMA with a 50 kV Ga^+ ion beam generated by a custom-built FIB system [37]. Sub-100 nm deposited metal patterns written by FIB interaction with precursor gas have been reported [40].

Ions are not expected to scatter as far as electrons in solids and the proximity effects due to elastically scattered electrons are not expected with ion-beam lithography. However, the key issues with ion-beam lithography are ion sources, ion optics, ion damage, and straggle in material. Current optics technology does not permit ions to be as finely focused as electrons and FIB systems typically have large tails on the Gaussian beam. In the laboratory, recent work has improved ion sources from a 6 eV energy spread to a 1 eV energy spread, which substantially reduces chromatic aberrations in the system [42]. However, ion straggle [43] in solids results in a Gaussian profile in depth and a graded error function profile laterally about an edge. This complicates the definition of the implanted patterns. In addition, ions are far more damaging to the underlying semiconductor material than other forms of radiation. This will be discussed in more detail below (in section 20.2.3.2).

20.2.2.3 X-ray lithography

In proximity x-ray lithography (XRL) a patterned mask is held close to an underlying resist coated wafer. The resist is exposed by a flood exposure of x-ray radiation through the mask. The x-ray source can either be a point source or a synchrotron. XRL is attractive as a high-speed method of printing nanostructures. Unlike the serial writing processes necessitated by the focused beam techniques, one can print an entire mask in one flood exposure, much in the manner of optical lithography. The shorter wavelengths of x-rays allow much greater resolution than conventional optical lithography. Lines of width 30 nm have been printed by contact proximity XRL, in which the mask was placed in contact with the wafer [44]. Furthermore, unlike the projection optical lithographies there is no lens system between the mask and the wafer and the mask pattern is proximity printed into the resist on the underlying wafer. The advantages of this lens-less system are no lens distortions and high lithographic resolution, without the loss of the depth of field encountered with conventional UV optical lithography. This allows for printing patterns on nonplanar surfaces. The large penetration depth of the x-rays and the large depth of field allow the use of much thicker (1 μm) resist than can be used with the other techniques of nanolithography. The major challenge of proximity XRL is that there is no pattern demagnification and 1$\times$ nanolithography requires severe tolerances in the mask fabrication.

The key issues in the application of the technique to nanolithography are diffraction limits and the compromise between high-resolution x-ray masks, and x-ray source wavelength. Ideally, the minimum feature size, P_{min}, is given by [45]

$$P_{min} = K\sqrt{\lambda G}$$

where K is a constant ≈ 1.6, λ is the wavelength of the x-rays (1 nm), and G is the distance between mask and substrate (5–40 μm). Obtaining high resolution is a trade-off between minimizing the gap, reducing the wavelength, and the ability to fabricate the mask. The gap is limited by how closely one can place the mask to the wafer with precision and without cross-contamination. It is not clear whether the technique can be pushed below the 30–50 nm feature sizes [45].

The fabrication of the x-ray masks is a critical factor in XRL. The mask material must be transparent to x-rays, which necessitates a membrane material. The absorber must produce high-contrast images and not place stress induced distortions on the membrane. The shorter the wavelength, the thicker the absorber that must be patterned on the x-ray mask, which is a nanofabrication challenge. For manufacturing purposes, one would like the critical dimension control, pattern placement, and overlay accuracy to be 10% of the critical dimension size. Sub-100 nm dimensions would require sub-10 nm precision in pattern placement and linewidth control (both along the length and in cross-section). The pattern is written in resist on a blank x-ray mask by electron-beam lithography. The absorber is most frequently formed by electroplating a $\approx$300–500 nm thick Au absorber in the resist pattern. This process requires electron-beam patterning of $\approx$0.75–1 μm thick resist layers, which limits the resolution of the x-ray mask [46]. Currently subtractive absorber schemes are under development, where the resist pattern is transferred into the absorber with a dry etch [47, 48].

To circumvent the problems associated with proximity 1$\times$ XRL, projection XRL is being developed [49]. The pattern demagnification (5–20$\times$) between mask and wafer relaxes the severe precision constraints required for 1$\times$ x-ray masks. The key problem areas are in the fabrication of the mirrors and mask materials and the thin resists required because of the short penetration depth of the 14 nm source. Both approaches to XRL are summarized in [45].

20.2.2.4 Proximal probe lithography

The newest and most innovative additions to the nanolithography arena are the proximal probes [51]. Although the demonstrations of the technique are still clearly restricted to the laboratory, proximal probes offer the possibility of true *nano*lithography. Lithographic patterns have been defined by proximal probes in resist, surface oxide modification, atomic manipulation, electrochemical etching, and chemical vapor deposition. In addition proximal probe source micromachined electron-beam columns have been envisaged for low-voltage electron-beam columns for electron-beam lithography [51]. Proximal probes offer the unique ability of *in situ* analysis while performing the lithographic modifications. This presents exciting scientific opportunities to study the lithographic process.

STM-based electron-beam lithography has been performed in resist on a substrate by several workers [14, 15, 52]. In this method the STM tip provides a small probe of low-voltage electrons to expose a wafer coated with a thin (< 80 nm) layer of resist. In all reports to date, the STM was operated with voltages above ~ 6 V, in the field emission mode rather than the tunneling mode. With the STM operated in the constant-current mode, the tip–sample separation is roughly proportional to the applied voltage. As the effective probe size is determined by the close proximity of the tip and sample, the probe size also scales with the tip–sample voltage. In this manner electron-beam lithography without electron scattering can be performed with a small low-voltage probe of electrons. The first demonstration of the improved resolution and the absence of proximity effects is shown in figure 20.5 [15]. Here are two lithographic patterns defined in the 50 nm thick Microposit SAL-601 resist on Si. On the left is a pattern defined with a tip–sample voltage of -25 V in the STM; on the right is a 100 nm line-to-line grating defined with a 15 nm diameter, 50 kV electron probe. The STM pattern shows both a smaller feature size (~ 40 nm) and closer line-to-line spacing (~ 55 nm) than the pattern written at 50 kV. The resist was processed identically in both cases. Using a lower tip–sample voltage and 30 nm thick resist, 23 nm lines have been written in the same resist. A limitation in this technique results from the tip–sample current being used to both expose the resist and drive the STM feedback loop to maintain a constant tip–sample current. As a result, the lithography is limited by the thickness and conductivity of the resist and the substrate surface. Some of these limitations can be circumvented by performing the lithography with a conducting atomic force microscope (AFM) tip [52]. In this case, the tip position is controlled by the tip–sample force, i.e. independently of the tip-sample current and voltage required to expose the resist.

To fully utilize the pattern resolution available with a proximal probe, novel resist and imaging layer schemes are an active area of research. Such resists include self-assembling monolayer (SAM) resists [53] and surface oxide [26, 54–56]. A robust process using organosilanes with end groups that can undergo reaction and bind to an Ni film was demonstrated by Marrian *et al* [53] and Perkins *et al* [57]. The Ni served as a robust etch mask and 15 nm lines were etched into the underlying Si [58]. The STM has also been used to directly pattern oxide on passivated Si and InGaAs substrates. In this case, a tip–sample voltage pulse of greater than 5 V induces desorption of the hydrogen passivation, thus allowing oxide to grow. The oxide patterns could be imaged *in situ* with the STM, operated at lower voltage, or with an AFM. The oxide pattern can be used as a mask for the sample during subsequent etching [54–57].

Simple electronic structures have recently been fabricated with proximal probe lithography and their transport properties measured, most notably by de Lozanne and co-workers at the University of Texas in Austin [59] and van

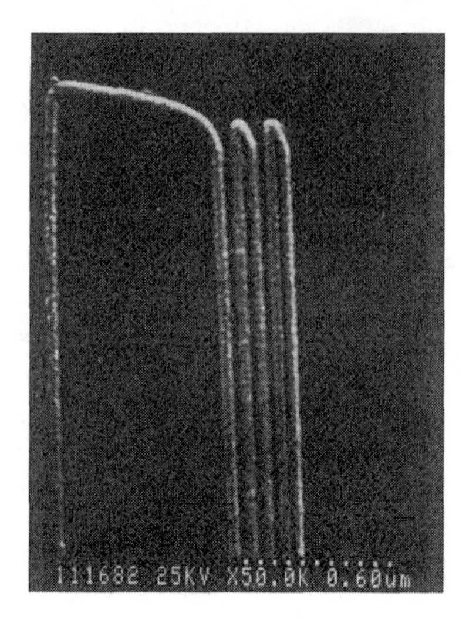

Figure 20.5. SEM micrographs of identically processed 50 nm films of Microposit SAL-601 (by Shipley Corporation) resist on Si. The pattern on the left was written with an STM at -25 V, the pattern on the right with a tightly focused (diameter $\approx$ 15 nm) 50 kV electron beam [15].

Haesendonck and co-workers at the University of Leuven in Belgium [60]. Using different approaches, thin metallic wires were written between optically defined pads. De Lozanne's group used STM stimulated decomposition of an Ni carbonyl precursor whereas van Haesendonck's group patterned a Langmuir–Blodgett film which acted as an etch mask for ion milling of the underlying Au film. This work demonstrates that the considerable problems of aligning a proximal probe lithography with previously defined structures can be overcome. In a similar vein, the latent exposure image in exposed polymeric resists has also been observed with both STMs [61] and AFMs [52] indicating the potential for implementing proximal probe lithography in mix and match lithographic schemes for practical definition of nanoelectronic structures. More recently, Campbell *et al* used an STM patterned oxide to etch a 30 nm poly-Si gate to form a working field-effect transistor (FET) device [61]. Lyding *et al* have demonstrated patterned oxide lines on Si that were two dimers wide [26], which are smaller than the smallest electron-beam patterned features. It is only a matter of time before working devices of those dimensions are made.

20.2.3 Pattern transfer

To fabricate a device, the facility to incorporate the resist pattern into the actual devices and circuits is crucial. This section describes several methods by which the lithographically defined pattern can be transferred to devices or nanostructures In the most straightforward approaches the resist pattern on a semiconductor or insulating layer is transferred to metal wires or gates or etch masks. In the case of an etch mask the pattern is etched into the underlying substrate. More subtle approaches involve implantation, clever growth techniques, or etching combined with regrowth.

20.2.3.1 Metal gates or wires by evaporation and lift-off

Patterned metal gates on top of a semiconductor FET-type material remain the most widely employed method to achieve patterning on the nanometer scale, as discussed in section 20.3.2.2. Metal patterns on the semiconductor surfaces are also used as etch masks, for etching of nanostructures into semiconductors, since thin resists, required for high-resolution lithography, offer little etch resistance. (Etching processes are discussed below.) The metal patterns are formed from the resist patterns with a 'lift-off' process. The lift-off technique is illustrated in figure 20.2. One begins by depositing metal (or other desired material) on top of a developed pattern in positive resist. The sample is dipped into a solvent, (i.e. acetone) which dissolves the resist and 'lifts off' the metal on top of the resist, leaving behind the metal pattern on the substrate (opposite tone to the resist pattern). Crucial to this technique is a discontinuity in the metal between that covering the substrate and that covering the resist. This implies sharp resist walls, undercut if possible, and a metal layer substantially thinner than the substrate layer (usually less than half the thickness of the resist). In nanolithography, the resist layer is most frequently 50–100 nm. For FET applications, this severely limits the conductivity of the gate. One method to increase the gate conductance is to make a T-gate or Γ-gate through lift-off with a tri-level resist [63]. For a lift-off process, the metal must be deposited anisotropically so that the sidewalls of the resist are not coated. Such anisotropy can be achieved by evaporation, if the source is sufficiently far from the wafer. Sputter coating and chemical vapor deposition are rarely as successful, since the processes tend to coat the resist sidewalls and heat the resist. Evaporated metals must have a low melting point so that the source does not radiatively heat the resist, and cause the sidewalls to flow. Furthermore, in nanometer-scale applications, the metal grain size must be less than the desired feature size. The technique has the advantage of little or no damage to the underlying semiconductor device.

The first low-dimensional device was formed by gate confined channels in a metal-oxide-semiconductor field effect transistor (MOSFET) structure [64]. Since then, several papers on mesoscopic physics have been reported on gate patterned modulation doped field effect transistor (MODFET) devices (reviewed in section 20.3.2). In the electric field confinement method, the sample is patterned with a series of gates across the top surface of layered material. A potential is applied to the gates, which depletes carriers in the material and effectively patterns the carrier channel beneath. The advantages of this method are simplicity and no bare semiconductor sidewall surfaces. The chief limitations of the technique are flexibility, softness of confinement potential, capacitances as the dimensions shrink, and gate conductivity. Furthermore, there are many compound semiconductors that cannot be gated. Hence the nanofabrication 'bag of tools' requires flexibility and the development of other processes.

20.2.3.2 Dry etching processes

Another widely employed method of pattern transfer is the selective removal of material. The desired characteristics of such a process include selectivity, anisotropy, and little damage to the electronic device. The consequences of low selectivity are erosion of the masking material and lack of etch stop control. The consequence of isotropic etching is that the horizontal etch rate is comparable to the vertical etch rate. In nanoscale pattern dimensions this means lack of lateral dimension control or determination. More catastrophic is the total loss of the feature due to lateral etching. Except for crystallographically selective etches, wet etching is isotropic. In contrast, accelerated ions are highly directional and are used almost exclusively in nanofabrication. The simplest process is to ion mill an electron-beam defined pattern in carbon contamination or cross-linked PMMA into an underlying material. This method is most frequently used for metals, because they sputter rapidly. The first demonstration of an Aharonov–Bohm [65] interference device was in a thin ion milled metal ring on top of a dielectric membrane [66]. The technique is limited by the erosion of the masking layer relative to the desired sputtering of the metal. Furthermore, ion milling is usually too damaging to be used in the fabrication of most semiconductor devices. To reduce the damage and increase the selectivity of the process chemically reactive (and selective) gases are used. Etching with reactive ions has been widely applied to semiconductors. Less success has been achieved in applying reactive gas etching to metals. Common reactive ion etching techniques are reactive ion etching (RIE), electron cyclotron resonance (ECR) reactive ion etching, chemically assisted ion-beam etching (CAIBE), and reactive ion-beam etching (RIBE).

RIE employs an ionized reactive gas plasma to etch the material. In this process, a radio frequency (RF) power is applied to the substrate electrode in a parallel-plate system. The RF field excites a plasma discharge. The key to anisotropic RIE is to use low enough gas pressures and high enough voltages that the mean free ion path is greater than the height of the dark-field region at the lower electrode. This means that the ion strikes the substrate before it has lost its directionality through elastic collisions with the gas molecules in the chamber. The lowest pressure at which an RF induced plasma is stable in standard RIE is in the 1 to 10 mT range. It operates at the highest pressure of the dry etching techniques, mentioned here, and undercutting can be a severe problem. The undercutting can be minimized by choosing a chemistry that passivates the sidewalls or low substrate temperatures. The ion energies, ranging from about a hundred volts to the kilovolt range, result from the RF induced d.c. bias on the base electrode and any applied d.c. bias. The ion energies in RIE are more difficult to control than in the other ion etching techniques. The technique has successfully produced optically active III–V nanostructures [67].

In ECR etching, the plasma is excited by a microwave source in resonance with large magnets which act to confine the plasma. As a result, a high-density plasma can be produced with operating pressures of 0.5 to 1 mTorr and ion energies as low as 20 eV. The ECR source creates a higher density of reactive radicals than RIE. The advantage of the technique is low ion damage to the semiconductor with a high etch rate. With identical gas chemistry and 1 mTorr pressure, the ECR has been found to etch at five times the rate of RIE [47]. Because of the low ion energies and high density of reactive radicals, undercutting can be a severe problem, as in the RIE case. Here, the undercutting can be minimized by the choice of chemistry, the sample position relative to the ECR plasma, and substrate bias. The technique has successfully etched nanostructures in HgCdTe, shown in figure 20.6 [68]. The application of ECR etching to nanofabrication is still newly emerging.

In the ion-beam techniques, a gas is leaked into an ion gun assembly (several types are commercially available) and a beam of ions is generated. The operating pressure ($\sim 10^{-4}$ Torr) is the lowest for the dry etching techniques and high ion energies (500 eV to 5 keV) are employed. Extremely directional ion etching can be achieved with aspect ratios as high as 40:1, as reported in the literature [69]. In the CAIBE configuration, an energetic beam of inert gas ions is generated and a neutral reactive gas is leaked into the vacuum chamber near the sample. In the RIBE case, a beam of energetic reactive ions is generated in the gun. Most work has been done with the CAIBE technique, due to corrosion of the ion gun from undiluted reactive gases in the RIBE case. Technology is producing more corrosion resistant guns and RIBE may become more widely used.

The major issues for ion etching are ion damage and the creation of a surface. The damage is the result of sputtering processes and is manifested as erosion of the etch mask and creation of traps for the carriers in semiconductors. Generally, the more chemistry and the less sputtering in the ion interaction, the greater the selectivity and the lower the ion damage [70]. This is accomplished by the use of appropriate chemically reactive gas and low ion energies. The amount of damage depends on the material, type of damage, and sensitivity of the measurement. Mobility and photoluminescence measurements of high-electron-mobility transistor (HEMT) and quantum-well material (undoped) have shown depletion of carriers 100 nm beneath the etched surface [71]. Raman and photoreflectance measurements in doped GaAs have shown the depletion of 10^{18} cm^{-3} carriers extending 30 nm beneath the surface [70]. Etched nanostructures possess a large surface-area-to-volume ratio and the electronic properties are determined both by surface science and etching damage. Even under ideal conditions, semiconductors such as GaAs and Si have surface states that trap carriers, pin the Fermi level, and can cause carrier depletion lengths comparable to the size of the nanostructure. In spite of this, excitation

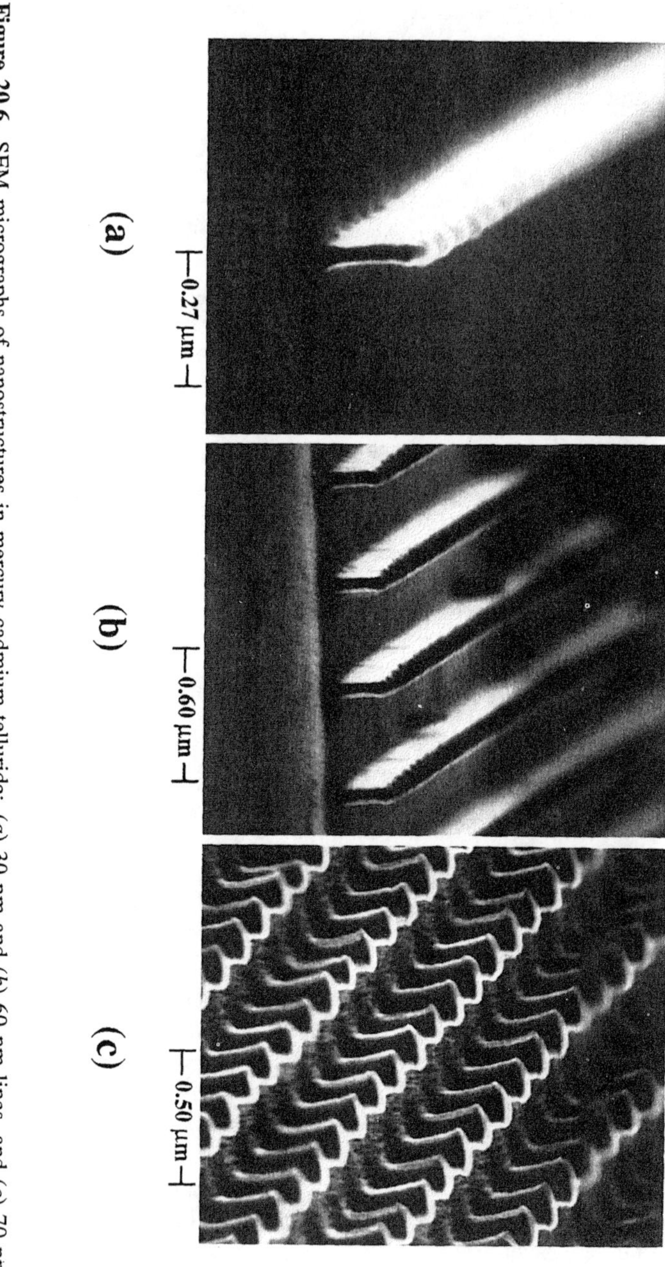

Figure 20.6. SEM micrographs of nanostructures in mercury cadmium telluride: (*a*) 30 nm and (*b*) 60 nm lines, and (*c*) 70 nm diameter dots. The patterns were written by electron-beam lithography and etched into the mercury cadmium telluride with a CH_4/H_2 ECR RIE process [68].

spectroscopy [67, 72] and photoluminescence [73] have shown that, in some instances, quantum wires and dots with dimensions below 50 nm still have carriers. Pattern replication is a critical area in nanofabrication.

20.2.3.3 Novel approaches to pattern replication

There have been many innovative approaches to the fabrication of nanodevices. The key idea behind theses methods is to embed laterally patterned nanowires (channels) or nanostructures in a layered heterostructure. This would achieve carrier confinement in two or three dimensions, rather than only one, as available with current epitaxial growth techniques. In this manner one can envisage three-dimensional heterostructural material and the creation of a free surface is eliminated. Ongoing research efforts include etching combined with surface passivation and/or regrowth [28, 29], interdiffusion patterning [39, 74], and growth of quantum dots and wires on stepped surfaces [75].

20.3 NANOELECTRONICS

As electronic device sizes decrease, conventional Si based MOS and bipolar IC technologies will reach a limit where the devices cease to function properly. For example, the definition of a p–n junction with sub-100 nm dimensions becomes questionable. In MOS devices with gate lengths below 100 nm, oxide thicknesses below 10 nm are required, which are difficult to grow with a suitable dielectric constant (i.e. the oxide is leaky) and uniformity. In dimensions below 100 nm, dielectric breakdown becomes a major issue for both p–n junctions and oxide in MOS. The fabrication of sub-100 nm complementary metal oxide semiconductor (CMOS) devices has recently been addressed by Wind *et al* [76].

On a more fundamental level, the device physics in conventional ICs obeys the law of large numbers and/or thermodynamic ensemble averages, neither of which apply in nanometer-scale dimensions. As conventional logic architecture becomes obsolete, science must examine new logic functions that exploit the newly available capabilities and produce devices that operate at room temperature. Conventional device dimensions are large compared to coherence lengths so that energy quantization, quantum interference, and discreteness of charge carriers do not play a significant role in device transport physics. Thus, in order for the down-scaling of electron devices to be realized in nanoelectronics one must consider quantum transport, tunneling and interference effects, and discreteness of electron charge in small semiconductor structures. Not only the down-sizing but also the new semiconductor materials available through MBE make the new physics attractive for device applications. The atomic-scale control of interfaces far exceeds that of oxides or p–n junctions. In addition, the layered semiconductor materials act as electron waveguides, having longer electron scattering lengths than more conventional semiconductor materials. The investigation of the properties of small structures is the common denominator between the research community from mesoscopic physics and nanoelectronics. Nanoelectronics research aims to apply the novel mesoscopic phenomena to the goals of continued down-scaling of IC components down to the atomic scale and the consequent up-scaling of computational complexity per chip [77]. Nanoelectronics is highly interdisciplinary and may eventually need cross-disciplinary approaches based on the mathematical, physical, chemical, and/or biological sciences to fully implement its goals [78].

The most useful device in solid-state electronics, which has brought the information revolution, is the transistor. The transistor is basically a three-terminal device which accomplishes two very important functions, namely, 'drive' and 'control'. The 'drive' function is accomplished by applying a large voltage difference across two terminals of the device, known as the source (emitter) and drain (collector). The 'control' function is implemented by simply controlling the current flow in a transport channel between the two terminals. Both functions are important in obtaining the 'gain' of the device. Electronic control of the current flow between the source (emitter) and drain (collector) is accomplished in two general ways:

(i) introduction of a control charge directly in the path of the current flow where the biasing charge voltage or current is applied to an electrode in contact with the 'base' region inside the transport channel;

(ii) depletion (or enhancement) of the charge carriers in the transport channel with an electric field generated by a voltage applied at the 'gate' electrode, electrically insulated from the transport channel.

Conventional transistors employing the first method above are called bipolar transistors and fall into two categories, i.e. the p–n–p transistor and the n–p–n transistor. In both cases, the biasing charge arises from the base–emitter current and has an opposite charge from the current carrying charge from the emitter to collector. Conventional transistors employing the second method are called unipolar transistors or FETs. Thus an electronic device which delivers voltage gain and current drive is basically a three-terminal device. Note that the role of the power supply is essential to obtain gain [79]. Because of the major role of lateral depletion regions in a conventional bipolar devices, these devices are less scaleable than field-effect devices. On the other hand, bipolar transistors are inherently high-speed devices compared to FETs by virtue of their low base–emitter capacitance. The prominence that three-terminal devices have in the electronics area is due to their ability to transmit and amplify (gain) a wide range of information signals and provide good isolation between input and output of a logic gate.

However, in the continued down-scaling of three-terminal device sizes and consequent up-scaling of complexity per chip, a 'wiring crisis' will result. As the number of devices per chip increases so do the number and net length of interconnections. Simultaneously, the cross-section of the wires will have to decrease to allow more communication paths per area in the chip. This will offset the benefits of faster device switching, since long connections and smaller wire conductivity will create longer delays. In fact, for non-ideal small metal wires, the resistance can go up exponentially as a function of its length. It is not yet clear whether research on novel computer architecture based on three-terminal devices or multi-valued logic will eventually solve this problem.

Research directed toward the development of a new general-purpose computational dynamics and novel computer architectures to avoid the wiring crisis are still at the embryonic stage [9, 77]. A direct approach to the interconnect problem is to scale down the interconnects as well, from the conventional architecture, by eliminating long interconnects per chip, through computerized circuit-layout optimization [7]. Another realistic approach may lie on the advances made in monolithic IC optical communication technologies, such as the one recently proposed by Yamanaka *et al* [80] using optoelectronic integrated circuit (OEIC) chipsets for chip-to-chip optical communications. Elimination of wiring interconnects, and/or interconnects whose number does not scale up with computational complexity per chip, is contained in various proposals which employ drastically different architectures. Texas Instruments has proposed the use of cellular automaton architecture [81] in which the devices are connected only to their nearest neighbor. What is intriguing about this architecture is that the coupling is not implemented by physical 'wires' but is accomplished through capacitive coupling between, say, neighboring quantum dots. However, the 'forces' and 'rules' of the new computational dynamics appropriate to a cellular automaton architecture that correspond to the two-valued logic in conventional transistor based computers are not entirely clear. Nevertheless, the ability to represent a bit of information by one electron in a quantum cellular automaton circuit is very intriguing, and clearly represents the ultimate efficiency in information representation. In any case, the quantum cellular automaton architecture idea should open up strong interest in information based mathematical, physical, and biological sciences to establish the necessary analytical knowledge base needed for a new general-purpose computational dynamics [9].

20.3.1 Nanodevices

This section focuses on the current research efforts in nanoelectronics, which are centered on nanodevices. These are essentially three-terminal nanodevices, or nanotransistors. The nanodevices, discussed below, involve multi-state switching phenomena, based on resonant tunneling, quantum interference (Bragg interference of coherent electron wavefunctions), and single-electron effects. The multivalued logic and negative transconductance distinguish nanoelectronics from conventional transistor electronics.

20.3.1.1 *Resonant tunneling devices (RTDs)*

Classically it is energetically forbidden for an electron to pass through an electronic barrier. However, if the barrier is sufficiently thin, there is a significant finite probability that an electron can pass through the barrier by the quantum mechanical process of tunneling. Consider a structure which has two potential energy barriers separated by a thin potential well, which has distinct energy levels. Here, there is a high probability (close to unity) that an electron can tunnel through the entire structure when it has an energy equal to an allowed energy state in the quantum well. Such a structure can be formed from a vertically layered, epitaxially grown, semiconductor material. Here the two barrier layers would consist of thin ($\sim$5 nm) layers of a higher-bandgap semiconductor material (for example $Al_{0.3}Ga_{0.7}As$, energy gap $\geq$ 1.9 eV) and the quantum well is a lower-bandgap semiconductor (such as GaAs or InGaAs, with energy gap $\leq$ 1.5 eV). The small vertical dimension of the quantum well ($\sim$ 10 nm) results in quantum confinement and the formation of discrete energy levels in the conduction band (and valence band) that can be occupied by electrons (holes). Between the quantized energy levels are gaps in the density of states. By changing the bias across the structure the relative positions of the energy levels can be changed with respect to the reservoirs of electrons on the outside of the barriers. The electrons tunnel into the quantum well when their 'longitudinal' energy outside the barriers matches an energy level in the quantum well. This is shown in figure 20.7(*a*) [82]. Thus, as the voltage bias across the double-barrier structure is varied, the current through the structure will pass through a series of maxima corresponding to the alignment of the energy levels of the quantum well with the energy of the source of electrons. Following a maximum, the current will decrease, as the energy of electrons in the reservoir outside the barriers lies in a minigap in the conduction band, as shown in figure 20.7(*b*). In this way a region of negative differential resistance is introduced into the device current–voltage (I–V) characteristic. Here we have a multi-state switch, in which current will flow (or not flow) at several resonant (nonresonant) voltages. It is an example of a device which cannot be described by thermodynamic Maxwell–Boltzmann statistics for a free electron gas.

Resonant tunneling phenomena have attracted the attention of the device community in the form of two-terminal GaAs/AlGaAs/GaAs diodes with significant voltages applied at the source and drain [83–85] (top and bottom of the layered semiconductor). An attractive feature is that resonant tunneling can operate at room temperature, whereas many quantum devices require cryogenic temperatures to operate. Furthermore RTDs, and other proposed quantum-based devices, exhibit an ultrasensitive response to voltage bias in going from the high-transmission state to the low-transmission state. This means that a very high transistor transconductance and ultrafast switching are obtainable. Indeed, numerical simulations of RTDs, to be discussed below, and microwave experimental results, indicate the intrinsic speed limit of RTDs to be in the

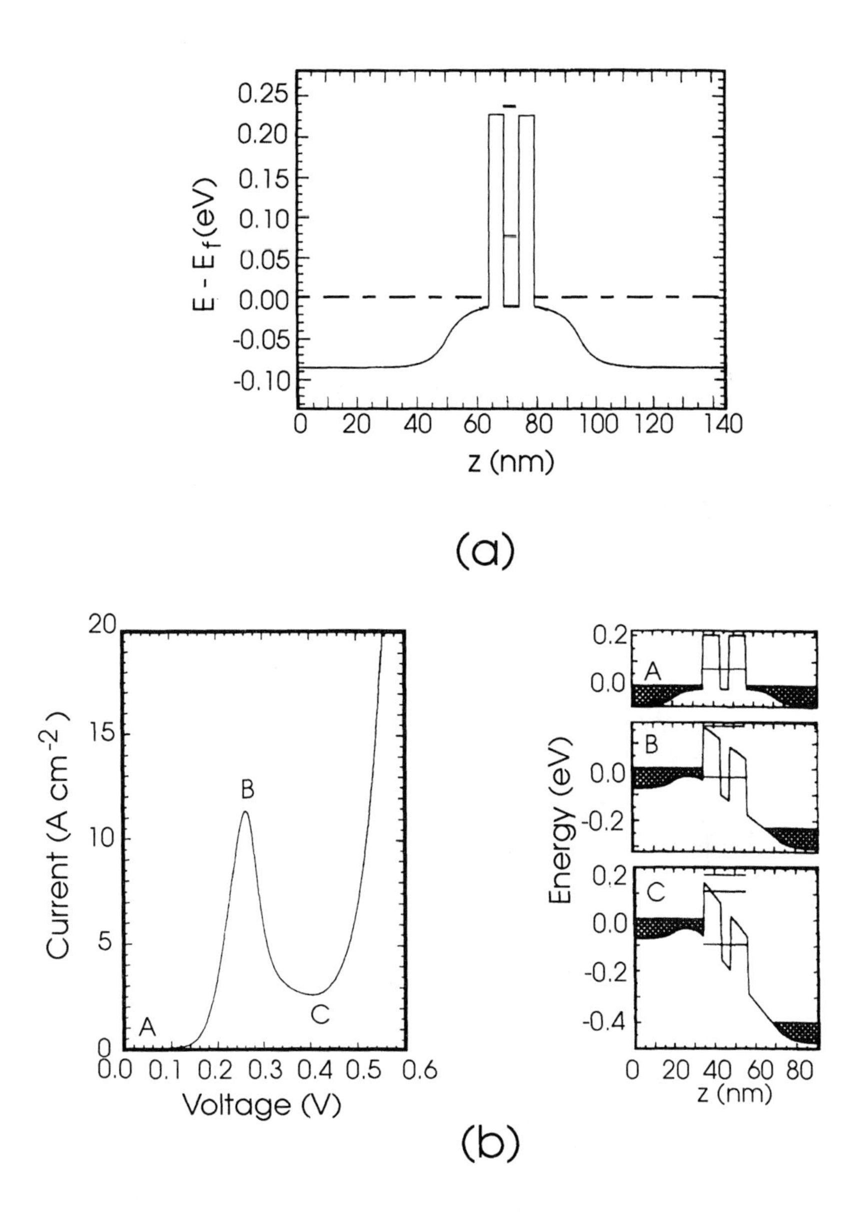

Figure 20.7. (*a*) Energy level diagram of the conduction band in a resonant tunneling structure. (*b*) Current (I) as a function of bias voltage (V) across a resonant tunneling structure is shown on left. On the right are energy level diagrams with bias applied corresponding to different points on the I–V curve (A, B, C). (From [82], with permission.)

terahertz range. This high sensitivity to bias can lead to a very high gain. Moreover, there is a strong indication that the gain can further be improved by appropriate design of the source and drain resistance [86].

The variability of input signal will become a critical issue, as pointed out by Landauer [87], for RTDs. Landauer noted that, unlike conventional transistors, the high-transmission conducting state in quantum based devices occurs only for particular values of the input signal and depends on the particular device structure. Tunneling current depends exponentially on the thickness of the barriers and high-transmission voltage depends on the thickness of the quantum well. Over time, the use of multi-valued logic over binary logic may, in practice, suffer serious signal noise tolerance and discrimination problems. Clearly, the high demands placed on manufacturing control are expected to be a challenging problem which will continue to confront researchers in quantum tunneling devices and ICs. We estimate that cascaded logic circuits with well fabricated RTDs, employing binary logic, will have about 20% of the total-voltage-swing signal tolerance, compared to about 50% of the total voltage swing in conventional CMOS devices.

To our knowledge, true RTD integrated circuits have, so far, not been tested. To date, RTDs have been experimentally demonstrated in the form of 'uncoupled' circuit configurations, i.e. as single-stage input/output circuit configurations [10]. Indeed, single-stage analog and digital logic circuits have satisfactorily been demonstrated throughout the world, e.g. as frequency multiplier/dividers, as exclusive NOR gates, as parity generators, as multi-state memories, and as analog-to-digital converters [10]. More recently, two-stage logic coupling has been demonstrated, i.e. two-stage XNOR logic gate combinations in the form of a fuller adder circuit [88]. It is critical, for future applications of RTDs to computer logic, that coupled and closed-loop arrays of input–output logic stages be successfully demonstrated.

20.3.1.2 Electron wavefunction effect devices

A second class of quantum-effect nanodevices is that based on the wavelike behavior of the electron. Here the electron wavefunction interferes constructively or destructively at an output terminal. The interference is modulated through an additional terminal [89]. Quantum interference between electron wavefunctions is the basis of the Schrödinger waveguide and the Aharonov–Bohm effect [65]. Three-terminal devices have been introduced, but with no demonstrated 'drive' and gain capability. The present research focuses on the novel controls, which are similar to electromagnetic waveguide devices. The 'stub', 'double constriction', and 'bend' are defined by the confining depletion-layer wall controlled by the gate voltage on the surface. For interference effect devices, the device size must be less or equal to the inelastic coherence length in order to operate. The devices so far reported operate only at very low temperatures. Furthermore, to provide gain, these devices must still be able to shut off at a high drain voltage (with the source at zero voltage). However, a high drain voltage will accelerate the carriers changing their wavelengths. This affects quantum interference in a complicated nonlinear manner, seriously limiting the gain.

20.3.1.3 Single-electron effect devices

In the nanometer regime, effects due to the motion of single electrons can be observed. Consider a nanocapacitor that is connected to a voltage supply. The capacitor can be a small-area tunnel junction, a small metal island between electrodes, or a small quantum well. The capacitance, C, is proportional to the cross-section of the junction or the size of the island and can be of the order of an attofarad (10^{-18} F) for nanometer-scale structures. The 'plate' of the capacitor acts as a reservoir of electrons or Coulomb island. The energy required to add an individual carrier of charge Q onto the capacitor is Q^2/C. If this energy is comparable to or larger than the thermal energy, kT (k is Boltzmann's constant and T is temperature), discontinuities or steps in the current–voltage characteristics of the device (or structure) corresponding to the addition (subtraction) of single electrons to (from) the island are observed [90–92]. The energy of charging by one electron can become significant for small areas and single-electron behavior begins to dominate the carrier flow across the tunnel junction or Coulomb island at temperatures below 10 K. Depending on the size of the island, this single-electron effect can be classical or quantum mechanical (the energy levels being quantized).

Consider the Coulomb island device shown in figure 20.8(*a*). This type of device can be implemented by patterning the metal gates on a GaAs HEMT device structure or a Si–SiO_2 MOSFET device structure. In a GaAs HEMT structure the 'gate' electrode can be applied to the heavily doped backside GaAs. One can add (remove) an electron from the island via the leads. The gate electrode induces a polarization and can continuously modify the charge on the island. The energy of the island is a parabolic function of the charge (or number of electrons (N)), as shown in figure 20.8(*b*). Through the polarization, the position of the minimum in the energy relative to charge can be shifted. If the energy of the minimum is at an integer number of electronic charges, as shown in figure 20.8(*b*), it will cost energy ($e^2/2C$) to add or remove an electron. Under these conditions no current will flow between the leads. This is the Coulomb blockade. Through the application of the gate voltage, one can modify the position of the minimum in the energy diagram. If the minimum is between two charge states, $N + \frac{1}{2}$, then the states N and $N + 1$ are degenerate, as shown in figure 20.8(*c*), and current can flow through the island (one electron at a time). This gives rise to periodic peaks in conductance as a function of gate bias, as seen in figure 20.9 [93].

If the size of the Coulomb island is further reduced to a quantum dot, then the electronic energy states become quantized in addition to the quantization imposed by the integer number of electrons. Groups from Massachusetts Institute

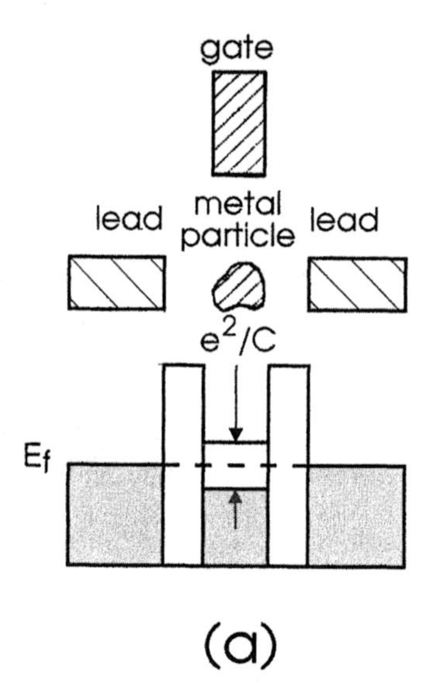

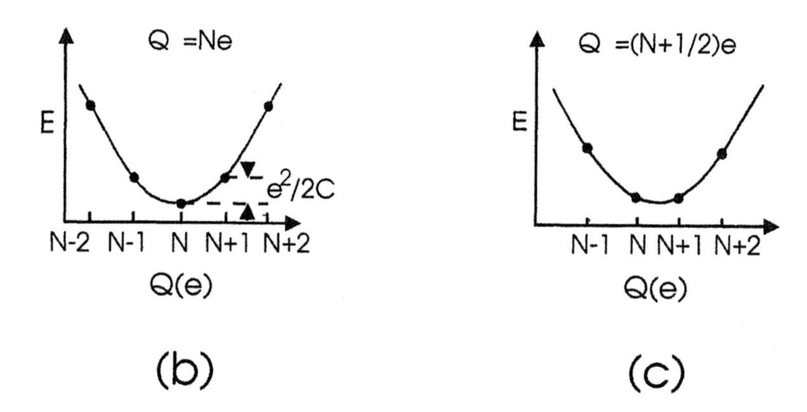

Figure 20.8. (*a*) A Coulomb island and corresponding energy levels. The Coulomb island has an energy gap in its tunneling density of states. (*b*) Charging energy versus polarization charge, Q, in a Coulomb island. Q_0 is the charge that minimizes the charging energy by varying the gate voltage. Because charge is discrete, only quantized values of the energy will be possible for a given Q_0. When $Q_0 = N_e$ there will be an activation energy for current to flow; this is the Coulomb blockade. (*c*) $Q_0 = (N + \frac{1}{2})e$. The states with $Q = Ne$ and $Q = (N + 1)e$ are degenerate, and the energy gap of the tunneling density of states vanishes. This results in the conductance with periodic sharp peaks as a function of the gate voltage, with period e/C. (Reprinted with permission from [91].)

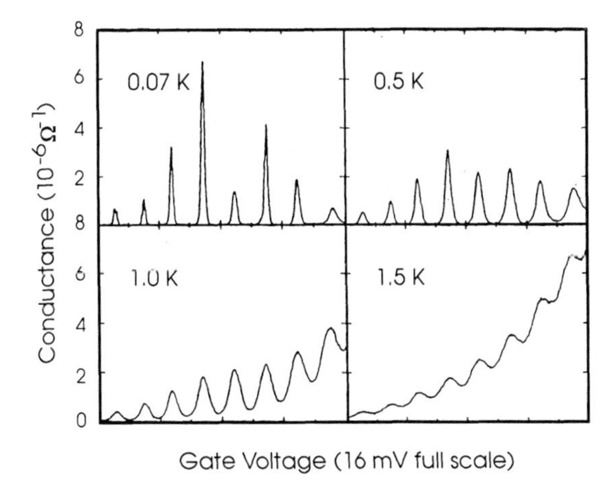

Figure 20.9. Conductance variation across a Coulomb island as a function of gate voltage. Each peak corresponds to the conduction condition as shown in figure 20.8(*c*). One can see that as the temperature increases the single-electron effects become less pronounced. (After [93], with permission.)

of Technology and IBM Thomas J Watson Research Center have observed Coulomb blockade in quantum dots [94]. The AT&T Bell Laboratories and the University of New York at Stony Brook have also obtained Coulomb blockade behavior using quantum wells [95].

The Coulomb blockade phenomenon, as applied to the concept of a 'single-electron dominated' transistor, is expected to have an attractive drive and gain capability. Single-electron transistors are anticipated as a result of 'lateral' scaling down of the present resonant tunneling transistors for ultrahigh-density ICs.

20.3.2 Nanotransistor designs

Through growth techniques, such as MBE, the layers can be grown with 0.2 nm (i.e. monolayer) precision and uniformity in the vertical direction. System and quantum effects in such materials, due to confinement in the vertical direction, can be readily observed. Thus, most experimental results for nanotransistors are based on vertical transport. Although there are many potentially attractive devices with nanoscale dimensions in one or two lateral directions, the lateral patterning cannot, at this time, achieve the resolution and precision that can be routinely grown in the vertical direction. For this reason laterally patterned devices can operate only at low temperatures, $\sim$4 K and below, and with heroic effects to obtain the required signal-to-noise ratio. The lateral fabrication still requires extensive development, and nanofabrication is an active field, as emphasized in section 20.2.

The different degree of fabrication control in lateral and vertical directions has resulted in a dichotomy of research effort into advanced electron devices. The planar-IC community efforts tend to be lateral quantum transport based and utilize the transport of carriers in the lateral direction, whereas a second group of researchers utilizes the quantum transport of carriers in the vertical direction, across heterojunction interfaces. Indeed, bandgap engineering of materials brought about by the advent of MBE and metalorganic chemical vapor deposition (MOCVD) was immediately employed by the second group to improve the performance of conventional bipolar and field-effect transistors with heterojunction transistors. Almost all of the lateral nanodevices and nanotransistors have been patterned by electric field confinement, produced by gates on the surface of high-electron mobility two-dimensional electron gas (2DEG) materials [2] (these are discussed in section 20.3.2.2).

20.3.2.1 Vertical transport nanotransistor designs

Several experimental nanotransistor devices are based on resonant tunneling, multi-barrier structures. Inherent in resonant tunneling electronic devices are the multivalued 'on' characteristics and negative transconductance as in the RTD case. In a transistor one must control the current by directly inserting a biasing charge into the current carrying region or by applying an external electric field that induces the current control. One obvious approach to control the emitter–collector resonant tunneling current is to connect the base directly to the quantum well as shown in figure 20.10. Here, the base bias alters the potential of the quantum well (and the resonant condition). However, in this configuration, the base electrode introduces carriers into the quantum well. If these carriers occupy the same quantized energy level in the quantum well as the tunneling electrons, between the emitter and collector, then the base–collector leakage current can become significant. This will short out the base and make transistor control impossible.

One way to keep the charge carriers introduced by the base separate from the ungated emitter to collector current is to assign one energy level (e.g. ground level) in the quantum well for the base signal. The next excited energy level is the current carrying channel between the emitter and collector. It is also desirable to 'hide' the ground level from both the emitter and collector Fermi level, to completely eliminate leakage current. These requirements are accomplished by choice of semiconductor material layers so that the quantum well has a narrower bandgap than the emitter and collector, as shown in figure 20.11. It should be emphasized that this design results in a unipolar transistor, i.e. current is carried only by one carrier, the electrons.

For high speed, one might wish to use bipolar devices. A bipolar, two-current carrier, resonant tunneling heterojunction transistor could be implemented by simply p-doping the quantum-well layer in the previous device, as shown in figure 20.12. Here too a narrow-gap quantum well is used to

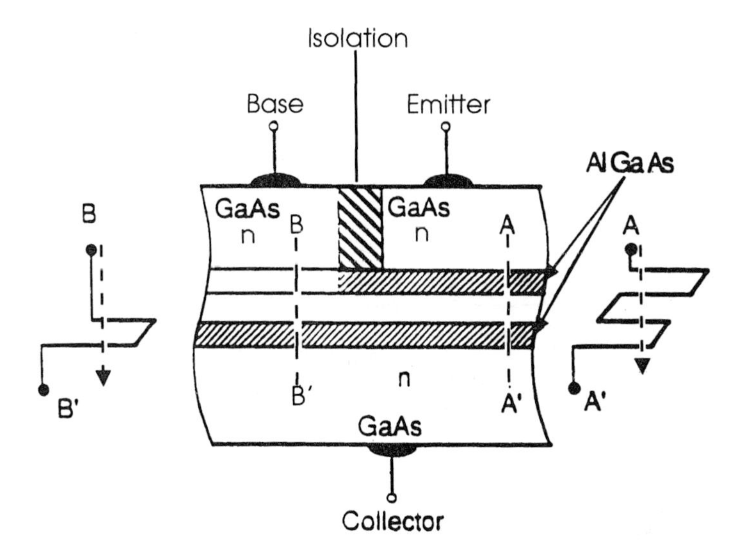

Figure 20.10. Unworkable unipolar resonant tunneling transistor (RTT) design. Section AA′ shows tunneling behavior, however BB′ shows a parasitic base current making transistor action impossible. (After [77], with permission.)

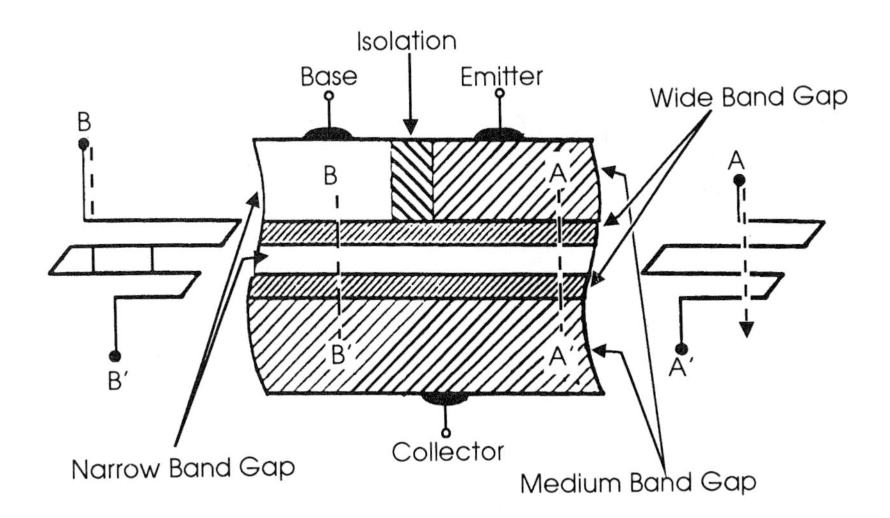

Figure 20.11. RTT design which eliminates the parasitic base current. The emitter–collector current channel is through the excited quantum-well state. Control charge carriers are introduced into the ground-state energy level by the base contact and prevented from flowing to the collector by the medium-bandgap collector region. (After [77], with permission.)

effectively 'hide' the hole energy level in the quantum well to prevent parasitic emitter–base hole currents. This design, a bipolar quantum resonant tunneling transistor (BiQuart), has been implemented at Texas instruments [8, 9].

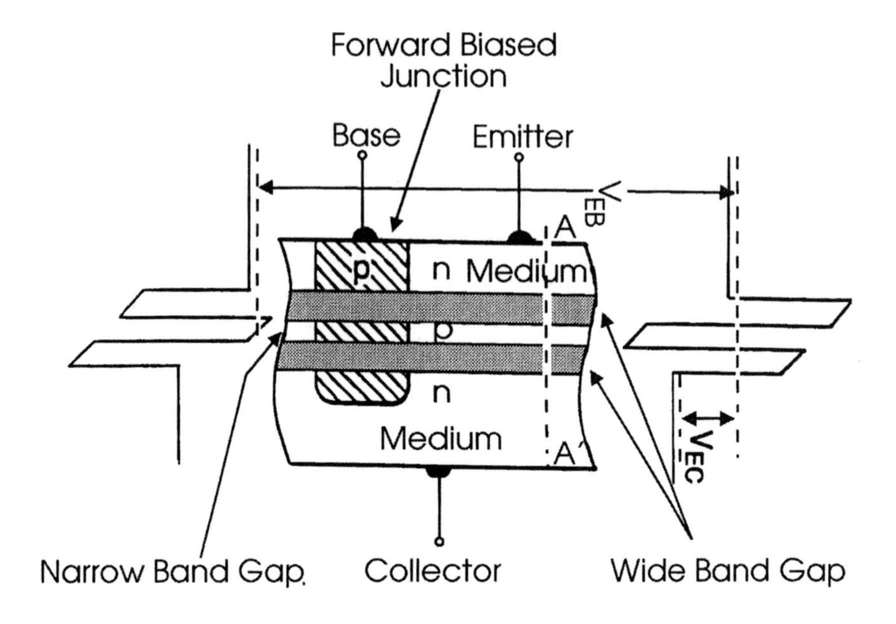

Figure 20.12. RTT design which eliminates p–n junction catastrophic parasitic current. A narrow-bandgap quantum well sandwiched between wide-bandgap barriers and medium-bandgap emitter and collector regions allows for a reduced forward bias which does not create catastrophic parasitic current across the p–n contact junction. (After [77], with permission.)

A Stark-effect transistor (SET) [85] can be made by interchanging the base and collector in the transistor in figure 20.11. In this new scheme, shown in figure 20.13(*a*), the base bias is introduced on the other side of the double barrier, creating an electric field, which controls the position of the energy levels in the well relative to the energy of the electrons in the emitter. The output current is extracted from the quantum well, i.e. the collector region. In this design, the two barriers are in general highly asymmetrical, with the second barrier many times larger than the first barrier, to eliminate tunneling leakage current from the 'base'. This design has a negligible base current and large current-transfer ratio. However, since the biasing charge is not directly introduced into the current channel but through the capacitive control of the 'base' electrode, this design also operates as a field-effect nanotransistor. The SET was proposed by Bonnefoi *et al* [85] and experimentally demonstrated at AT&T Laboratories [96]. Shown in figure 20.13(*b*) [96] are I–V measurements of such a device with different

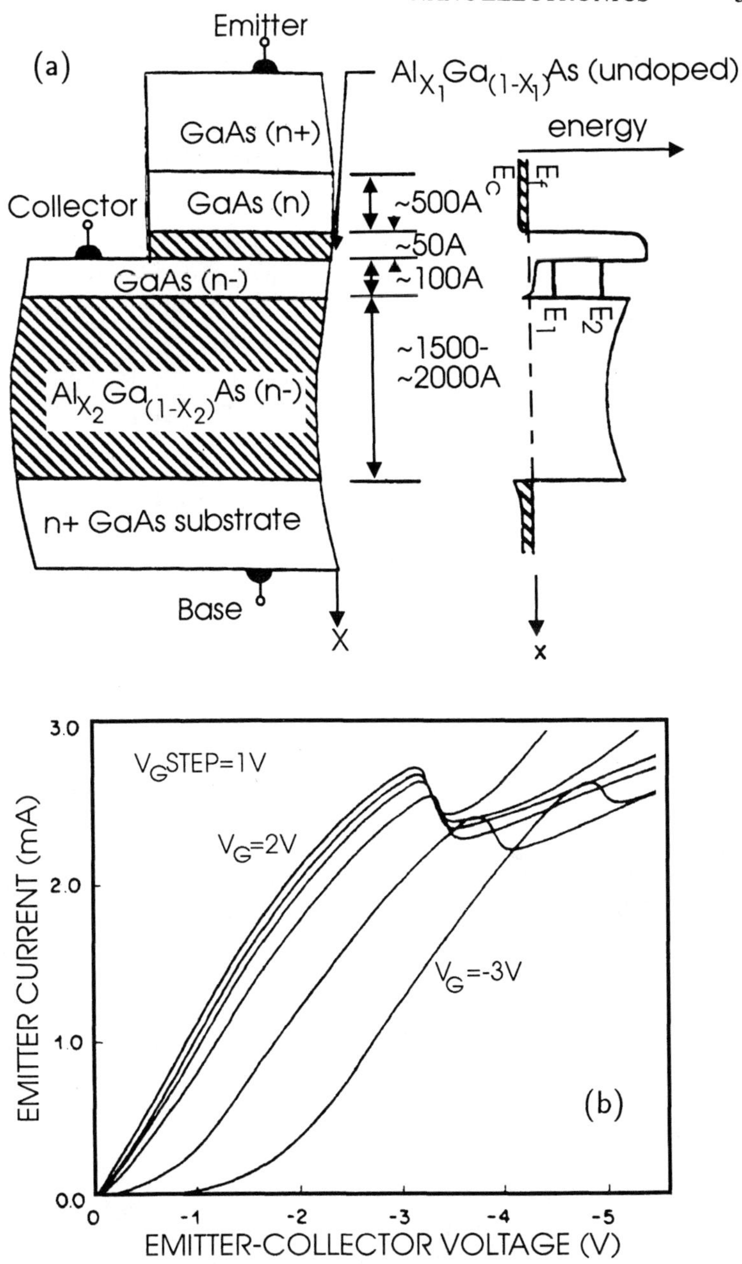

Figure 20.13. (*a*) SET design. The transistor is controlled by the polarization induced by the base electrode. (From [85], reprinted with permission.) (*b*) Measurement of emitter current versus emitter–collector voltage, at 7 K, with different gate biases. (From [96], reprinted with permission.)

gate voltages. The transistor switching action is controlled by the alignment of the energy level in the quantum well with the Fermi level in the emitter. One can see the emitter current increase as the emitter–collector voltage is increased until a resonant tunneling condition is reached. This is followed by a drop in the emitter current when the resonant tunneling condition is passed. The positions of the quantum well levels relative to the emitter Fermi level (and the resonant tunneling condition) are modulated by the gate voltage. For improved transistor design, one can add a potential step in the quantum well using a narrow-bandgap thin layer similar to the Texas Instruments approach. This will further decrease the base–collector leakage current, eliminate intervalley scattering, and enhance the current drive.

A vigorous effort aimed at developing multi-stage resonant tunneling transistors (RTTs) [97], in discrete mesa-device design configurations, has been undertaken at AT&T Bell Laboratories [98]. These are modifications of the transistor in figure 20.11, where multiple wells (and barriers) are placed in the base region [99] or multiple barriers are placed in the emitter region [100]. Each device has a multivalued transfer characteristic, having as many peaks as the number of additional layers. These devices have been demonstrated for potential multivalued logic applications [10].

The RTTs so far mentioned have been modifications of bipolar transistor approaches. A different approach based upon the modification of field-effect heterojunction transistors (e.g. the HEMT structure) has also been proposed at AT&T Laboratories [101]. The device is shown in figure 20.14. Here the device area is produced by a V-groove etching of the grown layered material. A thin AlGaAs layer is epitaxially grown on the etched groove. A gate metal is deposited on top of the barrier layer, and the structure is a HEMT structure. This fabrication technique and structure was also proposed earlier by Sakaki [102]. n^+ GaAs layers are on the top and bottom of the heterostructure stack forming the source and drain. A two-dimensional (2D) electron gas forms along the V-groove at the junction of the GaAs layers and the overgrown AlGaAs layer. The quantum well (thinnest GaAs layer) in the center imposes additional dimensional confinement in the original growth direction and the resulting electrons in the well occupy quantum-confined states of a one-dimensional (1D) wire. A positive gate voltage V_g induces electrons from the n-doped source and drain regions to occupy the delta-function width quantum wells formed by the GaAs/AlGaAs junctions along the V-groove. The novel feature of the proposed design is that tunneling transport occurs across a quantum 'wire' not across a planar layer as in previous nanotransistors [103]. The transport process is a tunneling between the 2D electron density of states and the 1D density of states in the quantum wire. Such a structure exhibits interesting characteristics. Because of large zero-point energy and transverse energy-level spacing in the 'quantum wire', there is a range of low V_g in which electrons are not induced in the wire. Applications of drain voltage V_d bring about a resonant tunneling current parallel to the AlGaAs (under the gate) interface. The number of tunneling electrons grows

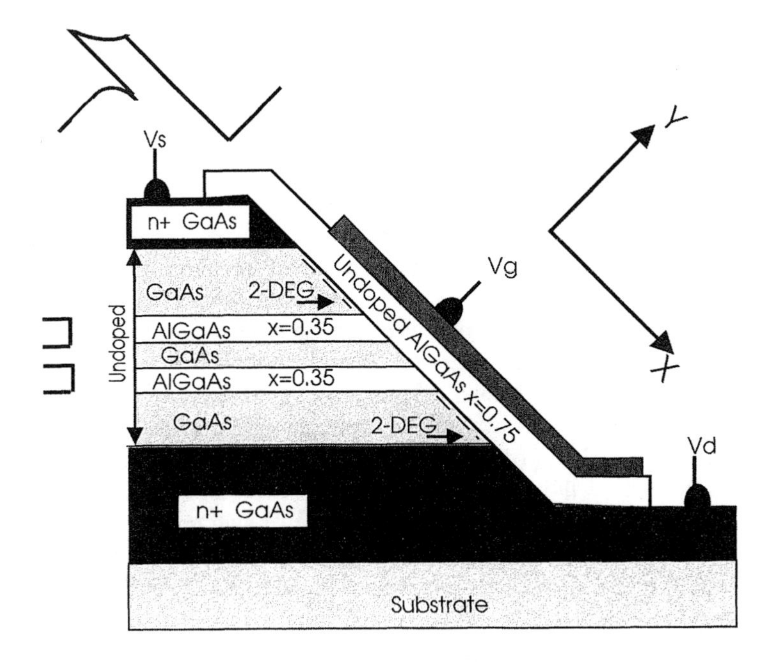

Figure 20.14. Unipolar surface quantum wire RTT and energy-band diagram along the gated 'surface' conduction channel. The thickness of the two undoped GaAs layers outside of the double-barrier region is sufficiently large to prevent parasitic resonant tunneling current in the bulk. (After [101], with permission.)

with V_d until there are no more electrons in the source with energy matching the quantized energy level in the wire. This situation gives rise to negative differential resistance, where the current decreases dramatically with increased voltage. The device can also be controlled by V_g rather than V_d. This implies the very interesting possibility of achieving negative transconductance with a unipolar transistor. Such a transistor could perform the function of, and hence replace, a p-channel transistor in silicon CMOS logic. This low-power inverter would find important applications in GaAs ICs. However, the success of the device depends on the development of reliable fabrication methods involving etching and regrowth [104, 105], as discussed in section 20.2.

20.3.2.2 Lateral transport nanotransistor designs

The last transistor design discussed above may be considered to be based on the combination of lateral transport and vertical transport: vertical transport, since current flows across heterojunction layers, and lateral transport, since current flows along the surface of the V-groove etching. Whereas most vertical transport based nanotransistor designs are essentially modifications of conventional heterojunction bipolar transistors, all lateral transport based nanotransistors designs, so far, are modifications of unipolar field-effect heterojunction transistors, HEMTs [106, 107]. The HEMT structures consist of a buried (50–100 nm below the surface) heterojunction between a higher-band-gap material (such as AlGaAs) and a lower-band-gap material (such as GaAs or InGaAs). The energy of the conduction bandedge as a function of depth exhibits a sharp, delta-function-type dip at the heterojunction. The higher-bandgap material (on top) is n-doped with the donors spatially placed 20–50 nm from the heterojunction. The electrons from the donors migrate to the sharp quantum well at the heterojunction and form a 2D electron gas (2DEG). Because the electrons are confined to the quantum well, they do not scatter with the ionized donors. Hence HEMT structures exhibit very high electron mobility. A FET is made by electrically contacting the 2DEG with the source and drain pads (usually by alloying contact pads to the material) and placing a gate(s) on the AlGaAs between the source and drain pads. Lateral transport based nanotransistor designs make use of patterned multi-gates rather than the single gate found in a MODFET-type structure [108]. This is achieved in a manner similar to the one proposed earlier for MOSFET structures [7]. Each gate creates an electric field, which induces a potential barrier in the 2DEG below which the carriers in the plane of the well are confined. The longer coherence length and mean free paths in the 2DEG transport channel in HEMT structures enable current lithographic techniques to fabricate lateral transport based nanotransistors. Lateral transport based designs offer a latitude of design parameters defined by the shape and geometrical pattern of the gates. This feature allows one to design nanotransistors whose function depends on the localization of charge carriers in more than one dimension, such as the use of 'quantum wires' and 'quantum dots' [109]. Here carrier scattering is further reduced due to multi-dimensional size quantization, thus enhancing the coherence length and low-field mobility.

Based on the modifications of the Schottky barrier gate structure of conventional heterojunction HEMTs (or MODFETs), a number of lateral transport based nanotransistors have been demonstrated [106, 107]. The various structural modifications are: (a) a dual gate or a split gate geometry, resulting in the so-called planar resonant tunneling field effect; (b) triple-gate geometry with an independently controlled middle gate (this has added features over (a) in that the quantum well depth and barrier heights are controlled independently); (c) 'grating gate' geometry, which leads to an induced array of quantum wires under the gate area, where lateral transport is across the wires and Bragg interference in this direction determines the overall transistor characteristics; and (d) 'grid gate' geometry, which lead to an induced array of quantum dots under the gate area; 2D Bragg interference (e.g. more bunching of states) determines the transistor characteristics.

At fixed drain bias, V_{ds}, as the gate bias, V_{gs}, is increased in quantum arrays, there are two correlated effects: (a) the quantum-well depth (non-depleted region) increases resulting in 'stronger' periodicity and a consequent larger negative-mass portion of the miniband; and (b) the electron concentration increases in each 'unit cell' resulting in 'filling' of the minibands. Thus as V_{gs} is increased in a continuous manner the Fermi energy is expected to pass through minibands and minigaps. This is illustrated from the experimental curves shown in figure 20.15, where characteristics of a HEMT structure with an array of 100 nm parallel wire gates on the surface are shown [110]. The source–drain current, (I_{DS}) and transconductance (g_m) are shown as a function of substrate voltage (V_{sub}) (applied from the backside). The drops in current (negative transconductance) are observed in the minigaps as the population of the different bands is changed with voltage.

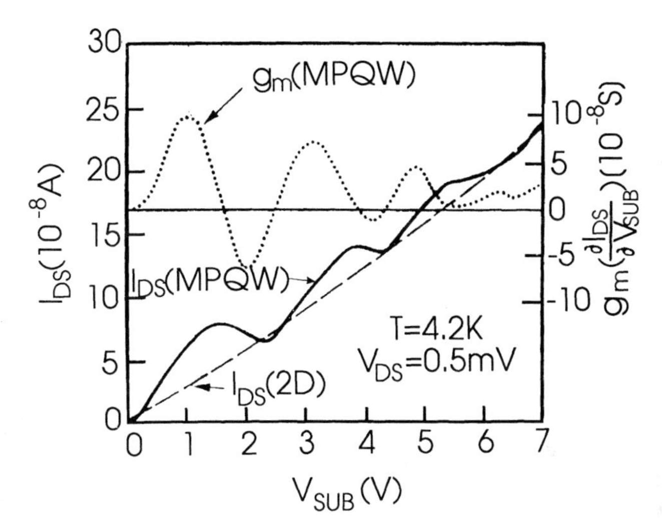

Figure 20.15. Source–drain current versus sample bias for an array of quantum wires in the conduction channel. The gates create a series of quantum wires in the 2DEG channel region of the HEMT, making 1D electronic states, which open minigaps in the conduction band. This is shown by the peaks in the source–drain current, which occur when the flowing electrons occupy allowed 1D energy levels. (Published with permission from [110].)

It was estimated that quantum wires (electrons confined in two dimensions), offer about two orders of magnitude improvement in low-field mobility, a significant higher saturation velocity, and longer coherence lengths compared to 2DEGs (electrons confined in only one dimension) [102]. Thus an obvious approach to further improvement of conventional field-effect high-speed devices based on the HEMT structure is to replace the 2DEG transport channel, controlled by the gate, by an array of quantum wires, say a few hundred parallel wires to maintain reasonable current drive capability. Indeed, size quantization by itself has immediate applications to conventional (i.e. nonquantum-transport-based) high-speed electronics and optoelectronic devices. One can expect an even further improvement if the 2DEG transport channel in HEMTs is replaced by an array of coupled quantum dots.

20.3.3 Nanoelectro-optics

Because size quantization effects are expected to have immediate real impact in optoelectronics, the optoelectronics community has a strong interest in nanostructure research. Lateral device array designs are the preferred choice since light can readily be coupled into lateral structures. The advantage of abrupt confinement across heterojunctions has led to vertical device designs in the form of a mesa device. The use of arrays of quantum wires and quantum dots, providing confinement of injected carriers in the active region of laser diodes, has been shown to dramatically improve the lasing characteristics. In the case of a quantum-dot array active region, the temperature dependence of the threshold current is virtually eliminated [111]. It was also shown theoretically that the electroabsorption and the associated electrorefraction in an array of quantum wires and quantum dots are greatly enhanced over those of 'planar' array of quantum wells by virtue of the multi-dimensional quantum-confinement Stark effect (i.e. bunching of quantum states). Such enhanced effects would lead to electroabsorptive and electrorefractive modulators and optical switches with even lower energy requirements than existing quantum-well devices [112]. Another interesting phenomenon in optoelectronics is the so-called carrier induced bleaching of optical absorption or the blue-shift of the absorption edge, primarily caused by band filling that quenches both exciton and band-to-band absorption in quantum-well arrays. The subsequent changes in refractive index (the refractive index in photon transport is akin to electrostatic potential in electron transport) have interesting applications. For example, the refractive index can be externally controlled by the injection of carriers or by a voltage applied at the gate in field-effect or HEMT configurations, where the 2DEG channel is replaced by the multi-dimensional confined arrays, so as to construct optical modulators and optical bistable switches [113]. Another example in the optoelectronics area is the self-electro-optic effect (SEED) which may have important applications in photonic switching [108] (see chapter 16).

20.3.4 Physical modeling and device simulation

A whole new approach to analyzing charge carrier transport in small structures is required for modeling the new regimes of operation and logic-gate characteristics of nanodevices. Nanodevice transport physics must incorporate all quantum effects by virtue of the wave nature of the charge carrier motion, all nonlinearities, all many-body effects, and all space and time-dependent nonuniformities. Indeed, device-simulation researchers are seriously turning to particle Monte Carlo [114] and various quantum approaches [115] to develop new computer aided research and device-characterization tools that will lead to novel computer aided design tools.

At present, we can identify two major thrusts in research in nonlinear, far from equilibrium, quantum transport physical modeling and simulation. The first is concerned with time-dependent, highly transient quantum transport in nanodevice physics. We believe that for 1D systems, these problems can be handled very well by the quantum distribution function (QDF) approach with a subsidiary boundary condition [116], including band-structure effects. Indeed, so far, this is the only approach that has produced realistic results for analyzing the time-dependent I–V characteristics for high-speed operation of RTDs [117, 118]. It can provide for the complete transition from basic quantum transport physics to an engineering computer aided design (CAD) tool for ICs. However, the success cannot be extended to multi-dimensional systems of arbitrary geometry with present-day supercomputers. We believe such devices might be best analyzed by coupling the self-consistent ensemble particle Monte Carlo (SEPMC) method, with a model of causal particle trajectory representation (CPTR) of quantum transport (e.g. the Wigner trajectory, Bohm trajectory, or a refined model quantum tunneling particle dynamics (MQTPD) [114]).

The second major direction is at the other end of the time scale spectrum, steady state. Time independence will result in a several orders of magnitude reduction in the computer simulation time. For 1D problems, this can be handled conveniently by the QDF pure quantum approach with a subsidiary boundary condition. For multi-dimensional problems of arbitrary geometry, the large-scale matrix solution for the non-equilibrium quantum correlation function $G(q, q', E)$, where q and q' represent lattice coordinates, should prove to be a very powerful quantum transport technique for real solid-state nanodevices. This is an area where one can start from the knowledge of atomic orbitals to account for surfaces, interfaces, multi-valley scattering processes, interband tunneling dynamics [119], and disorder effects in multi-dimensional finite systems where the energies of the charge carriers are not known *a priori* [120].

Advances in computational hardware and software are expected to alter the current research and simulation capability issues. It may be possible to perform a purely quantum transport simulation of a fully time-dependent, highly nonlinear, far-from-equilibrium multi-dimensional systems of 'atomic-orbital' real solid-state materials with arbitrary geometries.

20.4 SUMMARY

The quantum physics of small structures has evolved into a major field of condensed matter physics in the last 20 years [121]. Indeed, mesoscopic physics and nanoelectronics underwent a rapid development in the 1980s, and their study is expected to acquire more momentum as small structures (toward atomic sizes) become more generally accessible. Through current epitaxial growth techniques, layered materials with atomic control in interface uniformity and layer thickness can be routinely fabricated. Bandgap engineering of novel material systems, such as the InAs/GaSb/AlSb based electronic materials, is currently a very active field of research [122]. 'Vertical' heterojunction technology has clearly reached the atomic size range and has allowed abrupt 'atomic-size' confinement of electrons in solids. However, there is a strong need to confine electrons in solids in more than one dimension. Quantum based devices generally perform exceedingly well when the energy levels are well separated and the density of states per unit energy range is sharply defined [123].

Confinement in more than one dimension is currently being implemented by means of depletion barriers for digital applications, and by a 'mesa' device for optoelectronic and other applications. The inherent semiconductor volume required by the depletion layers inherently prevents scaling the device area for ultradense IC applications. Therefore, the development of abrupt lateral heterojunctions for device isolation and for lateral confinement would indeed set another milestone comparable to the development of vertical heterojunction MBE technology. The ability to fabricate lateral nanostructures is the major 'bottleneck' in the development of nanoelectronic ICs, at this time. The best fabrication techniques can produce semiconductor structures of lateral dimensions $\approx$ 20 nm, but atomic layer control is not available. Frequently the lateral patterning technique leaves the device region of the semiconductor depleted of carriers. Perhaps, developments in overgrowth techniques, advanced lithography, and other techniques using the principle of scanning tunneling microscopy [12, 13] may offer the needed giant step toward lateral heterojunction technology. Lateral and vertical heterojunction capabilities would foster the rapid development of coupled quantum-dot arrays and novel IC architectures.

In closing, once the lateral fabrication has been adequately developed, new frontiers of technological challenges will open. Although the interest in quantum-transport phenomena in high-speed, small systems is unprecedented, the present-day worldwide efforts in this area are negligible compared to other mainstream areas of transport research, such as classical and near-equilibrium quantum transport. This situation is expected to drastically change in the near future. The road to nanoelectronic ICs is a long winding road; nanofabrication is certainly one of its immediate greatest challenges. However, far more conceptual challenges await, such as: novel computer architecture, complete characterization and understanding of quantum-based IC components, ingenious routing of interconnects, partitioning of logic and processing blocks, and novel communication channels to make use of the capabilities of nanofabrication. There is no doubt that computer aided research and development, i.e. the development of software research tools, CAD, and R&D tools, will serve as the 'intelligence' behind the realization of a super computer on a chip.

REFERENCES

[1] Moore G E 1975 *Proc. IEEE IEDM* IEEE cat. No 75CH1023-1 ED, 11

[2] Pease R F W (ed) 1991 *Proc. IEEE* **79** number 8 special issue on nanoelectronics
[3] 1979 *Phys. Today* **32** October 17
[4] Anderson P W, Abrahams E and Ramakrishnan T V 1979 *Phys. Rev. Lett.* **43** 718
[5] Dolan G J and Osheroff D R 1979 *Phys. Rev. Lett.* **43** 721
[6] Krumhansl J S and Pao Y H 1979 *Phys. Today* November special issue on microscience
[7] Buot F A 1979 *Stanford Electronics Laboratory Report*; 1987 *Superlatt. Microstruct.* **3** 399
See also Buot F A 1986 *Phys. Rev.* A **33** 2544
For a review of theoretical and computational advances in mesoscopic physics see Buot F A 1993 *Phys. Rev.* **234** 73
[8] For an overview of mesoscopic physics, see, for example 1988 *IBM J. Res. Dev.* **32** No 3
[9] For an overview of nanoelectronics, see for example, 1989 *TI Tech. J.* **6** No 4 and reference [2]
[10] For an overview of the 'discrete' nanoelectronics work at AT&T Laboratories, see, for example, Capasso F, Sen S, Beltram F, Lunardi L M, Vengurlekar A S, Smith P R, Shah N J, Malik R J and Cho A Y 1989 *IEEE Trans. Electron Devices* **ED-36** 2065
[11] Yokayama N *et al* 1985 *Japan. J. Appl. Phys.* **24** L853
[12] Binnig G, Rohrer H, Gerber Ch and Weibel E 1982 *Phys. Rev. Lett.* **49** 57
[13] Eigler D M and Schweizer E I 1991 *Nature* **344** 524
Garcia R G 1992 *Appl. Phys. Lett.* **60** 1960
[14] McCord M A and Pease R F 1986 *J. Vac. Sci. Technol.* B **4** 86
[15] Dobisz E A and Marrian C R K 1991 *Appl. Phys. Lett.* **58** 2526
[16] Chen W and Ahmed H 1993 *Appl. Phys. Lett.* **62** 1499
[17] Dammel R 1993 *Diazonapthoquinone-Based Resists* (Bellingham, WA: SPIE)
[18] Berry A K, Graziano K A, Thomson S D, Taylor J W, Suh D and Plumb D 1991 *Proc. SPIE* **1465** 210
[19] Dobisz E A and Marrian C R K 1991 *J. Vac. Sci. Technol.* B **9** 3024
Umbach C P, Broers A N, Willson C G, Koch R and Laibowitz R B 1988 *J. Vac. Sci. Technol* B **6** 319
Yoshimura T, Nakayama Y and Okazaki S 1992 *J. Vac. Sci. Technol.* B **10** 2615
[20] Isaacson M and Murray A 1981 *J. Vac. Sci. Technol.* **19** 1117
[21] Hollenbeck J L and Buchanan R C 1990 *J. Mater. Res.* **5** 1058
[22] Clausen E M Jr, Harbison J P, Chang C C, Craighead H G and Florez L T 1990 *J. Vac. Sci. Technol.* B **8** 1830
[23] Allee D R and Broers A N 1990 *Appl. Phys. Lett.* **57** 2271
[24] Clausen E M Jr, Harbison J P, Florez L T and van der Gaag B 1990 *J. Vac. Sci. Technol.* B **8** 1960
[25] Harriott L R, Temkin H, Hamm R A, Weiner J and Panish M B 1989 *J. Vac. Sci. Technol.* B **7** 1467
[26] Lyding J W, Shen T-C, Hubacek J S, Tucker J R and Abein G C 1994 *Appl. Phys. Lett.* **64** 2010
[27] Dulcey C S, Georger J H, Krauthamer V, Fare T L, Stenger D A and Calvert J M 1991 *Science* **252** 551
[28] For a review see Ulman A 1991 *An Introduction to Ultrathin Organic Films: From Langmuir–Blodgett to Self-Assembly* (San Diego CA: Academic)

[29] Tiberio R C, Craighead H G, Lercel M, Lau T, Sheen C W and Allara D L 1993 *Appl. Phys. Lett.* **62** 476

[30] Schnur J M, Peckerar M C, Marrian C R K, Schoen P E, Calvert J M and Georger J H 1991/2 *US Patent* 5 077 085 and 5 079 600

[31] Calvert J M 1993 *J. Vac. Sci. Technol.* B **11** 2155

[32] Berger S D, Gibson J M, Camarda R M, Farrow R C, Huggins H A, Kraus J S and Liddle J A 1991 *J. Vac. Sci. Technol.* B **9** 2996

[33] Stengl G, Bösch G, Chalupka A, Fegerl J, Rischer R, Lammer G, Löschner H, Malek L, Nowak R, Traher C and Wolf P 1992 *J. Vac. Sci. Technol.* B **10** 2824

[34] Heidenreich R 1977 *J. Appl. Phys.* **48** 1418

[35] Dobisz E A, Marrian C R K, Salvino R E, Ancona M A, Perkins F K and Turner N H 1992 *J. Vac. Sci. Technol.* B **11** 2733

[36] Lee Y-H, Browning R, Maluf N, Owen G and Pease R F W 1992 *J. Vac. Sci. Technol.* B **10** 3094

[37] Kubena R L, Ward J W, Stratton F P Joyce R J and Atkinson G M 1991 *J. Vac. Sci. Technol.* B **9** 3079

[38] Ximen H, DeFreez R K, Orloff J, Elliott R A, Evans G A, Carlson N W, Lurie M and Bour D P 1990 *J. Vac. Sci. Technol.* B **8** 1361

[39] Kosugi T, Yamashiro T, Aihara R, Gamo K and Namba S 1991 *J. Vac. Sci. Technol.* B **9** 3099

[40] Blauner P G, Butt Y, Ro J S and Melngalis J 1989 *J. Vac. Sci. Technol.* B **7** 609

[41] Petroff P M, Li Y J, Xu Z, Beinstingl W, Sasas S and Ensslin K 1991 *J. Vac. Sci. Technol.* B **9** 3074

[42] Wilbert C, Miller T and Kalbitzer S 1994 *SPIE Symp. on Electron-Beam, X-Ray, and Ion-Beam Submicrometer Lithographies for Manufacturing IV (San Jose, CA, 1994)* (Bellingham, WA: SPIE)

[43] See for example Ziegler J F, Biersack J P and Littmark U 1985 *The Stopping and Range of Ions in Solids* (New York: Pergamon)

[44] Early K, Schattenburg M L and Smith H I 1990 *Microelectron. Eng.* **11** 317

[45] Peckerar M C and Maldonado J 1993 *Proc. IEEE* **81** 1249

[46] McCord M A, Wagner A and Donohue T 1993 *J. Vac. Sci. Technol.* B **11** 2958

[47] Dobisz E A *et al* 1994 *SPIE Proc.* **2194** 178

[48] Shirey L M *et al* 1994 *SPIE Proc.* **2194** 169

[49] Bjorkholm J E *et al* 1990 *J. Vac. Sci. Technol.* B **8** 1509

[50] For a recent review see Marrian C R K (ed) 1993 *The Technology of Proximal Probe Lithography* (Bellingham WA: SPIE)

[51] Chang T H P, Kern D P and Muray L P 1992 *J. Vac. Sci. Technol.* B **10** 2743

[52] Majumdar A, Oden P I, Carrejo J P, Nagahara L A, Graham J J and Alexander J 1992 *Appl. Phys. Lett.* **61** 2293

[53] Marrian C R K, Perkins F K, Brandow S L, Koloski T S, Dobisz E A and Calvert J M 1994 *Appl. Phys. Lett.* **64** 390

[54] Dagata J A, Schneir J, Haray H H, Evans C J, Postek M T and Bennett J 1990 *Appl. Phys. Lett.* **56** 2003

[55] Dagata J A, Tseng W, Bennett J, Dobisz E A, Schneir J and Haray H H 1992 *J. Vac. Sci. Technol.* A **10** 2105

[56] Snow E S, Campbell P M and McMarr P J 1993 *Appl. Phys. Lett.* **63** 749

[57] Perkins F K, Dobisz E A, Brandow S L, Calvert J M and Marrian C R K 1994 *J. Vac. Sci. Technol.* B **12** 3725

[58] Perkins F K, Dobisz E A, Brandow S L, Calvert J M and Marrian C R K 1996 *Appl. Phys. Lett.* **68** 550

[59] Ehrichs E E, Smith W F and de Lozanne A L 1992 *J. Ultramicrosc.* **42–44** 1438

[60] Stockman L, Heyvaert I, van Haesendonck C and Bruynseraede Y 1993 *Appl. Phys. Lett.* **62** 2935

[61] Dobisz E A, Marrian C R K and Colton R J 1990 *J. Vac. Sci. Technol.* B **8** 1754

[62] Campbell P M, Snow E S and McMarr P J 1995 *Appl. Phys. Lett.* **66** 1388

[63] Maile B E 1993 *J. Vac. Sci. Technol.* B **11** 2502

[64] Skocpol W J, Jackel L D, Hu E L, Howard R E and Fetter L A 1982 *Phys. Rev. Lett.* **49** 951

[65] Aharonov Y and Bohm D 1959 *Phys. Rev.* **115** 485

[66] Umbach C P, Washurn S, Laibowitz R B and Webb R A 1984 *Phys. Rev.* B **30** 4048

[67] Kash K, Scherer A, Worlock J M, Craighead H G and Tamargo M C 1986 *Appl. Phys. Lett.* **49** 1043

[68] Eddy C R, Dobisz E A, Hoffman C A and Meyer J R 1993 *Appl. Phys. Lett.* **62** 2362

[69] Lincoln G A, Geis M W, Pang S and Efremow N N 1983 *J. Vac. Sci Technol.* B **1** 1043

[70] Glembocki O J, Taylor B E and Dobisz E A 1991 *J. Vac. Sci Technol.* B **9** 3546

[71] Germann R, Forchel A, Bresch M and Meier H P 1989 *J. Vac. Sci. Technol.* B **7** 1475

[72] Gershoni D, Temkin H, Dolan G J, Dunsmuir J, Chu S N G and Panish M B 1988 *Appl. Phys. Lett.* **53** 995

[73] Maile B E, Forchel A, Germann R, Meier H P and Reithmaier J P 1989 *J. Vac. Sci. Technol.* B **7** 2030

[74] Dobisz E A, Marrian C R K, Craighead H G, Schwarz S A and Harbison J P 1989 *J. Vac. Sci. Technol.* B **7** 2053

[75] Nötzel R and Ploog K 1992 *J. Vac. Sci. Technol.* A **10** 617

[76] Wind S J, Taur Y, Lee Y H, Mii Y, Viswanathan R G, Bucchignano J J, Pomerene A T, Sicina R, Milkove K, Stiebritz J W, Hu C K, Manny M P and Chen W 1995 *J. Vac. Sci. Technol.* B **13** 2688

[77] Bate R T 1989 *Solid State Technol.* **32** 101

[78] Hopfield J J 1990 *Int. J. Quantum Chem.: Quantum Chem. Symp.* **24** 633

[79] Buot F A, Scott C, Mack I and Sleger K J 1988 *Molecular Electronic Devices* ed F L Carter, R E Siatkowski and H Wohltjen (New York: North-Holland) p 245

[80] Yamanaka N, Sasaki M, Kikuchi S, Takada T and Idda M 1991 *IEEE J. Selected Areas Commun.* **9** 689

[81] Lent C S, Tougaw P D and Porod W 1992 *Proc. Int. Workshop on Computational Electronics (Urbana-Champaign, IL, 1992)* p 163

[82] Luscombe J H, Randall J N and Bouchard A M 1991 *Proc. IEEE* **79** 1121

[83] Tsu R and Esaki L 1973 *Appl. Phys. Lett.* **22** 502
Chang L L, Esaki L and Tsu R 1974 *Appl. Phys. Lett.* **24** 593

[84] Sollner T C L G, Tannenwald P E, Peck D D and Goodhue W D 1984 *Appl. Phys. Lett.* **45** 1319

[85] Bonnefoi A R, Chow D H and McGill T C 1985 *Appl. Phys. Lett.* **47** 888

[86] The absolute value of the effective NDR was found to increase as a result of the voltage drop across the series resistors, creating a distortion of the *I–V* characteristic as calculated in [116].

[87] Landauer R 1989 *Nanostructure Physics and Fabrication* ed M A Reed and W R Kirk (Boston, MA: Academic) pp 17–30

[88] Takatsu M, Imamura K, Ohnishi H, Mori T, Adachihara T, Muto S and Yokoyama N 1992 *IEEE J. Solid State Circuits* **27** 1428; 1992 *IEEE Trans. Electron Devices* **ED-39** 2707

[89] For a review of quantum interference phenomena see chapter 4 of Kirk P and Reed M A (ed) 1992 *Nanostructures and Mesoscopic Systems* (New York: Academic)

[90] For an overview of single-electron phenomena in small area thin junctions see, for example, Averin D V and Likharev K K 1991 *Mesoscopic Phenomena in Solids* ed B L Altshuler, P A Lee and Webb R A (New York: Elsevier–North-Holland) p 173
See also Likharev K K and Claeson T 1988 *Sci. Am.* June 1992
Likharev K K 1988 *IBM J. Res. Dev.* **32** 144

[91] A brief review on single-electron phenomena in Coulomb islands is given by Kastner M A 1992 *Rev. Mod. Phys.* **64** 849
See also van Houten H 1992 *Surf. Sci.* **263** 442

[92] Harmans K 1992 *Phys. World* March 50

[93] Kastner M A, Foxman E B, McEuen P L, Meirav U, Kumar A and Wind S J 1992 *Nanostructures and Mesoscopic Physics* ed W P Kirk and M A Reed (New York: Academic) p 239

[94] McEuen P, Foxman E B, Meirev U, Kastner M A, Meir Y, Wingren N S and Wind S J 1991 *Phys. Rev. Lett.* **66** 1926

[95] Su B, Goldman V J and Cunningham J E 1992 *Science* **255** 313
See also Gueret P *et al* 1992 *Phys. Rev. Lett.* **68** 1896
Ashori R C *et al* 1992 *Phys. Rev. Lett.* **68** 3088

[96] Beltram F, Capasso F, Luryi S, Chu S N G and Cho A Y 1988 *Appl. Phys. Lett.* **53** 219

[97] Potter R C, Lakhani A A, Beyea D, Hier H, Hempling E and Fathimula A 1988 *Appl. Phys. Lett.* **52** 2163

[98] Capasso F (ed) 1990 *Physics of Quantum Electron Devices* (Berlin: Springer)

[99] Capasso F and Kiehl R A 1986 *J. Appl. Phys.* **58** 1366

[100] Capasso F, Sen S, Cho A Y and Sivco D L 1988 *Appl. Phys. Lett.* **53** 1056

[101] Luryi S and Capasso F 1985 *Appl. Phys. Lett.* **47** 1347

[102] Sakaki H 1980 *Japan. J. Appl. Phys.* **19** L735

[103] For an overview of resonant tunnelling between regions of different dimensionality see Tarucha S, Tokura Y and Hirayama Y 1992 *Nanostructures and Mesoscopic Systems* ed W P Kirk and M A Reed (New York: Academic) pp 153–63

[104] Clausen E M Jr, Craighead H G, Worlock J M, Harbison J P, Schiavone L M, Florez L and van der Gaag B P 1989 *Appl. Phys. Lett.* **55** 1427

[105] Mayer G, Maile B E, Germann R, Forchel A, Grambow P and Meier H P 1990 *Appl. Phys. Lett.* **56** 2016

[106] Ismail K E, Bagwell P F, Orlando T P, Antoniadis D A and Smith H I 1991 *Proc. IEEE* **79** 1106

[107] Chou S Y, Allee D R, Pease R F and Harris J S 1991 *Proc. IEEE* **79** 1131

[108] Buot F A 1987 *Int. J. Comput. Math. Electron. Elect. Eng.* COMPEL **6** 45

[109] See papers in Kirk W P and Reed M A (ed) 1992 *Nanostructures and Mesoscopic Systems* (New York: Academic)

[110] Ismail K E, Bagwell P F, Orlando T P, Antoniadis D A and Smith H I 1991 *Proc. IEEE* ed R F W Pease p 1112 (figure published with permission)

[111] Arakawa Y and Sakaki H 1982 *Appl. Phys. Lett.* **40** 639
Capasso F, Mohammed K, Cho A Y, Hull R and Hutchinson A L 1989 *Appl. Phys. Lett.* **47** 420

[112] Miller D A B, Chemla D S, Damen T C, Gossard A C, Wiegman W, Wood T H and Burrus C A 1984 *Appl. Phys. Lett.* **45** 13

[113] Sakaki H, Kato K and Yoshimura H 1990 *Appl. Phys. Lett.* **57** 2800

[114] Salvino R E and Buot F A 1992 *J. Appl. Phys.* **72** 5975

[115] Hess K, Leburton J P and Ravaioli U (ed) 1991 *Computational Electronics* (New York: Kluwer)
See also *Proc. Int. Workshop on Computational Electronics (Urbana-Champaign, IL, 1992)*

[116] Buot F A and Jensen K L 1990 *Phys. Rev.* B **42** 9429

[117] Jensen K L and Buot F A 1991 *Phys. Rev. Lett.* **66** 1078

[118] Buot F A and Jensen K L 1991 *Int. J. Comput. Math. Electron. Elect. Eng.* COMPEL **10** 241
Buot F A and Rajagopal A K 1993 *Proc. Int. Workshop on Computational Electronics (Leeds,1993)* ed C M Snowden (Leeds: University of Leeds) p 85

[119] Ting D Z-Y, Yu E T and McGill T C 1992 *Phys. Rev.* B **45** 3583

[120] Buot F A 1992 *Superlatt. Microstruct.* **11** 103

[121] Chang L L and Esaki L 1992 *Phys. Today* October 36

[122] Collins D A, Chow D H, Yu E T, Ting D Z-Y and Watson T J *Resonant Tunneling in Semiconductors* ed L L Chang, E E Mendez and C Tejedor (New York: Plenum) pp 515–28

[123] Sakaki H 1992 *Surf. Sci.* **267** 623

PART 9

ASSESSMENT OF TECHNOLOGICAL IMPACT

Chapter 21

Note from the editors

A S Edelstein and R C Cammarata

We thought it would be appropriate to close by presenting assessments of the present and likely future technological impact of nanomaterials and nanotechnology. The contributors of this part were chosen so that nearly all areas would be covered by people who had familiarity with those areas. In some fields, such as using nanomaterials as catalysts, there is no doubt that nanomaterials are already very important, whereas other areas are in a much earlier stage of development. Nevertheless, some discoveries have been put to use very quickly. It is impressive that even though giant magnetoresistance was only discovered in multilayer films in 1988, IBM has already made prototype read heads which use this phenomenon [1].

REFERENCE

[1] Baibich M N, Broto J M, Fert A, Nguyen van Dau F, Petroff F, Etienne P, Creuzet G, Friederich A and Chazeles J 1988 *Phys. Rev. Lett.* **61** 2472

Chapter 22

General overview

W M Tolles and B B Rath

The ability to fabricate nanostructures and to exploit their special properties is gaining widespread attention as a challenging frontier. Characterization and theory related to macroscopic and atomic/molecular behavior have received major emphasis for decades. The behavior of mesostructures (intermediate between macroscopic and molecular dimensions) exhibiting properties not readily explained by either macroscopic or atomic/molecular models has proved to be a recent research endeavor. Nobel Prize winning discoveries such as the scanning tunneling microscope [1] and the quantum Hall effect [2] provide new methods for examining mesostructures. The unexpected but readily understood quantization of conductance of mesostructures (the Landauer relationship [3]) represents another significant step in recognizing the unique properties of these materials. Mechanical parameters are dominated by the properties of the surfaces and interfaces when a significant fraction of a sample is represented by the interfacial region of nanostructures. New tools are being developed to observe and understand the behavior of mesoscopic structures, individually or in bulk. It is attractive to consider innovative material properties and the consequent exploitation for technological gain.

22.1 MECHANICAL PROPERTIES

The mechanical behavior of nanostructures are dominated by the nature of the interfaces in these structures. Properties such as strength, toughness, and crack initiation and propagation exhibit significantly different and often enhanced values compared to those for bulk solids [4]. Some attempts to use the improved behavior of nanomaterials have run up against prohibitive cost or technological barriers that impede the ability to make these materials in large quantities. Research to overcome the barriers offers an opportunity for innovation.

Materials designed and fabricated for their mechanical behavior are usually required in bulk quantities. Laboratory techniques used to demonstrate the behavior of milligram or gram samples are of some value to show the feasibility of these new materials. However, to be incorporated into technology, larger quantities of material must be produced at competitive cost. This points to a requirement to develop cost-effective methods of producing these materials in bulk. The physical and chemical methods for such preparation are mentioned below.

22.1.1 Physical methods

By controlling the thermomechanical processing history of a sample, the grain size may be modified to a surprising degree, with the expected consequences in the material properties. A notable example is CoWC [5] which demonstrates a readily obtained composite of nanostructures; this material exhibits a hardness making it attractive for machine cutting tools. Rapid solidification with controlled processing conditions and chemical composition produces grain structures over a wide range from being similar to glasses to submicron-size crystallites. Control of the annealing process to produce nanostructures is problematic. Rapid cooling rates may not always result in nanostructures, whereas slow cooling rates always produce relatively large grain sizes. Further, it appears as if nature conspires against an easy solution. When nanometer-sized grains have been obtained in metallic materials, grain growth, precipitation, phase transformations, and Ostwald ripening processes transform a desired nanomaterial into one with larger grain sizes. This occurs because of the thermodynamic properties that naturally increase the dimensions. Materials and processes must be carefully chosen. Nature has provided proof of this principle with attractive examples such as those listed by Dickson in chapter 18.

Rapid condensation of metallic vapor produces fine-grained structures. This attractive process is very flexible, and quite suitable for preparing small quantities of material in order to demonstrate desired properties in the laboratory. Research involving this approach to produce nanostructured materials must also recognize the above-mentioned coarsening processes. Thus there is a need to choose the materials and process variables carefully [6–8].

22.1.2 Chemical methods

Material transformation through reactions using chemical precursors can readily produce nanostructures in bulk quantities. Thermodynamic drivers for the production of certain chemicals are reasonably well understood, although the kinetic principles leading to a desired morphological structure from chemical reactions are less understood. This represents an interesting area for research. The attraction is the possibility that control of morphology may lead to highly desired properties.

The sol–gel process has successfully demonstrated the ability to make nanostructures [9]. In this process, glassy structures form in which nanometer-sized voids may contain other materials. Subsequent chemical and physical treatment of these sol–gels can lead to bulk quantities of material containing the gel, with or without inclusions, having the desired grain sizes.

Another approach is referred to as 'self-assembly'. This refers to the thermodynamic and kinetic processes that lead to a product with an ordered structure. The sol–gel process is but one of a wide variety of self-assembly processes. Another process of interest due to its unique approach is the use of vesicles (self-assembled lipid layers in spherical geometry) to cover nanostructures, hence preventing coagulation and ripening processes until the later stages of processing [10]. Understanding the complex interplay between thermodynamics and kinetics in order to take advantage of nature's processes is a challenge.

22.1.3 Composites

The admixture of nanostructures with a matrix material (organic, metallic, or ceramic) leads to a vast variety of composite materials having superior properties. The recent discovery of carbon tubules having nanometer dimensions [11] indicates the vitality and continued opportunity for unexpected discovery of new nanometer-sized materials. These nanotubes of carbon appear to have surprising mechanical and electronic properties. They may offer yet another improvement in material properties if researchers can successfully demonstrate cost-effective methods of synthesis and fabrication.

22.1.4 Material characterization

The conventional tools for characterizing materials include diffraction and microscopic examination. Instrumental developments along with automation have produced very powerful tools to better understand nanostructures. The newly developed and still advancing tools associated with proximal probes (scanning tunneling microscopy, atomic force microscopy, etc) are adding considerable opportunity to understanding the processes and properties of nanostructures. Surface techniques involving scattering along with the enhanced resolution available with these techniques offer yet additional instrumentation with which to understand nanostructures.

22.1.5 Properties and applications

Dominant mechanisms which control material strength are the elastic–plastic characteristics and failure. Ultimate failure results from crack nucleation and migration that frequently takes place along interfaces. By introducing barriers or inhibitors to delay or reduce the crack migration process, increased strength may be introduced into materials. It has been well established [4] that materials with smaller grain sizes (down to about 100 Å) are stronger. Composites that introduce barriers to slip and crack migration within a matrix operate on this principle. Nanostructured composites offer an increased density of inhibitors to slip crack migration, leading to enhanced mechanical properties. A large percentage of material is in the interfacial layer when the surface layer thickness is comparable to the dimensions of the particle. This interfacial layer has different properties than the bulk material within a grain. Thus, the inherent mechanical properties of nanostructured materials are different from those of the usually measured bulk properties. A number of research efforts have examined the properties of nanocrystalline materials and found them to be quite different [12] (factors of two are frequent, and some parameters change by as much as a factor of ten). Utilizing the properties of nanostructures for desired goals remains the challenge.

Another effect of interfaces in crystalline solids is their role in high-temperature creep. Since one of the predominant mechanisms of creep involves atomic transport along the grain boundaries, a material that is largely made up of grain boundaries will exhibit greater creep rates. The higher creep rates lead to superplasticity. This may or may not be desirable depending on the materials and the application. In the case of ceramics, superplasticity is desirable to enhance the formability of the material. However, metallic materials are often ductile even with large grain size and may have unacceptable levels of creep when made in nanocrystalline form, particularly if they are subjected to temperature excursions.

Additional properties that are inherent in nanostructures are the reduced variation in surface roughness and crystallographic texture. Whatever the material, particularly when grown from the vapor phase, a surface is uniform only to the dimensions of the grains or subunits from which it is composed. Materials with smaller crystallites can lead to improved machinability and surface finish, which for hard to machine materials provides a significant fabrication advantage [13]. This feature of the nanocrystalline aggregates is particularly important for diamond films. The recent discovery of new processes to prepare diamond without the usual application of high temperature/pressure has offered many new opportunities [14]. Nanostructures of diamond crystals adhering to each other and to a surface have become recognized as an important adjunct to modifying the properties of surfaces. Nanocrystalline diamond films can be used for many industrial applications without major polishing requirements. Research is under way to formulate advanced composites with diamond as a second-phase particulate.

22.2 ELECTRONIC, MAGNETIC, AND OPTICAL PROPERTIES

22.2.1 Semiconductors

The properties of structured semiconductors are one of the most intensely pursued high-technology areas today. There is no question of the importance attributed to these materials. Advances have enhanced the performance of semiconductor devices considerably, with improvements continuing through the exploitation of new properties engineered through material developments.

Beyond the silicon revolution are more specialized revolutions involving the two-dimensional structures formed from processes such as molecular beam epitaxy (MBE) and organo-metallic vapor phase epitaxy (OMVPE). Atomically smooth layers of only a few atomic dimensions are produced. Bandgap engineering through the fabrication of suitable materials and layer thicknesses represents a highly flexible approach to materials by design [15]. The term 'two-dimensional electron gas' has become commonplace as the carriers appear to have high mobility in the planes. Constraining the length in the second and third dimensions has resulted in many new properties beyond those obtained by simply scaling existing properties to smaller dimensions.

The production of micron or nanometer sizes in the second and third dimensions has usually taken place through the use of lithographic processes. Extending the lithographic process to dimensions smaller than the wavelength of the irradiating source has always been a major challenge. Phase contrast techniques applied to optical lithography [16] have provided surprising life to the continual trend of extending optical lithography to sub-micron dimensions. Anticipating an ultimate limit to optical techniques, shorter wavelengths such as those of x-rays are being exploited in anticipation of the requirements for smaller structures. Additionally, electron-beam projection techniques may give high-resolution exposure to resists with diffraction limits orders of magnitude below those associated with optical imaging.

Irradiation by electromagnetic or particle radiation is the current method for parallel production of microscopic structures. Parallel production is important (i.e. the production of many structures simultaneously) in order to obtain large numbers (billions) of transistors in a reasonable time period. Serial methods of producing structures through electron-beam exposure and lithographic development techniques are capable of making structures of 100 Å. However, currently these are only marginally reproducible at best and can be produced only serially, thus taking prohibitive amounts of time for mass production. Parallel techniques are quite likely to be the only way to fabricate large numbers of nanostructures for commercial applications.

A micro-/nanometer mask, produced by serial or parallel lithographic techniques, may serve as a pattern to produce images and subsequent nanostructures with the same basic pattern [17]. Less recognized for pattern production are processes able to use masks fabricated from other than lithographic processes. For example, it is found that by drawing concentric glass tubing over multiple cycles, and by carefully arranging the glass bundles in each draw, surprisingly regular arrays of holes in glass wafers can be fabricated. The dimensions of the holes in these wafers have been made as small as 30 nm, with a further reduction of these dimensions likely. Such laboratory materials can serve either as a template or mask in the subsequent production of quantum dots or quantum wires. Multiple exposure to different materials suggests nanostructures may be possible with these techniques that involve several different materials. The search continues for alternatives with innovative masking and lithographic techniques such as these.

Nature has provided a surprising number of nanostructures through thermodynamic and kinetic pathways. This may be referred to as 'self-assembly'. For example, a two-dimensional layer of a monomolecular film readily forms at the surface of a liquid for appropriate chemicals such as surfactants. Capturing this on a solid plate to produce Langmuir–Blodgett films is a familiar and popular technique. These films even demonstrate some degree of crystallinity and produce an ordered array of molecular dimensions. Chemisorption of molecules at a surface likewise is shown to produce an epitaxial molecular film with very few (perhaps one per cm^2) imperfections. Lithographically produced patterns along with subsequent development and processing offer great flexibility for producing sub-micron patterns of considerable complexity. Biological molecules having nanometer or even tens of nanometer dimensions also organize regularly. Transduction of information among such regular arrays represents the basic rudiments of communication. Harnessing these self-assembly methods for practical use represents a challenge that should be approached with the full recognition of the scientific opportunities and a careful definition of a pathway which may lead to scientific understanding or technological development.

22.2.2 Characterization

Characterization of the electronic and optical properties of nanostructures involves a great variety of instruments and techniques. Surface analysis represents a useful set of procedures, since a sizeable fraction of the atoms lies on or within a few monolayers of the surface. Spectroscopic methods such as x-ray photoelectron spectroscopy (XPS) and Auger microscopy provide valuable information about the chemical environment for selected species. Scattering techniques such as secondary ion mass spectroscopy (SIMS), and low-energy electron diffraction spectroscopy (LEEDS) provide near-surface chemical and structural information.

Of considerable importance to the field has been the development of proximal probes. The scanning tunneling microscope and the associated abilities to modify surface structures at the atomic level have proved to be remarkable tools for gathering information as well as for modifying the surface, through lithographic or other approaches. Structural damage has been shown to take place through displacements of atoms one or two atomic layers beneath the surface by electrons having only a few eV energy. The possibilities for examining and modifying materials seem to have expanded greatly with these tools. These procedures seem to be effective as a serial tool at present; the possibility that they may be applied in parallel seems attractive if the technological challenges involving control of multiple tunneling tips can be resolved satisfactorily.

22.2.3 Properties and applications

The electronic and optical properties of mesostructures do not resemble those extrapolated from larger dimensions. Tunneling and confinement effects change material properties considerably. Potential barriers along the transport direction or as boundary conditions introduce major perturbations to the solution of wave equations which change the properties of structures having dimensions of 10 nm or less.

Resonant tunneling transistors (RTTs) are high-speed devices with a useful negative transconductance [18], and are likely to become important new devices in the market place. These devices are fabricated by using the close tolerances of atomically smooth layers from MBE and are referred to as vertical structures (the transport is perpendicular to the plane of the surface). The negative transconductance exhibited with RTTs can be demonstrated with many alternative geometries, including those with fingers serving as a transistor gate, which produces an alternating potential energy field to the migrating charge carriers.

Confinement effects modify the energy levels much as the quantized energy levels for a particle in a box. The bandgap is typically increased with smaller structures, causing a blue-shift for resonances which are characteristic of most materials [19]. Confinement effects may serve to tune bandgaps for utility in semiconductor fabrication. This introduces interesting optical effects, including large nonlinear susceptibility coefficients as well as new lasing media with high gain. The dimensions of nanostructures are such that light is scattered minimally, hence optical transmission through composites is high, and they appear as transparent (or colored) materials.

Another phenomenon appears which is due to the discrete nature of the single elementary charge of an electron. As charge carriers migrate along nanostructures, they charge capacitors which are small, so small, in fact, that the presence of a single electron has a substantial influence on the voltage across the capacitance. This, in turn, influences the transport of additional charge carriers to that location in space, hence strong correlation effects are observed at low temperatures with these small structures [20]. There have been some reports of possible effects at room temperature [21], when charge migration across a small dielectric (such as a molecule between a tunneling tip and a surface) is observed.

Magnetic properties of nanostructures likewise provide a fertile ground for new discoveries. Thin layers of magnetic materials such as iron, in conjunction with chalcogenides in intervening layers, show evidence of high anisotropies and internal fields perpendicular to the plane instead of the parallel internal fields that usually occur. These materials show a large change in their resistance as the magnetic field is changed (magnetoresistive effect) [22], and appear to be promising for read heads. They also exhibit important properties needed for nonvolatile memory devices.

The ability to fabricate microstructures using lithographic techniques has been extended to the fabrication of machines consisting of moving structures, gears, levers, etc, characteristic of larger devices [23, 24]. A 10 μm gear, however, takes considerable expertise to fabricate and control. The community of micromechanical electromechanical machines (MEMS) is demonstrating a great deal of ingenuity for fabricating mechanical devices and sensors (on a chip) using lithographic techniques. This innovative field has demonstrated cost-effective alternatives for a number of sensors, and is likely to find introduction into the world of technology in many ways.

Catalysts have been used in the production of industrially important materials for decades. Of course, heterogeneous catalysis is highly dependent on two-phase reactions, and is enhanced by increasing the surface area exposed to a chemical reaction. As nanostructures together with the ability to make catalytically active surfaces and the mechanisms involving catalytic action are understood in greater detail, enhanced processing capabilities for the production of important and less costly materials should be discovered.

22.3 CONCLUSION

In summary, the advancing frontier of nanoscience and nanotechnology appears to offer exciting scientific and technological challenges. It is a world which brings together the macroscopic, yet increasingly smaller world of electronics and material behavior and the increasingly sophisticated world involving clusters and the behavior of large molecules. The interplay of new ideas in this frontier will be attractive for some time to come.

REFERENCES

[1] Binnig G and Rohrer H 1982 *Helv. Phys. Acta* **55** 726

[2] 1990 *High Magnetic Fields in Semiconductor Physics III: Quantum Hall Effect, Transport and Optics: Proc. Int. Conf. (Würzburg, 1990)* ed G Landwehr (New York: Springer)

[3] Smith H I and Craighead H G 1990 Nanofabrication *Phys. Today* February 24

[4] Chokshi A H, Rosen A, Karch J and Gleiter H 1989 *Scr. Metall.* **23** 1679

[5] McCandlish L E, Kear B H and Kim B K 1990 *Mater. Sci. Technol.* **6** 953

[6] Provenzano V, Louat N P, Sadananda K, Imam M A, Skowronek C J, Calvert J and Rath B B 1989 *Symp. on Surface Modification Technologies II* ed T S Sudarshan and D G Bhat pp 313–21

[7] Provenzano V, Louat N P, Imam M A and Sadananda K 1990 *Scr. Metall. Mater.* **24** 2065

[8] Provenzano V, Louat N P, Imam M A and Sadananda K 1993 *Nanostruct. Mater.* **1** 89

[9] Guizard C, Julbe A, Larbot A and Cot I 1992 *J. Alloys Compounds* **188** 8

[10] Liu H, Graff G L, Sarikaya M and Aksay I A 1991 *Materials Synthesis Based on Biological Processes* (Pittsburgh, PA: Materials Research Society) p 115

[11] Dresselhaus M S, Dresselhaus G and Eklund P C 1993 *J. Mater. Res.* **8** 2054

[12] Suryanarayana C and Froes F H 1989 *Nanocrystalline Metals: a Review, Physical Chemistry of Powder Metals Production and Processing (St. Mary's, PA, 1989)*
[13] Franks A 1987 Nanotechnology *J. Phys. E: Sci. Instrum.* **20** 1442
[14] Yarbrough W A 1992 *J. Am. Ceram. Soc.* **75** 3179
[15] Capasso F 1988 *Nucl. Instrum. Methods Phys. Res.* A **265** 112
[16] Levenson M D 1993 *Phys. Today* **46** 28
[17] Tonucci R J, Justus B L, Campillo A J and Ford C E 1992 *Science* **258** 783
[18] Sollner T C L G, Goodhue W D, Tannenwald P E, Parker C D and Peck D D 1983 *Appl. Phys. Lett.* **43** 588
[19] Wilcoxon J P, Williamson R L and Baughman R 1993 *J. Chem. Phys.* **98** 9933
[20] Averin D V and Likharev K K 1992 *Single-Electron Tunneling and Mesoscopic Devices (Springer Series in Electronics and Photonics 31)* ed H Koch and H Lübbig (Berlin: Springer)
[21] Nejoh H 1992 *Bull. Am. Phys. Soc.* **37** 188
[22] Hadjipanayis G C and Prinz G A (ed) 1991 *Science and Technology of Nanostructured Magnetic Materials* (New York: Plenum)
[23] Howe R T, Muller R S, Gabriel K J and Trimmer W S N 1990 *IEEE Spectrum* **27** 29
[24] Benson B, Sage A P and Cook G 1993 *IEEE Trans. Eng. Mgt.* **EM-40** 114

Chapter 23

Impact on chemistry and related technology†

James S Murday

23.1 INTRODUCTION

The chemical industry is one of a few components in the US industrial base which is sustaining a positive balance of trade [1]; innovative technology which sustains this strength is important. In a very real sense, chemistry has always been a materials science performed at the sub-nanometer scale; synthetic chemistry manipulates atoms and molecules which are that size. Chemistry has slowly been building the capability to handle the larger, more complex macromolecular structures and has recently utilized the term supramolecular chemistry to address structure in extended assemblies [2]. As Rolison points out in chapter 12, chemists have also been quick to utilize the unique properties of nanometer-scale materials, especially in catalysis where about 90% of all processes in the chemistry and petrochemical industry employ dispersed heterogeneous catalysts [3]. In contrast to chemistry, the physics and materials science communities have been developing the capability to understand and manipulate chunks of solid material in ever smaller pieces. In the last decade these three disciplines have all arrived at the nanometer scale in their level of sophistication—chemistry from below, physics/materials science from above. Not coincidentally, biology has also discovered its molecular/macromolecular/supramolecular basis.

The development of scanning tunneling microscopy in 1981 stimulated rapid progress in the analytical capability for measuring and manipulating nanometer-sized structures. The ensuing revolution in analytical capability may be discerned from the historical perspective in figure 23.1. Four classes of proximal probes have been established—tunneling, field emission, near field,

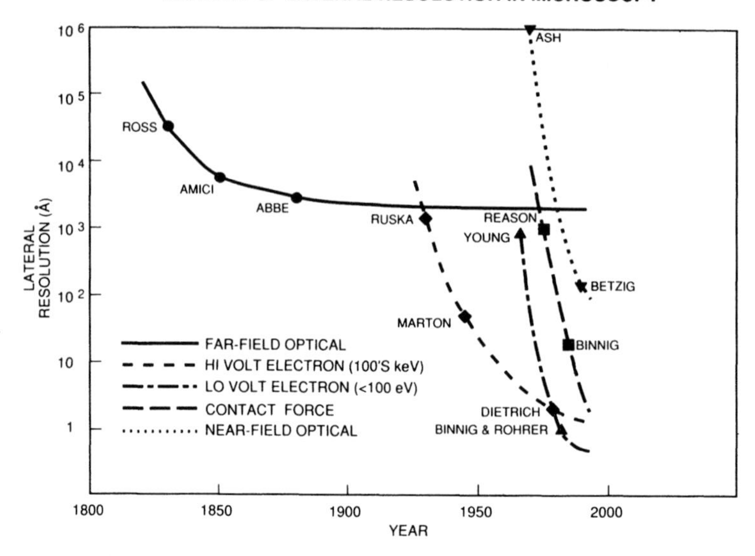

Figure 23.1. Historical perspective of resolution in microscopy.

and force—which derive their capabilities to measure localized properties at nanometer dimensions from the proximity between probe and surface [4–13]. All four classes have progressed in the last two decades into size scales previously the sole domain of electron microscopy. A multiplicity of microscopies is crucial to rapid progress in materials science and chemistry because each microscopy is limited in the properties it can measure—the proverbial story of blind men seeking to describe an elephant with each sensing only a limited part of the animal. The proximal probes are not limited to imaging; they can also directly measure chemical and physical properties at nanometer dimensions.

Further, the proximal probes enable new fabrication/manipulation techniques. These techniques can literally construct material structures atom by atom [14–16], write nanometer-dimensioned lithographic patterns [17–19], build clusters [16], and even interact with living cells [20–22].

The next section will briefly touch some areas of chemistry where the impact of nanometer-scale structures and proximal probes might be expected to have consequences for chemistry technology. References are provided for those who wish to pursue an area in more detail.

23.2 IMPACT ON CHEMISTRY AND RELATED TECHNOLOGY

23.2.1 Analytical chemistry: chemical sensors

The rapid improvements in computer capability in the last 50 years, especially in the last two decades, have provided significant enhancements in our ability to process and store data. Further dramatic progress is presently limited by software/display capability at the back end and by data acquisition rates (sensors) at the front end. Military and automotive needs have spearheaded the development of sensors for the detection of many phenomena—electromagnetic radiation, temperature, pressure as examples. Proximal probe techniques are being adapted to provide microfabricated variants of those sensors [23]. Environmental and medical concerns are compelling equivalent development of chemical/biological sensors [24]. There are hundreds of thousands of chemical species and, while any given environment will contain a substantially reduced subset, the demands on selectivity are clearly significant. The requirements on chemical sensors also have the usual demands for small size, low power, fast response, high sensitivity, and low cost. These latter requirements are the same ones as have driven electronic devices into nanometer dimensions.

Nanometer-scale materials will play at least two roles in the development of microfabricated chemical sensors. First, biology provides important lessons in the use of nanometer-sized structures to impart selectivity through the use of selective binding and topological conformation. These recognition phenomena must be better understood and applied to selective sensing; some initial research along this line is proving successful [25, 26]. In a demonstration of selectivity, coupled with the ultimate in sensitivity, Lee *et al* [27] have modified a force microscope and show evidence for the detection of a single streptavidin/biotin binding event and of complementary single-stranded DNA chains pairing.

The requirements for rapid response in a chemical sensor compel the second role for small structures. The acquisition of molecules from a fluid ambient requires transport in the fluid media. The transport distance must be kept to a minimum, especially in any condensed phases which are incorporated in the sensor. Small dimensions will be important for fast response times.

23.2.2 Lithography

The electronics industry has been reducing the size of electronic devices at approximately a factor of two every two years [24, 28]. By the year 2000, the device sizes will be approximately 0.1 micron and the spatial definition of their interfaces 1–10 nm. There are two very significant problems at these dimensions. The first is ignorance of the physical properties inherent to the nanometer-sized material structures. Many, if not most, of the phenomena we utilize in electronics structures have critical scale lengths in the nanometer domain [28–34]. The coincidence of structural size with: the electron wavelength indicates that quantum phenomena will become paramount; and the electron inelastic mean free path indicates that coherent effects will lead to quantum effects in transport. The small structures will mean that neither one electron nor large population statistics will be adequate to describe the relevant states. Other phenomena utilized in electronics, such as superconductivity [35], magnetism [36], and optics [37], will also have different properties. The second problem, the fabrication of the individual device and integrated circuit structures, will require substantial innovations in materials processing [24, 34, 38, 39]. The control of nucleation/growth (material addition) and etching (material removal) must be performed with near-atomic-scale precision. In order to do this, the chemical behavior of and around small structures must be accurately known.

The patterning of small structures by the lithographic process is accomplished by materials being exposed to a scanned focused beam or to a masked spread beam. As the patterned structure has grown smaller, higher-energy beam sources [19, 38, 39] have been necessary to maintain dimensional definition. Resist materials and materials processing have been developed to meet the need. Below 0.1 micron, present technologies have serious resolution problems and the high beam energy frequently also causes unwanted ancillary damage. Proximal probes can provide scanned, low-energy beams which have the required spatial resolution [16–19]. However, new chemistries will be required for pattern imprint and development; there are several recently published new chemical approaches to lithography utilizing low-energy electrons [19, 40–43].

23.2.3 Chemical vapor/electrochemical deposition of films/coatings: nucleation/growth

Electronic, magnetic, and electro-optic devices have large commercial markets and are strongly dependent on quality nucleation and growth processes [44]. In addition, many coatings such as metal coated plastics depend on inexpensive, yet reliable, deposition technology. In many of these applications chemical vapor or electrochemical deposition are the preferred modes. Nucleation and growth of solid phase materials from a fluid depend on localized atomic and electronic structure. The proximal probes are particularly powerful in this application because they probe the local surface properties which are exactly the properties governing the nucleation/growth phenomena. Previous surface science tools lacked two important qualities—the ability to probe the nanometer scale and the ability to work *in situ* (at the solid/fluid interface). There is a large and rapidly growing literature on nanometer-scale studies of nucleation/growth phenomena, especially for electronics materials [45, 46].

23.2.4 Maintainability and reliability: corrosion/adhesion/tribology

Corrosion, adhesive failure, friction, and wear are phenomena which have plagued human technology for millennia; estimates of their dollar cost to the US economy alone range into the tens of billions. There is a picture amongst the Egyptian hieroglyphs which shows attempts to reduce friction over three thousand years ago. Why are we still losing billions of dollars yearly to these scourges in spite of large amounts of scientific research? The demands of ever higher performance and system complexity are part of the answer. It is also true that we lack fundamental understanding of the nanometer-scale phenomena which dominate their chemical/physical processes.

The force of adhesion has several components [47, 48]—chemical bonding due to the sharing of electrons; physical bonding due to long-range forces associated with non-homogeneous distribution of charges; and mechanical bonding due to interlocking solid structures. Adhesion has long been promoted by surface treatments that are nanometer scale in dimension—generally a surface treatment to enhance chemical or surface forces. Surface preparations, adding 'tooth' to the surface or grading the mechanical properties, play major roles as well. It is likely that three-dimensional nanometer-scale materials will enhance adhesion, either by providing tailored valence electron states which promote chemical bonding between otherwise dissimilar materials, or by modifying the failure mechanisms through graded interface properties in the adhesion zone. Research to understand and tailor mechanical behavior on the nanometer scale is an active topic [48–54].

Tribology—the science of friction, wear, and lubrication—also depends critically on the mechanical behavior of structures in the nanometer scale [55–57]. It was recognized long ago that asperity contact dominated the phenomena in tribology. Asperities come in all scales, but contact will induce plastic/fracture deformations that result in reduced roughness. The mechanical properties of the asperities are clearly important. For more highly polished surfaces, either by virtue of fabrication or run in, potentially novel mechanical characteristics of nanometer-scale materials hold the promise for enhanced performance, especially for conditions of boundary lubrication. A new look at boundary lubricant fundamentals with a nanometer-scale perspective is under way [58–60].

Corrosion depends on the surface chemical reactivity and can be strongly dependent on localized perturbations in electronic states and/or structure [61, 62]. While it is not certain that nanometer-scale materials will dramatically reduce corrosion, it is conceivable they will be able to tailor the surface electronic states into a more passive configuration.

23.2.5 Macromolecules/biochemistry

Biotechnology has been identified by the US President's Office of Science and Technology Policy and by the Federal Coordinating Council for Science,

Engineering and Technology as a key to future US commercial competitiveness. The marriage of chemistry, physics, materials, and biology, coupled with nanostructural capability, is a key to rapid progress in biotechnology [63, 64].

The chain folding and the dynamical processes of macromolecules/supramolecules, and the recognition processes in biological systems (e.g. antibody/antigen recognition, enzymatic action, etc) depend on molecular structures with nanometer length scales. A recent review of the folding problem for proteins points out that this is one of the fundamental problems in biophysical science [65]. The advent of the proximal probes provides analytical capability which enable the definition of external structures and their properties [66, 67]. Those studies may be done in a fluid ambient, an extremely important consideration for the biological systems which are strongly influenced by waters of hydration.

Biochemical techniques have been developed which permit the construction of a great diversity of macromolecular structures [68–70]. This includes the synthesis of polymeric systems from nucleic acids, amino acids, and sugars. Each of these molecular building blocks is of nanometer size. The structure of these macromolecules (e.g. DNA, RNA, polypeptides, polysaccharides) is being investigated with some promise of being able to identify the polymer sequencing and secondary structure [71–73]. The synthesis of these deliberately sequenced polymers has important consequences for medical reasons, but it is also the subject of study for other functions [64].

Biological systems tend to construct macroscopic components by assembling molecules into the final form. In contrast, most materials fabrication is by cutting/machining large pieces into desired smaller conformations. The self-assembly approach has at least two conceptual advantages—thermodynamical driving forces lead to the desired end state with few flaws and only the minimum of material is necessary. Chemists are now examining this approach to the manufacture of new materials [47, 74–78].

23.2.6 Clusters

The importance of small structures in promoting catalytic chemical reactions is well documented [3]. The field and literature are far too large to address here. Suffice it to say that both the electronic structure and surface topology are known to be important. The progress in the fabrication and understanding of nanometer structures will significantly contribute to the development of this commercially important topic.

In addition to their catalytic potential, clusters may also serve as new reagents for chemical reactions [79]. The most prominent recent examples are the fullerenes [80] which contain 60 and more carbon atoms. The ellipsoidal fullerenes are the subject of many efforts to explore reactions which produce technologically interesting materials [81]; the most successful example has been demonstration of intermediate-temperature superconductivity [82]. The tubular fullerenes are of interest for their potential electrical conductivity [83, 84]. In another example of clusters as reagents, Weaver and Waddill [85] have demonstrated their use in the growth of electronic films with superior properties.

23.3 CONCLUSION

Research on nanometer-scale materials will have profound impact in many areas of chemistry and the proximal probes provide the 'eyes' and 'fingers' for that research [86]. Catalytic structures will continue to provide new, more economical ways of manufacturing materials. The generation of 21st century electronic devices and computers will almost certainly depend on nanometer structures. However, the strongest economic driver is likely to be biomaterials and medicine where the confluence of chemistry, physics, and materials science at the nanometer scale will provide the understanding necessary to unravel and control the processes of life.

REFERENCES

[1] 1993 *Chem. Eng. News* **71**(26) 38

[2] Bein T (ed) 1992 *Supramolecular Architecture: Synthetic Control in Thin Films and Solids (ACS Symposium Series 499)* (Washington, DC: American Chemical Society)

[3] Knözinger H 1991 *Fundamental Aspects of Heterogeneous Catalysis Studied by Particle Beams* ed H H Brongersma and R A van Santen (New York: Plenum) p 7

[4] Murday J S and Colton R J 1990 *Chemistry and Physics of Solid Surfaces VIII* ed R Vanselow and R Howe (New York: Springer) p 347

[5] Murday J S, Colton R J and Rath B B 1993 *Synthesis, Processing, and Modelling of Advanced Materials* ed F H Froes and T Khan (Brookfield, VT: Trans Tech) p 149

[6] Wickramasinghe H K (ed) 1992 *Scanned Probe Microscopy (AIP Conf. Proc. 241)* (New York: American Institute of Physics)

[7] Bonnell D A (ed) 1993 *Scanning Tunneling Microscopy and Spectroscopy* (New York: VCH)

[8] Stroscio J A and Kaiser W J (ed) 1993 *Scanning Tunneling Microscopy, Methods of Experimental Physics* vol 27 (New York: Academic)

[9] Behm R J, Garcia N and Rohrer H 1990 *Scanning Tunneling Microscopy and Related Methods* (Dordrecht: Kluwer)

[10] Sarid D 1991 *Scanning Force Microscopy* (New York: Oxford University Press)

[11] Chen C J 1993 *Introduction to Scanning Tunneling Microscopy* (New York: Oxford University Press)

[12] Frommer J 1992 *Angew. Chem. Int. Edn. Engl.* **31** 1298

[13] Wiesendanger R 1994 *Scanning Probe Microscopy and Spectroscopy, Methods and Applications* (New York: Cambridge University Press)

[14] Avouris Ph, Lyo I-W and Hasegawa Y 1993 *J. Vac. Sci. Technol.* A **11** 1725

[15] Avouris Ph (ed) 1993 *Atomic and Nanoscale Modification of Materials* (Dordrecht: Kluwer)

[16] Stroscio J A and Eigler D M 1991 *Science* **254** 1319

[17] Marrian C R K, Dobisz E A and Dagata J A 1992 *J. Vac. Sci. Technol.* B **10** 2877

[18] Schneir J, Dagata J A and Harary H H 1993 *J. Vac. Sci. Technol.* A **11** 754

[19] Marrian C (ed) 1993 *Technology of Proximal Probe Lithography* (Bellingham, WA: SPIE)

[20] Parpura V, Haydon P G, Sakaguchi D S and Henderson E 1993 *J. Vac. Sci. Technol.* A **11** 773

[21] Hansma H G, Vesenka J, Siegerist C, Kelderman G, Morret H, Sinsheimer R L, Elings V, Bustamante C and Hansma P K 1992 *Science* **256** 1180

[22] Häberle W, Hörber J K H, Ohnesorge F, Smith D P E and Binnig G 1992 *Ultramicroscopy* **42–44** 1161

[23] Kenny T W, Kaiser W J, Podosek J A, Rockstad H K, Reynolds J K and Vote E C 1993 *J. Vac. Sci. Technol.* A **11** 797

[24] 1991 *Office of Technology Assessment Report* OTA-TCT-514 (Washington, DC: US Government Printing Office)

[25] Grate J W and Abraham M H 1991 *Sensors Actuators* B **3** 85
Grate J W, Snow A, Ballantine D S Jr, Wohltjen H, Abraham M H, McGill R A and Sasson P 1988 *Anal. Chem.* **60** 869

[26] Göpel W, Hesse J and Zemel J N (ed) 1991 *Sensors: A Comprehensive Survey, Vols 2/3 Chemical and Biochemical Sensors* (New York: VCH)
Schierbaum K D and Göpel W 1993 *Synth. Met.* **61** 37

[27] Lee G U, Kidwell D A and Colton R J 1994 *Langmuir* **10**
Lee G U, Chrisey L A and Colton R J 1994 *Science* **266** 771

[28] Bate R T 1990 *Nanotechnology* **1** 1

[29] Luscombe J H 1993 *Nanotechnology* **4** 1

[30] Capasso F and Datta S 1990 *Phys. Today* **43**(2) 74

[31] Ferry D K and Grondin R O 1991 *Physics of Submicron Devices* (New York: Plenum)

[32] Kastner M A 1993 *Phys. Today* **46**(1) 24

[33] Kirk W P and Reed M A (ed) *Nanostructure and Mesoscopic Systems, Proc. Int. Symp.* (Boston, MA: Academic)

[34] Davies J H and Long A R (ed) 1992 *Physics of Nanostructures, Proc. 38th Scottish Universities Summer School in Physics* (Philadelphia, PA: Institute of Physics)

[35] Gubser D U *et al* (ed) 1980 *Inhomogeneous Superconductors* (New York: American Institute of Physics)
Kramer B, Bergmann G and Bruynseraede Y (ed) 1985 *Localization, Interaction and Transport Phenomena* (Berlin: Springer)

[36] Awschalom D D, DiVincenzo D P and Smyth J F 1992 *Science* **258** 414

[37] 1993 *Phys. Today* **46**(6)

[38] Fleming D, Maldonado J R and Neisser M 1992 *J. Vac. Sci. Technol.* B **10** 2511

[39] 1988 *IBM J. Res. Dev.* **32** 440

[40] Marrian C R K, Perkins F K, Brandow S L, Koloski T S, Dobisz E A and Calvert J M 1994 *Appl. Phys. Lett.* **64** 390

[41] Snow E S and Campbell P M 1994 *Appl. Phys. Lett.* **64** 1932
Snow E S, Campbell P M and Shanabrook B V 1993 *Appl. Phys. Lett.* **63** 3488

[42] Lyding J W, Shen T C, Hubacek J S, Tucker J R and Abeln G C 1994 *Appl. Phys. Lett.* **64** 2010

[43] Fay P, Brockenbrough R T, Abeln G, Scott P, Agarwala S, Adesida I and Lyding J W 1994 *J. Appl. Phys.* **75** 7545

[44] Granneman E H A 1993 *Thin Solid Films* **228** 1

[45] Lagally M G 1993 *Japan. J. Appl. Phys.* **32** 1493

[46] Hamers R J, Köhler U K and Demuth J E 1990 *J. Vac. Sci. Technol.* A **8** 195

[47] Israelachvili J 1992 *Intermolecular and Surface Forces* (New York: Academic)

[48] Creuzet F, Ryschenkow G and Arribart H 1992 *J. Adhesion* **40** 15

[49] Ducker W A, Senden T J and Pashley R M 1991 *Nature* **353** 239

[50] Burnham N A, Colton R J and Pollock H M 1993 *Nanotechnology* **4** 64
See also Bonnell D A (ed) 1993 *Scanning Tunneling Microscopy (AIP Conf. Proc. 241)* (New York: American Institute of Physics) ch 7

[51] Hartmann U 1992 *Ultramicroscopy* **42–44** 59

[52] Pethica J B and Oliver W C 1989 *Thin Films: Stresses and Mechanical Properties Symp.* ed J C Bravman, W D Nix, D M Barnett and D A Smith (Pittsburgh, PA: Materials Research Society) p 13

[53] Hues S M, Draper C F and Colton R J 1994 *J. Vac. Sci. Technol.* B **12** 2211

[54] Harrison J A, White C T, Colton R J and Brenner D W 1992 *Surf. Sci.* **271** 57

[55] Nieminen J A, Sutton A P, Pethica J B and Kaski K 1992 *Modelling Simul. Mater. Sci. Eng.* **1** 83

[56] Singer I L and Pollock H M (ed) 1992 *Fundamentals of Friction: Macroscopic and Microscopic Processes (NATO Series E)* vol 220 (Dordrecht: Kluwer)

[57] Harrison J A, Colton R J, White C T and Brenner D W 1993 *Wear* **168** 127

[58] Marti O, Colchero J and Mlynek J 1993 *Nanosources and Manipulation of Atoms under High Fields and Temperatures: Applications (NATO Series E)* ed Vu Thien Binh, N Garcia and K Dransfeld (Dordrecht: Kluwer)

[59] Meyer E, Overney R, Brodbeck D, Howald L, Luthi R, Frommer J and Guntherodt H-J 1992 *Phys. Rev. Lett.* **69** 1777

[60] Harrison J A, White C T, Colton R J and Brenner D W 1993 *J. Phys. Chem.* **97** 6573

[61] Bard A J and Fan F F 1993 *Scanning Tunneling Microscopy (AIP Conf. Proc. 241)* ed D A Bonnell (New York: American Institute of Physics) ch 9

[62] Bhardwaj R C, Gonzalez-Martin A and Bockris J O'M 1992 *J. Electrochem. Soc.* **139** 1050

[63] Goode M L (ed) 1988 *Biotechnology and Materials Science: Chemistry for the Future* (Washington, DC: American Chemical Society)

[64] 1992 Biology and materials synthesis, part I *MRS Bull.* **17**; 1992 Biology and materials synthesis, part II *MRS Bull.* **17**
Aksay I A, Baer E, Sarikaya M and Tirrell D A (ed) 1992 *Hierarchically Structured Materials, Mater. Res. Soc. Symp. Proc.* vol 255 (Pittsburgh, PA: Materials Research Society)
Viney C, Case S T and Waite J H (ed) 1993 *Biomolecular Materials, Mater. Res. Soc. Symp. Proc.* vol 292 (Pittsburgh, PA: Materials Research Society)

[65] Chan H S and Dill K A 1993 *Phys. Today* **46** 24

[66] Saraf R F 1993 *Macromolecules* **26** 3623

[67] Annis B K, Schwark D W, Reffner J R, Thomas E L and Wunderlich B 1992 *Makromol. Chem.* **193** 2589

[68] Gaber B P, Schnur J M and Chapman D (ed) 1988 *Biotechnological Applications of Lipid Microstructures, Advances in Experimental Medicine and Biology* vol 238 (New York: Plenum)

[69] Himmel M E and Georgiou G (ed) *Biocatalyst Design for Stability and Specificity (ACS Symposium Series 516)* (Washington, DC: American Chemical Society)

[70] Creel H S, Krejchi M T, Mason T L, Tirrell D A, McGrath K P and Atkins E D T 1991 *J. Bioact. Compat. Poly.* **6** 326

[71] Lindsay S M 1993 *Scanning Tunneling Microscopy (AIP Conf. Proc. 241)* ed D A Bonnell (New York: American Institute of Physics) ch 10
Lindsay S M, Lyubchenko Y L, Tao N J, Li Y Q, Oden P I, DeRose J A and Pan J 1993 *J. Vac. Sci. Technol.* A **11** 808

[72] Shaiu W-L, Vesenka J, Jondle D, Henderson E and Larson D D 1993 *J. Vac. Sci. Technol.* A **11** 820

[73] Amrein M and Marti O (ed) *STM and SFM in Biology* (London: Academic)

[74] Ulman A 1991 *An Introduction to Ultrathin Organic Films* (New York: Academic)

[75] Kuhn H and Möbius D *Physical Methods of Chemistry IXB* ed B Rossiter and R C Baetzold (New York: Wiley) p 375

[76] Charych D H and Bednarski M D 1992 *MRS Bull.* **17** 61

[77] Brandow S L, Turner D C, Ratna B R and Gaber B P 1993 *Biophys. J.* **64** 898

[78] Stenger D A, Georger J H, Dulcey C S, Hickman J J, Rudolph A S, Nielsen T B, McCort S M and Calvert J M 1992 *J. Am. Chem. Soc.* **114** 8435

[79] Reynolds P J (ed) *On Clusters and Clustering, from Atoms to Fractals* (New York: North-Holland)

[80] 1992 *Accounts Chem. Res.* **25** 98

[81] Diederich F, Ettl R, Rubin Y, Whetten R L, Beck R, Alvarez M, Anz S, Sensharma D, Wudl F, Khemani K C and Koch A 1991 *Science* **252** 548
Lamb L D, Huffman D R, Workman R K, Howells S, Chen T, Sarid D and Ziolo R F 1992 *Science* **255** 1413

[82] Hebard A F, Rosseinsky M J, Haddon R C, Murphy D W, Glarum S H, Palstra T T M, Ramirez A P and Kortan A R 1991 *Nature* **350** 600
Haddon R C 1992 *Accounts Chem. Res.* **25** 127

[83] White C T, Robertson D H and Mintmire J W 1993 *Phys. Rev.* B **47** 5485

[84] Gallagher M J, Chen D, Jacobsen B P, Sarid D, Lamb L D, Tinker F A, Jiao J, Huffman D R, Seraphin S and Zhou D 1993 *Surf. Sci. Lett.* **281** L335

[85] Weaver J H and Waddill G D 1991 *Science* **251** 1444

[86] Cohen M, Poate J and Silcox J (ed) 1993 *Report of NSF Panel on Atomic Resolution Microscopy* NSF 93–73

Chapter 24

Nanostructures in electronics

H G Craighead

It is clear that nanostructured materials will have increasing impact on electronics. The requirement for higher functionality, increased memory density, and higher speed is driving the need for smaller dimensions in electronics. The scaling down of conventional devices, such as field effect transistors, to channel lengths of around 100 nm is a clear path to the miniaturization of electronics with the related benefits. It is the hope of utilizing effects obtained with an additional ten times reduction in dimensions that drives the study of nanoscale devices based on new operating principles.

Essentially all conceivable electronic devices require highly organized, non-random structures. This is in contrast to a random distribution of particles or clusters. The ordered device and circuit structures are most easily thought of as being formed by an ultrahigh-resolution lithographic process. With x-ray proximity printing expected to be usable only to around 0.1 μm, charged particle lithography or projection x-ray lithography will be required for production of devices with nanometer dimensions. It is possible that highly periodic nanostructures can be employed in certain types of device, but the architectures for such systems have yet to be developed.

Quantum effect devices or single-electron devices are of potential utility for future electronic circuits. These would rely for their operation on the existence of quantized energy levels, the wave nature of the electron transport, or the effects of discrete electron charge.

One class of quantum based devices operates on interference or other wave phenomena of the electron transport. These devices tend to be analogs of microwave or optical devices such as waveguides, modulators, or interferometers. All of these devices rely on the elimination of scattering in order to preserve the phase of the electron wave. This requires a combination of highly ordered crystals, low temperatures, and small dimensions, with the cryogenic temperatures being the most restrictive in terms of wide-scale application. There is active research in these areas.

Controlled tunneling devices have the possibility to operate at higher temperatures if the characteristic energies of the barriers can be made greater than kT (thermal energy). The combination of the growth of epitaxial semiconductors and metals with fine lateral patterning makes this an approachable objective. More complex tunneling structures such as coupled systems of quantum dots are being considered as multifunctional logic devices or memory systems.

Devices based on the manipulation of single electrons have been demonstrated. Relying on quantized charge, rather than quantum transport, a variety of such devices and circuits have been described. However, the fabrication processes required for the mass production of the necessary small dimensions have yet to be developed.

Random nanostructured systems are useful for tailoring the magnetic and optical properties of materials. By the microscopic variation of composition, the intrinsic properties of the materials can be advantageously manipulated. In magnetic materials, for example, desired anisotropies can be introduced by the shape of microscopic inclusions while the materials are chosen to optimize the coercivity, Curie temperature, or other properties. This will be important for magnetic storage media. Linear and nonlinear optical properties of materials can be varied in desirable ways by using the microscopic structure variations as an additional degree of freedom in the material design. This is having an impact on optoelectronic devices and optical storage possibilities.

There is no doubt that the inclusion of nanostructures will be necessary for revolutionary advances in electronics and related technologies.

Chapter 25

Memories and electronic devices

R T Bate

The route to progress in memory technology is cost reduction and the development of ever faster, more compact, and less power consuming memory systems, with greater and greater storage capacity. Since all of these benefits can be obtained by reducing the size of the basic storage cell, it is natural to suppose that nanotechnology will eventually play a fundamental role. However, the system must also include a means for writing and reading the cells efficiently, and this turns out to be the most demanding aspect of memory design. For example, very compact memories based on scanning tunneling microscopy (STM) and related technologies, which employ storage cells approaching the size of a single atom, have been proposed, but the difficulty of designing sufficiently fast and reliable read/write mechanisms will be exceedingly formidable.

The most probable route by which nanostructure research could lead to a successful memory technology would be via nanoelectronics research resulting in the development of an ultrahigh-density very-large-scale-integration (VLSI) integrated circuit (IC) technology. In fact, memory would be an indispensable ingredient in such a nanoelectronic technology.

Conventional integrated circuit technology will probably reach a limit of functional density near the end of the century. The limiting process for complementary metal oxide semiconductor (CMOS) will be high-temperature sub-threshold leakage currents due to drain induced barrier lowering. If system requirements dictate that upper operating temperature limits for commercial direct access memories (DRAMs) be maintained at 80 °C, then the lower limit on transistor gate lengths for DRAMs will be about 0.2 μm.

Current research in quantum-confined structures such as quantum dots may lead to digital switching devices much smaller than 0.2 μm transistors. However, significant advances in materials and chip architectures would be required before these could form the basis for an ultradense memory technology. Experience has shown that quantum-confined structures for room-temperature operation must employ heterojunction barriers. For this reason, most of the higher-temperature work on quantum-confined structures has been done in GaAs related materials, where a well developed heterojunction technology exists. On the other hand, in order for the next generation of IC technology to continue the historical cost per bit experience curve, it seems inevitable that we must use a silicon technology.

Quantum-confined structures are an active area of research in silicon, but the confinement is usually achieved by means of electrostatic, as opposed to heterojunction barriers. Such structures seem to exhibit quantum confinement effects only at very low temperatures, which makes them primarily a subject of academic interest.

The current GaAs related heterojunction technology is sufficient for proof-of-principle demonstrations of new quantum devices and integrated logic chips, but it does not provide the low-leakage 'off' state required for large-scale memory applications. This difficulty can be traced back to the heterojunctions employed. These do not provide tunneling barriers sufficiently high to block all current flow in the 'off' state. Although higher barrier heights are available in the III–V system, the process technology is rudimentary. The way out of the above dilemma is the development of a new heterojunction with silicon which provides a barrier height of at least 1 eV. This could permit the realization of quantum-confined tunneling switching devices with sufficiently small leakage currents to permit integration of a trillion devices on a single silicon chip. This presents a challenge to the materials science and condensed matter physics communities which hold the key to much of the future of nanotechnology. The choice of materials is not at all obvious. Some of the most promising work has been done with silicon/calcium difluoride heterojunctions, but it is not at all clear that this is the optimal choice.

Another problem arises from the fact that the size of the basic storage cell is not the only limitation on the further integration of semiconductor memory. The technology has reached a level of integration such that further increases in circuit density are also limited by the size of interconnections. Many materials related innovations have been introduced into conventional DRAM technology in order to delay the onset of this limit. However, it appears that, by the end of the century, further progress will require a basic change in circuit architecture. A popular area of research stimulated by this need is the so-called 'cellular automaton' architecture in which individual cells are only locally connected to other cells, and communication with the array of cells takes place only at its periphery. It is not yet clear whether or not this approach will provide a suitable solution to the interconnection problem.

Chapter 26

Nanostructured materials for magnetic recording

Ami Berkowitz

26.1 INTRODUCTION

It is likely that in terms of economic value, the magnetic recording industry is the largest user of nanostructured materials. This situation will probably persist into the foreseeable future. Magnetic materials are employed in both the information storage media and in the 'write' and 'read' heads. At present, nanostructured materials dominate in both media and heads. In the future, it is likely that virtually all media and heads will be composed of nanostructured materials. In the following, the performance considerations leading to the use of nanostructured materials will be summarized, current media and heads will be discussed, and some future possible directions will be noted. This report is very brief and qualitative. A comprehensive review of all materials and systems aspects of magnetic recording is available [1], as is a set of review articles [2]. Recent shorter presentations of materials issues in magnetic recording have also been published [3, 4].

26.2 BACKGROUND

Magnetic recording essentially involves detecting changes in the direction of magnetization in the storage medium. Associated with these divergences of the magnetization are small magnetic fields immediately above the regions of magnetization changes. These stray fields are 'read' by producing either inductive signals or resistance changes in heads as the storage medium is moved past them. The magnetic transitions are 'written' with fields produced by an inductive head. These 'write' heads consist of a loop of high-permeability magnetic material with a gap at which the write field is produced by pulsing current through windings on the magnetic loop.

In order to store the magnetization transitions at high densities (e.g. $> 10^4$ cm^{-1}), several constraints are imposed on the storage media and on the write and read heads. Consider first the storage media. With a high storage density, the distance between magnetization reversals becomes very small. This produces strong demagnetizing fields on the stored 'bit'. For stable storage, the coercive force, H_c, must be high enough to withstand these demagnetizing fields. Another media consideration is a high remanence to ensure sufficient stray field for detection of the transition. This would suggest a high saturation magnetization, M_s, for the media; however, high M_s increases the demagnetizing field. Thus, a storage medium with both high H_c and M_s would seem to be required.

These media requirements must be matched by the head properties. The higher the H_c, the stronger the field needed from the write head to produce a transition. High permeability (for low write currents) and high data rates ($>$ 100 MHz) place additional constraints on the write head materials, e.g. eddy currents must be avoided. All these constraints on media and heads are met by nano-sized structures.

26.3 MEDIA

There are two types of storage medium: particulate (e.g. tapes), and thin films (e.g. disks). In both these configurations, the basic magnetic entities are single magnetic domains, i.e. regions in which the magnetization is essentially uniform. The coercive force is a measure of the energy required to reverse the magnetization direction in these single domains. The energy barriers to magnetization reversal usually arise from either the intrinsic magnetocrystalline anisotropy energy which favors specific crystallographic directions for the magnetization direction, or from an anisotropic shape of the domain, e.g. an acicular domain in which the long axis is the lowest-energy magnetization direction.

In particulate media, the current leading particles for high-density applications are composed of an acicular iron core passivated by an oxide surface which occupies about 70% of the particle volume. The particles are less than 150 nm long with an aspect ratio of about 7. H_c of these particles is about 130 kA m^{-1} and the remanent magnetization is about 250 kA m^{-1}. Obviously, the remanence would be more than doubled if the oxide coating were not required for chemical stability. Metallic particles have the advantage over oxides of much larger M_s values, but they must be protected from corrosion. A contender for high-density recording is, indeed, an oxide, i.e. barium ferrite particles doped with Co and Ti. These particles are platelets with diameters $\approx$ 50 nm and much smaller thicknesses. The chemical stability of these materials is satisfactory, but they have a remanence about half of that of the iron particles. The barium ferrites derive their coercive force from magnetocrystalline anisotropy. A lower performance, but very widely used particulate medium consists of elongated particles of slightly reduced γ-Fe_2O_3 particles with a chemisorbed surface of Co. These particles derive their H_c from a combination of shape and magnetocrystalline anisotropies. The γ-Fe_2O_3 core particles have lengths of ≈ 200 nm with aspect ratios of ≈ 8. H_c of these particles is ≈ 25 kA m^{-1}. This H_c is due principally to shape anisotropy. When a few monolayers of Co are chemisorbed on the particles' surfaces, H_c more than doubles. Co has a very high magnetocrystalline anisotropy which is certainly involved in this enormous increase in H_c, although the details of the mechanism are still obscure. This approach of creating high H_c by surface treatment probably has much more potential than has currently been realized.

26.3.1 Film media

For several reasons, metallic film media have superior performances to particulate media. The reasons include higher M_s, thinner and smoother films, and higher packing density of the magnetic single domains. These films are produced primarily by sputtering Co based alloys [2, 5]. The grains in these films are the single magnetic domains and they derive their coercive force from magnetocrystalline anisotropy. The smallest dimension of these grains is ≈ 20 nm and the magnetic film's thickness is less than 40 nm. Bit densities up to 3 bits μm^{-2} have been demonstrated from thin films [5, 6] and future needs will require even higher densities.

26.3.2 General media issues

For both particulate and film media, the magnetic single domains must become smaller to achieve higher densities with satisfactory signal-to-noise ratio [1, 3]. A question of limits arises because the magnetic energy barrier to magnetization reversal is proportional to the volume of the single domains. As the single-domain volumes decrease, the energy barriers may become of the order of $k_B T$, where k_B is the Boltzmann constant and T is temperature. In that case magnetization can be reversed by thermal activiation and stable magnetic storage ceases to exist. Higher energy barriers may be achieved utilizing magnetocrystalline anisotropy energy. However, magnetocrystalline anisotropy energy often has an unacceptable temperature dependence. One solution may be surface-treated particles and films, as in the case of the Co surface treated γ-Fe_2O_3 noted above. When Co is introducted into the volumes of the γ-Fe_2O_3 particles, H_c is larger, but so is its temperature dependence. Surface treatment minimizes the temperature dependence of H_c. Other surface treatments have also proven effective [7], and it seems appropriate to explore this approach to high density in both films and particles [8].

26.4 RECORDING HEADS

Write heads fall into the nanostructured category since the trend is toward utilizing thin magnetic films for high density [1, 2]. These films are high-permeability metallic alloys. Therefore, for high data rates, they often are laminated with insulators [2]. To prevent saturation in ferrite heads, magnetic alloy films are often incorporated in the gap [1].

The read heads of the future will employ films with significant magnetoresistance (MR). The signal to noise ratio is superior in thin-film MR heads to that of inductive heads at high data rates [1, 2]. Current MR heads utilize permalloy ($Ni_{81}Fe_{19}$) films whose resistance changes by about 2% as the recorded bits pass by. Recently, some truly nanocrystalline materials have demonstrated resistance changes more than an order of magnitude larger [9, 10]. These films consist of magnetic particles less than 10 nm in diameter dispersed in a nonmagnetic matrix. The potential of these materials has inspired a great deal of research and development.

26.5 SUMMARY

It is obvious that nanostructured magnetic materials dominate the magnetic recording industry. Future high-density information storage systems will require even smaller nanostructures. This important industry will therefore provide many opportunities for innovative developments in ultrafine structures.

REFERENCES

[1] Mee C D and Daniel E D (ed) 1990 *magnetic Recording Handbook* (New York: McGraw-Hill)

[2] For review articles on magnetic recording materials see 1990 *MRS Bull.* **15** 23–72

[3] White R M 1990 *J. Magn. Magn. Mater.* **88** 165

[4] Richter H J 1993 *IEEE Trans. Magn.* **MAG-29** 2185

[5] Yogi T, Tsang C, Nguyen T A, Ju K, Gorman G L and Castillo G 1990 *IEEE Trans. Magn.* **MAG-26** 2271

[6] Futamoto M *et al* 1991 *IEEE Trans. Magn.* **MAG-27** 5280

[7] Spada F E, Parker F T, Nakakura C Y and Berkowitz A E 1993 *J. Magn. Magn. Mater.* **120** 129

[8] Berkowitz A E and White R M 1989 *Mater. Sci. Eng.* B **3** 413

[9] Chien C L, Ziao J Q and Jiang J S 1993 *J. Appl. Phys.* **73** 5309

[10] Berkowitz A E *et al* 1993 *J. Appl. Phys.* **73** 5320

Chapter 27

Commercialization opportunities for nanophase ceramics: a small-company perspective

John C Parker

27.1 INTRODUCTION

Some of the more publicized property benefits achieved with nanophase or nanostructured materials include metals that exhibit increased hardness [1] and the emission of light from silicon [2]. For the most part, these materials have been a laboratory curiosity, but recently a handful of companies has tried to commercialize the property benefits associated with nanophase materials. In this article, we will present some of the commercial applications for nanophase materials as they relate to bulk consolidated structures with nanometer grains, porous structures with nanophase grains and pores, and dispersions of nanophase particles in solvents.

27.2 SYNTHESIS

The precursor to any nanophase material is a weakly agglomerated powder consisting of aggregates of 1–100 nm. Many high-surface-area powders exist today, but strong agglomeration creates particles comprised of several aggregates and it often require intense mechanical input to breakdown agglomeration. A variety of techniques have been cultivated to synthesize single- and multi-element nanophase powders, but few of these techniques have demonstrated the combination of commercial production and unique material properties. As a result there are only a few companies in the world that produce in volume inorganic materials that are within or near the nanophase regime [3]. Most of the 'ultrafine' (0.1–1 μm), inorganic products available today provide marginal benefits relative to conventional, less expensive ceramic materials. Marginal property enhancements do not often justify the higher price that is commonly associated with most ultrafine and *all* nanophase materials. Therefore, to commercially validate nanophase materials, processes must first be developed to produce the materials with demonstrable property benefits, in volume, and with competitive cost and pricing.

In our own experience we began with a laboratory-scale nanocrystalline material synthesis system based on gas phase condensation (GPC) [1]. The unit was ideal for synthesizing high-surface-area, nanophase materials (especially metals) because of the pristine chemical-free environment in which the materials were produced. However, this process has limited production capability (grams/day) largely due to the use of a natural convection flow current for inhibiting particle growth.

Figure 27.1. GPC produced nanophase metal oxide ceramics designed to have weak (right) and stronger (left) particle agglomeration that leads to varying nanometer-sized porosity.

The GPC process has demonstrated a unique level of microstructural control in fabricating bulk and porous nanophase materials. The photograph in figure 27.1 shows two room-temperature consolidated nanophase metal oxide samples; one that is transparent and one that is opaque. The transparent sample was designed to have pores *and* grains well below the wavelength of visible light [4]. The transparent sample was fabricated from GPC nanophase powders that have a weak degree of agglomeration. The opaque sample was designed to have a coarser microstructure due to a slightly harder particle agglomeration. A range of nanostructures between these two extreme cases can be fabricated by controlling the GPC particle growth process.

To maintain this level of uniqueness in synthesizing nanophase materials Nanophase Technologies Corporation (NTC) has developed a process analogous to GPC that allows control over the particle size and degree of agglomeration while sustaining a 'residue-free' surface chemistry. The process has been dubbed the physical vapor synthesis process (PVS); i.e. the synthesis of nanophase powders from elemental physical vapors. The scale-up of this process continues as it moves from pilot to full scale over the next few years. Starting with costs of several hundred dollars per gram at laboratory scale, NTC has maintained the unique quality and reduced prices to tens of dollars per pound at pilot scale. Most of this effort has been focused primarily on oxide ceramic materials. This enormous reduction in cost and increase in volume for these unique materials has enabled the development of significant commercial opportunities.

27.3 BULK CONSOLIDATED STRUCTURES

Beyond synthesis, the area of processing and application of nanophase materials has gained further attention. Many property benefits have been observed with consolidated nanostructure materials and many more have been suggested [1]. An application example for consolidated nanophase materials was developed during NTC's three-year effort funded by the Advanced Technology Program (ATP) of the US Department of Commerce. This program has been devoted to the superplastic net-shape forming of oxide ceramics. Similar work was recently sponsored in a joint collaboration between Japan and Europe [5].

In the ATP program, ceramic net-shaped parts are being formed in closed die configurations under strain rates comparable to the best commercial superplastic Al metal alloys ($0.2\text{–}1.0 \times 10^{-3}$ s^{-1}) [6]. The development of this plastic flow processing (PFP) presents a new and potentially more economic route to fabricating ceramic components, akin to the impact hot isostatic processing had several years ago.

The superplastic forming of ceramics is an obvious application for nanostructured materials because of the coupling of strain rate to grain size through the grain boundary sliding mechanism [6]

$$\dot{\varepsilon} \propto A \left(\frac{\sigma}{d^2}\right)^{3/2} \tag{27.1}$$

where d is the particle size and σ the applied stress. Thus, the rate of deformation, or in a more practical sense, the production rate of parts, becomes much more rapid as the grain size decreases. For some of the prototypical parts that have been made with this laboratory-scale process, a net-shaped ceramic can be achieved in as little as 15 min. Dimensional tolerances of 0.002 inches have been achieved due to the intimate replication of die tolerances in the final part. An example of a PFP net shaped part is shown in figure 27.2.

Figure 27.2. A plastically forged ceramic component fabricated from a nanostructured preform ring. The forging operation was performed at 1400 °C in less than 15 min. The estimated time for machining to net shape is 10 h.

For some applications, especially where cost to manufacture is dominated by machining, PFP can provide an incremental improvement in the cost to manufacture. Although machining is reduced or eliminated by the PFP the net cost reductions are only minor today because of the expense of the starting nanophase materials. As further developments continue in the scale-up and processing activities, it is expected that the quality of and cost to manufacture PFP parts will be significantly different from the more traditional fabrication routes. However, one should note there are some applications today for PFP that provide a unique solution to existing ceramic technologies, especially where machining is detrimental to the integrity of the part.

27.4 POROUS PLATFORMS

Porous ceramic platforms have been in use and under development for many years for a variety of applications [7]. These ceramic porous platforms have pore sizes ranging from ångström levels in zeolite materials to millimeter levels in reticulated ceramics. The extent of work in the nanophase regime has been dominated and well served by sol–gel chemical routes for synthesis, with applications ranging from functional gradient optics to catalyst supports and ultrafiltration membranes.

A survey of the technical literature has shown a repeated problem in consistently fabricating porous ceramic platforms for use as ceramic humidity sensors [8–10]. Conventional ceramic processing is often used to fabricate a porous platform from the sintering and microstructure evolution of a ceramic powder compact. By interrupting the sintering process, a pore distribution is obtained along with an adequate degree of strength to the platform. The objective for a chemi-resistor humidity sensor is to develop a nanostructure based in micro and meso pores in order to activate the capillary condensation of water vapor (humidity) from the air into the pores of the ceramic. The Kelvin equation for the critical diameter for condensation is expressed as

$$d = \frac{4\gamma V}{RT \ln(P/P_0)} \tag{27.2}$$

where γ is the surface tension, V the molar volume, R the gas constant, and P/P_0 is the supersaturation ratio (or in this case the percent relative humidity, RH) [10]. Figure 27.3 shows the Kelvin equation critical diameter as a function of the RH.

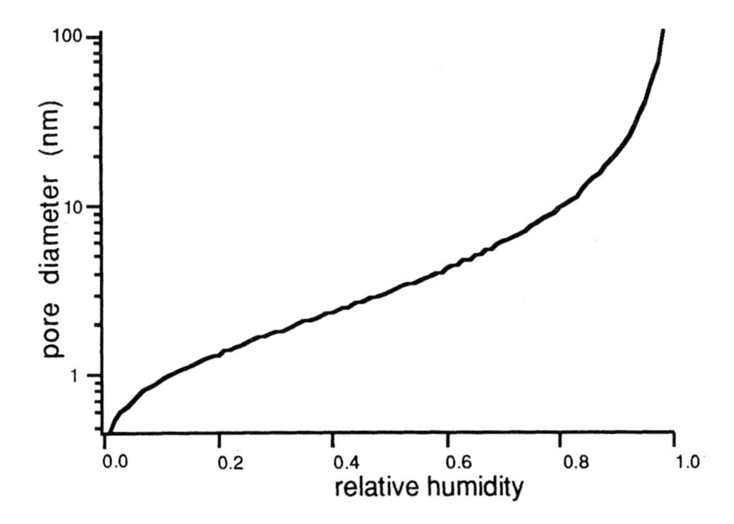

Figure 27.3. The critical pore diameter, determined from the Kelvin equation, versus the relative humidity. The limit for BET pore distribution analysis is indicated at 35 nm.

Considering the physical evidence of porosity control displayed in figure 27.1 it is clear that nanophase powders can be used as building blocks to form a porous platform that is capable of condensing (and detecting) the ambient moisture in an atmosphere. By coupling this with the considerable strength of green-body nanophase ceramics, it now becomes an economic and reproducible method for fabricating ceramic humidity sensors.

The performance of such a device is clearly demonstrated in figure 27.4 which shows the sensitivity (impedance versus relative humidity) of an as-

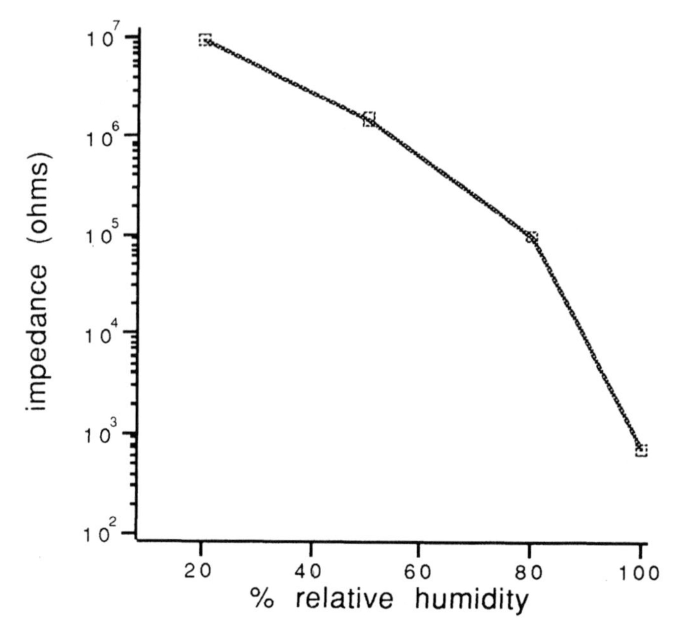

Figure 27.4. Impedance versus relative humidity of a nanostructured titanium oxide element. The element has 50 vol.% porosity.

made nanophase ceramic (0.5 in diameter, 1 mm thick element) that has open, porous Au electrodes on opposite faces. As can be seen, the impedance of the element varies over four orders of magnitude. The element was fabricated by uniaxially pressing the nanophase powders that have a particle size distribution and agglomeration state that leads to the maximum in sensitivity. The elements have excellent mechanical strength, and conductive paste electrodes can be easily applied by screen printing. In addition to sensitivity, the response times of these devices have also been measured and are comparable to commercially available devices.

This is one of a variety of applications that one can achieve with nanophase porous platforms. The future area of interest will be the development of the porosity ordering capability that is achievable with nanophase powders [11].

27.5 DISPERSIONS

The unique optical properties that are attainable with nanoparticles in dispersions has been studied for many years. The pioneering work of Brus *et al* in the area of nanoparticle quantum confinement [12] has demonstrated that unique optical properties can be obtained with semiconductor, nanoparticle dispersions. Also in this area, and of keen technological importance, is the recent development in the emission of light from porous silicon which has been shown to be a system of silicon nanoparticles dispersed in a silicon oxide matrix [2].

A great deal of commercial work has been focused on industrial coatings that alleviate the affects of solar ultraviolet (UV) radiation. Solar UV radiation effects range from hazards to the human skin to the color fading of some polymeric materials. For most of these systems, a suitable solution to this problem is the use of a protective coating that provides the maximum UV protection yet remains 'inert' to the system to which it is applied. Many coating material concepts have been attempted; however, the most recent developments have been the use of inorganic pigments to absorb *and* scatter solar UV radiation.

These inorganic pigment based UV attenuators, typically titanium oxide and zinc oxide, are highly attractive in this application because of their relative inertness and typically low cost when compared to most organic UV attenuating chemicals [13]. The drawback to inorganic pigment based coatings is the 'whiteness' they impart to the surfaces they are protecting. An obvious way to reduce or eliminate whitening is to shrink the particle size of the inorganic pigment and create a chemistry that allows the pigments to be highly dispersible in the UV attenuation coating vehicle.

Using the Rayleigh light scattering approximation (particle size $< \lambda/10$), the scattered light intensity (I_s) versus particle size (d) is expressed as

$$I_s \propto \omega^4 d^{-6} \tag{27.3}$$

where ω is the frequency of the electromagnetic radiation [14]. From this relationship it is evident that the smaller the nanoparticle the weaker the scattering in the visible spectrum. An optimal size is needed for this optical model in order to achieve the highest possible UV attenuation along with the highest degree of visible transparency.

In addition to particle size effects, the capacity for surface modification and encapsulation of the pigment (to allow for dispersability in a variety of solvents) needs to be as robust and reproducible as possible. In order for this to occur, the nanoparticle substrate needs to be as pure and uniform as possible. This can be a great challenge with nanophase powders because of their high specific surface areas which range from 20 to 200 m^2 g^{-1}, making them highly susceptible to surface contamination and process residue. This can be quite detrimental to the reproducibility of the particle coating technology, and thus challenge the optimal dispersibility of the nanoparticles. This is important to the UV attenuation coating application because an optimization of the dispersibility implies that less material is needed to obtain the UV protection.

Outstanding technical progress has been made with this type of UV attenuating pigment and as a consequence it is gaining a wider acceptance in commercial products. There are several development programs under way to expand this technology to polymer based materials, including the protection of fabrics and industrial plastics.

27.6 CONCLUSIONS

It is widely agreed that property benefits are achievable with nanophase materials, but it has been the practical demonstrations of the technology that have been key to establishing commercial viability. From our own experience the first stage of validation was to demonstrate that nanophase materials can be made reproducibly in volume at a reasonable cost. The definition of reasonable cost was obtained from the identification of applications where the benefits of nanophase materials are demonstrable, and material pricing is an important but secondary issue. The examples cited above are clearly within the bounds of this model.

As new and more versatile synthesis routes are developed, validated, and scaled *and* as the longer-term application development programs of today are completed, a much wider use of nanophase materials will occur within the next five years. As the environmental impacts of today's technologies come under closer scrutiny, nanophase materials will again demonstrate a benefit. The UV attenuation application discussed above is one such example, but many other nanophase material opportunities exist today including more efficient catalysts and CFC-free advanced refrigeration based on the magnetocaloric effect [15]. The combination of continued scientific and technical interest and short-term commercial success create a healthy prognosis for the future of nanophase materials

REFERENCES

[1] Siegel R W 1991 *Materials Science and Technology* vol 15, ed R W Cahn (Weinheim: VCH)
See also Andres R P *et al* 1989 *J. Mater. Res.* **4** 704

[2] Morisaki H, Ping F W, Ono H and Yazawa K 1991 *J. Appl. Phys.* **70** 1869

[3] Abraham T 1994 *Advanced Ceramic Powders and Nanosized Ceramic Powders—Processing, Technologies, New Developments, Markets, Players, and International Competition* (Norwalk, CT: Business Communications Co.)

[4] Skandan G, Hahn H and Parker J C 1991 *Scr. Metall.* **25** 2389

[5] 1995 *Am. Ceram. Soc. Bull.* **74** 22

[6] Wittenauer J 1995 *Superplasticity and Superplastic Forming 1995* ed A Ghosh and T R Bieier (Warrendale, PA: TMS)

[7] Komarneni S, Smith D M and Beck J S (ed) 1994 *Advances in Porous Materials MRS Symp. Fall (1994)* to be published

[8] Seiyama T, Yamazoe N and Arai H 1983 *Sensors Actuators* **4** 85

[9] Takami A 1988 *Ceram. Bull.* **67** 1956

[10] Kulwicki B M 1991 *J. Am. Ceram. Soc.* **74** 607

[11] Allen A J, Krueger S, Long G G, Kerch H M, Chen W, Malghan S, Parker J C and Skandan G 1995 *J. Nanostruct. Mater.* at press

[12] Brus L 1986 *IEEE J. Quantum. Electron.* **9** 1909

[13] Sayre R M, Kolloas N, Roberts R L and Baqer A 1990 *J. Soc. Cosmet. Chem.* **41** 103

[14] Jackson J D 1975 *Classical Electrodynamics* (New York: Wiley)

[15] Shull R D, Swartzendruber L J and Bennett L H 1991 *Proc. 6th Int. Cryocoolers Conf.* ed G Green and M Knox (Annapolis, MD: David Taylor Research Centre Publications) No DTRC-91/002 p 231

Subject index